Advances in Intelligent Systems and Computing

Volume 1226

AF738251

The series "Advances in Intelligent Systems and Computing" contains publications on theory, applications, and design methods of Intelligent Systems and Intelligent Computing. Virtually all disciplines such as engineering, natural sciences, computer and information science, ICT, economics, business, e-commerce, environment, healthcare, life science are covered. The list of topics spans all the areas of modern intelligent systems and computing such as: computational intelligence, soft computing including neural networks, fuzzy systems, evolutionary computing and the fusion of these paradigms, social intelligence, ambient intelligence, computational neuroscience, artificial life, virtual worlds and society, cognitive science and systems, Perception and Vision, DNA and immune based systems, self-organizing and adaptive systems, e-Learning and teaching, human-centered and human-centric computing, recommender systems, intelligent control, robotics and mechatronics including human-machine teaming, knowledge-based paradigms, learning paradigms, machine ethics, intelligent data analysis, knowledge management, intelligent agents, intelligent decision making and support, intelligent network security, trust management, interactive entertainment, Web intelligence and multimedia.

The publications within "Advances in Intelligent Systems and Computing" are primarily proceedings of important conferences, symposia and congresses. They cover significant recent developments in the field, both of a foundational and applicable character. An important characteristic feature of the series is the short publication time and world-wide distribution. This permits a rapid and broad dissemination of research results.

**** Indexing: The books of this series are submitted to ISI Proceedings, EI-Compendex, DBLP, SCOPUS, Google Scholar and Springerlink ****

More information about this series at http://www.springer.com/series/11156

Radek Silhavy
Editor

Applied Informatics and Cybernetics in Intelligent Systems

Proceedings of the 9th Computer Science On-line Conference 2020, Volume 3

Springer

Editor
Radek Silhavy
Faculty of Applied Informatics
Tomas Bata University in Zlín
Zlín, Czech Republic

ISSN 2194-5357 ISSN 2194-5365 (electronic)
Advances in Intelligent Systems and Computing
ISBN 978-3-030-51973-5 ISBN 978-3-030-51974-2 (eBook)
https://doi.org/10.1007/978-3-030-51974-2

This Springer imprint is published by the registered company Springer Nature Switzerland AG
The registered company address is: Gewerbestrasse 11, 6330 Cham, Switzerland

Preface

Modern cybernetics and computer engineering papers and topics are presented in the proceedings. This proceedings is a Vol. 3 of the Computer Science On-line Conference proceedings. Papers in this part discuss modern cybernetics and applied informatics in technical systems.

This book constitutes the refereed proceedings of the Applied Informatics and Cybernetics in Intelligent Systems section of the 9th Computer Science On-line Conference 2020 (CSOC 2020), held on-line in April 2020.

CSOC 2020 has received (all sections) more than 270 submissions from more than 35 countries. More than 65% of accepted submissions were received from Europe, 21% from Asia, 8% from Africa, 4% from America and 2% from Australia.

CSOC 2020 conference intends to provide an international forum for the discussion of the latest high-quality research results in all areas related to Computer Science.

Computer Science On-line Conference is held on-line, and modern communication technology, which are broadly used, improves the traditional concept of scientific conferences. It brings equal opportunity to participate for all researchers around the world.

I believe that you find the following proceedings exciting and useful for your research work.

April 2020 Radek Silhavy

Organization

Program Committee

Program Committee Chairs

Program Committee Members

Pavel Vaclavek	Faculty of Electrical Engineering and Communication, Brno University of Technology, Brno, Czech Republic
Miroslaw Ochodek	Faculty of Computing, Poznan University of Technology, Poznan, Poland
Olga Brovkina	Global Change Research Centre Academy of Science of the Czech Republic, Brno, Czech Republic, and Mendel University of Brno, Czech Republic
Elarbi Badidi	College of Information Technology, United Arab Emirates University, Al Ain, United Arab Emirates
Luis Alberto Morales Rosales	Head of the Master Program in Computer Science, Superior Technological Institute of Misantla, Mexico
Mariana Lobato Baes	Superior Technological of Libres, Mexico
Abdessattar Chaâri	Laboratory of Sciences and Techniques of Automatic Control and Computer Engineering, University of Sfax, Tunisian Republic
Gopal Sakarkar	Shri. Ramdeobaba College of Engineering and Management, Republic of India
V. V. Krishna Maddinala	GD Rungta College of Engineering & Technology, Republic of India
Anand N. Khobragade (Scientist)	Maharashtra Remote Sensing Applications Centre, Republic of India
Abdallah Handoura	Computer and Communication Laboratory, Telecom Bretagne, France

Technical Program Committee Members

Ivo Bukovsky
Maciej Majewski
Miroslaw Ochodek
Bronislav Chramcov
Eric Afful Dazie
Michal Bliznak
Donald Davendra
Radim Farana
Martin Kotyrba
Erik Kral
David Malanik
Michal Pluhacek
Zdenka Prokopova
Martin Sysel
Roman Senkerik
Petr Silhavy
Radek Silhavy
Jiri Vojtesek
Eva Volna
Janez Brest
Ales Zamuda
Roman Prokop
Boguslaw Cyganek
Krzysztof Okarma
Monika Bakosova
Pavel Vaclavek
Olga Brovkina
Elarbi Badidi

Organizing Committee Chair

Radek Silhavy — Tomas Bata University in Zlin, Faculty of Applied Informatics, email: radek@silhavy.cz

Conference Organizer (Production)

Silhavy s.r.o.
Website: https://www.openpublish.eu
Email: csoc@openpublish.eu

Conference Website, Call for Papers

https://www.openpublish.eu

Contents

Analysis and Verification of the Quality of the Structural Elements of a Bulgarian National Embroidery Information System

Desislava Baeva(✉), Kamelia Shoilekova, and Ivaylo Kamenarov

Angel Kanchev University of Ruse, 8 Studentska Street, 7000 Ruse, Bulgaria
dbaeva@uni-ruse.bg

Abstract. Intelligent technologies use models, software applications and practices of collection, integration, analysis and presentation of information necessary for automatic generation of knowledge. Integrating knowledge databases into the data processing would contribute to the creation of interactive and cognitive software solutions. Since historical research is a domain that provides interesting opportunities for the introduction of ontologies, not only computer scientists, but historians are also interested in popularizing this kind of data repositories. This article presents the analysis and verification of the functionality of a software system that allows working with factual information about Bulgarian national embroidery. As this task is the intersection of various disciplines and technological spaces, and it has not been covered by any specific software system it poses a number of problems, the resolution of which is the main objective of the working team.

Keywords: Applied ontologies · Information system

1 Introduction

Intelligent technologies are aimed at the development of contemporary methods of analysis and synthesis of complex intelligent information and management systems, based on dynamic ontologies, multi-agent systems and soft sensing.

They use models, software applications and practices of collection, integration, analysis and presentation of information necessary for automatic generation of knowledge. The instruments applied by the information systems encompass both traditional forms of query, response and online analytical processing and extraction of knowledge from data, and provide suitable visualization.

Graph structures and semantic networks are successfully used as a visual and universal means of information presentation in various subject matters. Semantic networks are a type of representation of the informal knowledge in the form of a directed network in which the nodes are conceptual objects – concepts or entities, and the directed arches between them express various semantic relations (properties) linking the respective pair of objects [5]. Based on this definition, resources that were not designed as such but acquired similar characteristics as a result of their extensive analysis can be assigned to the semantic networks.

R. Silhavy (Ed.): CSOC 2020, AISC 1226, pp. 1–9, 2020.
https://doi.org/10.1007/978-3-030-51974-2_1

The semantic networks organise the objects from a given area of the human knowledge in a method reflecting the relationships between them, and hence allowing the drawing of various types of conclusions about the respective objects.

Practically, the real applications oriented towards the representation of the human knowledge and its organisation in the semantic memory combine various types of semantic networks.

The focus of this presentation is placed on the national embroideries and the representation of the historical knowledge about them through the formalism of the semantic networks. To this end, a definitional or taxonomic network is used, characterised by the fact that it organised the concepts using the genus-species relation. The subject matter of this article is the description of the software system and the analyses of its functioning.

The reasons that necessitated this study are as follows:

- The text data about the cultural heritage are mostly art catalogues, photographs, etc. in digital format, from which it is difficult to extract information for analyses and comparisons.
- It is a challenge to establish connections and coincidences between the elements set out in the graphic images from various geographic districts.
- The specific information is unsystematised and inaccessible to the general public of users.

This necessitates the detailed study of the embroidery symbolism which will underlie the building of an information system.

As this task is the intersection of various disciplines and technological spaces, and this task has not been covered by any specific software system, it is necessary to develop a new specialised instrument to cover this case. It is necessary that the team include an engineer specialising in databases, a software engineer, a programmer and an ethnologist and historian, as the specificity of the embroidery is that it keeps information about pre-Christian faiths. It is a kind of encoded information borne from the most ancient times [2].

2 Conceptual Framework

The main advantage of the experimentally built semantic network with subject matter the various historical data about the Bulgarian national embroidery is its visualisation and modularity of presentation of the relations, the inheritance of the properties and its flexibility.

The building of an information system based on databases uses a theoretical concept whose composition consists of three sub-systems – for entry of the data, for acquisition of knowledge, and for analysis and recommendations.

The sub-system for acquisition of knowledge serves to form the initial structure of the subject matter, the model of description of the embroidery as an element of the general knowledge base.

Any task for knowledge acquisition is referred to the knowledge base engineer's expertise and the description of the subject matter. The main objective of the knowledge base engineer's interface is to determine the connection between the given concepts in the specific context, as well as the definition of the semantic relationships between them.

The system allows for structuring and formalising knowledge from the application area in order to complete the knowledge base, and also provides the users with a wide range of search options. The created knowledge base is presented and described in the Micro proceedings [1].

The web information system suggested herein is founded on a conceptually-based semantic network. This type of network allows the user to easily obtain the desired information from an autonomous database without preliminary knowledge of the design by using the Sprawl query semantic language. [7] The full and good description of the data also suggests a multiaspect search in the database, which is undoubtedly the greatest achievement of the software product (Fig. 1).

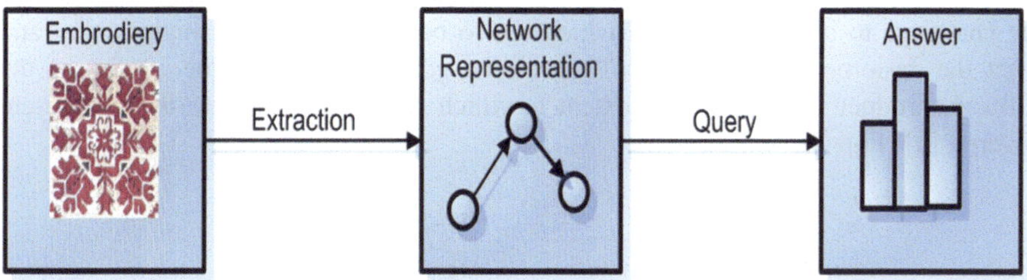

Fig. 1. Conceptual scheme for the presentation of semantic knowledge

3 Data Management

In various sources, the process of verification of the quality of the knowledge is given different terms. The two most common variants are verification and validation. It is more appropriate to use the term verification when a general analysis of the characteristics of the obtained knowledge base is performed for compliance with the requirements which are applied to the quality of the ontological bases and the compliance of the entered knowledge with the specificity of a given area of application is not checked. The very evaluation of the quality does not aim at proving the official compliance with all requirements of the process of building ontology, but to identify the weaknesses and the erroneous semantic entities for the purpose of their further improvement. The problem of accuracy and completeness of the entry of the concepts is one of the most important in creating ontology, and is rooted in ontology's universal nature.

The verification of the ontologies means an exceptionally official inspection of the building process correctness, but a successful validation can be discussed only after an analytical inspection by experts, administrators and users. It is also necessary to perform periodic checks of the weakest ontology areas through automatic evaluations of the functionality [6].

The testing of the knowledge base poses two main problems: 1) how big the full set of tests should be, and 2) how small the necessary number of tests should be. These problems have no solution as a whole, and therefore such solutions should be found in specific cases: either through limiting the classes of specifications and realisations under consideration, or through the assumption for the presence of additional opportunities for testing, or through a combination between these two.

4 Testing of the Information System

The quality of software is a complex characteristic which summarises the effect of the impact of a number of various factors on the process of development and realisation of the software product. The performance of a quantitative evaluation of the software product quality is important for the effective management and improvement of the processes in software engineering. The contemporary software engineers are faced with the challenge to develop software which has to be of exceptionally high quality and meet the customers' requirements. One of the goals of ensuring the quality of the software product is to check the extent to which the requirements to the system are covered [9] (Fig. 2).

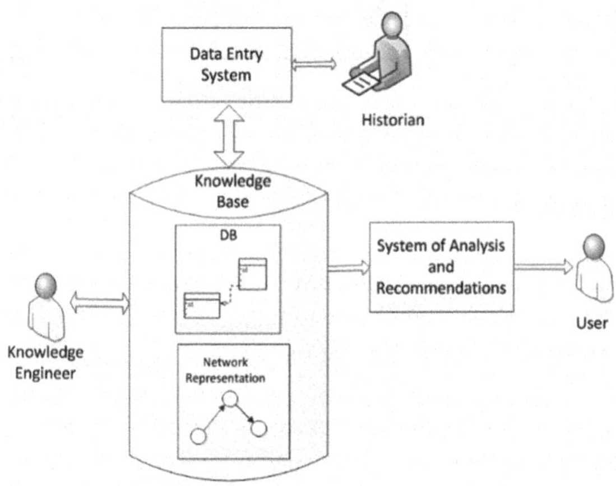

Fig. 2. Architecture of a knowledge-based system

The aim of the functional testing is to check whether certain functionalities and characteristics of the system function in accordance with the specification. The tests are conducted through the graphic interface of the system and an analysis of the output data is performed based on the submitted input data. It is necessary to use both valid and invalid input data in order to establish the following:

- whether the expected results are generated in case of valid input data;
- whether in case of invalid input data the system generates a suitable error message and the user is denied access to the next levels of the system. (In the described

information system, in the event the actor Historian fails to enter a correct username and/or password, the system will not allow him to manipulate the information in it);
- the navigation between the various windows, hyperlinks and fields is correct.

During the functional testing, the attention is focused on the correctness of the realised processes and their compliance with the functional requirements to the system.

The reliability of the system can be defined as its capacity to perform the functions required from it under certain conditions for a given period of time. The reliability testing is an attempt to cover all functionalities without including complex functionalities, as they are the subject matter of the functional testing. The reliability testing aims at confirming that the system will not stop functioning [4].

In testing the reliability, the attention is focused on:

- identification of the vulnerable points allowing ill-intentioned people to gain access to the system. To solve this problem, instruments that provide the opportunity to prevent possible breaches are used, which identify possible threats and suggest control measures for management of the vulnerabilities before the ill-intentioned hackers succeed in taking advantage of them. A test is performed, aiming at checking whether the confidential information is encrypted with sufficiently effective algorithms, so that an unauthorised person will not be able to modify it.

In the Semantics of the Bulgarian embroidery information system, one of the vulnerabilities is the entry into the system of the actor Historian. To solve this problem, a decision was made to give the actor Historian's password a high level of complexity and be of the type passphrase, in which additional protection such as a digit or a special character are included for the purposes of increasing the security level.

- hacker attacks of the DoS type. This attack aims at making the customer's infrastructure collapse and in particular to disrupt the offered service. This type of attacks is implemented through generation of an enormous amount of traffic, which floods the capacity of the network channels. This starts processes which are performed infinitely and consume the resources of the hardware devices.

To solve this problem, a stable online environment was selected and a choice was made of a correct hosting solution. The addition of security to the hosting was a key factor for accessibility to the site and guaranteeing the uninterruptibility of the online presence.

As a whole, each individual element of the process of developing databases and information systems has been discussed extensively. However, there are still reserves in the less studied aspects, in the application of combinations of successful methods and solutions on the borderline between the various areas or in search of new aspects of application of already employed practices [3] (Fig. 3).

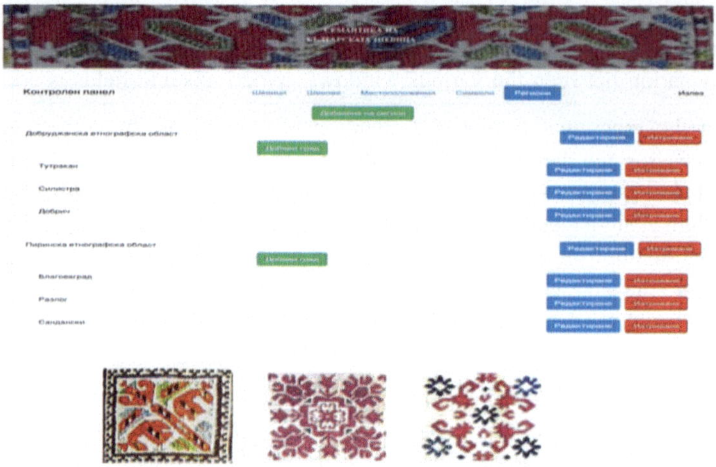

Fig. 3. Information system working desktop.

5 Interface Testing

The effectiveness is an external metric for validation of the information systems that reflects the degree of the users' satisfaction from the manufactured products or rendered services. It shows the extent to which the system has fulfilled its purpose.

Wixom, Todd et al. discussed the characteristics of the quality software product and classify them in several groups: quality of the system, quality of the information, quality of the service, utility, ease of use, expected results, and organisation factors [9]. They pointed out that these specific conditions were obtained from integration of the factors listed in the literature about the users' satisfaction.

Although these characteristics have general applicability, the importance of each of them depends on a specific system and direction of application [2].

To identify the degree of user satisfaction from the software system, an anonymous online survey was developed and positioned on a social media platform. The questionnaire was compiled by groups generated according to the criterion for the qualities of the system which refer to: the characteristics of the software system; its functional capacity; and the access to specialised information.

The survey consisted of five questions, and the main rule in the formulation of the questions was that they had to be short and clear for the respondents as well as unambiguous. The development of the survey was based on the following two stages:

Stage 1: A system of indicators was created, through which the research hypotheses would be verified. At this stage, the content aspect of the indicators was determined and one of the main goals here was the formulation of questions whose answers were expected to be evaluations, opinions and recommendations about the system. Based on this stage, the following three groups of questions were formulated:

The first group of questions determined the users' profile; the second group of questions evaluated the characteristics of the software system and the access to specialised information presented through links from the software system;

The third group included only one question, to which the user had to answer expressing an opinion on the software system.

Stage 2: Determination of the type of questions:
According to the content – questions about evaluations and degree of their intensity.

Two of the five questions involved the expression of evaluation. The constructs were formulated in a manner that they required a level of agreement, evaluated using a 5-point Likert scale, which varied from 1 (very dissatisfied) to 5 (very satisfied).

According to the form – open-ended, closed-ended, mixed questions

The survey contained mixed questions: 4 questions of the closed-ended type and 1 questions requiring an open-ended response. Two out of the four questions were formulated and contained in their composition four more questions, which were also of the closed-end type.

According to the function – main and additional; filter questions.

Two of the questions were not closely related to the system specificity, they were filter questions applied for the purpose of establishing which users were interested in testing the system and what version of the system they preferred – mobile or desktop version. The other three questions were main questions.

The survey was conducted in October 2019; it was planned to conduct it annually, as a feedback instrument in the process of management and integration of the designer solutions in building the specialized interface.

The data obtained from the survey were as follows:

54 users were surveyed. They were divided into 3 categories – professionals and non-professionals with or without any interest in the field. It could be noted that those interested in testing the system were mainly non-professionals, those with interest in the field - 70%. The most critical of the software functionality were the professionals, who comprised 12% of the surveyed users. Regardless of the good functionalities of the mobile version of the system, the users' preferences were still directed towards its use via desktop computers - 62%.

The users from all categories shared their impressions from the service. The percentage of their satisfaction is presented in Fig. 4.

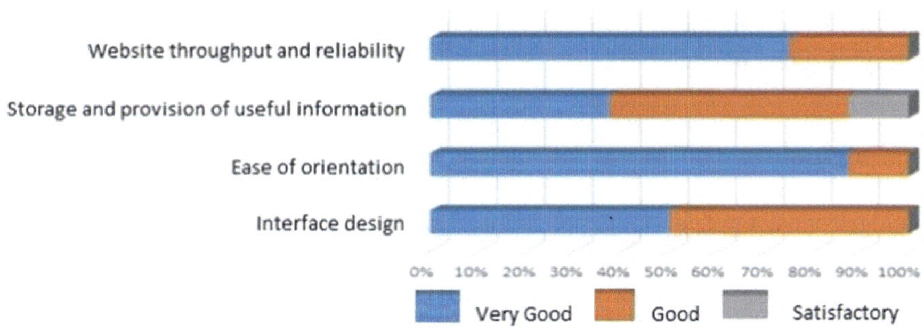

Fig. 4. The users impressions from the service.

How will you evaluate the access to specialised information provided through links from the software system? (Fig. 5)

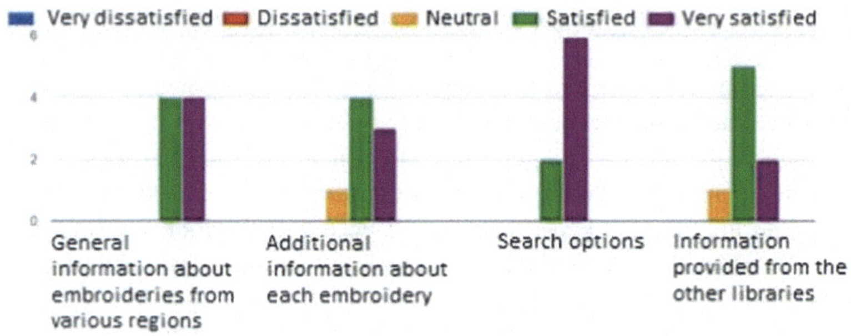

Fig. 5. Excerpt from responses of the respondent groups on a 5-point likert-type opinion scale.

The summarised opinions of the respondents were an objective prerequisite for improving the characteristics of the metrics with the lowest (end) values. The prevalent opinion was "very satisfied" (73–77% of the respondents) about the opportunities provided by the presented software.

According to the analysis of the groups of respondent's answers, the following generalised conclusions can be drawn:

- A large part of the respondents believed that the well-organised interface and the easy access to the system facilitated its operation and were a prerequisite for good service.
- The majority of the users of the PhD Student system were pleased with the operation of the software resource. They reported the use of the electronic administrative services offered by it as an advantage.
- The attitude to the opportunity for online access to current information – regulatory documents, study resources, etc. was definitely positive.

The further development of the system will take into consideration the recommendation made by the respondents to pay attention to the opportunity to compare embroideries containing similar elements from the folklore areas that are already in the territory of different countries.

6 Conclusion

Ontologies have a potential for applications that provide a glossary of a subject matter designed for users and software systems. With their assistance, intelligent interfaces can be designed, which, based on the underlying rules, personalise the dialogue with the user based on simple conclusions from the available restrictions on certain properties.

The creation of new ontological resources gives rise to new types of data collections, which are oriented towards the representation of a complex abstract model of the human knowledge. The dominating tendency is for these united resources to combine a conceptually modelled description, characteristic of the ontologies, with the large volume of facts in the knowledge bases.

The theoretical concepts underlying the building of the software system for Semantics of the Bulgarian national embroidery, their practical implementation and the opinion of the users showed that the web-based platform thus designed and built was a successfully functioning software resource that reflected and met the requirements of the users from the target groups and the stakeholders to a great extent.

Acknowledgements. This research was supported by the Bulgarian Science Fund, with the grant for financial support for projects of junior researchers and postdocs, agreement DM12.1/2017 "Develop and explore a modern approach to conceptual software presentation of the semantic information stored in the Bulgarian national embroidery".

References

1. Baeva, D., Baev, B.: Semantic approach in encoding of the meanings in Bulgarian Folklore embroidery in digital libraries. In: 2019 42nd International Convention on Information and Communication Technology, Electronics and Microelectronics (MIPRO), pp. 1082–1086. IEEE (2019)
2. Ivanova, G., Kozov, V., Zlatarov, P.: Gamification in software engineering education. In: 2019 42nd International Convention on Information and Communication Technology, Electronics and Microelectronics (MIPRO), pp. 1445–1450. IEEE (2019)
3. Kaloyanova, K.: Information systems analysis and design course with projects based on real customers requirements. In: Eighth Mediterranean Conference on Information Systems, Verona, Italy (2014)
4. Krippendorff, K.: Content Analysis: An Introduction to Its Methodology, 2nd edn. Sage Publications, Thousand Oaks (2004)
5. Lehmann, F.: Semantic networks. Comput. Math Appl. **23**(2–5), 1–50 (1992)
6. Nikonenko, A.A.: Podkhody k verifikatsii znaniy v lingvisticheskikh ontologiyakh, Nau-kovi pratsi Donets'kogo natsional'nogo tekhnichnogo universitetu. Series: Informatika, kibernetika ta obchislyuval'na tekhnika, vol. 16, pp. 192–201 (2012)
7. Ogiela, L., Ogiela, M.R.: Advances in Cognitive Information Systems, vol. 17, pp. 17–18. Springer Science Business Media (2012)
8. Kalushkov, T., Valcheva, D., Markova, G.: A model for pseudo-cloud hosted e-learning module for collaborative learning. In: 2018 2nd International Symposium on Multidisciplinary Studies and Innovative Technologies (ISMSIT), pp. 1–5. IEEE (2018)
9. Valcheva, D., Todorova, M., Kalushkov, T.: Structuring multimedia scenarioes according to the different learning modalities. In: Proceedings of the 2009 Euro American Conference on Telematics and Information Systems: New Opportunities to increase Digital Citizenship, p. 23. ACM (2009)

Implementing Sticky Bit Generators Based on FPGA Carry-Chains for Floating-Point Adders

I. V. Ushenina[1(✉)] and E. V. Chirkova[2]

[1] Penza State Technological University, Penza, Russia
ivl23@yandex.ru
[2] Penza State University, Penza, Russia

Abstract. When performing addition or subtraction in the floating point format, the sticky bit is used in some rounding modes. To generate the sticky bit, the multiple-input OR gate is required. If addition or subtraction is implemented in FPGA, the sticky bit generation can slow down the work of the adder. This paper proposes the sticky bit generators based on FPGA carry-chains. The proposed generators are intended for those adders which work with normalized numbers of single and double precision (according to IEEE 754 standard). The specific feature of the generators is that they work simultaneously but are not integrated with shifters involved in the alignment of the summands. This paper assesses resource utilization and performance of the proposed sticky bit generators; evaluates the contributions to the sticky bit generation delay of FPGA logic elements and routing resources used in the generators' circuits.

Keywords: Floating-point adder · Rounding · Sticky bit · FPGA

1 Introduction

Nowadays real-time floating-point arithmetic operations are demanded in many digital signal processing applications, when performing high-accuracy scientific calculations, in math co-processors, etc.

In contrast to the fixed-point format, the advantages of the floating-point format are high accuracy of calculations, robustness to quantization errors, and wide dynamic range [1]. The disadvantage of the floating-point format is complexity of hardware implementation of calculations.

One of the platforms for implementing calculations in the floating-point format is FPGAs. Their strong point is the possibility of simultaneous calculations. One of the most demanded and at the same time the most complex for implementation floating-point arithmetic operations is addition (subtraction) [1–3]. When implementing adders on FPGA resources, the most complex is implementation of shifters and the sticky bit generators [4, 5].

The sticky bit is generated in the course of alignment of the mantissas of the two operands before their addition or subtraction. The operand mantissa with the least exponent is shifted to the right until the exponents of the two operands are aligned.

R. Silhavy (Ed.): CSOC 2020, AISC 1226, pp. 10–20, 2020.
https://doi.org/10.1007/978-3-030-51974-2_2

The sticky bit is the result of logical addition of all the bits rejected during the shift except the guard bit and the round bit which follow the least significant bit. Guard, round and sticky bits are required if, when rounding addition or subtraction results, the modes "round to nearest even", "round to positive infinity, round to negative infinity" are used [1, 6].

The sticky bit generator can be represented as a multi-input OR gate, the mantissa of the least of the operands being applied to its input as a data bus **d[23:0]** (Fig. 1). To calculate the sticky bit, the data bus bits must be used which will be rejected as the result of alignment. That is, as a control bus of the sticky bit generator, **sv[7:0]**, it is necessary to use the exponent difference, and the width of the control bus equals the width of the exponents. The sticky bit generator presented in Fig. 1 corresponds to the single-precision adder operating according to IEEE 754 standard.

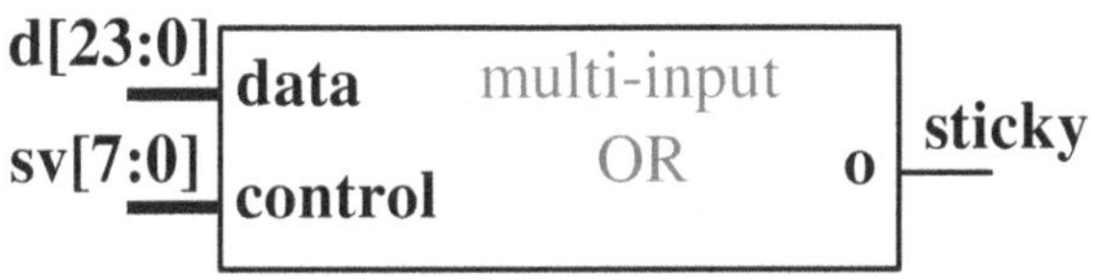

Fig. 1. Sticky bit generator as a multi-input OR gate with capability of controlling input selection.

Among FPGA resources there are no hardware multi-input OR gates. It means that the sticky bit generator must be implemented on FPGA general-purpose configurable logic blocks. This paper considers the potential of using FPGA carry-chains for implementing the sticky bit generators. It provides circuits of sticky bit generators intended for adders operating with single-precision and double-precision normalized numbers (according to IEEE 754 standard). It also researches the dependence of sticky bit generators' performance on the following factors:

- precision of representing numbers in the adder (single and double precision options are considered);
- FPGA placement settings of sticky bit generators.

Besides, this paper assesses resource utilization of sticky bit generators for number adders with single and double precision.

The research has been carried out in Xilinx FPGA Artix 7 [7]. Design of sticky bit generators and research of their characteristics has been done in Vivado 2017.2 IDE.

2 Related Work

Carry chains had been used to generate the sticky bit in adders before FPGAs became widely spread. For example, [8] proposes the sticky bit calculation circuit in the form of a precharged Manchester carry chain. [9] Presents a floating-point arithmetic processor utilizing the circuit proposed in [8].

In [5, 10–12] FPGA carry-chains are used for generating the sticky bit. Sticky bit generators considered in these works are integrated with shifters performing the alignment of mantissas before addition or subtraction. Shifters consist of several stages, and each stage is joined with part of the sticky bit generator doing logical addition of the bits rejected by this stage. The figure demonstrating this approach can be seen in [10].

[5] considers the option of supplementing FPGA slices with specialized hardware units designed for aligning mantissas before addition and for calculating the sticky bit. In the hardware units proposed, sticky bit generators are also integrated with shifters.

This paper proposes sticky bit generators using mantissa of the least of summands and the exponent difference as initial data. It provides for simultaneous work of the sticky bit generator and the shifter. That is, sticky bit generators are not integrated with shifters, and do not use intermediary results of the shift obtained at particular stages of shifters.

3 Sticky Bit Generators

The advantage of FPGA carry-chains is that the pathways between them cause the minimal delay to signal propagation [13]. This paper proposes sticky bit generators implemented on look-up tables (LUTs) and carry chain multiplexers (MUXCYs) of FPGA [7].

A sticky bit generator is a series of separate units. As shown in Fig. 2, each unit contains one LUT and one MUXCY. The unit's LUT implements a given logical function by processing from 1 to 6 input signals and forms a control signal applied to the MUXCY input **s**. If logic 1 is applied to the MUXCY input **s**, its output **co** is linked with the input **ci**; if logic 0 is applied to the MUXCY input **s**, its output **co** is linked with the input **di**. The MUXCY input **ci** is linked with the output **co** of the previous MUXCY in the carry-chain. The constant logic 0 or logic 1 is applied to the MUXCY input **di**.

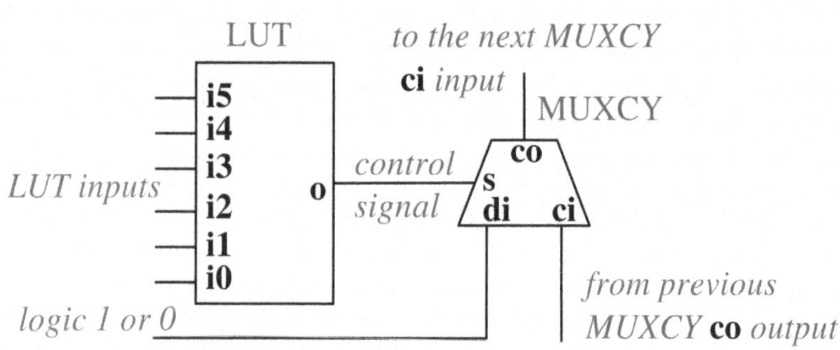

Fig. 2. Sticky bit generator unit implemented on the LUT and MUXCY of FPGA.

Figure 3 demonstrates the sticky bit generator for a single precision adder, that is for the case when the mantissa is represented by 24 bits whereas the exponent by 8 bits.

The first unit LUT (implementing **ftest** function) checks three most significant bits of the control bus. If at least one of the bits **sv[7:5]** equals one, the mantissa is shifted more than by 24 bits at any values of **sv[4:0]**. That is, most significant mantissa bit which always equals one for normalized numbers will take part in the sticky bit generation. **Ftest** function implemented by LUT is a 3-input NOR. If **sv[7:5]** > 0, logic 0 appears at the LUT output of the first unit; it gives a pass to logic 1 to the first unit output.

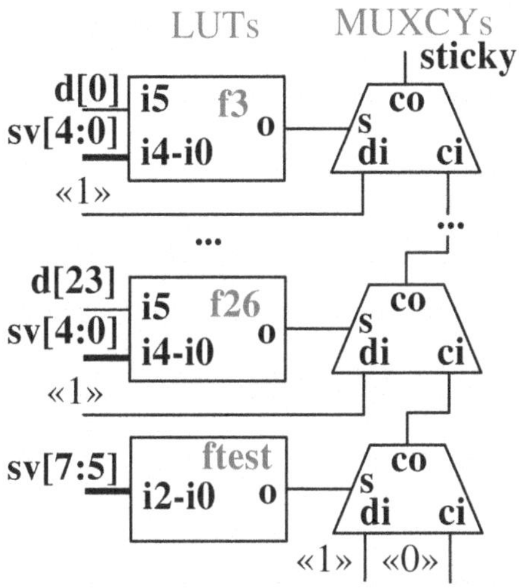

Fig. 3. Sticky bit generator for a single precision adder.

Interaction between LUTs and MUXCYs in the rest of the units is organized in the following way: LUT takes **sv[4:0]** and (**j** − **3**)-th bit of the data bus and implements the function **fj** (**j** = 3..26). If **sv[4:0]** $\geq$ **j**, the choice of MUXCY input connected with LUT depends on **d[j** − **3]**. When **d[j** − **3]** = 1, the MUXCY output **co** is linked with the input **di**, logic 1 being applied to it. When **d[j** − **3]** = 0, and also when **sv[4:0]** < **j**, the MUXCY is included into the chain and gives a pass to its output to the signal formed by previous units.

The circuit similar to the one presented in Fig. 3 can be applied for generating the sticky bit in a double precision adder. However, there emerges a problem of the limited number of inputs in LUTs. For generating the sticky bit from 53 mantissa bits, at least 6-bit control bus is necessary. Consequently, **sv[5:0]** and **d[j** − **3]** cannot be applied to 6-input LUT similar to the way which is presented in Fig. 3.

The sticky bit generator for a double precision adder is shown in Fig. 4. The generator can be divided into two parts, one of which is used if **sv[5]** = 1, and the other if **sv[5]** = 0. Checking of **sv[5]** bit is carried out by the generator unit with **check1** function. This unit separates the two parts of the generator.

In the first part of the generator logical addition of the data bus bits **d[28:0]** is carried out without checking the control bus as it is supposed that **sv[5]** = 1.

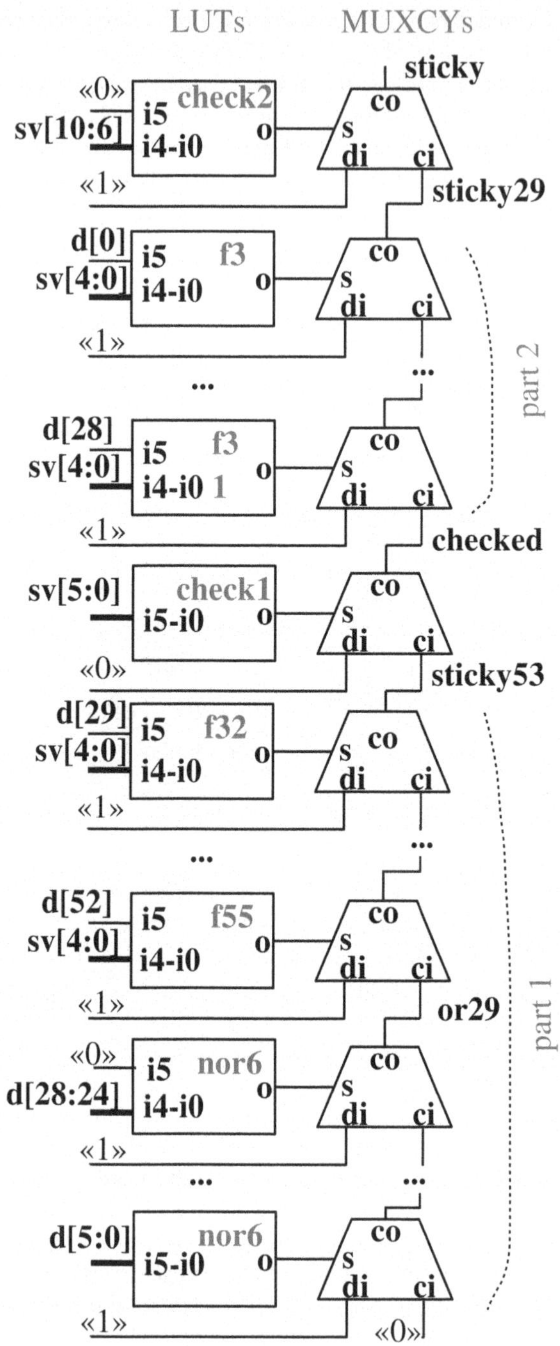

Fig. 4. Sticky bit generator for a double precision adder.

Logical addition of **d[28:0]** bits is done by five lower units of the generator. LUTs of these units operate as 6-input NOR gate (implement **nor6** function). The result of **d [28:0]** bits logical addition is marked **or29**.

The rest of the units of the generator's first part, which perform **fj** (**j** = 32..55) functions, allow adding most significant bits of the data bus **d[52:29]** to **or29**. Each of these units operates on condition that **sv[5]** = 1. In each of the units, functions of LUTs **fj** control MUXCYs: if **sv[5:0]** $\geq$ **j** and **d[j − 3]** = 1, the unit output is linked with the MUXCY input **di**, to which logic 1 is applied. In all other cases, the MUXCY output is linked with the input **ci** and gives pass to the next unit to the signal generated by previous units.

The **sticky53** signal generated by the first part of the generator is applied to the unit controlled by the **check1** function. If **sv[5:0]** $\leq$ 31, that is **sv[5]** = 0, the **sticky53** signal is pushed back, and the sticky bit calculation starts again by the second part of the generator under the control of the **sv[4:0]** bus. As the maximum shift value, corresponding to **sv[4:0]**, equals 31, the second part of the generator processes the data bus bits **d[28:0]**. The functions **fj** (**j** = 3..31) performed by unit LUTs of the generator's second part are similar to the functions **fj** of LUTs of the generator's first part. The output signal of the second part of the generator is **sticky29**.

The last unit of the generator checks five most significant bits of the control bus **sv [10:6]**. If at least one of them equals one, during the alignment all the mantissa bits are shifted to the right of the round bit position, and the mantissa most significant bit, which always equals one for normalized numbers, participates in the sticky bit generation. The **check2** function performed by the last unit's LUT allows equating sticky with logic 1, if **sv[10:6]** > 0. If **sv[10:6]** = 0, the MUXCY controlled by the **check2** function gives pass to **sticky29** signal to its output.

4 Methods

Within the framework of the research conducted, sticky bit generators for single and double precision adders were described using the VHDL language in a structural style, that is with specification of FPGA resources used and connections between them. For LUTs, codes (hexadecimal values) were specified, which correspond to the functions implemented by them. Based on VHDL descriptions, projects of generators were made in Vivado 2017.2 IDE. Both projects were intended to be implemented on FPGA XC7A200T-3.

Owing to the fact that generators are described in the structural style, the projects' resource utilization had been assessed before implementation. The assessment results were confirmed by the reports on the placement and routing of the projects generated in Vivado.

The performance assessment of sticky bit generators was based on timing analysis of the projects. To launch the timing analysis, constraints were imposed for the period of clock signals which synchronize generators [14]. The sticky bit generators themselves are combinational, but in order to assess their performance they were synchronized with the clock signal for the period on which the constraints were imposed. The timing analysis was carried out for several values of the clock signal period; as a result, the ranges of values of the maximum clock frequency were obtained for each generator.

The performance of devices implemented on FPGA is also influenced by the method of their placement [15]. Assessment of sticky bit generators performance was carried out for two methods of their placement: automatic placement and placement on the released area of the chip (Pblock). The size of the chip area was determined by the resource utilization of the implemented devices.

On the results of the timing analysis carried out for each of the generators, contributions were defined which were made by FPGA logic elements and interconnections between them to total delay of the sticky bit generation. For this purpose, the longest chains of each generator were considered. As it is shown in Fig. 5, the longest generator chain starts at one of the input D flip-flops connected with the LUT of the first unit, goes through all the units and finishes at the input of D flip-flop which receives the generated sticky bit. Figure 5 has the following notations: 1 – input D flip-flop receiving the control bus bit; 2 – LUT of the generator's first unit; 3 – intermediate units of the generator (four in each FPGA slice); 4 – output D flip-flop receiving the generated sticky bit; 5 – FPGA slice. Analysis of delays in the sticky bit generation was performed under automatic placement of generators and under time constraint which provides maximum clock frequency.

5 Results and Discussion

The results of assessing resource utilization of the sticky bit generators are demonstrated in Table 1. When assessing the resource utilization, D flip-flops, included in the projects in order to perform the timing analysis, were not taken into consideration. Table 1 shows that the sticky bit generators for both types of adders occupy less than 0.05% of all the slices available on FPGA.

Tables 2 and 3 show ranges of values of generators' maximum clock frequencies. Ranges of clock period values, specified in constraints, are also presented in Tables 2 and 3.

Tables 2 and 3 lead us to the conclusion, that the clock period given in constraints, as well as the method of project placement (automatic and Pblock) influence the performance of the sticky bit generator insignificantly. This is due to the fact that carry-chain units are always located by the place and root tool strictly above each other in neighboring slices independently of implementation settings of projects. Nevertheless, if necessary it is possible to increase the generator's performance by a few per cents by means of time constraints and to keep the placement result obtained for usage in the project of the adder.

Restricting the chip area available for placement leads to reducing the sticky bit generators' performance up to 6.6%. It is connected with restricting the amount of variants for placing and routing. Nevertheless, in big projects each block has the area of the chip occupied by it, and insignificant reduction in performance under restricting the chip area is the advantage of the generator.

Table 4 demonstrates the results of assessing contributions made by logic elements of sticky bit generators and interconnections between them into total delay of the sticky

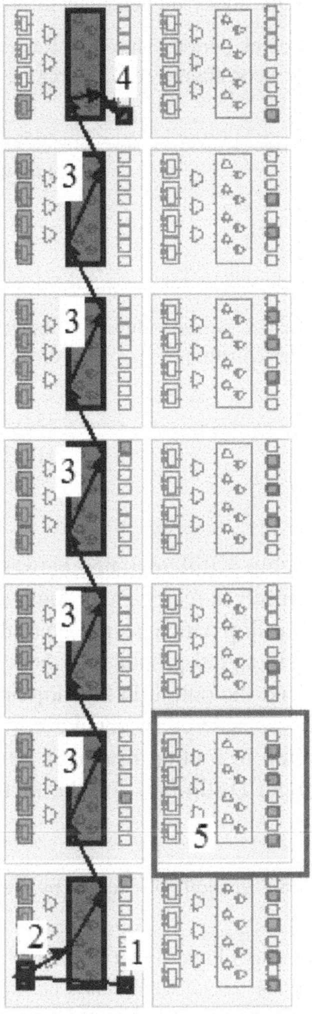

Fig. 5. Placement results of the sticky bit generator for a single precision adder. For analysis, the longest chain which contains all generator units was chosen.

Table 1. Resource utilization of sticky bit generators.

Width of the generator's data bus	Resource utilization	
	LUTs (MUCXYs)	Slices
24 bits	25(25)	7
53 bits	60(60)	15

Table 2. Performance of the sticky bit generator for a single precision adder.

Method of the generator's placement	Range of clock period values given under constraints, ns	Range of the generator's maximum clock frequency values, MHz
Automatic	1.5–2.6	465–488
Pblock	1.5–2.6	444–483

Table 3. Performance of the sticky bit generator for a double precision adder.

Method of the generator's placement	Range of clock period values given under constraints, ns	Range of the generator's maximum clock frequency values, MHz
Automatic	2–3.5	332–350
Pblock	2–3.5	310–345

bit generation (total delay). Table 4 differentiates between the delay caused by the sticky bit generators' units themselves (carry-chain logic delay) and total delay of the generator's units and input D flip-flop which keeps one of the bits of data bus or control bus (total logic delay). Delays in signal propagation between the generator's units as well as between the final unit and the output D flip-flop are so insignificant that in the timing analysis report they are considered to equal zero. Thus, in each of the generators, delay in signal propagation in routing resources (net delay) equals the delay in interconnection between the input D flip-flop and the first unit (Fig. 5).

Table 4. Delays in the sticky bit generation caused by generators' logic elements and interconnections between them.

Width of the generator's data bus	Carry-chain logic delay, ns	Total logic delay, ns	Net delay, ns	Total delay, ns
24 bits	1.124	1.465	0.351	1.816
53 bits	1.764	2.105	0.613	2.718

From reports on timing analysis it also follows that the largest delay (about 0.45 ns) in signal propagation in the chains under research is caused by the generators' first units where signals from D flip-flops go through LUTs to MUXCY control inputs (Fig. 5). Contribution of intermediate units which form multiplexer chains is significantly smaller: each four MUXCYs cause about 0.09 ns delay. Contribution of final units, generating the sticky bit, depends on which MUXCY of FPGA slice the output signal comes from. Contributions from final units are smaller than the ones from the first units, and can be larger or equal to contributions of intermediate units.

6 Conclusion

The advantages of the proposed circuits of sticky bit generators are: 1) independence from shifters performing the alignment of mantissas before addition and simultaneous operation with the shifters, 2) maximum usage of internal pathways of FPGA slices and due to it – the least delays of propagation in interconnections between logic elements, and 3) predictable placement results on the chip. The results obtained during the study allow us to conclude that the proposed sticky bit generators can keep high performance of the adders which use them.

References

1. Ercegovac, M.D., Lang, T.: Digital Arithmetic. Elsevier (2004)
2. Malik, A., Ko, S.B.: A study on the floating-point adder in FPGAs. In: 2006 Canadian Conference on Electrical and Computer Engineering, pp. 86–89. IEEE (2006)
3. Shirke, M., Chandrababu, S., Abhyankar, Y.: Implementation of IEEE 754 compliant single precision floating-point adder unit supporting denormal inputs on Xilinx FPGA. In: 2017 IEEE International Conference on Power, Control, Signals and Instrumentation Engineering (ICPCSI), pp. 408–412. IEEE (2017)
4. Shirazi, N., Walters, A., Athanas, P.: Quantitative analysis of floating point arithmetic on FPGA based custom computing machines. In: Proceedings IEEE Symposium on FPGAs for Custom Computing Machines, pp. 155–162. IEEE (1995)
5. Beauchamp, M.J., Hauck, S., Underwood, K.D., Hemmert, K.S.: Architectural modifications to enhance the floating-point performance of FPGAs. IEEE Trans. Very Large Scale Integr. (VLSI) Syst. **16**(2), 177–187 (2008)
6. Nannarelli, A.: Tunable floating-point adder. IEEE Trans. Comput. **68**(10), 1553–1560 (2019)
7. Xilinx 7 series FPGAs Configurable Logic Block User Guide (2019). https://www.xilinx.com/support/documentation/user_guides/ug474_7Series_CLB.pdf. Accessed 12 Dec 2019
8. Peng, V., Bowhill, W.J., Gavrielov, N.M.: U.S. Patent No. 4,864,527. Washington, DC: U.S. Patent and Trademark Office (1989)
9. Benschneider, B.J., Bowhill, W.J., Copper, E.M., Gavrielov, M.N., Gronowski, P.E., Maheshwari, V.K., Peng, V., Pickholtz, J.D., Samudrala, S.: A pipelined 50-MHz CMOS 64-bit floating-point arithmetic processor. IEEE J. Solid-State Circuits **24**(5), 1317–1323 (1989)
10. Lee, B., Burgess, N.: Parameterisable floating-point operations on FPGA. In: Conference Record of the Thirty-Sixth Asilomar Conference on Signals, Systems and Computers, vol. 2, pp. 1064–1068. IEEE (2002)
11. Paschalakis, S., Lee, P.: Double precision floating-point arithmetic on FPGAs. In: Proceedings. 2003 IEEE International Conference on Field-Programmable Technology, pp. 352–358. IEEE (2003)
12. Hemmert, K.S., Underwood, K.D.: Fast, efficient floating-point adders and multipliers for FPGAs. ACM Trans. Reconfigurable Technol. Syst. (TRETS) **3**(3), 1–30 (2010)
13. Chapman, K.: Multiplexer design techniques for datapath performance with minimized routing resources. Xilinx Application Note (2019). https://www.xilinx.com/support/documentation/application_notes/xapp522-mux-design-techniques.pdf. Accessed 12 Dec 2019

14. Xilinx Inc.: Timing closure user guide (2019). https://www.xilinx.com/support/documentation/sw_manuals/xilinx14_7/ug612.pdf. Accessed 12 Dec 2019
15. Xilinx Inc.: Floorplanning methodology guide (2019). https://www.xilinx.com/support/documentation/sw_manuals/xilinx14_7/Floorplanning_Methodology_Guide.pdf. Accessed 12 Dec 2019

Hardware Realization of GMSK System Using Pipelined CORDIC Module on FPGA

Renuka Kajur[1(✉)] and K. V. Prasad[2]

[1] Department of Electronics and Communication Engineering, PESIT, Bengaluru, India
renukakajur@gmail.com
[2] Department of Electronics and Communication Engineering, BIT, Bengaluru, India

Abstract. GMSK system is used in many diverse communication systems and which provides compact spectral Bandwidth and high spectral efficiency for next-generation communication systems. In this manuscript, the cost-effective GMSK system is designed using a Pipelined CORDIC model and realized on the FPGA hardware system. The GMSK system mainly has NRZ Encoder-Decoder, Integrator, and differentiator, Gaussian filter, FM modulator and demodulator along with Channel. The FM modulator and demodulator is designed Using a pipelined CORDIC model and Digital Frequency Synthesizer (DFS). The 8-bit Pipelined CORDIC model is used for In-phase and Quadrature-Phase (IQ) generation along with DFS for arbitrary waveform generation for the formation of IQ modulation. The complete GMSK system is designed using Verilog-HDL on Xilinx ISE 14.7 environment and prototyped on FPGA. The GMSK system implementation results are compared with existing approaches with better improvement in hardware constraints like chip area (Slices) and operating frequency.

Keywords: GMSK · Cordic algorithm · Pipelined architecture · FPGA · Verilog-HDL · DFS · FM modulation and FM demodulation

1 Introduction

Gaussian-Minimum-Shift-Keying (GMSK) system is one of the established and widely used modulation systems with proven spectral efficiency. The main key element of GMSK is Gaussian filtering using a low pass filter (LPF), which controls the Inter-system interference (ISI) without affecting the system performance. The MSK is a popular modulation scheme in Continuous phase modulation (CPM), which offers constant envelope modulation (CEM) with better phase continuity. This MSK Modulation is designed using nonlinear amplifiers, which are mainly used wireless communication applications. The MSK modulation signals are pre-filtered using a narrow band Gaussian filter to improve the MSK spectral efficiency. The GMSK with Bandwidth-time (BT) is applied in many applications [1–3]. The BT product is set to 0.3 for GSM communications, 0.25, and 0.5 for space mission applications, which are standardized by the Space data system (SDS) committee [4]. The GMSK Modem is

R. Silhavy (Ed.): CSOC 2020, AISC 1226, pp. 21–31, 2020.
https://doi.org/10.1007/978-3-030-51974-2_3

used for an automatic identification system (AIS), which includes a GMSK modulator for wireless communication between vessels. The GMSK modulation is the best option in wireless and telecommunication systems because of high immunity to interference and noise, consumed compact power spectrum utilization in mobile Radio, and constant envelope features [5]. The MSK type of modulation is used in the non-aided feed-forward synchronization algorithm to provide better the signal to noise ratio, and symbol timing and provides the robust operation in digital circuits [6]. The GMSK mixed signals meet the separation conditions by using Blind source separation receiving technology (BSSRT) along with Euclidean distance (ED) for the better system performance [7]. The GMSK systems are implemented on low-cost, reliable FPGA for real-time applications [8–10]. In this manuscript, an efficient-cost effective GMSK system is designed using pipelined CORDIC architecture. Section 2 elaborates on the existing approaches towards the GMSK system and CORDIC techniques with its findings. The pipelined CORDIC principles and Hardware architecture are discussed in Sect. 3. Section 4 explains the detailed GMSK hardware architecture. The results and analysis of the Pipelined CORDIC and GMSK system are evaluated with different resource constraints in Sect. 5. The overall works with improvements are concluded in Sect. 6 with future scope.

2 Literature Review

The review of the existing methods towards the GMSK system and are discussed in this section, along with CORDIC techniques. Udawant et al. [11] present the GMSK Technique for image transmission along with other digital modulation techniques like BASK, BFSK, and QPSK. The work analyses the GMSK in GSM at BT product 0.3 and error rate improvement. The radio communication-based GNU radio platform, and GMSK Modulation is explained by Luo et al. [12]. This design GMSK Modulation is used to improve the spectrum utilization and communication quality. It also tests the reliability, compatibility, and accuracy experiment on GNU Radio-HACK RF devices. Zaidi et al. [13] present the GMSK Very high frequency (VHF)/Ultra high frequency (UHF) transceiver system for satellite applications. This work analysis the simulation modeling, thermal analysis transmitter, and receiver characterization with constraint improvements. Lijun-Ge et al. [14] explain the GMSK based Frequency hopping modulation method on FPGA, which includes DFS based carrier generator using Frequency hopping technique for phase generation, along with phase accumulation and CORDIC module for cosine generation. Zhu et al. [15] present the GMSK system using LDPC coded Non-Recursive continuous Phase encoder (CPE) for the space transmission system. It includes GMSK Modulator using Non-recursive CPE (NRCPE) and memory less modulator and GMSK Demodulator and also analyze the error rate between Recursive and Non-recursive CPE.GMSK based Viterbi demodulator is designed for Satellite- automatic Identification system (AIS) by Li et al. [16], which includes 1-bit differential detection using GMSK for easy access of AIS in real-time scenarios. It also analyses the bit error rates for different GMSK demodulation with improvement in differential Viterbi based GMSK demodulator.

The Pipelined CORDIC module for 3-DOF inverse kinematics calculation by Evangelista et al. [17] on the FPGA platform is designed. The pipelined CORDIC is designed based on state machines to achieve low latency and high throughput. The architecture looks complex, which affects the overall computational performance. Meenpal et al. [18] present the multiplexor based CORDIC for signal processing application on FPGA, which includes three and four-stage iterations that are replaced by Multiplexors to improve the critical path delay and operating frequency. The Diwei-Li et al. [19] present the Out-phasing Modulator based Unrolled and pipelined CORDIC architectures are designed with hardware constraint's improvements in Area efficient, low power, and high throughput. The low-latency pipelined CORDIC module is designed by Lakshmi et al. [20]. The CORDIC Module acts as Rotator and which predicts the direction and final coordinates between numbers of iterations. The sign pre-computation block is designed using the ROM block, which consumes more chip area in real-time implementations. Changela et al. [21] present the Asynchronous pipelined CORDIC architecture on FPGA, which provides high throughput and consumes low power utilization on real-time FFT applications.

3 Pipelined CORDIC Architecture

In general, the CORDIC algorithm is used to compute trigonometric sine, cosine, and Tangent values along with hyperbolic values. The conventional CORDIC algorithm consumes more iteration stages for the sine and cosine generation, and it consumes more chip area and power. The pipelined CORDIC model speeds up the computations by inserting the pipelined registers between the iterative stages.

In general, the model has Vector 'V' and having a coordinate's x_i and y_i is rotated by an angle θ for each stage is

$$\begin{aligned} x_R &= x_i \cos\theta - y_i \sin\theta \\ y_R &= x_i \sin\theta + y_i \cos\theta \end{aligned} \tag{1}$$

For the initial iteration stage, the $xi = 1$ and $yi = 0$, then $x_R = cos\theta$ and $y_R = sin\ \theta$. simply the Eq. (1) for the generation of sine and cosine angle values in rotation mode and it is represented as

$$\begin{bmatrix} x_R \\ y_R \end{bmatrix} = \cos\theta \begin{bmatrix} 1 & -\tan\theta \\ \tan\theta & 1 \end{bmatrix} \begin{bmatrix} x_i \\ y_i \end{bmatrix} \tag{2}$$

As per Eq. (2), for one rotation computation, it needs a minimum of 4 multipliers along with some adders and subtractors. The small arbitrary angles are introduced $\tan\theta_i = 2^{-i}$ for $i = 0, 1, \ldots n$, for replacing the multiplication operation. The ith stage iterations are calculated by using the following Eq. (3), and it represented as

$$
\begin{aligned}
x_{i+1} &= x_i - S_i 2^{-i} y_i \\
y_{i+1} &= y_i + S_i 2^{-i} x_i \\
z_{i+1} &= z_i - S_i \tan^{-1}(2^{-i})
\end{aligned}
\tag{3}
$$

In our pipelined CORDIC, the 'i' is set to six-stage iteration is processed in rotation mode. The Sign '*Si*' is set to –1 for clockwise and +1 for anticlockwise direction. The hardware architecture of the 8-bit pipelined CORDIC is represented in Fig. 1. It has pipelined registers (X, Y, and Z), adder/sub tractors, shifters, and counter for constant tangent values. When the clock signal is activated and reset is high, set the Initial stage (i = 0), to $x_0 = 1$ and $y_0 = 0$ and $z_0 = \theta$, to starts the pipelined CORDIC Process. The results of sine and cosine values initial stage are discrete in form, and it is challenging to realize the on hardware. So tune the X_i and Y_i values to avoid the fractional values by dividing the 2^6 by scaling factor 1.647 for next stage iterations. Where 2^6 is the maximum number of iterations is a CORDIC process. The initial values are stored in pipelined registers and shift by i-bits (where i = 0 to 5). The shifter is used to replace the division of X_i and Y_i by 2^i. The intermediate stage process the new vector values, which rotates iteratively to generate the desired angle Z_i. The sign S_i of the addition and subtraction is decided by the Z_i. If the $Z_i$is less than zero, then sign S_i set to -1 for subtraction else $Z_i$is more significant than zero, then sign S_i set to +1 for addition. The shifted outputs and pipelined register outputs are inputs to the addition/subtraction in iterative stages. The tangent values are processed based on countering operations, and the values are generates based on Tan^{-1} (2^i). Addition/subtraction feedback outputs generate each stage sine and cosine output values along with inputs.

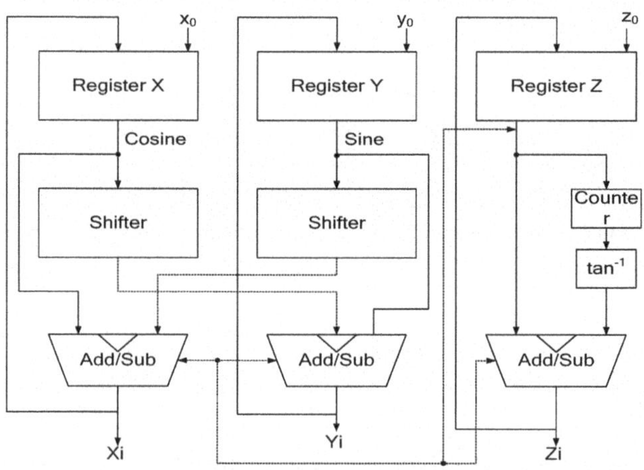

Fig. 1. Hardware architecture of the pipelined CORDIC model

4 Proposed GMSK System

The GMSK is a popular modulation technique and mainly used in GSM for high spectral efficiency. The complete hardware architecture of the GMSK system is represented in Fig. 2. It mainly consists of GMSK Modulator, Channel, and demodulator. The GMSK Modulator has mainly NRZ (Non-Return-to-Zero) Encoder integrator, Gaussian filter, followed by FM Modulator. Similarly, the GMSK demodulator has NRZ Decoder, differentiator, followed by FM Demodulator. The FM modulation and demodulation are modeled using Pipelined CORDIC and Digital Frequency Synthesizer (DFS).

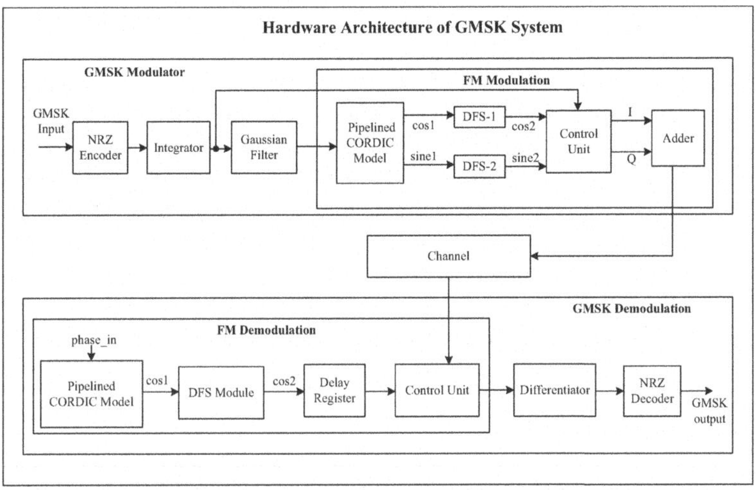

Fig. 2. The complete hardware architecture of the GMSK system

The GMSK Modulator receives the 1-bit GMSK input sequentially in binary form and input to NRZ encoder, which encodes based on NRZ Logic. NRZ encoder has an XOR module and Data flip-flop (D-FF) module. The XOR module receives GMSK input along with feedback D-FF output and stores the XOR output temporarily in D-FF either as high '1' or low '0' value, until next data transaction and mainly used for slow speed communications. The integrator receives 1-bit NRZ encoded data and has four 1-bit temporary registers. The fourth register output added with NRZ encoder output to generate a 1-bit Integrator output. The integrator output is input to the Gaussian filter and used in FM Modulator for controlling the modulated signals. The Gaussian filters are used to reduce the group delays in the GMSK system, and it is considered as 8- tap FIR Filter in design. The 8-tap FIR Filter has 8-multipliers, 15-Delay elements, and 7-adders along with 8-bit Gaussian co-efficient values. The integrator output is multiplied with 8 Gaussian coefficient values parallelly and stored in temporary delay elements. The present delay is added with the next delay element, and these outputs will be added again with the next delay element, and it continues until the last Gaussian coefficient. The Gaussian filter output is input to the FM Modulator.

The FM Modulator has a Pipelined CORDIC Model for In-phase (I) and Quadrature (Q) Phase Generation and two DFS for IQ modulation. The Pipelined CORDIC model is explained in Sect. 3, and it receives the Gaussian filtered output as a phase angle and generates the sine and cosine signals, which are input to two separate DFS architectures for modulation (IQ) signals. The control unit works based on the integrator signal, if the signal is set high '1', then cosine signal is generated as in-phase modulated data else signal is low '0', sine signal is generated as Quadrature-phase-modulated data. The IQ modulated signals are added by adder for the formation of the GMSK Modulated signal.

The Random number sequences are generated using standard linear feedback shift register (LFSR) in Channel. These are similar to AWGN channel generation in real-time hardware scenarios. The LFSR is generated based on Galois field polynomial P (x) = $1 + X^2 + X^5$. The Random number sequences are XORed with GMSK modulated signals for the generation of a corrupted modulation signal. The GMSK demodulation receives the corrupted modulation data for recovery using FM demodulation and Differentiator, followed by NRZ decoder. The FM demodulator has pipelined CORDIC for cosine signal (cos1) generation followed by DFS for waveform signal generation (cos2). The delay unit synchronizes the DFS output signal with a corrupted GMSL modulated signal. The differentiator is the reverse process of the integrator, and instead of addition, subtraction has used the process. The NRZ decoder decodes the differentiator data and generates the original GMSK signal as an output.

5 Results and Analysis

The proposed GMSK system using Pipelined CORDIC is synthesized and implemented on a low-cost FPGA device. The simulation analysis is carried out using Modelsim 6.3f simulator. The present work is designed using Verilog-HDL on Xilinx 14.7 ISE environment. In this results section, the pipelined CORDIC model and Proposed GMSK transceiver system, synthesis, and implementation results are analyzed along with comparative analysis of proposed work with existing approaches on the same FPGA device are represented with improvements.

5.1 Pipelined CORDIC Model Results

The Pipelined CORDIC Model synthesis results are analyzed in terms of Area (Slices), and Time using Different FPGA Devices like Spartan -3E, Artix-7, and Zynq are tabulated in Table 1 and graphical representation in Fig. 3. The low-end Spartan -3E FPGA operates at 90 nm Technology, whereas high-end Artix-7 and Zynq FPGA devices are operate using 28 nm. The Resources like Slice Registers, LUT's, and LUT-FF pairs are consumed more chip Areas in Spartan-3E than High-end Artix-7 and Zynq FPGA devices. The Artix-7 FPGA utilizes 146 Slice registers, 331 slice LUTs, and 124 LUT-FF pairs. The pipelined CORDIC model in the Zynq device operates at

162.084 MHz, whereas on Spartan -3E and Artix-7 operate at 68.03 MHz and 157.58 MHz respectively.

Table 1. Resource Utilization of Pipelined CORDIC model on different FPGA devices

Resource utilization	Spartan 3E	Artix-7	Zynq
Device	XC3S250E-5 FT256	XC7A100T-3 CSG324	XC7Z020-1 CLG484
Area			
Slice registers	412	146	146
Slice LUTs	201	331	331
LUT-FF pairs	789	124	129
Time			
Minimum period (ns)	14.698	6.346	6.17
Max.Frequency (MHZ)	68.034	157.588	162.084

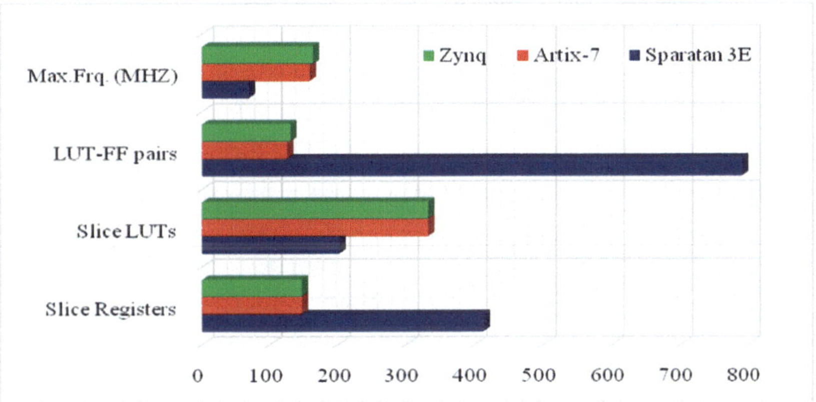

Fig. 3. Graphical representation of resource utilization of pipelined CORDIC model

The performance of Pipelined CORDIC is measured in terms of Latency and Throughput. The latency is calculated in terms of Clock cycles and for the pipelined CORDIC uses 8-clock cycles. 1 clock cycle is set to 10 ns. The throughput of Pipelined CORDIC is 157.588 Mbps on the Artix-7 FPGA device.

The resource utilization of pipelined CORDIC model in terms of Slice registers, Maximum frequency (MHz), and power are compared with an existing similar Approach [22] with improvements on the same Artix-7 FPGA device are tabulated in Table 2. The Proposed pipelined CORDIC utilizes 75.25% less area (Slices) overhead, maximum operating frequency improved 40.5%, and Utilizes 46.7% less power overhead than the existing similar CORDIC model [22].

5.2 GMSK System Results

The GMSK system synthesis and implementation results are represented in terms of Area (Slices), Time and Power using Different FPGA Devices like Spartan-3E, Artix-7, and Zynq are tabulated in Table 3. The GMSK System consumes more Chip Area in terms of Slice Registers, LUT's, and LUT-FF pairs in Spartan-3E than High-end Artix-7 and Zynq FPGA devices. The GMSK System utilizes the almost same amount of chip area in Artix-7 and Zynq because both these devices are processed on 28 nm Technology. The Artix-7 FPGA utilizes 248 Slice registers, 581 slice LUTs, and 215 LUT-FF pairs. The GMSK system operates at 163.666 MHz in the Zynq device, whereas on Spartan -3E and Artix-7 operates at 68.97 MHz and 158.88 MHz, respectively. The GMSK system utilizes a minimum amount of total power 0.053 W in Spartan-3E, 0.092 W in Artix-7, and 0.123 W in the Zynq FPGA device.

Table 2. FPGA resources comparison of proposed CORDIC with previous [22] CORDIC model

Resources	CORDIC [22]	Proposed CORDIC
Slice registers	590	146
Max.Frequency (MHz)	94	157.588
Power (mW)	169.35	90
Device	Artix-7	Artix-7

Table 3. Resource utilization of GMSK system on different FPGA devices

Resource utilization	Spartan 3E	Artix-7	Zynq
Device	XC3S250E-5 FT256	XC7A100T-3 CSG324	XC7Z020-1 CLG484
Area			
Slice registers	936	248	248
Slice LUTs	542	581	581
LUT-FF pairs	1780	215	217
Time			
Minimum period (ns)	14.499	6.294	6.11
Max.Frequency (MHZ)	68.972	158.887	163.666
Power			
Dynamic power (W)	0.001	0.01	0.01
Total power (W)	0.053	0.092	0.123

The GMSK system graphical representation of resource utilization is shown in Fig. 4 for different FPGA devices. The Overall GMSK system utilizes more area (Slices) in Spartan -3E than then the other two devices. The proposed GMSK system is compared with existing similar GMSK systems [23, 24] with improvement, and it is tabulated in Table 4. The proposed GMSK system utilizes 71.3%less overhead for slice registers and 31.4% overhead for Slice-LUT's than the previous GMSK approach [23] on Zynq FPGA. Similarly, the GMSK System utilizes 69.2% less overhead for slice registers, 67.5% overhead for Slice-LUT's, and 60.6% less overhead for LUT-FF pairs than previous GMSK approach [24]. The DSP Multipliers are used around 17 in GMSK [23] and 38 in GMSK [24]. The Proposed GMSK system is not utilized any DSP multipliers.

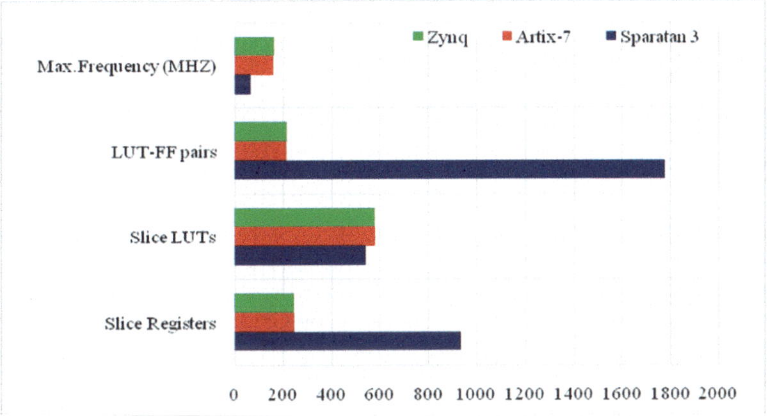

Fig. 4. Graphical representation of resource utilization of GMSK system

Table 4. FPGA resources utilization comparison of proposed GMSK with existing systems [23, 24]

Resources	GMSK [23]	GMSK [24]	Proposed GMSK
Slice registers	867	807	248
Slice LUTs	847	1793	581
LUT-FF pairs	168	552	217
BUFG/BUFGCTRLs	2	3	1
DSP multipliers	17	38	0
Power (mW)	109	NA	114
Device	Zynq	Virtex-5	Zynq

6 Conclusion and Future Work

The GMSK system using pipelined CORDIC and DFS is designed and implemented on a Low-cost Artix-7 FPGA device. The GMSK system used for high spectrum efficiency in GSM communications. It mainly contains a GMSK modulator and a demodulator along with the channel. The FM modulation and demodulation are modeled using Pipelined CORDIC for IQ generation and DFS for IQ Modulation signals. The pipelined CORDIC improves the latency of the overall GMSK system. The GMSK system and Pipelined CORDIC models are synthesized individually for hardware constraint analysis. The hardware constraints like Area (Slices), Time, and Power utilization are evaluated. The pipelined CORDIC model is compared with the existing CORDIC with an improvement of 40.5% in frequency and 46.7% in power utilization. The GMSK system is compared with the previous GMSK system with better improvements in a resource like Slice registers, Slice LUT's, and LUT-FF pairs. In the future, the GMSK system is designed using an Optimized CORDIC model for better improvements in Chip area and other performance computations.

References

1. Yang, R.H.-H., Chern, S.-J., Shiu, G.-C., Lee, M.-T.: Space-time coded GMSK for wireless communication. In: 2005 International Symposium on Intelligent Signal Processing and Communication Systems, pp. 413–416. IEEE (2005)
2. Bauer, I., Gordan, S., Borivoj, M.: Modeling of GMSK communication systems for educational purposes. In: Proceedings ELMAR 2006, pp. 267–272. IEEE (2006)
3. Hietala, A.W.: A quad-band 8PSK/GMSK polar transceiver. IEEE J. Solid-State Circ. **41**(5), 1133–1141 (2006)
4. Puengnim, A., Thomas, N., Tourneret, J-Y., Guillon, H.: Classification of GMSK signals with different bandwidths. In: 2008 IEEE International Conference on Acoustics, Speech and Signal Processing, pp. 2013–2016. IEEE (2008)
5. Souissi, M.G., Grati, K., Ghazel, A., Kouki, A.: Software efficient implementation of GMSK modem for an automatic identification system transceiver. In: 2008 Canadian Conference on Electrical and Computer Engineering, pp. 000601–000606. IEEE (2008)
6. Gudovskiy, D.A., Chu, L., Lee, S.: A novel nondata-aided synchronization algorithm for MSK-type-modulated signals. IEEE Commun. Lett. **19**(9), 1552–1555 (2015)
7. Sen, D., Yang, Y., Cui, P., Guo, B.: Research on separability of GMSK mixed signals based on modulation parameters. In: 2015 IEEE 5th International Conference on Electronics Information and Emergency Communication, pp. 175–177. IEEE (2015)
8. Babu, K.M.N., Vinaymurthi, K.K.: GMSK modulator for GSM system, an economical implementation on FPGA. In: 2011 International Conference on Communications and Signal Processing, pp. 208–212. IEEE (2011)
9. Lee, J.F.M., Montenegro, J.F.P., Morales, C.M., Parrado, A.L., Gutiérrez, J.J.G.: Implementation of a GMSK communication system on FPGA. In: 2011 IEEE Second Latin American Symposium on Circuits and Systems (LASCAS), pp. 1–4. IEEE (2011)
10. Jhaidri, M.A., Laot, C., Thomas, A.: Nonlinear analysis of GMSK carrier phase recovery loop. In: 2016 International Symposium on Signal, Image, Video and Communications (ISIVC), pp. 230–235. IEEE (2016)

11. Udawant, S.R., Magar, S.S.: Digital image processing by using GMSK. In: 2016 International Conference on Automatic Control and Dynamic Optimization Techniques (ICACDOT), pp. 695–698. IEEE (2016)
12. Luo, J., Chai, S., Wang, Y., Hu, Z., Zhang, B., Cui, L.: A maritime radio communication system based on GNU Radio_HackRF platform and GMSK modulation. In: 2018 IEEE 18th International Conference on Communication Technology (ICCT), pp. 711–715. IEEE (2018)
13. Zaidi, Y., van Zyl, R.R., Fitz-Coy, N.G.: A GMSK VHF-uplink/UHF-downlink transceiver for the CubeSat missions. Ceas Space J. **10**(3), 453–467 (2018). https://doi.org/10.1007/s12567-018-0217-5
14. Ge, L., Li, X., Wang, X.: Design of GMSK frequency hopping modulation scheme based on FPGA. Am. Sci. Res. J. Eng. Technol. Sci. (ASRJETS) **47**(1), 28–38 (2018)
15. Zhu, H., Xu, H., Zhang, B., Xu, M., Zhu, S.: Design of efficient LDPC coded non-recursive CPE based GMSK system for space communications. IEEE Access **7**, 70654–70661 (2019)
16. Li, S., Chen, L., Zhao, Y.: GMSK viterbi demodulation for satellite-AIS. In: 2018 IEEE 3rd International Conference on Signal and Image Processing (ICSIP), pp. 327–331. IEEE (2018)
17. Evangelista, G., Olaya, C., Rodríguez, E.: Fully-pipelined CORDIC-based FPGA realization for a 3-DOF hexapod-leg inverse kinematics calculation. In: 2018 WRC Symposium on Advanced Robotics and Automation (WRC SARA), pp. 237–242. IEEE (2018)
18. Meenpal, T.: Efficient MUX based CORDIC on FPGA for signal processing application. In: 2019 IEEE International Conference on Electrical, Computer and Communication Technologies (ICECCT), pp. 1–6. IEEE (2019)
19. Li, D., Zhao, D.: High-throughput low-power area-efficient outphasing modulator based on unrolled and pipelined radix-2 CORDIC. IEEE Trans. Very Large Scale Integr. (VLSI) Syst. **28**(2), 480–491 (2019)
20. Lakshmi, B., Dhar, A.S.: Low latency pipelined CORDIC-like rotator architecture. Int. J. Electron. **104**(1), 64–78 (2017)
21. Changela, A., Zaveri, M., Lakhlani, A.: FPGA implementation of asynchronous mousetrap pipelined radix-2 cordic algorithm. In: 2018 International Conference on Current Trends towards Converging Technologies (ICCTCT), pp. 252–258. IEEE (2018)
22. Ramadoss, R., Kermani, M.M., Azarderakhsh, R.: Reliable hardware architectures of the CORDIC algorithm with a fixed angle of rotations. IEEE Trans. Circ. Syst. II: Express Briefs **64**(8), 972–976 (2016)
23. Kajur, R.R., Tejas, S.P., Prasad, K.V.: Efficient hardware design of single carrier GSMK modulator and demodulator for next generation communication using flexible and optimal sub-modules. Int. J. New Innov. Eng. Technol. J. **8**(2), 10–23 (2018)
24. Adarsh, M.A., Adithya, K., Nimbal, A.: FPGA implementation of space-based AIS. In: 2018 International Conference on Wireless Communications, Signal Processing and Networking (WiSPNET), pp. 1–5. IEEE (2018)

Actor-Network Method of Assembling Intelligent Logistics Terminal

Yury Iskanderov[1(✉)] and Mikhail Pautov[2]

[1] The St. Petersburg Institute for Informatics and Automation of RAS, 39, 14-th Line, St. Petersburg, Russia
iskanderov_y_m@mail.ru

[2] Foscote Group, 23A Spetson St., 102A, Mesa Geitonia, 4000 Limassol, Cyprus
mikhail.pautov@foscote.net

Abstract. Actor-network theory based approach was used to assemble a model of logistics terminal viewed as a network of interacting heterogeneous actors. Elements of applied semiotics and logics of action were utilized to formalize some concepts of the actor-network theory. The system of fuzzy indicators was introduced to describe statuses and dynamics of the model system. A virtual automatic device to support coordination and decision making between the heterogeneous actors in the multimodal logistic process (including intelligent material products) was suggested. The paper aims at bringing attention of the AI researchers, MAS theorists, human-machine systems engineers, ergonomists, knowledge engineers, logistics specialists and broader research community to the actor-network paradigm and its applied potential in socio-technological systems research.

Keywords: Actor-network theory · Heterogeneous actors · Applied semiotics · Logic of action · Intelligent logistics · Logistics terminal

1 Introduction

The core principle of the actor-network theory (ANT) is based on the idea that the actions of any human or nonhuman agent (or "actor" in terms of ANT) are mediated by the actions of a set of other heterogeneous actors. This set is informally defined as "actor-network" [16]. The origins of any actor-network are considered a "rhizome" in Deleuze-Guattarian understanding of this term, i.e. a self-organizing multiplicity with totally decentralized structure and no hierarchical relations between its heterogeneous elements having only an initial affinity with each other. It can hardly be identified as a system since it lacks order and never inherits any order from its predecessors or constituents, however the initial affinity between its heterogeneous members helps establish functional links between them [1]. Heterogeneity of actors traditionally understood in the actor-network theory as their belonging to one of the two opposite worlds (human/social or

R. Silhavy (Ed.): CSOC 2020, AISC 1226, pp. 32–44, 2020.
https://doi.org/10.1007/978-3-030-51974-2_4

nonhuman/natural/technological) and ability to form steady networks of human-nonhuman (socio-technological) hybrids (or quasi-objects) needs to be redefined with the recent advent and coming out of the blue of the artificial intelligence. The pendulum of the actor-network theory which was earlier oscillating in the two-dimensional field between the two opposite poles (human and nonhuman) leaving after every sway the two-dimensional hybrids, now acquires the third pole and the third dimension: nonhuman intelligent objects-subjects in their interactions with humans and non-intelligent nonhuman actors [16]. Hence the emerging demand for integration of the actor-network paradigm with the intelligent systems research, multi-agent systems studies, knowledge engineering, ergonomics and other related fields of research and applications [16]. As it is stated in [16], we foresee the trajectory of evolution of the actor-network approach from the descriptive theory created (and further revised) by Latour, Callon, Law and other ANT protagonists, through its formalization and integration with other relevant methods of AI and agent-based research, toward its eventual conversion into a full-fledged applied tool for modelling and simulation of socio-technological systems. Starting paving this way we have demonstrated that the actor-network theory provides new semantics for some core concepts of the multi-agent systems theory [2], and suggested to use the elements of applied semiotics and logics of action (TI, SAL) to formalize some basic concepts of the actor-network theory [16]. Semiotics is considered the ideological nucleus of ANT, the whole theory being viewed as the newest phase of evolution of semiotics toward its object-orientedness. Therefore, in our move to assemble basic formal definitions in the next section we follow the semiotic route of conceptualization: sign – actor – actor-network.

2 Signs, Actors, Actor-Networks

Sign s can be formally defined as a set of the four components: $<n, p, m, a>$, where n is the name of sign s; p is the portrait of sign s corresponding to node $w_p(s)$ of the causal network on the portraits (W_p); m is the meaning of sign s corresponding to node $w_m(s)$ of the causal network of meanings (W_m); a is the ascription (individual sense or attribution) of sign s corresponding to node $w_a(s)$ of the causal network on the ascriptions (W_a). R_n defines the relations on the set of signs; Θ defines the operations on the set of signs based on the fragments of the causal networks where belong relevant sign components. Tuple of the five elements $<W_p, W_m, W_a, R_n, \Theta>$ represents the semiotic model of actor A_1 [3]. We assume that actor A_1 has a predetermined goal $G_1(A_1)$ not achievable by his own efforts due to the existing obstacle(s). In this situation actor A_1 can either abandon the goal or try to achieve it by taking an alternative route (*detour*) through involvement of other actor(s) A_2, A_3… (human or nonhuman) [4, 5]. Together they can either strive to achieve the initial goal (G_1) or choose alternative goals (G_2, G_3…). Return to the initial goal G_1 is only one virtual scenario in a set of alternative scenarios (Fig. 1) [5].

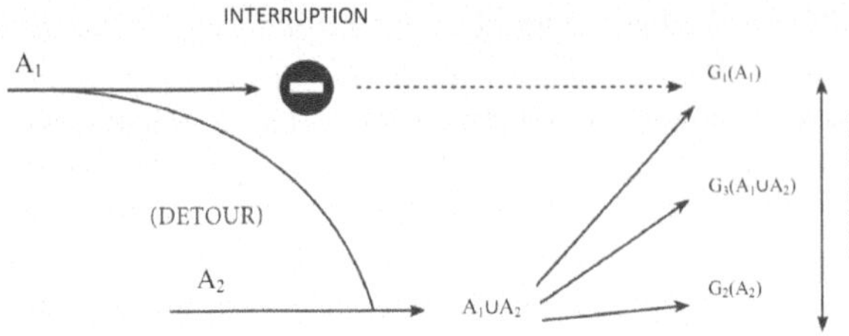

Fig. 1. Translation [5].

Thus actor A_1 together with the mediating actors A_2, A_3... accepted by A_1 after negotiation and transformation form a network which, in turn, is transformed by A_1 [6]. Such network of heterogeneous actors is called actor-network (AN). In terms of ANT network is understood as a work done by actors, i.e. by entities who act or undergo an action [7]. ANT rejects the dualism that tends to separate the social from the material. Under the fundamental ANT principle of generalized ontological symmetry heterogeneous actors share the same capacity for agency [8]. Thus actor-network AN can be formally presented as: $AN = AN_H \cup AN_{AI} \cup AN_{NH}$, where AN_H is the set of human actors, AN_{AI} is the set of AI actors, AN_{NH} is the set of other natural and artificial (technological) actors [2]. Operation of translation in actor-networks is defined as a delegation of powers of representation from a set of actors (actor-networks) to any particular [black-boxed] actor or actor-network in a particular programme of actions: $A = T(A_1,\ldots,A_n)$, where T is the translation of actors $A_1,\ldots,A_n$ to A. In other words, actions of actors $A_1,\ldots,A_n$ (translants) are brought into being or expressed through representative A acting on behalf of the entire actor-network. Prescription index $P(A) \in [0,1]$ of actor A is a fuzzy estimate of possible actions of actor A from the viewpoint of other actors in actor-network AN [2]. More formally, the more complete and determined knowledge actors $AN\backslash A$ of actor-network AN have in regard to actor A the higher the value of $P_{AN\backslash A}(A)$. The less prescribed actors are more easily translatable in the interest of others, than more rigidly prescribed ones [9]. Hence the theorem [2]: *For any actors A_1 and A_2: $P(A_1) < P(A_2) \Rightarrow \tau(A_1) > \tau(A_2)$, where $\tau(A_i) \in [0,1]$ is a quantitative metric to measure ability of actor A_i to be translated.* For the purposes of this ongoing research we find it more logical and convenient to change the expression for $\tau(A_i)$ appearing in [2] to read: $\tau(A_i) = 1 - P(A_i)$.

3 Assembling an Intelligent Terminal

3.1 Terminal Logistics as an Actor-Network

In this paper we reflect on the critical role played by logistics operators in planning, coordination and partial automation of logistic processes aiming at the smoothest passage of people (passengers) and materials (cargoes, mail) through terminals considered as the points of high logistic resistance distributed across transport network. Logistics terminals are the critical nodes (hubs) of all transport networks. We will consider a model multimodal terminal as a scene of interactions between logistics operator and other actors involved in transfer of people and materials. Without going deep into the terminal specifics (considering a model terminal an unopened black-box) we can determine five major actors participating in the process of transition of people and materials through the terminal ($N_{act} = 5$). Those are: Terminal (TERM) combining human (administration, personnel) and nonhuman (buildings and constructions, equipment, technologies) actors; Local Transport (LT) used for inward and outward transfer of people and materials; Long Distance Transport (LDT) represented by ocean, air or rail carriers; Controls (CNTR) represented by a set of controlling organizations (inspections) and Logistics Operator or n-th party logistics provider (nPL, where $n \geq 3$) playing the role of planning actor. We further consider two main stages of the intra-terminal transfer process ($N_{ops} = 2$): the preparatory stage and the main operations stage. We adopt the actor-network approach combined with the methods discussed in [10] for the purposes of our research. Semiotic model of actor-network is used as a basic tool of presentation of knowledge by every participating actor for individual and collective planning and further implementation of the created plans [3]. The planning procedure is based on the actors' vision of the target situation the actors strive to achieve. Knowledge of the potential of the planning actor and other actors is represented by so called causal matrices (semantic networks) on the ascription networks of sign "nPL" (the planning actor) and the signs of other involved actors accordingly. This knowledge helps estimate the prescription indices (and, hence, translatability) of actors as defined in the previous section. The actors create plans aiming to achieve the target situation (goal), distribute their roles based on the criteria of feasibility of actions performed by various actors. Like any other sign "nPL" comprises ascriptions of the planning actor, his portrait and meaning. The meaning of sign "nPL" and the meanings of signs of other actors represent generalized scenarios (actions) where an actor may act as a subject either directly (without mediation) or mediated through other actors. All actions performed by an actor are represented in his ascriptions [3]. The ascription network of the planning actor includes the action matrices as described below in the following section. Signs of other actors are subject to the "class-subclass" relation with the abstract actor-network sign "Others". Signs of the actors contain the knowledge and understanding of other actors based on the general description of the planned and performed tasks. The causal matrix on the network of meanings of the abstract sign "Others" contains references to the signs of other actors [3]. Signs "nPL" and "Others" are symbolically depicted as pyramids with vertices *n, p, m, a* (see Fig. 2 in the next section).

3.2 Actor-Network Model of Multimodal Logistic Process

Logistics Operator nPL playing the role of the planning actor in our case-study mobilizes other involved human and nonhuman actors (represented by sign "Others") to overcome logistic resistance of the system and thus to achieve the target situation (goal) of the logistic process by bringing it to its optimum. The goal of the multimodal logistic process can be formulated in terms of smooth passage of people/materials with minimum stops and delays (logistic effect), minimum cost for nPL and Others (economic effect), minimum adverse impact on the environment (ecological effect), minimum human, material and informational loss or damage (asphalistic effect). For the purposes of this paper we consider a model material (cargo) traffic (**T**) controlled by one general Trading Operator (TO) sourcing cargo from a set of shippers (SH) and distributing it on a set of consignees (C) [10]. TO hires nPL to arrange for logistics operations: cargo transportation, intermodal transhipment at terminal, passage of all necessary controls, and thus appears in our model as the sixth major actor in addition to the five defined earlier. TO's functions comprise: transmission of data and information (orders, instructions, hard/soft documents) $D = \{d_1 \ldots d_u\}$ to nPL necessary for the successful transit of cargo through the terminal; transmission of funds (nPL and third party fees). Also, TO informs nPL of the readiness of Shippers $SH = \{sh_1 \ldots sh_p\}$ to ship their cargoes, and readiness of Consignees $C = \{c_1 \ldots c_m\}$ to receive the inbound cargoes at their premises respectively. The starting point of the logistic process is the transmission of the order ($d_{or} \in D$) by TO to nPL. This order necessarily contains shipping plan for a specific period of time and detailed specifications of cargoes. Also, TO provides nPL with the full set of hard and soft documents (D) required by nPL to organize terminal handling and other logistic operations. Fuzzy indicator $F(D, s) \in [0,1]$ expresses nPL's degree of readiness to start the logistic process. $F(D, s) = 1$ when all the necessary data $d_i \in D$ and funds $s = s_f + s_r + s_p + \sum_i s_i$ are transmitted by TO to nPL (here s_f is the freight cost, s_r is the local transport cost, s_p is the total of terminal fees, s_i is the fee payable to the i-th control). Otherwise $F(D,s) < 1$ and nPL is subject to the K-criterion of the logic of action [11]:

$$[p]nPL \supset \tilde{M}[\sim p]nPL, \tag{1}$$

i.e. actor nPL can either commit action [p] or refrain from committing action [p], where [p] is the launch of the logistic process. Apart from fuzzy indicator F(D,s) in this model we have to take into account another critical factor: the values of translatability factors τ("Others") of other actors which affect the likelihood of an actor to be translated/mobilized by nPL to commit necessary actions and thus achieve the target situation. nPL will most likely refrain from committing [p] in cases when F(D,s) is significantly less than 1, and when nPL cannot ally with other actors involved in the process through *detour*. The bigger the values of F(D,s) and τ(TO,D,s |nPL) the sooner the logistic process is likely to start (here notation "|nPL" means that TO, D, s are translatable toward nPL). Or, in terms of the logic of action:

$$\tilde{M}[p]nPL \;\&\; \sim[\sim p]nPL \supset [p]nPL, \text{ or :} \tag{2}$$

$$\tilde{M}[p]nPL \supset [p]nPL \vee [\tilde{p}]nPL, \tag{3}$$

i.e.: if actor nPL can commit action [p] (i.e. launch the logistic process) and does not refrain from committing this action, he commits it. As it will be shown further these same criteria of the logic of action are applicable to other (fuzzy) indicators used in this model. Based on the TO's orders nPL includes the expected cargoes in the terminal operations plan and mobilizes the Local Transport to have the necessary number of cargo carrying units to move the planned cargo volumes to/from the terminal. Also, the funds received by nPL from TO are distributed to other actors through payment of: freight charges (s_f), terminal operations and service costs (s_p), local transportation fees

(s_r), control fees (s_i). By the moment when planned long-distance freight carrying units (ships, aircraft etc.) arrive at terminal and get ready for discharge/loading, the shippers must have their cargoes delivered to the terminal (directly or through nPL) and transmit to nPL all shipment related data and documents. Consignees must be prepared in all respects for the receipt of their cargoes. Degree of preparedness of shipper sh_i and consignee c_j is characterized by fuzzy indicators $r(sh_i) \in [0,1]$ and $r(c_j) \in [0,1]$ respectively, where: $sh_i \in SH$, $c_j \in C$. Obviously, in case of full preparedness of all shippers and consignees: $r(sh_i) = 1$, $r(c_j) = 1$ for all *i* and *j*. Otherwise, the following situations are possible:

$$\begin{aligned} & r(sh_i) < 1 \vee r(c_j) < 1 \supset [\sim READY]T([READY]I[\sim READY]) \\ & r(sh_i) < 1 \vee r(c_j) < 1 \supset [READY]T([READY]I[READY]) \\ & r(sh_i) < 1 \vee r(c_j) < 1 \supset [\sim READY]T([\sim READY]I[\sim READY]) \end{aligned} \tag{4}$$

Here we use the language of TI-formalism [11, 12] to indicate the status shift of the actor-network in the following standard format:

$$[A]T([B]I[C]), \tag{5}$$

where [A] is the initial status of actors, [B] is the next status when translation of actors is successful, [C] is the next status when translation fails. It is taken for granted that the shippers, consignees, cargoes, funds, documents and shipping data are initially pre-translated to TO, meaning that the value of translatability factor $\tau(SH,C,Cargoes, Funds,D,Shipping\_Data\,|TO)$ allows for such translation. Preparatory stage prior to the handling operations at terminal encompasses transfer of relevant data and funds to the Terminal, Controls and Local Transport expressed by the respective fuzzy indicators: $P(D_p,s_p)$, $S(D_i,s_i)$ and $R(D_r,s_r)$. Interaction between the actors involved in the multimodal logistic process is schematically depicted on Fig. 2 below.

Let us introduce other relevant fuzzy indicators (we use the syntax of the logic of action in the below definitions):

a) Indicator of inclusion of cargo in the terminal operations plan/schedule:

$$PP([incl\_plan]TERM) = \begin{cases} 1, [incl\_plan]TERM \\ \tau(nPL,Cargo,LDT,Shipping\_Data,D_p,s_p|TERM), \\ \quad \tilde{M}[incl\_plan]TERM \\ 0, \sim\tilde{M}[incl\_plan]TERM \end{cases} \tag{6}$$

For transport and terminal safety reasons the next two indicators do not admit any fuzziness and must unequivocally confirm or reject the readiness of the long-distance transport and the terminal for the cargo operations.

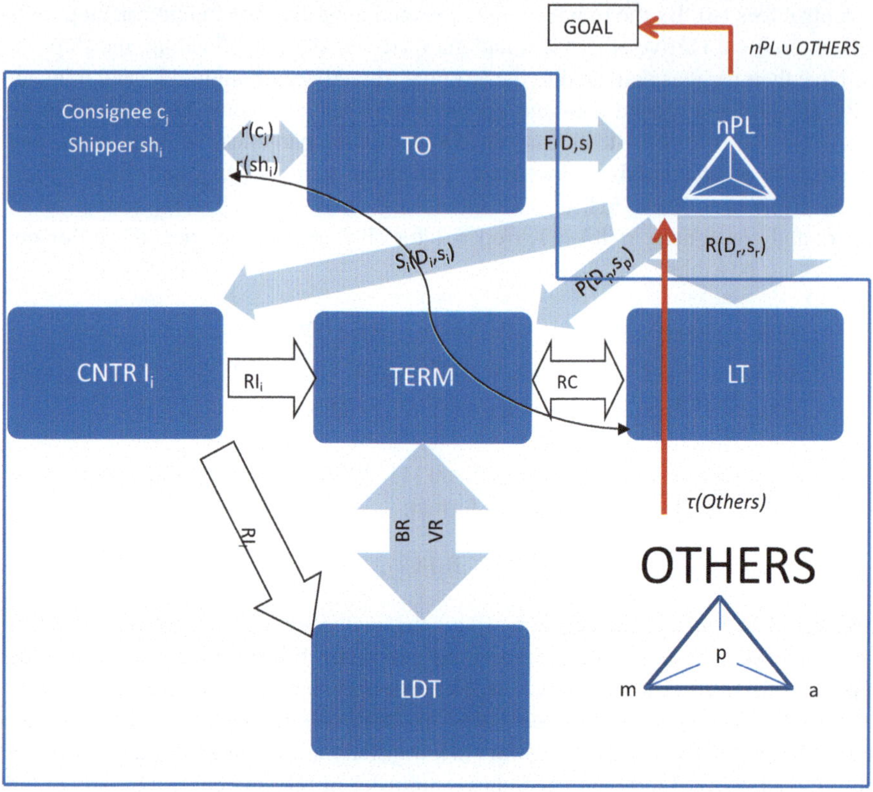

Fig. 2. Interaction of actors involved in multimodal logistic process at terminal.

b) Indicator of LDT readiness for cargo operations at the terminal:

$$VR = \begin{cases} 1, [NOR]LDT \\ 0, [\sim NOR]LDT \end{cases} \tag{7}$$

where [NOR] is the action of tendering the arrival notice by LDT to the terminal (e.g. ship's notice of readiness sent to port authorities, landing report of aircraft commander to the traffic control).

c) Indicator of the terminal readiness for cargo operations:

$$BR = \begin{cases} 1, PP = 1 \ \& \ P(D_p, s_p) > 0 \\ 0, PP < 1 \vee P(D_p, s_p) = 0 \end{cases} \tag{8}$$

The next indicator cannot be fuzzy for the legal reasons and must unambiguously point to successful or failed passage of Cargoes through the i-th control:

d) Indicator of cargo passage through the i-th control:

$$\mathrm{RI_i} = \begin{cases} 1, [release]I_i \\ 0, [\sim release]I_i \end{cases} \tag{9}$$

where [$\sim$ release] means any action by i-th control (I_i) other than cargo release.

e) Indicator of readiness of the local transport (export scenario):

$$\mathrm{RC} = \begin{cases} 1, R(D_r, s_r) = 1 \;\&\; r(sh_i) = 1 \\ \tau(nPL, Shipper, Cargo, Shipping\_Data, D_r, s_r|LT), R(D_r, s_r) < 1 \vee r(sh_i) < 1 \\ 0, R(D_r, s_r) = 0 \;\&\; r(sh_i) = 0 \end{cases} \tag{10}$$

f) Weather indicator:

$$\mathrm{W} = \tau(\mathrm{Weather}\;|\mathrm{Others} \cup \mathrm{nPL}) \tag{11}$$

g) Indicator of Force Majeure circumstances:

$$\mathrm{FM} = \tau(\mathrm{Force\_Majeure}|\mathrm{Others} \cup \mathrm{nPL}) \tag{12}$$

The string $w_T = <$ F(D,s),($r(c_j) \vee r(sh_i)$),$R(D_r,s_r)$,$P(D_p,s_p)$,$S_I(D_i,s_i)$,PP,VR,BR,RC, RI_i,W,FM $>$ reflects the status of the logistic process at moment T. Table 1 [10] shows actor statuses in specific logistic process where TERM = Port, LDT = Vessel, LT = Inland Transport.

As it was demonstrated above:

$$\mathrm{F(D,s)} = 0 \;\&\; \tau(\mathrm{TO, D, s}\;|\mathrm{nPL}) = 0 \supset\; \sim\tilde{M}[\mathrm{p}]\mathrm{nPL} \tag{13}$$

$$0 < \mathrm{F(D,s)} < 1 \vee 0 < \tau(\mathrm{TO,D,s}\;|\mathrm{nPL}) < 1 \supset \tilde{M}[\mathrm{p}]\mathrm{nPL} \;\&\; \tilde{M}[\sim \mathrm{p}]\mathrm{nPL} \tag{14}$$

$$\mathrm{F(D,s)} = 1 \;\&\; \tau(\mathrm{TO,D,s}\;|\mathrm{nPL}) = 1 \supset\; \sim\tilde{M}[\sim \mathrm{p}]\mathrm{nPL} \supset [\mathrm{p}]\mathrm{nPL}, \tag{15}$$

i.e. nPL launches the logistic process permanently coordinating actions with Others based on the changing values of the indicators. Using the language of SAL logic introduced in [13] for the description of multi-agent techno-societies the situation where PP = 1 can be narrated as follows:

$$\mathbf{Agree}_{\{\mathrm{nPL,Port}\}}\mathbf{Sat}_{\{\mathrm{nPL,Port}\}}\mathbf{Do} < \mathrm{nPL} : \left(\left[\mathrm{Transfer}\left(\mathrm{D_p}\right)\right], \left[\mathrm{Pay}\left(\mathrm{s_p}\right)\right]\right), \mathrm{Port} : [\mathrm{incl\_plan}] > \mathrm{T}, \tag{16}$$

Table 1. Statuses of actors of the multimodal logistic process [10]

Symbol	Value	STATUS				
		Vessel	Port	CNTR I_i	Inland transport	sh_i/c_j
F(D,s)	0	STANDBY	STANDBY	STANDBY	STANDBY	STANDBY
	(0,1)	READY → LOADING → SAILING	STANDBY	STANDBY	STANDBY	STANDBY
	1	WAIT FOR: VR = 1; BR = 1	WAIT FOR: $P(D_p,s_p) = 1$; PP = 1; BR = 1; VR = 1	WAIT FOR: $S_I(D_i, s_i) = 1$; BR = 1; VR = 1	WAIT FOR: $(r(sh_i) = 1 \vee r(c_j) = 1)$; $R(D_r,s_r) = 1$	STANDBY
$r(sh_i)$ $r(c_j)$	[0,1)				WAIT FOR: $(r(sh_i) = 1 \vee r(c_j) = 1)$; $R(D_r,s_r) = 1$	STANDBY
	1				WAIT FOR: $R(D_r,s_r) = 1$	READY
$R(D_r, s_r)$	[0,1)				WAIT FOR: $R(D_r,s_r) = 1$	READY
	1				WAIT FOR: BR = 1; $RI_i = 1$	READY
$P(D_p, s_p)$	[0,1)		WAIT FOR: $P(D_p,s_p) = 1$; PP = 1; BR = 1; VR = 1			READY
	1		WAIT FOR: PP = 1; BR = 1; VR = 1			READY
$S_I(D_i, s_i)$	[0,1)			WAIT FOR: $S_I(D_i, s_i) = 1$; BR = 1; VR = 1		READY
	1			WAIT FOR: BR = 1; VR = 1		READY
PP	[0,1)		WAIT FOR: PP = 1; BR = 1; VR = 1			READY
	1		WAIT FOR: BR = 1; VR = 1			READY

(*continued*)

Table 1. *(continued)*

Symbol	Value	STATUS				
		Vessel	Port	CNTR I_i	Inland transport	sh_i/c_j
VR	0	WAIT FOR: BR = 1; VR = 1	WAIT FOR: BR = 1; VR = 1			READY
	1	WAIT FOR: BR = 1	WAIT FOR: BR = 1	WAIT FOR: BR = 1		READY
BR	0	WAIT FOR: BR = 1	WAIT FOR: BR = 1	WAIT FOR: BR = 1		READY
	1	BERTH → WAIT FOR: RI_i = 1	BERTH → WAIT FOR: RI_i = 1;	READY → ACT		READY
RC	[0,1)				WAIT FOR: RC = 1	
	1				READY	READY
RI_i	0	WAIT FOR: RI_i = 1	WAIT FOR: RI_i = 1			READY
	1	READY FOR CARGO OPERATIONS	VESSEL DISCHARGE/LOADING	CARGO RELEASE	CARGO DELIVERY	READY
W	0	SUSPENSION OF CARGO OPERATIONS	SUSPENSION OF CARGO OPERATIONS			
	(0,1]	RESTART/CONTINUE CARGO OPEARTIONS	RESTART/CONTINUE CARGO OPERATIONS			
FM	0	FULL STOP	FULL STOP	FULL STOP	FULL STOP	FULL STOP
	1					

i.e. nPL and Port agree that the situation in which nPL transfers data and documents (D_p) and funds (s_p), and Port includes the cargo in its operations plan is satisfactory for both actors. This matrix reflects possible situations of Port readiness/non-readiness for the cargo operations:

$P(D_p,s_p)$	PP	BR
[0,1)	[0,1)	0
1	[0,1)	0
1	1	0
1	1	1

For successful berthing of Vessel the following criteria must be simultaneously satisfied:

VR = 1, BR = 1, W = 1, FM = 1

For flawless Vessel discharge/loading the following criteria must be satisfied:

VR = 1, BR = 1, $S_I(D_i,s_i) = 1\&RI_i = 1$ (for all involved controls I), W = 1, FM = 1

For the dispatch of import cargoes from Port to Consignees (C) the following criteria must be satisfied: $r(c_j) = 1$ for all $c_j \in C$, RC = 1, $RI_i = 1$ for all relevant import controls. Table 2 [10] shows a summary of indicators reflecting functional interactions between the actors involved in the logistic process as specified in Table 1:

Table 2. Indicators of functional interactions between the actors in multimodal logistic process [10].

	TO	nPL	Port	Inland transport	Vessel	Control I_i	Shipper/Consignee	Environment
TO		F(D, s)						FM
nPL	F(D,s)		P, PP	R		S_i	$r(sh_i),r(c_j)$	FM
Port		P, PP		RC	VR, BR	RI_i		W, FM
Inland transport		R	RC			RI_i	$r(sh_i),r(c_j)$,RC	FM
Vessel			VR, BR			RI_i		W, FM
Control I_i		S_i	RI_i	RI_i	RI_i		RI_i	FM
Shipper/Consignee		$r(sh_i)$, $r(c_j)$		$r(sh_i)$,r (c_j)		RI_i		FM
Environment	FM	FM	W, FM	FM	W, FM	FM	FM	

A virtual automatic device supporting coordination and decision making between the actors in the multimodal logistic process at every particular moment T generates string w_T = <$F(D,s),(r(c_j)\vee r(sh_i)),R(D_r,s_r),P(D_p,s_p),S_I(D_i,s_i)$,PP,VR,BR,RC,$RI_i$,W,

FM > reflecting the current system situation. Also, a set of messages $\{msg_i(w_T)\}_i$ as functions of w_T is generated by the same virtual device at the same moment T. Thus, string w_T = < 1 1 1 1 1 1 1 1 1 0 0.2 1 > generates the following messages (here for the sake of simplification we reduce the total number of controls to 1):

$msg_1(w_T)$ = '*CARGO IS PENDING PASSAGE THROUGH CONTROL*'. This message provides a link to status details uploaded into the system by the control.

$msg_2(w_T)$ = '*WEATHER CONDITIONS ARE LIKELY TO SUSPEND CARGO OPERATIONS*'. This message provides a link to a reliable online weather monitor recognized by all actors.

4 Conclusions

This paper describes a multimodal terminal logistic process as an actor-network, where interacting heterogeneous actors are presented as four-component signs as it was suggested in [3]. If we go one level up, this same approach can be applied to the logistics networks where the individual terminals are the nodes (hubs). The method introduced in this paper represents actor-network theoretical motherboard with formal elements of applied semiotics and action logics "welded" on it. This theoretical assemblage was used to upgrade the approach earlier suggested in [10]. Our method which is still *in statu nascendi* already tends to proliferate from its modest status of a descriptor for logistics systems towards a fully functional tool for future study of techno-info-societies. More information on the actor-network theory underlying the suggested approach can be found in our earlier papers [2, 16]. The suggested virtual automatic device to support decision-making and coordination between the actors is sought to be materialized on the platform of the Internet of Things used for real-time mobile interactions between the actors, including intelligent (smart) cargoes: material products linked to information and rules governing the way they are intended to be stored, prepared or transported which enables these products to be the actors supporting such operations [14, 15]. Our paper aims at bringing attention of the AI researchers, MAS theorists, human-machine systems engineers, ergonomists, knowledge engineers, logistics specialists and broader research community to the actor-network paradigm and its applied potential in socio-technological systems research.

References

1. Deleuze, G., Guattari, F.: A Thousand Plateaus. University of Minnesota Press, Minneapolis (1993). ISBN 0-8166-1402-4
2. Iskanderov, Y., Pautov, M.: Actor-network approach to self-organisation in global logistics networks. In: Kotenko, I., Badica, C., Desnitsky, V., El Baz, D., Ivanovic, M. (eds.) IDC 2019. SCI, vol. 868, pp. 117–127. Springer, Cham (2020). https://doi.org/10.1007/978-3-030-32258-8_14
3. Kiselev, G.A., Panov, A.I.: Sign-based approach to the task of role distribution in the coalition of cognitive agents. In: SPIIRAS Proceedings 2018 Issue 2(57). ISSN 2078-9181 (print), ISSN 2078-9599 (online), pp. 161–187 (2048). https://doi.org/10.15622/sp.57.7

4. Shirokov, A.A.: The politics of explanation and strategy of description of Bruno Latour: how to write infra-reflexive texts. Russ. Sociol. Rev. **18**(1), 186–216 (2019)
5. Latour, B.: On technical mediation: philosophy, sociology, genealogy. Common Knowl. **3** (2), 29–64 (1994)
6. Callon, M.: Techno-economic networks and irreversibility. In: Law, J., (ed.) A Sociology of Monsters, pp. 132–161. Routledge, London (1991)
7. Latour, B.: On actor-network theory. A few clarifications plus more than a few complications. Soziale Welt. **47**, 369–381 (1996)
8. Balzacq, T.: A theory of actor-network for cyber-security. Eur. J. Int. Secur. **1**, 11–23 (2016). https://doi.org/10.1017/eis.2016.8
9. Cordella, A., Shaikh, M.: Actor-network theory and after: what's new for IS research. In: European Conference on Information Systems, 2003-06-19–2003-06-21 (2003)
10. Pautov, M.: Model of transshipment of imported grain cargoes through seaports of Russia. Sci. Technol. Transp. **2**, 55–61 (2015). ISSN 2074-9325
11. Blinov, A.L., Petrov, V.V.: Elements of Logic of Action. NAUKA Publishers (1991). ISBN 5-02-008150-7
12. Von Wright, G.H.: The logic of action – a sketch. In: Rescher, N., (ed.) The Logic of Decision and Action, Pittsburgh, pp. 121–139 (1967)
13. Lorini, E., Verdicchio, M.: Towards a logical model of social agreement for agent societies. In: Padget, J., Artikis, A., Vasconcelos, W., Stathis, K., da Silva, V.T., Matson, E., Polleres, A. (eds.) COIN -2009. LNCS (LNAI), vol. 6069, pp. 147–162. Springer, Heidelberg (2010). https://doi.org/10.1007/978-3-642-14962-7_10
14. McFarlane, D., Giannikas, V., Wong, A.C., Harrison, M.: Product intelligence in industrial control: theory and practice. Ann. Rev. Control **37**(1), 69–88 (2013)
15. Meyer, G.G., Främling, K., Holmström, J.: Intelligent products: a survey. Comput. Ind. **60** (3), 137–148 (2009)
16. Iskanderov, Y., Pautov, M.: Agents and multi-agent systems as actor-networks. In: Rocha, A., Steels, L., van den Herik, J. (eds.) Proceedings of the 12th International Conference on Agents and Artificial Intelligence ICAART 2020, 22–24 February 2020, vol. 1, pp. 179–184 (2020)

A Framework to Enhance ICT Security Through Education, Training & Awareness (ETA) Programmes in South African Small, Medium and Micro-sized Enterprises (SMMEs): A Scoping Review

Mvelo Walaza(✉), Marianne Loock, and Elmarie Kritzinger

University of South Africa, Pretoria, South Africa
mvelo.walaza@gmail.com

Abstract. ICT security has proven to be one of the fundamental needs of organizations across the world. There has been a lot of research that has been conducted on the subject of ICT security. There has also been a lot of initiatives that have been implemented to improve ICT security education, training and awareness in South Africa. With all these efforts in place, there is still an increase in ICT-related crime in the country. ICT-related crime occurs almost on a daily basis in small enterprises in South Africa. There is a need for the enhancement of ICT security in South African organizations. The South African Small, Medium and Micro-sized Enterprises (SMMEs) play a vital role in improving the economy. Literature has shown that SMMEs create a number of employment opportunities thereby contributing to the country's economy. Therefore it is important that these institutions run effectively, are profitable and sustainable. Many small organizations (SMMEs) have been victims of ICT-related crime in South Africa. Most of these crimes happen in these institutions because they do not necessarily have the resources to protect themselves. In some cases, these crimes occur because, due to issues such as financial constraints and lack of skills, ICT security is not one of their priorities. This research aims to explore the literature review that has been conducted; in an effort to propose a framework that will use of Education, Training & Awareness (ETA) programmes to improve ICT security in South African small enterprises.

Keywords: ICT · ICT security · Education · Training & Awareness (ETA) · Framework · SMME · South Africa

1 Introduction

ICT Security is a term that is used to describe the ability to be safe when using resources such as computers, laptops, mobile phones, and many others and to prevent unauthorized access, use, destruction, and modification of information stored in them [1]. Many people across the world make use of these resources for various reasons. Some people use computers and laptops for work-related activities and some use them for personal and entertainment purposes such as gaming and social media. Whatever

R. Silhavy (Ed.): CSOC 2020, AISC 1226, pp. 45–58, 2020.
https://doi.org/10.1007/978-3-030-51974-2_5

the use of these resources may be, there is still a great need for ICT security awareness, training, and education.

Many ICT-related incidents of crime over the past couple of years have shown that ICT security is of great importance. Some researchers have even suggested that it be included as part of the schools' curriculum because literature indicates that there is still a lack of integration of ICT into the South African education system [2, 3]. This means that schools learners are still not taught about ICT security when they are still at school. Many scholars have written papers and published research at academic conferences emphasizing the importance of this field in society. This reiterates the importance of the field of ICT security in modern society.

The crimes that occur as a result of ICT vary from one incident to another depending on what the perpetrators want to achieve. According to Griffiths [4], the advent of ICT over the past number of years has had advantages as well as many disadvantages. Unfortunately many criminals have used the increase in the usage of the internet for their own selfish benefit [5]. The increase in the usage of the internet has not been on par with the awareness, training and education of users regarding ICT security [4]. This leaves many users vulnerable to criminality by perpetrators who are always looking for ways to take advantage of ICT security uneducated and unaware individuals.

The Small Enterprise Development Agency (SEDA) defines a Small, Medium and Micro-sized Enterprise (SMME) as a broad range of companies – some formally registered, informal and non-VAT registered [6]. SEDA further state that these companies range from medium-sized (employing over one hundred people) to informal micro-enterprises. There are many organizations of this nature in South Africa and they are the main focal point of this research.

Government plays an overarching role when it comes to ICT security in their countries, therefore they need to publish appropriate policies that promote ICT security education, training and awareness. National key assets become the most vulnerable in the event of ICT-related crimes like a cyber war. SMMEs are some of these national key assets – in that they play a pivotal role in the country's economy and they are key drivers of economic growth and job creation [6].

Like many other countries around the world, South Africa needs to be prepared for events of such nature. South Africa does not have a national cyber strategy. South Africa is more reactive when it comes ICT-related crime rather than proactive and the national cyber security policy framework is not clear on how to deal with ICT-related crime [7]. This paper depicts the initial scoping literature review that has been conducted for this research.

2 Method

The method that has been used in this research for the literature review process has been scoping reviews. By their nature, scoping reviews are ongoing and they provide a way to interrogate and explore the body of literature. They include work in progress, practitioner reports, credible websites, and case studies [8].

2.1 Search Strategy

In an effort to identify papers relevant to the research, databases such as the IEEE Explore, ACM digital library, Scopus and ScienceDirect were utilized. The papers that were considered relevant were those that have topics related to ICT security, frameworks and models, education, training and awareness in SMMEs. A search using these keywords was also done on Google Scholar filtering it to papers published in the past five years.

2.2 Inclusion and Exclusion Criteria

The initial search criteria that has been used so far in this research included keywords such as "ICT Security", "Frameworks", and "Education, Training & Awareness". Then as literature review was being narrowed, the search criteria included keywords such as "SMMEs", "small businesses", and "South Africa".

The papers that were included were those that are relevant to the topic of this research. These papers either contained a part of the keywords in the heading or were in some way closely related to the angle of the research. The papers that were excluded were those papers that were more generic about ICT security and those that are not written in English. These are papers that are written about ICT security in general and not specifically about any of the keywords.

3 Literature Review

3.1 ICT in South Africa

ICT can be used to increase economic growth and reduce poverty in Africa [9]. South Africa is one of the leading countries when it comes to adoption of ICT on the African continent [10]. Many technology companies have even expanded their businesses to explore avenues in some of the countries in the continent [11]. This is evidence of the good standing of the ICT sector in South Africa.

The ICT sector has seen an increase in the usage of online tools such as for online trading and e-commerce in South Africa [12]. However, Goga, Paelo & Nyamwena [13] state that this increase has been tampered by the lack of adequate network infrastructure, skills and knowledge in this sector. Even though there is a visible digital divide between rural and urban areas, ICT remains one of the sectors that can be used for economic growth in South Africa.

The usage of ICT in South Africa has increased significantly over the last couple of years - there has been an up rise in the use of technologies such as mobile phones, computers, laptops and many more among the citizens of South Africa [10, 12]. This is the reason why large companies such as Huawei, EOH, Amazon and many more are coming to invest in the ICT environment in South Africa. Sutherland [14] states that it is important for the South African government and the private sector to influence the practices of businesses, individuals and households when it comes to ICT security.

3.2 ICT Security

ICT security has become significantly important in organizations because its violation has tremendous consequences. Many organizations value information as a great asset and they go out of their way to ensure that it is protected [15]. ICT security refers to the broader protection of accessing and the usage of resources via a network with the aim of preserving integrity, confidentiality and availability [16]. These resources can be accessed and used using various devices such as mobile phones, computers, laptops, iPads, ATMs, and many more.

The governance of ICT security should be one of the top priorities of any government, simply because they are usually the custodians of their citizens' personal information and therefore cannot afford to be compromised [16]. As the demand increases for government systems to be automated, so is the risk of unauthorized access – hence it becomes vital that there is proper governance (strategies, policies, initiatives) in place to ensure ICT security awareness among the people.

3.3 The Importance of ICT Security

Studies have shown that ICT security plays an important role in both individuals and organizations' perspectives. As a result, Moneer et al. [15] advises that there should be effective information security education, training, and awareness (SETA) so that they can be equipped to deal with ICT related crime. This attests to the importance of ICT security in combating ICT-related crime.

It has become vital to manage ICT security risk in organizations for various reasons, some of those being that a violation – damages the organizations' reputation, the attacks disrupt operations and can be expensive for organizations to endure [17]. Some of the reasons for the continued increase in ICT security violations are that – the sophistication of the attacks has increased at a faster pace, many of the devices that are connected to the internet, the availability of the tools for hacking and online resources, and poor patch management [18].

3.4 Common ICT Security Attacks

ICT security attacks occur in various forms and formats – ranging from social engineering, malware and virus attacks, unauthorized access to data – to the defacing of important websites. There are many different types of ICT security attacks that perpetrators have used. Table 1 outlines some of the common ICT security attacks, as specified by Melnick [19]:

According to Williams, Maharaj & Ojo [20] many of the ICT security attacks occur as a result of perpetrators gaining access to ICT systems through employees. They emphasize the importance of employee behaviour in an effort to curb ICT security attacks in an organization. Most attacks occur as a result of employee behaviour within an organization [21]. Employee behaviour is vital in ensuring that ICT security attacks are kept to a minimum [21]. This reiterates the importance of effective ICT security awareness initiatives in organizations.

Table 1. Common ICT security attacks [19]

ICT security attack	Brief explanation
Malware	- is a software (mostly viruses) attack that gets installed into a computer without the owner's consent
Phishing	- is when attackers send malicious emails to unsuspecting recipients pretending to be people from trusted sources with the intent of acquiring their personal information
DoS	- is when a system's resources are flooded so much that the system is unable to respond to service requests
Man in the Middle (MitM)	- This is when a perpetrator manages to intercept communication between client and server
SQL Injection	- This occurs when an SQL query is executed to the database from client to server in plain text

3.5 ICT Security in South Africa

The responsibility of ICT security within a country lies with the government of that particular country [22]. The South African government has demonstrated a level of interest in the field of ICT security. In 2016, they gazetted a cybercrimes and cyber-security bill which aims to assist, among other things, assist in the prosecution of perpetrators that are involved in cybercrime. Some of the aspects that the bill covers is to impose penalties on those that are involved in cybercrime and to criminalize various actions that perpetrators use during their modus operandi [23].

The South African institutions are constantly under attack from criminals illegally trying to access its online resources [24]. As a result they have adopted the National Cyber Security Policy Framework (NCPF) to assist the government by conducting audits and assessments that will determine the states' readiness to curb ICT-related crimes [25].

Because of the increase in the usage of computers and technology in general, ICT security related crime is on the increase in South Africa [26, 27]. The private sector has played a significant role when it comes to ICT security in South Africa. There are companies (such as ISG-Africa and SABRIC) whose sole mandate is to curb the scourge of ICT security in the country. These companies specialise in ICT security and pride themselves in making sure that ICT security threats are minimized.

In some cases there has been partnerships between government and the private sector in South Africa with the aim to assist with ICT security. Institutions like the Council for Scientific and Industrial Research (CSIR) are funded by both the South African government as well as the private sector to conduct research and come up with solutions to curb ICT-related crime in South Africa. These institutions play a significant role in the field of ICT security in South Africa.

One of the ways that can assist to improve ICT security awareness in South Africa is to include it in the schools' curriculum [28]. Research has shown that when school learners learn about something at school from a young age, they tend to remember and apply it in their everyday lives. This is one of the recommendations made by Kritzinger et al. [29] where they state that ICT security awareness should be considered to be included in the South African schools' curriculum. This research further highlights the importance of stringent ICT security education, training and awareness initiatives.

Even though ICT is seen as a solution to many problems in South Africa, issues of ICT security have not been dealt with adequately [30]. According to Griffiths [4], ICT security has become one of the problems that are threatening national security in South Africa. As a result of the ever-increasing adoption of ICT in Africa and the risks and barriers that are often associated with this adoption [31], there are many companies whose primary core business is ICT security in South Africa. Many of these private companies have explicitly expressed concern over the state of ICT security in the country. Many commercial banking institutions in South Africa are affiliated to the South African banking Risk Information Centre (SABRIC) – which prides itself as a vanguard against banking scams and online banking fraud in South Africa.

The South African private sector plays a significant role when it comes to ICT security in South Africa. Institutions such as ITWeb often host events and conferences where the state of ICT security in the country is discussed [31, 32]. The speakers and key stakeholders in these events are usually individuals and companies with vast knowledge and experience about the ICT security industry. The information that is shared and discussed in these conferences gets published on the internet in various reputable websites and publications. This is but one of the ways that the South African private sector is contributing to the ICT security awareness and education in South Africa.

It is important that the government's systems are secure because they carry critical information about the citizenry of South Africa [24]. The South African government institutions have experienced a number of ICT-related crime over the last decade. There has been a considerable increase in ICT-related crime syndicates that are operating in South Africa. The perpetrators of these crimes are not only targeting privately owned organizations, but they are also targeting government institutions.

South Africa is considered vulnerable when it comes to cybercrime – as a result, there has been a considerable number of ICT-related crime that has occurred in the country in recent times [32]. The scourge of ICT related crime is not only a problem to the South African government, but it is a problem to ordinary citizens as well. An article on BusinessTech showed that an individual was scammed of R140 000 from his bank account through a 'Microsoft scam' [33]. This and many other cases is an indication that there is still a lot that needs to be done when it comes to ICT security in South Africa.

3.6 ICT Security in Other Countries

ICT security awareness, training and education is a phenomenon that should be preached in other countries as well. Even though many organizations have invested significantly in ICT security, there is still an increase ICT-related crime [34]. Global organizations such as Facebook have been victim of syndicates that have attempted to breach their security and access their data [35].

Many government institutions and multi-national organizations host large amounts of critical data – such as personal health information, taxes, and many more [24]. As a result of the growing increase in the usage of ICT across the world, governments are realizing that ICT security is a major concern. According to Sutherland [14], many governments are adopting various strategies in order to combat the scourge of ICT related crime.

There is a shortage of properly trained ICT security professionals that will advocate for ICT security culture around the globe [36]. It is important that organizations employ ICT security professionals that have adequate skills to prevent cybercrime [36]. These professionals should possess enough education and competency to be able to deal with cyber threats and be able to keep up with the pace of the ever-changing technology.

3.7 Dimensions of This Research

The dimensions that have been identified in this research are Awareness, Training, and Education in ICT security for South African SMMEs. This section dissects each of these dimensions in detail.

Education. ICT security education refers to the structured formalised dissemination of information towards the participants – this usually involves a curriculum or the content is accessed online. Grobler *et al.* [26] state that because of the ever-increasing reported incidents of ICT security related crime in the world, more ICT security training has become necessary. There is a need for ICT security education towards all farmers, but they state that there is a greater need for emerging farmers [37].

Many academic institutions in South Africa have conducted ICT security education initiatives in order to address the scourge of ICT-related crime [26]. The education dimension of this research is necessitated by the gap that was identified by these institutions. This dimension ensures that the participants (in this case SMMEs) are well-educated and informed about ICT security.

Training. ICT security training refers to the structured formalised dissemination of information and knowledge towards the participants – the training is usually offered by experts. During their investigation, Spaulding & Wolf [37] found that there is a great need for ICT security training for emerging farmers in the United States of America (USA). ICT security awareness plays one of the roles in the life of an SMME.

This dimension ensures that all stakeholders in SMMEs (employees, management, suppliers and customers) are given sufficient training about ICT security. During their research, Grobler *et al.* [26] have identified places in South Africa where ICT security training is needed; and the CSIR and the University of Venda then collaborated and provided such training. This kind of training would be beneficial for South African SMMEs especially since it has been proven that there is a spike in ICT-related crimes among them.

Awareness. In the context of this research, awareness refers broadly to the establishment of ICT security skills and knowledge towards users in order to protect physical and intellectual property [38]. Awareness is a form of security training with the aim of inspiring, stimulating and building skills and knowledge to system users [35]. The goal of an awareness initiative should be to impart ICT security knowledge to the participants.

There are many ways that ICT security awareness initiatives can be conducted [35]. For these initiatives to be successful, they must be well-planned and well-executed – taking into consideration factors such as the targeted audience, the covered topics, the delivery methods and many more. This research focuses on awareness initiatives that are relevant for SMMEs in South Africa.

Table 2. Analysis of literature

Author (Reference)	**Culture**	**Knowledge (risk assessment, mgmt., …)**	**Awareness (Education & Training)**	**Strategy, Policy, Bill**	**Human Behaviour**	**Technology**	**Governance**	**Frameworks and Models**	**Cyber Attacks (Viruses, Phishing, …)**	**People (school learners, employees, …)**	**Institutions (Private & Public)**	**Challenges**
Bruijn, H. De, & Janssen, M. (2017)			X	X								
Burbidge, M. (2019)									X			
Chandarman, R., & Niekerk, B. Van. (2017)			X								X	
Department of Telecommunications and Postal Services. (2019)		X										
Dlamini, Z., & Modise, M. (2012)			X									
Griffiths, J. L. (2016)		X										
Gundu, T., & Flowerday, S. V. (2013)			X		X							
Kaušpadienė, L., Ramanauskaitė, S., & Čenys, A. (2019)								X			X	
Kelley, D. (2018)					X							
Kreicberga, L.					X						X	

(2010)												
ada, M., & Nurse, J. R. C. (2017)			X							X		
Magalla, A. (2013)									X			
Minister of Justice and Correctional Services. Cybercrimes and Cybersecurity Bill. , 94 § (2017)				X					X			
Minne, D. (2018)									X			
Moneer, A., Humza, N., Atif, A., & Maynard, S. B. (2019)			X		X							
NTIS. (2014)												X
Patrick, H., Van Niekerk, B., & Fields, Z. (2016)											X	
Shaaban, H. K. (2014)							X					
Staff Writer. (2019)									X			
Sutherland, E. (2017)							X					
van Heerden, R., von Soms, S., & Mooi, R. (2016)									X			
Vegter, I. (2019)									X			
Walaza, M., Loock, M., & Kritzinger, E. (2015)			X					X				
Wangen, G. B. (2017)		X						X				
Whitman, M. E. (2018)												
Williams, A. S., Maharaj, M. S., & Ojo, A. I. (2019)					X							
Walaza, M., Loock, M., & Kritzinger, E.			X					X				

(2015)												
UNICEF (2012)						X						
Melnick, J. (2018)									X			
Mashiane, T., Dlamini, Z. and Mahlangu, T. (2019)			X	X								
Banerjee, C., Banerjee, A. and Murarka, P. D. (2013)			X					X				
Spaulding, A. D. and Wolf, J. R. (2018)			X							X		
Barraco, R. A. (2013)			X								X	
Grobler, M. et al. (2011)			X									
Small Enterprise Development Agency (2016)											X	
McKane J. (2019)												
Molosankwe, B (2019)												

4 Analysis of Literature

Table 2 provides an analysis of the literature that has been used in this research. The literature will be categorized according to the various aspects associated with ICT security.

The first column of Table 2 lists the authors of the publication that has been referenced. The rest of the columns are some of the high level dimensions (building blocks) that have been reviewed in this research. The cross (X) that is in the rows indicates whether the publication reviewed (on the left) falls within that particular dimension. At times a publication may fall under more than one dimension. The scoping review table depicts the exact number of publications that have been referenced in this research.

Table 3. Source dates and types ranges

Period of Publication	Conference Proceeding	Book	Thesis/Dissertation	Web site	Journal	Magazine	Newspaper	Speech	Report	Policy Brief	Bill
No Date											
Before 1990											
1990 – 2000											
2001 – 2005											
2006 – 2010		1	2	1	1						
2011 – 2015	4	1	3		2				2		
2016 – 2020	6		3	8	7				2	1	1

5 Source Dates and Types Ranges

Table 3 depicts the review of literature that has been used in this research. It is based on the analysis of literature table in Sect. 3.

The first column of Table 3 lists the year of publication in ranges of five. The rest of the columns are the different types of publications that have been used in this research. The digit shown in the rows specifies the number of publications that have been referenced which fall on that particular year range.

6 Critical Discussion

Literature has shown that the increase in the usage of ICT in the world has brought advantages as well as disadvantages. According to UNICEF [10] and Barraco [39], the increase in the usage of ICT has opened an opportunity for risks associated with its usage. This has prompted and necessitated the study of ICT security and reiterates the importance of ICT security education, training and awareness [40].

There has been a rise in ICT-related crime in South Africa in recent years in both private and public sector. In 2019, it was reported that there had been a DDoS attack in one of the South Africa's major banks [41] as well as the City of Johannesburg

(CoJ) municipality [42]. This rise in ICT-related crime elevates the need for ICT security education, training and awareness.

The aim of this research is to initiate the discussion of the education, training and awareness needs of South African SMMEs. This research will propose a framework that will be used to improve ICT security using ETA initiatives in South African organizations.

7 Gap Analysis

The literature review depicted the current usage of ICT in South Africa. It showed that there is a significant increase in the adoption of ICT in South Africa – but there is still a concern when it comes to ICT security education, training and awareness, especially in South African SMMEs. An extensive literature review was conducted and a gap was identified. The process that was followed in order to identify this gap was to do an analysis of the various spheres within the field of ICT security. This analysis of literature is depicted in Table 2 of this research.

From the analysis of literature where the various spheres were grouped in Table 2 – a few building blocks were identified. The identified building blocks were Frameworks, Education, Training and Awareness (ETA), and Institutions (SMMEs). Literature also guided the decision to select SMMEs as the institution that will be used in this research. All this resulted in the proposal of a framework to enhance ICT security through ETA programmes in South African SMMEs.

8 Conclusion

The aim of this research was to provide a scoping review of the literature that has been conducted related to the topic. This method has been discussed in Sect. 2 of this paper. Section 3 depicts the literature that has been conducted in this research. By its nature, a scoping review examines the nature of research that has been conducted in a field. This examination has been done in Sect. 4 where the analysis of literature that has been used is depicted in Table 2.

The scoping review is work in progress so this research shows the literature review that has been conducted up to so far. The dates of the sources that have been used in this research have been depicted in Sect. 5 in Table 3. A critical discussion about the research that has been conducted is done in Sect. 6, and lastly, the gap analysis found is depicted in Sect. 7 of the research.

References

1. van Heerden, R., von Solms, S., Mooi, R.: Classification of cyber attacks in South Africa. In: IST-Africa 2016 Conference Proceedings 2016, pp. 1–16 (2016)

2. Draper, K.: Understanding science teachers' use and integration of ICT in a developing country context. University of Pretoria (2010). http://upetd.up.ac.za/thesis/available/etd-02032011-132142/unrestricted/thesis.pdf
3. Mdlongwa, T.: Information and Communication Technology (ICT) as a means of enhancing education in schools in South Africa : challenges, benefits and recommendations. Africa Institute of South Africa, Pretoria, pp. 1–8 (2012)
4. Griffiths, J.L.: Cyber security as an emerging challenge to South African National Security. University of Pretoria (2016)
5. Kelley, D.: Investigation of attitudes towards security behaviors. McNair Res. J. SJSU. **14**, 124–139 (2018)
6. Small Enterprise Development Agency. The Small, Medium and Micro Enterprise Sector of South Africa (2016)
7. Vegter, I.: We're horrifically exposed, expert tells CISOs (2019). https://www.itweb.co.za/content/kYbe9MXxax3MAWpG?utm_source=enews&utm_medium=email. Accessed 5 Jun 2019
8. Weiss, M., Botha, A., Herselman, M.: Coming to terms with telemetry: a scoping review. Commun. Comput. Inf. Sci. **933**, 206–222 (2019)
9. Africa Partnership Forum. ICT in Africa: Boosting Economic Growth and Poverty Reduction. 10th Meeting of the Africa Partnership Forum, Tokyo (2008)
10. UNICEF. South African mobile generation. Study on South African young people on mobiles [Internet],pp. 1–47 (2012). http://www.unicef.org/southafrica/SAF_resources_mobilegeneration.pdf
11. Prior, B.: Microsoft and Amazon data centres in South Africa – The latest (2019). https://mybroadband.co.za/news/cloud-hosting/297398-microsoft-and-amazon-data-centres-in-south-africa-the-latest.html. Accessed 3 Jun 2019
12. Kreutzer, T.: Assessing cell phone usage in a South African Township School. Int. J. Educ. Dev. Inf. Commun. Technol. **5**, 43–57 (2009). http://emerge2008.net. Cape Town: e/merge
13. Goga, S., Paelo, A., Nyamwena, J.: Online retailing in South Africa: an overview (2019)
14. Sutherland, E.: Governance of cybersecurity – the case of South Africa. Afr. J. Inf. Commun. 83–112 (2017)
15. Kreicberga, L.: Internal threat to information security - countermeasures and human factor within SME. Lulea University of technology (2010)
16. Shaaban, H.K.: Enhancing the Governance of information security in developing countries: the case of Zanzibar. University of Bedfordshire (2014)
17. Kaušpadienė, L., Ramanauskaitė, S., Čenys, A.: Information security management framework suitability estimation for small and medium enterprise. Technol. Econ. Dev. Econ. **25**, 979–997 (2019)
18. Wangen, G.B.: Cyber security risk assessment practices core unified risk framework. Norwegian University of Science and Technology (2017)
19. Melnick, J.: Top 10 most common types of cyber attacks (2018). https://blog.netwrix.com/2018/05/15/top-10-most-common-types-of-cyber-attacks/. Accessed 25 Sep 2018
20. Williams, A.S., Maharaj, M.S., Ojo, A.I.: Employee behavioural factors and information security standard compliance in Nigeria Banks. Int. J. Comput. Digit. Syst. **8**, 4 (2019)
21. Gundu, T., Flowerday, S.V.: Ignorance to awareness: towards an information security awareness process. SAIEE Afr. Res. J. **104**, 69–79 (2013)
22. Bruijn, H.D., Janssen, M.: Building cybersecurity awareness : the need for evidence-based framing strategies. Gov. Inf. Q. **34**, 1–7 (2017). http://dx.doi.org/10.1016/j.giq.2017.02.007
23. Minister of Justice and Correctional Services. Cybercrimes and Cybersecurity Bill. Republic of South Africa, p. 88 (2017). http://www.dhet.gov.za/Gazette/DHET Research Agenda 19 Aug 2014 Final edited [1].pdf

24. Patrick, H., Van Niekerk, B., Fields, Z.: Security-information flow in the South African public sector. J. Inf. Warf. **15**, 68–85 (2016)
25. Department of Telecommunications and Postal Services. Official website of the South African National CSIRT (2019). https://www.cybersecurityhub.gov.za/. Accessed 2 May 2019
26. Grobler, M., Flowerday, S., von Solms, R., Venter, H.: Cyber awareness initiatives in South Africa: a national perspective. In: Proceedings of South African Cyber Security Awareness Work, Gaborone (2011)
27. Dlamini, Z., Modise, M.: Cyber security awareness initiatives in South Africa: a synergy approach. In: 7th International Conference on Information Warfare Security, Seattle, USA, Academic Conferences International, pp. 62–83 (2012). http://hdl.handle.net/10204/5941
28. Walaza, M., Loock, M., Kritzinger, E.: A pragmatic approach towards the Integration of ICT security awareness into the South African education system. In: Second International Conference on Information Security Cyber Forensics, Cape Town, pp. 35–40 (2015)
29. Kritzinger E, Bada M, Nurse JRC. A Study into the Cybersecurity Awareness Initiatives for School Learners in South Africa and the UK. IFIP WG 118 -10th World Conf Inf Secur Educ. 2017
30. NTIS. Tackling the challenges of cybersecurity in Africa. Economic Commission for Africa, Addis Ababa, pp. 1–6 (2014)
31. Pankomera, R., van Greunen, D.: Opportunities, barriers, and adoption factors of mobile commerce for the informal sector in developing countries in Africa: a systematic review. Electron. J. Inf. Syst. Dev. Countries **85**(5), 1–18 (2019)
32. Burbidge, M.: How vulnerable is South Africa to cyber attack? ITWeb Africa (2019). https://www.itweb.co.za/content/4r1ly7RoW3kMpmda?utm_source=enews&utm_medium=email. Accessed 20 May 2019
33. Staff Writer. This South African lost R140,000 to fraud because his bank didn't act fast enough (2019). https://businesstech.co.za/news/banking/317084/new-case-deals-with-south-african-who-lost-r140000-in-microsoft-scam/. Accessed 20 May 2019
34. Moneer, A., Humza, N., Atif, A., Maynard, S.B.: Toward sustainable behaviour change: an approach for cyber security education training and awareness. In: Twenty-Seventh European Conference on Information Systems, Stockholm-Uppsala (2019)
35. Mashiane, T., Dlamini, Z., Mahlangu, T.: A rollout strategy for cybersecurity awareness campaigns. In: 14th International Conference on Cyber Warfare and Security (2019)
36. Whitman, M.E.: Industry priorities for cybersecurity competencies. J. Colloq. Inf. Syst. Secur. Educ. **6**, 1–21 (2018)
37. Spaulding, A.D., Wolf, J.R.: Cyber-security knowledge and training needs of beginning farmers in Illinois. In: 2018 Agricultural & Applied Economics Association Annual Meeting, Washington (2018)
38. Banerjee, C., Banerjee, A., Murarka, P.D.: An improvised software security awareness model. JIMS 8i-Int. J. Inf. Commun. Comput. Technol. **I**, 43–48 (2013)
39. Barraco, R.A.: Information Security Awareness and Training for Small Businesses. Ferris State University (2013)
40. Walaza, M., Loock, M., Kritzinger, E.: A framework to integrate information and communication technology security awareness into the south african education system. University of South Africa (2017)
41. McKane, J.: South African banks hit by massive DDoS attack. MyBroadband (2019). https://mybroadband.co.za/news/banking/324881-south-african-banks-hit-by-massive-ddos-attack.html?source=newsletter. Accessed 28 Oct 2019
42. Molosankwe, B.: City of Joburg hacking: how it happened (2019). https://www.iol.co.za/the-star/news/city-of-joburg-hacking-how-it-happened-35889367. Accessed 28 Oct 2019

Development of the Pattern Recognition Theory for Solving the Tasks of Object Classification and Yard Processes

Nikolay Lyabakh, Anna Saryan(✉), Irina Dergacheva, Aleksandr Nebaba, Tatyana Lindenbaum, and Victor Panasov

Department of "Informatics", Rostov State Transport University, Rostov-on-Don, Russian Federation
liabakh@rambler.ru, {saryan83,nebaba72}@mail.ru, ira.dergacheva@gmail.com, {saryan83,nebaba72}@mail.ru

Abstract. The necessity and trends of developing the methods of the pattern recognition theory for solving nonstandard tasks of train sorting at the marshalling yard have been substantiated. The examples of similar tasks from other spheres of production and social life have been given. We have proposed to expand the idea of class standard (the standard- line is being discussed in particular) and to introduce the notion of class limits. The method of evaluating the class structure, the mechanism of constructing the standard- line and the model of presenting the compact class limits have been developed.

Keywords: Sorting process · Classification of objects and processes · The type of cut-runner · Theory of pattern recognition · Method of standard · Class structure

1 Introduction

The pattern recognition theory (PRT) has found a wide application in various systems of intellectual functioning. Systems of automated train dismissal at marshalling yards are no exception [1]. The spectrum of applying PRT approaches and methods is rather wide: classification of cuts according to runners' types, choice of the shunter's arrival track to normalize the dismissal process, selection of the braking stage on retarders, choice of a cutter' siftings track with irregular sliding regime, etc. Numerous works [2–5] are devoted to solving the above-mentioned and many other similar tasks. At the same time the formalization of these tasks by means of the PRT is restrained by lacking theoretical models appropriate to practical situations.

The approach using a point as a class standard [2, 3] is well known. However, this approach doesn't suit K_1 and K_2 classes represented on Fig. 1. Let's substantiate the limited possibilities of the existing PRT approaches and methods in the multidimensional feature space (FS) by the following examples:

1. The task of controlling the cut rolling down the hump presupposes the cut to keep the property of a "good" runner, with lower weight (feature x_2) but improved slipping in the bearing (feature x_1) - Fig. 1. That is., the property of the given cut

R. Silhavy (Ed.): CSOC 2020, AISC 1226, pp. 59–68, 2020.
https://doi.org/10.1007/978-3-030-51974-2_6

type is distributed in the feature space FS without concentrating in the point-modelled local area.

2. While controlling the cuts on the retarder, the braking result will be the same (belonging to one class), if increasing the pressing force (feature x_1), one will reduce the time of applying this force to the actuation mechanism (feature x_2).
3. The similar situation can be observed in economy:

- the demand for the product is stable when the higher price is accompanied by the improved quality of the produced goods;
- profit from product selling is constant when the lower product price is accompanied by the higher sales volume.

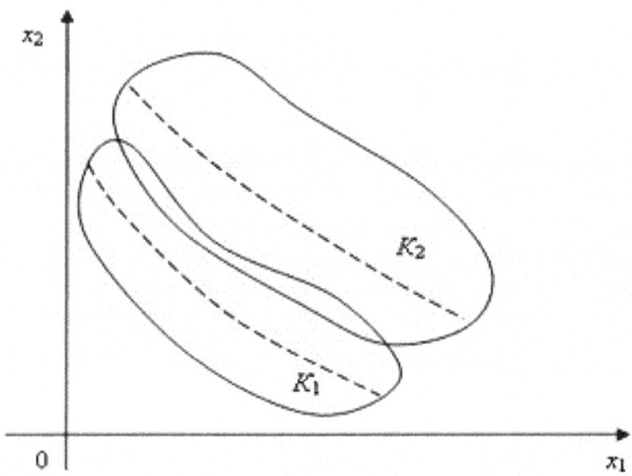

Fig. 1. Graphic representation of poor K_1 and good K_2 runner classes in two-dimensional feature space: x_1 – a bearing type and x_2 – cut's mass

In both the first and the second examples the standards turn to be not points but some lines (shown by a dotted line in Fig. 1). Moreover, the researched classes occupy the restricted FS domain. Beyond its bounds the objects (cuts, products, etc.) begin to possess other properties or they are not characterized by these parameters. In other words, setting the class standard (in any form) appears to be not enough. It is necessary to limit the FS domain where the points of the researched class are concentrated.

In the case of the "distributed" standard it is natural to suppose that the class standard is a certain surface in the multi-dimensional FS. This approach was first proposed in the work [6], and the piecewise-linear approximation was suggested to simplify the standard construction. However, this led to some technical difficulties in identifying the object class belonging (Fig. 2).

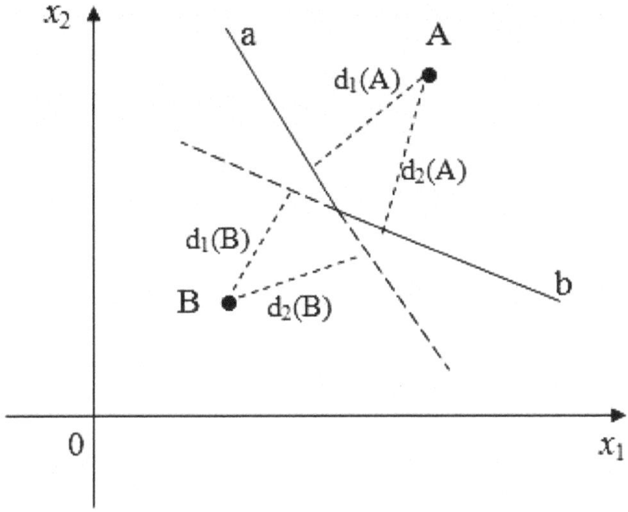

Fig. 2. Graphic illustration of decision-making on the point class belonging in piecewise-linear approximation of line-standard

In fact, the Fig. 2 introduces a certain class standard by means of two broken lines *a* and *b* (continuous sections of appropriate lines). Two researched points *A* and *B* are given. Everything is evident for point *A*: we calculate distances $d_1(A)$ и $d_2(A)$, the minimum one being taken as the distance to the standard. But it is impossible to take the distance from the point *B* to *a* and *b* lines as $d_1(B)$ and $d_2(B)$ present the distances to the standard line continuation (dotted sections), not to the standard itself.

2 Development of PRT

As follows from the given analyses the PRT development in the research under consideration is supposed to be carried out in two directions:

- Broadening the notion of a standard.
- Developing the technology of describing class limits.

2.1 Development of PRT Standard Method

We are going to approximate the line-standard with an arbitrary curve using the self-organization methods in determining its structure (in this particular case one can get a straight line as well) [2, 7]. In this case one can easily construct the aggregate of different structure regressive dependences, choose the optimal one according to a certain criterium. This very dependence will be the class standard. The criteria of line-standard identification can be classified by different features:

- designation of the classified model (development forecast, determining the essence of the researched process and its controlling);

- method of calculating the model parameters (least-squares procedure (LSP), minimizing the sum of theoretical value absolute deviations from the really observed ones, etc.)

Classifying the model in our task one should reflect the properties of the examined class. Thus, the criterium of the unknown dependence classification is to be the requirement of the model unbiasedness [7].

The principle of self-organization for unbiased models is realized in the following way:

- Data retrieval introduced for the unknown dependence identification is divided into two peer parts.
- The aggregate of regressive dependences with different model structure (linear, quadratic, ...) is constructed for every training set.
- The degree of mismatching one-type (of the same structure) models built on different retrievals are found. (The coefficients of the corresponding models are compared).
- The model which is optimal according to a certain comparison test (i.e. providing maximum model coefficient coincidence) is chosen. This model will be the class standard.
- The optimal structure model is recalculated across the data retrieval.

In every case the choice of the model parameter computing method is determined individually on the basis of data error analysis (as a rule, stationary, normally distributed error envisages the choice of LMS (least mean square). It is this very method that is used as a criterium for parameter estimation.

Provided the class element scatter round the line-standard is negligible (class K_1 in Fig. 1), the correlation coefficient $RK_1(x_1, x_2)$ between the corresponding variables will be higher modulo (the sign doesn't determine the variation direction: increase or decrease). In our case $|RK_1(x_1, x_2)| > |RK_2(x_1, x_2)|$. The correlation coefficient decreasing, the standard-line transforms to the standard-point. Thus, the matrix of correlations between the variables of the corresponding multidimensional feature space can testify to the class structure.

Let the illustrated example for three-dimensional case be considered. The data are set by the Table 1 (the first line shows the numbers of points A, the 2nd, 3d and 4th lines give their coordinates x_1, x_2 и x_3 correspondingly).

Table 1. Class of objects in the three-dimensional feature space

i	1	2	3	4	5	6	7	8	9	10	11	12	13	14
x_1	7	10	5	2	7	3	4	12	8	4	5	9	7	5
x_2	4	1	5	7	3	6	6	1	2	4	7	3	5	3
x_3	7	11	6	3	7	2	5	11	8	5	6	8	6	6
$d(Q, A_i)$	0.9	6.6	1.7	6.3	1.4	5.9	3.3	7.9	3.1	2.7	3.2	3.3	1.3	1.7

Let the block $R(x_i, x_j)$ be found:

$$R(x_i, x_j) = \frac{\overline{(x_i - \overline{x_i}) \cdot (x_j - \overline{x_j})}}{\sigma_{x_i} \cdot \sigma_{x_j}} \tag{1}$$

From (1) and Table 1 it goes that
$R\ (x_1, x_2) = -\ 0{,}84$; $R(x_1, x_3) = 0{,}95$; $R(x_2, x_3) = -\ 0{,}85$.
The structure of the researched class will be set by the matrix

$$S = \{R(x_i, x_j)\}. \tag{2}$$

In our case it is

$$S = \begin{Bmatrix} 1; & -0,84; & 0,95; \\ -0,84; & 1; & -0,85; \\ 0,95; & -0,85; & 1. \end{Bmatrix}$$

The matrix analysis shows that the points of the researched class are located relative to the line identified with the standard in the following way:

- In $x_1 0 x_2$ and $x_2 0 x_3$ planes data scattering is approximately the same and is less than in the $x_1 0 x_3$ plane.
- In the first two cases the projections to the corresponding planes represent the decreasing functions, in the third case – the increasing one.

In other words, the researched class presents a "flattened" ellipsoid of revolution.

Let our class center coordinates be found – its "center of gravity", as normal point coordinates of the training set (points of Table 1):

$$Q(x_{1avg}, x_{2avg}, x_{3avg}) = Q(6{,}3;\ 4{,}1;\ 6{,}5).$$

The distances to all points A_i from the class center Q (6,3; 4,1; 6,5) ate given in line 5 of the Table 1.

The maximum distance is: $d_{max} = d(Q, A_8) = 7{,}9$. This distance limits the section of the class occupied space. The average distance is: $d_{avg}\ (Q, A_i) = 3{,}5$.

Then we turn to constructing the standard models. The priori information being reliable, the form of the sought dependence can be set a priori. In more complex cases one should resort to special procedures [2, 7].

In terms of the two-dimensional case set by the Tables 2 and 3 which are synthesized on the basis of Table 1, we can find straight line-standards of K_1 и K_2 classes.

The comparison of Tables 1 and 2 shows that class K_1 is the ellipsoid projection (Table 1) to the $x_1 0 x_2$ plane. According to calculations this ellipsoid projection to the $x_2 0 x_3$ plane will be analogous: $R(x_1, x_2)$ is equal to $R(x_2, x_3)$ approximately.

If it is known that one of the variables is set correctly or has a small tracking error, the regression is formed with the argument coinciding with this variable. Let x_1 be a variable in our case. Then we construct dependences $x_2 = fK_1(x_1)$ on the basis of

Table 2. Class K_1 in two-dimensional x_1 и x_2 feature space

i	1	2	3	4	5	6	7	8	9	10	11	12	13	14
x_1	7	10	5	2	7	3	4	12	8	4	5	9	7	5
x_2	4	1	5	7	3	6	6	1	2	4	7	3	5	3

Table 3. Class K_2 in the two-dimensional x_1 и x_2 feature space

i	1	2	3	4	5	6	7	8
x_1	1	5	1	4	3	1	6	2
x_2	7	1	5	2	2	3	1	4

Table 2 and $x_2 = fK_2(x_1)$ on the basis of Table 3. That is, according to Tables 2 and 3 data we construct regressive (x_2 – function, x_1 – argument) dependences and plot them in Fig. 3.

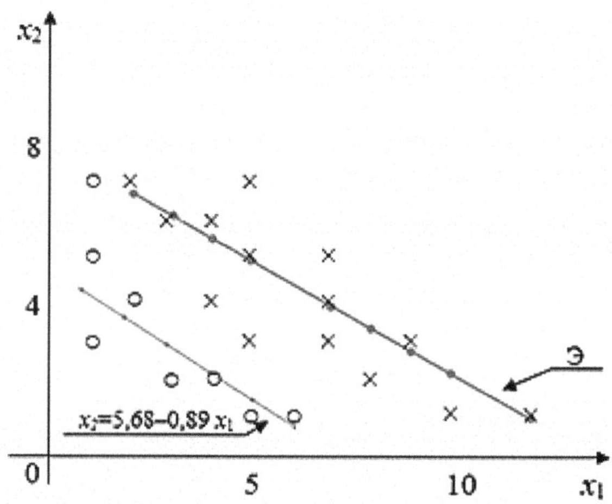

Fig. 3. Graphical task illustration of class K_1 (crisscrosses) and class K_2 (circles) set by Tables 2 and 3 (straight lines-standards)

$$x_2 = fK_1(x_1), x_2 = 7,88 - 0,6x_1. \tag{3}$$

In normal form it looks like:

$$0,52x_1 + 0,85x_2 - 6,74 = 0. \tag{4}$$

Herewith, the approximation error calculated by the formula $\varepsilon = (\sum(x_{2Hi} - x_{2Ti})^2)^{0,5}/n$, (where x_{2Hi} – tracked values (from table), x_{2Ti} – corresponding theoretical values (calculated by formula (3) and n – the number of points) equals ε = 0,279 in the training sample.

$$x_2 = fK_2(x_1), x_2 = 5,68 - 0,89x_1.$$

The error of this approximation is ε = 0,389. This comparatively high error can result from:

- Large scattering of class points. This is an objective class characteristic and it requires no deforming.
- Incorrect type of approximating line. To eliminate this effect, one should apply the above-described logics of constructing complex dependences (choosing the optimal structure model).

Taking into account this procedure recommendations, the model is refined. Quadratic dependence $x_2 = fK_2(x_1)$ gives the expression:

$$x_2 = 0,153\,x_1^2 - 1,891\,x_1 + 6,778, \tag{6}$$

with approximation error ε = 0,365 (Fig. 4).

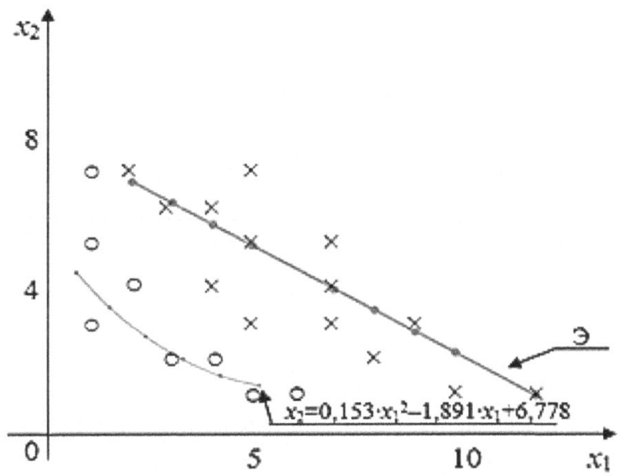

Fig. 4. Graphical task illustration of class K_1 and class K_2 set by Tables 2 and 3 (with quadratic line-standard)

If an equally significant error is tracked in both variables one should apply the method based on minimizing the sum of distances to the sought line [8].

To describe the class for the researched space with more than 2 dimensions one should set several dependences whose crossing determines the class standard. In this case the suitable parametric task of the line-standard will be:

$$x_i = \phi_i(t). \tag{7}$$

In some cases, one of the FS variables is possible to be taken as a t parameter.

2.2 Development of the Class Limit Describing Method

In a number of cases the sufficient class model may present an ellipse. For the form shown in Fig. 5 its equation is set by the ratio:

$$\begin{cases} x_1 = a + m \cdot \cos t, \\ x_2 = b + n \cdot \sin t. \end{cases} \tag{8}$$

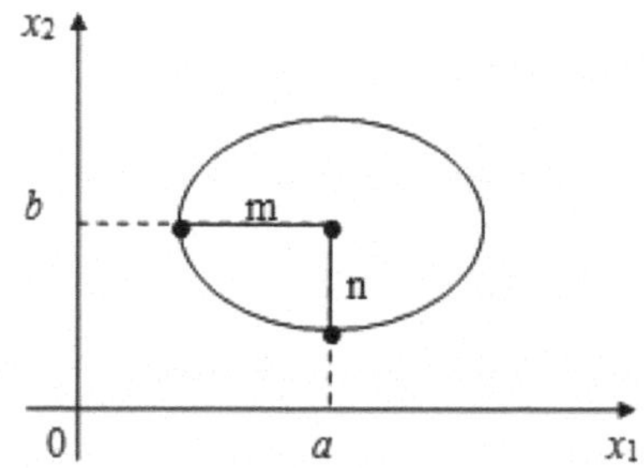

Fig. 5. Ellipse as a class limit model

Excluding parameter t from (8) we have the task of the ellipse represented in Fig. 5 in x_1, x_2 coordinates [9]:

$$\left(\frac{x_1 - a}{m}\right)^2 + \left(\frac{x_2 - b}{n}\right)^2 = 1. \tag{9}$$

The class being situated as it is shown in Fig. 6, the model reduction to a canonical view requires only rotating the axes and transferring the coordinate origin.

This is performed according to formulas [9]:

$$\begin{cases} x_1 = x_1' \cos\alpha - x_2' \sin\alpha + a, \\ x_2 = x_1' \sin\alpha + x_2' \cos\alpha + b. \end{cases} \tag{10}$$

If the axes of the Cartesian coordinate system are moved parallel to a values on Ox_1 axis and to b values on Ox2 axis, and, moreover, they are turned by a angle, the coordinate transformation formulas expressing old coordinates via new ones correspond to the system changes.

Let the ellipse modelling K_1 class in the previous case be considered as an example. The ellipse center is presented by a and b coordinates. We have already calculated: a = 6,3; b = 4,1: O_1 (6,3; 4,1). The value m for the canonical form is calculated

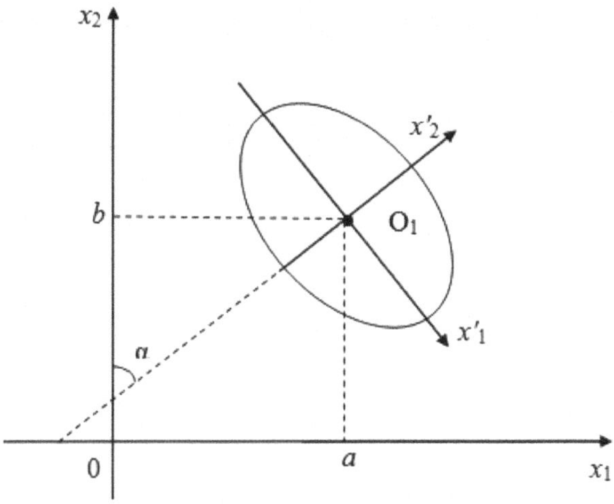

Fig. 6. Coordinates transformation for random class-ellipse

according to the distance between O_1 center and the most distant point A_8: $m = d(O_1, A_i) = 6{,}49$. The n parameter is calculated according to the most distant point from the standard straight line Э set by the ratio (4). This point is $A_{14}(5;\ 3)$: $n = d(Э, A_{14}) = 1{,}59$. Consequently, the ratio (9) in this particular case will be as follows:

$$\left(\frac{x_1^{'}}{6,49}\right)^2 + \left(\frac{x_2^{'}}{1,59}\right)^2 = 1. \tag{11}$$

Or

$$\begin{cases} x_1^{'} = 6,49 \cdot \cos t, \\ x_2^{'} = 1,59 \cdot \sin t. \end{cases} \tag{12}$$

To find the line-standard angle of inclination the ratio (3) will be used. Thus, it follows that tg$\alpha = 0{,}3$. Then $\alpha = 16{,}7°$. The angle tangent will allow define the corresponding sine and cosine functions. We will get sin$\alpha = 0{,}29$; cos$\alpha = 0{,}96$.

These values substituted in (10), $x_1^{'}$ and $x_2^{'}$ are expressed through x_1 и x_2. The final class limit equation will look like:

$$\left(\frac{0,96x_1 + 0,29x_2 - 7,237}{6,527}\right)^2 + \left(\frac{-0,29x_1 + 0,96x_2 - 2,109}{1,599}\right)^2 = 1. \tag{13}$$

3 Conclusions

The following suggestions have been made:

- the class standard is to be a certain surface but not a point of the feature space;
- the class extraction in the feature space requires its limits calculation.

The method of class structure evaluation, the mechanism of constructing regressive standard-line model and the models of presenting compact class limits in the form of a ellipse have been developed.

References

1. Shabelnikov, A., Ivanchenko V., Kovalev S., Lyabakh, N., Sokolov V., Odikadze V., Sachko V.: Hump yard automatization systems on the basis of up-to-date computer Technologies. Textbook for Railway High Schools. In: Shabelnikov, A. (ed.) NIIAS, Rostov State Transport University, Rostov-on-Don, p. 436 (2010)
2. Lyabakh, N., Shabelnikov A.: Engineering Cybernetics on Railway Transport. Rostov State Transport University, SKNTcVSh, Rostov-on-Don (2002)
3. Adadurov S., Gapanovich V., Lyabakh, N., Shabelnikov A.: Railway Transport: Towards Intelligent Management. SSC RAS, Rostov-on-Don (2010)
4. Lyabakh, N., Saryan A.: Automatic adviser for the predicted maintenance of mobile railway objects. Transport: Science, Engineering, Control, № 8 (2018)
5. Shabelnikov A.N., Liabakh N.N., Gibner Ya.M., Saryan A.S.: Automatic adviser on mobile objects status identification and classification. J. Phys.: Conf. Ser. (JPCS). Russian Federation, Tomsk, **1015** (2018)
6. Lyabakh N.N., Gibner Ya.M.: Development of the intelligent system for monitoring the operating conditions of marshalling facilities and objects and the maintenance automatic adviser. In: The 5th International Scientific and Practical Conference "Intellectual Control Systems on Railway Transport. Computer and Mathematical Modeling, ISUZhT -2016, Moscow (2016)
7. Ivanchenko, A.: Inductive method for self-organizing complex system models. Naukova Dumka, Kiev, p. 296 (1981)
8. Orlov, A.: Decision-making. Theory and development of managerial decisions. Engineering and Consulting Center "Mart", Rostov-on-Don, p. 496 (2005)
9. Privalov, I.: Analytic Geometry: Textbook for High Schools. 40th ed. Bachelor and Specialist, ISBN 978-5-534-01262-0, Publishing house, Moscow "URAIT", p. 233 (2018)

A Framework for the Assessment of Information Security Risk, the Reduction of Information Security Cost and the Sustainability of Information Security Culture

S. G. Govender(✉), E. Kritzinger, and M. Loock

University of South Africa (UNISA), Pretoria 0001, Gauteng, South Africa
32393113@mylife.unisa.ac.za,
{kritze,loockm}@unisa.ac.za

Abstract. Information security data breaches are becoming larger and more frequent. Incorporating information security into the culture of the information technology (IT) staff members that support these technologies is a key function that must be considered in parallel to improved security technology. The framework proposed in this paper considers focusing on cost reducing products, services and structures while building the correct behaviour and values in IT staff members, and strengthening their ability to improve information security assessment capabilities in the organization to better support information security management.

Keywords: Information security culture · Information security assessment · Information security cost · Information security evaluation · Information security · The ARCS security framework

1 Introduction

The value of IT has revolutionized the enablement of business and thus brought many advantages to organizations. However, with the expansion of IT there has been significant concern for the information security of these technologies [1]. The information security of any business system and their people continues to remain a top priority for most organizations. Those systems and people present a significant cybersecurity risk [2]. However the inclusion of organizational security objectives, such as information security risk assessments, sustainability of information security cultural and behavioral traits, and reducing cost overruns of information security products and services, is deficient in general strategic planning [3].

Information security breaches have received significant publicity and become more frequent [4]. In the past three years there have been more than six thousand five hundred reported data breaches that have exposed more than nine billion information records [4–6]. Studies conducted by IBM and Kaspersky indicate that even though the cost per breach has been reduced incrementally, with the adoption of better technology,

R. Silhavy (Ed.): CSOC 2020, AISC 1226, pp. 69–84, 2020.
https://doi.org/10.1007/978-3-030-51974-2_7

organizational structures and awareness, the number of breaches has risen [7–9]. This implies that the overall cost to protect an organization has and will continue to increase.

The remediation factors to the major data breach vectors are a combination of social (structural and awareness) and technical (product and service acquisition). With the increased likelihood of breaches, technical solutions are increasingly employed to reduce risk [10]. However, these solutions incur significant cost for the solutions themselves as well as the scarce IT human resources required to manage, maintain, monitor and administer these solutions [11, 12].

In previously published articles the authors of this article have developed three models related to information security assessment, information security culture and information security cost reduction. These are discussed in section three below and are the foundations for the framework discussed in this paper.

2 Information Security Assessments

Security practices in large organizations are notoriously difficult to assess. The challenge only increases when organizations turn to third parties to provide technology and business services, which typically require tight network integration and sharing of confidential data, potentially increasing the organization's attack surface [13]. The security maturity of an organization describes how well it mitigates known risks and responds to new threats. Maturity is typically assessed with audits and questionnaires, which are difficult to quantify, lack objectivity and may not reflect current threats [13].

The underlying goal of an information security risk assessment is to identify and quantify risks related to the organization's information assets [14]. This may include the risk to business functions, image, reputation, organizational assets and people.

The main justification for conducting an information security assessment is so that organizations can take a proactive and repetitive approach to addressing information security concerns. Additionally legal, compliance and regulatory requirements aimed at protecting sensitive or personal data, as well as general public security requirements, create an expectation for companies of all sizes to devote the utmost attention and priority to information security risks. The information gained during an assessment assists an organization in determining the best ways to treat the risk identified.

Further key justifications for conducting an information security assessment have been adapted from [13–15] and are described in Table 1.

Security assessments are conducted so that risks identified can be understood, resolved or otherwise accepted as part of the information risk management lifecycle [15]. The traditional approach to information security assessments is to conduct a security audit, vulnerability test or penetration test [14]. These may or may not be aligned to best practices such as the Centre for Internet Security benchmarks [13] or standards such as ISO 27002 audits [16]. Irrespective of the approach to the assessment, a general process should be adhered to conduct the assessment.

The following information security assessment process described in Table 2. is adapted from [17].

Table 1. Key justifications for conducting information security assessments

Justification	Description
Cost Justification	As expenditure on information security does not provide a positive business case from an earning perspective, a security assessment will advise senior management on the critical technology risks within the organization, which will allow them to focus security spend on these key areas for remediation. In this paper a link is described between key cost reduction drivers and information security assessment in Sect. 3
Productivity	As risks are addressed with technology and systems, there will be greater potential for uptime of these systems. The net effect will be higher productivity for the organization as systems will be functioning for longer periods of time
Breaking barriers	Information security assessment expose risks in systems, technology and processes. People in the IT department are responsible for the technology and systems and people are the key drivers of processes. The risk exposed during an information security assessment breaks down the barriers of assigning risk solely to technology but also to IT staff as well. In this paper, the relationship between information security assessments, and human values and behavior is described in a model developed and documented in Sect. 3
Self-analysis	An information security assessment allows for introspection by the IT team. The risks identified allow the people in this team an opportunity to reflect on their knowledge and expertise. This gives the IT employees an opportunity to take ownership of the information security concerns of their environments. The self- analysis concept and the imbibing of a common thought process is a key feature of the information security culture enhancement model described in Sect. 3

A comprehensive security risk assessment also helps determine the value of the various types of data generated and stored across the organization. Without valuing the various types of data in the organization, it is nearly impossible to prioritize and allocate technology resources where they are needed the most. To accurately assess risk, management must identify the data that are most valuable to the organization, the storage mechanisms of said data and their associated vulnerabilities.

IT security risk assessments is therefore a key method in order for an organization to evaluate its security posture. It allows for organizations to assess, identify and modify their overall security posture and to enable security, operations, organizational management and other personnel to collaborate and view the entire organization from an attacker's perspective. This process is required to obtain senior management's commitment to allocate resources and implement the appropriate security solutions.

Table 2. Information security assessment process

Process step	Description
Support	Any security assessment initiative must be proactive and repetitive. It must be fully supported by senior management and leadership so that the security management team can undertake investigations without intervention from teams being evaluated. The initiatives must also supported by management so that the inputs, processes and outputs are communicated effectively across the IT department
Scope	When conducting an information security assessment all environments within an organization must be scrutinized and assessed with the equal rigor. IT staff may already understand where inherent risks are and try to exclude those environments from the assessment. The correct approach is to assess all internal and external systems. People and processes should also be part of assessments conducted
Testing	Testing must be conducted in a consistent manner using a specific testing benchmark, guideline or best practice. Testing outputs must be repeatable. Key performance Indicators (KPIs) must be defined for the resolution of the risks found, before testing
Reporting	As tools that conducts automated security tests produce large technical output reports, it is necessary to create simpler, clear and concise reports that define the high, medium and low risks found and the potential resolution mechanisms. Reports must easy to understand and highlight key risk areas that of priority for resolution, taking the context of systems and business into account
Resolution	Resolution must be defined by clear action plans that contain the resolution action, the person or team responsible for resolution and the time or date by which this action will be resolved. Resolution is the key action in this process as this is what ultimately assist in reducing information security risk
Oversight	The security team must ensure that security management is ongoing between the routine of conducting security assessments. Controls, policies and procedures must be adhered to during these periods. As the security landscape changes on a daily basis, it important try to improve security incident sand alert management incrementally rather than trying to attain perfection in security. Additionally management must be kept engaged so that compliance and contractual obligations are adhered to

3 Information Security Culture and Cost Reduction

The three models discussed in this section emanate from previously published work by the authors of this paper and are related to information security assessment, information security culture and information security cost reduction.

Studies have shown that the establishment of an information security culture in the organization is necessary for effective information security [18]. However, the security culture cannot be assessed in isolation from the overall corporate culture [19].

In organizations, the information security culture is usually managed through basic awareness and training programs. These usually deal with simple issues and are generally not assessed to provide an honest view of the actual learning taking place. Furthermore, these awareness programs are usually focused on general staff, without a focus on management and, more importantly, the technical IT staff. A security culture is far more important amongst IT staff in an organization compared to general staff, as these IT staff members manage and implement security controls. Sub-cultures are formed within the IT department, reducing the concentration of security efforts but rather focusing on the technical specifics of each environment. In addition, technical staff do not consider security as a primary concern even though technical actions that support information security management, such as patching, updates, application enhancement, and upgrade equipment lifecycle management, are implemented by them. It has been established that technical security measures are key to information security management success [20, 21].

Considering that technical staff have a greater influence on factors that support information security management, it would be reasonable to contend that enhancing the information security behavior and values of these technical staff members would improve information security [22, 23].

A model to enhance information security culture was developed and described by [24] and is depicted below in Fig. 1

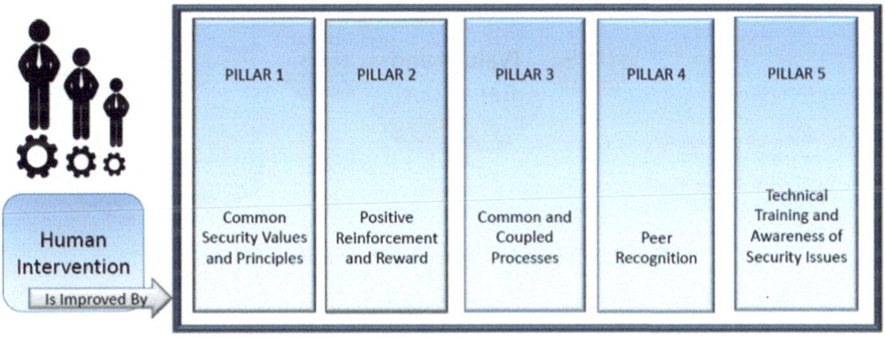

Fig. 1. Five pillar model to enhance information security culture [24]

The economics of information security are of significant interest to senior management and leadership in organizations [25]. In all businesses, cost-benefit justifications for information security investments are important [26]. There has always been a difficulty in quantifying an organization's cost for information security in a comprehensive and comparable way, and information security professionals need to be able to articulate their value in economic terms [27]. Using economic and business administration principles helps to quantify information security management costs [26].

Literature on information security budgeting, indicates that the cost-benefit analysis for information security is difficult and applying traditional IT budgeting techniques is counter-productive [25, 28, 29]. Furthermore, investment is focused on reducing security breaches through technical means [30]. The caveat to these types of technical solutions is that the products are relatively expensive, the expertise to implement and maintain these is scarce and therefore expensive, and most importantly the processes and periodic analysis of information and remediating activities as a result of information generated by these products is dependent on the human intervention, skill and motivation.

In the study conducted by Ponemon, twenty factors were considered as remediation for security breaches [7]. Of these factors, twelve were considered to have an effect of decreasing the cost of a breach whereas eight were considered to increase the cost.

Based on the cost-reducing factors, six of which are socially influenced (human, managerial or structural) and six of which are technically influenced, a model describing these factors and the factors relationship to culture was developed [24] and is described in Fig. 2. Therefore, concentrating the information security management effort on these twelve factors will provide the best information protection at the lowest cost.

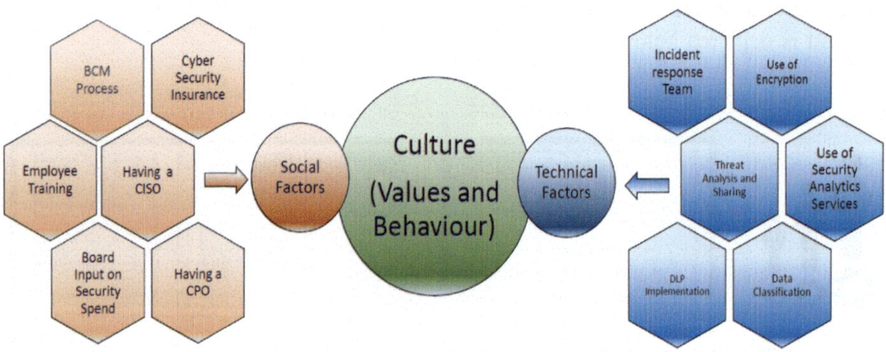

Fig. 2. Cost reducing remediation factors [24]

In Sect. 2 it was established that organizations determine the information security landscape using several methods such as ISO 27001 assessments, risk assessments, or longer-term strategy defined by a security architecture programme. The outcomes of these assessments or programmes are supported by the selection of products and/or solutions that fit the information security and business needs. Information security products and solutions are complex in nature and require many technical inputs to function appropriately or effectively within the organization. Products and solutions purport to be autonomous and self-running but in reality, require significant human input and intervention in order to function in a valuable way for the business. Figure 3 developed by [24] describes the relationship between generalized common information security evaluation methods and the reliance on human resources to run, manage, monitor and maintain information security systems that are identified through these methods.

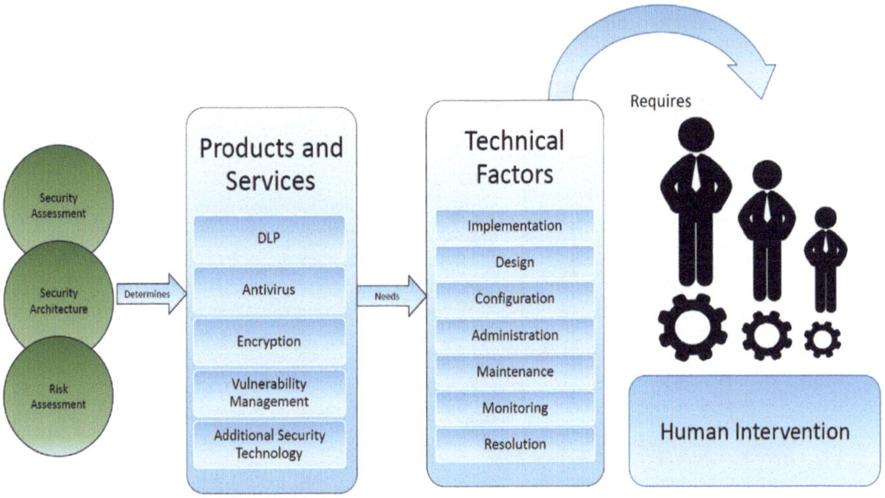

Fig. 3. Information security assessment and human intervention [24]

Security tools are not always managed by the security function within the organization and staff that do manage these solutions are from alternative functional areas within the IT department, i.e. application development, infrastructure, end user computing or networks. The motivation and behavior for these IT staff members to consider security first is generally incongruent with their motivation for their primary job responsibilities. The effect of what staff consider additional work to their primary job responsibility is lower motivation to consider their information security responsibilities as important.

4 The ARCS Security Framework

In Sects. 2 and 3 above it was established that the key drivers to understanding, managing, supporting and changing information security culture are driven by understanding the current state of information security, investing in specific initiatives that help reduce cost while also reducing information security risk, and creating a sustainable approach to managing and improving information security culture.

The framework discussed below combines these three drivers to establish a structured evaluation method, which will assist an organization in addressing and improving information security management.

The framework extracted from the models described in Figs. 1, 2 and 3 is structured into three features (F1, F2 and F3) called **A**ssessment of Information Security Risk (F1), **R**eduction of Information Security **C**ost (F2) and **S**ustainability of Information Security Culture (F3) and is referred to as the ARCS Security Framework as described in Fig. 4.

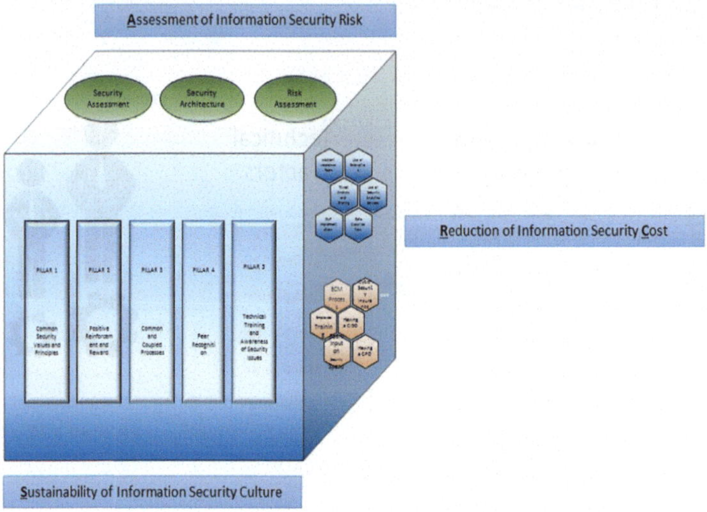

Fig. 4. ARCS security framework

Each feature is broken down into an evaluation area. Each evaluation area is given a shortened tagged description and is broken down further into questions related to that evaluation area as described in Fig. 5.

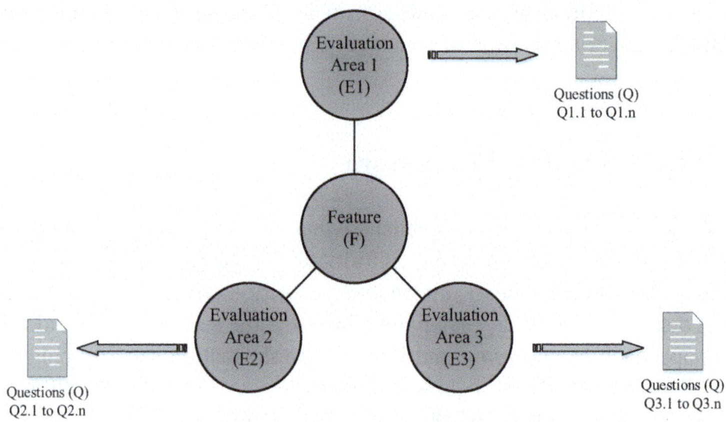

Fig. 5. ARCS framework reference model

The framework compromises of three features, twenty-one evaluation areas and eighty-three questions, which is described in Table 4.

Table 4. Description of the structure of the ARCS framework

Feature name (F)	Evaluation area (E)	Questions (Q)
F1 ASSESSMENT OF INFORMATION SECURITY RISK	**E1.1** - SECASSESS	**Q**1.1.1 to **Q**1.1.6
	E1.2 - SECARCH	**Q**1.2.1 to **Q**1.2.6
	E1.3 - RSKASSESS	**Q**1.3.1 to **Q**1.3.5
F2 REDUCTION OF INFORMATION SECURITY COST	**E2.1** - BCM	**Q**2.1.1 to **Q**2.1.6
	E2.2 - CSI	**Q**2.2.1 to **Q**2.2.4
	E2.3 - EIST	**Q**2.3.1
	E2.4 - HACISO	**Q**2.4.1 to **Q**2.4.2
	E2.5 - BIOSS	**Q**2.5.1 to **Q**2.5.3
	E2.6 - HACPO	**Q**2.6.1 to **Q**2.6.2
	E2.7 - IRT	**Q**2.7.1 to **Q**2.7.4
	E2.8 - UOE	**Q**2.8.1 to **Q**2.8.3
	E2.9 - TAS	**Q**2.9.1 to **Q**2.9.4
	E2.10 - SAS	**Q**2.10.1 to **Q**2.10.4
	E2.11 - DLP	**Q**2.11.1 to **Q**2.11.7
	E2.12 - DC	**Q**2.12.1 to **Q**2.12.2
	E2.13 - ISIC	**Q**2.13.1 to **Q**2.13.6
F3 SUSTAINABILITY OF INFORMATION SECURITY CULTURE	**E3.1** - CSVP	**Q**3.1.1 to **Q**3.1.4
	E3.2 - PRR	**Q**3.2.1 to **Q**3.2.4
	E3.3 - CCP	**Q**3.3.1 to **Q**3.3.2
	E3.4 - PREC	**Q**3.4.1 to **Q**3.4.3
	E3.5 - TTSA	**Q**3.5.1 to **Q**3.5.5
3 FEATURES	**21 EVALUATION AREAS**	**83 QUESTIONS**

The features F1 to F3 and the respective evaluation areas are described below.

Feature 1 (F1) – Assessment of Information Security Risk

The Assessment of Information Security Risk feature supports the evaluation and understanding of the current information security landscape, the risks faced by the organization and the key success factors to help an organization address information security management. The objectives of Feature 1 also assists with creating short, medium and long term goals in driving information security management and reducing risk in the organization.

The function of Feature 1 of the ARCS framework aims to evaluate:

- An organizations ability to assess its current information security environment.
- The depth and breadth of the assessment methodology used.
- The alignment of the assessment(s) to best practice or standards.
- The frequency of assessment(s).
- The structural level to which the assessment is communicated.

Feature 1 is structured into three evaluation areas as depicted in Fig. 6 and contains a total of seventeen questions.

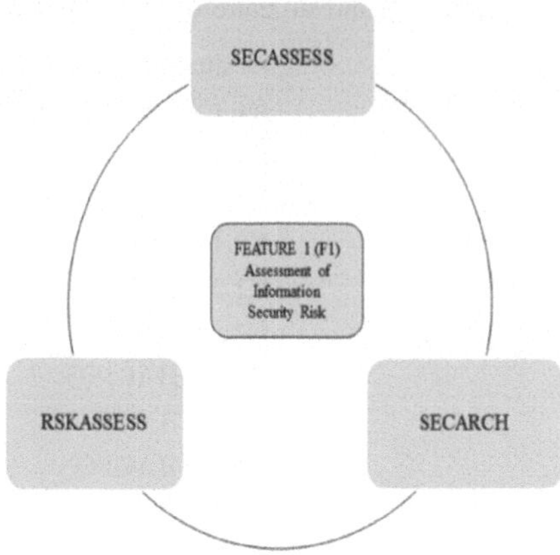

Fig. 6. Structure of Feature 1 (F1)

In Feature 1 the evaluation areas are described in Table 5.

Table 5. Description of components of Feature 1 (F1)

Evaluation Area	Description
E1.1 Security Assessment (SECASSESS)	This evaluation area focuses on general security assessments that may or may not be aligned to best practices or standards. Questions that relate to vulnerability management and penetration testing are weighted higher as these are linked to cost reduction factors
E1.2 Security Architecture (SECARCH)	This evaluation area focuses on the structured implementation of a security architecture programme. Questions in this area are rated higher as these are linked to sustainability factors
E1.3 Risk Assessment (RSKASSESS)	This evaluation area focuses on general information security risk assessments that may or may not be aligned to best practice or standards. No questions within this area are weighted higher

Feature 2 (F2) - Reduction of Information Security Cost

The Reduction of Information Security Cost feature evaluates the remediation factors noted in Fig. 2. These factors have been established to help information cost reduction in an organization. Feature 2 focuses on whether and to what extent the cost reduction factors are implemented within an organization.

This function of Feature 2 of the ARCS framework aims to evaluate:

- How many of the cost reduction remediation factors are implemented.
- The extent to which these factors are implemented.
- The alignment of the implementation(s) to best practices or standards.

- The structural level that is involved in the decision making of these implementations.
- The extent to which input cost factors are evaluated and measured.

The Reduction of Cost feature is structured into thirteen evaluation areas as depicted below in Fig. 7 and contains forty eight questions in total.

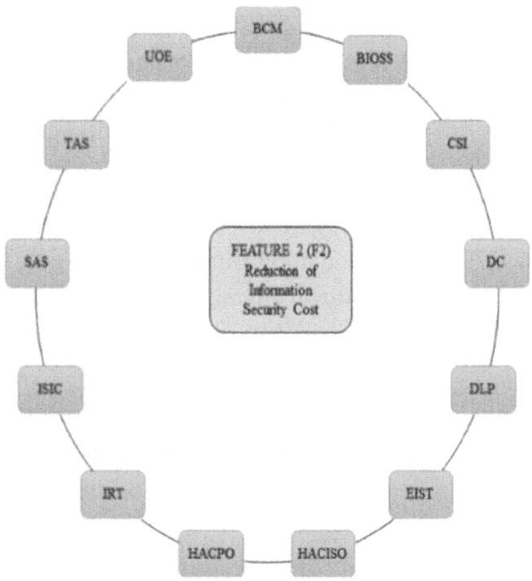

Fig. 7. Structure of Feature 2 (F2)

In Feature 2 the evaluation areas are described in Table 6.

Table 6. Description of components of Feature 2 (F2)

Evaluation area	Description
E2.1 Business Continuity Management (BCM)	This evaluation area focuses on whether the organization conducts a business continuity management programme, follows a business continuity process and is aligned to best practice or standards. It also evaluates how effective this process is and to what structural level the results of this process are communicated
E2.2 Cyber Security Insurance (CSI)	This evaluation area focuses on whether the organization invests in cyber security insurance
E2.3 Employee Information Security Training (EIST)	This evaluation area focuses on whether the organization implements regular and focused information security awareness and training, to what structural level and the evaluation thereof
E2.4 Having a CISO (HACISO)	This evaluation area focuses on whether the organization has a Chief Information Security Officer (or equivalent) and the relevant supporting structure

(*continued*)

Table 6. (*continued*)

Evaluation area	Description
E2.5 Board Input on Security Spend (BIOSS)	This evaluation area focuses on the input the board or highest managerial structure has on information security spending
E2.6 Having a CPO (HACPO)	This evaluation area focuses on whether the organization has a Chief Privacy Officer (or equivalent) and the relevant supporting structure
E2.7 Incident Response Team (IRT)	This evaluation area focuses on whether the organization has a security incident response team, the scope of their work and how they are measured
E2.8 Use of Encryption (UOE)	This evaluation area focuses on whether the organization uses encryption as a standard
E2.9 Threat Analysis and Sharing (TAS)	This evaluation area focuses on whether the organization conducts threat analysis activities, to what extent and whether this information is shared with peer organizations
E2.10 Security Analytics Services (SAS)	This evaluation area focuses on whether the organization runs a Security Operations Centre (SOC), if security analytics are conducted and if so what is done with that analysis
E2.11 Data Loss Prevention (DLP)	This evaluation area focuses on whether the organization runs a comprehensive DLP programme
E2.12 Data Classification (DC)	This evaluation area focuses on whether the organization manages and classifies data
E2.13 Employee Information Security Training (EIST)	This evaluation area focuses on whether the organization evaluates and budgets for information security input costs

Feature 3 (F3) – Sustainability of Culture

The Sustainability feature evaluates the readiness, behavior and culture of the organization in respect of information security. Feature 3 focuses on long term human behavior in relation to information security. The feature helps an organization determine its current cultural and behavioral fit from an information security management perspective and assists in developing a longer term sustainable road map to imbibe information security culture into the organization rather than a traditional cycle of information security analysis and addressing gaps through technology products and services.

The function of Feature 3 of the ARCS framework aims to evaluate:

- An organizations ability to assess whether common security values and principles exist within the organization and to what extent these are entrenched.
- An organizations ability to assess whether positive reinforcement and reward is applied for good information security behavior.
- An organizations ability to assess whether the organization implements common and couple security processes across the IT/IS disciplines.
- An organizations ability to assess whether peer recognition is conducted by employees for employees when good information security behavior is identified.

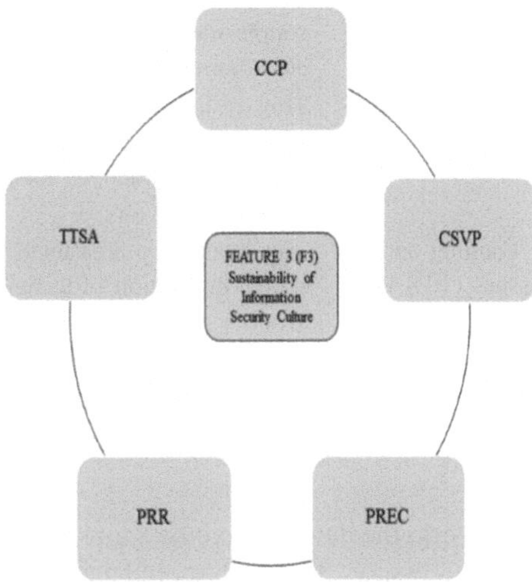

Fig. 8. Structure of Feature 3 (F3)

- An organizations ability to assess whether technical training and awareness of security issues is conducted within the organization.

The Sustainability feature is structured into five evaluation areas as depicted below in Fig. 8 and contains eighteen questions in total.

In this feature the evaluation areas are described in Table 7.

The questions that are part of the evaluation framework will be described in detail in future papers. The questions focus on the quality and efficacy of the state of information security assessments, cost reduction and culture within the organization.

Table 7. Description of components of Feature 3 (F3)

Evaluation area	Description
E3.1 Pillar 1- Common Security Values and Principles (CSVP)	This evaluation area focuses on the identification, communication and evaluation of common security values and principles
E3.2 Pillar 2- Positive Reinforcement and Reward (PRR)	This evaluation area focuses on the communication, method and process of positive reinforcement and reward for good information security behavior
E3.3 Pillar 3- Common and Couple Processes (CCP)	This evaluation area focuses on the communication and evaluation of common and couple security processes in the IT/IS environment
E3.4 Pillar 4- Peer Recognition (PREC)	This evaluation area focuses on the evaluation, structural approach and visibility of peer recognition initiatives for good information security behavior
E3.5 Pillar 5- Technical Training and Awareness of Security Issues (TTSA)	This evaluation area focuses on the approach, scope, evaluation of technical training and awareness of information security

5 Future Work

In order to apply the framework, the development of a tool to assist with an evaluation against the framework, is necessary. The tool should be able to link common and critical areas of the framework and evaluate responses from organizations. The output should give organizations critical management insight into the current state of the organizations information security assessment maturity, the drivers to reduce information security cost and the state of information security culture.

6 Conclusion

Information security data breaches are becoming larger and more frequent. The reporting on such events is part of mainstream media reports, making people and organizations much more aware of these threats. Organizations have focused their security management efforts on technology at significant cost and complexity. Organizations have however, neglected the consideration that the number of devices and therefore the amount and variety of threats is far greater than ever before. Incorporating information security into the culture of the IT staff members that support these technologies is a key function that must be considered in parallel to improved security technology. Understanding the threats, security landscape and gaps in information security is also key. Additionally managing cost efficiency in deploying key security solutions, including the human resources to manage these solutions is important. The framework proposed in this paper considers focusing on cost reducing products, services and structures, while building the correct behaviour and values in IT staff members, and strengthening their ability to improve information security assessment capabilities in the organization, to better support information security management.

References

1. Safa, N.S., Maple, C., Watson, T., Von Solms, R.: Motivation and opportunity based model to reduce information security insider threats in organisations. J. Inf. Secur. Appl. **40**, 247–257 (2018). https://doi.org/10.1016/j.jisa.2017.11.001
2. Mukherjee, S.: Overview of the importance of corporate security in business. SSRN Electron. J. (2019). https://doi.org/10.2139/ssrn.3415960
3. Dhillon, G., Torkzadeh, G., Chang, J.: Strategic planning for IS security: designing objectives. In: Chatterjee, S., Dutta, K., Sundarraj, R.P. (eds.) DESRIST 2018. LNCS, vol. 10844, pp. 285–299. Springer, Cham (2018). https://doi.org/10.1007/978-3-319-91800-6_19
4. Lord, N.: The history of data breaches. https://digitalguardian.com/blog/history-data-breaches. Last Accessed 07 Oct 2019
5. Winder, D.: Data breaches expose 4.1 billion records in first six months of 2019. https://www.forbes.com/sites/daveywinder/2019/08/20/data-breaches-expose-41-billion-records-in-first-six-months-of-2019/#7fe19849bd54. Last Accessed 07 Oct 2019
6. Targett, E.: Global data breaches 2018: 4.5 billion records compromised thus far. https://www.cbronline.com/news/global-data-breaches-2018. Last Accessed 07 Oct 2019
7. Ponemon Institute: cost of data breach study (2018)
8. Verizon: 2017 data breach investigation report, 10th edition (2017)
9. Kaspersky Lab: damage control: the cost of security breaches, IT Security Risks Special Report Series (2016)
10. Chong, J.: How security technology improves business operations. https://www.securitymagazine.com/articles/89706-how-security-technology-improves-business-operations. Last Accessed 07 Oct 2019
11. Debar, H.: Cybersecurity: high costs for companies. http://theconversation.com/cybersecurity-high-costs-for-companies-110807. Last Accessed 07 Oct 2019
12. Wilczek, M.: Cybercrime is increasing and more costly for organizations. https://www.cio.com/article/3386417/cybercrime-is-increasing-and-more-costly-for-organizations.html. Last Accessed 07 Oct 2019
13. Edwards, B., Jacobs, J., Forrest, S.: Risky business: assessing security with external measurements. arXiv Prepr. arXiv1904.11052 (2019)
14. Schmittling, R.: Performing a security risk assessment. https://www.isaca.org/Journal/archives/2010/Volume-1/Pages/Performing-a-Security-Risk-Assessment1.aspx. Last Accessed 10 Oct 2019
15. Mayer, N., Aubert, J., Grandry, E., Feltus, C., Goettelmann, E., Wieringa, R.: An integrated conceptual model for information system security risk management supported by enterprise architecture management. Softw. Syst. Model. **18**(3), 2285–2312 (2018). https://doi.org/10.1007/s10270-018-0661-x
16. Kosutic, D.: ISO 27001 checklist: 16 steps for the implementation. https://advisera.com/27001academy/knowledgebase/iso-27001-implementation-checklist/. Last Accessed 11 Sep 2019
17. Beaver, K.: Best practices for an information security assessment. https://searchsecurity.techtarget.com/tip/Best-practices-for-an-information-security-assessment. Last Accessed 10 Oct 2019

18. Eloff, M.M., Von Solms, S.H.: Information security management: an approach to combine process certification and product evaluation. Comput. Secur. **19**, 698–709 (2000). https://doi.org/10.1016/S0167-4048(00)08019-6
19. Ruighaver, A.B., Maynard, S.B., Chang, S.: Organisational security culture: extending the end-user perspective. Comput. Secur. **26**, 56–62 (2007). https://doi.org/10.1016/j.cose.2006.10.008
20. Crossler, R.E., Johnston, A.C., Lowry, P.B., Hu, Q., Warkentin, M., Baskerville, R.: Future directions for behavioral information security research. Comput. Secur. **32**, 90–101 (2013). https://doi.org/10.1016/j.cose.2012.09.010
21. Hsu, J.S.C., Shih, S.P., Hung, Y.W., Lowry, P.B.: The role of extra-role behaviors and social controls in information security policy effectiveness. Inf. Syst. Res. **26**, 282–300 (2015). https://doi.org/10.1287/isre.2015.0569
22. D'Arcy, J., Hovav, A.: Deterring internal information systems misuse. Commun. ACM **50**, 113–117 (2007). https://doi.org/10.1145/1290958.1290971
23. Vance, A., Benjamin Lowry, P.: A new approach to the problem of access policy violations: increasing perceptions of accountability through the user interface. MIS Q. **39**, 345 (2015)
24. Govender, S.G., Loock, M., Kritzinger, E.: Enhancing information security culture to reduce information security cost: a proposed framework. In: Castiglione, A., Pop, F., Ficco, M., Palmieri, F. (eds.) CSS 2018. LNCS, vol. 11161, pp. 281–290. Springer, Cham (2018). https://doi.org/10.1007/978-3-030-01689-0_22
25. Mercuri, R.T.: Analyzing security costs. Commun. ACM **46**, 15–18 (2003). https://doi.org/10.1145/777313.777327
26. Brecht, M., Nowey, T.: A closer look at information security costs. In: Böhme, R. (ed.) The Economics of Information Security and Privacy. LNCS, pp. 3–24. Springer, Heidelberg (2013). https://doi.org/10.1007/978-3-642-39498-0_1
27. Scholtz, T.: Articulating the Business Value of Information Security. Gart. Inc., Stamford (2011)
28. Bodin, L.D., Gordon, L.A., Loeb, M.P.: Evaluating information security investments using the analytic hierarchy process. Commun. ACM **48**, 78–83 (2005). https://doi.org/10.1145/1042091.1042094
29. Schatz, D., Bashroush, R.: Economic valuation for information security investment: a systematic literature review. Inf. Syst. Front. **19**(5), 1205–1228 (2016). https://doi.org/10.1007/s10796-016-9648-8
30. Asen, A., Bohmayr, W., Deutsche, S., Gonzalez, M., Mkrtchian, D.: Are you spending enough on cybersecurity? https://www.bcg.com/publications/2019/are-you-spending-enough-cybersecurity.aspx. Last Accessed 26 Mar 2019

The Improvement of the Stylometry-Based Cognitive Assistant Performance in Conditions of Big Data Analysis

E. V. Melnik[1], I. S. Korovin[2], and A. B. Klimenko[2](✉)

[1] Federal Research Centre the Southern Scientific Centre of the Russian Academy of Sciences, 41, Chekhov Street, 344006 Rostov-on-Don, Russian Federation

[2] Scientific Research Institute of Multiprocessor Computer Systems of Southern Federal University, 2, Chekhov Street, 347928 Taganrog, Russian Federation

anna_klimenko@mail.ru

Abstract. The current research is devoted to the problem of individual's victim behavior in the Internet. The victim behavior is a complex one and can be characterized by numerous markers from the classical victimology point of view. Yet, there are no general approaches to identify the victim behavior in the Internet due to the novelty of this communication environment. In this paper the marker of numerous false pseudonyms is presented to identify the individual's victim behavior, and the cognitive assistant is proposed as a facility to deal with this marker. The cognitive assistant is based on the stylometry analysis, yet this approach has some scalability and velocity issues under the big data conditions. So, the technique for cognitive assistant performance improvement is presented. The basic estimates are made to prove the efficiency of the proposed complex technique, based on the fog-computing and the distributed ledger technology.

Keywords: Cognitive assistant · Distributed ledger · Fake accounts · Stylometry · Distributed computing

1 Introduction

Nowadays the Internet has become a global communication environment. Yet, the lack of physical representation and the possibility to create the desired "personality" implemented in profile/account provokes the multiple "victim-perpetrator interactions". In terms of classical victimology, the individual's victim behavior takes place, which is the paramount of numerous crimes [1–3].

The problem of victim behavior in the Internet is complex and sophisticated because of the following:

- there are no general approaches in contemporary victimology, which allow to identify the potential victims in the Internet;
- the organizational strategies and technical facilities, which exist in this area, are not sufficient, due to the lack of general approach to the victim behaviour identification.

R. Silhavy (Ed.): CSOC 2020, AISC 1226, pp. 85–99, 2020.
https://doi.org/10.1007/978-3-030-51974-2_8

Turning to the victim behavior identification, the following must be mentioned: two main victim behavior classes there exist, including active and passive victim behavior. The passive victimizing is the carelessness, naïve behavior, chatting with strangers, sharing the individual private information, online dating and so on. The active victim behavior presupposes very aggressive, rather amoral behaviour. Both of the behavioural patterns lead to crimes.

As far as it is known, the victim potential can be identified through the questionnaire procedure [4], but it is obvious that such an approach is efficient only in case of individual's conscent to be tested. Also, it is problematic to organize the questionnare-based victim behavior identification in the Internet. The main reasons for this, again, the possible lack of conscent to be tested and the insufficiency of classical psychological markers to identify the victim behavior in the Internet, as far as the Internet is the relatively new and unexplored communication environment. At the same time, there are lots of precedents when the individuals who had multiple fake accounts, acted as a victim [5–9].

For example, considering the cyberbullying as sub-area of victim-perpetrator interaction, the victim creates new accounts to restore the reputation, or to affect the discussion (passive victim). At the same time the bully creates new accounts to bully the victim more and more, while the victim can use the social networks facilities to block him. It must be mentioned that the bully also features the victim behaviour according to the victim behaviour definition. Another example of victim behavior objectification is fake accounts of those individuals, who share private, perhaps, amoral information with the network communities though the fake accounts. Also, individuals with active victim behavior use fake accounts while being involved in some destructive and even terroristic communities.

So, considering multiple cases of the fake accounts usage to objectificate the victim behaviour in the Internet, the new victim behavior marker is proposed. It is the multiple fake accounts, which belong to the individual or other people around him. Yet, the problem is not to state that the account does not belong to a real individual only, but to create the linkability graph. In other words, the problem is to answer the following question: if the account considered is one of the existing fakes and which person they belong to.

In this field the stylomenty-based approach proves to be sufficient. Authorship identification techniques are proved as usable and efficient [10–13]. Yet, the conditions of big data make the stylometry approach to be of a low sufficiency due to their focus on the limited data volumes [14, 15].

So, the problem of fake accounts identification from the linkability point of view can be formulated as a problem of stylometry-based false pseudonyms search and identification under the big data conditions and the geographically distributed data sources. As we determined the multiple fake accounts as a victim behaviour marker, the individual can be checked from this point of view. It is more prospective in terms of a parental control, because the parental control facilities are limited, not agile and include the standard set of functions, such as: site blocking, black lists, content filters, etc.

In the current paper the technique of the stylometry-based cognitive assistant performance improvement is proposed and described. Also, a brief estimation of the technique proposed is presented.

2 Victim Behavior Identification, False Network Pseudonyms Detection and the Functional Structure of a Cognitive Assistant

As was mentioned above, the person with multiple pseudonyms is supposed to have a potentially victim behavior. The relation between the victim behavior and the multiple pseudonyms is showed in the following publications [16–20]. Some cases are devoted to the efforts to restore the reputation in the social networks, some cases, in contrary, contain the indirect relation between multiple pseudonyms and being a bully. Also multiple fake accounts are used, when individuals are involved in destructive and religious communities. Another case of active victim behavior is aggressive blogging with pseudonyms usage, or trolling, which stimulates the victimizing of the individuals and affects their mental health.

Of course, the fake account marker is not sufficient to cover all types of a victim potential, yet it can be used to identify numerous cases, including the identity theft cases (in this case the victim's pseudonyms can be found, or, on the contrary, the pseudonyms of a perpetrator).

The existing facilities counteract the selected aspects of the victim behavior, such as, for example, aggression in messages, amoral content. Also a wide range of anti-fake facilities exists.

The strategies to counteract the victim behavior are as follows:

Parental control facilities (NetKids, Kindergate Parental Control, Kaspersky Internet Security Component, etc.);

Social networks politics (destructive content prevention, fake-account blocking). As an example, here the brief review is presented. For VK social network: 7900000 content units are blocked, 1800000 accounts have been blocked, 11300 suicidal communities are identified, 300000 content units, 8600 pages, 500 communities related to the cyberbulling are identified. Yet, it must be mentioned, that the strengthening of the social networks policies can lead to the auditory outflow (the case have taken place once and mentioned in [21].

Numerous specialized facilities for de-anonymizing, linkability graph creation and fake account detection, but such systems are corporate usage-oriented and, in general, the objective is not a victim behavior at all.

So, there is a lack of agile, individual-oriented facilities on the market. Issues listed are covered with such facility as an individual cognitive assistant (CA) – adaptive software, which aims the enhancement of the individual's intelligence and computing capabilities.

In our case the objective of the CA is to identify and prevent the individual's victim behavior, with the usage of the fake-accounts marker.

Within the fake account detection approaches the following ones are researched and used:

- String-based matching [22, 23];
- Stylometric matching [24–26];
- Time-profile-based matching [27, 28];
- Social network-based matching [29].

The stylometric approach is proposed to be efficient from the false pseudonym detection point of view, yet it has some issues in conditions of big distributed data, where there is a need to gather and analyze the information from the heterogeneous sources, to form the text features vectors and to compare them with the text feature vector given as a reference. So the problem arises to create a methodology, which would improve the performance of software with the functions, presented in the Fig. 1.

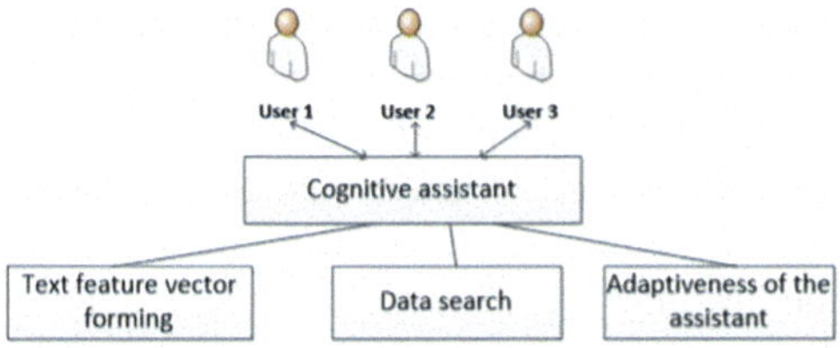

Fig. 1. The general functions of the stylometry-based CA.

The functional listed is implemented in the following functional algorithmic scheme, as is shown in the Fig. 2.

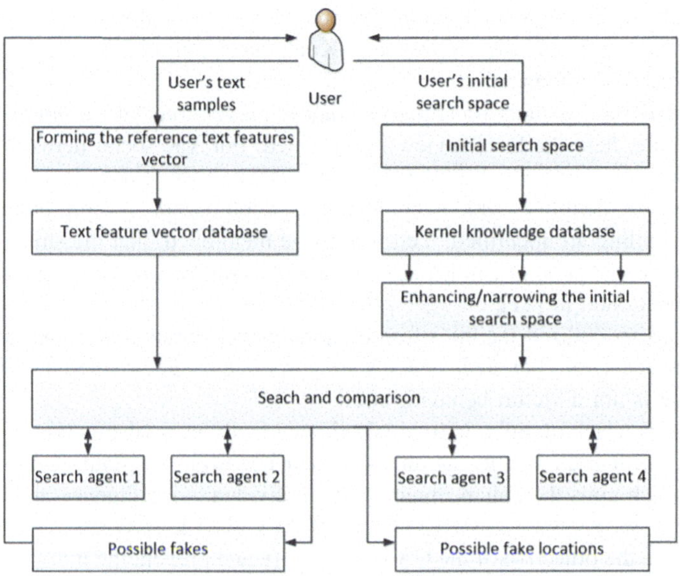

Fig. 2. The generic algorithmic structure of the CA

Here "Text feature vector forming" operation presupposes the creation of a text feature vector from the set of text instances according to one of the stylometry approaches. Text samples are provided by user, who searches the pseudonyms.

"Data search" process includes the stages of information retrieving and processing, when the accounts and the text sets are retrieved from the Internet resources. The output of the data search is the sets of pairs "pseudonym-text sample". Then the comparison stage takes place: the reference text features vector is compared to all vectors, formed the retrieved data. It is obvious, that this operation is non-trivial due to the geographical distribution of the data and its volume.

"Adaptiveness" of the assistant is the key for the operation time consuming, the operational quality and the possibility of an assistant to adapt to the environment. In this research the adaptiveness is implemented in history-based search-space correction and in adapting the text sample size.

The next section of the paper contains the description of the main issues of the CA operation and the description of the complex technique to cope with.

3 The Issues of the Cognitive Assistant Operation Under the Big Data Conditions and the Technique to Improve Its Performance

To outline the main issues of the stylometry-based CA, consider the following scheme (Fig. 3).

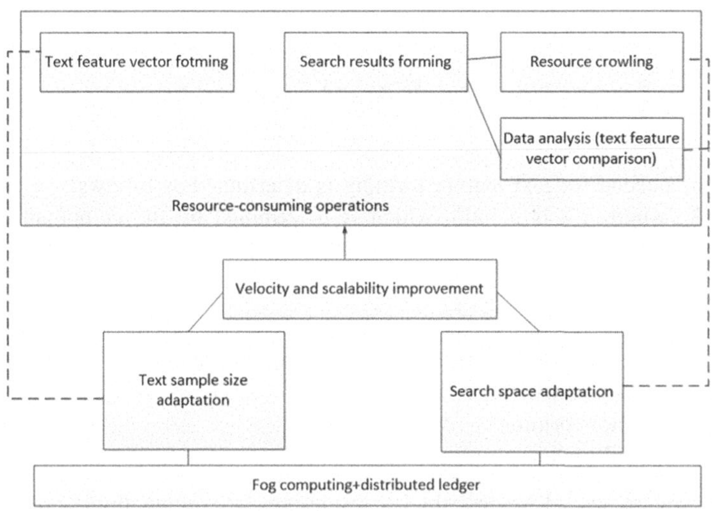

Fig. 3. The main resource-consuming operations and the technique to improve the performance

As is shown in Fig. 3, the following operations are the most time-consuming:

Text feature vector forming, the time needed depends on the text features selected and the size of the text samples;

Search results forming, the time needed depends on the volumes of the information to process, the number of pseudonyms to form text samples and transfer them to the text feature vector forming subsystem, and the set of the resources to be crawled.

As a solution for the time-consuming decrease problem the following technique is proposed: to improve the velocity and scalability of the CA in question, the cooperated application of the adaptive mechanisms of the assistant, fog computing concept elements and the distributed ledger technologies elements are needed.

Leaving the fog-computing concept [30] and the distributed ledger technologies [31, 32] description outside of this paper's body, we has made some simple estimates to prove the expediency of the technique proposed.

4 The Adaptive Mechanisms of the CA and Their Efficiency

Consider a model estimating the relation between the text sample size and the time needed for the text feature vector forming. Text sample size is an element of the adaptive cognitive assistant mechanisms proposed and depends on the user's feedbacks or the types of data sources.

Consider c as a vector size (feature number), i – the feature identifier. Then the computational complexity of the text processing for the i-feature is as follows:

$x_i = \xi_i V_t$, where V_t – is the size of text sample (in words), ξ_i - is a ratio, which determines the complexity of text processing for the x_i forming.

Then the complexity of the text feature vector forming is:

$$X = \sum_{i=1}^{c} x_i = \sum_{i=1}^{c} \xi_i V_t \tag{1}$$

The time needed for text feature forming is determined as follows:

$T_c = \alpha X$, where α - is a ratio, which is determined by the computational node performance.

Full time of the text features vector forming is a sum of the following elements, assuming the text sample is ready:

- the time of text sample transfer to the computational node, where the text sample processing is conducted;
- the time of vector forming;
- the time of result delivery to the user.

The procedure should be described, considering the computational node location, which is supposed to form the text features vector, in relation to the data source, as is shown in equation below:

$$T_0 = V_t k + T_c + k\eta c \tag{2}$$

where T_0 – the full time of text feature vector delivery to user;

k – the ratio, which determines the velocity of data transferring though the network in dependency of the transitory network fragments number;
ηc – the value of the text features vector.

As is seen in the Fig. 4, text sample size affects the time of text feature vector forming and providing, so, it is expedient to use this dependency for time consuming decrease.

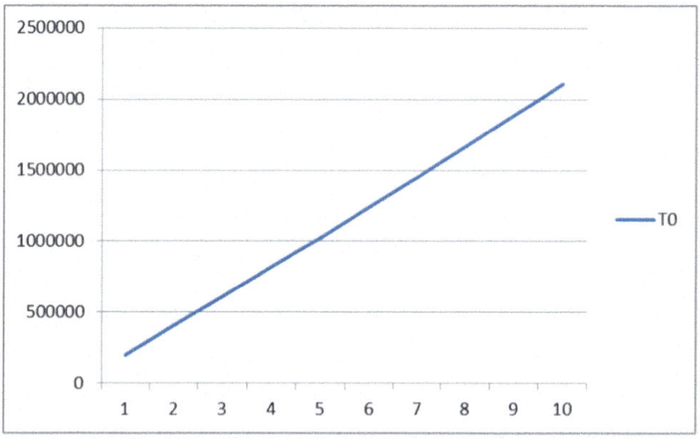

Fig. 4. The dependency between the time of the text feature vector forming and the text sample value

Further, consider the effect of the cognitive assistant search space adaptivity.
Consider the following parameters:

T_s – the time of one search session, when for the text features vector given the search of pseudonyms is conducted.
W_i – the data value to be crawled.
μ – the crawler velocity;
σ – the velocity of the pseudonyms search assuming the crawler data received, including the text feature vector forming

Then one session of the fake pseudonyms search with the text feature vector given on the resource I is:

$$T_{si} = \mu W_i + \sigma P_i + \theta(P_i) \quad (3)$$

where P_i - the full number of pseudonyms in the resource i,
$\theta(P_i)$ - the time of the text features vectors comparison.
For m resources to be processed, the following expression is conducted:

$$T_s = \sum_{i=1}^{m} \mu W_i + \sigma P_i + \theta(P_i) \quad (4)$$

So, it is seen that the adaptive features of the cognitive assistant does affect the time of system reaction. The possibility to correct the search space is expedient and shorten the time of system reaction (Fig. 5).

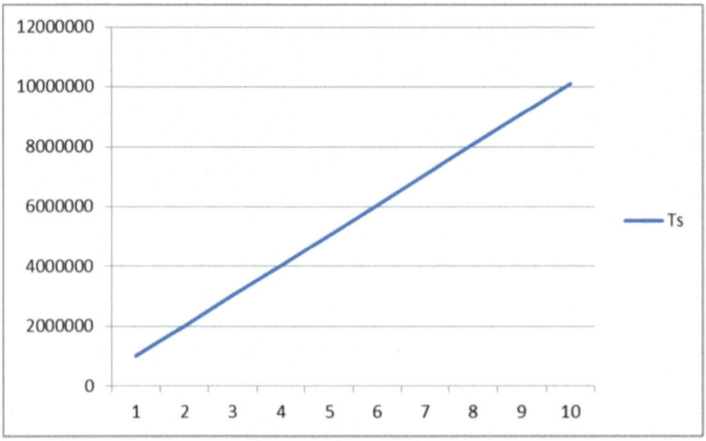

Fig. 5. The search time dependency on the number of resources to be processed

5 The Parallel Information Retrieving and Processing in the Fog-Layer

Consider the following cases of search agents location:

- service agent is located at the same computational node as data customer service;
- service agent is located at the same computational node as Data Source and preprocess the information;
- service agent is located in the fog-layer of the network near the Data Source node and preprocess the information retrieved.

To estimate roughly the cases in terms of time of information processing, consider the following model parameters:

n_1 - is the distance between data customer service and the data source in network hops;
n_2 - is the distance between the data source node and the fog nodes on which the search agent is located (in network hops);
n_3 - is the distance between the search agent and the data customer service computational node;
γ - is the percentage of data processing volume of the overall data volume V;
v - is the speed of information processing/analysis of the agent;
V_{req} - is the volume of request for information block from a search agent to the data source;
ξ - is a ratio between time and number of network hops.

With the simple models the cases of search agents location can be presented in following way:

$$T_{agent\_in\_house} = \xi n_1 V_{req} + \xi n_1 V + \frac{V}{v}, \tag{5}$$

where $T_{agent\_in\_house}$ - is the time, which is needed for search agent on the FADCA node to process the information with the volume V.

$$T_{agent\_in\_place} = \xi n_1 (1 - \gamma) V + \gamma \frac{V}{v}, \tag{6}$$

where $T_{agent\_in\_place}$ is the time needed to process the information, when the search agent is located at the same node with the data source, assuming the possible information preprocessing.

$$T_{agent\_in\_fog} = \xi n_2 V_{req} + \gamma \frac{V}{v} + \xi n_3 (1 - \gamma) V, \tag{7}$$

where $T_{agent\_in\_fog}$ is the time needed to process the information, when the search agent is located on the fog-node with the partial information preprocessing.

In the Fig. 6, 7 one can see that with the fixed distance between the data source and data customer service the best time is got when the search agent is located on the same node with the data source.

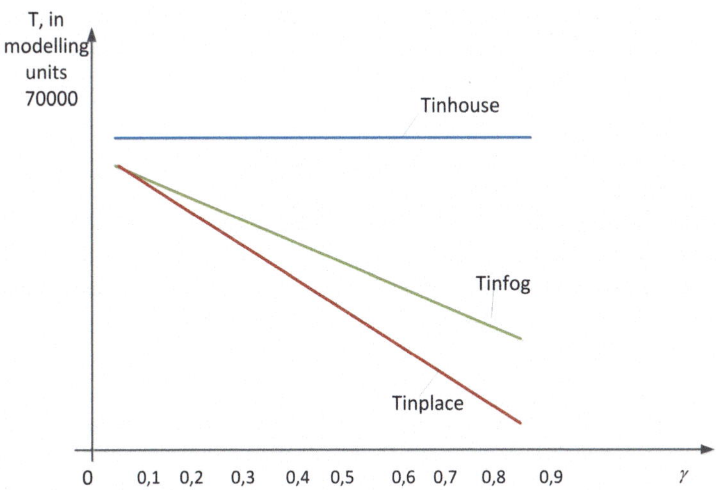

Fig. 6. Simulation results with the variables values: n1 = 10; n2 = 1; n3 = 9; $\gamma = 0..0.9$

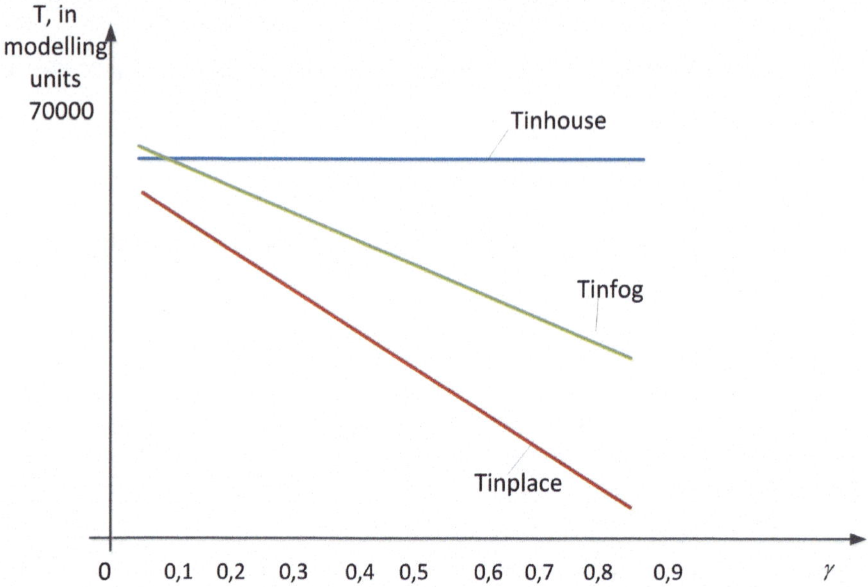

Fig. 7. Simulation results with the variables values: n1 = 10; n2 = 3; n3 = 7; $\gamma = 0..0.9$.

With the increase of distance between data source and the search agent the time continues to decrease, up to the time equal to that of an agent located on the data customer node.

To estimate the expediency and efficiency of the parallel data search and processing consider two situations:

- when crawlers and data processing agents locate on the server (in a datacenter);
- when crawlers and data processing agents locate in the fog-layer and provide simultaneous data crawling and processing.

Assume the following:

V – the value of data to be crawled and processed,
v_1 – the centralized crawler data crawling velocity,
v_2 – the centralized service data processing velocity,
N – the accounts number to be processed,
v_3 – the crawling velocity for the "collective" crawler, $v_3 < v_1$, assuming that the fog-nodes performance is less than the central server's one,
v_4 – the fog-node located agent data processing velocity, $v_4 < v_2$

The lower estimate of the time needed for the data crawling and processing, can be got for the case, when it is mandatory to get all data crawled before the processing. Then, for the centralized solution with one data processor agent and one crawler the following estimate is provided:

$$T_{lc} = \frac{V}{v_1} + \frac{V}{v_2} \tag{8}$$

In case when the data processing within the datacenter can perform simultaneously, the lower estimate is as follows:

$$T_{lc} = \frac{V}{v_1} + \frac{V}{kv_2} \tag{9}$$

The lower estimate for the simultaneous crawling and processing in the fog-layer is as follows:

$$T_{lf} = \frac{V}{mv_3} + \frac{V}{mv_4}, \tag{10}$$

where m is the fog-layer located agents number.

In case when the data processing can be provided as soon as the first (not large) portion of data received (Fig. 8), the following higher estimates can be provided:

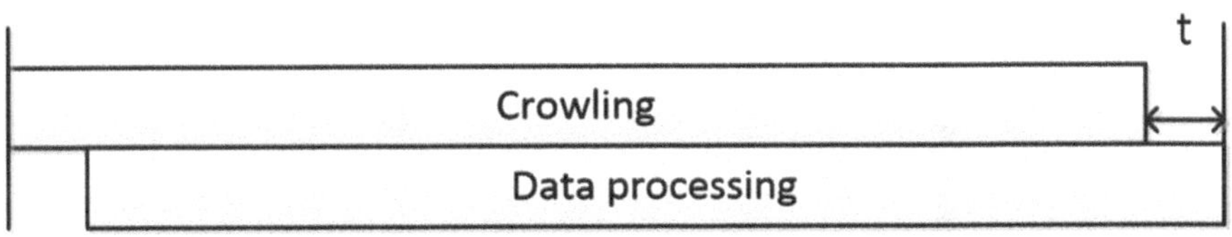

Fig. 8. The simultaneous crawling and processing

$T_{lc} = \frac{V}{v_1} + t$ - for the centralized case,

$T_{lf} = \frac{V}{mv_3} + t$ - for the simultaneous case.

With the decrease of initial data portion size, $t \to 0$.

So, the higher and lower estimates allow to conclude the efficiency of simultaneous fog-located data crawling and processing agents. It is seen that the efficiency improvement is possible to two-times for the lower and higher estimates (Figs. 9 and 10).

6 The Distributed Ledger-Based Data Storage for the Adaptive Assistant Behaviour

The considered above adaptive mechanism of the cognitive agent, which implements the correction and adaptation of the search space, functions as following:

- Every precedent of the fake pseudonym identification is put into the database, reflecting the resources links.

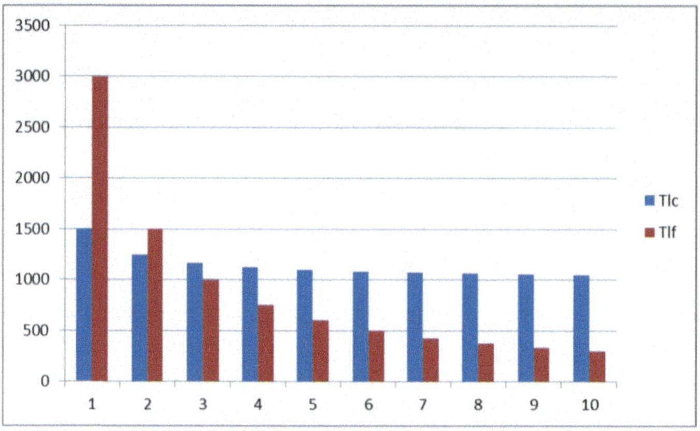

Fig. 9. The lower estimate of the time needed for data crawling and processing (with the k and m increase).

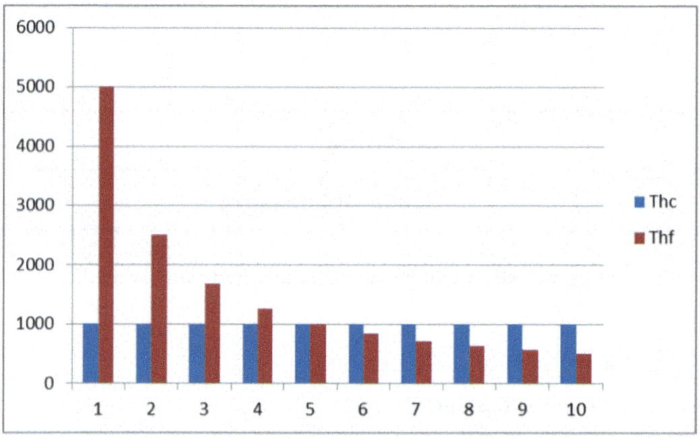

Fig. 10. The lower estimate of the time needed for data crawling and processing (with the k and m increase).

- Every new search session begins with the database analysis. User provides the initial search space, and the precedent database analysis constructs the cognitive map of resource linking. It is explained by the multiple auditory overlapping cases, so, the cognitive map is a mechanism, which allows the search to be adapted to the historical data of pseudonyms identification. As the different users have the different initial search points, their cognitive maps would not be equal, as the cognitive map is a directed graph.
- In case of search space decrease as a result of cognitive map analysis, the time of search decreases, otherwise, the resources added on the basis of cognitive map allow to increase the accuracy of the search.

As every user initiates every new search session with the cognitive map building, the information about the Internet resource linking must be taken from the data storage. The situation of massive read-operations takes place, while write-operations are rare enough and conduct just in cases of pseudonyms identification. With the numerous read- and relatively seldom write- operations it is expedient to store the historical data in a distributed ledger structure, when every user has its own ledger copy. This optimizes the velocity and scalability of data analysis operations, as far as implements the fog-computing concept, reducing the distance between the data source and user.

So, the general structure of the cognitive assistant considered is presented in the Fig. 11.

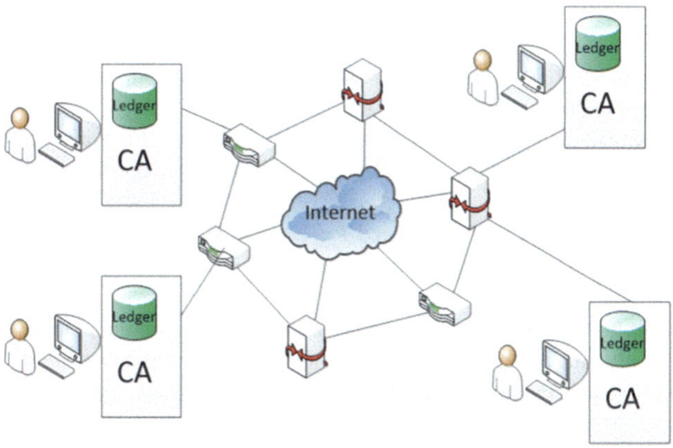

Fig. 11. Distributed ledger-based cognitive assistant structure

7 Conclusions

A problem of individual's victimization in the Internet is quite topical nowadays. The main issues in this field are the lack of general approaches for the victim behavior identification, the lack of methods and facilities, which could identify the victims in time. The marker of multiple fake accounts usage allow to suspect the individual in a victim potential, so, as a technical facility, the stilometry-based cognitive assistant is proposed.

The main issue of such software agent is the need to gather and process the information from the geographycally distributed data sources. So, special techniques and algorithms must be developed to decrease the time of agent's reaction.

In the current paper the new technique for the stylometry-based cognitive assistant performance improvement is proposed and described. The paramount of the technique is that the adaptive mechanisms in cooperation with fog-computing and distributed ledger technologies affect the system performance.

Some estimates of the technique proposed has been made and presented in the paper. The results prove the efficiency of the technique proposed.

Acknowledgements. The paper has been prepared within the RFBR projects 18-29-22093.

References

1. Abramenkova, V.V., Avanesov, V.S., Ageev, V.S.: Short Psychological Vocabulary. Politizdat, Moscow (1990)
2. Victimology and its main categories and concepts in native and foreign science. http://obrazovanie9.ru/articles/345-victimology.html. Accessed 10 Dec 2019
3. Psychological factors for adolescent's victimization. https://www.dissercat.com/content/psikhologicheskie-faktory-vozniknoveniya-viktimnogo-povedeniya-podrostkov. Accessed 10 Dec 2019
4. Victimization identification. http://psihologn.org/index.php/diagnostika-sklonnosti-k-viktimnomu-povedeniyu. Accessed 10 Dec 2019
5. Zhang, L., Liu, B.: Sentiment analysis and opinion mining. In: Sammut, C., Webb, G.I. (eds.) Encyclopedia of Machine Learning and Data Mining, pp. 41–48. Springer, Boston (2017)
6. Not an obscene racist after all: 4 flagged for Facebook fake. https://arstechnica.com/tech-policy/2009/09/that-obscene-racist-may-be-fake-4-sued-for-profile-prank/. Accessed 10 Dec 2019
7. Girl, 13, commits suicide after being cyber-bullied by neighbour posing as teenage boy. https://www.dailymail.co.uk/news/article-494809/Girl-13-commits-suicide-cyber-bullied-neighbour-posing-teenage-boy.html/. Accessed 10 Dec 2019
8. Man faces cyber-bullying charge in ex-girlfriend's fake Adult-DateProfile. https://5newsonline.com/2013/11/07/man-faces-cyber-bullying-charge-in-ex-girlfriends-fake-adult-date-profile/. Accessed 10 Dec 2019
9. Hinduja, S., Patchin, J.: Bullying, cyberbullying, and suicide. Arch. Suicide Res.: Offic. J. Int. Acad. Suicide Res. **14**(3), 206–221 (2010)
10. Afroz, S., Caliskan Islam, A., Stolerman, A., Greenstadt, R., McCoy, D.: Doppelgänger finder: taking stylometry to the underground. In: 2014 IEEE Symposium on Security and Privacy, pp. 212–226. IEEE, Danvers (2014)
11. Almishari, M., Kaafar, D., Oguz, E., Tsudik, G.: Stylometric linkability of tweets. In: WPES 2014, pp. 205–208 (2014)
12. Juola, P., Vescovi, D.: Analyzing stylometric approaches to author obfuscation. In: Peterson, G., Shenoi, S. (eds.) Advances in Digital Forensics VII. Digital Forensics 2011. IFIP Advances in Information and Communication Technology, vol 361, pp. 15–125. Springer, Heidelberg (2011)
13. Afroz, S., Brennan, M., Greenstadt, R.: Detecting hoaxes, frauds and deception in writing style online. In: Proceedings of the IEEE Symposium on Security and Privacy, pp. 461–475 (2012)
14. Bhargava, M., Mehndiratta, P., Asawa, K.: Stylometric analysis for authorship attribution on twitter. In: Bhatnagar, V., Srinivasa, S. (eds.) Big Data Analytics. BDA 2013. Lecture Notes in Computer Science, vol. 8302, pp. 37–47. Springer, Cham (2013)
15. Tempestt, N., Kalaivani, S., Aneez, F., Yiming, Y., Yingfei, X., Damon, W.: Surveying stylometry techniques and applications. ACM Comput. Surv. **50**, 1–36 (2017)

16. Dealing with fake profiles, https://www.thecyberhelpline.com/guides/fake-profiles. Accessed 10 Dec 2019
17. Students create fake online profiles to bully peers. https://www.edweek.org/ew/articles/2012/04/04/27facebook.h31.html. Accessed 10 Dec 2019
18. Fake Instagram accounts being used by kids to 'destroy reputations', eSafety Commissioner says. https://www.abc.net.au/news/2018-03-16/children-using-fake-instagram-accounts-to-bully-others/9553548. Accessed 10 Dec 2019
19. Experts warn teenagers are creating fake Instagram accounts to torment their peers in growing 'finsta' trend - with a worrying 133 per cent spike in cyber-bullying in just two weeks. https://www.dailymail.co.uk/news/article-5508891/Students-using-fake-Instagram-accounts-bully-children.html. Accessed 10 Dec 2019
20. How to safeguard children against cyberbullying. https://www.nytimes.com/2019/09/05/well/family/how-to-safeguard-children-against-cyberbullying.html. Accessed 10 Dec 2019
21. The main Facebook's problems: sex, hate, fakes. https://texterra.ru/blog/glavnye-problemy-facebook-seks-nenavist-feyki.html. Accessed 10 Dec 2019
22. Levenshtein, V.I.: Binary codes capable of correcting deletions, insertions and reversals. In: Doklady Akademii Nauk SSSR, vol. 163, no. 4, pp. 845–848 (1965)
23. Thor, A.: Toward an adaptive string similarity measure for matching product offers. In: GI Jahrestagung, no. 1, pp. 702–710 (2010)
24. Stamatatos, E.: A survey of modern authorship attribution methods. J. Am. Soc. Inf. Sci. Technol. **60**, 538–556 (2009)
25. Abbasi, A., Chen, H.: Writeprints: a stylometric approach to identity-level identification and similarity detection in cyberspace. ACM Trans. Inf. Syst. (TOIS) **26**(2), 1–29 (2008)
26. Juola, P.: Authorship attribution. Found. Trends Inf. Retr. **1**(3), 233–334 (2006)
27. Johansson, F., Kaati, L., Shrestha, A.: Timeprints for identifying social media users with multiple aliases. Secur. Inform. **4**, 7 (2015)
28. Faouzi Atig, M., Cassel, S., Kaati, L., Shrestha, A.: Activity profiles in online social media. In: 2014 IEEE/ACM International Conference on Advances in Social Networks Analysis and Mining (ASONAM 2014), pp. 850–855. IEEE, Danvers (2014)
29. Cataldi, M., Di Caro, L., Schifanella, C.: Emerging topic detection on twitter based on temporal and social terms evaluation. In: MDMKDD, p. 4. ACM (2010)
30. Yi, S., Li, C., Li, Q.: A survey of fog computing: concepts, applications and issues. In: Workshop on Mobile Big Data, pp. 37–42 (2015)
31. Distributed Ledger Technology: beyond block chain: a report by the UK Government Chief Scientific Adviser. https://assets.publishing.service.gov.uk/government/uploads/system/uploads/attachment_data/file/492972/gs-16-1-distributed-ledger-technology.pdf. Accessed 10 Dec 2019
32. Burkhardt, D., Werling, M., Lasi, H.: Distributed ledger definition & demarcation. IEEE (2018)

A Peer-to-Peer Crowdsourcing Platform for the Labeled Datasets Forming

E. V. Melnik[1] and A. B. Klimenko[2(✉)]

[1] Federal Research Centre, The Southern Scientific Centre of the Russian Academy of Sciences, 41, Chekhov Street, 344006 Rostov-on-Don, Russian Federation

[2] Scientific Research Institute of Multiprocessor Computer Systems of Southern Federal University, 2, Chekhov Street, 347928 Taganrog, Russian Federation

anna_klimenko@mail.ru

Abstract. Training dataset forming is quite labor intensive and frequently is of high costs. Also the cost overheads depend on the policies of the service, which implements the crowdsourcing approach to data labeling. In this paper a new peer-to-peer data labeling platform concept is presented, as well as the framework of the decentralized labeling approach is described briefly. The architecture proposed allows to avoid the intermediary labeling service and to perform the crowdsourcing-based data labeling by the computational facilities of users involved. Besides, the additional consensus procedure improves the quality of the labeled data by means of the voting procedure.

Keywords: Training dataset · Machine learning · Distributed ledger · Peer-to-peer service

1 Introduction

A labeled dataset creation is a topical problem of machine learning field. Despite the existence of numerous labeled datasets, e.g., Imagenet, MS-COCO, CIFAR-10 [1–3] and others, the uniqueness of machine learning projects demands the unique labeled datasets frequently. This leads to the questions of training dataset forming, data cleansing, choosing the way how to deliver the labeled data and so on.

There are some well-known approaches to the data labeling, including the following [4]:

- In-house labeling;
- Outsourcing;
- Crowdsourcing;
- Involvement of special outsourcing companies;
- Automated labeling, including synthetic labeling and data programming approaches.

Among these approaches the crowdsourcing should be emphasized because of its relatively low costs. The main disadvantage of the crowdsourcing is the possibly low quality of labeling, which affects the quality and the velocity of the neuronet training.

R. Silhavy (Ed.): CSOC 2020, AISC 1226, pp. 100–107, 2020.
https://doi.org/10.1007/978-3-030-51974-2_9

Another important feature of the crowdsourcing in data labeling is that all data are labeled with the involvement of a central element – server, which provides the workers with the data to be labeled and then collects the data and transfers the labeled content to the customer. This leads to the customers dependency on the crowdsourcing service suppliers and their prices, as well as their hardware dependability.

In this paper we propose the new crowdsourcing platform architecture based on the principles of distributed ledger technologies. The platform includes the Orders ledger to deal with content, and the Index ledger to provide the data search procedures. Also the architecture proposed presupposes the facilities to improve the quality of the labeled data by means of implying the consensus procedure on the stage of data labeling.

2 Centralized Solution for the Crowdsourcing-Based Data Labeling

As is known, the contemporary data labeling services have the centralized architecture. The examples are: Amazon Mechanical Turk, Click Worker, MicroWork for all and others [5, 6]. The basic architecture of such services is shown in Fig. 1.

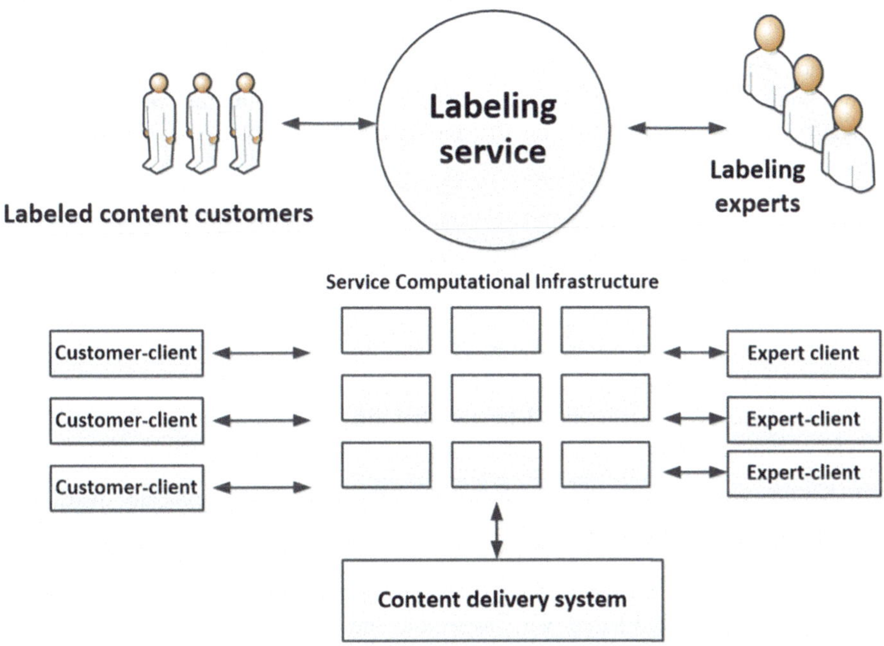

Fig. 1. Centralized labeling service basic architecture

In this paper we presuppose that, besides customers and labeling experts, there is a subsystem, possibly automated, of content delivery. This subsystem extracts data from the Internet resources according to the keywords given, and provides the experts with data to be labeled. Optionally, the content delivery subsystem functions can be incorporated into the customers' requests, i.e., to contain the piece of data to be labeled as a part of request.

So, as is shown in the Fig. 1, the centralized architecture of crowdsourcing data labeling system implements the following:

- Customers leave their requests in the labeling service;
- The data to be labeled is provided;
- The labeling experts label the elements of content and return the results to the labeling service;
- The labeling service transfers the results to the customers.

As was mentioned earlier, the critical disadvantage of such architecture and the interaction model is the dependency between labeling cost, labeling velocity and the labeling service prices and dependability.

The alternative to such a centralized model is the decentralized one. Decentralization tendency is quite topical nowadays, and the reasons for this are following:

- The need of scalability under the big data conditions;
- The need of data processing velocity;
- The need to offload the network infrastructure;
- And, of course, the need to avoid the intermediators in the business processes.

In the next section the distributed ledger -based data labeling platform architecture is presented and described, leaving the distributed ledger definitions and description outside of this paper body. Yet, the distributed ledger technology description can be found in details in [7–9].

3 Peer-to-Peer Crowdsourcing Platform for the Labeled Datasets Forming

Aiming the avoidance of the intermediators, the distributed-ledger-based architecture has been developed. This architecture of the system is presented in Fig. 2.

Consider the peer-to-peer data labeling platform.

The paramount terms of this platform's framework are as following:

- Labeling expert community, which is the community of geographically distributed users subscribed to the same order.
- Orders ledger, which contains order requests and the labeled content as transactions.
- Index ledger, which implements the maintenance of the search index, which is stored in a distributed ledger way, as well as the orders.
- Interaction agent, which implements the particular functions of users' interaction and the coordination of the user-ledger operations.

It must be mentioned, that the first stage of system functioning takes place on the users level, as well as the orders and index ledgers are the basic elements of the system.

The Orders ledger contains the customers requests and the experts answers as transaction analogues.

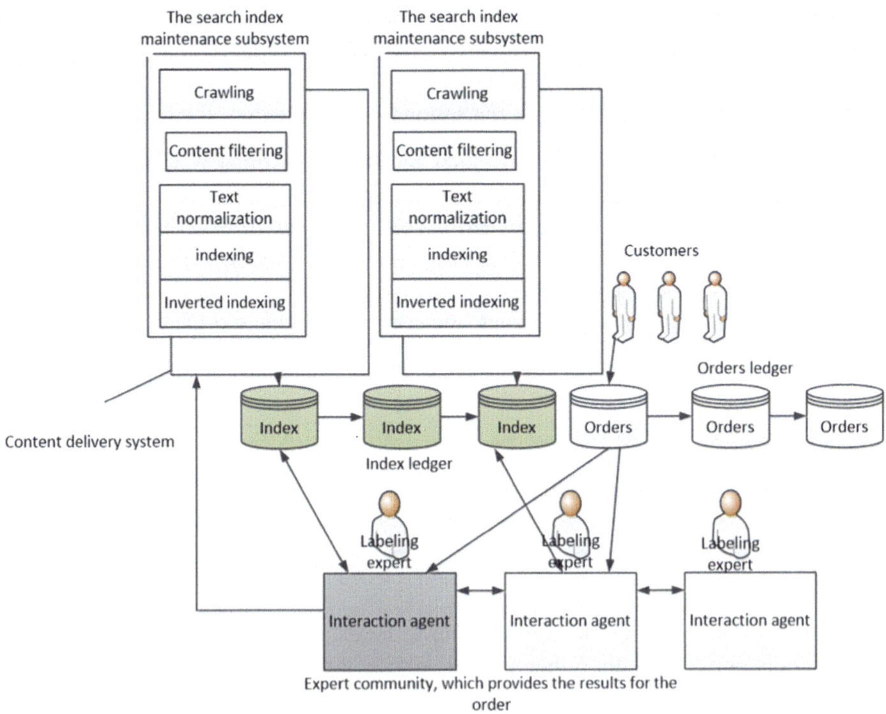

Fig. 2. The peer-to-peer data labeling platform

Transactions are transferred through the platform nodes according to the gossip algorithms (for example, as is implemented in BitCoin). Then, depending on the implementation method, blocks are formed and added to the ledger according to the consensus method selected. Also every transaction can be put into ledger separately and definitely depending on the consensus type.

Consider the basic function priciples of the platform.

A customer put the transaction into the Orders ledger. The transaction contains the Data Labeling Order description, possibly informal. In case of gossip algorithms usage the transaction reaches all ledger copies in time O log N, where N – the number of Orders ledger nodes. Then the experts make a decision about their involvement to the order, and the new transactions are added, which are called "subscription transactions".

The interaction agent represents the labeling expert in the expert community.

This agent interacts with the Orders ledger, retrieves the subscription transactions and so gets the information about the experts, which desire to perform this order. So the community is created. Also the interaction agent requests the content from the content delivery subsystem.

The interaction agent implements functions of labeling, interaction with ledgers and consensus within the expert community.

The search index maintenance subsystem implements the functions of inverted index updating, including content filtering, document lemmatization and stemming, if it is needed. The search index maintenance subsystem can be located at the same nodes as the Index ledger is. Such decision is expedient because of the intensive data exchange between crawlers and indexes.

The basic data structures contain the following information:

- Order transaction. Contain the customer identifier, the order description (subject area, the number of labeled elements and so on).
- Subscription transaction is the expert's guarantee that he takes the order. This transaction type contains the order transaction and the expert identifier.
- The labeled content transaction is a part of the order. It contains the order transaction identifier, the expert identifier, the content element and the content class.

The Index Ledger contains the inverted index fragments, which are replicated through all the nodes.

4 The Expert Communities Level Consensus

Within the system the expert is represented by the interaction agent, which implements the following:

- Ledger monitoring, with the aim to control the expert's status about the orders;
- The cooperation with the other experts within the community, which is formed according to the expert's subscription;
- The interaction with the content delivery subsystem.

The Orders ledger monitoring is performed to monitor the order status and so determines if there is a need to request the content for labeling. The interaction with the content delivery subsystem takes place if there is a need to label a piece of content.

The interaction with other experts takes place if the redundancy is presupposed to improve the labeling quality.

Experts interaction is optional but highly desirable due to the possibility to minimize the labeling noise level [10].

The labeling noise is anything that obscures the relationship between the features of an instance and its class [11]. The main sources of labeling noise are as follows [10]:

- the information which is provided to the expert may be insufficient to provide reliable labeling;
- the errors can occur in the expert's labeling process itself;
- the labeling task is subjective, as, for example, in medical applications and image data analysis;
- the presence of data encoding errors and communication problems, which is caused by insufficient level of dependability.

Generalizing, one can say that there are two main causes for labeling noise: the human-based and the infrastructure-based ones.

At least four major techniques for labeling noise elimination are described in literature:

- label noise-robust methods; [12–14]
- probabilistic label noise-tolerant methods [15, 16];
- data cleansing methods [17, 18];
- model-based label noise-tolerant methods [19–21].

It must be mentioned that besides the approaches exist, the problem of labeling noise is relevant and is not solved. The labeling noise worsens the neuronet training time and the neuronet-based classifier efficiency.

In the current paper the additional consensus procedure is proposed, which is implemented between experts and proves the labeled data element. Such an approach minimizes the number of labeling errors and so shortens the time of neuronet training as well as improves the efficiency of classifier.

The content verification stage, implemented with the consensus mechanism, can be performed as following: every expert receives some content, labeled by the other experts. He must classify this additional content too. As the asynchrony is assumed, and delays can take place, the consensus procedure can be performed as every expert receives the content from the others.

The alternative way to implement the labeled content verification is to send the same piece of content to every expert in the group of more than one expert. So, when the consensus on the labeled data is reached, the labeled piece of content is sent to the Orders ledger.

For the implementation of procedure mentioned above, the following steps are needed:

- to form the experts groups within the experts community, which are presupposed to receive the equal content elements;
- to deliver the content to every member of the group;
- to perform the consensus operation within the group.

As the groups are presupposed to be relatively small, the PBFT consensus [22], or the simple majority voting can be used. So, the algorithm of the content verification is as presented below:

1. To perform the leader election within the expert community.
2. A leader determines the groups of experts who receive the equal content.
3. A leader requests the content in the Content delivery subsystem for the community.
4. A leader transfers the content to groups in the borders of the community.
5. A leader performs the labeled content gathering and chooses that labels that are proved by the majorities in the groups.
6. Leader sends the labeled content to the Orders Ledger.
7. The new leader election (optionally).

Here step 7 is optional in dependence of those events, which are the conditions for leader change. For example, the leader can be elected every verification round (as it is implemented in PBFT consensus algorithm), or just in case of node crash (as it is implemented in View Stamped Replication) [22].

In the Fig. 3 the main stages of content verification are shown.

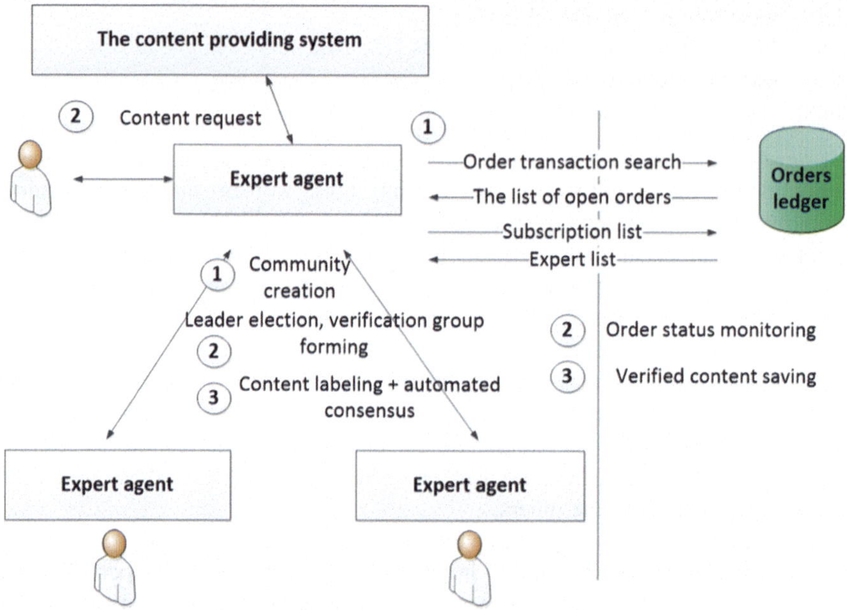

Fig. 3. The expert communities-level consensus on labeled data

So, the procedure described is supposed to improve the quality of training dataset, which is the issue of crowdsourcing labeling frequently.

5 Conclusions

In this paper the new approach to the training data sets forming is proposed. It presupposes the usage of crowdsourcing for data labeling, yet, differs from the existing approaches in terms of decentralization. The new architecture is supposed to be fully decentralized and designed on the basics of distributed ledger technologies. Such a design allows to avoid the intermediary service to form the labeled content, and so to decrease the labeling cost and dependency on the intermediator service. The platform architecture and its basics are presented and described, as well as the algorithm of additional consensus on the level of experts groups is proposed to improve the quality of the labeled data and to minimize the labeling noise of the training dataset.

Acknowledgements. The paper has been prepared within the RFBR projects 18-29-22086 and 18-29-03229.

References

1. ImageNet. http://www.image-net.org. Accessed 12 Dec 2019
2. Common objects in context. http://cocodataset.org/#home. Accessed 12 Dec 2019

3. The CIFAR-10 dataset. https://www.cs.toronto.edu/~kriz/cifar.html. Accessed 12 Dec 2019
4. How to organize data labeling for machine learning: approaches and tools. https://www.kdnuggets.com/2018/05/data-labeling-machine-learning.html. Accessed 12 Dec 2019
5. Clickworker: Data labeling service. https://www.clickworker.com/crowdsourcing-glossary/data-labeling/. Accessed 12 Dec 2019
6. Amazon Mechanical Turk. https://www.mturk.com/. Accessed 12 Dec 2019
7. Natarajan, H., Krause, S., Gradstein, H.: Distributed Ledger Technology (DLT) and Blockchain. FinTech note, no. 1. Washington, D.C., World Bank Group (2017)
8. Walport, M.: Distributed ledger technology: beyond blockchain. UK Government Office for Science (2016)
9. Crosby, M., Nachiappan, P.P., Verma, S., Kalyanaraman, V.: Blockchain technology explained. Sutardja Center for Entrepreneurship & Technology Technical Report (2015)
10. Frénay, B., Verleysen, M.: Classification in the presence of label noise: a survey. IEEE Trans. Neural Netw. Learn. Syst. **25**(5), 845–869 (2014)
11. Hickey, R.J.: Noise modeling and evaluating learning from examples. Artif. Intell. **82**(1–2), 157–179 (1996)
12. Manwani, N., Sastry, P.S.: Noise tolerance under risk minimization. IEEE Trans. Cybern. **43**(3), 1146–1151 (2013)
13. McDonald, A., Hand, D.J., Eckley, I.A.: An empirical comparison of three boosting algorithms on real data sets with artificial class noise. In: Proceeding of 4th International Workshop Multiple Classifier Systems, Guilford, UK, pp. 35–44, June 2003
14. Abellán, J., Masegosa, A.R.: Bagging decision trees on datasets with classification noise. In: Link, S., Prade, H. (eds.) Foundations of Information and Knowledge Systems. FoIKS 2010. Lecture Notes in Computer Science, vol. 5956, pp. 248–265. Springer, Heidelberg (2010)
15. Joseph, L., Gyorkos, T.W., Coupal, L.: Bayesian estimation of disease prevalence and the parameters of diagnostic tests in the absence of a gold standard. Am. J. Epidemiol. **141**(3), 263–272 (1995)
16. Perez, C.J., Giron, F.J., Martin, J., Ruiz, M., Rojano, C.: Misclassified multinomial data: a Bayesian approach. Revista de la Real Academia de Ciencias Exactas, Físicas y Naturales. Serie A, Matemáticas **101**(1), 71–80 (2007)
17. Brodley, C.E., Friedl, M.A.: Identifying mislabeled training data. J. Artif. Intell. Res. **11**, 131–167 (1999)
18. Gamberger, D., Boskovic, R., Lavrac, N., Groselj, C.: Experiments with noise filtering in a medical domain. In: Proceeding of 16th International Conference on Machine Learning, Bled, Slovenia, pp. 143–151. Springer, San Francisco, June 1999
19. Krauth, W., Mezard, M.: Learning algorithms with optimal stability in neural networks. J. Phys. A: Gen. Phys. **20**(11), L745 (1987)
20. Clark, P., Niblett, T.: The CN2 induction algorithm. Mach. Learn. **3**(4), 261–283 (1989)
21. Cantador, I., Dorronsoro, J.R.: Boosting parallel perceptrons for label noise reduction in classification problems. In: Proceedings of First International Work-Conference on the Interplay Between Natural and Artificial Computation, IWINAC 2005. Lecture Notes in Computer Science, vol. 3562, pp. 586–593. Springer, Berlin (2005)
22. Liskov, B.: From viewstamped replication to Byzantine fault tolerance. In: Charron-Bost, B., Pedone, F., Schiper, A. (eds.) Replication. Lecture Notes in Computer Science, vol. 5959. Springer, Heidelberg (2010)

Assessing the Relationship Between BPM Maturity and the Success of Organizations

Joana Pinto(✉) and Vitor Duarte dos Santos

Universidade Nova de Lisboa - Information Management Systems School, Lisbon, Portugal
m20170209@isegi.unl.pt

Abstract. For the past decades, organizations have been investing heavily in BPM projects in the hope of improving their competitive advantage in an increasingly complex environment. However, although it is believed that the higher the level of BPM maturity the greater the success of the organization, experience shows that this relationship is not always possible to prove.

The purpose of this study is to help clarify the relationship between the level of BPM maturity and the success of an organization. This was done through the implementation of a case study-based research within a global company, focusing on the shared services organization.

An analysis of the existing BPM maturity models and its level of coverage of BPM core areas was conducted to select the most suitable BPM maturity model to conduct the assessment of the current BPM maturity level. It was also established a framework to characterize the success of an organization. These two inputs, along with information gathered to understand implemented process improvements, were the basis for conducting the research.

Results show a successful organization, with a high maturity level according to the BPM OMG maturity model, that has been investing in continually improving its processes with a strong focus on digital transformation. The identified benefits from a high level of BPM maturity, namely improved productivity, cost reduction, error & risk prevention, higher agility, employee upskilling and knowledge retention, were shown to have a positive influence in the majority of the dimensions used to characterize the success of the organization.

Keywords: Business process management maturity · Success of organizations · Shared services

1 Introduction

1.1 Background

Since the first industrial revolution, companies have been focusing on continuously improving their productivity by introducing new technical innovations, as well as by (re)organizing the work in a way to become more effective and efficient [1]. In this context, over the past decades, companies became aware on the benefits of being

R. Silhavy (Ed.): CSOC 2020, AISC 1226, pp. 108–126, 2020.
https://doi.org/10.1007/978-3-030-51974-2_10

process oriented as a way to create competitive advantage and respond to the fast pace changing environment [2].

Along with that, the concept of Business Process Management (BPM) has been evolving in the past decades. It has its roots on the concept of workflow management (WFM), inspired by the production processes in the manufacturing industry, that is mainly focused on the automation of business processes [1]. BPM, on the other hand, has a wider scope and aims to improve business processes, possible without the use of new technologies [1].

The BPM institute describes BPM as the "definition, improvement and management of a firm's end-to-end enterprise business processes in order to achieve three outcomes crucial to a performance-based, customer-driven firm: 1) clarity on strategic direction, 2) alignment of the firm's resources, and 3) increased discipline in daily operations" [3]. In fact, BPM can be understood as the efforts of an organization to analyse and continually improve its fundamental activities [4].

In the past decades, many maturity models have emerged claiming to guide an organization towards its competitive advantage and, therefore, to its success [5]. The BPM maturity models have the purpose to outline the stages of maturation paths and provide a tool to assess the as-is situation, identify gaps, define improvement initiatives and control the progress [6]. Research appoints that higher levels of maturity in a business process results, among others, in greater effectiveness to reach the defined goals and improving management ability to propose new and higher targets for performance [6].

Among several models, it is possible to highlight the BPM Maturity Model (BPMMM) [7, 8], BPR Maturity Model (BPRMM) [9], Business Process Maturity Model (BPMMFisher) [10], Process Management Maturity Assessment (PMMA) [11], BPO Maturity Model (BPOMM) [5], Process and Enterprise Maturity Model (PEMM) [12], Process Maturity Ladder (PML) [13], Business Process Maturity Model (BPMMOMG) [14] and the Business Process Maturity Model (BPMMLee) [6, 15].

In spite of the fact that many empirical researches indicate that there is a positive correlation between process management and business success [4], and that organizations can improve their performance by adopting a business process orientation [5], research also suggests that only few organizations were able to obtain efficiency gains from the process-oriented transformation programs they have implemented [16].

In fact, there is a common understanding among organizations that the greater the process orientation is, the larger its success. This premise has been guiding many of the investment decisions of organizations in the past years that expect to enhance their competitive advantage by investing significantly in a business process-oriented approach. However, experience shows that this relationship is not always straightforward once there are other variables that might influence the availability of the organization to incorporate the benefits of BPM, as well as its success.

1.2 Objectives

The purpose of this research is to provide a contribution to clarify the relationship between the level of BPM maturity and the success of an organization. Specifically, the research aims to answer the following question: Is there a relationship between the level of business process management maturity of an organization and its success?

This was done by conducting a qualitative research through a case study within an organization with high perceived BPM maturity level to assess its current maturity and its level of success, as well as to understand the evolution done in the past years regarding process improvements. The data collected was the basis for analysing the results and providing insights about the relationship between the two areas.

2 Methodology

To achieve the desired outcome, the work was structured in three phases: 1) conceptualization, where the goal was to, through a literature review, identify the problem and research questions that will guide the study, as well as to define the basis for the research; 2) implementation, where the focus was the implementation of the research according to the inputs provided in the first phase and 3) analysis, where the results were systematized, analysed and discussed in order to answer the research question.

The first step was to conduct an extensive literature review to understand the key concepts related with business process management, business process management maturity and success of an organization.

Once there are many BPM maturity models that can be applied to assess the BPM maturity level of the organizations, a selection of one model was done. For that, an analysis of the existing BPM maturity models and its level of coverage of BPM core areas was conducted. Along with that, and based on the literature review, a conceptual model to measure the success of the organization was defined.

These two inputs were the basis to conduct a qualitative research, implemented through a case study. The selection of this approach and method was done taking into consideration several factors such as the research problem, purpose of the study and its specific goals. The case study was implemented within a shared services organization of an international company who operates in Portugal and has a high perceived process maturity level. This was done through in-depth interviews and documentation analysis.

After collecting all the data, results were systematized in order to bring inputs to support the answer to the research question. These results were analysed, focusing on ensuring the accomplishment of the study objectives, as well as identifying limitations and recommendations for further research.

2.1 Selection of the BPM Maturity Model

The literature review has shown a multiplicity of business process maturity models. Although most of them are based on the CMMI model, they represent different variations of a solution to answer some fundamental questions: (i) how mature are the processes of an organization and (ii) how mature is the organization to manage those processes. There is no best answer to these questions. The key is to understand the needs of the organization and how can the evaluation of its BPM maturity contribute to achieve positive outcomes. Based on these factors, organizations should select the model it is considered to best fit their purpose [17].

Even though several BPM maturity models could be used in this research, that would increase significantly the complexity of the assessment tool, which might diminish the company's willingness to participate. Therefore, one of the critical success factors is the selection of one maturity model that best address the purpose of the study.

The selected model should be applicable to any company, despite its dimension and business sector. It is also important that the model is aligned with the key areas of business process management, providing a holistic approach on the maturity of the organization for this subject. To address these issues, an analysis of the coverage of the BPM maturity models through the BPM core areas was undertaken.

The starting point were the models analysed in depth during the literature review phase. Those models were studied in depth in order to understand to which extent they refer to the practices contained within each BPM core area.

A BPM maturity model was considered to have coverage of one BPM core area if they have a dimension, a factor or similar, according to its specific structure, that clearly matches the description and principles of the core area. On the opposite, if no match was possible, the model was considered as having no coverage of the BPM core area. In some situations, where the match was not direct or totally clear, or if it only covers some components of the core area, the model was considered has having partial coverage. The results are presented in Table 1.

With the conducted analysis, it is possible to conclude that there is no BPM maturity model that covers all the BPM core areas. However, some provide greater coverage such as the BPM Maturity Model (BPMMM) of Rosemann & de Bruin [7, 8, 18] and the Business Process Maturity Model (BPMMOMG) of the OMG Group [14]. While the first model lacks on the coverage of the enterprise process management (EPM) core area, the second one lacks on the information systems/ information technology (IS/IT) core area.

Once it is intended to evaluate the level of maturity of an organization and not specifically the level of maturity of a process, it is important that the selected model covers the core areas that are under the enterprise perspective. In this context, and despite its limitations, the BPM OMG maturity model was selected.

2.2 Selection of the Measures of Organizational Success

The literature review has shown that defining the success of an organization is not straightforward and, in some extent, there is some confusion with the term.

Based on the assumption that the level of success represents in what extent can an organization fulfils its objectives [19], and if we consider that an organization is a complex system composed by several dimensions with specific objectives, a multidimensional approach to the success of an organization should be considered.

In this context, a 5-dimension model to characterize the success of an organization is proposed.

Table 1. Results of the analysis of level of coverage of BPM maturity models through the BPM core areas

		BPM core areas								
		Process perspective							Enterprise perspective	
Model	References	Modelling	Analysis	Design	Implementation	Performance Management	Transformation	IT/IS	Organization	EPM
BPM Maturity Model (BPMMM)	[7, 8, 18]	C	C	C	C	C	C	C	C	NC
Business Process Maturity Model (BPMMFisher)	[10]	NC	NC	NC	NC	C	NC	C	PC	C
Process Management Maturity Assessment (PMMA)	[11]	C	NC	C	NC	C	C	C	C	PC
Process and Enterprise Maturity Model (PEMM)	[12]	C	C	C	NC	C	NC	C	C	C
Process Maturity Ladder (PML)	[13]	PC	NC	PC	NC	C	C	NC	PC	NC
BPO Maturity Model (BPOMM)	[5, 20]	C	NC	NC	NC	C	NC	C	C	MC
Business Process Maturity Model (BPMMOMG)	[14]	C	C	C	C	C	C	NC	C	C

C – Covers; PC – Partially Covers; NC – No Cover

Fulfilment of Mission. Mission statements are one of the cornerstones of strategy definition, providing a sense of purpose and establishing the direction of an organization [21]. Considering that the mission statement of an organization represents its ultimate goal, it cannot be disregarded the fulfilment of the mission as a dimension to measure organizational success.

Financial Performance. Financial performance has been one of the most traditional perspectives to measure the success of an organization. Although the fact that other perspectives have been included in measuring organizational performance and success, the financial perspective is still widely used in the literature, as well as in business practice, and cannot be discarded [22–26].

Market Performance. Along with financial performance, market performance has been widely considered in measuring organizational performance, providing an external point of view [22, 23, 26]. In fact, besides the importance of having a financially well performed company, it is also relevant to consider how the company is performing in the market against its competitors, as well as to which extent it is satisfying its customer needs [23, 27, 28].

Internal Performance. Internal performance refers to the internal aspects such as processes, human capital, corporate culture as well as other internal practices that allow companies to develop and sustain its competitive advantage. The Balanced Scorecard [22] introduces this variable through the internal and the learning & growth perspective. The DMP approach also reflects the internal performance through its people development and process dimensions [23]. Also, Pickle & Frielander [29] refer to one of the most important stakeholders within a company: the employees. Therefore, internal performance is considered to be related with human capital management issues, quality issues as well as cultural issues.

Sustainability. Another view on company's success relates to the sustainability of the organization. Organizations perform their activities in a given environment, influencing it. In this context, the concept of corporate sustainability has been gaining importance [30–32]. However, the sustainability concept within the business context does not have a single definition. The study conducted by Lankoski [33] identifies different usages of the concept highlighting the concept of sustainability as the long term viability of the business and sustainability as corporate social responsibility. Both of these approaches to the sustainability concept will be considered in this study.

For each of the five dimensions proposed to characterize organizational success, some categories were defined based on the literature review. The purpose of defining categories is to provide a more consistent logical framework for measuring success. Also, defining categories helps with the process of selecting metrics for measurement, allowing it to be more focused and straightforward. In this context, for each category, one or two measures were selected, as presented in Table 2. The number of measures was not too extensive, so the research would not be highly complex, increasing willingness to participate from the company. Besides ensuring that the proposed metrics were aligned with the dimension to be evaluated, three key aspects were considered when selecting the metrics for each category: the obtainability of quantitative or quantifiable data, availability of information in the company and relevance of metrics.

Furthermore, the metrics should be simple, understandable and useful as a tool [34].

Table 2. Dimensions, categories and metrics to measure organizational success

Dimension	Category	Metric	References
Fulfilment of mission	–	Perception of managers	[21]
Financial performance	Growth	Income growth rate	[34]
	Profitability	Net profit margin	[28, 35]
	Efficiency	Return on equity	[23, 28, 35]
		Return on assets	[23, 28, 35]

(*continued*)

Table 2. (*continued*)

Dimension	Category	Metric	References
Market performance	Market position	Market share	[23, 27, 28, 35]
		Sales growth rate	[23, 28, 35]
	Customer satisfaction	Level of customer satisfaction	[27]
Internal Performance	Human capital	Turnover rate	[35]
		Employee growth rate	[25, 28, 35]
	Organizational culture	Perception of managers on how strong the culture is	[36]
	Quality & risk management	Implementation of operational internal audits	[37]
		Implementation of risk management procedures	[37]
Sustainability	Longevity	Number of years company is in business	[35]
	Corporate social responsibility	Investment in CSR initiatives on total revenue	[38]
	Innovation	Investment in R&D and self-innovation	[38]

2.3 Case Study Implementation

Selecting the best approach and methodology is usually one of the most challenging aspects when implementing a research.

As previously mentioned, there is a lack of research between the relationship between BPM maturity and the success of the organizations and the purpose of this research is to provide insights and to better understand this relationship, which could be a starting point for further research.

According to Darke, Shanks, & Broadbent [39], case study has been one of the most used methods to implement a qualitative research in information systems (IS) research. It assumes particular relevance when it is used to investigate a complex phenomenon that rely on several evidences. Hughes & McDonagh [40] also refer its popularity in IS research and refer it as a powerful method when implementing an exploratory research as it allow to get higher richness of data. Trkman [4] also refers that case study methods can be used when we are still trying to understand or discover a problem within an emerging area.

In spite of its limitations [39, 40], it is considered as suitable to be used when trying to relate information systems innovation and organizational contexts [39], which fits the scope of this research. In this context, an exploratory research based on a qualitative approach with case study was the selected method.

The research was implemented within a global company that operates in the energy sector and has an operation in Portugal. The starting point to identify a company was the list of the five hundred (500) bigger companies in Portugal, published annually by Exame magazine, a magazine leader in the business sector. Within the list, the criteria applied to select the company were the high perceived process maturity level, the fact that has an operation in Portugal and its willingness to cooperate.

The selected company operates in Portugal for one hundred and twelve (112) years in several areas such as energy, infrastructures and mobility. It is also an important hub for providing services to other companies and locations within the same business group through its shared services organization.

Considering the existent body of knowledge and literature available in this area, it is possible to understand that business process management practices have been assuming particular importance in the shared services area.

Shared services could be understood as the support functions from different departments that are consolidated into a specific organizational entity that aims to provide these services in an efficient and effective way. This services usually include areas such as human resources, finance and procurement [41, 42].

Establishing a shared service organization has been an approach that many companies have been following in the past decades as a way to allow them to focus on its business, improve quality, reduce costs and develop new capabilities, among other benefits [42].

Considering the benefits of a high business process management maturity and business process orientation, along with the expected benefits from implementing a shared services area, it is possible to identify similarities regarding efficiency, quality and continuous improvement which raises the importance of conducting this study within a shared service organization.

3 Results

3.1 BPM Maturity Analysis

The assessment of the current BPM maturity level, based on the BPM OMG Maturity Model, allow to portrait the current situation of the organization regarding its process management practices. This can be used to identify constraints and improvement initiatives that can be undertaken to ensure that the organization captures and incorporates the benefits of having a high process orientation.

Each maturity level, except the first one, have a set of defined process areas in a total of thirty (30) process areas. The maturity level is achieved when the goals of the process areas are accomplished. Process areas can be related with four main cornerstones: organization, work unit, product & service offering and improvements [14].

Based on the interview conducted with the process manager, as well as from the analysis of the documentation provided, it was possible to establish what was the level of fulfilment of each process area hence each maturity level. To do that, and due to the fact that a process area can only be considered as accomplished if its specific goals are achieved [14], an analysis per specific goal was undertaken. For each specific goal, a score between 0 and 1 was attributed. The accomplishment level of each process area is given by the average of the scoring of each specific goal.

Results suggest that the first three maturity levels – initial, managed and standardized – are achieved once each process area and its related specific goals are fulfilled.

The fourth maturity level – predictable – is partially achieved with an accomplishment level of 66%. This is mainly due to the fact that the quantitative management of the processes is still done in an incipient way without robust and statistically managed approaches to establish predictable results. In fact, the process area with lower score (0,33 out of 1,00) is the quantitative process management (QPM), followed by the quantitative product and service management (QPSM) (0,50 out of 1,00).

The fifth and last maturity level – innovating – is also not completely achieved, with an accomplishment level of 82%. Although the organization has established several practices for innovation and continuous improvement, there is still a lack of understanding of which are the direct quantitative impacts of each improvement implemented. These outcomes are comprehensible if we consider the results from the fourth maturity level where one of the least rated areas is associated with the lack of a quantitative management approach.

Also, in spite of the quality management system in place that promotes a high process management maturity, the focus on individual process performance and workgroup process performance is limited as there is no specific practices in place that address these components. This is reflected in the continuous capability improvement area (CCI) that has the lowest score on this maturity level (0,25 out of 1,00).

As a conclusion, it is possible to state that the studied organization has a high BPM maturity level. As the model states that an organization can only achieve a maturity level if it meets the requirements associated with the previous maturity level [14], the maturity of the studied organization is established at level 4 – predictable.

3.2 Process Improvements and Impacts

The second area of the study is focused on understanding what improvements have been made on the organization's processes, in the past years, and what were the results of those improvements, in a qualitative and, if possible, quantitative way.

Shared services organizations seek to provide cost savings, increased efficiency and the development of new capabilities in the environment they are established [42]. To leverage their ability to fulfil its goals, they have been heavily investing in information systems and technology such as enterprise resource planning (ERP), self-service portals and automation tools [41]. This has been one of the drivers of the organization, that has been investing on the digital transformation of the value chain including areas such as master data, customer order management, purchase to pay and cash management. This is aligned with the trends of the shared service organizations such as digital transformation and service automation as appointed by Lacity & Willcocks [41].

3.3 Characterizing the Success of the Organization

The third component of the study aims to characterize the success of the organization. This was done according to the proposed model that is structured in five areas: (i) fulfilment of mission, (ii) financial performance, (iii) market performance, (iv) internal performance, and (v) sustainability.

Fulfilment of Mission. Fulfilment of mission is evaluated through the perception of the manager regarding to which extent is the mission of the organization being accomplished. This input was collected in the interview conducted to the process manager and the result is that it is confirmed that the organization is accomplishing the purpose for what it was created for.

Financial Performance. Although a shared service organization does not pursue higher profits but rather efficiency and optimization, this variable could not be disregarded from analysis.

In spite of the fact that financial performance has not been improving, according to growth and profit metrics, the organization is stable from the financial point of view as it provides significant income every year. Profit rate is not very high but, if we consider that shared services organization do not pursue profit, these values are as expected.

Two additional metrics for measuring financial performance were proposed in the model: return on equity and return on assets. However, this data was not possible to obtain from the organization.

If we consider the specific context of the organization, an important metric arises: savings. As previously mentioned, a shared services organization pursues optimization which, in the end, might be reflected as savings. This metric has been evolving positively in the past years, demonstrating the focus on efficiency and optimization.

Market Performance. To asses this variable, two perspectives are appointed in the model: one related with competition and one related with customer satisfaction.

The studied organization does not operate in an intensive competitive market as the shared service area provides business services to other organizations within the same business group. Therefore, market share is not a relevant metric in this context.

On the other hand, sales growth metric allows to understand how the organization has been evolving in terms of providing more or new services to its customers. According to the data provided, this has been evolving positively in the last years, which shows a positive trend regarding service provision to customers.

Customer satisfaction is also a very important metric. The results show a high level of customer satisfaction throughout the time, with the lowest score of 8,59 out of 10 achieved in 2017. The score has been consistently higher in the Portuguese organization when comparing to other locations that provide similar services.

Internal Performance. The defined model to measure the success of the organization also considers the internal perspective concerning human capital, corporate culture and quality management practices. The first metric, turnover rate, was between 10%–16% in the last three years. These values are aligned with the industry practices [43].

The number of employees has been increasing in the past years, with special attention between 2018 and 2019 where a big increase was verified, along with the increase of the service volume

Organizational culture was evaluated through the perception of manager regarding on how strong the culture is. The organization shows a well-established culture, with a set of principles, values and guidelines transmitted to the employees that ensure their implementation in their day-to-day work.

Regarding quality, two metrics were analysed: the implementation of internal audits and the implementation of risk management procedures. Both are visible in the company.

Sustainability. This component, along with the internal performance, allows to complement the traditional perspective on the success of an organization, providing insights on how the company lives in an increasingly complex environment.

The first indicator of sustainability is the longevity of the organization, meaning the number of years that it is performing its business. The studied organization has more than a decade in business and has been growing significantly during this period in terms of clients, employees and business model which shows its consolidation as a shared service organization.

The second indicator refers to the capacity of the organization to give back to the community and is focused on the investment in corporate social responsibility initiatives. Corporate social responsibility has been an area of importance in the organization which can be seen by the investment done in the past years.

The third indicator is related with the investment in R&D and self-innovation. Due to the specificity of the organization and its governance, it was not possible to obtain this information focused only in the shared services organization.

4 Discussion

The results from the BPM maturity assessment allow to understand that the organization has established itself on the maturity level 4, even though it had already accomplished some process areas from maturity level 5.

Although the organization does not have implemented a BPM model, it is able to achieve a good BPM maturity level due to the existence of a high focus on quality management. In fact, through the analysis conducted, it is possible to conclude that the high level of BPM maturity is based on nine cornerstones:

1. The existence of a solid and consolidated quality management system that was implemented since the organization was established. The quality management system includes a systematic assessment and auditing process to ensure the compliance with the defined guidelines. The assessment provides a score that reflects the maturity of the quality management practices which allows organization to define and implement a roadmap, as well as to compare itself with other similar organizations within the same business group. To show the importance of this quality management system to executive management, one of the annual performance targets defined by the organization is the score achieved in these assessments.

2. The implementation, within the scope of the quality management system, of several approaches and methodologies to identify and treat the non-conformances as well as to establish process improvement initiatives. In this context it is important to highlight the 8D tool and FMEA (failure mode and effect analysis). 8D tool allows the organization to have a reactive response to non-conformances and issues that arise in their operation, allowing it to be registered, tracked, solved and communicated. FMEA analysis, on the other hand, allows the organization to have a proactive response to potential risks, as well as to identify and implement improvement activities in the processes.
3. The fact that processes and work instructions are documented and available in the organization, reflecting the work that needs to be conducted by each work unit according to the requirements established in the internal service agreements (ISAs).
4. The establishment of internal service agreements (ISAs) with the customers that reflect the requirements for the service, as well as the key performance indicators (KPIs) that represent the performance agreements. They are also an important instrument to conduct the capacity planning process to ensure the alignment between demand and resources.
5. The application of a change request process that documents the changes that are requested to a service, analyses them with the support from a business case and an impact analysis and implements them in a structured way, agreed by the relevant stakeholders.
6. The definition of organizational targets, drilled down to the different organizational levels, from top management to individuals, through the performance management program, that promote the alignment of business activities and drives the implementation of improvement activities.
7. The promotion of an open error culture, where errors are accepted and treated with a positive approach, as well as a continuous improvement system, that promotes innovation, best practice sharing and the implementation of improvement measures.
8. The recent transformation to an organization focused on end-to-end processes that promote coordination, alignment and synergies throughout the work units and organizational levels.
9. The high technological maturity, once most of the processes are supported by information systems and technology to promote efficiency and accountability.

This is aligned with the findings from Feldbacher, Suppan, Schweiger, & Singer [44] that establish that service companies reach an high maturity level than manufacturer ones. Also, companies that are active in the IT sector were shown to have a higher maturity level than companies from other industries.

The next steps to climb in the maturity ladder, according to the BPM OMG maturity model [14] are related with increasing the focus on managing the individual processes and workgroup processes in a more quantitative and statistically way, in order to achieve predictable results.

The organization is aware of the steps to be done and has already started to implement tools to improve the maturity level as it recognizes that it will allow to increase efficiency, reduce costs and improve client satisfaction. This is aligned with the findings from several authors that refer that organizations constantly seek to improve their maturity level as they believe greater maturity leads to greater performance [8, 10, 11].

Organization is also conscious of the importance of technology as a catalyst to promote their process management maturity. Although the technology maturity can be perceived as high, there is still a path to pursue towards a digital transformation of the operations. In this context, the organization has been highly investing in projects that focus on process automation, robotization and data analytics solutions.

The projects already implemented in this context have been providing significant results in terms of improved productivity, end-to-end integration, increased data quality, manual effort reduction with impact in the occurrence of errors, cycle time reduction and increased customer satisfaction. It also allowed the organization to improve its monitoring activities, establishing a more robust and accountable continuous improvement system, as well as to implement workforce upskilling programs. This was key to establish the organization as a successful shared service provider and is coherent with the appointed benefits of adopting a BPM approach.

Looking to the dimensions defined to characterize the success of the organization, they can be impacted by some of the appointed benefits from a high BPM maturity.

The main benefits from a high BPM maturity, leveraged by the usage of information systems and technology, that were identified through the study, can be grouped in six building blocks: (i) improve productivity; (ii) reduce costs; (iii) error and risk prevention; (iv) upskilling; (v) high agility; and (vi) knowledge retention.

According to the conducted literature review, a high BPM maturity allow to increase productivity and efficiency. If the organization improves productivity, this might lead to customer satisfaction, especially in a shared services organization where the customers pursue an increase in efficiency. Higher customer satisfaction will promote the consolidation of the organization in the market as it will allow a growth of sales. It is also a driver of growth as higher customer satisfaction might lead to income increase.

One of the impacts noted from a high BPM maturity is also cost reduction. This might influence several dimensions of the success of the organization. If we reduce costs, we are able to increase our net profit margin. On the other hand, as shared services organizations aim to provide cost effectiveness to their customers, this will also impact the customer satisfaction, hence contributing to strengthen the market position and possible revenue increase. Although this relationship is not always possible to verify from the literature review, this was pointed as a consequence of the investment of the organization in several process improvement projects, especially due to the context it operates.

Also, with cost reduction, organizations might have higher availability to invest in corporate social responsibility activities hence contributing to the sustainability component of the success of an organization.

Error and risk prevention are also appointed as impacts from a high BPM maturity. In fact, if the organization has a process-oriented culture in terms of people, processes and technology, it is more likely to have a proactive posture towards risk and potential errors. This directly impacts the quality of the delivered products and services hence influencing customer satisfaction in a positive way. Customer satisfaction, on the other hand, might lead to a stronger market position and revenue growth. Also, if the organization manages issues and risks properly, the negative impact of unwanted events might diminish, helping the organization deal with unexpected costs, hence impacting profit.

Another appointed benefit relates with the upskilling of employees. With process improvements focused on automation, employees do not need to perform routine tasks on a regular basis, which allow them to conduct higher value activities. This might represent an increase in the available capacity that could allow organization to gather new customers and increase its revenue.

Also, the available capacity could be shifted to conduct new tasks, more challenging and rewarding, that might lead to higher employee retention and commitment or even employee growth, hence lower turnover rate. Companies with less employee rotation have shown to have strengthen organizational cultures [45, 46] so it is possible to consider that lower turnover rate impacts positively the organizational culture. Employee turnover, on the other hand, is shown to affect growth, profitability and customer satisfaction [46].

High agility is also appointed as one of the benefits of a high BPM maturity has streamlined processes and less efforts on error detection and solving allow organizations to respond faster to customer needs. This improves customer satisfaction, strengthen market position, as well as increase potential income.

Knowledge retention is also an important point when it comes to BPM maturity. With documented processes, requirements and changes, the organization is promoting its knowledge retention which leads to a strengthened organizational culture and employee retention. A lower turnover rate, on the other hand, has impacts related with growth, profitability and customer satisfaction.

Besides the direct relations between the BPM maturity benefits and the success of the organization, other relationships should be highlighted in a broader sense. Experience shows that an organization that is growing, that is profitable and that has a well-established position in the market is likely to be sustainable and have greater longevity. This will impact the fulfilment of the mission as an organization, to accomplish the purpose for which was created, needs to have these four pillars well established and solid. These relationships are summarized in Fig. 1.

In fact, the benefits from a high BPM maturity influence, in a positive way, the key dimensions of growth, profitability, market position, customer satisfaction, human capital, organizational culture, quality & risk management and corporate social responsibility that, on their turn, contribute to the longevity of the organization and, consequently, to the fulfilment of its mission. On the other hand, the success of an organization is something complex and cannot be explained only by the high BPM maturity as there are other variables, internal and externals, that influence the ability of an organization to be successful.

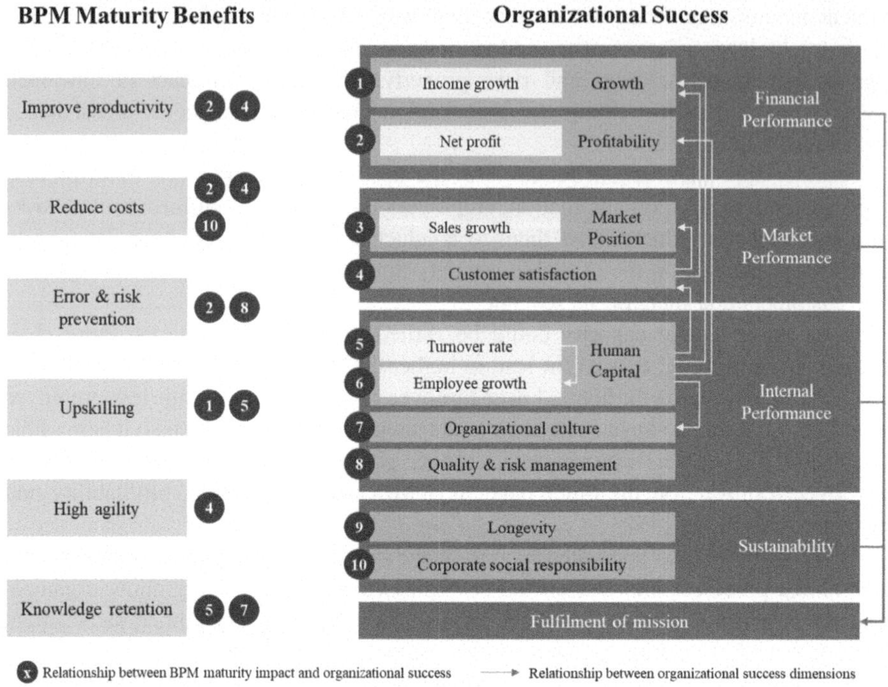

Fig. 1. Established relationships between BPM maturity and organizational success

5 Conclusion

Business process management has been assuming greater importance in the past years, as organizations continuously seek to achieve greater efficiency and effectiveness to leverage its competitive advantage in an increasingly complex environment.

The purpose of the study was to understand if there is a relationship between the level of business process management maturity and the success of an organization.

The research started with a literature review to understand the key concepts related with business process management, business process management maturity and success of an organization.

Once there are many BPM maturity models that can be applied to assess the BPM maturity level of the organizations, a selection of one model was done through an analysis of the existing BPM maturity models and its level of coverage of BPM core areas. Along with that, and based on the literature review, a conceptual model to measure the success of the organization was defined.

These two inputs were the basis to conduct a qualitative research, implemented through a case study, grounded on in-depth interviews and documentation analysis, within a shared services organization of an international company who operates in Portugal and has a high perceived process maturity level. The case study was structured in three areas: (i) assessment of the current BPM maturity level, based on the

BPM OMG maturity model; (ii) process improvements and its impacts; and (iii) characterization of the success of the organization.

The results from the BPM maturity assessment allow to understand that the organization has established itself on the maturity level 4, even though it already had fully accomplished some process areas from maturity level 5. Although the organization does not have implemented a BPM model, it is able to achieve a good BPM maturity level due to a strong focus on quality management and a high technological maturity.

The organization is aware of the steps to be done to climb the maturity ladder as it recognizes that it will allow to increase efficiency, reduce costs and improve client satisfaction. It is also conscious of the importance of technology as a catalyst to promote their process management maturity. In this context, the organization has been highly investing in projects that focus on process automation, robotization and data analytics solutions.

The projects already implemented, have been providing significant results can be grouped in six building blocks: (i) improve productivity; (ii) reduce costs; (iii) error and risk prevention; (iv) upskilling; (v) high agility; and (vi) knowledge retention.

This building blocks were shown to affect some dimensions of organizational success such as customer satisfaction, net profit, turnover, innovation, quality & risk management, income growth, organizational culture, among others. Those dimensions are key to establish a good financial, market and internal performance, as well as to promote the sustainability of the organization allowing it to fulfil its mission.

In this context, results suggest a positive relationship between the BPM maturity and the success of the organization, from a qualitative perspective. In fact, the research shows that an organization that has a high BPM maturity and has a high focus on process improvement is able to incorporate several benefits that leverage its ability to be more successful. However, the success of an organization is something complex and cannot be explained only by the high BPM maturity as there are other variables, internal and externals, that influence the ability of an organization to be successful.

Nevertheless, the research provides insights on how the relationship between process maturity and performance and success of an organization might be established which is important for the decision-making process of organizations when it comes to invest in BPM projects.

5.1 Limitations of the Study and Recommendations for Further Research

In spite of the fact that the research goals were achieved, there are some limitations of the study that should be pointed, along with some recommendations for future research.

The research was conducted through a case study within a shared service organization. Although this methodology provides deep insights on how organization has established its BPM practices, it lacks on comparability with other organizations. A suggestion for further research could be to expand the range of studied organizations. Other suggestion could be to study, with a quantitative approach, several organizations that operate in different business sectors allowing to collect statistical data that could corroborate the assumption that higher maturity levels lead to greater success.

The BPM maturity assessment itself also has some limitations to be pointed. The assessment was designed to ensure the willingness to participate from the organization and, therefore, was conducted up to the level of the specific goals within each process area and did not include the specific practices.

The fact that the BPM assessment was done through in-depth interviews also presents a limitation. By collecting data through interviews, we are, in some extent, exposed to the subjectivity of the answers from our interviewed. Although this risk was reduced by conducting a documentation analysis, some subjectivity might still exist.

Regarding definition of success, the framework that was established was intended to be the most generic as possible, in order to be applied to different organizations regardless of their business and context. However, by doing this, some of the specificities of the organization might have been disregarded.

Concerning the results of the study, two major limitations are ought to be appointed. First, the established relationship between BPM maturity and the success of the organization is qualitative and lacks quantitative data to corroborate the findings. Furthermore, it does not allow to establish the strength of the identified relationships. Future research should rely on establishing a quantitative relationship between these two dimensions in order to validate the achieved results. Second, the established relationships only consider the dimensions of success that were possible to analyse within the company. Additional measures related with efficiency and market position should be considered in further researches.

References

1. Van Der Aalst, W.M.P., La Rosa, M., Santoro, F.M.: Business process management: don't forget to improve the process! Bus. Inf. Syst. Eng. **58**, 1–6 (2016). https://doi.org/10.1007/s12599-015-0409-x
2. Willaert, P., Van Den Bergh, J., Willems, J., Dirk, P.: The process-oriented organisation: a holistic view developing a framework for business process orientation maturity. In: BPM 2007. LNCS, vol. 4714, pp. 1–15 (2007). https://doi.org/10.1007/978-3-540-75183-0_1
3. Rock, G., Dwyer, T.: What is BPM anyway? Business Process Management Explained. http://www.bpminstitute.org/
4. Trkman, P.: The critical success factors of business process management. Int. J. Inf. Manag. **30**, 125–134 (2010). https://doi.org/10.1016/j.ijinfomgt.2009.07.003
5. McCormack, K., Willems, J., van den Bergh, J., Deschoolmeester, D., Willaert, P., Indihar Štemberger, M., Škrinjar, R., Trkman, P., Bronzo Ladeira, M., Paulo Valadares de Oliveira, M., Bosilj Vuksic, V., Vlahovic, N.: A global investigation of key turning points in business process maturity. Bus. Process Manag. J. **15**, 792–815 (2009). https://doi.org/10.1108/14637150910987946
6. Röglinger, M., Pöppelbuß, J., Becker, J.: Maturity models in business process management. Bus. Process Manag. J. **18**, 328–346 (2012). https://doi.org/10.1108/14637151211225225
7. Rosemann, M., de Bruin, T., Hueffner, T.: A Model for Business Process Management Maturity. In: ACIS 2004 Proceedings, Paper 6, pp. 1–7 (2004)
8. Rosemann, M., De Bruin, T.: Towards a business process mangement maturity model. In: ECIS 2005 Proceedings of Thirteen European Conference on Information Systems, pp. 26–28 (2005). https://doi.org/10.1109/EUROMICRO.2007.35

9. Maull, R.S., Tranfield, D.R., Maull, W.: Factors characterising the maturity of BPR programmes. Int. J. Oper. Prod. Manag. **23**, 596–624 (2003). https://doi.org/10.1108/01443570310476645
10. Fisher, D.M.: The business process maturity model: a practical approach for identifying opportunities for optimization. Bus. Process Trends **9**, 11–15 (2004)
11. Rohloff, M.: Process management maturity assessment. In: AMCIS 2009 Proceedings, pp. 631–642 (2009)
12. Hammer, M.: The process audit (2007)
13. Harmon, P.: Evaluating an organization's business process maturity. Bus. Process Trends **2**, 1–11 (2004)
14. Weber, C., Curtis, B., Gardiner, T.: Business Process Maturity Model (BPMM) Version 1.0 (2008)
15. Lee, J., Lee, D., Kang, S.: An overview of the business process maturity model (BPMM), vol. 4537 (2007). https://doi.org/10.1007/978-3-540-72909-9
16. Alibabaei, A., Bandara, W., Aghdasi, M.: Means of achieving business process management success factors. In: Proceedings 4th Mediterranean Conference on Information Systems, 25–27 September 2009, pp. 25–27, Athens University Economics Business, Athens (2009)
17. Van Looy, A., De Backer, M., Poels, G., Snoeck, M.: Choosing the right business process maturity model. Inf. Manag. **50**, 466–488 (2013). https://doi.org/10.1016/j.im.2013.06.002
18. Rosemann, M., de Bruin, T.: Application of a holistic model for determining BPM maturity. In: Proceedings of the AIM Pre-ICIS Workshop on Process Management and Information system, pp. 46–60 (2004)
19. Miller, D., Peter, H.: Archetypes of strategy formulation. Manag. Sci. **24**, 921–934 (1978)
20. Lockamy, A., Childerhouse, P., Disney, S.M., Towill, D.R., McCormack, K.: The impact of process maturity and uncertainty on supply chain performance: an empirical study. Int. J. Manuf. Technol. Manag. **15**, 12–27 (2008). https://doi.org/10.1504/IJMTM.2008.018237
21. Pinho, C., Silva, A.M., Macedo, I.M.: Revisiting the link between mission statements and organizational performance in the non-profit sector: the mediating effect of organizational commitment. Eur. Manag. J. **34**, 36–46 (2016). https://doi.org/10.1016/j.emj.2015.10.003
22. Kaplan, R.S., Norton, D.P.: The stategy-focused organization: how balanced scorecard companies thrive in the new business environment. Harvard Business School Press (2001)
23. Maltz, A.C., Shenhar, A.J., Reilly, R.R.: Beyond the balanced scorecard: refining the search for organizational success measures. Long Range Plan. **36**, 187–204 (2003). https://doi.org/10.1016/S0024-6301(02)00165-6
24. Crumpton-Young, L.L., Ferreras, A.M.: The measure of success. Ind. Manag. **55**(3), 27–30 (2013)
25. Leković, B., Marić, S.: Measures of small business success/performance – importance, reliability and usability. Industrija **43**, 7–26 (2015). https://doi.org/10.5937/industrija43-7209
26. Keerthika, D., Alagarsamy, S.: A multiple measure of organizational performances and its effect on distinctive marketing competencies: an empirical study of india and the Maldives. J. Manag. Mark. Rev. **3**, 129–142 (2018)
27. Clark, B.: Measuring marketing performance: research, practice and challenges. In: Neely, A. (ed.) Business Performance Measurement: Unifying Theories and Integrating Practice, 2nd edn. pp. 36–63. Cambridge University Press (2007)
28. Richard, P.J., Devinney, T.M., Yip, J.S., Johnson, G.: Measuring organizational performance: towards methodological best practice †. J. Manag. **35**, 718–804 (2009). https://doi.org/10.1177/0149206308330560
29. Pickle, H., Frielander, F.: Seven societal criteria of organizational success. Pers. Psychol. **20**, 165–178 (1967). https://doi.org/10.1111/j.1744-6570.1967.tb02277.x

30. Linnenluecke, M.K., Griffiths, A.: Corporate sustainability and organizational culture. J. World Bus. **45**, 357–366 (2010). https://doi.org/10.1016/j.jwb.2009.08.006
31. Alexandre, A., De Francisco, A.C.: Organizational sustainability practices: a study of the firms listed by the corporate sustainability index. Sustainability **10** (2018). https://doi.org/10.3390/su10010226
32. Asemah, E.S., Okpanachi, R.A., Edegoh, L.O.N.: Business advantages of corporate social responsibility practice: a critical review. New Media Mass Commun. **18** (2013). https://doi.org/10.4028/www.scientific.net/MSF.879.1019
33. Lankoski, L.: Alternative conceptions of sustainability in a business context. J. Clean. Prod. **139**, 847–857 (2016). https://doi.org/10.1016/j.jclepro.2016.08.087
34. Székely, F., Knirsch, M.: Responsible leadership and corporate social responsibility: metrics for sustainable performance. Eur. Manag. J. **23**, 628–647 (2005). https://doi.org/10.1016/j.emj.2005.10.009
35. Murphy, G.B., Trailer, J.W., Hill, R.C.: Measuring performance in entrepreneurship. J. Bus. Res. **2963**, 15–23 (1996)
36. Garmendia, J.A.: The impact of corporate culture on company performance (2004)
37. Shin, W.S., Dahlgaard, J.J., Dahlgaard-park, S.M., Kim, M.G.: A quality scorecard for the era of industry 4.0. Total Qual. Manag. Bus. Excell. 29, 1–18 (2018). https://doi.org/10.1080/14783363.2018.1486536
38. Kao, E.H., Yeh, C., Wang, L., Fung, H.: The relationship between CSR and performance: evidence in China. Pac.-Basin Finan. J. **51**, 155–170 (2018). https://doi.org/10.1016/j.pacfin.2018.04.006
39. Darke, P., Shanks, G., Broadbent, M.: Successfully completing case study research combining rigour, relevance and pragmatism. Inf. Syst. J. **8**(4), 273–289 (1998)
40. Hughes, J., McDonagh, J.: In defence of the case study methodology for research into strategy practice. Irish J. Manag. **36**, 129–145 (2017). https://doi.org/10.1515/ijm-2017-0013
41. Lacity, M., Willcocks, L.: Robotic process automation: transformation lever for shared services. Credit Finan. Manag. Rev. **22**, 16–45 (2016)
42. Fielt, E., Bandara, W., Miskon, S., Gable, G.: Exploring shared services from an is perspective: a literature review and research agenda. Commun. Assoc. Inf. Syst. **34**, 1001–1040 (2014)
43. Deloitte: Global Shared Services 2017 Survey Report (2017)
44. Feldbacher, P., Suppan, P., Schweiger, C., Singer, R.: Business process management : a survey among small and medium sized enterprises (2016)
45. Bosomtwe, T.E., Obeng, B.: The link between organizational culture and turnover intention among employees in Ghana. Int. J. Contemp. Res. Rev. **9**, 20951–20958 (2018). https://doi.org/10.15520/ijcrr/2018/9/08/566
46. Mutua, C.M., Simba, F.T.: Effect of organisational culture on employee turnover: a case study of technical university of Mombasa. J. Manage. **4**, 1–14 (2017)

An Approach for Creating Data Structure of a Hierarchical Competency Model

Ondrej Pektor[1], Bogdan Walek[1(✉)], and Ivo Martinik[2]

[1] Department of Informatics and Computers, University of Ostrava, Ostrava, Czech Republic
{ondrej.pektor,bogdan.walek}@osu.cz

[2] Department of Computer Science, VSB-Technical University of Ostrava, Ostrava, Czech Republic
ivo.martinik@vsb.cz

Abstract. The article deals with the approach for creating, storing and manipulating a hierarchical competency model of a company. This approach is implemented in the ASP.Net Core web application environment. The article also describes the data model of the hierarchical competency model and the process of mapping to/from the JSON format. Techniques such as reflection, recursion, or dynamic data structures are used in the proposed approach. The article also includes an experimental verification of the proposed approach. The hierarchical competency model is used to create a general system for evaluating professional competencies of employees.

Keywords: Hierarchical competency model · Competency model · Competencies · System for evaluating competencies · Job applicants

1 Introduction

Nowadays, the area of recruiting suitable and competent staff to job positions is a very current issue. It is crucial for all companies to find workers who will not only be loyal and reliable but will, above all, have all the competencies required for the job.

In companies, it is emphasized that job positions are occupied by competent employees. Competencies are defined as knowledge, skills and attributes that distinguish an expert from a layman. Another definition is that competency is a construction which helps define the level of skill and knowledge [1, 2]. A person competent to take up a particular job must meet the given competencies to a minimum. This information is called the competency model. It can then be successfully used to define which employee skills are crucial for the job position [3, 4].

In our research we focus on creating a system for evaluating competencies. The system is useful for evaluating the suitability of job applicants or employees in a company that is able to determine its hierarchical competency model. The main goal of this system is to bring efficiency and to decrease the degree of subjective evaluation of job applicants and employees in job tenders or regular evaluation [5, 6].

The most important component of this system is the hierarchical competency model. This model has to be structured both in the relational database and in the

R. Silhavy (Ed.): CSOC 2020, AISC 1226, pp. 127–136, 2020.
https://doi.org/10.1007/978-3-030-51974-2_11

graphical interface of the system. The model is represented as a traditional tree structure in a relational database. The class representation of the hierarchical competency model is shown in Fig. 1.

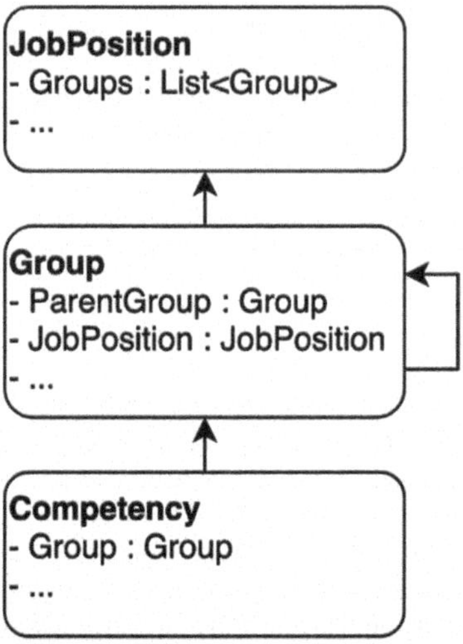

Fig. 1. Class diagram of database entities in the hierarchical competency model

The hierarchy shown in Fig. 1 is advantageous for several reasons. On one hand, when it is necessary to evaluate one person in several positions, it is sufficient to evaluate the same competencies only once. Another situation is when it is necessary to compare persons across different job positions. In this case, it is possible to compare persons in common competencies that are in a general part of the hierarchy.

2 Problem Formulation

The class model described above is necessary to implement the system. But the problem is how to provide an efficient and ergonomic management and usability of this hierarchical structure to the end users (company employees or HR managers). It is necessary for users to be able to modify the properties of individual elements in the hierarchy (competencies, competency groups, assigning groups to the job position, etc.) and also to change the representation of these elements in the hierarchy.

3 Problem Solution

Our proposed hierarchical competency model and the system for evaluating competencies is implemented in the ASP.NET environment as a web application. We assume that the best solution is to visualize the hierarchical structure with the minimum amount of information of each element [7]. Another function should be editing any element data through a dialog box. And the last necessary feature is the possibility to drag & drop the elements in the hierarchy. All these properties must be functional in JavaScript because it is a web application.

The process of creating a hierarchical competency model will be described in the following sections.

3.1 Data Visualization

Currently, there are several data-driven libraries for visualizing and manipulating JavaScript data, such as GoJS [8]. This library provides work with diagrams in a web application environment. We use this library to display a hierarchical tree diagram in several parts of the system. The library is used to visualize the hierarchical competency model of a company or a job position and also to visualize the hierarchical evaluating of a person in a job position. Next, the diagram of the hierarchical competency model will be discussed. After creating a basic diagram in a web page, we need to provide the data source. The data for the visualized tree diagram is represented by an array of objects where each object has a unique identifier and a parent element identifier. With this type of data structure, it is possible to realize a generic tree, where each parent element has a finite number of children elements. For using in a specific web application, we only need to customize the layout of the information that appears in each element and from which attributes of the model objects are to be populated. An example of a tree diagram of the hierarchical competency model is shown in Fig. 2.

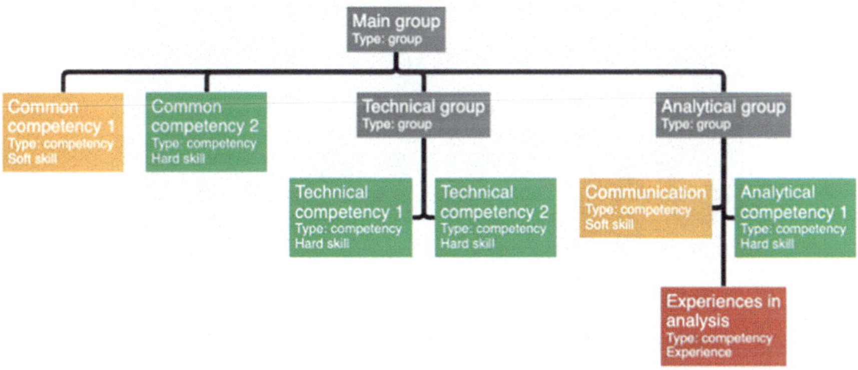

Fig. 2. Generated tree diagram of the hierarchical competency model

The hierarchical model consists of a few types of elements. The groups of competencies are represented by grey rectangles. The competencies are represented in the following way:

- Soft skills – orange rectangles,
- Hard skills – green rectangles,
- Work experience – red rectangles.

As we can see in Fig. 2, there is one main group of competencies called Main group. This group has one common soft skill competency, one common hard skill competency and two groups of competencies. The group called Technical group has two hard skill competencies. The group called Analytical competencies has one hard skill competency, one soft skill competency and one work experience competency.

3.2 Data Source

In next part, it is necessary to define the data source. In the ASP.NET Core environment, a data source means creating an endpoint in some way. After calling the endpoint, the requested data is in the response. In our system, we created the REST method which accepts the company ID as an input parameter and its output is a hierarchical structure in an open-standard file format JSON, which is easy to process in JavaScript.

Sample data for a diagram in the JSON format is shown in Fig. 3.


```json
[
    {
        "key": "ef644ae8-cd9a-4a79-a31a-61408e8459f6",
        "name": "Main group",
        "type": "GROUP"
    },
    {
        "key": "1d1dbfac-86c2-4296-945b-04e03bfff578",
        "parent": "ef644ae8-cd9a-4a79-a31a-61408e8459f6",
        "name": "Common competency 1",
        "type": "COMPETENCY",
        "competencyType": "SOFT_SKILL"
    },
    {
        "key": "43d1f6c0-3a42-4930-8663-e6f6f48435d5",
        "parent": "ef644ae8-cd9a-4a79-a31a-61408e8459f6",
        "name": "Technical group",
        "type": "GROUP"
    }
]
```


Fig. 3. Example of hierarchical model data in JSON format

The figure shows selected objects with selected attributes. The example represents three arrays which consist of objects. The first object is the main group (tree root), which does not have a parent object identifier. The second object is a competency called “Common competency 1”, which has the previously mentioned “Main group” set as the parent element. The last object is another group named “Technical group”, whose parent object is, again, a group named “Main group”.

3.3 Data Manipulation

For all data manipulation, a local copy is used on the client side. After each change, the local copy is confirmed as a model of the displayed diagram and after pressing the "Save model" button, the data is sent to the server.

Element Management
Double click function is used to access the editing of individual elements. This function invokes a modal window into which information from the selected element is loaded. When this information is saved, it is overwritten in a local copy of the data that is stored within the web page. The dialog box for editing a specific competency is shown in Fig. 4.

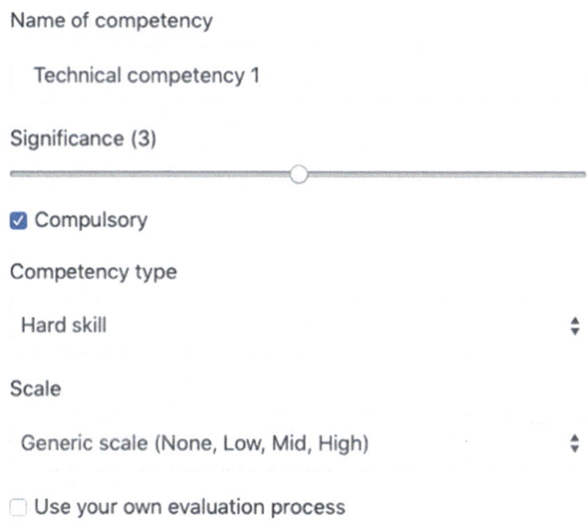

Fig. 4. Dialog for competency editing

Adding a New Element
To add a new element, a context menu is created that offers options to add a group of competencies or a competency to the hierarchy below the selected element. After selecting this option, a new element is added to the local data copy. The element is pre-filled with generic values and is ready for editing – see the description in the previous section.

Removing an Element
The removal of elements function is realized using a separate array of objects that have the "removed" flag. The element is first copied from the local data copy to this array, then the flag is set and then the element is removed from the local copy. The last step is changing the diagram model by a local copy of the data. In the process of saving a model to the database, the content of the array to delete is added to existing elements

from the local copy, and the server-side method deletes the required records. Figure 5 shows the context menu for the group of competencies.

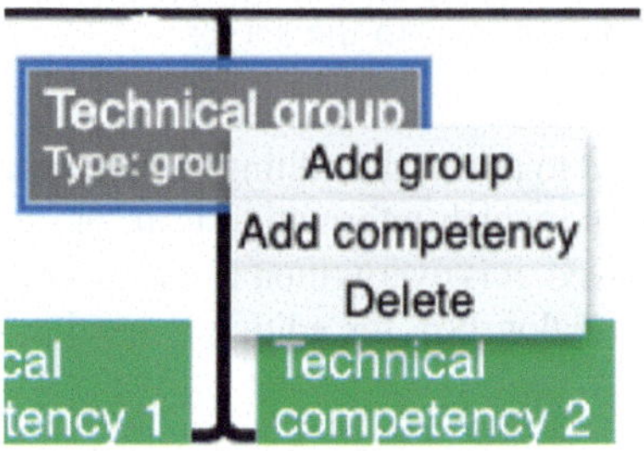

Fig. 5. Context menu for group

Structural Changes

The term Structural changes represents the process of moving an element (group of competencies or competency) in a hierarchy under another element (group of competencies). The drag & drop function is used to activate this process. We can use the left mouse button to grab an element and drag it to another element. There is no change in model if the target element is a competency. In the case of a group of competencies, the parent element identifier of the held element changes to the target element (group of competencies) identifier in the local data copy, and the local data copy is retrieved as a diagram model again. After this step, the diagram is arranged in a well-arranged tree with modified data. The child elements of the moved element remain unchanged and the hierarchy under the moved element is also unchanged. Figure 6 shows a diagram during editing in progress in structure and Fig. 7 shows a diagram after the change.

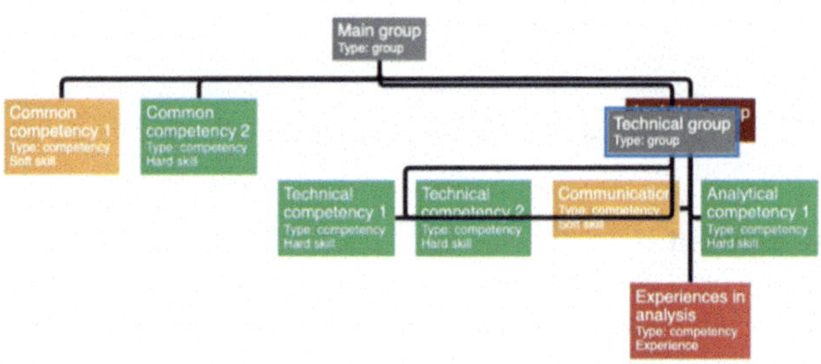

Fig. 6. Diagram during editing in progress

Data Persistence

In this step, the data persistence of the hierarchical model will be described. The data of modified model on a client side (possibilities are described in the previous subsections) is sent to the server and stored in the database for further use. The example of model data in the JSON format to be sent to the server is shown in Fig. 8. The structure of the data is similar to the structure presented in Fig. 3.

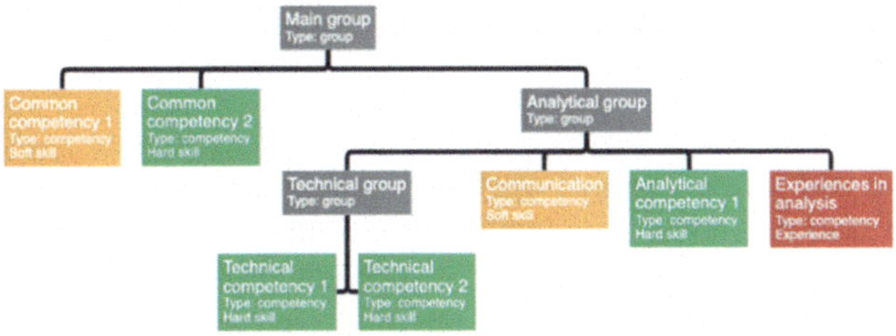

Fig. 7. Diagram after editing


```
[
    {
        "key": "1d1dbfac-86c2-4296-945b-04e03bfff578",
        "parent": "ef644ae8-cd9a-4a79-a31a-61408e8459f6",
        "name": "Common competency 1",
        "type": "COMPETENCY",
        "state": "REMOVED"
    },
    {
        "key": "ef644ae8-cd9a-4a79-a31a-61408e8459f6",
        "name": "Main group",
        "type": "GROUP"
    },
    {
        "key": "43d1f6c0-3a42-4930-8663-e6f6f48435d5",
        "parent": "ef644ae8-cd9a-4a79-a31a-61408e8459f6",
        "name": "Technical group",
        "type": "GROUP"
    },
    {
        "name": "New competency",
        "type": "COMPETENCY",
        "parent": "ef644ae8-cd9a-4a79-a31a-61408e8459f6",
        "key": "new0"
    },
    {
        "name": "New group",
        "type": "GROUP",
        "parent": "ef644ae8-cd9a-4a79-a31a-61408e8459f6",
        "key": "new1"
    }
]
```


Fig. 8. Example of hierarchical model data in JSON format to be sent to the server

The difference is that the competence called "Common competency 1" is removed. In this first object, the "state" attribute is set to "REMOVED". This will delete this competency from the database after sending it to the server. Then, a new competency called "New competency" and a new group "New group" are added. These objects have the string "new" in the "key" attribute and the sequence number of the new object. This identifier is on the server side replaced by the (UUID) database identifier. If a children are assigned to the new group of competencies, the same identifier is assigned to the parent attribute.

Dynamic Saving
For comfortable and safe interactive work with this hierarchical competency model, it is necessary to save the model after important operations. To avoid data inconsistency (multiple values stored, etc.), it is necessary to reload the data from the server after each save. To minimize the number of hits, it is possible to return data for the diagram in the save request response. The disadvantage of such an approach is that the data does not contain information about the graphical coordinates of the individual objects, so the tree diagram may be rearranged.

4 Verification

The verification of our hierarchical model will be demonstrated on several models of specific companies.

Figure 9 shows a hierarchical competency model of an accounting company which has common competencies, competencies for the accounting department, for the assistant department, and for lawyers. Each department is represented by one group of competencies.

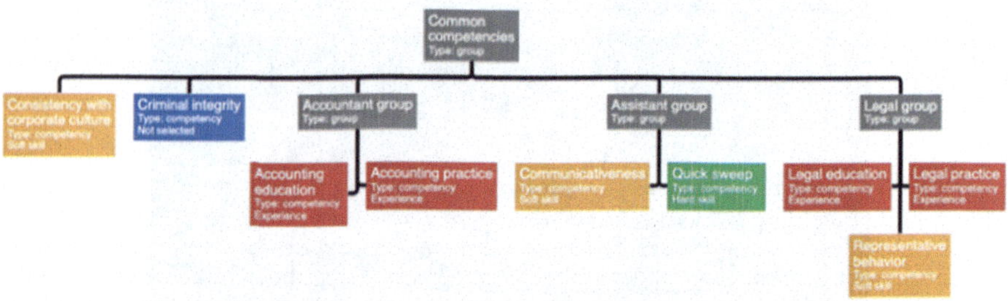

Fig. 9. Accounting company competency model

Figure 10 shows the hierarchical competency model for a software development company. We can see several groups of competencies which represent several divisions in company.

Figure 11 shows visualization of the hierarchical model for a manufacturing company. The competencies are divided to the two main groups: Product development and Product manufacturing.

As we can see, our proposed hierarchical allows to create models of various companies with various structure.

The advantage of our approach is the possibility to group competencies into groups according to the organizational structure of the company. In addition, the model allows to create common competencies which are the same for various job positions.

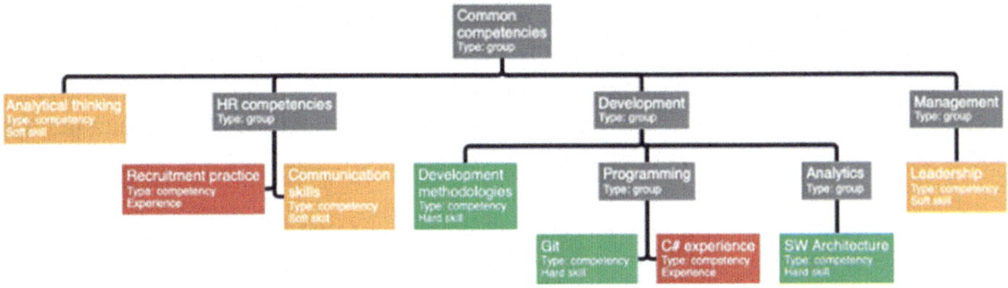

Fig. 10. Software development company competency model

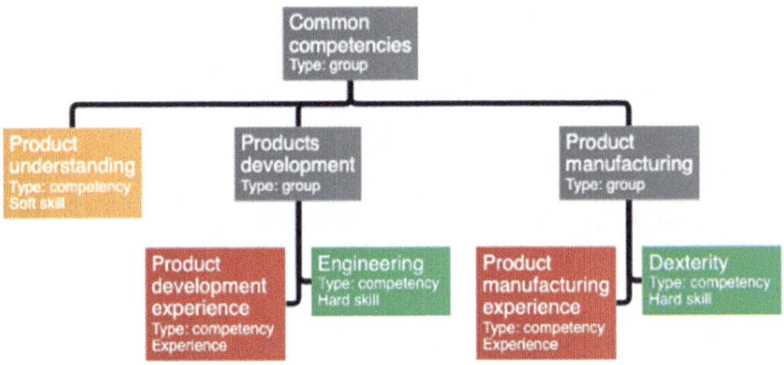

Fig. 11. Manufacturing company competency model

Therefore, if a person applies for more than one position, common competencies need to be filled only once.

Another advantage of this approach is that this hierarchical structure can be modified based on company needs. We can relocate or create new competencies or a group of competencies.

Furthermore, it is possible to evaluate groups of competencies separately and then obtain an overall evaluation based on the evaluation of groups and their different significance for the job position.

5 Conclusion

In this article, the approach for creating a data structure of a hierarchical competency model was proposed. We introduced how to represent and manipulate a hierarchical competency model of a company in a web application environment (ASP.NET Core). The JSON format is used to transfer data between the server and the front-end application. The GoJS library is used to visualize the tree diagram. Furthermore, the behavior of all important functionalities necessary for the work with the hierarchical competence model was described. The created hierarchical model was verified on several examples.

The proposed approach is used to create an expert system for evaluating competencies, which is developed as part of our research. In addition, the approach is usable for creating a hierarchical competency model of a company or an organization.

In future work, we will focus on creating other functionalities to the hierarchical model and on verifying the model on other companies.

Acknowledgement. This work was supported by a Student Grant SGS02/UVAFM/2020 with student participation, supported by the Czech Ministry of Education, Youth and Sports.

References

1. Shippmann, J.S., et al.: The practice of competency modelling. Pers. Psychol. **53**(3), 703–740 (2000)
2. Harzallah, M., Vernadat, F.: IT-based competency modelling and management: from theory to practice in enterprise engineering and operations. Comput. Ind. **48**(2), 157–179 (2002)
3. Campion, M.A., et al.: Doing competencies well: best practices in competency modelling. Pers. Psychol. **64**(1), 225–262 (2011)
4. García-Barriocanal, E., Sicilia, M.A., Sánchez-Alonso, S.: Computing with competencies: modelling organizational capacities. Expert Syst. Appl. **939**(39), 12310–12318 (2012)
5. Walek, B., Pektor, O., Farana, R.: Evaluating suitable job applicants using expert system. In: International Conference on Soft Computing-MENDEL, pp. 186–196. Springer, Cham (2016)
6. Pektor, O., Walek, B., Martinik, I., Jaluvka, M.: Hierarchical system for evaluating professional competencies using Takagi-Sugeno rules. In: Computer Science On-line Conference, pp. 318–325. Springer, Cham (2018)
7. Introduction to ASP.NET Core. Microsoft Docs. Redmond: Microsoft Corporation. https://docs.microsoft.com/cs-cz/aspnet/core/. Accessed 06 Mar 2020
8. GoJS: Diagrams for JavaScript and HTML, by Northwoods Software. New Hampshire: Northwoods Software. https://gojs.net/. Accessed 06 Mar 2020

Document Clustering Using Hybrid LDA- Kmeans

Oi-Mean Foong(✉) and Alia Nabila Ismail

Computer and Information Sciences Department,
Universiti Teknologi PETRONAS, Perak, Malaysia
{foongoimean, a.nabila_20790}@utp.edu.my

Abstract. This paper presents a Hybrid Latent Dirichlet Allocation – Kmeans (HLDA-Kmeans) Algorithm for document clustering. The overload information has became a challenge for users due to the existence of abundance information and heterogeneous nature of the Web. Researchers such as academician as well as people who are involved in text analytics have encountered challenges to analyze documents because of ambiguity in keywords/keyphrases. Hence, the objective is to perform document clustering analysis using HLDA - Kmeans algorithm to discover the clusters among the unlabelled text data, classify the keyphrases based on topics and visualize the clustering results. Online news from Oil and Gas is used as a dataset for training and testing using 70%–30% split. The system performance of the proposed HLDA - Kmeans algorithm was assessed using Precision, Recall and F-Score Formulas. Experimental results show that the proposed HLDA - Kmeans has achieved clustering results satisfactorily.

Keywords: Latent Dirichlet Allocation · Document clustering · Information overload

1 Introduction

In the technology era, the World Wide Web (WWW) is incessantly booming with unbounded information and it giving users with a lot of information. Due to this exponential growth of data on the Internet, users are inundated with infinite information yet have limited time for comprehension. Thus, clustering the search results may help find the desired information quickly and accurately. Clustering is use when there is no prior knowledge about the data [1]. It also known as unsupervised or unlabelled data. There are a lot of research papers, news and articles that have been uploading in the Internet which may have different topics. In this paper, it can reduce the effort of reading each document by doing document clustering to form group of retrieved documents into list of useful categories of topics without having the user to suffer from cognitive overload.

The challenges faced by researchers and academicians are: (1) It is very difficult to browse through relevant topics and information because search engines return thousand of search results and quite often the search result is not related to the information that they desired. (2) Time is highly precious in today's rat race world but with this problem

R. Silhavy (Ed.): CSOC 2020, AISC 1226, pp. 137–146, 2020.
https://doi.org/10.1007/978-3-030-51974-2_12

it will waste their time to find the relevant data that they need and they have limited time to digest all the information. In addition, the Internet is exploding with huge volume of text documents and this will make it harder to retrieve the relevant data. Researchers find it hard to search for the relevant keyphrases because there is no grouping of similar documents together. Thus, there exists a need to cluster text documents according to topics.

Moreover, there is less data visualization to visualize the results of the clustering analysis. Researchers that do not have any technical background might find it hard to analyze the keyphrases in the topics without visualization aid. Thus, the need for data visualization in analyzing the results of the document clustering is highly needed to make it easy for users to understand. Accordingly, the research objectives two-folds: (1) To apply Hybrid LDA – Kmeans algorithm for documents clustering based on topics detected. (2) To visualize the clustering results by sorting them according to topics clusters.

2 Related Works

Vast amounts of new information and data are generated everyday through economic, academic and social activities, much with significant potential economic and societal value. Techniques such as text mining and analytics are required to exploit this potential. Text mining is usually refers to text data mining or text analytics is a process of extracting high quality information from the text and make the information accessible to the various data mining algorithms such as machine learning.

The information that have been extracted usually use to derive the summaries of the documents based on the words contain in the documents as well as analyze words. There are many applications of text mining in today's world such as analyzing open-ended survey response, automatically process the email, analyzing customer satisfaction and many more. Text analyses also involve information retrieval, pattern recognition, visualization and predictive analysis.

In this research, clustering is used to group the set of objects into clusters that are different from each other. According to Xiao [1, 2], Google currently uses clustering methods to match a specific website with query since the website is represented as a collection of topics while query itself is viewed as topic or combination of several topics. Document clustering is crucial and important in organizing and separating documents into groups to be used in document organization, summarization, classification, browsing and information retrieval.

Topic modeling comprises of algorithm that extract latent topics inside a text corpus [3]. Topic modeling is used for unsupervised machine learning and it is working by running a collection of texts or documents through an algorithm that group a words in the corpus into topics [4]. The output displayed is the list of candidate phrases and its probabilities that belong to that particular topic. One of the popular techniques is Latent Dirichlet Allocation (LDA) Algorithm. The LDA was introduced to overcome the overfitting problem of PLSA algorithm. Overfitting is the result of analysis that contains more parameters that fail to fit the additional data and predict future observations correctly. According to Jelodar et al. [5], LDA is a generative probabilistic model for collection of discrete data and it is used for unsupervised learning. Each of the group is represented as a random mixture of latent topics whereas each topic is a discrete distribution over the vocabulary collection.

Semantic discovered by topic models will provide the accurate similarity measure that is helpful in gaining coherent clusters [6, 7]. The LDA algorithm was adopted to uncover the topics present in the corpus. The keyphrases that have the highest probabilities in particular topics will form the main topics of the documents. By using LDA, it generated the latent topics to find the relationship between the document and keyphrases. LDA is often used for unsupervised learning for unclassified text or data. As LDA is based on 3-tier Bayesian model, potential topic nodes would be repeatedly sampled and multiple topics would be predicted as illustrated in Fig. 1. The LDA presumes the generative process of document w in a corpus D [8]:

I. The $N \sim$ Poisson(x) is chosen.
II. The $\theta \sim$ Dir (α).
III. For each of the N words w:
   1. Topic, z is chosen.
   2. Word, w from $p(w \mid z)$ is chosen where it is a multinomial probability conditioned on the topic, z.

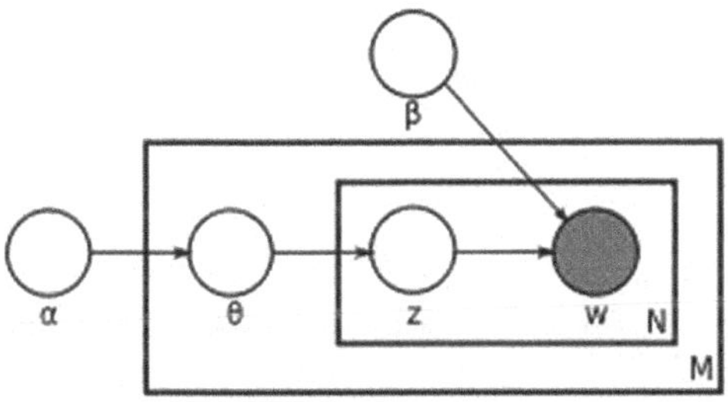

Fig. 1. Graphical representation of LDA

Based on Bartohome et al. [9], if both of topic modeling and document clustering is used separately, the accuracy of the result is not high. Thus, to improve the accuracy of clustering, topic modeling was implemented using LDA with document clustering. The semantics that have been discovered from LDA can enable the accuracy of similarity measure which can facilitate in discovering coherent clusters and topics.

3 Proposed HLDA-Kmeans Algorithm

The proposed HLDA-Kmeans is illustrated in Fig. 2.

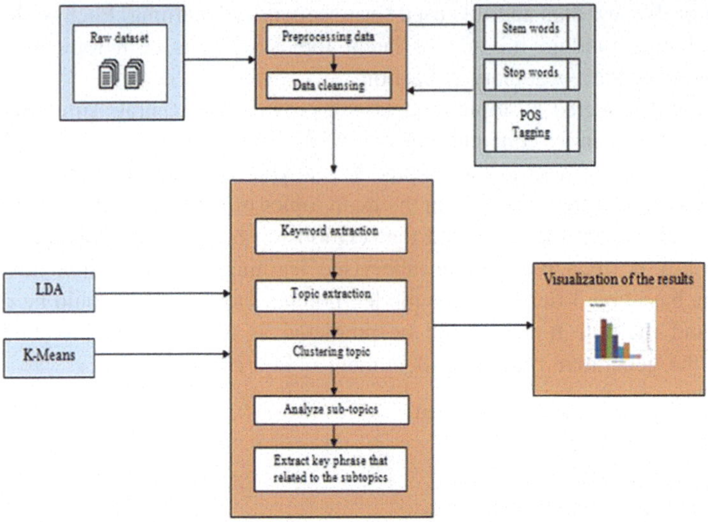

Fig. 2. Proposed HLDA-Kmeans algorithm

3.1 Preprocessing

The preprocessing phase of the system starts by tokenization which will separate or split text into meaningful words, phrases and other elements called tokens.

3.1.1 Tokenization

During the tokenization process, some of the characters such as exclamation marks and punctuation marks are discarded. This process is done on raw dataset of the articles by parsing the text into single word and eliminates any character.

3.1.2 Stemming

Stemming is a process dealing with prefixes and suffixes of each word in the sentences. The input of stem word process is taken from the file produced by tokenization process. The stem word process will execute each word to remove the suffixes of each word and keep the root word only in the new file. For example, the word 'furthered', 'furthering' and 'furthers' are having the same root word which is 'further'. The purpose of word stemming is to have only one root word which is written for more than one occurrence in a different type of format. By having different format of the same words, it will lead to different meaning. So, stemming will only keep the root words to make it easy for the system to process.

3.2 Latent Dirichlet Allocation

The first step of the LDA algorithm is to randomly assign each word in the document after go through each document [10]. It assigns the words to *k* topics. After that, word-topic matrix will be created. It is a count of each word that is being assigned to each topic. Next, document-topic matrix will be creating to assigned number of words in the topic for each of the document. These two matrices will be use throughout the algorithm process. After that, the probability of the words is calculated and used the result to analyze which topic does each document belongs to and possible words that are related with each topic. In azure machine learning, LDA used Bayesian theorem to determine which topic might be associated with each words, *P*(word|topic) [11]. Bayesian methods works better with smaller dataset to avoid overfitting of data.

Table 1 contains the transformed dataset, based on the dataset above. The output contains the input text, and a specified number of discovered categories, together with the scores for each category.

Table 1. Classification of documents according to probability of topics

Doc_Id	Document	Topic 1	Topic 2
1001	Profit plunge for Malaysia Petronas for low oil price	0.001109	0.998891
1002	Malaysia Petronas post profit drop on low oil price	0.998891	0.001109
1003	Malaysia success key to attract foreign investment	0.998575	0.001425
1004	Strength in Malaysia economy drawing factor for global investment	0.001109	0.998891
1005	PETRONAS attract investment to increase economy in Malaysia	0.998753	0.001247

In this study, the number of topics is initialized to 2. The LDA algorithm creates two topics. The module also assigns a score to each entry for each of the five categories that represent topics. A score indicates the probability that the row should be assigned to a particular category. In this case, the dataset is normalized as a conditional probability of topic given documents, *P*(topic|doc). The sum of each row is 1. The second output is the topic matrix where it contains the keywords as the features. It also has score for each of the categories in Topic 1 and 2. In this case, the sum of each column is equal to 1. The result of the computation is shown in Table 2:

Table 2. Extraction of important feature words

Feature	Topic 1	Topic 2
Increase	0.041556	0.041295
Global	0.041479	0.041258
Economy	0.042005	0.041152
Investment	0.041801	0.041400
Foreign	0.041328	0.041136
Attract	0.041283	0.041526
Key	0.041789	0.043470
Success	0.04152	0.041449
Price	0.041303	0.041165

(*continued*)

Table 2. (*continued*)

Feature	Topic 1	Topic 2
Oil	0.041369	0.042140
Drop	0.041413	0.041169
Malaysia	0.041675	0.041563
Petronas	0.042352	0.041319
Plunge	0.04152	0.041295
Profit	0.042201	0.04214

Based on observation, it can be concluded that the topics are about investment and oil based on the probability of words/phrases that are inferred from the documents (Table 3).

Table 3. Classification of keywords according to topics

Topic 1		Topic 2	
Word	Probability	Word	Probability
Increase	0.041556	Strength	0.043759
Global	0.041479	Success	0.041449
Economy	0.042005	Oil	0.04214
Investment	0.041801	Profit	0.04214

3.3 Integrating LDA with K-means

In this phase, K-means algorithm will be used to cluster the document that has the same topic together in one cluster [7, 12]. Document clustering enables us to extract local topics specific to each document cluster and global topics shared across clusters. The first step in document clustering is specifying *k*-centroids in the model. The centroid acts as a representative of each cluster. K-means algorithm starts with the initial set of randomly chosen centroid when processing the training data that serve as the starting points for each cluster. Then, it will iteratively refine the locations of the centroids and will stop refining when the centroid is stabilized which is when the cluster assignment no longer changes and the algorithm has found the solution. As illustrated in Table 4, it displays the clustering results with two clusters.

Table 4. Summary of clustering results

Doc Id	Text	Cluster_Id	Keyphrases
1001	Profit plunge for Malaysia Petronas on low oil price	0	Malaysia Petronas, low oil price
1002	Malaysia Petronas post profit drop on low oil price	0	Malaysia Petronas, port profit, low oil price
1003	Malaysia success key to attract foreign investment	1	Malaysia success key
1004	Strength in Malaysia economy drawing factor for global investment	1	Malaysia economy, factor
1005	Petronas attract investment to increase economy in Malaysia	1	Investment, increase economy

From the tables above, Table 4 contains two clusters with five different documents. Cluster 0 contains document 1001 and 1002 while cluster 1 contains document 1003, 1004 and 1005. It also represents the key phrases that have been extracted from the documents present in that particular cluster. Based on Table 5, the cluster size is the total number of documents presents in that particular cluster. For cluster 0, it has cluster size 2 which means that it has document 1001 and 1002 while for cluster 1, the cluster size is 3 which contains document 1003, 1004 and 1005. The keyword had been extracted from the documents in each cluster. In cluster 1, it has keyword ‘increase’, ‘economy’, ‘malaysia’, ‘success’, ‘key’ and ‘investment’. For cluster 0, it has keywords ‘petronas’, ‘malaysia’, ‘low’, ‘oil’, ‘price’, ‘post’, ‘profit’.

3.4 Labeling Clusters

After the documents have been clustered, we need to label the cluster. Labeling the cluster is arbitrary because clustering is an unsupervised learning. Human can group the objects together based on our own perception and view but computer cannot do that. Computer categorize data based on their feature terms. The standard clustering algorithm where in this case is K-means usually do not produce any labeling of the cluster. The goal of cluster labeling is to find the labeling after analyzing the content of the document and summarize the topic present in each cluster. For this study, title label will be used to label the clusters. Title label work by finding the document within the cluster that has the smallest Euclidean distance to the centroid and after that used the topic words as a label for the cluster. Based on the results in Table 5, cluster 0 will be labeled as ‘oil’ whereas cluster 1 as ‘investment’ by topic inferencing from document contents [13–15]. A visual clustering of documents is shown in Fig. 3.

Table 5. Distance to cluster

Cluster	Cluster size	Doc_ID	Distance to cluster 0	Distance to cluster 1
0	2	1001	0.000230	1.410630
		1002	0.000019	1.410880
1	3	1003	1.410863	0.000000
		1004	1.411077	0.000214
		1005	1.410863	0.000000

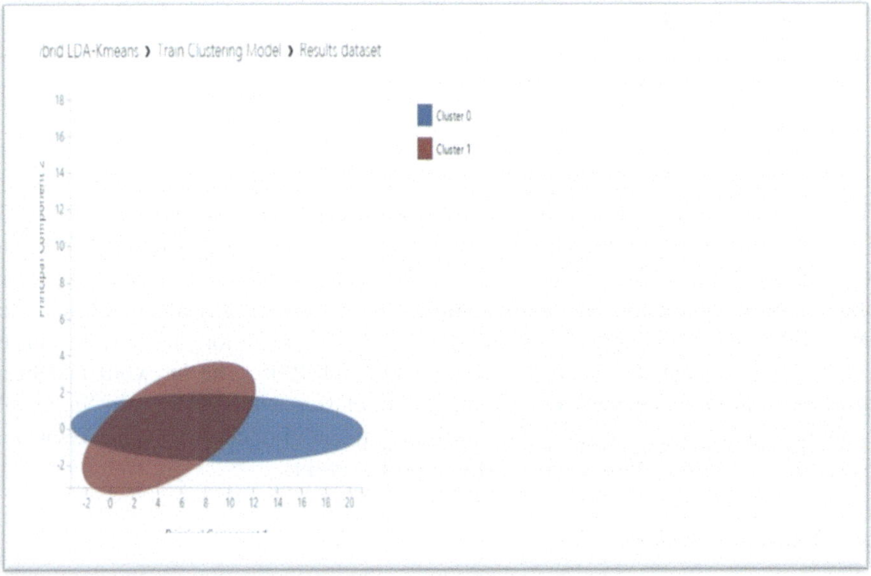

Fig. 3. Clustering document based on similarity of topics

4 Result and Discussion

The results generated by both HLDA-Kmeans algorithm and LDA were obtained after extracting the hidden topics in the document and listing down the words/keyphrases that are represented in the topics. The results will be compared with human-generated data. The model was evaluated using Recall-Oriented Understudy for Gisting Evaluation (ROUGE) [16]. The Recall, Precision & F-score were shown in Table 6. Evaluation of all the results by the system is based on the overlapping keywords/keyphrases found in the particular topics in the corpus. The average F-score for HLDA - Kmeans algorithm outperforms LDA with F-score 0.735 and 0.682 respectively.

Based on observation, document 1005 talks about economy and oil and gas industry. The system extracted the words with the highest probability to represent the document. In document 1003, the word 'investment' has the highest probability when compared to other words so we can predict that the document elaborates on investment.

The next process is clustering the document together based on the similarity of the topics present in the specific documents. The topic generated by LDA algorithm is being used to label the cluster of document.

Table 6. Precision, Recall and F-score

HLDA-Kmeans algorithm				LDA algorithm (Baseline)		
Doc_ID	Precision	Recall	F-score	Precision	Recall	F-score
5	0.666	0.666	0.666	0.636	0.636	0.636
10	1.000	0.687	0.814	0.620	0.909	0.740
15	0.833	0.595	0.694	0.561	0.741	0.638
20	0.681	0.750	0.714	0.615	0.631	0.623
25	0.800	0.678	0.733	0.666	0.555	0.606
30	0.754	0.689	0.721	0.700	0.573	0.630
35	0.808	0.687	0.743	0.714	0.735	0.724
40	0.763	0.725	0.743	0.731	0.769	0.750
45	0.772	0.734	0.753	0.743	0.698	0.720
50	0.747	0.764	0.755	0.747	0.755	0.751
		Average	0.734		Average	0.682

5 Conclusion and Recommendation

In conclusion, the HLDA- Kmeans algorithm outperforms the LDA algorithm for document clustering based on topics. We could also visualize the clustering results based on the topic lebels or keyphrases/keywords. These keyphrases will be used in the document indexing and document searching From the keywords found in the document, it can predict the topics that used to label the clusters of documents. For future work, different machine learning algorithms can be considered to enhance the clustering results.

Acknowledgement. The authors would like to thank YUTP (Cost Center 015LC0-173) for the financial support in the research.

References

1. Xiao, Y.: A survey of document clustering techniques & comparison of LDA and moVMF (2010)
2. Thakare, P., Karche, R.K.P., Gaikwad, S., Khaladhar, M.: Data analysis using document clustering. Int. J. Eng. Comput. Sci. **4**(4), 11267–11271 (2015)
3. Hoffman, T.: Probabilistic latent semantic analysis. In: Proceedings of the 15th Conference on Uncertainty in Artificial Intelligence, pp. 289–296 (1999)
4. Li, Z.X., Shi, Z.P., Li, Z.Q., Shi, Z.Z.: Automatic image annotation by fusing semantic topics. J. Softw. **22**(4), 801–812 (2011)

5. Jelodar, H., Wang, Y., Yuan, C., Feng, X., Jiang, X., Li, Y., Hao, L.: Latent Dirichlet Allocation (LDA) and topic modeling: models, applications, a survey. Multimed. Tools Appl. **78**(11), 15169–15211 (2019)
6. Foong, O.-M.: Swarm LSA-PSO clustering model in text summarization. Int. J. Adv. Soft Comput. Appl. **8**(3), 88–99 (2016)
7. Liu, L., Tang, L., Dong, W., Yao, S., Zhou, W.: An overview of topic modeling and its current applications in bioinformatics. SpringerPlus **5**(1), 1–22 (2016)
8. Blei, D.M., Ng, A.Y., Jordan, M.I.: Latent dirichlet allocation. J. Mach. Learn. Res. **3**, 993–1022 (2003)
9. Bartolome, A., Islam, M., Vundekode, S.: Clustering and topic analysis. Information Storage and Retrieval (2016)
10. Blei, D.M., Lafferty, J.: A correlated topic model of science. Ann. Appl. Stat. **1**(1), 17–35 (2007)
11. Chappell, D.: Introducing Azure machine learning. A guide for technical Professionals (2015)
12. Jelodar, H., Wang, Y., Yuan, C., Feng, X.: Latent Dirichlet Allocation (LDA) and Topic modeling: models, applications, a survey (2017)
13. Ian. W.: Recent advances and applications of probabilistic topic models. In: AIP Conference Proceedings, vol. 1636, no. 1, p. 124 (2014)
14. Christophe, D.: Inference and applications for topic models. Machine Learning. PSL Research University (2017)
15. Lirong, Q., Jia, Y.: CLDA: an effective topic model for mining user interest preference under big data background (2018)
16. Lin, C.Y.: ROUGE: a package for automatic evaluation of summaries. In: Proceeding of Workshop on Text Summarization ACL 2004, pp. 74–81 (2004)

Performance Evaluation of RISC-Based Memory-Centric Processor Architecture

Danijela Efnusheva(✉)

Computer Science and Engineering Department,
Faculty of Electrical Engineering and Information Technologies,
Skopje, Republic of North Macedonia
danijela@feit.ukim.edu.mk

Abstract. Recent technological advances in computer hardware and software industry resulted in a wide range of single- and multi-core processors, operating systems, compilers and applications, with the ultimate goal to increase the overall performances of the computing system. Although the number of transistors that can fit on a chip has been doubling every 18 months, there is still difficult to improve the performances of sequential single-core processors, and even of parallel many-processor and multi-core shared-memory systems. The main issue resides in the computer's hardware and software inability to alleviate the occurrence of processor-memory bottleneck and to provide parallel instruction execution at peak level. Therefore, this paper proposes a memory-centric approach of parallel processing in a distributed system, which includes several RISC-based processor cores that integrate on-chip memory and provide direct access to it, without using registers and cache memory. Furthermore, the execution of several algorithms with different arithmetical intensity is evaluated and the achieved improvements are discussed.

Keywords: Computer architecture · Memory-centric computing · Parallel programming · RISC

1 Introduction

The constant race of various computer technologies resulted in a wide range of processor architectures, including: CISC (complex instruction set computing), RISC (reduced instruction set computing), [1], Vector, [2], Superscalar, [3], VLIW (very long instruction word), [4] and EPIC (explicitly parallel instruction computing), [5]. Despite the great advances in processor architectures and the evolvement of parallel multi-core processors, current computer systems are still dominantly based on the classical Von Neumann's model, [6]. This means that the memory and the processing units in the computer system are strictly separated, in such a way that the memory is used to store data and programs, while the CPU is purposed to read and decode the program code, load and store data to and from the cache memory or/and the registers and to execute

R. Silhavy (Ed.): CSOC 2020, AISC 1226, pp. 147–163, 2020.
https://doi.org/10.1007/978-3-030-51974-2_13

arithmetical-logical operations over the loaded operands. This organization suffers from limited bandwidth (bottleneck), [7] on the processor-memory interface, caused by the memory and processor speeds discrepancy. This dissimilarity leads to many wasted processing cycles since the memory system is not capable to provide instant response to each CPU request. Therefore, superscalar and EPIC processors, which are capable to execute several instructions per cycle, did not manage to achieve the expected success. Furthermore, multi-core processors, [8], introduced even more intensive work with the memory, which resulted in an increased number of slow off-chip memory accesses.

The existing model of processor-centric computing allows performance to scale only if these two conditions hold: the processor has sufficient work to do, so it can mitigate the cache miss latencies, and the processor has enough bandwidth to load the changes into the cache without excessive delay, [9]. This means that the performances of the computer system are not only related to the CPU speed but are also dependant on the memory system behavior, expressed with its bandwidth and latency, [1]. In order to improve these memory parameters several approaches have been proposed, [10], including branch prediction, speculative and re-order instructions execution, data and instruction pre-fetching, multi-threading, wider and faster memory connections, and multi-level cache memory. Each of these techniques can serve as a temporal solution but is still not able to completely overcome the occurrence of a processor-memory bottleneck.

Cache memory is one of the most popular methods for alleviating the processor-memory bottleneck. For example, Intel processors use up to 40% of the chip die area [11,12], to implement on-chip caches. Although cache memory can significantly reduce the memory access time, it still adds extra hardware resources for storing redundant data and also requires complex mechanisms for maintaining memory consistency, which causes up to 40% increase of the energy consumption in the system, [13]. Besides that, cache memory is not very suitable for real-time applications, because it brings unpredictability in the timing of the program.

Contrary to the standard model of processor-centric computing, some researchers have proposed several memory-centric approaches of computing which suggest integrating or approaching the memory and the processing elements into the same chip die or closer to each other. This research has resulted with several proposals, including: register-less PERL processor, [14], high-speed Scratchpad memory, [15], various processing in memory chips (computational RAM, Mitsubishi 32R/D, DIVA, Terasys, intelligent RAM, parallel processing RAM, and DataScalar), [16], and intelligent memory systems (ex. active pages model, [17]). The given memory-logic merged chips are characterized with high internal memory bandwidth, low latency, and high power efficiency, which makes them very suitable for applications that require stride memory accesses and high data throughput, like Fast Fourier Transform (FFT), multimedia, and network processing.

The aim of this paper is to propose a novel approach of parallel processing in a distributed system which utilizes memory-centric RISC-based processing cores. These RISC-based processing cores implement on-chip memory and provide direct access to it, without the use of cache memory and general-purpose registers (GPRs), which is different from the other memory-logic merged chips that mostly use the standard memory hierarchy model. Besides that, the proposed processing cores operate in a pipeline mode, without explicit LOAD and STORE instructions, allowing every (arithmetical-logical, branch or control) instruction to be completed in a single cycle. Each processing core allows concurrent execution of different program segments that can be parallelized and executed simultaneously.

The paper is organized into five sections. Section 2 presents the state of the art and discusses a variety of alternative memory-centric approaches of computing. Section 3 describes the architecture of the proposed RISC-based memory-centric processing cores and presents their application in a multi-core system with the distributed organization. The next section presents the results of the performance evaluation of the proposed distributed system when executing algorithms with different arithmetical intensity (according to the Roofline model, [1]) in parallel. The paper ends with a conclusion, stated in section five.

2 Current State

Nearly all modern computer systems are based on the stored-program Von Neumann model, [18], which describes a system with a separated memory, used for data and programs store and a dedicated central processing unit purposed to perform control, branch and arithmetical/logical operations. Assuming that the processor has the main role in this processor-centric organization, computer designer's research was aimed at improving the processor architectures, and innovating in their internal organization, instruction set and the abilities of program code parallelization at instruction, data or thread level.

Besides the great advances in processor technologies, computer designers also suggested innovations in memory technologies, which resulted in embedding more than 10 MB eDRAM memory as on-chip cache into the processor, [19]. This was due to the increased number of processing cores into the chip, which brought even greater requirements to the memory system. Actually, it is well known that the increase in computing resources without a corresponding increase in the on-chip memory leads to an unbalanced system.

The performances of a computing system mainly depend on the communication between the processor and the main memory. This behavior is defined with the parameters: memory latency and bandwidth, [9]. Memory latency is the time between the initiation of a memory request, and its completion, while the memory bandwidth is the rate at which data can be transferred to or from the memory. Perceiving the complex relationship between memory latency and bandwidth, computer architects have proposed the use of "smart memories" that will be able to simultaneously provide latency reduction and memory bandwidth increase. Thus, several memory-centric approaches of integrating or placing the memory near to the processing elements were introduced, [16,20].

There are certain proposals, which present some modifications into the common multi-level memory hierarchy and provide nonstandard faster access to the main memory. For example, an approach of register-less computing that is given in [14] describes a processor, named PERL, which uses only cache memory (inside and outside the processor) to communicate with the main memory. Additionally, the authors of [15] suggest the use of Scratchpad memory as a small software-managed on-chip memory that is separated from the cache memory, but accessible in a single cycle.

Several proposals of processing in memory (PIM) chips, including Sun PIM, Mitsubishi 32R/D, DIVA, Terasys, intelligent RAM, parallel processing RAM, DataScalar, and active pages intelligent memory system, are described in [20–23]. These merged memory/logic chips implement on-chip SRAM or embedded DRAM memory which allows high internal bandwidth, low latency, and high power efficiency, eliminating the need for expensive, high-speed inter-chip interconnects, [16]. In general it can be concluded that processing in/near memory brings latency and bandwidth improvements, but still, the merged system has to operate with redundant data and to perform many moves and copy operations into the on-chip memory, caches and GPRs. In addition to that, the processing speed, the on-chip memory size, and the cost of the chip is limited to the production process and the implementation technology. Besides that, it is even a greater challenge to develop a compiler that will recognize the program parallelism and will provide efficient utilization of the internal memory bandwidth (ex. wide data path in DIVA, [22]).

Considering that modern processors are lately dealing with both technical and physical limitations, while the memory capacity is constantly growing, it seems that now is the right moment to reinvestigate the idea of placing the processor close to the memory in order to overcome their speed difference. A promising approach that targets this problem is presented by the Hewlett Packard international information technology company that suggests novel computer architecture, called the Machine, [24], which utilizes non-volatile memory as a true DRAM replacement. Recent proposals of near-memory and processor in-memory computing architectures are discussed in [25] and [26]. Our research will continue into the direction of developing a parallel system with shared virtual memory, composed of many RISC-based memory-centric processors, similar to PERL.

3 Design of RISC-Based Memory-Centric Processor Architecture

Assuming the adjustments of the standard memory hierarchy model, presented in some of the previously discussed approaches (ex. PERL, Scratchpad, Machine), we can say that an extension or revision of their work can be a promising direction for our research. Therefore, we are developing a RISC-based memory-centric processor core which replaces the top memory hierarchy levels (GPRs and cache) with an on-chip memory that is directly accessible. For that purpose, we are modifying a single-cycle pipelined MIPS processor core, presented by D. A. Patterson and J. L. Hennessy in [1]. This RISC-based processor core is also used as a basis for the design of the PERL register-less processor core, discussed in [14].

MIPS implements a load-store architecture, which is characterized with: fix-length instruction format, simple addressing modes, memory accesses with explicit load and store instructions, hardwired control unit, large GPR set and pipeline operation in five stages (fetch, decode, execute, memory access and write back). This RISC-based processor core is still an industry-standard leader, which is mostly applicable to the embedded industry.

Figure 1 presents the pipelined load-store architecture of a MIPS processor, which includes several building blocks: Program counter - PC, Instruction Register - IR, 32 GPRs, 32-bit arithmetical and logical unit - ALU, pipeline registers, separated instruction and data cache memory, control unit (marked with blue), and other selecting and control logic (extenders, decoders, multiplexers, adders etc.). According to Fig. 1, the MIPS processor operates only on operands found in the local GPRs, which requires a frequent transfer of data between the memory and the processor's registers, via load and store instructions. In order to provide easier access and manipulation of memory data, this paper proposes an extension and modification of the original pipelined RISC architecture and creation of a novel pipelined MEMRISC (Memory Access Reduced Instruction Set Computing) processor architecture, shown in Fig. 2.

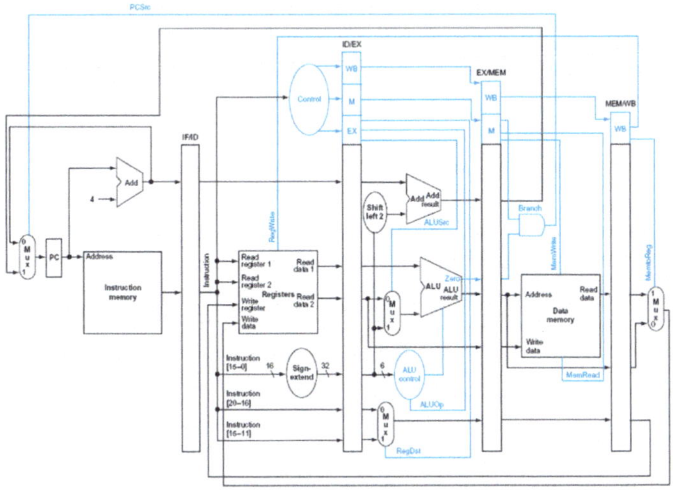

Fig. 1. Pipelined RISC architecture of MIPS processor, [1]

As shown in Fig. 2, the proposed processor with MEMRISC architecture uses separated on-chip local data and program memory, instead of GPRs and on-chip

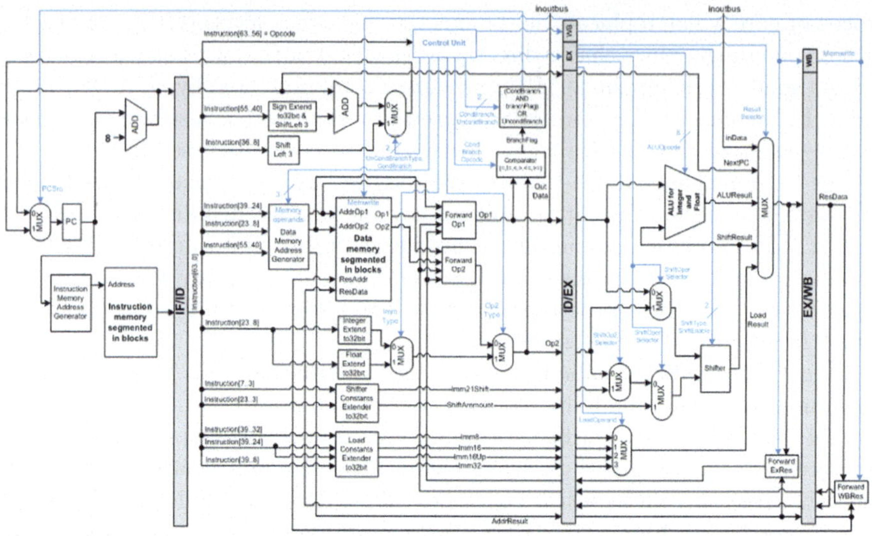

Fig. 2. Pipelined MEMRSIC architecture of MIMOPS processor

cache memory. This means that the given processor executes all the operations on values found in the local on-chip memory, avoiding the unnecessary and redundant data copying and movements, which are performed in the MIPS processor, during (load/store) data transfers. Therefore if the RISC-based MIPS processor is able to execute a million instructions per second, then the proposed processor with MEMRISC architecture would be able to execute a million instructions on memory operands per second, which is the reason why it is called MIMOPS processor in continuation.

The proposed MIMOPS processor excludes the GPRs and the cache memory from the memory hierarchy and thus allows direct and simultaneous access to two sources and one result operand, specified in the MIMOPS instruction. These operands are selected by a specialized memory address generator unit that is used to perform the translation of the input virtual addresses into physical addresses of the paged on-chip memory. Once the operands are read from the local on-chip data memory, the operation is executed and then the result is written back to the local on-chip data memory. In fact, the MIMOPS processor operates in a 4-stage pipeline (instruction fetch, instruction decode, execute and write back), excluding the MEM phase, and allowing every (arithmetical, logical, branch or control) MIMOPS instruction to be completed in single tact cycle. The instructions that are supported by the proposed MIMOPS processor are MIPS alike, but the way of their interpretation and execution is slightly different.

Unlike the MIPS processor given in Fig. 1, the proposed MIMOPS processor operates directly with the on-chip memory and thus simplifies the access to the operands, the execution of the instructions (pipelining without MEM phase) and the instruction set (removes explicit LOAD/STORE instructions).

This way of operation of the MIMOPS processor is managed by a specialized control unit (marked with blue on Fig. 2), which provide supports for several addressing modes (ex. direct, immediate and base addressing). Generally, the memory operands are addressed directly, while the translation of the virtual addresses to physical addresses is performed via specialized hardware support for virtual memory that is implemented inside the MIMOPS processor. This refers to segmentation of the local on-chip memory and filling it with virtual pages, and implementation of page translation tables and page replacement mechanisms (ex. FIFO). In addition to that, the MIMOPS processor also implements control mechanisms for dealing with interrupts and exceptions (ex. incorrect result) and as well provides support for communication with I/O devices.

The proposed MIMOPS processor can operate with 32-bit virtual address space (4GB capacity), divided into equal-sized virtual pages. Although the most common applied size of a virtual page is 4KB, the study given in [27] recommends the use of larger virtual pages in the range of 16KB to 256KB. Therefore, this paper proposes segmentation of the MIMOPS processor's on-chip memory in 64KB physical frames (for virtual pages). It is assumed that the application of 64KB page(frame) size provides low miss rate for most of the analyzed benchmark scenarios given in [27] and thus it is expected that it would decrease the number of off-chip accesses, providing an increase in the performances of the proposed MIMOPS processor. In addition to that, the MIMOPS processor implements special type of small capacity inverted page table (in fetch and decode stages), [10], that has one entry for each physical frame, instead of each page.

Regarding pipelining, it can be noticed that both MIPS and MIMOPS processors provide overlapping of the execution of the instructions, by implementing pipeline registers for every inter-phase (ex. instruction fetch/instruction decode). Besides these similarities, the MIMOPS processor differs from the MIPS processor in many ways, since it allows: reducing of the pipeline stages number by one, finishing the execution of conditional and unconditional branches in the decode pipeline stage and support of data forwarding for overcoming data hazards during parallel instructions execution. Additionally, the MIMOPS processor implements a separate shifter logic that is purposed to generate a second flexible operand for the arithmetical-logical unit (ALU). This is achieved by shifting the second operand by a specific constant value before it is being used by the ALU (this is similar to the ARM - Advanced RISC Machine architecture, [28]). Therefore, the ALU of the MIMOPS processor is able to perform operations over two integer or floating-point input numbers, where the second operand might be previously shifted.

The recent trend in processor design indicates that more and more of the chip area will be devoted to memory. Actually, a huge portion of many modern processors is already dedicated to the top of the memory hierarchy i.e. GPRs and one or two levels of cache. Assuming this trend continues, and as the gap between on-chip and off-chip memory latency grows, it may become desirable to integrate the entire memory on-chip, leaving off-chip accesses for references that are treated more like page faults than cache misses. Even if the complete memory cannot be included inside a single processor chip, a multi-processor system with distributed memory might be built. This way, many merged memory-logic chips could create a homogeneous distributed system that would be easily extendable and upgradable.

When it comes to distributed computing, the proposed MIMOPS processor could be used in a system with distributed shared virtual memory, shown in Fig. 3, and discussed in [29]. In such a system, the virtual memory is a single address space that is shared by a number of processors, so that any processor can directly access any memory location in the address space. According to Fig. 3, the mapping between the local memories and the shared virtual memory is performed via mapping managers, which besides mapping are also responsible to keep the address space coherent. Every mapping manager views its local memory as a large cache of the shared virtual memory address space for its associated processor. In fact, the shared virtual memory exists only virtually, like the traditional virtual memory.

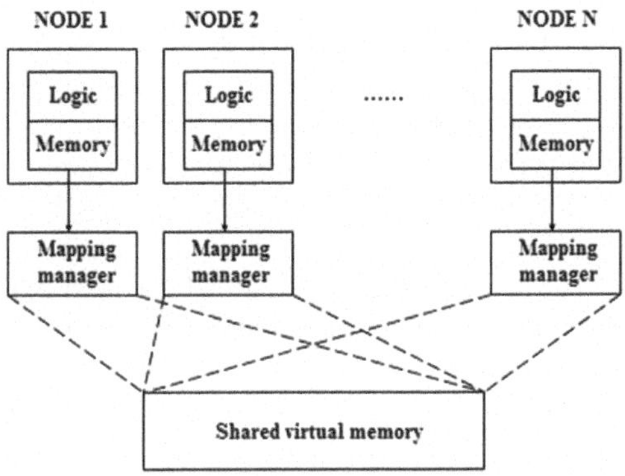

Fig. 3. System with distributed shared virtual memory, [29]

The shared virtual memory address space is partitioned into pages. The pages that are marked as read-only can simultaneously have copies in the physical memory of many processors. On the other hand, the pages that are marked as write can reside in only one processor's physical memory. According to that, memory reference can cause a page fault, when the page containing the memory location is not in the processor's current physical memory. In such a case, the mapping manager retrieves the page either from the disk or from the memory of another processor. However, if the memory reference results with writing, the mapping manager must guarantee atomicity of the operation. For that purpose, after a write, all copies are invalidated or updated.

In a system with shared virtual memory, the model of a parallel program is a set of processes (or threads) that share a single virtual address space. In fact, one of the main goals of the shared virtual memory is to allow processes (or threads) of a program to execute on different processors in parallel. According to that, the performances of a parallel program in a system with shared virtual memory would mainly depend on the number of parallel processors and the degree of data sharing (i.e. contention). Generally, the performances improve as the number of parallel processor increases and also if the program exhibits locality of reference (low level of contention), [29]. Although memory references in parallel programs differ from those in sequential ones, a single process is still a sequential program that should have a high degree of locality. However, the contention among parallel processes would mostly depend on the type of algorithm that is executed.

4 Performance Evaluation

The performance evaluation that is given in the continuation analyzes the behavior of the proposed MIMOPS processor while executing test-programs that are characterized with different arithmetical intensity (dense matrices multiplication, FFT, and partial differential equation solver), according to the Roofline model, [1]. Additionally, a parallel approach of memory-centric computing for some of the test programs is proposed and the achieved results are compared with other similar processors, like MIPS, PERL and intelligent RAM.

The performance analysis of the proposed MIMOPS processor is made with an instruction-level simulator that is specially developed for that purpose, (described in [30]). This simulator allows parallel execution of separate programs or program segments in different MIMOPS processor cores. For that purpose, it implements a scheduler that executes one instruction in every active processor core at a time. If the instruction operates with memory operands that are not available inside the data blocks, then the scheduler puts the new page in a free data block. If all the blocks are busy, the scheduler picks a victim block by looking at the time stamp and then replaces its content with the new page. In general, the MIMOPS simulator is purposed to examine the processor's state and to present the number of processor cycles, required for the execution of diverse test programs.

For the purposes of the comparative analysis, a publicly available instruction-level MARS simulator, [31], of MIPS processor is used, and other results from comparable experiments on similar processor architectures are referenced. It is considered that the proposed MIMOPS processor includes on-chip memory that is segmented into 64KB physical blocks and is with a total capacity that equals to the amount of the on-chip cache memory into the initial MIPS processor (128KB L1 and 2M L2 cache). Both processors make use of pipelining, allowing each instruction to be executed in single cycle, excluding the MIPS's memory instructions that generate a miss into L1 or L2 cache. In that case, it is considered that the MIPS processor operates with associative cache consisted of 128-word blocks, where L1 hit time is 1 cycle, L2 hit time is 21 cycles and L2 miss penalty is 272 cycles.

The first test program is characterized with largest arithmetical intensity and performs multiplication of dense matrices with different dimensions (8×8,

16×16, 32×32, 64×64, 128×128, accordingly). This program executes many repetitive arithmetical operations over large data sets (768B, 3072B, 12288B, 49152B, 196608B, accordingly).

Figure 4 presents the execution time (in processor cycles) of multiplication of dense matrices with different sizes (8×8, 16×16, 32×32, 64×64 and 128×128) on the initial MIPS processor and the proposed MIMOPS processor. According to the results, it can be concluded that the MIMOPS processor reduces the execution time for each of the given matrices size, in comparison to the general-purpose MIPS processor. Additionally, the results that are shown in Fig. 5 present the percentage speedup of the execution time of dense matrices multiplication (for sizes: 8×8, 16×16, 32×32, 64×64, 128×128, accordingly) that is achieved by the proposed MIMOPS processor. According to the results, it can be noticed that as the problem size is increased, the proposed MIMOPS processor accelerates the program execution by 12%, 15%, 25%, 39%, 99%, accordingly, in comparison to MIPS. This is due to the constant increase in the number of misses in the MIPS processor's cache, which introduces further delay during the program execution.

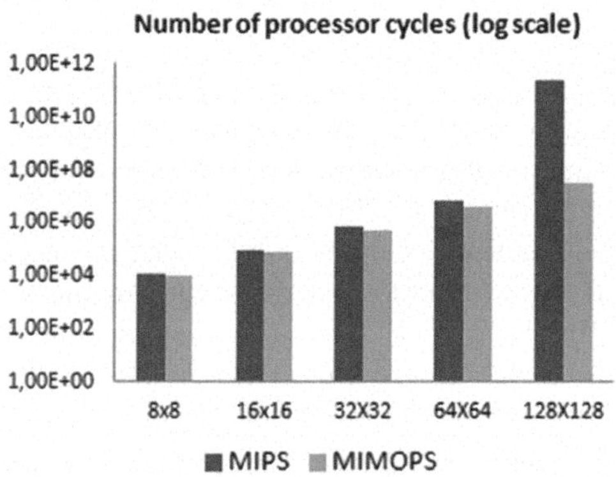

Fig. 4. Execution time of multiplication of dense matrices with different sizes on MIPS and MIMOPS

Considering that the time complexity of matrices multiplication program is $O(N^3)$, the next text describes a "divide and conquer" method which is used to parallelize the multiplication of two square matrices with size $N \times N$ (A and B). This method includes two steps:

1. Divide matrices A and B in four sub-matrices of size $N/2 \times N/2$ as shown below;
2. Calculate the following values: ae + bg, af + bh, ce + dg and cf + dh;

$$\underset{A}{\begin{bmatrix} a\ b \\ c\ d \end{bmatrix}} X \underset{B}{\begin{bmatrix} e\ f \\ g\ h \end{bmatrix}} = \underset{C}{\begin{bmatrix} ae + bg\ af + bh \\ ce + dg\ cf + dh \end{bmatrix}}$$

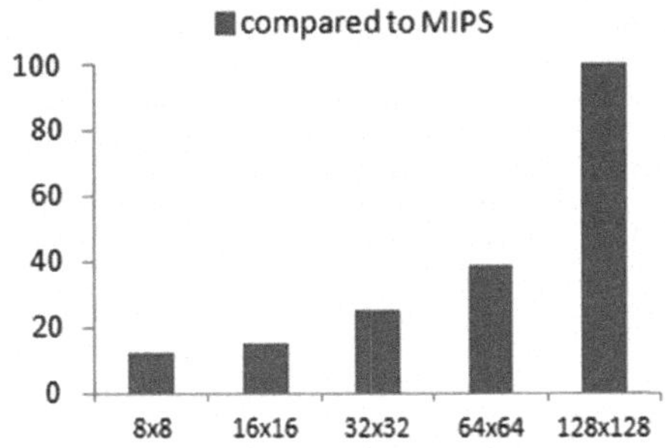

Fig. 5. Percentage speedup of execution time of multiplication of dense matrices with different sizes on MIMOPS

In the above method, 8 multiplications and 4 additions for matrices of size N/2 × N/2 are performed. Addition of two matrices takes $O(N^2)$ time, so the time complexity of the method is $T(N) = 8T(N/2) + O(N^2)$. This method is applied in a 4-core parallel implementation of a MIMOPS processor, which executes multiplication of two 32 × 32 dense matrices. In that case, each MIMOPS core operates on 16 × 16 sub-matrices and therefore performs two multiplications and one addition over the given sub-matrices. It is also considered that the required sub-matrices are copied in the appropriate MIMOPS cores (ex. a is stored in cores: 1 and 2) that should operate with them. The results of 32 × 32 dense matrices multiplication are shown in Fig. 6 and Fig. 7, where Fig. 6 presents the execution time of the test program run on four different processors, including MIPS, PERL, [14], MIMOPS, and 4-core parallel MIMOPS, while Fig. 7 illustrates the improvement that is achieved by the parallel MIMOPS processor. Referring to these results, it can be concluded that the parallel MIMOPS processor provides an improvement of 79.8% (4.96 times), 77.8% (4.52 times) and 73.1% (3.72 times) in comparison to MIPS, PERL and MIMOPS processors, accordingly.

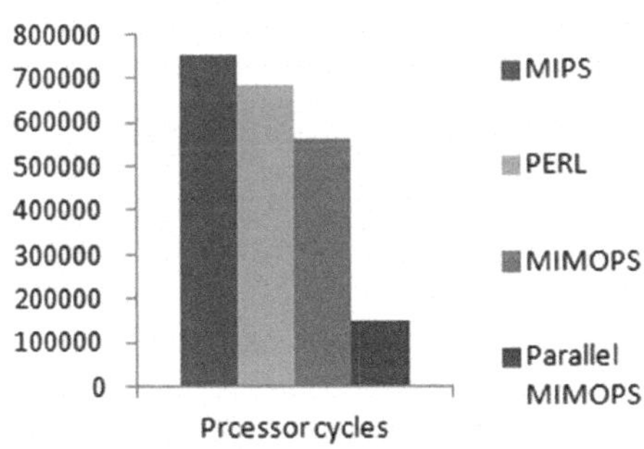

Fig. 6. Execution time of 32 × 32 matrices multiplication on four different processors: MIPS, PERL, MIMOPS and parallel MIMOPS

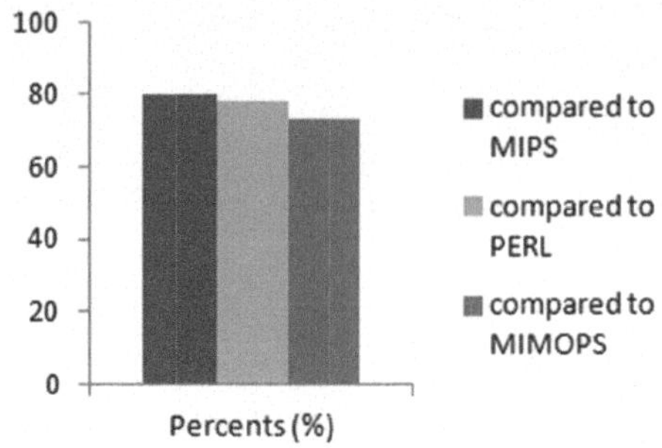

Fig. 7. Percentage speedup of execution time of 32×32 matrices multiplication on parallel MIMOPS

The second test program is characterized with medium arithmetical intensity and it implements a spectral method problem. This program executes a Fast Fourier Transform according to the well known radix-2 Cooley–Tukey algorithm, [32]. In general, FFT is an efficient and fast algorithm for converting signals from time to frequency domain, defined with Eq. 1.

$$X(k) = \sum_{n=0}^{N-1} x(n) W_N^{nk}, k = 0, 1, ..., N-1 \quad (1)$$

In Eq. 1, X(k) represents frequency output for the k-th spectral point, where N is the number of points and x(n) is the n-th sample obtained in the process of sampling. The inputs and outputs of the FFT algorithm are complex or floating-point numbers. The variable W is called a Twiddle factor and is defined in Eq. 2.

$$W_N^{nk} = e^{\frac{-j2nk\pi}{N}} \quad (2)$$

The radix-2 Cooley–Tukey algorithm is based on the "divide and conquer" paradigm, so it performs FFT computation of the even-indexed and odd-indexed FFT inputs separately and then combines them to produce the FFT outputs of the whole sequence. This approach allows recursive execution of the algorithm in $\log_2 N$ phases, whereas each phase performs parallel computations, organized in "butterfly" processes. Every "butterfly" process executes one multiplication with twiddle factor and two additions of complex numbers.

Figure 8 presents the execution time (in processor cycles) of FFTs with different number of points (8, 16, 32, 64, 128, 256, 512, 1024, 2048, 4096, 8192 and 16384) on the initial MIPS processor and the proposed MIMOPS processor. According to the results, it can be concluded that the MIMOPS processor reduces the execution time for each of the FFTs with a different number of points, in comparison to the general-purpose MIPS processor. Additionally, the results that are shown in Fig. 9 present the percentage speedup of the execution time of FFTs with 8, 16, 32, 64, 128, 256, 512, 1024, 2048, 4096, 8192 and 16384 points, which is achieved by the proposed MIMOPS processor. According to the results, it can be noticed that there is a decrease, and then an increase of the speedup that is accomplished by the MIMOPS processor. This is because the MIMOPS processor raises the number of instructions that are executed, by increasing the extent of the problem (FFT size), in comparison to MIPS. Still, as a result of the increased number of cache misses that occur into the MIPS processor when calculating FFTs with over 500 points, the MIMOPS processor instead of deceleration it introduces a speedup of the execution time.

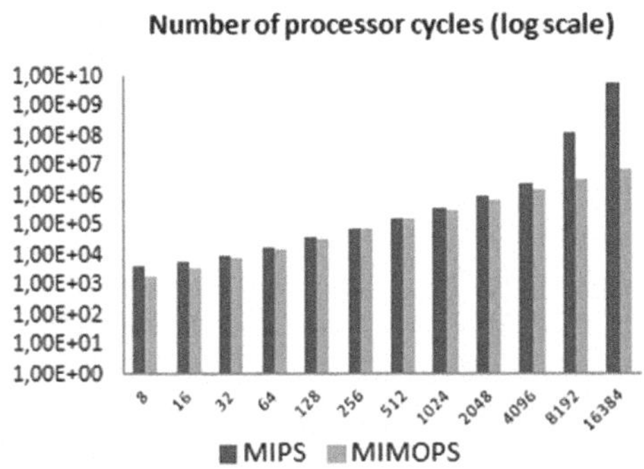

Fig. 8. Execution time of FFTs with different sizes, on MIPS and MIMOPS

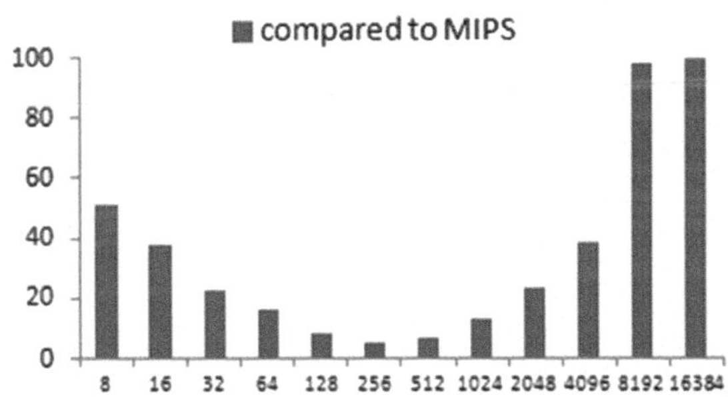

Fig. 9. Percentage speedup of execution time of FFTs with different sizes on MIMOPS

In order to parallelize the execution of 1024-point FFT, an approach that allows computing of 1024-point FFT as a 2D structure made of two 32-point FFTs is used. This approach follows the rule that each FFT module with N points can be represented as a decomposition of other two FFT modules with p and q points, in such a way that $N = p^*q$ and assuming that $p, q \geq 2$ (in this specific case $1024 = 32^*32$), [32]. In order to compute 1024-point FFT as a 2D structure made of two 32-point FFTs, the following five-step algorithm is executed:

1. Arrange 1024-point input vector in a 32×32 matrix, in rows/columns major order;
2. Calculate 32-point FFT for each column/row of the matrix and store the obtained results in the appropriate column/row;
3. Multiply every (i, j) element of the matrix by conjugated twiddle factor, whereas i = [0–31], j = [0–31];
4. Calculate32-point FFT for every row/column of the matrix and store the results in the corresponding row/column;
5. The result is acquired by outputting the derived 32×32 matrix, in columns/rows major order;

This method is applied in an 8-core parallel implementation of MIMOPS processor that executes 1024-point FFT. For that purpose, each MIMOPS core performs 32-point FFT on 4 rows/columns of the input 32×32 matrix. It is considered that the required data are distributed to the other MIMOPS cores, after the third step of the algorithm. The execution time of the 1024-point FFT on an 8-core parallel MIMOPS processor and other digital signal processors, presented in [33], is given in Table 1. These results show that the 8-core parallel MIMOPS processor outperforms TMS3206701 and ADSP21160, while it achieves very close results to VIRAM and TigerSHARC digital signal processors.

Table 1. Execution time of 1024 point FFT on various digital signal processors

Processor	Execution time (μs)
Pathfinder-1	22.3
Wildstar	25
Pulsar	27.9
Parallel MIMOPS	35.2
VIRAM	37
TigerSHARC	41
ADSP21160	91
TMS3206701	124.3

The third test program is characterized with small arithmetical intensity and it implements a PDE (partial differential equation) solver, [34], like a significant problem from the structure grids area. The appropriate PDE is defined with Eq. 3.

$$\begin{cases} \frac{\partial^2 u(x,t)}{\partial x^2} = \frac{\partial u(x,t)}{\partial t} \\ u(0,t) = u(1,t) = 0 \\ u(x,0) = \sin \pi x \end{cases} \tag{3}$$

This program includes a small data set (128B) and performs repetitive operations over it by calling two functions. As a result, the number of MIPS's cache misses is only 2, causing just two blocks to be transferred from the main memory to the two-level cache memory. The results that are shown in Fig. 10 present the PDE execution time, measured in processor cycles, for the MIPS and MIMOPS processors. According to that, it can be concluded that MIMOPS slows down the program execution time by 14.7%, in comparison to MIPS. Assuming that the proposed MIMOPS processor is not applicable to problems with low arithmetical intensity, a parallel implementation of MIMOPS is not further purposed for accelerating the execution of this test program.

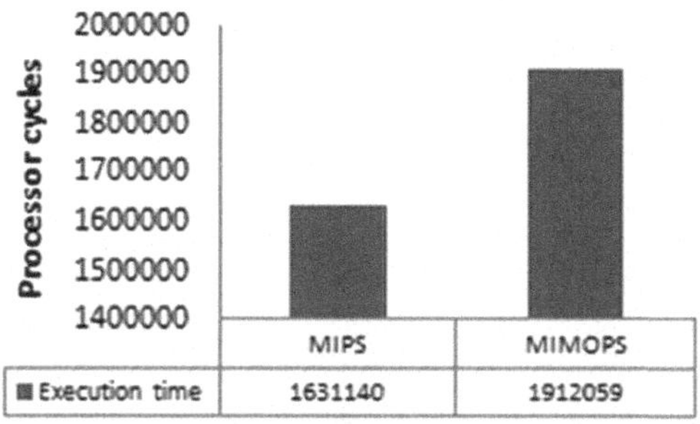

Fig. 10. Execution time of partial differential equation solver, on MIPS and MIMOPS

5 Conclusion

This paper proposes a novel RISC-based memory-centric (MEMRISC) processor architecture that integrates processor and memory on the same chip die, and removes GPRs and cache (in and out of the processor chip) from the memory hierarchy. As a result of these abilities, the proposed MIMOPS processor provides: fast and simple access to the on-chip memory (without explicit load/store operations and pipeline phase), high internal memory bandwidth, decrease of the capacity of redundant on-chip memory resources (GPRs and cache memory), avoidance of copy operations and transfers of redundant data and blocks into GPRs and cache memory, and removal of complex mechanisms for cache memory management. Furthermore, this paper also proposes a distributed approach of parallel processing with several MIMOPS processor cores, where every MIMOPS core views its local memory as a large cache of the shared virtual memory address space for its associated core.

The performance evaluation that is presented in this paper shows that the proposed single-core MIMOPS processor is suitable for the execution of programs that are characterized by medium and high arithmetical intensity, according to the Roofline model. This is basically because it provides direct access to the memory, avoiding the time-consuming stalls which occur during data exchanges between GPRs and caches in a processor that operates with standard memory hierarchy. In addition to that, it is shown that a 4-core parallel MIMOPS processor achieves 4.96 (79.8%), 4.52 (77.8%) and 3.72 (73.1%) times better results than MIPS, PERL and single-core MIMOPS processors, accordingly, while executing 32×32 matrices multiplication, as a program characterized with high arithmetical intensity. Furthermore, an 8-core parallel implementation of MIMOPS achieves comparable results, like IRAM, and even better results than other DSPs (ADSP21160 and TMS3206701), while executing 1024-point FFT, as a program characterized with medium arithmetical intensity.

References

1. Patterson, D.A., Hennessy, J.L.: Computer Organization and Design: The Hardware/Software Interface. Elsevier, Amsterdam (2014)
2. Kozyrakis, C., Patterson, D.: Vector vs. superscalar and VLIW architectures for embedded multimedia benchmarks. In: 35th International Symposium on Microarchitecture, Turkey (2002)
3. Silc, J., Robic, B., Ungerer, T.: Processor Architecture: From Dataflow to Superscalar and Beyond. Springer, Germany (1999)
4. FitzRoy-Dale, N.: The VLIW and EPIC processor architectures. Master Thesis. University of New South Wales, Australia (2005)
5. Smotherman, M.: Understanding EPIC architectures and implementations. In: ACM Southeast Conference (2002)
6. Hennessy, J.L., Patterson, D.A.: Computer Architecture: A Quantitative Approach. Morgan Kaufmann, Burlington (2012)
7. Yan, Y., Brightwell, R., Sun, X.: Principles of memory-centric programming for high performance computing. In: Workshop on Memory Centric Programming for HPC, USA (2017)
8. Jakimovska, D., et al.: Modern processor architectures overview. In: XVIII ICEST Conference, Bulgaria, pp. 239–242 (2012)
9. Carvalho, C.: The gap between processor and memory speeds. In: ICCA 2002, Portugal (2002)
10. Machanick, P.: Approaches to addressing the memory wall. Technical Report. University of Queensland Brisbane, Australia (2002)
11. Borkar, S., Chien, A.A.: The future of microprocessors. Commun. ACM **54**(5), 67–77 (2011)
12. Intel Corporation: New Microarchitecture for 4th Generation. Intel Core Processor Platforms. Product Brief (2013)
13. Bao, W., Tavarageri, S., Ozguner, F., Sadayappan, P. PWCET: power-aware worst case execution time analysis. In: International Conference on Parallel Processing Workshops (2014)
14. Suresh, P.: PERL - a register-less processor. Ph.D. thesis. Indian Institute of Technology, Kanpur (2004)
15. Venkataramani, V., Chan, M.C., Mitra, T.: Scratchpad-memory management for multi-threaded applications on many-core architectures. ACM Trans. Embedded Comput. Syst. **18**(1), 1–28 (2019)
16. Efnusheva, D., Cholakoska, A., Tentov, A.: A survey of different approaches for overcoming the processor-memory bottleneck. Int. J. Comput. Sci. Inf. Technol. **9**(2), 151–163 (2017)

17. Oskin, M., Chong, F.T., Sherwood, T.: Active pages a computation model for intelligent memory. In: 25th Annual International Symposium on Computer Architecture, pp. 192–203 (1998)
18. Eigenmann, R., Lilja, D.J.: Von Neumann computers. In: Wiley Encyclopedia of Electrical and Electronics Engineering, pp. 387–400 (1998)
19. Hamzaoglu, F., Arslan, U., Bisnik, N., Ghosh, S., Lal, M. B., Lindert, N., Meterelliyoz, M., Osborne, R. B., Park, J., Tomishima, S., Wang, Y., Zhang, K.: A 1GB 2GHz embedded DRAM in 22nm tri-gate CMOS technology. In: IEEE International Solid-State Circuits Conference (2014)
20. Azarkhish, E., Rossi, D., Loi, I., Benini, L.: Design and evaluation of a processing-in-memory architecture for the smart memory cube. In: 29th International Conference Architecture of Computing Systems, Germany (2016)
21. Vermij, E., Fiorin, L., Jongerius, R., Hagleitner, C., Van Lunteren, J., Bertels, K.: An architecture for integrated near-data processors. ACM Trans. Architect. Code Optimiz. **14**(3), 1–25 (2017)
22. Draper, J., et al.: A prototype processing-in-memory (PIM) chip for the data-intensive architecture (DIVA) system. J. VLSI Sig. Process. Syst. **40**(1), 73–84 (2005)
23. Gebis, J., et al.: VIRAM1: a mediaoriented vector processor with embedded DRAM. In: 41st Design Automation Student Design Contest, San Diego (2004)
24. Hewlett Packard Labs: The Machine: The future of technology. Technical Paper (2016)
25. Singh, G., Chelini, L., Corda, S., Javed Awan, A., Stuijk, S., Jordans, R., Corporaal, H., Boonstra, A.: A review of near-memory computing architectures. In: 21st Euromicro Conference on Digital System Design (2018)
26. Ghose, S., Hsieh, K., Boroumand, A., Ausavarungnirun, R., Mutlu, O.: The processing-in-memory paradigm: mechanisms to enable adoption. In: Beyond-CMOS Technologies for Next Generation Computer Design. Springer (2019)
27. Weisberg, P., Wiseman, Y.: Virtual memory systems should use larger pages rather than the traditional 4KB pages. Int. J. Hybrid Inf. Technol. **8**(8), 57–68 (2019)
28. ARM: ARM architecture reference manual. Technical Paper (2005)
29. Li, K.: IVY: a shared virtual memory system for parallel computing. In: International Conference on Parallel Processing (1988)
30. Dokoski, G., Efnusheva, D., Tentov, A., Kalendar, M.: Software for explicitly parallel memory-centric processor architecture. In: Third International Conference on Applied Innovations in IT, Germany (2015)
31. Vollmar, K., Sanderson, P.: MARS: an education-oriented MIPS assembly language simulator. In: 37th SIGCSE Technical Symposium on Computer Science Education (2007)
32. Efnusheva, D., Tentov, A.: Hardware implementation of 64 point FFT module in FPGA. In: XI International Conference ETAI, North Macedonia (2013)
33. Thomas, R., Yelick, K.: Efficient FFTs on IRAM. In: 16th International Parallel and Distributed Processing Symposium (2002)
34. Cheney, W., Kincaid, D.: Numerical Mathematics and Computing. Thomson Brooks, San Francisco (2008)

Technical and Behavioural Training and Awareness Solutions for Mitigating Ransomware Attacks

Abubakar Bello(✉) and Alana Maurushat

School of Social Sciences, WSU, Sydney, NSW, Australia
{A.Bello,A.Maurushat}@westernsydney.edu.au

Abstract. Over the years, there has been a significant increase in cyber security risks and vulnerabilities with one of the most severe threat being ransomware attacks. Ransomware, a variant of malware, encrypts files, data, and often locks computer systems, and retains the decryption key until victims pay a ransom. Current method of ransomware mitigation is the analysis and classification of the ransomware and its variants to propose solution for detection and prevention. This mitigation approach omits technology users as part of the solution especially given their role in falling prey to ransomware by means of social engineering attack vectors. The purpose of this qualitative study was to highlight current and emerging ransomware vectors, and to identify cyber security awareness and education solutions that can be applied to mitigate socially engineered ransomware attacks. A semi-structured interview with executives and managers from several financial, technology, construction, transportation, education, and health industries revealed the lack of current awareness and training approaches to mitigate against socially engineered ransomware attacks. This study recommends some specific cybersecurity training and awareness approaches to consider in order to enable technology users resist and mitigate against ransomware attacks.

Keywords: Ransomware · Malware · Social engineering · Cybersecurity · Mitigation

1 Introduction

Online technologies have become interwoven in the daily lives of individuals, businesses, and governments due to tremendous benefits and opportunities. However, sophisticated security threats exist and continue to rise with little or no awareness about these risks and how they could be mitigated [1]. Technology users, particularly organisations, are constantly at risk of encountering a cyberattack such as ransomware, and having their confidential information stolen or destroyed affecting business continuity.

Ransomware is a new extortion business criminal model in form of a cyberattack. The threat is propagated using malware or malicious codes, where data and confidential information is held ransom until a sum of money is paid. In the last quarter of 2018 alone, 1,493 cases of ransomware extortion valued at about $3.6 million not including

R. Silhavy (Ed.): CSOC 2020, AISC 1226, pp. 164–176, 2020.
https://doi.org/10.1007/978-3-030-51974-2_14

loss of business, time, document or files were reported in the US [2]. Ransomware victims still pay the ransom to cybercriminals despite warnings and advice not to pay [3]. The rapid rise and spread of ransomware suggest that it is highly profitable for cybercriminals as a significant number of technology users lack appropriate cybersecurity awareness and education against socially engineered ransomware attacks. A socially engineered ransomware attack involves the use of psychological manipulation, influence or deceiving of individuals to install a malware that infect and encrypts data on their device for digital extortion.

The security stakes are increasing for both public and private sectors regarding ransomware attacks. Recent attacks target all user categories, businesses and government departments and critical infrastructures. Therefore, mechanisms for security empowerment, awareness and education against socially engineered ransomware attacks is crucial. Modern methods of training and awareness are often generic and limited to learning hardware and software skills. However, technology users need psychological training to enable them to identify and assess social engineering tactics employed by cybercriminals.

This study aims to identify the current and potential social-engineered ransomware attack methods, as well as cyber security awareness and education solutions that can be used to mitigate the risks for socially engineered ransomware attacks. The study begins by providing a succinct overview of ransomware, followed by the method of inquiry and analysis to aid findings on the current and emerging socially engineered ransomware attack techniques. Ransomware control gaps across organisations, including the identification of system and user centric mitigation approaches based on effective training and awareness strategies are also presented.

2 Ransomware Anatomy

2.1 Overview

Ransomware first appeared in the late 1980s deployed as PC CYBORG where victims were socially engineered to install an infected floppy disk received by mail [4]. The crypto malware was triggered when victims' systems reboot counts to 90, allowing the malware to spread during the dormancy period. Victims were required to post ransom cheque to recover the lost files and directories. This paved the way for the sophisticated socially engineered ransomware attacks reported today.

Traditionally, cryptography aim to provide protection, privacy and security to users; however, its mutation to crypto-virology by cyber criminals allowed cryptography to be used offensively [5]. Crypto-virology allows criminals to install viruses and malicious codes that could mount extortion-based attacks when deployed on a host system. Ransomware strains today are more sophisticated and operate within this dynamic ecosystem, with the ability to adapt and evolve when deployed.

2.2 Lifecycle of Ransomware

The lifecycle of a ransomware begins from deployment of a malicious code or application, propagation, and concludes in most cases when the attacker financial demands are met. Several actions between the victim's device and the attacker's Command-and-Control (C2C) server take place during this lifecycle for the hijack to be successful [6]. The entire ransomware process involves the four stages illustrated in Fig. 1, followed by discussion.

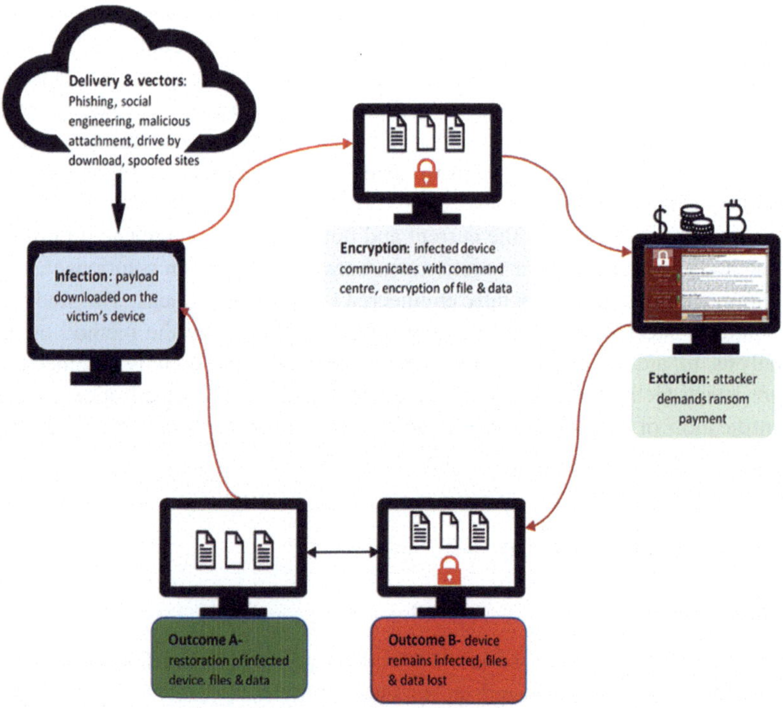

Fig. 1. Ransomware lifecycle

Infection: There are over 4 billion internet users worldwide and about 7 billion internet-connected devices, with additional 100 billion devices or more expected to be connected to the internet due to the Internet of Things [7–9]. A cyber criminal may employ social engineering technique to lure a victim into clicking and download a malicious email attachment, redirect the victim to a spoofed site or use a drive-by download to disseminate the ransomware payload into a victim's device. Security researchers often obtain samples of an infection code for analysis and development of possible detection solutions [10].

Encryption: Once the ransomware infection is successful, the cybercriminal employs the cryptographic process to search and encrypt the victim's files or lock down the infected device. The ransomware code usually targets and encrypt the victim's files and documents, critical system files, backup files, attached devices and shared network drives [6, 9]. Encryption is the core process necessary for the extortion to proceed to the next stage in the lifecycle [11, 12].

Extortion: In this stage, the victim is made aware of the infection by the cybercriminal. The ransomware software displays a prompt to the victim that files have been encrypted, and timely remittance of the ransom will ensure the decryption of files or release of locked devices, otherwise, files will be deleted permanently. As previously mentioned, ransomware is profitable for cybercriminals as a result of electronic payments and anonymous payment channels such as MoneyPak, DarkCoin or BitCoin [13].

Outcome: The ransomware lifecycle is completed in this final stage based on the actions of the victim. There are three possible outcomes in this stage:

a) The victim's is able to eliminate the ransomware code from their devices and restore all encrypted files.
b) The victim succumb to the extortion crime and remit the ransom via the anonymous channels for restoration of files and devices, or
c) No payment is made, the ransomware is not eliminated and the victim's files are lost or destroyed.

Ransomware attacks could be socially engineered using phishing techniques as point of entry in the propagation of malicious payloads. The result of a socially engineered ransomware attack is devastating to the victim, from loss of sensitive data, extortion, and denied access to the infected device. For example, in 2017 more than 300 computer systems across the world were infected with the WannaCry ransomware, costing the UK government millions of pounds [14]. Ransomware attack victims whose computer systems were infected by the WannaCry ransomware paid the ransom but were not guaranteed restoration of their data and computer systems [15].

3 Methodology

This study used a qualitative methodology to address the issue of socially engineered ransomware attacks. The qualitative data for the study was collected via semi-structured interviews from 30 senior management employees across financial, technology, construction, transportation, education, and health organisations to understand the reasons why users are susceptible to socially engineered ransomware attacks, as well as the mechanisms to resist these attacks. The data collected were transcribed, and the content of the transcripts analysed using thematic content analysis technique to identify themes and patterns from the responses.

4 Findings

The themes that emerged from the interviews are discussed in the following sections, including knowledge and understanding of current and emerging ransomware attack methods.

4.1 Perceived Understanding of Ransomware

The findings follow the theme of a general lack of understanding, awareness, and how to mitigate ransomware. These understandings are explored below.

Lack of Knowledge on Ransomware: Unfamiliarity with ransomware attacks was a strong theme identified from the interview. Majority of the participants are not familiar with media reports or conference discussions on ransomware attacks and have vague or no actual knowledge on what ransomware attacks entails. Only 11% of the interviewed participants had a decent understanding of ransomware and how it could potentially impact them.

Lack of Awareness on Socially Engineered Ransomware Attack Vectors: Participants had no very little knowledge or information on the various types of socially engineered ransomware attack vectors. Although Phishing emails was considered as an attack vector by 2 participants, further probe indicated no knowledge of how a phishing email could spread a ransomware payload.

Assumption of Miscreant Behaviour: Acommon misconception identified was participants assume that miscreant behaviour causes victims to fall prey to ransomware attacks. Analysis of ransomware attacks shows that cyber criminals may exploit network vulnerabilities, use exploit kits or brute force attacks to propagate ransomware payload to any user [16].

Equating IT Savviness with Susceptibility to Ransomware Attacks: Information technology (IT) savviness was a resounding theme that participants equated to socially engineered ransomware insusceptibility. Other studies have shown individuals who perceive themselves to be IT savvy are more likely to be vulnerable to ransomware attacks [17, 18].

Training and Awareness: Training and awareness on ransomware threats which was reported by participants as lacking in their organisations, was referred to as a crucial method that could enable individuals recognise and resist socially engineered ransomware attacks. However, participants also emphasised that basic training approaches alone may not increase an individuals' insusceptibility to socially engineered ransomware attacks. Pragmatic approach to training and awareness ensuring that emerging ransomware threats are encapsulated in cyber security education, and incorporating practical methodologies is crucial for mitigating socially engineered ransomware attacks.

4.2 Current and Emerging Ransomware Attack Vectors

Ransomware research is very recent, and academic publications have increased in the last three years, improving the general knowledge on ransomware attacks, strains and vectors. However, as technology advances, cybercriminals consistently adapt new techniques to propagate ransomware codes.

Ransomware vectors are the attack transmission channel that enables the propagation of malicious ransomware code [19]. Attack vectors vary and differ in their degree of stealthiness, effectiveness and complexity [9, 13, 19]. Some common and evolving attack vectors used in socially engineered ransomware attacks are explored below.

Email: Phishing is a persisting cyber security issue, and the majority of socially engineered ransomware utilise various forms of phishing, especially emails [20]. Email remains popular as it affords cybe rcriminals a cheap and easy avenue to reach millions of unsuspecting users [9, 21, 22]. A cybercriminal only needs to send out a malicious email that contains the ransomware payload. When a victim opens the mail, the payload is triggered and begins the ransomware lifecycle process. An analysis of ransomware email showed that victims fall prey to phishing emails because of social engineering techniques used in framing the subject lines, employing a persuasive characteristic and urgency [22].

Man-in-the Middle (MITM): MITM attack occurs when an attacker or malicious third party covertly takes control of a seemingly protected and encrypted channel of communication [23]. The cybercriminal within the compromised network can intercept sensitive data, but also direct the victim to the spoofed website where the ransomware payload will be deployed. MITM is usually executed using various communication channels such as Near Field Communication (NFC), Bluetooth, Long-Term Evolution (LTE) and Wi-Fi.

Drive-by-Download: This attack vector allows a malware to infect a user's devices by exploiting simple security flaws. Attackers place the ransomware application often in compromised website, the ransomware automatically downloads and installs when the victim access the website [24]. Drive-by-download could be embedded in malicious emails and attachments.

Remote Desktop Protocol (RDP): RDP is a service that allows users to connect to a server and access applications from another location. Ransomware attacks via RDP have been on the rise since 2016 due to the administrative privilege it may give cyber criminals if they successfully gain access. RDP as an attack vector could be used to propagate ransomware on a large scale, by installing malicious files on all devices within the network [25].

Internet of Things (IoT): IoT ransomware attacks is a growing trend and likely to increase as more devices are incorporated into an IoT environment. IoT devices span from Fitbit wearables to smart home and cars, and they are constantly accessing the internet gathering and distributing data [21]. A malicious IoT device as a ransomware vector could cause severe disruptions to networks, surveillance systems, critical infrastructures and manufacturing in the future [7]. For example, the Android.Lockdroid.E ransomware that accessed IoT devices and locked smart TVs [26].

Application Programming Interfaces (APIs): An API is a set of procedures and functions that allows two or more software applications to communicate with each other. As an attack vector, a cyber criminal could obfuscate an API call and stealthily deploy a ransomware code within the network to infect the connected devices [27, 28].

5 System and User Centric Ransomware Mitigation Approaches

Existing ransomware research focuses on the development of detection, prevention, prediction and possible recovery of ransomware attack. The trend from previous studies shows that security professionals usually obtain samples of ransomware infection code, analyse and classify them based on attack attributes to provide possible solutions. However, these solutions do not present a silver bullet for ransomware attacks. For example, security analysts are still seeking solutions to disable the NotPetya ransomware after it gains access to a victim's device, no fixed solution has been found [29].

Ransomware prevention solutions requires a holistic approach. Unfortunately, studies often focus on preventive solutions for ransomware codes, and regard social engineering techniques as a silo event. This study makes a case for research to focus on a holistic preventive solution for socially engineered ransomware attacks. The ransomware prevention techniques presented in this section are categorised into system/network and user levels. Two general approaches emerge: technical and user/behavioural.

5.1 Technical

System/Network Level: The studies that examine ransomware prevention and detection usually propose solutions from a network or system perspective. Some studies propose platform-based detection solutions to deal with ransomware [6, 30, 31]. While others focus on specific ransomware such as crypt-ransomware [3, 32, 33]. At a technical level, the techniques for ransomware prevention should be focused on neural network, machine learning, or data centric approaches.

Neural Network Approach: Homayoun et al. [34] projected a Deep Ransomware threat Hunting and Intelligence System (DRTHIS) by combining Long Short-Term Memory (LSTM) and Convolutional Neural Network (CNN) to detect ransomware and their families within seconds as the most effective way to address ransomware. The research model analysed large samples of crypto-ransomware families and achieved a precision and recall (F-measure) value of 0.996 in detecting ransomware with a true positive rate of 0.972 and a false positive of 0.027 [34]. Azmoodeh et al. [35] on the other hand studied ransomware detection based on energy consumption within IoT networks to halt the effect of a ransomware attack within a network. They analysed four classifiers: k-Nearest Neighbor (KNN), Neural Network (NN), Support Vector Machine (SVM) and Random Forest (RF) on power usage. The results from their experiment showed that KNN using Dynamic Time Wrapping (DTW) achieved an optimal performance and achieved a detection rate of 95.5% and precision rate of 89.19% within an IoT network [35]. Neural network, from a technical perspective, presents a brilliant effective solution in detecting ransomware and their families.

Machine Learning Approach: Machine learning (ML) techniques enables a defined set of algorithms to continually monitor, search and analyse computer networks, systems and devices for abnormal processes that might initiate suspicious alteration activities [36]. In recent studies, ML is proposed as an effective technique in detecting ransomware resulting from an intelligent algorithm that learns and classify ransomware variants and families. For example, Sgandurra et al. [37] proposed EldeRan, a dynamically based approach used to test, analyse and classify 582 ransomware belonging to 11 families. The result from the study showed that EldeRan achieved a detection rate of 96.3%. Similarly, Zhang et al. [38] collected 1787 ransomware samples from eight families, using N-gram sequence and Term Frequency-Inverse Document Frequency (TF-IDF), their result achieved an accuracy rate of 91.43%. ML detection techniques have proved effective for malicious programs; however, some ransomware families may not be adequately analysed and classified. Additionally, new ransomware strains may incorporate logic bomb techniques (waiting for a user action before carrying out an attack) or require a shorter encryption time which may allow the ransomware to escape detection via ML techniques. Nevertheless, further development on this technical approach could see ransomware threats become obsolete in the future.

Data-Centric Detection Approach: Data-centric approach monitors user's resources that are subject to attack rather than the malicious process that carries out the ransomware attack. Azmoodeh et al. [35] proposed a detection approach that monitors ransomware energy consumption to distinguish ransomware from non-malicious applications. The detection technique creates energy consumption log files of the most popular Android applications, and ransomware. Analysis of the log files are retrieved and labelled according to the application type (goodware or ransomware) [35]. Scaife et al. [32] proposed a technique focused on monitoring the real-time change of user data, over time the accumulated observations allows an analyst to examine multiple suspicious indicators rather than relying on one isolated indicator to detect ransomware activities.

5.2 User/Behavioural

Security researchers often develop conceptual models and propose solutions to a ransomware attack. For example, Dion et al. [39] developed a prototype to gamify the procedure of a ransomware attack. Lika et al. [29] developed a conceptual preventive solution intended to educate users on NotPeyta attack through a gamification process. These studies proposed highly conceptual solutions that have not been empirically tested despite the increasing rate of ransomware attacks. Socially engineered ransomware preventive solutions require more user-focused approach involving awareness and training programs since humans are the primary vectors used in socially engineered ransomware attacks.

Awareness and Training Solutions for Mitigating Socially Engineered Ransomware: Since 2007, Luo and Liao [40] proposed a ransomware prevention framework focused on policy regulation, access control and management, awareness training and education [40] which has received little attention and now outdated. Academic literature is now less focused on the human aspect in ransomware prevention, and more recently, Ferreira [22] highlighted the dearth of ransomware research focused on user-related solutions. One way to ensure the prevention of ransomware attacks is to incorporate users as part of the ransomware solution through the development of serious games, simulations, virtual labs, apps, and gamification based training and awareness programs.

Serious games is beneficial for individuals to understand the concept of socially engineered ransomware attacks using real time scenarios to test their knowledge to implement mitigation strategies (see Table 1). Serious game technique of training is based on virtual scenario formation that enables a trainee to learn different attack situations and how to deal with them [41].

Table 1. Serious games training methods

Serious game training methods/platforms	Source
TATA Interactive Systems (e-learning based)	[42]
PEX as a web-based serious gaming training program using (virtual threats)	[43]
Serious games comprising of reward systems	[44]
HATCH Training (Hack and Trick Capricious Humans)	[45]
CYBERCIEGE serious gaming platform	[41]

Virtual labs are also quite similar to serious games in that they help to mitigate against cyber security threats by confining them to a virtual environment [41]. Virtual labs could be set up remotely enabling the trainees learn about security threats such as ransomware, social engineering and phishing along with virtual solutions to the threat scenario [46]. Table 2 shows the different virtual labs training methods that can be adapted for education on ransomware.

Table 2. Virtual labs training methods

Virtual lab training methods/platforms	Source
Cyber Zone (remote virtual lab environment)	[42]
Virtual communications & collaboration training (social networks)	[47]
DECAMP (virtual online platform)	[46]

Simulation training is another approach that utilises virtual labs. In a simulated training session, a model of a real ransomware threat can be deployed to conduct experiments solely for understanding the behaviour of the trainees or evaluation of various ransomware mitigation strategies [48]. Simulations training are often used in large organisations with complex networks and multi-layers of operations (Table 3).

Table 3. Simulation training methods

Simulation training methods/platforms	Source
Cyber Arrest (real time challenges based on issues faced by employees i.e. using simulated phishing)	[42]
(Testing, Training, Case Phishing, and Evaluation) simulation method	[49]

Gamification training can focus on assessing the behavior of hypothetical ransomware victims using gaming strategies in non-game activities. The gamification process should include psychological manipulation of ransomware threats [50]. Table 4 shows some current methods of gamification training that can be adapted for education on ransomware.

Table 4. Gamification training methods

Gamification training methods/platforms	Source
Phishing Game (physical games organised with respect to real time)	[42]
Modulated Light Intensity using a 3D sensor and Sun Small Programmable Object Technology	[50]
Communication threats competitions	[49]
Information gathering and leaking competitions	[52]

Lastly, software applications are currently becoming a key method of security awareness and training of individuals especially in large organisation. Some organisations build their own defense solutions and framework or purchase commercial solutions based on the security threats they need to address [51]. Software-based ransomware training solutions can be very effective in raising awareness about ransomware threats, however, but may require a safety model of training individuals not working virtually.

6 Conclusion

Socially engineered ransomware is a growing menace and currently affect individuals, businesses and government based institutions that house critical economic infrastructure and control systems. Psychological manipulation techniques coupled with sophisticated ransomware codes allows an attacker to gain access into organisations and encrypt, steal or destroy sensitive data. There are various methods and techniques that can be employed to overcome ransomware threats ranging from threat intelligence and detection technologies, to implementing security standards, policies and procedures. However, these mitigation methods may be less efficient because developing an effective training and awareness solution plays a major role in defending against socially engineered ransomware attacks. Cyber criminals are constantly changing their attack methodologies; hence, training approaches also need to change, whereby organisations employ virtual real life scenarios and simulations to learn about attacks and threats, and the strategies to mitigate them.

References

1. Kostopoulos, G.K.: Cyberspace and Cybersecurity, Second. Taylor and Francis (2017)
2. FBI. 2018 Internet Crime Report. Federal Bureau of Investigation (2018)
3. Moore, C.: Detecting ransomware with honeypot techniques. In: 2016 Cybersecurity Cyberforensics Conference (CCC), pp. 77–81 (2016)
4. Wilding, E.: Popp goes the weasel. Virus Bull. 20 (1992)
5. Young, Y., Yung, M.: Cryptovirology: extortion-based security threats and countermeasures. In: Proceedings of the IEEE Symposium Security Privacy, pp. 129–140 (1996)
6. Hampton, N., Baig, Z.A., Zeadally, S.: Ransomware behavioural analysis on windows platforms. J. Inform. Secur. Appl. **40**, 44–51 (2018)
7. Bello, A., Mahadevan, V.: A cloud based conceptual identity management model for secured internet of things operation. J. Cyber Secur. Mobility **8**(1), 53–74 (2019)
8. Lueth, K.L.: State of the IoT 2018: Number of IoT devices now at 7B - Market accelerating. IOT Analytics (2018). https://iot-analytics.com/product/state-of-the-iot-02–2017/. Accessed 14 May 2019
9. O'Kane, P., Sezer, S., Carlin, D.: Evolution of ransomware. IET Netw. **7**(5), 321–327 (2018)
10. Hull, G., John, H., Arief, B.: Ransomware deployment methods and analysis: views from a predictive model and human responses. Crime Sci. **8**(1), 2 (2019)
11. Bajpai, P., Sood, A.K., Enbody, R.: A key-management-based taxonomy for ransomware. In: APWG Symposium on Electronic Crime Research, pp. 1–12 (2018)
12. Zimba, A., Mulenga, M.: A dive into the deep: demystifying wannacry crypto ransomware network attacks via digital forensics. Int. J. Inform. Technol. Secur. **4**(2), 57–68 (2018)
13. Salvi, M.H.U., Kerkar, M.R.V.: Ransomware: a cyber extortion. Asianj J. Convergence Technol. 2 (2016)
14. Gibbs, S.: Wannacry: Hackers Withdraw £108,000 Of Bitcoin Ransom| Technology| The Guardian (2017). https://www.theguardian.com/technology/2017/aug/03/wannacry-hackers-withdraw-108000-pounds-bitcoin-ransom. Accessed 28 Oct 2019
15. Bistarelli, S., Parroccini, M., Santini, F.: Visualising bitcoin flows of ransomware: wannacry one week later. In: ITASEc, p. 8 (2018)
16. Liao, K., Zhao, Z., Doupe, A., et al.: Behind closed doors: measurement and analysis of cryptolocker ransoms In: Bitcoin. APWG Symposium on Electronic Crime Research, pp. 1–13 (2016)
17. Ament, C., Jaeger, L.: Unconscious on their own ignorance: overconfidence. Inform. Secur. 13 (2017)
18. Wang, J., Li, Y., Rao, H.R.: Overconfidence in phishing email detection. J. Assoc. Inform. Syst. **17**(11) (2016)

19. Zimba, A., Wang, Z., Chen, H.: Reasoning crypto ransomware infection vectors with bayesian networks. In: IEEE International Conference Intelligence and Security Informatics, pp. 149–151 (2017)
20. Gallegos-Segovia, P.L., Bravo-Torres, J.F., Larios-Rosillo, V.M., et al.: Social engineering as an attack vector for ransomware. In: CHILEAN Conference on Electrical, Electronics Engineering, and Informatics and Communication Technologies, pp. 1–6 (2017)
21. Al-rimy, B.A.S., Maarof, M.A., Shaid, S.Z.: Ransomware threat success factors, taxonomy, and countermeasures: a survey and research directions. Comput. Secur. **74**, 144–166 (2018)
22. Ferreira, A.: Why ransomware needs a human touch. In: International Carnahan Conference on Security Technology, pp. 1–5 (2018)
23. Conti, M., Dragoni, N., Lesyk, V.: A survey of man in the middle attacks. IEEE Commun. Surv. Tutorials **18**(3), 2027–2051 (2016)
24. Provos, N., Mavrommatis, P., Rajab, M.A., et al.: All Your iFRAMEs Point to Us, p. 15 (2008)
25. Wang, Z., Wu, X., Liu, C., et al.: Ransomtracer: exploiting cyber deception for ransomware tracing. In: IEEE Third International Conference Data Science Cyberspace, pp. 227–234 (2018)
26. Symantec: Android.Lockdroid.E| Symantec (2015). https://www.symantec.com/security-center/writeup/2014–103005–2209-99. Accessed 21 May 2019
27. Chen, Z.G., Kang, H-S., Yin, S.N., et al.: Automatic ransomware detection and analysis based on dynamic API calls flow graph. In: RACS, pp. 196–201 (2017)
28. Sheen, S., Yadav, A.: Ransomware detection by mining API call usage. In: International Conference Advance Computing Communication Information, pp. 983–987 (2018)
29. Lika, R.A., Murugiah, D., Brohi, S.N., Ramasamy, D.: NotPetya: cyber attack prevention through awareness via gamification. In: International Conference on Smart Computing and Electronic Enterprise, pp. 1–6 (2018)
30. Alzahrani, A., Alshehri, A., Alharthi, R., et al.: An overview of ransomware in the windows platform. In: Conference on Computer Science and Computational Intelligence, pp. 612–617 (2017)
31. Maiorca, D., Mercaldo, F., Giacinto, G., et al.: R-PackDroid: API package-based characterization and detection of mobile ransomware. In: SA, pp. 1718–1723 (2017)
32. Scaife, N., Carter, H., Traynor, P., et al.: CryptoLock (and Drop It): stopping ransomware attacks on user data. In: IEEE 36th. International Conference Distributed Computing Systems, pp. 303–312 (2016)
33. Al-rimy, B.A.S., Maarof, M.A., Prasetyo, Y.A., et al.: Zero-day aware decision fusion-based model for crypto-ransomware early detection. Int. J. Integr. Eng. **10**(6), 82–88 (2018)
34. Homayoun, S.: DRTHIS: deep ransomware threat hunting and intelligence system at the fog layer. Future Gener. Comput. Syst. **90**, 94–104 (2019)
35. Azmoodeh, A., Dehghantanha, A., Conti, M., et al.: Detecting crypto-ransomware. In: Iot Network Based Energy Consumption Footprint, vol. 9, no. 4 (2018)
36. Torres, P.E.P., Yoo, S.G.: Detecting and neutralizing encrypting ransomware attacks by using machine-learning techniques: a literature review. Int. J. Appl. Eng. Res. **12**(18), 7902–7911 (2017)
37. Sgandurra, D., Muñoz-González, L., Mohsen, R., et al.: Automated dynamic analysis of ransomware: benefits, limitations and use for detection arXiv:1609.03020 [cs] (2016)
38. Zhang, H., Xiao, X., Mercaldo, F., et al.: Classification of ransomware families with machine learning based on n-gram of opcodes. Future Gener. Comput. Syst. **90**, 211–221 (2019)
39. Dion, Y.L., Joshua, A., Brohi, S.N.: Negation of ransomware via gamification and enforcement of standards. In: Proceedings of the 2017 International Conference on Computer Science and Artificial Intelligence, pp. 203–208 (2017)

40. Luo, X., Liao, W.: Awareness education as the key to ransomware prevention. Inform. Syst. Secur. **16**(4), 195–202 (2007)
41. Raman, R., Lal, A.K.: Achuthan: serious games based approach to cyber security concept learning: indian context. In: 2014 International Conference on Green Computing, Communication and Electrical Engineering (ICGCCEE), Coimbatore, India, pp. 1–5 (2014)
42. Aldawood, H., Skinner, G.: An academic review of current industrial and commercial cyber security social engineering solutions. In: Proceedings of the 3rd International Conference Cryptography, Security and Privacy. ICCSP 2019, Kuala Lumpur, Malaysia, pp. 110–115 (2019)
43. Tillmann, N., de Halleux, J., Xie, T.: Pex for fun: engineering an automated testing tool for serious games in computer science, p. 9 (2011)
44. Micallef, N., Arachchilage, N.: Involving users in the design of a serious game for security questions education. arXiv:1710.03888 [cs] (2017)
45. Beckers, K., Pape, S., Fries, V.: HATCH: hack and trick capricious humans – a serious game on social engineering. In: Proceedings of the 30th International BCS Human Computer Interaction Conference (2016)
46. Soceanu, A., Vasylenko, M., Gradinaru, A.: Improving cybersecurity skills using network security virtual labs. In: International MultiConference Engineering Computer Science, Hong Kong, vol. 2, p. 6 (2017)
47. Krombholz, K., Hobel, H., Huber, M., et al.: Advanced social engineering attacks. J. Inform. Secur. Appl. **22**, 113–122 (2015)
48. Veksler, V.D., Buchler, N., Hoffman, B.E., et al.: Simulations in cyber-security: a review of cognitive modeling of network attackers, defenders, and users. Front. Psychol. 9 (2018)
49. Crouse, P., Farmer, R.: Information security awareness: a course module using simulated spear-phishing. Inform. Secur. Educ. J. **3**(1), 21–25 (2016)
50. Alarcon-Ramirez, A., Martinez-Pabón, M., Kim, C.: Design and prototype of a wireless tailgate detection system using sun spot platform. Int. J. Embed. Syst. Appl. 14 (2011)
51. Airehrour, D., Nair, N.V., Madanian, S.: Social engineering attacks and countermeasures in the new zealand banking system: advancing a user-reflective mitigation model. Inform. **9**(5), 110 (2018)
52. Social-Engineer LLC.: The 2018 Social Engineering Capture the Flag. Report (2018)

CollaborativeHealth: Smart Technologies to Surveil Outbreaks of Infectious Diseases Through Direct and Indirect Citizen Participation

Óscar Apolinario-Arzube[1,2], José Antonio García-Díaz[3(✉)], Sheila Pinto[4], Harry Luna-Aveiga[2], José Jacinto Medina-Moreira[2], Juan Miguel Gómez-Berbis[4], Rafael Valencia-Garcia[3], and José Ignacio Estrade-Cabrera[5]

[1] Facultad de Ciencias Matemáticas y Físicas, Universidad de Guayaquil, Cdla. Universitaria Salvador Allende, Guayaquil, Ecuador
oscar.apolinarioa@ug.edu.ec

[2] VIAMATICA S.A., Edif. San Francisco 300, Córdova y Av. 9 de Octubre, 090313 Guayaquil, Ecuador
{harry.lunaa, jose.medinamo}@ug.edu.ec

[3] Computer Science Department,Universidad Carlos III de Madrid, Madrid, Spain
{joseantonio.garcia8,valencia}@um.es

[4] Facultad de Informática, Universidad de Murcia, Campus de Espinardo, 30100 Murcia, Spain
100346103@alumnos.uc3m.es, juanmiguel.gomez@uc3m.es

[5] DANTIA Tecnología S.L., Parque Empresarial de Jerez 10, Calle de la Agricultura, 11407 Jerez de la Frontera, Cádiz, Spain
jiestrade@dantia.es

Abstract. Early warning systems are essential to mitigate the consequences of outbreaks of infectious diseases, which causes millions of deaths every year. Surveillance systems collect data on epidemic-prone diseases to trigger prompt public health interventions. In recent years, these systems have improved significantly thanks to infodemiology, a recent research field that promotes the use of health-data collected from the Internet. However, early warning systems can be improved regarding the interpretability and confidentiality of the compiled evidences. In addition, they still rely heavily on human intervention to distinguish among confident and non-confident evidences. To solve these concerns we present CollaborativeHealth, an infodemiology platform that compiles evidence from (1) social networks, (2) public reports, and (3) voluntary citizen participation; and makes use of deep-learning technologies to extract knowledge regarding infectious diseases, their symptoms, or poor environment conditions what promote the propagation of these diseases. Finally, all the compiled evidence is available to health-professionals in real-time through a configurable dashboard. The validation of CollaborativeHealth was performed with a real use-case about monitoring infectious disease cases related to Zika, Dengue, Chikungunya, and Influenza in Ecuador.

R. Silhavy (Ed.): CSOC 2020, AISC 1226, pp. 177–190, 2020.
https://doi.org/10.1007/978-3-030-51974-2_15

Keywords: Infodemiology · Information-retrieval · Opinion-mining · Blockchain · Progressive web app · Ontologies

1 Introduction

Although outbreaks of infectious diseases have been appearing since ancient times, still in the 21st century they are one of the biggest challenges we face. Only in the last two decades we have suffered the Severe Acute Respiratory Syndrome (SARS) outbreak in China; several dengue outbreaks in Asia, Africa and America; a global influenza pandemic; and the Ebola outbreak in Africa among other cases. Moreover, three infectious diseases concerning lower respiratory infections, diarrheal diseases, and tuberculosis were ranked in the top ten causes of death worldwide in 2016 by the World Health Organization (WHO) [36].

The consequences of infectious diseases go beyond the physical and psychological suffering of direct victims. Some infectious diseases can be transmitted from parents to children causing congenital problems. The Zika virus, for example, causes on newborns severe microcephaly, and damage to the back of the eye among other defects and disabilities [28]. Regarding the social dimension, outbreaks and epidemics collapse local health services which also contributes to the panic spreading [10] that causes, on the medium and long term, loss of vaccine efficiency due to the indiscriminate use of vaccines [39]. Moreover, the mismanagement of recent outbreaks has lead to discredit and distrust of citizens towards governments and institutions. Therefore, the struggle against infectious diseases threat has become one of the main objectives of the majority of research initiatives, such as Horizon 2020 in Europe [21].

The International Health Regulations (IHR) require countries to develop effective early warning mechanisms to report public health threats which may constitute a potential public health emergency of international concern [37]. The first early warning systems consisted of controls in high traffic places, such as airports or hospitals. Currently, Infodemiology has emerged as a novel research field that exploits health data on the Internet with the ultimate goal of informing public health and public policies [17]. Healthmap [6] is an example of new tools capable of performing infodemiology studies. Healthmap analysis feed news sites like Google News, and different data-sources specialized in the monitoring of infectious diseases, such as GeoSentinel [8] or ProMED Mail [5]. Healthmap was used to monitor Rio de Janeiro during the 2016 Olympics due to the risk of getting gastrointestinal diseases due to pathogens in the water and infections transmitted by mosquitoes such as Dengue, Chikungunya or Zika virus in an worldwide event that involved nearly one thousand athletes and half a million fans from all over the world. In addition, diseases could have cross borders when people returned to their countries of origin.

Early warning systems based on infodemiology presents potential challenges: (1) an inaccurate interpretation of health status [11]; (2) privacy issues related to confidentiality [11]; and (3) there is great dependence on human intervention when validating the evidence found [6,29]. To solve these drawbacks, advances in modern technologies can play a crucial role identifying relevant content, and improving security and confidentiality by means of distributed databases. Therefore, we propose CollaborativeHealth, a platform for the early warning of outbreaks of infectious diseases that makes use of semantic technologies for a deep understanding of texts written in social networks and the usage of automatic-based filtering that relies in the network confidence to discard false evidences and to prioritize relevant ones. From a geographical area, and a list of infectious diseases, the platform compiles health-data from three sources: (1) public social networks, (2) official sources, and (3) evidences reported directly by citizens through an Progressive Web App. Then, semantic technologies are applied to identify concepts related to the target infectious diseases, such as symptoms, or their transmission methods; then, all the evidences are filtered and prioritized by using a trust filter based on blockchain technology, in order to discard false evidences. Finally, health-professionals can monitor in real-time the evidences identified through a configurable dashboard. This information will help to define strategies to stop the propagation of infectious diseases as soon as possible. The validation of CollaborativeHealth is performed with a real use-case about monitoring infectious disease cases related to Zika, Dengue, Chinkungunya, and Influenza in Ecuador.

The rest of the paper is organized as follows: Sect. 2 contains some background information related to infodemiology. Section 3 shows the architecture of the platform and describes each module. Finally, Sect. 4 describes the current status of the project and describes the following tasks to perform.

2 Background Information

This section contains background information regarding the smart technologies involved in CollaborativeHealth: (1) Infodemiology (see Sect. 2.1), (2) Blockchain (see Sect. 2.2), and (3) Semantic Deep Learning (see Sect. 2.3).

2.1 Infodemiology

Infodemiology, also known as infoveillance, is the science that focuses on exploit data on the internet related to health [17]. Despite its novelty, there are already works that have explored the reliability of extracting data from the Internet towards meningitis, herpes virus, or various sexually transmitted diseases such as AIDS or Syphilis.

Infodemiology studies have two major data-sources: (1) search queries, and (2) social networks. On the one hand, search queries analyse queries from search engines to discover trends about health-behaviour [4,16]. On the other hand, exploit data from social networks can provide more fine-grained information. In this sense, Ofoghi et al. [24] conducted a study to measure the efficiency of social network analysis for the early warning of diseases by analyzing topics related to public domain health. The information was also studied giving special priority to texts produced in the place of origin rather than messages in surrounding areas.

Twitter is a popular micro-blogging social network where health-related information can be found. Twitter is very popular for infodemiology due to the size limit of their posts (also known as Tweets) that restricts users to be concise about what they think and write. In this sense, Haruna et al. [20] applied text analysis in search of contraindications and risks in certain market-based products, such as cosmetic products or food products. Chew and Eysenbach [9], collected about 2 million tweets related to H1N1 virus to observe how the trend of topics and keywords varied. Berkelman et al. [3] created a public health surveillance tool that estimates the state and behavior of the population's health so that governments can carry out more effective interventions. In their work, private laboratory networks related to certain types of flu, active population surveys and laboratory supervision were analyzed.

2.2 Blockchain

Blockchain is the underlying data structure of Bitcoin that provides decentralization, transparency, and immutability in order to provide incorruptible transaction records delivering health-data securely. Therefore, blockchain technology is suitable in environments where network nodes do not necessarily trust each other [13,38]. It is possible to use this technology in order to evaluate the trustworthiness of the evidence based on the reputation of the author's profile and other meta data information.

2.3 Semantic Deep Learning

Semantic Deep Learning refers to the use of deep learning algorithms guided by formal knowledge representation [35]. Deep learning is a field of Artificial Intelligence (IA) composed by a set of machine-learning algorithms that attempt to imitate the cognitive abilities of human beings. However, prior to emulate these human abilities, it is needed to capture, represent, and perform critical features on data. An ontology can be defined as "a formal, explicit specification of a shared conceptualization" [32]. Ontologies have been applied in different domains obtaining promising results, such as, recommender systems [19], opinion mining based at aspect-level [27], software engineering [22], or security [18]. On the other hand, semantic Data Mining is the process of mining information supported by semantic information [14]. Semantic Annotation consists in label parts of the speech of a document with concepts of a formal knowledge representation.

Due to the length restriction of tweets, there are cases where the main topic of the post does not appear explicitly. The use of ontologies has proven to be effective in order to identify hidden, or not obvious, knowledge. The hypothesis is that concepts related to the explicit terms on the document are also related but in a minor degree. The TF-IDF-E formula (see Eq. 1) is an extension of the TF-IDF that measures how a document is related to the concepts in a specific ontology taking into account concepts that are at certain distance from the explicit concept [30]. For example, if we have a tweet where an user states that he is suffering an illness, this tweet is also related to the symptoms of that disease.

$$\sum_{j=1}^{n} \frac{tf - idf}{e^{dist(i,j)}} \tag{1}$$

3 Architecture

The architecture of CollaborativeHealth platform is shown in Fig. 1. In a nutshell the behaviour of the system can be described as follows: First, the platform collects evidence from: (1) social networks, (2) official reports, and (3) evidence directly reported by citizens through a multi-device application. Second, all the compiled documents are semantically annotated by using an ontology of infectious diseases to classify the evidence concerning what diseases, what symptoms, and what causes are the documents referring to. In addition, geographical information is labelled. Third, evidence are filtered and prioritized. Finally, health-professionals can monitor all this information in real-time through a configurable control panel where they indicate which infectious diseases they want to monitor and within what area, with the possibility of generating alerts when new cases are detected. In the following subsections each module is described in detail.

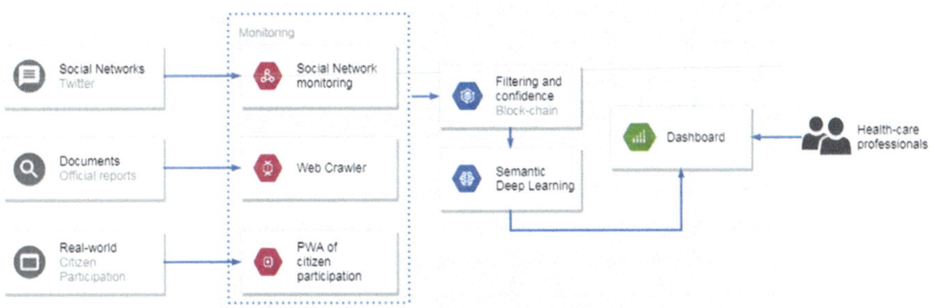

Fig. 1. Architecture of CollaborativeHealth

3.1 Ontology of Infectious Diseases

An ontology of infectious diseases has been developed to conceptualize all the concepts regarding infectious diseases, symptoms, risks, or transmission methods. The infectious diseases concepts were extracted from the Disease Ontology (DO) [31] and from the Infectious Disease Ontology (IDO) [12]. Risks were modeled from WHO reports, such as [7,34]. To keep the ontology small, we selected only the relevant concepts. However, to keep interoperability, we included the SNOMED-CT code for each concept [2] and the DOID (Disease Ontology ID).

The ontology was developed following an iterative process by using the tool Protegé [23]. A screen capture of the ontology is shown in Fig. 2.

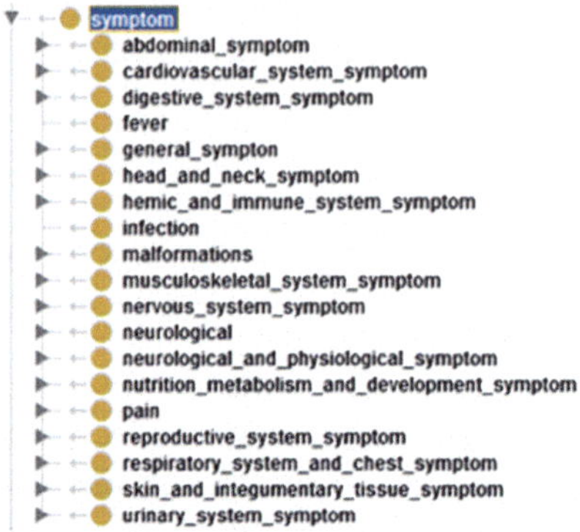

Fig. 2. CollaborativeHealth ontology. Examples of diseases

3.2 Social Network Monitoring Module

This module extracts posts regarding infectious diseases from Twitter. Specifically, we are compiling tweets from trusted Twitter accounts such as the *Ministerio de Salud Pública del Ecuador* (@Salud_Ec) and from public hospitals. Moreover, we are compiling in Ecuador (see Fig. 3) tweets that contains the following keywords and hashtags: *zika*, *dengue*, or *chinkungunya*.

Fig. 3. Area where tweets were collected

The compiled tweets were semantically annotated and its sentiment was extracted. This module is based on previous works of the research team [25,26].

However, prior to Opinion Mining, we manually Labelled the tweets with the help of 40 volunteers, divided into two periods of 2 months each. The volunteers classified each tweet into: *very positive*, *positive*, *neutral*, *negative*, *very-negative*, and *out-of-domain*. In order for volunteers to carry out manual labeling, a web tool was created to help the volunteers in their task. Once a volunteer logs in, the tool chooses among the tweets that the volunteer has not classified, randomize them and offers him the options mentioned above. The graphic interface of this tool, with a random tweet, is shown in Fig. 4. In addition, the classification tool can choose the same tweet to different volunteers so, at the end of the labelling process, we have several ratings for the same tweet and we can discard those tweets without consensus, that is, those Tweets that were labelled differently by the volunteers. The labelling process was reviewed periodically, to ensure that each volunteer tagged tweets on a regular rate and to ensure that all volunteers collaborated uniformly, in order to avoid bias.

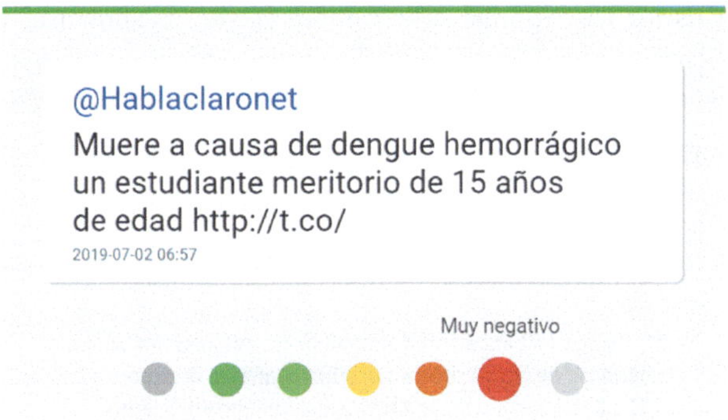

Fig. 4. Example of tweet

We created three machine-learning binary classifiers based on the corpus. First, from *out-of-domain* tweets and the rest of the tags, to detect non-related tweets. Second, from *neutral* tweets against non-neutral tweets, to distinguish among tweets that contains objective or subjective information. Third, from all positive tweets and all negative tweets, to distinguish among tweets that provide positive subjective from negative subjective information.

In order to detect what are the best linguistic features we tried with different information retrieval models. (1) General-purpose models, such as Bag of Words (BoW), and (2) char-grams were applied in conjunction with Machine-Learning Classifiers for creating a supervised classifier. The benchmark of this process is based on the work presented in [1]. These models will be used to build a sentiment-analysis at aspect-level in order to evaluate the positive, neutral and negative statements concerning infectious diseases [15].

3.3 Crawler Module for Extraction from Official Sources

This module is responsible for extracting data related to infectious diseases from official organizations from the communication channels they use, such as web pages, annual reports, syndicated news or social networks. This module makes use of a web-crawler to extract all documents from specific sites that have been previously introduced by the administrators. Then, according to the format of the resource, the module selects which is the best strategy that allows to extract information of the asset. In this sense, it is possible to extract content from web-pages and other types of documents such as PDF, Word, XML, or JSON files.

CollaborativeHealth makes use of the ontology of infectious diseases (see Sect. 3.1) to classify documents. Each concept of the ontology includes a comprehensive list or regular expressions with synonyms and related terms to identify each concept in a document. In addition, this expressions can be translated to different languages. For example, the regular expression for matching synonyms and related terms for the Spanish word *fiebre* (fever) is shown in Eq. 2.

$$(fiebres?|decimas|calenturas?|hipertermia|temperatura) \tag{2}$$

We can see an example of an annotated tweet in Fig. 5.

Fig. 5. Semantic Annotation of tweets

3.4 Multi-device Citizen Participation App

This module consists of an application for citizen participation, which allows citizens to collaborate directly with the collection of evidence. To do this, a progressive, lightweight and easy-to-use web application was created that allows citizens to report cases they see directly. An screen capture of this app is shown in Fig. 6. The information collected by the citizens is automatically complemented by the device. If the user's device allows it, the application will include accurate geo-location information where the user has taken the sample, although it will give him the opportunity to modify it. In addition, it will allow the user to work without an Internet connection and resend the sample once the connection is restored.

Image analysis (or computer vision) is understood as the set of techniques that allow the extraction of significant information from digital images. The most modern computer vision techniques, based on edge detectors and neural networks are inspired by human visual perception models for the recognition of high-level concepts. In this sense, there are techniques that allow obtaining a set of semantically relevant categories from applying machine-learning techniques trying to identify patterns in the photos. This task can simplify and assist citizens when reporting evidence. In addition, it can be used to quantify the categories of labels and their relevance to a specific domain from the use of ontologies of a specific domain. Google API Vision is a Google tool that contains a set of general domain machine learning models. Through this system it is possible to catalog an image among a set of predefined categories. In addition, this technology detects objects and faces, reads printed and handwritten text, and achieves valuable meta data for user's image catalog. Google is not an Open-Source application, and currently applies the Google subscription model, which is from Software as a Service.

3.5 Blockchain Trust and Filtering Module

This module aims to create a system of filtering sources based on trust and that allows prioritizing or discarding among the evidence and documents collected through the application of citizen participation, data from official sources, and data from the social networks.

This system will be based on blockchain technology creating a public and transparent list of registered sources. The long-term objective is to have a distributed system of trust that is able to distinguish between false positives and prioritize valid sources. To do this, this system will extract information from the meta data obtained, in order to obtain information from the sources and will also feed on the manual reports that users download.

3.6 Semantic Deep Learning for Intelligent Data Analysis, Led by Ontologies

Deep-learning is a set of learning algorithms that model high-level abstractions using computational architectures that support iterative data transformations.

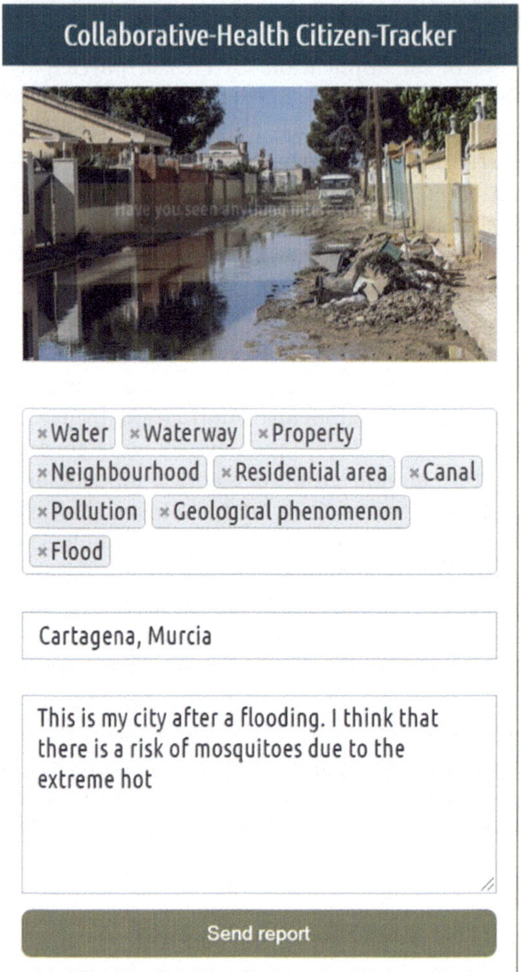

Fig. 6. Citizen Participation PWA

Since the ability to reason about a set of data is essential for the compression of information, Deep-learning technologies with ontologies go one step beyond rule-based systems or systems based on symbol logic. Therefore, this module trains a set of neural networks to design a model capable of learning to reason and infer from an ontological model based on the analysis of social network sources, official sources and the evidences identified by the users that have been filtered and sorted according to urgency by the trust module based on blockchain.

This module combines Semantic Web techniques with Deep Learning classifiers. The nexus of these technologies is the use of the infectious diseases ontology (see Sect. 3.1), as it provides meaning. We use ontology population techniques to transform data sources into instance data of previous works of our research group [33].

3.7 Real-Time Intelligent Disease Monitoring

This module is a web dashboard where health-professionals can manage projects and monitor the data compiled and analyzed from the previous modules under a Semantic Deep-Learning approach. The graphic interface is a customizable dashboard where professionals can shape the vision they want to have from a series of configurable indicators (or KPIs). In addition, the health-professionals can send messages to the citizens that use the citizen app (see Sect. 3.4).

4 Future Work

CollaborativeHealth is scheduled to be complete in mid 2020. We successfully have developed the extraction modules: (1) social networks, (2) the web crawler, and (3) the citizen participation PWA. During the remaining time the technologies of intelligent analysis of data guided by ontologies will be developed in which different concepts of the ontologies will be selected in order to only perform the intelligent analysis on the data related to those concepts.

As future work, it is scheduled to ease process to adapt CollaborativeHealth to other languages. Although CollaborativeHealth was designed following internationalization principles, the process of translate and identifying synonyms of the ontology is time-consuming. Therefore, we are considering the use of automatic translation tools, such as Google Translation API, and available thesaurus in order to provide an automatic preliminary list of the translation. Moreover, the confidence filter based on blockchain requires more available data to perform real validation. Regarding the crawler, we will improve the evidence identification in multimedia files, such as videos.

Acknowledgments. This work is being funded by CDTI and the European Regional Development Fund (FEDER/ERDF) through project CollaborativeHealth IDI-20180989.

References

1. Apolinardo-Arzube, O., García-Díaz, J.A., Medina-Moreira, J., Luna-Aveiga, H., Valencia-García, R.: Evaluating information-retrieval models and machine-learning classifiers for measuring the social perception towards infectious diseases. Appl. Sci. **9**(14), 2858 (2019)
2. Benson, T.: Principles of Health Interoperability HL7 and SNOMED. Springer, Heidelberg (2012)
3. Berkelman, R.L., Sullivan, P., Buehler, J.W., et al.: Public health surveillance. Oxford textbook of public health, Volume 2: the methods of public health 2(Ed. 5), pp. 699–715 (2009)

4. Bragazzi, N.L., Bacigaluppi, S., Robba, C., Nardone, R., Trinka, E., Brigo, F.: Infodemiology of status epilepticus: a systematic validation of the Google trends-based search queries. Epilepsy Behav. **55**, 120–123 (2016)
5. Breit, N., Allen, T., Arnold, B., Huff, A., Madoff, L., Pollack, M.: 20.142 evaluation of promed-mail global surveillance capability. Int. J. Infect. Dis. **53**, 140 (2016)
6. Brownstein, J.S., Freifeld, C.: Healthmap: the development of automated real-time internet surveillance for epidemic intelligence. Euro Surveillance: Bull. Europeen sur les maladies transmissibles= Eur. Commun. Dis. Bull. **12**(11), E071129–5 (2007)
7. Checchi, F.: Principles of Infectious Disease Transmission: Short Course on Infectious Diseases in Humanitarian Emergencies. World Health Organization, London (2009)
8. Chen, L.H., Leder, K., Barbre, K.A., Schlagenhauf, P., Libman, M., Keystone, J., Mendelson, M., Gautret, P., Schwartz, E., Shaw, M., et al.: Business travel-associated illness: a geosentinel analysis. J. Travel Med. **25**(1), tax097 (2018)
9. Chew, C., Eysenbach, G.: Pandemics in the age of Twitter: content analysis of tweets during the 2009 H1N1 outbreak. PloS One **5**(11), e14118 (2010)
10. Chingwaru, W., Vidmar, J.: Culture, myths and panic: three decades and beyond with an HIV/AIDS epidemic in Zimbabwe. Glob. Public Health **13**(2), 249–264 (2018)
11. Choi, J., Cho, Y., Shim, E., Woo, H.: Web-based infectious disease surveillance systems and public health perspectives: a systematic review. BMC Public Health **16**(1), 1238 (2016)
12. Cowell, L.G., Smith, B.: Infectious disease ontology. In: Infectious Disease Informatics, pp. 373–395. Springer, Heidelberg (2010)
13. Dedeoglu, V., Jurdak, R., Putra, G.D., Dorri, A., Kanhere, S.S.: A trust architecture for blockchain in IoT. arXiv preprint arXiv:1906.11461 (2019)
14. Dou, D., Wang, H., Liu, H.: Semantic data mining: a survey of ontology-based approaches. In: Proceedings of the 2015 IEEE 9th International Conference on Semantic Computing (IEEE ICSC 2015), pp. 244–251. IEEE (2015)
15. Dragoni, M., Poria, S., Cambria, E.: Ontosenticnet: a commonsense ontology for sentiment analysis. IEEE Intell. Syst. **33**(3), 77–85 (2018). https://doi.org/10.1109/MIS.2018.033001419
16. Eysenbach, G.: Infodemiology: tracking flu-related searches on the web for syndromic surveillance. In: AMIA 2006, American Medical Informatics Association Annual Symposium, Washington, DC, USA, 11–15 November 2006 (2006). http://knowledge.amia.org/amia-55142-a2006a-1.620145/t-001-1.623243/f-001-1.623244/a-049-1.623601/a-050-1.623598
17. Eysenbach, G.: Infodemiology and infoveillance: framework for an emerging set of public health informatics methods to analyze search, communication and publication behavior on the internet. J. Med. Internet Res. **11**(1), e11 (2009)
18. García-Crespo, Á., Gómez-Berbís, J.M., Palacios, R.C., Alor-Hernández, G.: Securontology: a semantic web access control framework. Comput. Stand. Interfaces **33**(1), 42–49 (2011). https://doi.org/10.1016/j.csi.2009.10.003
19. García-Sánchez, F., Colomo-Palacios, R., Valencia-García, R.: A social-semantic recommender system for advertisements. Inf. Process. Manag. **57**(2), 102153 (2020)
20. Isah, H., Neagu, D., Trundle, P.R.: Social media analysis for product safety using text mining and sentiment analysis. CoRR abs/1510.05301 (2015). http://arxiv.org/abs/1510.05301

21. Lane, D.C., Pala, Ö., Barlas, Y.: Health, demographic change and well-being: the European union's horizon 2020 programme and system dynamics. Syst. Res. Behav. Sci. **32**(4), 407–413 (2015)
22. Lopez-Lorca, A.A., Beydoun, G., Valencia-García, R., Martínez-Béjar, R.: Supporting agent oriented requirement analysis with ontologies. Int. J. Hum. Comput. Stud. **87**, 20–37 (2016). https://doi.org/10.1016/j.ijhcs.2015.10.007
23. Noy, N.F., Sintek, M., Decker, S., Crubézy, M., Fergerson, R.W., Musen, M.A.: Creating semantic web contents with protege 2000. IEEE Intell. Syst. **16**(2), 60–71 (2001)
24. Ofoghi, B., Mann, M., Verspoor, K.: Towards early discovery of salient health threats: a social media emotion classification technique. In: Biocomputing 2016: Proceedings of the Pacific Symposium, Kohala Coast, Hawaii, USA, pp. 504–515 (2016). http://psb.stanford.edu/psb-online/proceedings/psb16/ofoghi.pdf
25. del Pilar Salas-Zárate, M., Alor-Hernández, G., Sánchez-Cervantes, J.L., Paredes-Valverde, M.A., García-Alcaraz, J.L., Valencia-García, R.: Review of English literature on figurative language applied to social networks. Knowl. Inf. Syst. 1–33 (2019). https://doi.org/10.1007/s10115-019-01425-3
26. del Pilar Salas-Zárate, M., Paredes-Valverde, M.A., Rodríguez-García, M.Á., Valencia-García, R., Alor-Hernández, G.: Automatic detection of satire in Twitter: a psycholinguistic-based approach. Knowl. Based Syst. **128**, 20–33 (2017). https://doi.org/10.1016/j.knosys.2017.04.009
27. del Pilar Salas-Zárate, M., Valencia-García, R., Ruiz-Martínez, A., Palacios, R.C.: Feature-based opinion mining in financial news: an ontology-driven approach. J. Inf. Sci. **43**(4), 458–479 (2017). https://doi.org/10.1177/0165551516645528
28. Rasmussen, S.A., Jamieson, D.J., Honein, M.A., Petersen, L.R.: Zika virus and birth defects-reviewing the evidence for causality. N. Engl. J. Med. **374**(20), 1981–1987 (2016)
29. Roccetti, M., Marfia, G., Salomoni, P., Prandi, C., Zagari, R.M., Kengni, F.L.G., Bazzoli, F., Montagnani, M.: Attitudes of crohn's disease patients: infodemiology case study and sentiment analysis of Facebook and Twitter posts. JMIR Public Health Surveill. **3**(3), e51 (2017)
30. Rodríguez-García, M.Á., Valencia-García, R., García-Sánchez, F., Zapater, J.J.S.: Creating a semantically enhanced cloud services environment through ontology evolution. Future Gen. Comput. Syst. **32**, 295–306 (2014). https://doi.org/10.1016/j.future.2013.08.003
31. Schriml, L.M., Arze, C., Nadendla, S., Chang, Y.W., Mazaitis, M., Felix, V., Feng, G., Kibbe, W.A.: Disease ontology: a backbone for disease semantic integration. Nucleic Acids Res. **40**(D1), 940–946 (2012). https://doi.org/10.1093/nar/gkr972
32. Studer, R., Benjamins, V.R., Fensel, D.: Knowledge engineering: principles and methods. Data Knowl. Eng. **25**(1–2), 161–197 (1998). https://doi.org/10.1016/S0169-023X(97)00056-6
33. Valencia-García, R., Ruiz-Sánchez, J.M., Vicente, P.J.V., Fernández-Breis, J.T., Martínez-Béjar, R.: An incremental approach for discovering medical knowledge from texts. Expert Syst. Appl. **26**(3), 291–299 (2004). https://doi.org/10.1016/j.eswa.2003.09.001
34. Viliani, F., Edelstein, M., Buckley, E., Llamas, A., Dar, O.: Mining and emerging infectious diseases: results of the infectious disease risk assessment and management (IDRAM) initiative pilot. Extr. Ind. Soc. **4**(2), 251–259 (2017)
35. Wang, H.: Semantic deep learning. University of Oregon, pp. 1–42 (2015)
36. WHO, W.: The top 10 causes of death. World Health Organization (2017)

37. Zeraatkar, K., Ahmadi, M.: Trends of infodemiology studies: a scoping review. Health Inf. Libr. J. **35**(2), 91–120 (2018)
38. Zhang, P., White, J., Schmidt, D.C., Lenz, G.: Applying software patterns to address interoperability in blockchain based healthcare apps. arXiv preprint arXiv:1706.03700 (2017)
39. Zuzek, L.G.A., Buono, C., Braunstein, L.A.: Epidemic spreading and immunization strategy in multiplex networks. CoRR abs/1507.01837 (2015). http://arxiv.org/abs/1507.01837

Mathematical Modeling of Dynamics of Crime Indicators in the Field of Computer Information

A. S. Bogomolov[1,2](✉), A. F. Rezchikov[3], V. A. Kushnikov[1,2,4], V. A. Ivashchenko[1], L. Yu. Filimonyuk[1], N. V. Yandybaeva[8], D. S. Fominykh[1], N. N. Kovaleva[6], O. L. Soldatkina[6,7], E. V. Berdnova[5], and E. Yu. Kalikinskaya[4]

[1] Institute of Precision Mechanics and Control of RAS, 24 Rabochaya street, Saratov 410028, Russia
alexbogomolov@ya.ru
[2] Saratov State University, 83 Astrakhanskaya street, Saratov 410012, Russia
[3] V. A. Trapeznikov Institute of Control Sciences of RAS, 65 Profsoyuznaya street, Moscow 117997, Russia
[4] Yuri Gagarin State Technical University of Saratov, 77 Politechnicheskaya street, Saratov 410054, Russia
[5] Saratov State Vavilov Agrarian University, 1 Teatral'naya pl. street, Saratov 410012, Russia
[6] Saratov State Law Academy, 1 Volskaya Street, Saratov 410028, Russia
[7] Saratov Branch of the Institute of State and Law of RAS, 135 Chernyshevskogo street, Saratov 410028, Russia
[8] Balakovo Branch, Russian Presidential Academy of National Economy and Public Administration, 107 Chapaeva street, Balakovo 413865, Russia

Abstract. The actual task of analyzing and predicting the dynamics of the main crime indicators in the field of computer information is investigated. The registered values of the key parameters from the official statistics of the Ministry of the Interior are considered as indicators. To solve the problem, mathematical software is proposed based on the principles of system dynamics. The proposed complexes of models include causal matrices reflecting direct and inverse relationships between the variables in question, and systems of nonlinear differential equations. Causal matrices are based on the study of the correlation of the values of the variables in question, given from statistics. Systems of differential equations are built on the basis of cause-effect matrices and principles of system dynamics. The solution of these systems of differential equations by approximate methods makes it possible to analyze trends and make predictions of the values of the variables in question at different time intervals with fairly high accuracy. Experiments with the developed models will allow us to quantify the effect of various factors on the growth or decrease in crime in the field of computer information. The use of the developed software is shown in an example that compares the effect of the number of sentences and new legislation on reducing the number of computer crimes. The results are intended to support decision-making on combating crime in the field of computer information.

R. Silhavy (Ed.): CSOC 2020, AISC 1226, pp. 191–200, 2020.
https://doi.org/10.1007/978-3-030-51974-2_16

Keywords: Crime · Computer · Information · Internet · Hacker attack · Hacking · Causal link · System dynamics model · Forrester model

1 Introduction

According to the Russian Criminal Code, crimes in the sphere ofcomputer information (IT) include: illegal access to computer information (article 272 of the Criminal Code), the creation, use and dissemination of harmful computer programs (article 273 of the Criminal Code), and violations of the rules governing the use of storage, processing or transmission of computer information and information and telecommunications networks (article 274 of the Criminal Code). Responsibility for crimes committed with the use of information and telecommunication technologies has been established in a number of other articles of the Criminal Code of the Russian Federation [1].

Currently, the number of crimes in the field of IT in the world is estimated at tens of thousands per year and tends to grow rapidly. These tendencies are conditioned by high rates of informatization, increasing role of the information sphere in the life of the society and, consequently, increasing interest of criminals in this sphere [2–4].

Crimes in the field of IT are committed by both organized groups and individuals who are involved in criminal activities under the influence of external factors, against the background of insufficient legal regulation of the Internet space. At the same time, according to experts, the main problem in the fight against such crimes is their high latency – more than 80%. The main reasons are anonymity of users, instability or absence of material traces of the crime, unwillingness of publicity on the part of the victim.

According to the information of the Department "K" of the Ministry of Internal Affairs of the Russian Federation, 11,000 crimes in the field of telecommunications and IT were registered in Russia in 2014, in 2016 their number amounted to 65,949, in 2017 – 90,587 [5].

In these circumstances, it is necessary to quantify the resources used to combat IT crime, to assess the consequences of the impacts taken. Therefore, the development of mathematical software that will allow for the quantification and prediction of computer crime indicators, taking into account the many complex causal links that arise between these indicators, is necessary.

2 Task Setting

The objective of the study is to develop and use a set of mathematical models to quantify and predict the dynamics of crime indicators in the field of IT. Such models will help to identify trends and key factors affecting the growth of crime in IT and to specify measures to reduce that growth.

3 Problem-Solving Method

To solve the problem, it is proposed to develop a mathematical model based on system dynamics [6–12]. Mathematical models of system dynamics were previously used by domestic scientists in the study of crime indicators [13] and processes in information security [14, 15]. To build a model that allows to analyze the dynamics of crime indicators in the field of IT, it is necessary to take into account the current level of theoretical concepts and the results of special empirical research in this field.

For the solution of the problem it is offered:

- to identify a set of indicators that quantify the state of crime in the field of IT and its control, as well as the factors that influence these indicators;
- to develop a causal scheme of the selected indicators and the factors that influence them;
- to construct differential equations of system dynamics and to define algorithms of their solution on the basis of the established dependencies;
- to determine the values of variables of the system at given intervals of time as a result of solving the system of differential equations;
- to perform the analysis of the adequacy of the built model based on the retrospective data and to correct the model if necessary;
- to analyze the influence of various factors on the indicators that quantitatively characterize the state of crime in the field of IT and its control, and to model the dynamics of computer crime indicators depending on different conditions;
- to make conclusions and form recommendations to the decision-maker based on the results of numerical experiments.

4 Selection of System Variables

On the basis of studying statistics, research of mechanisms of occurrence and investigation of computer crimes [16, 17], the indicators have been identified, which, as assumed in the creation of the model, quite fully reflect the state of crime in the field of IT and its control. Among these indicators there are the numbers in the country per year:

$X_{att}(t)$ – of hacking into computer systems (hacking is illegal access to computer information);

$X_{msw}(t)$ – of the use of malicious software;

$X_{frd}(t)$ – of cases of fraud in the field of IT and fraud committed using computer and telecommunication technologies;

The following indicators have also been identified for the forces aimed at fighting computer crime (the number in the country per year):

$X_{reg}(t)$ – registered crimes in the field of IT;
$X_{unc}(t)$ – uncovered IT crimes;
$X_{conv}(t)$ – the convictions for IT crimes;
$X_{leg}(t)$ – adopted legislation against crimes in the field of IT, as well as the number of shares and events in the country per year:
$X_{safe}(t)$ – the promotions of safe online behavior;
$X_{ctz}(t)$ – the forces against computer crime by citizens and organizations, as well as the amount of $X_{cmb}(t)$ money (rubles per year) allocated by the state for the development of the sphere of combating computer crime.

In addition, indicators have been identified whose relationship to computer crime development is more complex and diverse – the number:
$X_{sci}(t)$ – of scientific developments in the field of information technologies per year;
$X_{spec}(t)$ – of IT specialists graduated by universities;
$X_{us}(t)$ – of internet users in the country, as well as $X_{dsw}(t)$ – share of domestic software and information technologies in the market and $X_{mon}(t)$ – amount of money in the accounts of the population (average estimate, rubles per year).

The above indicators, which have an impact on the number of computer crimes in a country, will also be referred to as system variables. In addition to the system variables, there are external factors in the model that are assumed to be independent of the system variables but that affect them themselves:

$Y_{unem}(t)$ – the number of unemployed in the country;
$Y_{sanc}(t)$ – the extent of foreign sanctions against the country;
$Y_{doll}(t)$ – the ruble-dollar exchange rate.

The choice of such a set of external factors is due to the high role of these indicators in the formation of the social environment and economic realities, which significantly affect the financial situation of citizens and their motivation to commit certain types of crimes. It can be argued that a number of other indicators related to the socio-economic situation in the country have the same importance, but most indicators are largely linked to the selected ones, so their addition to the list of external factors would be redundant.

5 Development of a Cause-Effect Model

As the regressive and correlation analysis and expert opinions have shown, there are cause-effect links between the listed variables and factors, which are reflected in the cause-effect Table 1 below. The presence of a plus sign (or a minus sign) in the cell at the intersection of the line corresponding to variable *A* with a column corresponding to another variable or external factor *B* means that the positive (or negative) influence of variable or external factor B on the growth of variable A is considered in the proposed model (the designation of the time argument t in Table 1 and system (1) is omitted for a more compact form).

Table 1. Matrix of interaction of variables and external factors of the system

	X_{reg}	X_{unc}	X_{conv}	X_{dsw}	X_{sci}	X_{spec}	X_{safe}	X_{ctz}	X_{leg}	X_{mon}	X_{cmb}	X_{us}	X_{att}	X_{msw}	X_{frd}	Y_{unem}	Y_{sanc}	Y_{doll}
X_{reg}		+	+					+	+									
X_{unc}	+					+	+				+							
X_{conv}	+	+					+	+	+		+							
X_{dsw}						+	+				+						+	+
X_{sci}						+					+						+	+
X_{spec}									+		+							
X_{safe}				+	+	+		+	+		+		–	–	–	–		
X_{ctz}		+	+				+				+		–	–	–			
X_{leg}							+	+										
X_{mon}							+						–		–	–	–	–
X_{cmb}									+	+		+					–	–
X_{us}							+										+	+
X_{att}	–	–	–	–	–	–	–	–	–			+		+		+	+	+
X_{msw}	–	–	–	–	–	–	–	–								+	+	+
X_{frd}	–	–	–			–	–	–	–			+		+		+	+	+

The absence of a value in the cell of the table means that the corresponding dependence is considered to be quite small and is not taken into account in the model.

This table is used to compose differential equations of system dynamics.

6 Differential Equations of System Dynamics

Based on Table 1 of cause-effect relationships and principles of system dynamics [6], a differential equation is constructed for each system variable.

Since the variables and the external factors have different dimensions and numerical scales, dependencies and equation systems have each value taken in relation to the same value in the base year in which 2015 was taken. One year was taken as the unit of time. After the system of differential equations was solved and the results of the simulation were obtained, the normalized values were converted back to the original dimension by multiplying by the corresponding value of year 2015.

The system of differential equations has a general form (1).

$$\begin{cases} dX_{att}/dt = F_{att-us}(X_{us})F_{att-msw}(X_{msw})(Y_{unem} + Y_{sanc} + Y_{doll}) \\ -F_{att-reg}(X_{reg})F_{att-unc}(X_{unc})F_{att-conv}(X_{conv})F_{att-dsw}(X_{dsw})F_{att-sci}(X_{sci}) \\ \times F_{att-spec}(X_{spec})F_{att-safe}(X_{safe})F_{att-ctz}(X_{ctz})F_{att-leg}(X_{leg}) \\ dX_{msw}/dt = (Y_{unem} + Y_{sanc} + Y_{doll}) - F_{msw-reg}(X_{reg})F_{msw-unc}(X_{unc}) \\ \times F_{msw-conv}(X_{conv})F_{msw-dsw}(X_{dsw})F_{att-sci}(X_{sci})F_{msw-spec}(X_{spec})F_{msw-safe}(X_{safe})F_{msw-ctz}(X_{ctz}) \\ dX_{frd}/dt = F_{frd-us}(X_{us})(F_{att-frd}(X_{frd})(Y_{unem} + Y_{sanc} + Y_{doll}) - F_{frd-reg}(X_{reg}) \\ \times F_{frd-unc}(X_{unc})F_{frd-conv}(X_{conv})F_{frd-safe}(X_{safe})F_{frd-ctz}(X_{ctz}) \\ \times F_{frd-ctz}(X_{ctz})F_{frd-leg}(X_{leg}) \\ dX_{reg}/dt = F_{reg-unc}(X_{unc})F_{reg-conv}(X_{conv})F_{reg-ctz}(X_{ctz})F_{reg-leg}(X_{leg}) \\ dX_{unc}/dt = F_{unc-reg}(X_{reg})F_{unc-spec}(X_{spec})F_{unc-safe}(X_{safe})F_{unc-cmb}(X_{cmb}) \\ dX_{conv}/dt = F_{conv-reg}(X_{reg})F_{conv-unc}(X_{unc})F_{conv-safe}(X_{safe})F_{conv-ctz}(X_{ctz}) \\ \times F_{conv-leg}(X_{leg})F_{conv-cmb}(X_{cmb}) \\ dX_{leg}/dt = F_{leg-safe}(X_{safe})F_{leg-ctzc}(X_{ctz}) \\ dX_{safe}/dt = F_{safe-dsw}(X_{dsw})F_{safe-sci}(X_{sci})F_{safe-spec}(X_{spec})F_{safe-ctz}(X_{ctz}) \\ \times F_{safe-leg}(X_{leg})F_{safe-cmb}(X_{cmb}) - F_{safe-att}(X_{att})F_{safe-msw}(X_{msw})F_{safe-frd}(X_{frd})Y_{unem} \\ dX_{ctz}/dt = F_{ctz-unc}(X_{unc})F_{ctz-conv}(X_{conv})F_{ctz-safe}(X_{safe})F_{ctz-cmb}(X_{cmb}) \\ -F_{ctz-att}(X_{att})F_{ctz-msw}(X_{msw})F_{ctz-frd}(X_{frd}) \\ dX_{cmb}/dt = F_{cmb-leg}(X_{leg})F_{cmb-mon}\left(X_{\prod c}\right)F_{cmb-us}(X_{us}) - (Y_{sanc} + Y_{doll}) \\ dX_{sci}/dt = F_{sci-spec}(X_{spec})F_{sci-cmb}(X_{cmb})(Y_{sanc} + Y_{doll}) \\ dX_{spec}/dt = F_{spec-leg}(X_{leg})F_{spec-cmb}(X_{cmb}) \\ dX_{us}/dt = F_{us-safe}(X_{safe})(Y_{sanc} + Y_{doll}) \\ dX_{dsw}/dt = F_{dsw-spec}(X_{spec})F_{dsw-safe}(X_{safe})F_{dsw-cmb}(X_{cmbe})(Y_{sanc} + Y_{doll}) \\ dX_{mon}/dt = F_{mon-safe}(X_{safe}) - F_{mon-att}(X_{att})F_{mon-frd}(X_{frd})(Y_{sanc} + Y_{doll} + Y_{unem}) \end{cases} \quad (1)$$

To model external factors, such as the growth in the number of unemployed persons without jobs $Y_{unem}(t)$, the volume of foreign sanctions against Russia $Y_{sanc}(t)$ and the exchange rate of the US dollar $Y_{doll}(t)$, trends corresponding to the forecasts of the Ministry of Social and Economic Development of Russia were used.

The model assumptions assume that the dependencies, which appear in the above system of equations, are close to linear, which is confirmed by the establishment of correlations between them. After establishment of a concrete kind of these dependences and expressions for mathematical modeling of dynamics of external factors the system of nonlinear differential equations has been received which decision of methods Runge-Kutta of an order 4 of accuracy gives an admissible error in relation to available statistics. In particular, the results of the solution for the variable $X_{frd}(t)$ – the number of cases of fraud in the field of computer information and fraud committed using computer and telecommunications technologies (Fig. 1, the curve for the case of term $k_{unc} = 1$, $k_{conv} = 0.4$) differ from the values of this indicator from the statistics of the Ministry of Internal Affairs of 2014–2016 by no more than 20%. On this basis, we assume that the results of numerical experiments with the constructed model will provide certain grounds for conclusions regarding the nature of the dynamics of computer crime and its dependence on various factors.

7 Mathematical Modeling and Discussing the Results of the Experiment

A series of computational experiments was carried out with the developed model, fragments of which are used in the following example.

Of the system variables related to computer-related crime, we will consider $X_{frd}(t)$ – the number of cases of computer-related information fraud and computer- and telecommunications-related fraud. As noted by V.A. Minayev and V.N. Sablin, the significant factors reducing the effectiveness of the fight against computer crime include "…the lack of a well-functioning system of legal and organizational-technical support for the legitimate interests of citizens, the state and society in the field of information security" [18]. Let us consider the impact on the dynamics of the $X_{frd}(t)$ indicator of the variables $X_{unc}(t)$ – the number of convictions for crimes in the field of IT – and $X_{leg}(t)$ - the number of legislative norms adopted against crimes in the field of IT. As can be seen from Table 1, the model assumes that both indicators reduce the number of computer-related fraud cases. Additional information about this effect can be obtained by varying the first term k_{conv} and k_{leg} coefficients for these indicators in the system of differential equations of the form (1). The above decision with the forecast of $X_{frd}(t)$ dynamics for the nearest years taking into account the existing tendencies has been received at the term $k_{conv} = 1$, $k_{leg} = 0.4$. In the course of the computational experiment, it was found that with the increase k_{conv} from 1 to 1.5 (k_{leg} remains equal to 0.4), the number of cases of fraud in the field of IT calculated by the model decreases significantly by 2019 than with the increase $k_{leg}$from 0.4 to 0.9, Fig. 1.

The growth of k_{conv} with little change in the k_{leg} corresponds to a situation in which the number of convictions under existing articles increases. In the context of the constant growth in the number of cases of fraud in the $X_{frd}(t)$ network, such a tendency should be justified and sufficiently high $X_{conv}(t)$ values corresponding to the current level of $X_{conv}(t)$ values are evidence of strengthening the work of law enforcement agencies. However, the number of convictions is a clear deterrent to potential criminals.

In the second case study, k_{conv} remains constant with the increase of k_{leg}, i.e. the number of imposed convictions $X_{conv}(t)$ does not change much with the increase in the number of new legal provisions. Given that the number of cases of online fraud is increasing, the authors believe that this situation requires a review of a number of issues relating to the legislation itself, new and existing regulations, and their implementation. At the same time, the increase in the number of new legal provisions restrains potential violators less than the increase in the number of actual convictions. In the opinion of the authors, this is explained by the fact that information about the new laws is communicated to the public very slowly. In addition, new legal provisions, as a rule, do not work at once in full force, and some time is necessary for the development of judicial practice for their implementation. Imperfection of the provisions also sometimes leads to impossibility of their implementation in practice, and in some cases, it requires a number of specific conditions, which are not always provided for. This latter circumstance is also a rather complicated problem, the solution of which

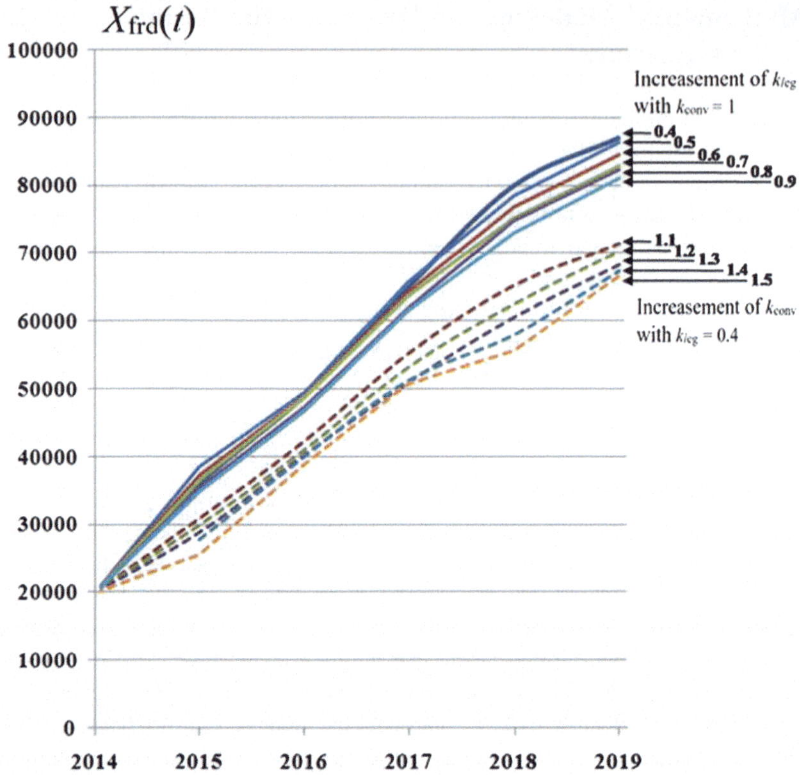

Fig. 1. Dependency of growth of cases of fraud in the network $X_{frd}(t)$ on changing k_{leg} and k_{conv} coefficients for the number of adopted legislative norms against IT crimes and the number of convictions for IT crimes and, accordingly, in the type of system (1).

requires efforts in several directions at once, one of which is the use of the developed mathematical tools for modeling and analyzing the dynamics of crime indicators in the field of computer information.

8 Conclusion

A complex of mathematical models has been developed, which allows to carry out modeling and analysis of the dynamics of indicators, quantitatively characterizing the state of crime in the sphere of computer information and fighting it. The matrix of cause-effect links established between the indicators of committing, investigation and prevention of these crimes and the system of differential equations of system dynamics based on this matrix are constructed.

The developed complex of mathematical models makes it possible to predict the dynamics of indicators of crime and computer crime detection depending on the various factors and to establish the specificity of influence on these indicators of measures adopted by the state. The example of practical application of the developed models for forecasting and comparative analysis of the influence of the growth of the number of adopted legislative provisions and convictions on crimes in the sphere of computer information on the growth of the number of cases of fraud in this sphere is given.

The results are intended for decision makers in the field of computer crime control at various levels.

References

1. Code of Criminal Procedure of Russian Federation. Omega-L, Moscow, p. 288 (2016)
2. Chirkov, D.K., Sarkisjan, A.Zh.: Crimes in the field of high technology: trends and prospects. Natl. Secur. **2**, 160–181 (2013)
3. Bykov, V.M., Cherkasov, V.N.: Crimes in the field of computer information: criminological, criminal law and criminalistic problems: a monograph. Jurlitinform 328 (2015)
4. Efremova, M.A.: Criminal liability for crimes committed using information and telecommunication technologies: monograph. Jurlitinform 200 (2015)
5. Right.ru. http://pravo.ru/news/view/122040. Accessed 16 Sept 2018
6. Forrester, D.: Global Dynamics, LLC, p. 379. AST Publishing House, Moscow (2003)
7. Spiridonov, A.Yu., Rezchikov, A.F., Kushnikov, V.A., et al.: Prediction of main factors' values of air transportation system safety based on system dynamics. In: Proceedings of International Conference on Information Technologies in Business and Industry, Tomsk Poly-technic University. Series: Journal of Physics Conference Series, Tomsk, vol. 1015, p. 032140 (2018)
8. Tikhonova, O.M., Kushnikov, V.A., Fominykh, D.S. et al.: Mathematical model for prediction of efficiency indicators of educational activity in high school. In: Proceedings of International Conference on Information Technologies in Business and Industry, Tomsk Poly-technic University. Journal of Physics Conference Series, Tomsk, vol. 1015, p. 032143 (2018)
9. Rezchikov, A.F., et al.: The dynamical cause-effect links' presentation in human-machine systems. News Saratov Univ. (N. S.). Math. Mech. Inform. **17**(1), 109–116 (2017)
10. Rezchikov, A.F., Kushnikov, V.A., Ivashchenko, V.A., Bogomolov, A.S., Filimonyuk, LYu.: Models and algorithms of automata theory for the control of an aircraft group. Autom. Remote Control **79**(10), 1863–1870 (2017)
11. Bogomolov, A.S.: Analysis of the ways of occurrence and prevention of critical combinations of events in man-machine systems. Izv. Saratov Univ. (N. S.), Math. Mech. Inform. **17**(2), 219–230 (2017)
12. Akinfiev, V., Tsvirkun, A.: Managing development of companies with a complex structure of assets. In: 2018 Eleventh International Conference Management of large-scale system development (MLSD), pp. 1–3 (2018). https://doi.org/10.1109/mlsd.2018.8551762
13. Minaev, V.A., Kurushin, V.D., Zaharov, D.V.: Mathematical modeling of regional criminological processes. In: Minaeva, V.A. (ed.) Russian Academy of Science, Siberian Branch, Novosibirsk Computing Center, p. 159 (1992)
14. Minaev, V.A., et al.: Application of methods of system-dynamic modeling to solve the problems of information security. In: Modern Problems and Challenges of Information Security Collection of Articles of the International Scientific-Practical Conference, pp. 177–183 (2017)

15. Minaev, V.A., Nikiforov, O.D.: Modeling of interactions in system of counteraction to information threats to computing complexes. In the collection: Modern problems and tasks of information security provision. In: Makarova, O.A. (ed.) Proceedings of the International Scientific and Practical Conference ISS – 2014, pp. 204–210 (2014)
16. Guzeeva, O.S.: Criminal policy in respect of crimes committed in the Russian segment of the internet. Russ. Laws: Experience Anal. Pract. **6**, 74–77 (2014)
17. Garbatovich, D.A.: Problems of the effectiveness of norms criminalizing computer information crimes. Forensic Libr. **5**(10), 6–14 (2013)
18. Minaev, V.A., Sablin, V.N.: Main problems of combating computer crimes in Russia. Economy and production, p. 56. Svet, Moscow (2009)

Computer Visualization of Optimality Criterion's Weighting Coefficients of Electromechanical System

Nikita S. Kurochkin[1], Vladimir P. Kochetkov[1], Maksim V. Kochetkov[2](✉), Mikhail F. Noskov[3], and Aleksey V. Kolovsky[1]

[1] Khakass Technical Institute – Branch of the Siberian Federal University, Shchetinkina str. 27, 655017 Abakan, Russian Federation
nikita-kurochkin@yandex.ru

[2] Norilsk State Industrial Institute, 50 years of October str. 7, 663310 Norilsk, Russian Federation
m-kochetkov@yandex.ru

[3] Sayano-Shushensky Branch of the Siberian Federal University, Svobodnyi str. 79, 660074 Krasnoyarsk, Russian Federation

Abstract. Computer visualization of optimality criterion's weighting coefficients is solved by modelling tools of MATLAB SIMULINK. The excavator's AC drive is studied as an electromechanical system. Thus the actual practical problem is studied – improvement of mining excavator's rotary drive performance. Asset target is designed by differential equations. Optimality criterion is measured as a result of minimization of squared deviation location and controlled impacts. Algorithm of optimal control is presented according to Pontryagin's maximum theory. Choosing of weighting coefficients resides into finding the intersection of the permissible values surfaces of the elastic moment and the possible values of the gap in the mechanical part of the rotary mechanism's electric drive during the transition process. The time of the transition process is directly related to the performance of the excavator, and the nature of the transition process - with the safety of the equipment. The nature of the transition process depends on the dynamic loads determined by the gap in the tooth zone of the rotation mechanism at the initial moments of rotation. Computer visualization of the weighting factor selection process as a result of using MATLAB SIMULINK tools was demonstrated. This method of choosing the optimality criterion coefficients helps to reduce the maximum value of the elastic moment throws while reducing the time of the transition process by taking into account the operating conditions of a particular excavator. #CSOC1120

Keywords: Optimality criterion's weighting coefficients · Tools for computer visualization · Automated electric drive · Optimal control algorithms

R. Silhavy (Ed.): CSOC 2020, AISC 1226, pp. 201–209, 2020.
https://doi.org/10.1007/978-3-030-51974-2_17

1 Introduction

Problems of productivity gain and reliability for single-bucket excavators stay relevant for the present moment [1–3]. Productivity is mostly determined by amount of time for turning activity. The turning drive of excavators is a complex electromechanical system which works in constant start-stop cycles. This system should be limited from dynamic loads exceeding 2–3 times the possible static values.

Service reliability increase and increase of excavator's performance can be obtained by updating control system or by enhancement of electric drive's mechanical part. This paper studies possible ways of enhancement for electromechanical system.

Electric drive's enhancement in the control system can be done by conventional or unconventional control systems. There are two types of conventional control systems [4, 5]. Firstly, incremental correction systems or secondary control systems and secondly, optimal control systems [6–8]. Above mentioned systems have their own advantages and disadvantages which lead to presence of different basis control systems [1, 9–11].

Correctional optimal control system (COCS) is one of the promising system based on combination of two above mentioned systems. Acknowledged combination provides an opportunity to change feedback coefficients. Secondary control systems do not have such opportunity. COCS saves benefits of both secondary control systems and optimal control systems upon realization. Mentioned aspect of COCS provides accomplishment of the specific claim to the electromechanical system of excavator's turning mechanism: decreasing elastic torque shots while reducing the time of transition.

COCS synthesis take place as a result of electric drive's consecutive correction of internal coordinates and optimal adjustment of external coordinates. Synthesis of regulators' consecutive correction achieves according to "technical minimum" method while synthesis of regulators' external coordinates achieves according to Pontryagin's maximum theory. It helps to find the best solution for the whole range of possible gaps in the excavator's turning mechanism.

Mine excavators are one of the excavator's varieties, such as mine excavators EKG-8I. Mine excavators are used in the high dynamic conditions. These excavators are in the list of the highly valuable mechanisms during open mining of rock formations. Open mining is more efficient, than underground mining.

Turning mechanism acquire most of the excavator's cycle time. There is a choice of gap in a crown gear, which is the location of the considerable part of the equipment, after turning electric drive start-up. Oscillatory process of elastic moment (My) begins after the end of the gap choice. Value change increases 1,5–2 times the allowed static ones which is determined by specified transition time (ttt). Smooth choice of the gap helps to get rid of that moment. However, it leads to increase of cycle time and reduction of efficiency as a result.

Optimization of turning drive is designed to reduce dynamic loads and specified transition time ttt as well. Optimal control is attached to explanation of optimality criterion's weighting coefficients. In the case of several key characteristics that affect the transition processes, the choice of weighting coefficients is always quite a difficult task, which is associated with specific operating conditions in specific production conditions. As for excavator's electric drive, the task is complicated by the fact that excavator's fleet in Russia is worn off for 70%. This lead to the fact that real values of My can be 2–3 times higher than the calculated ones.

Computer visualization of weighting coefficient choice is a milestone to increase work efficiency of electromechanical systems which is especially true when it comes to excavator's electric drive.

Obviously, application of specialized software products allowing to carry out a reliable assessment of dynamic properties of the projected mechanical system is expedient in that case. Programs like VEHICLESIM, ADAMS AUTO, SIMPACK are successful in completing of such tasks. However, their mathematical models and algorithms are closed for users.

Matlab Simulink has developed a library of typical Simscape blocks to model objects of different physical nature. It makes possible to compile models of dynamic objects in the form of schematic diagrams, elements and compounds, real physical quantities, taking into account units of measure [12].

Matlab Simulink tools model dynamic processes based on standardized block diagrams. The block diagrams are relatively simply obtained from differential equations.

2 Problem Statement

Computer visualization of weighting coefficient choice for optimal criterion as an example of excavator's electric drive control system is an actual theoretical and practical problem. The theoretical aspect of the problem solution is determined by the mathematical description of the electromechanical system, which would allow modeling of the control system and visualization of optimality criterion's weighting coefficients. The practical aspect of solving a problem is associated with a rational choice of the control system weights, taking into account the production features of the specific equipment exploitation. In the case of an excavator's electric drive, these are equipment wear, dynamic loads of mining mechanisms, as well as dynamic loads of the turning electric drive, which are caused by the requirements for excavator's start-stop cycles. Computer visualization of the weighting coefficients resides in determining the intersection of the permissible values surfaces of the control object coordinates and the time of the transition process for the electric drive.

3 Methods

The paper describes the AC electric drive. The engine is considered either in the classical form [1, 4], or in the form of a dual power machine for driving mining equipment [9, 10], or in some other forms when it comes to mathematical description.

The paper presents the synthesis of the electromechanical system control algorithm which was carried out on the basis of the following principles and methods: the

Pontryagin's maximum theory, Kessler's "technical optimum" sequential correction method; V.P. Kochetkova's correction method of optimal control systems [4].

The catalog data of the EKG-8I excavator and the 5AM280s4e, 110 kW asynchronous motor, were used for the simulation of the control object. We used catalog data of EKG-8I excavator to simulate the mechanical part of the turntable drive,

There are a few parameters that have been taken into account while choosing electromechanical systems computer simulation method in order to visualize the necessary parameters: ease of programming, the availability of proven mathematical methods and the simplicity of the simulation results representation. The MATLAB package satisfies the designated criteria that are implemented in the structure of the visual modeling tool - SIMULINK.

4 Results

4.1 Theoretical Results

The basis for the synthesis of AC electric drives with the vector control principal is a mathematical description of an asynchronous motor which is powered by current source:

$$\begin{cases} k_R r_R i_{sx} = \frac{1}{T_R}\psi_{Rx} + p \cdot \psi_{Rx} - (\omega_1 - p_n \omega)\, \psi_{Ry}, \\ 0 = \frac{1}{T_R}\psi_{Ry} + p \cdot \psi_{Ry} + (\omega_1 - p_n \omega)\, \psi_{Rx}, \\ M = 1.5 p_n k_R(\psi_{Rx} i_{sy} - \psi_{Ry} i_{sx}), \\ J_{\sum} p\, \omega = M - M_c, \end{cases}$$

where $k_R = x_m/x_R$; $T_R = x_R/r_R$; x_m, x_R, r_R are coefficients in relative units; p is complex operator; p_n is number of electric poles pairs for a drive; ψ_{Rx}, ψ_{Ry} are flux linkage for x and y axes; i_{sx}, i_{sy} are stator currents for x and y axes; M, M_c are static and electromagnetic moments; ω, ω_1 are angle speeds for rotor turning and magnetic field respectively; $J_{\sum}$ is joint inertia.

Let us assume the regulate structure of rotor's flux linkage as single-loop. According to the system calculating method for "technical optimum", we determine the transfer function of the rotor flux linkage regulator along the axis "x", equating the desired and real transfer functions of the open loop flux linkage. The desired transfer function is specified as $W^d_{p.k} = \frac{1}{2T_\mu p(T_\mu p + 1)}$, where T_μ is short time constant not compensable by regulator.

The obtained transfer function of the flux linkage regulator along the "x" axis has the form of a PID-controller:

$$W_{p\psi}(p) = \frac{(T_s p + 1)(T_R p + 1)}{2T_\mu \frac{k_n}{R_s} L_m k_{o\psi} p} = \frac{T_s + T_R}{T_{\text{in}1}} + \frac{1}{T_{\text{in}1} p} + \frac{T_s T_R}{T_{\text{in}1}} p,$$

where $T_{\text{in}1} = 2T_\mu \frac{k_n}{R_s} L_m k_{o\psi}$ and $\frac{T_s + T_R}{T_{in}}$ are proportional part, $\frac{1}{T_{\text{in}1} p}$ is integral part, $\frac{T_s T_R}{T_{\text{in}}} p = Tp$ is ideal differential part of the flux linkage regulator.

In practice, the operation of differentiation carried out with a filter, which has time constant is smaller by an order of magnitude than the time constant of ideal differentiation.

Disregard the uncompensated time constant of the closed loop of the stator flux linkage in the "x" channel. The mathematical description of external coordinates is represented by differential equations system in relative units:

$$\begin{cases} \frac{di_{sy}^*}{dt} = -\frac{1}{T_S} i_{sy}^* - \frac{k_R \omega_\partial}{L_S' I_\partial} \omega_1^* + \frac{u_\partial}{L_S' I_\partial} u_{sy}^*, \\ \frac{d\omega_1^*}{dt} = \frac{3 p_n k_R I_\partial}{2 J_{\sum} \omega_\partial k_{o\psi_x}} i_{sy}^*. \end{cases}$$

For stator current along the "y" axis, engine's speed and control impact which are coordinates of controlled process, the following dependencies are true:

$$i_{sy}^* = i_{sy}/I_{s\partial},\ \omega^* = \omega/\omega_\partial,\ u_{sy}^* = u_{sy}/u_\partial.$$

Add legend to the system:

$$\begin{gathered} i_{sy}^* = x_1,\ \omega^* = x_2,\ u_{sy}^* = u_y,\quad a_{11} = 1/T_s,\ a_{12} = k_R \omega_\partial / L_S' I_\partial, \\ b_k = u_\partial / L_S' I_\partial,\ a_{21} = 3 p_n k_R I_\partial / 2 J_\Sigma \omega_\partial k_{o\psi_x}. \end{gathered}$$

Control object can be described by the following system of equations in the formalized form:

$$\begin{cases} \dot{x}_1 = -a_{11} x_1 - a_{12} x_2 + b_\kappa u_y \\ \dot{x}_2 = a_{21}^0 x_1. \end{cases} \tag{1}$$

Let's consider the regulation object in the form of a single-mass electromechanical system described by a system of Eq. (1).

Synthesis of excavator's electric drive control systems based on popular part of optimal control theory based on making analytically designed optimal regulator (ADOR). The minimization of the quadratic deviations of the coordinates and the control action is chosen as the optimality criterion:

$$J = \frac{1}{2} \int_0^\infty \left(\mathrm{k}_1 x_1^2 + \mathrm{k}_2 x_2^2 + u^2 \right) dt.$$

We use the computational procedure of the Pontryagin's maximum principle theory.

Then make the Hamiltonian:

$$H = -\frac{1}{2}(x_1^2 + x_2^2 + u^2) + \psi_1(-a_{11} x_1 - a_{12} x_2 + b_k u) + \psi_2 a_{21}^0 x_1.$$

Equating to zero the partial derivative of the Hamiltonian with respect to the control and assuming that the optimal control is negative as for the specifying action, we obtain the optimal control form:

$$u^0 = -b_k \psi_1 \tag{2}$$

Then take the partial derivatives of the H function on the coordinates and make a system of associated equations:

$$\begin{cases} \frac{d\psi_1}{dt} = -a_{11}\psi_1 + a_{21}^0\psi_2 - \mathrm{k}_1 x_1, \\ \frac{d\psi_2}{dt} = -a_{12}\psi_1 - \mathrm{k}_2 x_2. \end{cases} \tag{3}$$

Regulator structure is built on the basis of Eqs. (2) and (3). We place the ADOR in the direct channel of the drive speed control. The structural scheme of the speed regulation of an ADOR drive in a forward channel can be represented as a two-loop subordinate control system (SCS). SCR consists of an internal stator current contour along the "y" axis with an aperiodic regulator and an external drive speed contour with an integral regulator. There is also internal feedback from the stator current controller to the drive speed controller (see Fig. 1).

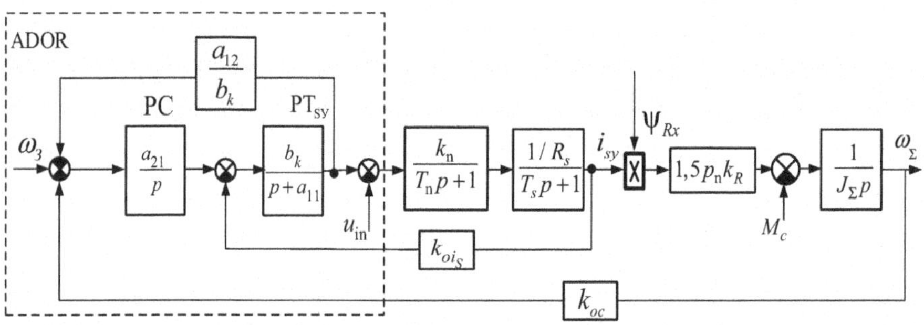

Fig. 1. Structural scheme of electromechanical system.

4.2 Experimental Results

Initial modelling foundation is based on structural scheme on the Fig. 1.

The following typical blocks are used in structural scheme building in Matlab Simulink: Constant (sets a constant signal level), Product (multiplies current signal values), Integrator (integrates input signal), Transport Delay (provides input signal delay for a specified time), Derivative (performs numerical differentiation of the input signal), Gain (performs multiplication of current signals with a constant coefficient), Sum (performs addition of current values of signals), Signal Bulder (forms a linear signal of arbitrary shape using a graphical user interface).

Modelling results are shown on the Fig. 2. The objectivity of the obtained modelling results can be indirectly verified by the coincidence of the processes described by the model with the real characteristics of the turning mechanism of the mining excavator. Figure 2 shows that an increase in the weighting coefficient for the stator current leads to a decrease in the maximum value change of Mu with an increase in the transient time. An increase of the weighting coefficient of the drive speed leads to the following consequences: a decrease in the maximum value change of the elastic moment (Mmax) while decreasing the transient time (t). Epy average transient time for stop-start cycles of a mining excavator is 10 s (see Fig. 2).

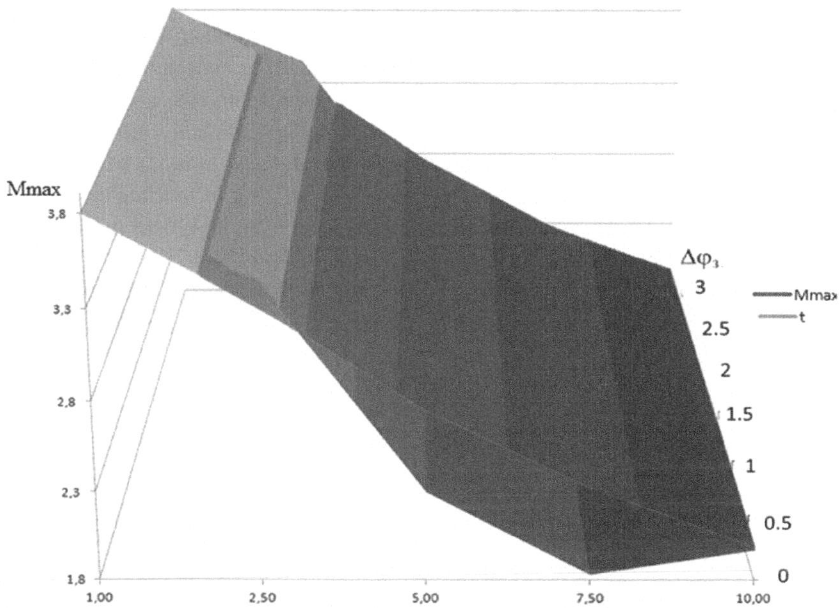

Fig. 2. Computer visualization results.

Figure 2 shows two intersecting surfaces: elastic moments and transient time in an electromechanical turning system as function of optimality criterion's weighting coefficient. Each surface depends, on the optimality criterion's weighting coefficients (shows on abscissa axis of Fig. 2) and on the size of the gap ($\Delta\varphi$) in the gear connection of the crown gear. Value of elastic moment and transition time show on ordinate axis of Fig. 2. Analysis of the intersection of these surfaces just provides information for choosing the optimality criterion's weighting coefficients of the turning mechanism control system.

That way, we are able to adjust the system to the required parameters of transient processes in the context of the operating conditions for a particular mining excavator, as well as the state of its equipment worn-out.

5 Discussions

Particularity of computer visualization with MATLAB SIMULINK in regard to electromechanical system based on ability of computer program to work with structural scheme directly. In other words, there is no need in direct solving of differential equations with software. Set of standard library blocks of Simulink can build structural scheme instead.

There is also no need in creating a mathematical model to study transient processes while solving a similar problem using Matlab Simulink/Simscape. Only a schematic diagram is needed. The following typical blocks can be used when building a scheme with Matlab Simulink/Simscape: Ideal Translation Motion Sensor (simulates a motion sensor in mechanical systems), Simulink-PS Converter (converts a Simulink input signal into a physical signal), Ideal Force Source (simulates an ideal source mechanical energy, which generates in proportion to the input signal), Mass (models mass in mechanical systems), Translational Damper (models damping in mechanical systems), Solver Configuration (determines modelling information from all blocks using physical signals and sets the parameters for solving a simulation problem), Lever (simulates a lever in mechanical systems), Mechanical Translational Reverence (simulates a control point, or a frame for mechanical transfer ports), Translational Spring (models flex pattern in mechanical systems).

6 Conclusion

The application of the described method of the optimality criterion's weighting coefficients choosing for the control system presented in this work makes possible to adjust the drive operation in terms of speed and limitation of the elastic moment in the best possible way by finding the permissible values surfaces intersection for the coordinates of control object and this make it possible to find optimality criterions. However, this increases the relevance of computer visualization of the optimality criterion's weighting coefficients. Simulating dynamic processes with Matlab Simulink/Simscape is relatively simple. Difficulties can be associated only with the correct development of a basic block scheme of an electromechanical system.

Designing of transient models with a help of Matlab Simulink and Matlab Simulink/Simscape tools provide an opportunity for a rational choice of the optimality criterion's weighting coefficients depending on the working conditions of a particular excavator and state of its worn-out. We determine the value of the weighting coefficients, which makes it possible to determine the value of the elastic moment over the entire range of possible gaps in the gears of the gearbox while finding the permissible values surfaces intersection for the coordinates of control object. Under these values, the control system will ensure the speed of operation of the electromechanical system within the acceptable limits of the electric drive's elastic moment.

References

1. Kurochkin, N.S., Kochetkov, V.P., Platonova, E.V., Glushkin, E.Y., Dulesov, A.S.: Combined Optimal Control System for excavator electric drive. In IOP Conference Series: Materials Science and Engineering, vol. 327, pp. 1–6 (2018). https://doi.org/10.1088/1757-899X/327/1/011002
2. Strelnikov, A., Markov, S., Rattmann, L., Weber, D.: Theoretical features of rope shovels and hydraulic backhoes using at open pit mines. In: International Innovative Mining Symposium, E3S Web of Conferences, vol. 41, p. 01003. EDP Sciences (2018). https://doi.org/10.1051/e3sconf/20184101003
3. Demirel, N., Taghizadeh, A., Khouri S., Tyuleneva, E.: Optimization of the excavator-and-dump truck complex at open pit mines – the case study. In: International Innovative Mining Symposium, E3S Web of Conferences, vol. 41, p. 01006. EDP Sciences (2018). https://doi.org/10.1051/e3sconf/20184101006
4. Kochetkov, V.P.: Fundamentals of non-traditional control theory. Abakan (2013). 215 p. (in Russian)
5. Klassen, S.V., Klassen, T.S., Shtein, D.A., Volkov, A.G., Dubkova, R.Yu., Luft, S.V.: Study of a dual-loop subordinate control system for a DC-DC converter with galvanic isolation. In: 19th International Conference of Young Specialists on Micro/Nanotechnologies and Electron Devices (EDM), Erlagol, Russia, pp. 6403–6410 (2018). https://doi.org/10.1109/edm.2018.8435034
6. Ross, I.M., Proulx, R.J., Karpenko, M., Gong, Q.: Riemann-Stieltjes optimal control problems for uncertain dynamic systems. J. Guidance Control Dyn. **38**(7), 1251–1263 (2015). https://doi.org/10.2514/1.G000505
7. Mardanov, M.J., Melikov, T.K.: On the theory of singular optimal controls in dynamic systems with control delay. Comput. Math. Math. Phys. **57**(5), 749–769 (2017). https://doi.org/10.1134/S0965542517050086
8. Balasubramaniyan, S., Srinivasan, S., Kebraei, H., Subathraa, B., Balas, V.E., Glielmo, L.: Stochastic optimal controller design for medium access constrained networked control systems with unknown dynamics. Intell. Decis. Technol. **11**(2), 223–233 (2017). https://doi.org/10.3233/IDT-170290
9. Ostrovlyanchik, V.Yu., Popolzin, I.Yu., Kubarev, V.A., Marshev, D.A.: Mechanical characteristics of a double-fed machine in asynchronous mode and prospects of its application in the electric drive of mining machines. In IOP Conference Series: Earth and Environmental Science, vol. 84, pp. 1–9 (2017). https://doi.org/10.1088/1755-1315/84/1/012030
10. Ostrovlyanchik, V.Yu., Popolzin, I.Yu.: Equivalent model of a dually-fed machine for electric drive control systems. In IOP Conference Series: Materials Science and Engineering, vol. 354, pp. 1–7 (2018). https://doi.org/10.1088/1757-899X/354/1/012017
11. Vostrikov, A.S., Prokhorenko, E.V., Filosov, P.: Multi-criteria synthesis of multidimensional control of alternating current electric drive. Nauchnyy vestnik Novosibirskogo gosudarstvennogo tekhnicheskogo universiteta [Sci. Bull. Novosibirsk state tech. Univ.] **1**, 39–50 (2018). (in Russian)
12. Liu, X., Zuo, Y.: CompPD: a MATLAB package for computing projection depth. J. Stat. Softw. **65**(2), 1–21 (2015). https://doi.org/10.18637/jss.v065.i02

The Mechanism of Microdroplet Fraction Evaporation in the Plasma of the Cathode Region of a Low-Pressure Arc Discharge

A. V. Ushakov[1,2(✉)], I. V. Karpov[2], A. A. Shaikhadinov[2], and E. A. Goncharova[2]

[1] Federal Research Center Krasnoyarsk Scientific Center, Siberian Branch of the Russian Academy of Sciences, 660036 Krasnoyarsk, Russia
sfu-unesco@mail.ru

[2] Siberian Federal University, 660041 Krasnoyarsk, Russia
karpovsfu@mail.ru

Abstract. A model of the motion of the microdrop fraction in a high-temperature plasma of the near-cathode region of a vacuum arc is presented. Mathematically, the model is presented in the form of a system of nonlinear differential heat equations of the second order. The solution of the presented system of equations was carried out by the finite element method in two coordinates in numerical form using the COMSOL Multiphysics program. The solution showed that a drop with a diameter of up to 10 μm evaporates efficiently during the time the micro-droplet fraction is in the plasma of the near-cathode region of the low-pressure arc discharge #CSOC1120.

Keywords: Vacuum arc · Cathode spot · Numerical simulation

1 Introduction

Based on the analytical review of the literature [1–7], one of the main problems hindering the use of a vacuum arc discharge to obtain nanopowders is the presence of a microdrop fraction in cathode erosion products. However, most of the eroded mass of the cathode is carried away precisely in the form of molten microdrops with a size of 10 nm to 50 μm. Let us evaluate the interaction of large droplets with a heated buffer gas in the near-cathode region of size r_p.

2 Methods

Consider a droplet of radius R of a spherical shape that moves in an infinite medium consisting of an inert gas. Since the drop is spherically symmetric, it is possible to solve the problem in spherical coordinates and take into account the dependence of temperature on the radial component. For definiteness, we assume that the atomized cathode is made of a M1 grade copper alloy, and high-purity argon is used as an inert gas. To simplify the problem, we neglect the process of rapid evaporation of the

R. Silhavy (Ed.): CSOC 2020, AISC 1226, pp. 210–215, 2020.
https://doi.org/10.1007/978-3-030-51974-2_18

droplet, although taking into account saturated vapors on the surface of the droplet significantly increases the evaporation time. By excluding the evaporation process from consideration, we neglect the heat consumption for cooling the droplet.

Since we consider rather large droplets far from the dew point, the influence of surface tension can be neglected in our model, since the critical radius weakly depends on the free energy of the interface [8]. To calculate the flow of molecules and heat, we use the continuous medium approximation, which is applicable for sufficiently large drops. At the same time, we consider all the thermophysical characteristics of the environment known. Spectroscopic measurements showed [8] that the gas in the cathode region has a temperature of the order of 1.5–2 eV. We assume that the droplet is heated due to heat exchange with the environment.

The terms of the heat equation associated with diffusion and temperature are considered time-dependent. The heat conduction equations for a drop and the environment, taking into account the movement of the drop, are as follows.

$$\rho_1 C_{p1}\frac{\partial T_1}{\partial t} + \rho_1 C_{p1}\vec{u}\nabla T_1 = \nabla(k_1\nabla T_1) + Q_1$$

$$\rho_2 C_{p2}\frac{\partial T_2}{\partial t} + \rho_2 C_{p2}\vec{u}\nabla T_2 = \nabla(k_2\nabla T_2) + Q_2$$

$$\rho_2\frac{\partial \vec{u}}{\partial t} + \rho_2\vec{u}\nabla\vec{u} = -\nabla p + \mu\Delta\vec{u} + \vec{F}$$

$$\rho_2\frac{\partial \vec{u}}{\partial t} + \nabla(\rho_2\vec{u}) = 0$$

where $\rho_{1,2}$, $C_{p1,2}$, $k_{1,2}$ is the density, specific heat and coefficient of thermal conductivity of the drop and gas, μ is the kinematic viscosity coefficient of the gas, p is the gas pressure. To set the boundary conditions, we assume that the flow of heated gas around the droplet is laminar in nature, and the droplet is heated as a result of heat conduction. In addition, since we are only interested in estimating the time a droplet is heated to boiling point, the latent heat of vaporization can be ignored. A characteristic feature of our problem statement is the temperature jump at the drop-gas interface. The existence of such jumps was first indicated by Langmuir in 1915. We write the boundary condition in the form:

$$-\vec{n}k\nabla T = h(T_2 - T_1)$$

where $T_2 - T_1$ is the temperature jump at the drop-gas interface, k is the coefficient of thermal conductivity of the drop, which generally depends on the direction, is a tensor. However, by virtue of symmetry, we can assume k = const. In addition, since it is believed that droplet evaporation is negligible, the change in thermal conductivity at the droplet boundary is neglected. h heat transfer coefficient. For a buffer gas, the boundary condition associated with the conductivity at the drop-gas interface is similar. The solution of the presented system of equations was carried out in two coordinates in numerical form using the COMSOL Multiphysics program. Figure 1 shows a two-dimensional model for computing.

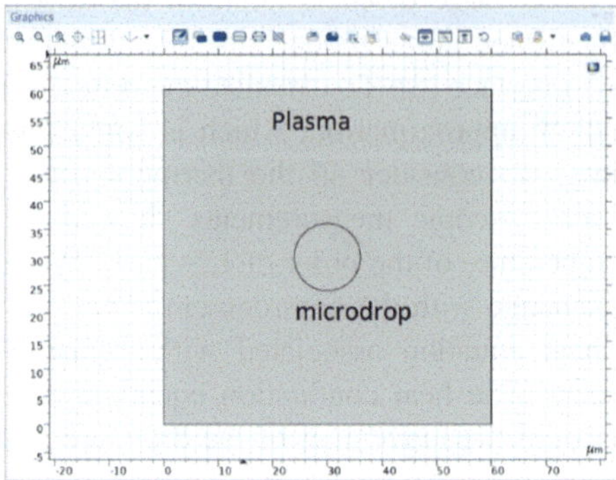

Fig. 1. Two-dimensional model for computing.

3 Results and Discussions

Figure 2, Fig. 3, Fig. 4 and Fig. 5 show the results of calculations. As can be seen from Fig. 2, Fig. 3, Fig. 4 and Fig. 5, a droplet motion creates an uneven distribution of the velocity field and gas pressure around the droplet, which has a significant effect on the temperature distribution. In the zone of high pressure, the drop heats up much faster. In addition, the disadvantage of this model associated with the intense evaporation of a drop is leveled out by intense interaction with the external environment. The diffusion rate of the evaporated material is added to the speed of the buffer gas, and the evaporation of the evaporated material of the droplet is more intense.

Figure 5 shows that for complete evaporation of a sufficiently large drop, the time required is significantly shorter than the duration of the drop in the zone of radius r_ρ.

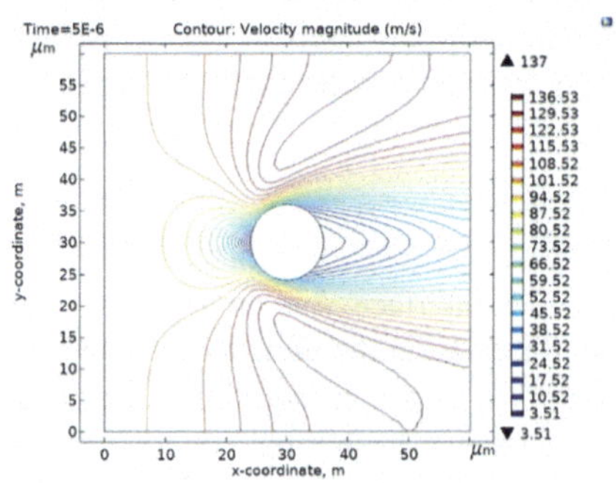

Fig. 2. Contour field of buffer gas velocities in the near-cathode region of a low-pressure arc discharge around a copper drop in 5 µs.

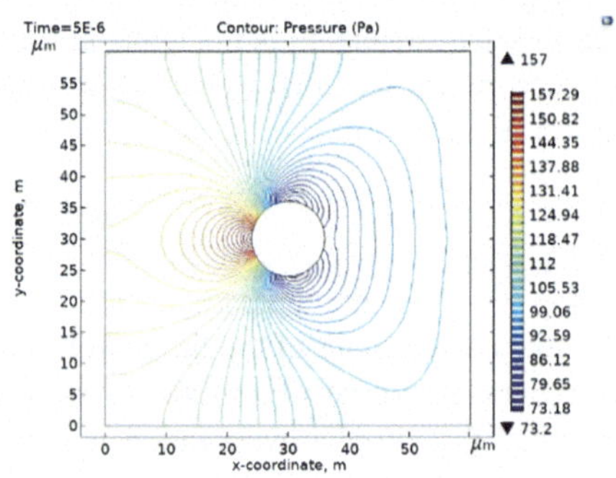

Fig. 3. The contour field of the buffer gas pressure near the droplet in the near-cathode region of the low-pressure arc discharge.

For the first time, the mechanism of droplet formation and crushing in vacuum arc discharges was proposed by McClure [9] and Boxman [10]. In his works, he suggested that the intense steam flows in the erosion products of the cathode are nothing more than an evaporated microdrop fraction. According to various estimates, for various materials, the cathode erosion in the form of a neutral vapor is from 10 to 25 wt%.

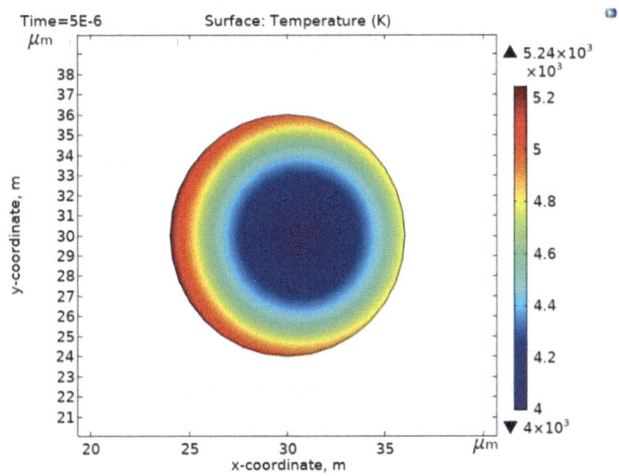

Fig. 4. Profile distribution of temperature inside a drop.

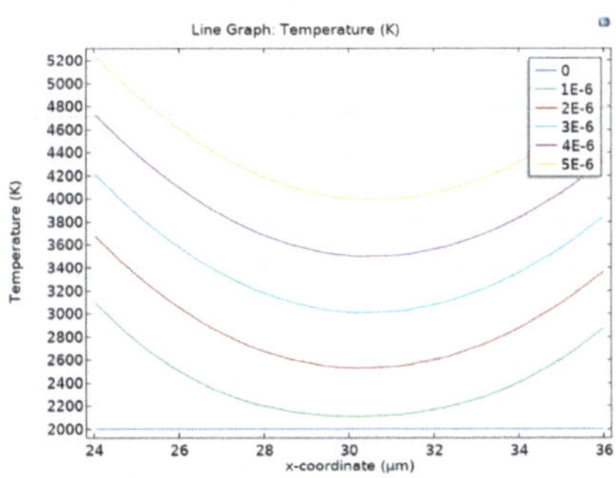

Fig. 5. Profile dependence of droplet temperature on time. The x axis passes through the diagonal of the drop.

Since the calculations are evaluative in nature, this model does not take into account a number of effects that significantly accelerate the heating of the droplet. The process of droplet interaction with charged particles is not taken into account, but, as is known [11], the Coulomb interaction in strongly ionized media can play a key role in the processes of droplet decay. This is indicated by coating technology using electric arc metallizers, powerful plasmatrons, where the process of droplet decay is undesirable.

The Coulomb effect of droplet fragmentation becomes especially noticeable in strong electric fields [11]. This effect explains such phenomena as the decay of liquid jets of metal during an electric explosion of conductors [12], the dispersion of melts [13], and the production of dispersed media in the form of artificial mists and emulsions. The limitation of the model is also associated with the hydrodynamic consideration of the motion of a droplet in a buffer gas. Only the laminar regime is considered, although the explosive nature of the cathode spot implies a wide variety of flows, both of a buffer gas and of metal and gas plasmas. Taking these currents into account would allow one to estimate the time of the drop's stay in the cathode region. All these additional effects compensate for unaccounted for energy losses, such as cooling a droplet as a result of evaporation, accompanied by a decrease in its temperature relative to the ambient temperature, surface tension, which play a significant role in decreasing droplet size, the formation of refractory films on the surface of droplets when using chemically active gases.

4 Conclusion

Thus, the microdroplet fraction, which is the main product of erosion in the cathode spot of a low-pressure arc discharge, effectively evaporates in the cathode region due to heat exchange with the surrounding heated gas, the most optimal pressure being in the region of 100 Pa.

Acknowledgments. The reported study was funded by Russian Foundation for Basic Research, Government of Krasnoyarsk Territory, Krasnoyarsk Regional Fund of Science and «Seysmik-Lab», project number 20-48-242906.

References

1. Davis, W.D., Miller, H.C.: Analysis of the electrode products emitted by DC arcs in a vacuum ambient. J. Appl. Phys. **40**(5), 2212–2221 (1969)
2. Boxman, R.L., Goldsmith, S.: Macroparticle contamination in cathodic arc coatings generation, transport and control. Surf. Coat. Technol. **52**(1), 39–50 (1992)
3. Karpov, I.V., Ushakov, A.V., Lepeshev, A.A., Fedorov, L.Y.: Plasma-chemical reactor based on a low-pressure pulsed arc discharge for synthesis of nanopowders. Tech. Phys. **62** (1), 168–173 (2017)
4. Lepeshev, A.A., Ushakov, A.V., Karpov, I.V.: Low-temperature magnetic behavior of nanostructured ferrite compositions prepared by plasma spraying. J. Appl. Phys. **122**(10), 104103 (2017)
5. Ushakov, A.V., Karpov, I.V., Lepeshev, A.A., Fedorov, L.Y.: Copper oxide of plasma-chemical synthesis for doping superconducting materials. Int. J. Nanosci. **16**(4), 1750001 (2017)
6. Ushakov, A.V., Karpov, I.V., Lepeshev, A.A.: Peculiarities of magnetic behavior of CuO nanoparticles produced by plasma-arc synthesis in a wide temperature range. J. Supercond. Novel Magn. **30**(12), 3351–3354 (2017)

7. Ushakov, A.V., Karpov, I.V., Lepeshev, A.A.: Plasma-chemical synthesis of Fe_3O_4 nanoparticles for doping of high-temperature superconductors. J. Supercond. Novel Magn. **30**(2), 311–316 (2017)
8. Green, H.L., Lane, W.R.: Particulate Clouds: Dusts, Smokes and Mists, 2nd edn. Van Nostrand, Princeton (1964)
9. McClure, G.W.: Plasma expansion as a cause of metal displacement in vacuum-arc cathode spots. J. Appl. Phys. **45**(5), 2078–2084 (1974)
10. Boxman, R.L., Goldsmith, S.: The interaction between plasma and macroparticles in a multicathode-spot vacuum arc. J. Appl. Phys. **52**(1), 151–159 (1981)
11. Ermolaev, Y.L., Gorokhov, M.V., Kozhevin, V.M., Yavsin, D.A., Gurevich, S.A.: Dispersion of metal microdrops exposed to an electron beam with dynamical retention in an electrostatic trap. Tech. Phys. Lett. **40**(1), 32–35 (2014). https://doi.org/10.1134/S1063785014010040
12. Kotov, Y.A., Beketov, I.V., Demina, T.: Characteristies of ZrO_2 nanopowders produced by electrical explosion of wires. J. Aerosol Sci. **28**(1), 905–906 (1995)
13. Ilyuschenko, A.F., Savich, V.V., Syroezhko, G.S.: The innovations in the technology and manufacture of components by the methods of powder metallurgy in Belarus. In: Proceedings of the Euro Powder Metallurgy Congress and Exhibition, Euro PM 2007, pp. 39–44, Toulouse (2007)

Cyber Safety Awareness Framework for South African Schools to Enhance Cyber Safety Awareness

Dorothy Scholtz[1(✉)], Elmarie Kritzinger[1], and Adele Botha[2]

[1] University of South Africa, Pretoria, South Africa
{scholid, kritze}@unisa.ac.za
[2] CSIR, Pretoria, South Africa
abotha@csir.co.za

Abstract. Nowadays Information Communication Technology (ICT) is so entwined in our daily lives that the cyber user finds it second nature to use the Internet to accomplish their daily activities. The Internet has not only opened many opportunities to the cyber users but also introduced the cyber users to cyber risks and cybercrimes. It is important for cyber users to become cyber safety aware to protect themselves. However, there is currently a definite lack in South Africa for cyber safety awareness and cyber education. It is of utmost importance that South Africa as a developing country start creating and implementing cyber safety awareness policies, procedures, measures and initiatives for the education of all cyber users within South Africa. To include cyber safety awareness in the South African school curriculum is one strategy for improving online safety among cyber users. A literature study was conducted, where the research used searches for phrases with 'Cyber Safety Awareness', 'Education', 'Schools', 'Policies' and 'Frameworks'. The aim of this research is to determine what policies, procedures and measures a school needs to have in place to enhance cyber safety awareness of the school learners within a South African school environment. This research then proposed a theoretical framework that can be used to create and enhance cyber safety awareness.

Keywords: Cyber safety awareness · Education · Schools · Policies · Frameworks

1 Introduction

Nowadays Information Communication Technology (ICT) is so entwined in our daily lives that the cyber user finds it second nature to use the Internet to accomplish their daily activities. The Internet has opened up many opportunities to cyber users like online banking, communication and socializing. On the other hand, the Internet also opened up a world where the cyber user is exposed to cyber risks and cybercrimes. Not all cyber users are aware of the cyber risks and cybercrimes and thus leaving the cyber user unprotected to numerous forms of cybercrimes and the cybercrimes consequences. It is important for cyber users to become cyber safety aware when they are active in

R. Silhavy (Ed.): CSOC 2020, AISC 1226, pp. 216–223, 2020.
https://doi.org/10.1007/978-3-030-51974-2_19

cyber space. Numerous countries have developed and implemented cyber safety awareness and education measures to educate and to make the cyber users aware of the cyber risks and cybercrimes. However, there is currently a definite lack in South Africa for cyber safety awareness and cyber education [1].

In the 2019 State of Malware Report, education was found to be at the top ten industries to be targeted by cybercriminals. However when looking at the major cybercrimes in 2018, that included information stealing and ransomware attacks, education (schools) were listing in the top five industries [2]. Stating that schools is at high risk and targeted by cybercriminals, it is of utmost importance that schools are protected. The main reason why schools are not protected, especially in a developing country such as South African is that the schools do not have cyber policies in place for protection [3]. Furthermore, the teachers are ill equipped to understand and offer assistance in cyber safety awareness. The teachers and the schools have a limited budgets and resources within the school environment, thus making it is even more difficult for teachers to education the school learners on cyber safety awareness. Further, hardly any African governments are providing any governmental support in attempting to raise the levels of cyber safety awareness amongst school learners. Thus, cyber safety awareness among the youth in most African countries is becoming a growing problem [4].

It is of utmost importance that South Africa as a developing country start creating and implementing cyber safety awareness policies, procedures, measures and initiatives for the education of all cyber users within South Africa. To include cyber safety awareness in the South African school curriculum is one strategy for improving online safety among school learners [5].

The aim of this research is to determine what policies, procedures and measures a school needs to have in place to enhance cyber safety awareness of the school learners within a South African school environment. This research then proposed a theoretical framework that can be used to create and enhance cyber safety awareness.

2 Background

Schools are data rich gold mines in terms of stored data, from student and employee records to sensitive financial information that can be obtained by cybercriminals. Three main threats impacting schools are data breaches, phishing and ransomware [6]. In the State of K-12 Cybersecurity: 2018 Year in Review, 122 publicly disclosed cybersecurity incidents were discussed that were affecting 119 public K-12 education agencies across 38 states in 2018. Some of the examples of the cybersecurity incidents were: data security incidents, malware attacks and ransomware attacks [7]. Lazzarotti [7] indicates some steps that schools can take to address the cybersecurity incidents:

- Educate the community (includes learners and parents)
- Appoint a data protection officer
- Develop data security and privacy policies
- Consider privacy and security at the outset of any new technology initiative
- Establish a vendor management program

- Provide training for administrators, teachers, staff, and others
- Develop an incident response plan

The research could not found clear information of cyber safety awareness within the South African school system. Thus it was important that Mungadze [8] reported from Threat 2019 the cyber security conference, that e-education in South Africa has become a threat due to cybercrime. Mungadze also indicated that 42% of South African organisations experienced a ransomware attack in the past 12 months, compared to 23% in the prior 12 months [8]. At the Threat 2019 conference the delegants were urged to participate in the president of South Africa – Cyril Ramaphosa's commission on the fourth industrial revolution. Also at the Threat 2019 conference the government was advise on policies and focus was placed on enhancing Africa's cyber security posture [8].

3 Method

A literature study was conducted, where the research used searches for phrases with 'Cyber Safety Awareness', 'Education', 'Schools', 'Policies' and 'Frameworks'. Relevant work for consideration were identified by means of using the following databases: IEEE Xplore, Scopus and ScienceDirect. In the research work in progress, practitioner reports and credible websites were also included. The search period had a date limit from 2008–2019. South Africa have a lack in centralize data sources on the topic cyber safety awareness in schools and thus must be based on internal academic work. Based on the literature studied, the research concluded that cyber safety awareness must urgently be implemented in South Africa schools. The research proposes a cyber safety awareness and education framework, in the form of an artefact. The proposed framework and a discussion on the proposed framework will follow in the next section.

4 Proposed Framework for Cyber Safety Awareness for South African Schools

4.1 Introduction

This section provides information on the Cyber Security Framework that was studied and adopted to be used in the research. The National Institute of Standards and Technology (NIST) published in 2014 and revised during 2017 and 2018, the Cyber Security Framework. The Cyber Security Framework contains standards, guidelines, and best practices to manage cybersecurity-related risk. The Cyber Security Framework can be used for identifying, assessing, and managing cybersecurity risk. The Cyber Security Framework consists of three parts: the Framework Core, the Implementation Tiers and the Framework Profiles. A sort summary of the three parts are given in Table 1.

Table 1. Cyber security framework – core, implementation tiers and profiles.

Core	Set of cyber security activities, desired outcomes and applicable references that are common across critical infrastructure sectors. Consists of five functions: identify, protect, detect, respond and recover. Identifies underlying key categories and subcategories, which are discrete outcomes for each function
Implementation Tiers	Provide context on how an organization views cyber security risk and the processes in place to manage that risk
Profiles	Represents the outcomes base on the organization needs

The advantages of using the Cyber Security Framework are as followed:

- The Framework enables organizations to apply the principles and best practices of risk management to improving security and resilience
- The Framework offers a flexible way to address cybersecurity, including cybersecurity's effect on physical, cyber, and people dimensions
- The Framework can assist organizations in addressing cybersecurity as it affects the privacy of customers, employees, and other parties
- The Framework is aimed at reducing and better managing cybersecurity risks.

It is indicated that organizations outside the United States may also use the Cyber Security Framework to strengthen their own cybersecurity efforts [9]. The Cyber security framework will thus be used as a guideline for this research study.

4.2 Proposed Framework for Cyber Safety Awareness for South African Schools

As indicated in the previous section the Cyber Security Framework can be used to strengthen the cybersecurity efforts of an organization, and in this research study, the organization will be South African schools. Out of the research done in the literature study it was clearly stated that schools are not protected. Schools in South Africa do not have cyber policies in place for protection [3]. This is the main reason why the research have been conducted. Although the Cyber Security Framework can be used, the framework is not a one-size-fits-all approach to manage cyber security. For this reason, other resources were also used to complete the research framework. The other resources were obtained from schools and experts with cyber knowledge. The resource references can be found within the research framework. The research framework embraced the functions of the Cyber Security Framework. Some of the categories and sub categories that were appropriated to cyber-safety awareness, of the Cyber Security Framework were implemented within the research framework. Another reason why other resources were used is to indicate that the components in the research framework is valid. The proposed research framework is given in Table 2.

Out of the proposed Framework for Cyber Safety Awareness for South African Schools, it is clear that the following categories, listed below needs to be addressed within the schools:

Table 2. Proposed framework for cyber safety awareness for South African Schools.

Functions	Categories and references	Sub categories
1. Identify	**1.1 Asset Management** The data, personnel, devices, systems, and facilities that enable the school to achieve day-to-day activities are identified and managed [2, 9]	1.1.1 Physical devices and systems within the school are inventoried 1.1.2 Hardware and software platforms and applications within the school are inventoried 1.1.3 Personnel and data within the school are inventoried 1.1.4 Student data should be backed up and encrypted end-to-end in storage 1.1.5 Keep all software and hardware updated regularly
	1.2 Governance The policies, procedures, and processes to manage and monitor the school regulatory, legal, risk, environmental, and operational requirements are understood Data security and privacy policy in place Cyber safety policy in place [2, 9–11]	1.2.1 Data protection officer identified for the school to be responsible for implementing security and privacy policies 1.2.2 Cyber safety roles and responsibilities for the school are established 1.2.3 Legal and regulatory requirements regarding cyber safety, including privacy are understood and managed 1.2.4 Data security and student privacy a fundamental part of on boarding new school employees
	1.3 Risk Assessment The school understands the cyber safety risk to school operations [9]	1.3.1 Asset vulnerabilities are identified and documented 1.3.2 Cyber threat intelligence is received from information sharing forums and sources 1.3.3 Threats, both internal and external, are identified and documented 1.3.4 Threats, vulnerabilities, likelihoods, and impacts are used to determine risk 1.3.5 Risk responses are identified and prioritized
	1.4 Risk Management Strategy The schools priorities, constraints, risk tolerances, and assumptions are established and used to support operational risk decisions [9]	1.4.1 Risk management processes are established, managed, and agreed to by school management 1.4.2 Risk tolerance is determined and clearly expressed
	1.5 Identity Management, Authentication and Access Control Access to physical and logical assets and associated facilities is limited to authorized users, processes, and devices, and is managed consistent with the assessed risk of unauthorized access to authorized activities and transactions [2, 9]	1.5.1 Identities and credentials are issued, managed, verified, revoked, and audited for authorized devices, users and processes 1.5.2 Physical access to assets is managed and protected 1.5.3 Remote access is managed 1.5.4 Access permissions and authorizations are managed, incorporating the principles of least privilege and separation of duties 1.5.5 Network integrity is protected 1.5.6 Filters are in place 1.5.7 Password are regularly changed 1.5.8 Use password managers 1.5.9 Install security software on all devices in the school environment

(*continued*)

Table 2. (*continued*)

Functions	Categories and references	Sub categories
	1.6 Awareness and Training The schools personnel and partners are provided cyber safety awareness education and are trained to perform their cyber safety-related duties and responsibilities consistent with related policies, procedures, and agreements [2, 7, 9–11]	1.6.1 All users are informed and trained 1.6.2 All users understand their roles and responsibilities 1.6.3 School have a system to educate parents about cyber safety (potential risks and safety protocols) 1.6.4 School have a system to educate community about cyber safety (potential risks and safety protocols) 1.6.5 School use and create a cyber safety awareness curriculum for school learners, to be included in the school curriculum 1.6.6 Knowing what to teach students and teachers alike about cyber safety 1.6.7 Establish a cyber safety awareness culture
	1.7 Data Security Information and records (data) are managed consistent with the schools risk strategy to protect the confidentiality, integrity, and availability of information [9]	1.7.1 Data-at-rest is protected 1.7.2 Data-in-transit is protected 1.7.3 Protections against data leaks are implemented 1.7.4 Integrity checking mechanisms are used to verify software, firmware, and information integrity
	1.8 Information Protection Processes and Procedures Security policies processes, and procedures are maintained and used to manage protection of information systems and assets [2, 9–11]	1.8.1 Backups of information are conducted, maintained, and tested 1.8.2 School create backups of valuable information and store them separately from the central server to protect against ransomware attacks 1.8.3 Data is destroyed according to policy 1.8.4 Protection processes are improved 1.8.5 Response plans and recovery plans are in place and managed 1.8.6 Response and recovery plans are tested
2. Detect	**2.1 Anomalies and Events** Anomalous activity is detected and the potential impact of events is understood [9]	2.1.1 Detected events are analyzed to understand attack targets and methods 2.1.2 Impact of events is determined
	2.2 Security Continuous Monitoring The information system and assets are monitored to identify cyber safety events and verify the effectiveness of protective measures [9]	2.2.1 The network is monitored to detect potential cyber safety events 2.2.2 Malicious code is detected 2.2.3 External service provider activity is monitored to detect potential cyber safety events 2.2.4 Vulnerability scans are performed

(*continued*)

Table 2. (*continued*)

Functions	Categories and references	Sub categories
3. Response	**3.1 Response Planning** Response processes and procedures are executed and maintained, to ensure response to detected cyber safety incidents [2, 7, 9–11]	3.1.1 Develop an incident response plan 3.1.2 Cyberattack response plan is develop 3.1.3 Response plan is executed during or after an incident
	3.2 Communications Response activities are coordinated with internal and external stakeholders [9]	3.2.1 Personnel know their roles and order of operations when a response is needed 3.2.2 Reporting cyber crimes and risks
4. Recover	**4.1 Recovery Planning** Recovery processes and procedures are executed and maintained to ensure restoration of systems or assets affected by cyber safety incidents [9]	4.1 1 Recovery plan is executed during or after a cyber safety incident
	4.2 Improvements Recovery planning and processes are improved by incorporating lessons learned into future activities [9]	4.2.1 Recovery plans incorporate lessons learned
	4.3 Communications Restoration activities are coordinated with internal and external parties [9]	4.3.1 Public relations are managed 4.3.2 Reputation is repaired after an incident 4.3.3 Recovery activities are communicated to internal and external stakeholders as well as executive and management teams

- The data, personnel, devices, systems, and facilities that enable the school to achieve day-to-day activities are identified and managed
- All policies, procedures, and processes to manage and monitor the school regulatory, legal, risk, environmental, and operational requirements are understood
- The school understands and managed the cyber safety risk to school operations
- Management, authentication and access control
- Awareness and training
- Data Security
- Response processes and procedures are executed and maintained
- Response activities are coordinated with internal and external stakeholders
- Recovery processes and procedures are executed and maintained
- Recovery planning and processes are improved by incorporating lessons learned into future activities.

Furthermore, that needs to be address within the schools are the corresponding sub categories of the listed categories that can be found within the research framework. With the research framework in place, the schools will be able to address cyber safety awareness and thus keeps the school and the school environment safe for cyber risks and cybercrimes.

5 Conclusion

It is important for cyber users to become cyber safety aware when they are active in cyber space. Numerous countries have developed and implemented cyber safety awareness and education measures to educate and to make the cyber users aware of the cyber risks and cybercrimes. However, there is currently a definite lack in South Africa for cyber safety awareness and cyber education. Schools are data rich gold mines in terms of stored data that can be obtained by cybercriminals. The aim of this research is to determine what policies, procedures and measures a school needs to have in place to enhance cyber safety awareness of the school learners within a South African school environment. A literature study was conducted, where the research used searches for phrases with 'Cyber Safety Awareness', 'Education', 'Schools', 'Policies' and 'Frameworks'. The research then proposed a theoretical framework that can be used to create and enhance cyber safety awareness. The proposed research framework is given in Table 2. Out of the proposed Framework for Cyber Safety Awareness for South African Schools, it is clear that the functions, categories and sub categories needs to be addressed within the schools. With the research framework in place, the schools will be able to address cyber safety awareness and thus keeps the school and the school environment safe for cyber risks and cybercrimes.

References

1. Kortjan, N., Von Solms, R.: A conceptual framework for cyber-security awareness and education in SA. S. Afr. Comput. J. **52**(1), 29–41 (2014)
2. Zamora, W.: What K-12 schools need to shore up cybersecurity - Malwarebytes Labs—Malwarebytes Labs, 26 February 2019. Accessed 18 Sept 2019
3. Kritzinger, E.: Short-term initiatives for enhancing cyber-safety within South African schools. S. Afr. Comput. J. **28**(1), 1–17 (2016)
4. Von Solms, S., Von Solms, R.: Towards cyber safety education in primary schools in Africa. In: Proceedings of the 8th International Symposium on Human Aspects of Information Security & Assurance, HAISA 2014, vol. 3, no. HAISA, pp. 185–197 (2014)
5. Kritzinger, E.: Online safety in South Africa-a cause for growing concern. In: 2014 Information Security for South Africa - Proceedings of the ISSA 2014 Conference (2014)
6. Sarang, R.: School of Cyberthreats: 3 Attacks Impacting Today's Schools—McAfee Blogs, 23 July 2019. Accessed 02 Sept 2019
7. Lazzarotti, J.: K-12 Schools Data Security Risks for Students and Parents, 23 August 2019. Accessed 02 Sept 2019
8. Mungadze, S.: Minister: Cyber crime limits govt, business potential—ITWeb, 28 June 2019. Accessed 17 Sept 2019
9. Framework for Improving Critical Infrastructure Cybersecurity, Version 1.1, Gaithersburg, MD, April 2018
10. Birdsong, T.: 7 Questions to Ask Your Child's School About Cybersecurity Protocols—McAfee Blogs, 31 August 2019. Accessed 02 Sept 2019
11. Miller, M.: Cyberattacks find easy target in nation's schools—The Hill, 8 September 2019. Accessed 09 Sept 2019

Devices Control and Monitoring on the Production Level Using Wonderware Platform

Dmitrii Borkin, Martin Barton(✉), Martin Nemeth, and Pavol Tanuska

Faculty of Materials Science and Technology in Trnava, Institute of Applied Informatics, Automation and Mechatronics, Slovak University of Technology in Bratislava, Trnava, Slovakia
{dmitrii.borkin,martin.barton,martin.nemeth, pavol.tanuska}@stuba.sk

Abstract. In our paper we present the method for cross-platform control and monitoring of a device on a production level. Hardware and software inconsistency makes it often difficult to build a consistent and robust control and monitoring environment for a production process. For the purpose of demonstrating this inconsistence and application of our method, we have chose a motor with the frequency converter made by Siemens and the Wonderware platform as the supervisory control and monitoring software environment. In this paper we present the results in a form of proposed hardware setup scheme and software application for supervisory control and monitoring of the device, developed in the Wonderware environment.

Keywords: Archestra · Wonderware · Telegram · Frequency converter

1 Introduction

The inconsistency of hardware and software in production processes can be caused by many factors. For example various upgrades and new features can be the motivation for switching to the third-party software for creating application on SCADA and MES control level.

The ideal case is to use hardware on production level and software to control the hardware from one manufacturer. The reason for this is the guarantied compatibility and support from the manufacturer. Often times however, the third-party software developers offer more features and new solutions, which can lead to applying software applications from different manufacturer than is the current used hardware.

In our paper we present the method for applying software solution for supervisory control and monitoring from one manufacturer to the existing device and control hardware solution from another manufacturer. For the purpose of presenting the method we have chose to design, create and apply the supervisory application based on Wonderware platform on to the motor controlled by a frequency converter by Siemens.

R. Silhavy (Ed.): CSOC 2020, AISC 1226, pp. 224–229, 2020.
https://doi.org/10.1007/978-3-030-51974-2_20

The application was design in the Wonderware environment, which allows monitoring and controlling the engine by Siemens. The advantage of this solution is, that the Wonderware Platform represents the industrial software platform mainly created for Supervisory HMI, SCADA [3, 5, 8].

This platform also contains the integrated set of services and the extensible data model to manage plant control. System Platform supports both the supervisory control layer and the manufacturing execution system (MES) layer, presenting them as a single information source [4–6].

The motor used in this example is using a G120CU frequency converter by Siemens. The native control environment for this device is TIAPortal also by Siemens. This platform contains native modules that will ensure the proper communication and control of the frequency converter with the use of a Siemens PLC.

The Fig. 1. shows the default environment for controlling the frequency converter. In this environment we can basically monitor only 2 parameters, which are speed in rpm and 2 parameters which we can choose from list from parameters like output voltage. This example shows, that the default software environment does not provide sufficient amount of parameters that can be controlled and monitored. On the other hand with the use of third-party software platform, like for example Wonderware we can achieve more detailed control and monitoring of given device.

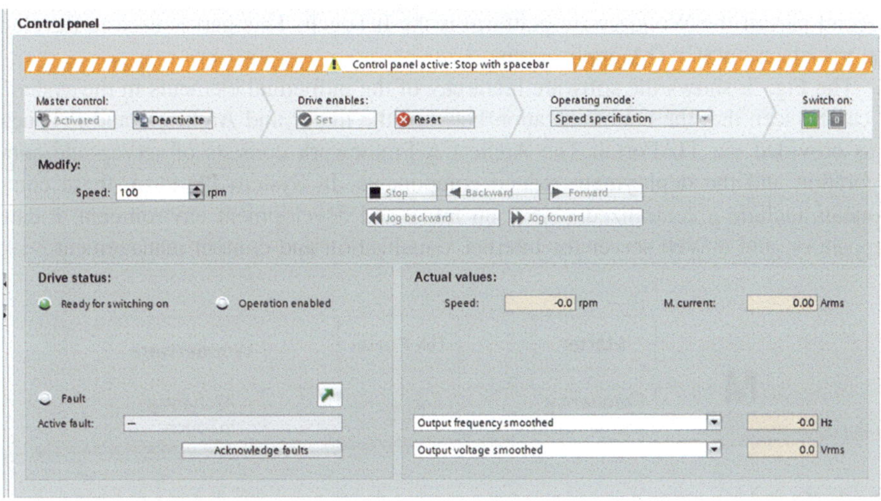

Fig. 1. Control panel in TIAPortal, which allows monitoring 2 parameters

2 Proposal of Communication

The proposed method is based on the hardware configuration and interconnection between the individual elements. The hardware connection scheme is shown in the Fig. 2. The frequency converter is connected to the motor it controls. However, the frequency converter cannot be connected directly to the Wonderware based application.

This is due the direct incompatibility of used software and hardware elements. The Wonderware platform is however capable to communicate with the Siemens PLC's. In our case we have decided to use the Siemens PLC as the middle element between our application and frequency converter.

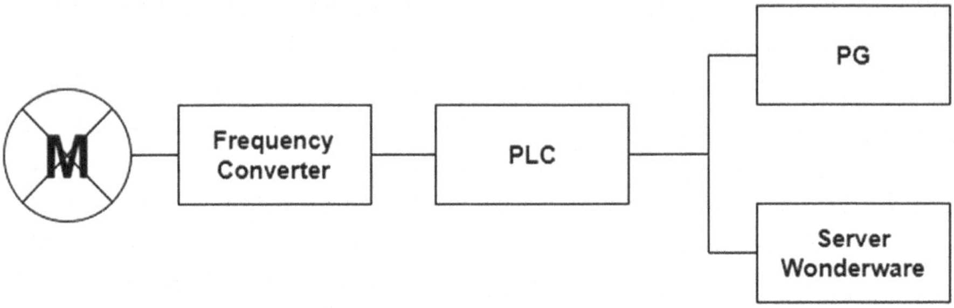

Fig. 2. The scheme of hardware engagement of individual elements in the system

The Wonderware platform consists of several standalone elements. First element is the ArchestrA IDE, which can be understood as an application server. We can define the architecture of our setup and also various functional scripts in the ArchestrA. Second part of the Wonderware platform is the InTouch. This part is used to develop the visualization/SCADA applications.

The Fig. 3. shows the software hierarchy of the individual elements in the system. It can be seen that the communication between the motor and Archestra and InTouch was provided via TIAPortal. The ArchestrA Framework consists of server-side configuration and the deployment related components. In System Platform, these components include a centralized object, an integrated development environment, a data repository, and a Web server for Internet visualization and content management.

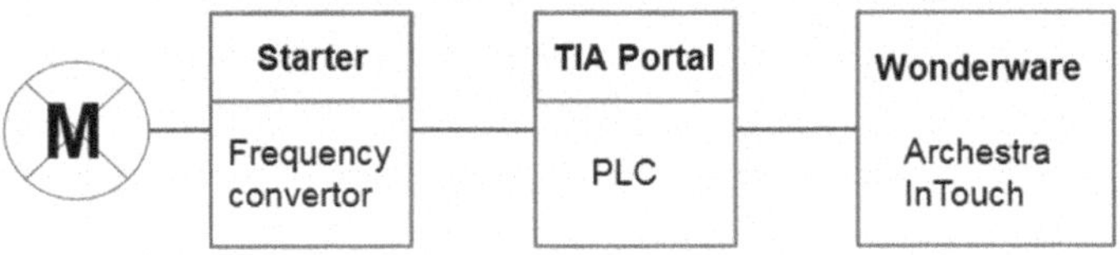

Fig. 3. The scheme of hardware and software engagement of individual elements in the system

The standard telegram is used to communicate between the PLC and the frequency converter. It's the standard way to send and receive information. Depending on the type of telegram, a distinction is made between how many different words (how many data) can be sent and received at the same time, differentiated between Actual Value and Setpoint. These two can have different word lengths. (Standard Telegram 1–2 words both, Standard Telegram 20-6 words Actual Value and 2 words Setpoint, Siemens Telegram 352-6 words both). In our case we use Standard Telegram 1 with QW 256 and QW258 outputs. With QW256, we set the engine to turn on, off, and change direction. The QW258 adjusts the engine speed (Fig. 5).

Table 1. Overview of used standard telegram 1 in tiaportal environment

Drive object	Telegram	Length	Extension		Partner	Partner data area
actual value	Standard telegram 1	2 WORDS	0 words	CD	PLC_2	I 260…263
Setpoint	Standard telegram 1	2 words	0 words	CD	PLC_2	Q 260…263

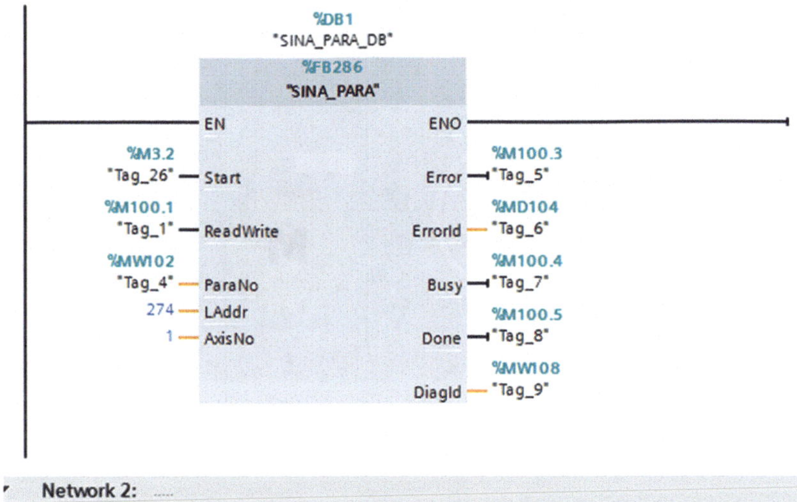

Fig. 4. Data block used for read parameters from G120CU

2.1 Proposal of the Application

The application displays five parameters that we are able to read using the PLC. The parameters we can read from the frequency converter are rpm, temperature, torque, absolute current smoothed and output voltage smoothed. The PLC Connection status shows the status of the telegram that is in the 16th set. The M2.1, M2.2 and M2.3 are the gauges we use to control the frequency converter. The application displays their status in the real time.

$$Rmp = speed * 0.09155 \tag{1}$$

From the Fig. 5. we see that the value of output voltage smoothed and absolute actual current smoothed degree when we increase the engine speed. The value of *Torgue* varies from 1 to-1. If the motor is turned in the forward direction, the value is negative and if the backward direction is on, the value is positive. The speed can be set through the speed, while the *rpm* is calculated via constant.

In Table 1. are markers which we used in PLC program to control G120CU. M2.1 is used for stop the motor, M2.2 is used for start motor and turn motor direction to turn right and M2.3 is used for change the motor direction to left. The forbidden condition is condition when the motor is constantly switching between these states and it can cause problems with frequency converter or motor itself (Table 2).

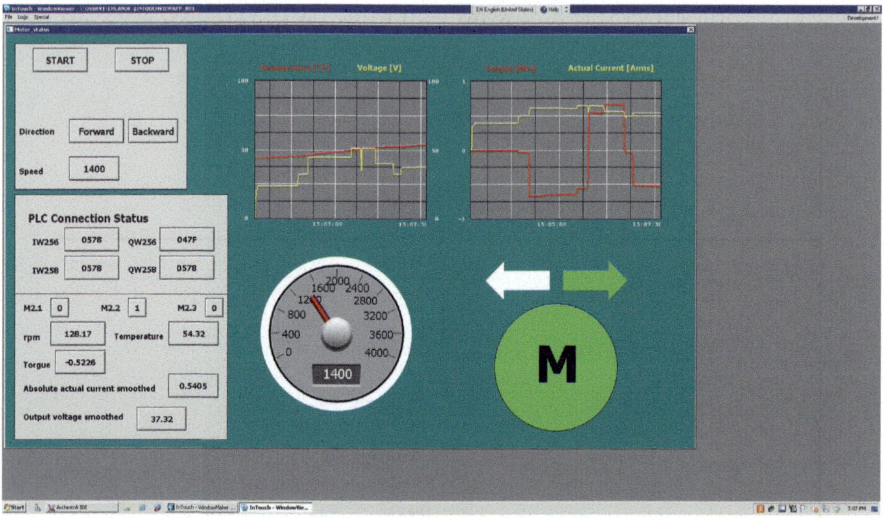

Fig. 5. The main window of the application created in wonderware in touch

Table 2. Overview of motor states using following markers

M2.1	M2.2	M2.3	
0	0	0	Previous state
0	0	1	Turn left
0	1	0	Motor start, turn right
0	1	1	Forbidden condition
1	0	0	Motor Stop
1	0	1	Forbidden condition
1	1	0	Forbidden condition
1	1	1	Forbidden condition

3 Conclusion

In our paper we have presented the designed application for the control and monitoring of frequency converter using the Wonderware software package. The main aim of this paper was to showcase the methodology of controlling and monitoring devices on the production control level with inconsistent control software and hardware. In this paper we dealt with establishing stable connection between the device (frequency converter) and third-party software for control and monitoring purposes. This approach is beneficial for those types of industry companies, which attempt to upgrade their software side of production, while they keep the existing control hardware and devices. Our methodology is based on the use the Wonderware platform as a control and monitoring platform. We chose this platform because of its capabilities in establishing communication with various control hardware manufacturers.

Acknowledgment. This contribution was written with the financial support of the VEGA agency in the frame of the project 1/0272/18: "Holistic approach of knowledge discovery from production data in compliance with Industry 4.0 concept."

References

1. Zhong, R.Y., et al.: RFID-enabled real-time manufacturing execution system for mass-customization production. Robot. Comput.-Integr. Manuf. **29**(2), 283–292 (2013)
2. Ahmed, M.E., Orabi, M., Abdelrahim, O.M.: Two-stage micro-grid inverter with high-voltage gain for photovoltaic applications. IET Power Electron. **6**(9), 1812–1821 (2013)
3. Edelman, E.C.: Command and control system and method for multiple turbogenerators. U.S. Patent No 6,169,334 (2001)
4. Stemmler, H., Guggenbach, P.: Configurations of high-power voltage source inverter drives. In: 1993 Fifth European Conference Power Electronics Applications IET, pp. 7–14 (1993)
5. Cwikla, G.: Methods of manufacturing data acquisition for production management-a review. In: Advanced Materials Research, pp. 618–623. Trans Tech Publications (2014)
6. Mackay, R.: Power generating system. U.S. Patent No 4,754–607 (1988)
7. Persson, E.: Transient effects in application of PWM inverters to induction motors. IEEE Trans. Ind. Appl. **28**(5), 1095–1101 (1992)
8. Yoshida, Y., et al.: Frequency inverter. U.S. Pat. Appl. No 29/026,178 (1995)
9. Gulko, M., Medini, D., Ben-yaakov, S.: Inductor-controlled current-sourcing resonant inverter and its application as a high pressure discharge lamp driver. In: Proceedings of 1994 IEEE Applied Power Electronics Conference and Exposition-ASPEC 1994, pp. 434–440. IEEE (1994)

Models and Algorithms for Improving the Safety of Oil Refineries of the Republic of Kazakhstan

A. A. Dnekeshev[4](✉), V. A. Kushnikov[1,2,4], A. F. Rezchikov[3], V. A. Ivashchenko[1], A. S. Bogomolov[1,2], L. Yu. Filimonyuk[1], and O. N. Dolinina[4]

[1] Institute of Precision Mechanics and Control of RAS, 24 Rabochaya Street, Saratov 410028, Russia

[2] Saratov State University, 83 Astrakhanskaya Street, Saratov 410012, Russia

[3] V. A. Trapeznikov Institute of Control Sciences of RAS, 65 Profsoyuznaya Street, Moscow 117997, Russia

[4] Yuri Gagarin State Technical University of Saratov, 77 Politechnicheskaya Street, Saratov 410054, Russia

dnekeshev1991@gmail.com

Abstract. The article discusses mathematical models and algorithms used in automated control systems to increase the safety of oil refineries in the Republic of Kazakhstan and reduce the likelihood of accidents due to a critical combination of events. The main attention is paid to the mathematical apparatus, which allows identifying and preventing the occurrence of critical combinations of events, each of which individually does not significantly affect the safety of the operation of the oil refinery. For this purpose, along with the traditionally used fault trees, it is proposed to use the tree in the form of a dynamic graph, which will allow us to take into account when preparing and making decisions to improve the safety of functioning also the rapidly changing cause-effect relationships between the modeled variables. An example of such a fault tree describing the development of a fire in the tank farm of an oil refinery is considered in detail in the article. For this fault tree, for the five-element minimum section of the state graph of the functioning process, a list of minimum sections is defined that formalize the critical combinations of events consisting of 5 independent events. A general algorithm has been developed for solving the problem of identifying and preventing critical combinations of events. It is based on the apparatus of Markov chains, which allows, under certain limitations imposed on the process of occurrence of critical combinations of events during the operation of an oil refinery, to consider it as a random process satisfying the Markov property with natural filtration. This assumption allows us to use when determining the probability of occurrence of critical combinations of events of the Kolmogorov-Chapman system of equations, the solution of which is carried out in the article using the fourth-order accuracy numerical Runge-Kutta method. The developed software will be used during the modernization of company group "Condensate" of the Republic of Kazakhstan.

Keywords: Oil · Control · Accident · Critical combination of events · Fault tree

R. Silhavy (Ed.): CSOC 2020, AISC 1226, pp. 230–239, 2020.
https://doi.org/10.1007/978-3-030-51974-2_21

1 Introduction

At present, in connection with the modernization of the existing oil refineries of the Republic of Kazakhstan [1–4] and the simultaneous increase of their production capacities, the issue of increasing the fault tolerance of the technical systems of these enterprises has risen on the agenda. Logical-probabilistic methods [5–12], methods of the theory of automata [13–15] and development control methods [16] are effectively used to determine the potential danger of human-machine and technogenic systems.

Modern oil refining enterprises have systems and resources that can reduce the risk of critical situations occurring both individually and simultaneously [8]. But in most cases, the accident is caused by several unfavorable combinations of events differing in their properties and types, to some extent, distributed in time and sequence of occurrence. Critical combinations of events are understood to mean complexes of events that can be separately parried by the system, but together lead to an accident [8]. The concept of emergency event combinations is used to model the development of such combinations [17, 18].

The occurrence of critical combinations of events at oil refineries contributes to an increase in emergency situations. This problem requires a mathematical solution through the use of formal models and algorithms for the analysis of emergency combinations of events [8].

2 The Statement of the Problem

The content consistently consists of the selection of control actions that reduce the likelihood of their most dangerous critical combinations by parrying individual adverse events. The formal statement of the problem can be represented as follows.

Let a list of emergency and catastrophic situations $A = \{A_1, \ldots, A_n\}$ be given, each of which arises as a result of adverse combinations of events characteristic of a given type of oil refinery. Based on the analysis of the accidents that have occurred, an fault tree [19–21] has been built for each catastrophic situations from A characterizing the process of the occurrence and development of their adverse combinations. The set of all fault trees $D = \{D_1, \ldots, D_m\}$ is denoted by, $E = \{e_1, \ldots, e_k\}$ – the set of their terminals.

Let the intensities $\lambda_i(t)$, $i = 1, \ldots, k$, the occurrence of events from the set E, be also known. Let denote $\mu_i(t)$ the intensities of the flows of parrying events from the set E. We believe that each value $\mu_i(t)$ corresponds to a set of instructions $Q(\mu_i(t))$ – intended for operators and other decision-makers. Let $x(t)$ be the state of the environment.

It is necessary to do:

- on a modeling complex in preparation for system functioning for any moment, determine the probabilities $p_i(\lambda(t), \mu(t), x(t))$ characterizing the possibility of the most common accidents or catastrophes from A;
- for accidents and catastrophes, from the set determine the vector $\mu(t)$ of actions for which, at a given time interval for all permissible environmental conditions $x(t)$ and system resource requirements the condition is satisfied:

$$\forall i \in \{1,\ldots,n\} p_i(\lambda(t),\ \mu(t),\ x(t)) \le \varepsilon_i \quad (1)$$

where ε_i are the given non-negative numbers, the limiting values of the probabilities of accidents and catastrophes.

3 Proposed Approach to Solving

To represent the processes of development of critical combinations of events, it is proposed to use fault trees in which "fault" are understood as manifestations of technical defects in the system, as well as external natural phenomena, operator errors, and errors in computer programs. Critical combinations of events are represented by minimal sections of such trees. The probability of realization of the minimum cross section depends on the intensity (frequency) of the occurrence of various adverse events and on the intensities of parrying such events. Thus, it is required to construct algorithms that allow one to determine such parry intensities of adverse events at which the probabilities of the development of critical combinations of events will not exceed a predetermined relatively safe value.

4 Used Mathematical Models

In solving of this problem, it is proposed to use the minimum sections of fault trees and event graphs to represent critical combinations of events and their development. In determining the probability of the development of critical combinations, the Kolmogorov-Chapman differential equations are used [22–24]. When we find a solution of the problem, the considered period of time for the operation of the automatic telephone exchange is divided into parts in such a way that, on each of these parts $\lambda_i(t)$ are considered as constants, which makes it possible to use the Markov model. The number of such parts depends on the operating conditions of a particular system.

4.1 Fault Tree

To determine the critical combinations of events leading to an accident at an oil refinery, it is proposed to use them in the form of complexes of fault tree. A variant of the upper levels of the fault tree leading to a fire at the refinery is shown in Fig. 1 [8].

The following symbols are adopted in Fig. 1 [8]: – logical operator AND; – logical operator OR; F – the fire at tank farm; 1 – oil product spill, oil product overflow; 2 – ignition source; 3 – torch burning; 4 – oil product ignition, explosion of gas-air mixture; 5 – fiery sphere; 6 – strait fire; 7 – oil product ignition, explosion; 8 – lack of level control devices; 9 – low production discipline of staff; 10 – ignition source; 11 – formation of steam-gas-air cloud; 12 – explosion of gas-air mixture; 13 – flash vapor-air mixture; 14 – formation of explosive gas-air environment in the tank farm square, release of light hydrocarbons, vaping in a fuel oil tank, overheating of sub-standard fuel oil; 15 – ignition source; 16 – long product expiration; 17 – ignition source; 18 – instant

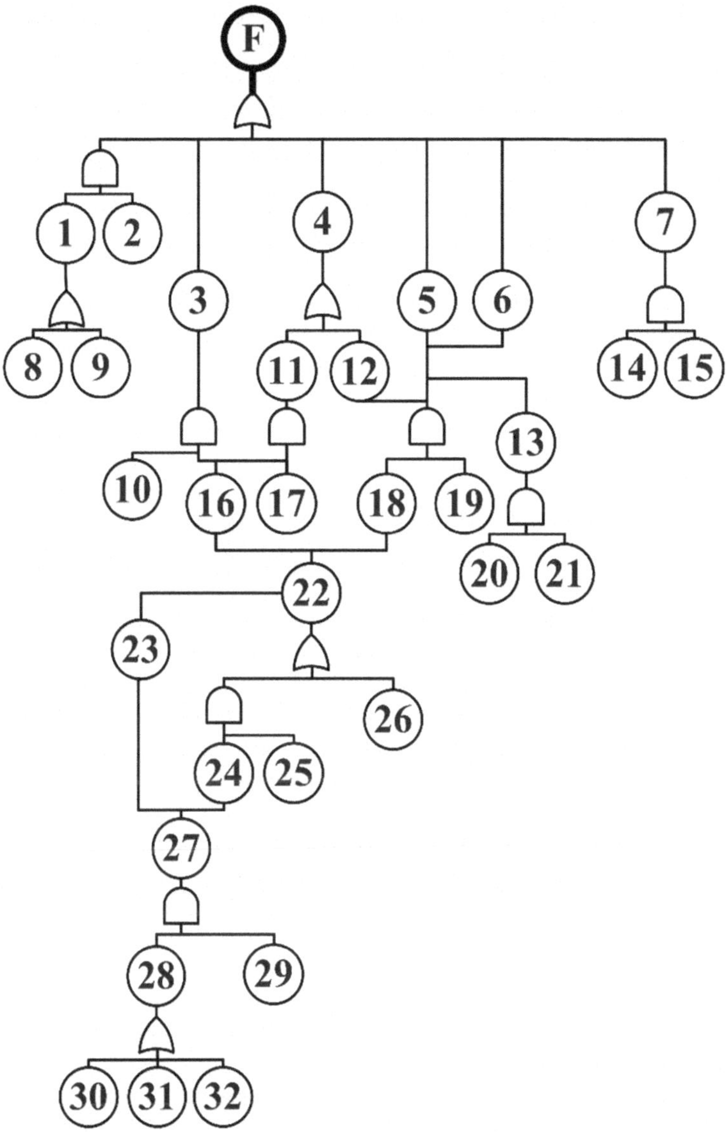

Fig. 1. Possible upper levels of the fault tree leading to a fire at the refinery [8].

depressurization; 19 – ignition source; 20 – formation of explosive vapor-air mixture, gas contamination of the space adjacent to the reservoir, the lower manhole of the reservoir is in the open position for a long time, violation of instructions during operation, the pumping of "dead gasoline ; 21 – synthetic clothing of one of the employees; 22 – destruction of the tank of equipment with product release; 23 – oil product fire in the tank; 24 – the explosion inside the tank; 25 – no protective weak layer between the roof and the tank wall; 26 –possible deformation of the tank, collapse of the floating roof, jamming of the floating roof, flow of oil over the surface of the floating roof, exit of oil through the ring seal of the floating roof; 27 – the flash inside the tank; 28 – ignition source; 29 – presence of gas-air mixture in the tank in dangerous concentrations, absence of ventilation peripheral openings in the roof or on the walls of the tank (breathing valves); 30 – incorrect performance of fire works, employees of the contractor did not pass the internship before carrying out fireworks; 31 – lightning; 32 – applied electrical equipment of non-explosion-intrinsic safety design, spark in switching equipment of portable pump, explosion-intrinsic safety of portable pump does not meet the condition of works performance [8].

Terminal vertex combinations-critical event combinations correspond to the minimum sections of the fault tree, the execution of which, regardless of the execution of the other end vertices, leads to the root event [8].

Critical combinations of events are represented in the form of combinations of end vertices of the tree, during the implementation of which the root event (vertex *F*) occurs according to a given logical scheme, regardless of the implementation of the other end vertices. For example, one of the possible scenarios for the development of a critical combination of events can be presented as follows. In a case of a thunderstorm, two lightning strikes occur (event 31), the first of which destroys the air terminal, and the second enters the surface of the oil tank. At this time, an increased concentration of oil vapor is observed in the reservoir, which is a consequence of blockage of the respiratory valves (event 29). Therefore, oil vapor ignites in the free space of the tank (event 28) and an explosion inside the tank (event 27, 24). An explosion destroys the roof of the tank in the absence of a protective weak layer between the roof and the wall of the tank (event 25). This leads to the destruction of the fire fighting equipment of the tank (event 22), which, together with the prolonged outflow of oil (event 16), leads to the explosion of a vapor-gas cloud and a fire in the tank farm.

An example of a fault tree (Fig. 1) and a scenario of the situation leading to a backward refinery is developed taking into account adverse situations arising at production sites in the Republic of Kazakhstan.

4.2 Event Graph

Let the vertices $e_1, \ldots, e_k$ form the minimum section of the fault tree [8]. The process of occurrence of the root event is analyzed using the event graph G_k, 2^k vertices of which correspond to the occurrence of all combinations of events $e_1, \ldots, e_k$. For events e_1, e_2, e_3, e_4, e_5 of the graph G_5, arising with intensities $\lambda_1, \lambda_2, \lambda_3, \lambda_4, \lambda_5$, which counteracts with intensities $\mu_1, \mu_2, \mu_3, \mu_4, \mu_5$ respectively, is shown in Fig. 2 [8].

The correspondence of the vertices of the graph G_5 to the occurrence of combinations of events from the set $\{e_1, e_2, e_3, e_4, e_5\}$ is displayed in Table 1 [8]. Vertex 0—absence of events e_1, e_2, e_3, e_4, e_5, vertex 31—their combination, other vertices are intermediate combinations of events [8].

Following the principle used in Table 1 for numbering the vertices G_5, in the general case, numbers for numbering the vertices denoting all combinations of $k = 0, \ldots, n$ events can be taken according to the following rule:

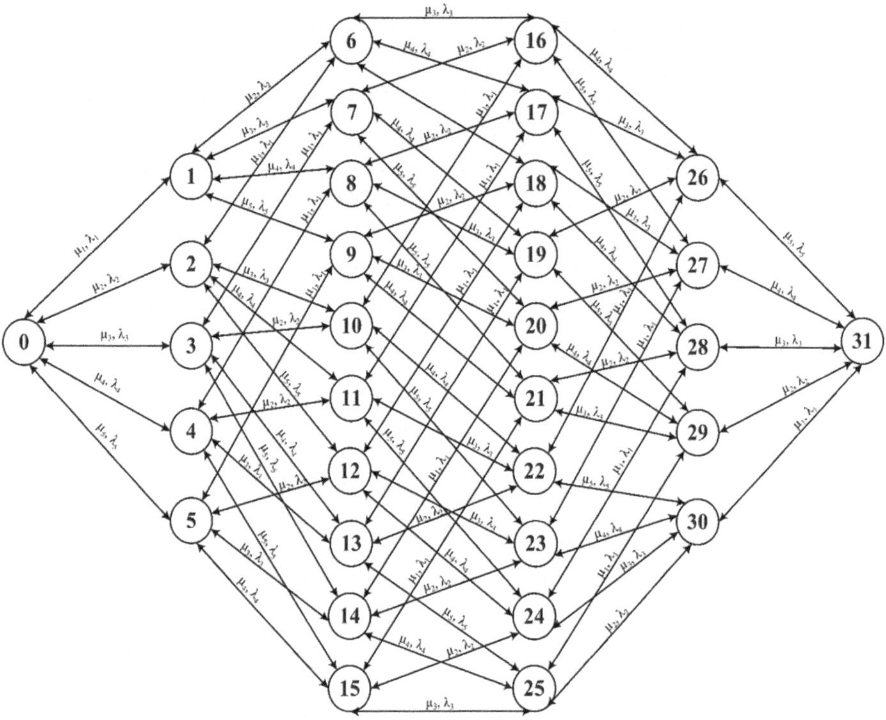

Fig. 2. Graph G_5 operational status for five-element minimum sections [8].

- the vertex corresponding to the absence of events e_1, …, e_k is numbered 0;
- the vertices denoting the appearance of all possible combinations of k events ($n \geq k > 0$) are numbered by natural numbers C_n^k following after those used in the previous iteration.

Table 1. The correspondence of the vertices of the graph G_5 to the occurrence of combinations of events from the set $\{e_1, e_2, e_3, e_4, e_5\}$ [8].

Top of the graph G_4	Events
0	–
1	e_1
2	e_2
3	e_3
4	e_4
5	e_5
6	e_1, e_2
7	e_1, e_3

(*continued*)

Table 1. (*continued*)

Top of the graph G_4	Events
8	e_1, e_4
9	e_1, e_5
10	e_2, e_3
11	e_2, e_4
12	e_2, e_5
13	e_3, e_4
14	e_3, e_5
15	e_4, e_5
16	e_1, e_2, e_3
17	e_1, e_2, e_4
18	e_1, e_2, e_5
19	e_1, e_3, e_4
20	e_1, e_3, e_5
21	e_1, e_4, e_5
22	e_2, e_3, e_4
23	e_2, e_3, e_5
24	e_2, e_4, e_5
25	e_3, e_4, e_5
26	e_1, e_2, e_3, e_4
27	e_1, e_2, e_3, e_5
28	e_1, e_2, e_4, e_5
29	e_1, e_3, e_4, e_5
30	e_2, e_3, e_4, e_5
31	e_1, e_2, e_3, e_4, e_5

5 General Algorithm for Solving the Problem

The beginning of the algorithm [8].

1. Determine the set of accidents elements of the refinery $A = \{A_1, \ldots, A_n\}$.
2. Construct a set of fault trees $D = \{D_1, \ldots, D_m\}$ corresponding to accidents from set A, with a set of elementary events $E = \{e_1, \ldots, e_n\}$.
3. For each tree from D define all minimum sections.
4. To construct for each class $L^i(D)$, $i = 1, \ldots, k$, of system of differential equations in the form of Kolmogorov-Chapman for determination of probability of occurrence of emergency situations at the oil refining enterprise [8].
5. Determine the input parameters of the mathematical model, allowing to solve the system of differential equations Kolmogorov-Chapman [8].
6. Solve a system of differential equations to determine the probability $P_i(t)$, $i = 1, \ldots, k$, of accidents at the refinery [8].

7. Using the method of a computational experiment, determine the sets of values of $\mu(t)$ for which condition (1) is satisfied [8].
8. Analyze and interpret the results of solving systems of differential equations.
9. Taking into account the resources of the system and the requirements for its normal functioning, determine the realized value of $\mu'(t)$ at which condition (1) is satisfied.
10. To develop recommendations for operators and decision makers on the implementation of actions $Q(\mu'(t))$.

End of the algorithm [8].

6 Discussion

The developed software allows you to create a promising safety management circuit for the operation of oil refineries, which allows you to timely identify and quickly prevent the occurrence of critical combinations of events, each of which individually does not significantly affect the safety of the functioning of the refinery.

The currently used critical warning systems (Honeywell, Siemens, ABB, Emerson, Bently Nevada) do not fully take into account combinations of events that are relatively harmless separately, but lead to an accident in aggregate.

Such events are of a different nature, and therefore are usually not taken into account in a single mathematical model. However, in many cases, the occurrence of earlier events leads to the fact that the system is deprived of resources for a regular response to later adverse events. In this case, there is a process of development of a critical combination of events. The proposed concept is aimed at the development of models, methods and algorithms to stop the development of such processes leading to accidents.

The idea of the approach is to fix the occurring adverse events and determine on this basis the probability of the development of their critical combinations. The composition of critical combinations of events is determined by the fault trees, and the probability, in particular, from the solution of the Kolmogorov-Chapman equations in the case.

If the probability determined by this turns out to be greater than the permissible value ε_i, the prospective security system offers the personnel a series of actions from the lists $Q(\mu_i(t))$ aimed at counteracting certain adverse events in order to reduce the likelihood of their combinations.εIf the probability determined by this turns out to be greater than the permissible value. Information for compiling fault trees, determining the frequency probabilities of adverse events, lists of specific actions $Q(\mu_i(t))$ will have to be taken from the structure of the refinery, including statistical data, the characteristics of the personnel's work, as well as the climatic and geographical features of the area.

The prospect under development is the development of the proposed approach in terms of accounting for events arising in critical combinations.

Mathematical software was developed, which are based on logical-probabilistic models and numerical methods that allow to identify and prevent critical combinations of events in the process of functioning of oil refineries. This software will be used in the modernization of company group "Condensate" of the Republic of Kazakhstan.

References

1. Kapital. https://kapital.kz/gosudarstvo/70924/k-koncu-goda-kazahstan-budet-polnostyu-obespechen-nefteproduktami.html. Accessed 28 Nov 2019
2. Pomfret, R.: Kazakhstan's economy since independence: does the oil boom offer a second chance for sustainable development? Eur.-Asia Stud. **57**(6), 859–876 (2005)
3. Kaiser, M.J., Pulsipher, A.G.: A review of the oil and gas sector in Kazakhstan. Energy Policy **35**(2), 1300–1314 (2007)
4. Heim, I., Kalyuzhnova, Y., Li, W., Liu, K.: Value co-creation between foreign firms and indigenous small- and medium-sized enterprises (SMEs) in Kazakhstan's oil and gas industry: the role of information technology spill overs. Thunderbird Int. Bus. Rev. **61**(6), 911–927 (2019)
5. Dhillon, B.S., Chanan, S.: Engineering Reliability: New Techniques and Applications. Wiley, New York (1981)
6. Guck, D., Spel, J., Stoelinga, MIA.: DFTCalc: reliability centered maintenance via fault tree analysis (tool paper). In: International Conference on Formal Engineering Methods (ICFEM). LNCS, vol. 9407, pp. 304–311 (2015)
7. Ruijters, E., Reijsbergen, D., de Boer, P.-T., Stoelinga, M.I.A.: Rare event simulation for dynamic fault trees. Reliab. Eng. Syst. Saf. **186**, 220–231 (2019)
8. Dnekeshev, A.A., et al.: Models and algorithms for improving the safety of oil refineries in the Republic of Kazakhstan. Natl. Tech. Sci. **7**, 145–150 (2019). (in Russian)
9. Fominykh, D.S., Rezchikov, A.F., Kushnikov, V.A., Ivashchenko, V.A., Bogomolov, A.S., Filimonyuk, L.Y., Dolinina, O.N., Kushnikov, O.V., Shulga, T.E., Tverdokhlebov, V.A.: Problem of quality assurance during metal constructions welding via robotic technological complexes. In: International Conference on Information Technologies in Business and Industry, vol. 1015, no. 3, p. 032169 (2018)
10. Bogomolov, A.S.: Integrated resource control of complex man-machine systems. Izvestiya Saratovskogo Universiteta Novaya Seriya – Matematika Mekhanika Informatika **13**(3), 83–87 (2013)
11. Spiridonov, A.Y.; Rezchikov, A.F., Kushnikov, V.A., Ivashchenko, V.A., Bogomolov, A.S., Filimonyuk, L.Y., Dolinina, O.N., Kushnikova, E.V., Shulga, T.E., Tverdokhlebov, V.A., Kushnikov, O.V., Fominykh, D.S.: Prediction of main factors' values of air transportation system safety based on system dynamics. In: International Conference on Information Technologies in Business and Industry, vol. 1015, no. 3, p. 032140 (2018)
12. Radosteva, M., Soloviev, V., Ivanyuk, V., Tsvirkun, A.: Use of neural network models in the market risk management. Adv. Syst. Sci. Appl. **18**(2), 53–58 (2018)
13. Rezchikov, A.F., Kushnikov, V.A., Ivashchenko, V.A., Bogomolov, A.S., Filimonyuk, L. Yu.: Models and algorithms of automata theory for the control of an aircraft group. Autom. Remote Control **79**(10), 1863–1870 (2018)
14. Zheng, Y., Jia, B., Li, X.-G., Zhu, N.: Evacuation dynamics with fire spreading based on cellular automaton. Phys. A: Stat. Mech. Appl. **390**(18–19), 3147–3156 (2014)
15. Alexandridis, A., et al.: Wildland fire spread modelling using cellular automata: evolution in large-scale spatially heterogeneous environments under fire suppression tactics. Int. J. Wildland Fire **20**(5), 633–647 (2011)
16. Baranov, V.V., Tsvirkun, A.D.: Development control: structural analysis, problems, stability. Autom. Remote Control **79**(10), 1780–1796 (2018)
17. Bogomolov, A.S.: Analysis of the ways of occurrence and prevention of critical combinations of events in man-machine systems. Izv. Saratov Univ. (New Ser.) Ser. Math. Mech. Inform. **17**(2), 219–230 (2017)

18. Filimonyuk, L.: The problem of critical events combinations in air transportation systems. In: Advances in Intelligent Systems and Computing, vol. 573, pp. 384–392. Springer (2017)
19. Dugan, J.B., Bavuso, S.J., Boyd, M.A.: Fault trees and sequence dependencies. In: Proceeding of Annual Reliability and Maintainability Symposium, pp. 286–293. IEEE (1990)
20. Guck, D., Spel, J., Stoelinga, MIA.: DFTCalc: reliability centered maintenance via fault tree analysis (tool paper). In: Proceeding 17th International Conference Formal Engineering Methods (ICFEM). LNCS, vol. 9407, pp. 304–311 (2015)
21. Ruijters, E., Guck, D., Drolenga, P., Stoelinga, M.I.A.: Fault maintenance trees: reliability centered maintenance via statistical model checking. In: Proceedings of IEEE 62nd Annual Reliability and Maintainability Symposium (RAMS) (2016)
22. Siergiejczyk, M., Rosinski, A.: Reliability and exploitation analysis of power supply of alarm systems used in railway objects. In: 26th Conference on European Safety and Reliability (ESREL), Glasgow, Scotland (2016)
23. Laskowski, D., Lubkowski, P., Pawlak, E., Stanczyk, P.: Anthropo-technical systems reliability. In: Safety and Reliability: Methodology and Applications, pp. 399–407 (2015)
24. Pilo, E.: Power Supply, Energy Management and Catenary Problems, pp. 1–191. WIT Press, Ashurst (2010)

Logical-Probabilistic Models of Complex Systems Constructed on the Modular Principle and Their Reliability

Gurami Tsitsiashvili(✉)

Institute for Applied Mathematics, Far Eastern Branch of Russian Academy Sciences, Vladivostok, Russian Federation
guram@iam.dvo.ru

Abstract. The aim of this paper is to define a system built on a modular principle and to construct a recursive algorithm for calculating its reliability. For this purpose, the system is described using a monotone Boolean function that characterizes its performance. The oriented tree is mapped to a Boolean function (which characterizes operability). From any leaf of this tree there is a (single) path to its root. We assume that all the edges of this path are directed towards the root. A recursive algorithm for calculating the Boolean function and the reliability of the system, described by this function, is constructed. It is proved that the number of arithmetic operations for calculating a Boolean function, built on the modular principle, and the reliability of the system depend linearly on the number of leaves in the tree. Then how for systems of general form of this dependence is exponential. #CSOC1120.

Keywords: Boolean function · Recursive definition · Reliability · Modular principle.

1 Introduction

The paper [1] describes the basics of logical-probabilistic modelling of complex systems used in reliability problems. The concept of the probabilistic logic is briefly described as an extension of inductive logic. A clear distinction is made between the probabilistic logic and logical-probabilistic calculus – the branch of mathematics that defines the rules of computation and operation with two-valued (true and false) statements. The logical-probabilistic calculus is based on the algebra of logic and rules of substitution of logical arguments in functions of the algebra of logic by their truth probabilities, and logical operations by arithmetical operations.

Logical-probabilistic calculus is closely related to the fundamental work on the theory of reliability, one of the main tasks of which is to improve the reliability of technical systems. So, in the fundamental monograph on the theory of reliability [2] for this purpose, it is proposed to use the theorem on the advantage of separate redundancy over block redundancy.

R. Silhavy (Ed.): CSOC 2020, AISC 1226, pp. 240–246, 2020.
https://doi.org/10.1007/978-3-030-51974-2_22

But recently, there are more and more technical designs based on the modular principle. This principle is used in the design of computer equipment [3], machine tools with software control and industrial robots [4], in the modelling of industrial production processes, in particular, in petrochemistry [5], etc. The modular principle is widely used in the design of large technical systems for responsible use. An important application of the modular principle is the ability to design highly reliable technical systems. However, there is no mathematical equivalent to the concept of the modular principle and its connection with reliability analysis of the systems constricted on this principle.

In this paper, the definition of a system built on the modular principle is given. For this purpose, the system is described using a monotone Boolean function that characterizes its performance. The oriented tree is mapped to a Boolean function (which characterizes operability). From any leaf of this tree there is a single path to its root. We assume that all the edges of this path are directed towards the root. A recursive algorithm for calculating the Boolean function and the reliability of the system described by this function is constructed. It is proved that the number of arithmetic operations for calculating a Boolean function built on the modular principle and the reliability of the system depends linearly on the number of leaves in the tree. Then how for systems of general form this dependence is exponential.

2 Methods

Let the health of the system is determined by the Boolean function $A(x_1, \ldots, x_n)$, where $x_1, \ldots, x_n$ – Boolean variables that characterize the health of individual elements of the system: if $x_k = 1$, then the element k is in working condition, otherwise the element k fails, $k = 1, \ldots, n$. Assume that the Boolean variables $x_1, \ldots, x_n$ are independent random variables, $P(x_1 = 1) = p_1, \ldots, P(x_n = 1) = p_n$.

It is natural to assume that the Boolean function A, describing the state of the system, is monotonous. In other words, for any sets of Boolean variables $(x_1, \ldots, x_n)$, $(x_1^*, \ldots, x_n^*)$ from the relations $x_1^* \leq x_1, \ldots, x_n^* \leq x_n$ follows the inequality $A(x_1^*, \ldots, x_n^*) \leq A(x_1, \ldots, x_n)$. It is known that any monotone Boolean function $A(x_1, \ldots, x_n)$ can be defined using the superposition of conjunction and disjunction operations [6].

Let's say that a monotone Boolean function $A(x_1, \ldots, x_n)$, represented as a superposition of conjunction and disjunction operations, obeys the modular principle if it is defined by a disjunction or conjunction of Boolean functions only from disjoint sets of Boolean variables

$$A_1(x_k : k \in K_1),\ A_2(x_k : k \in K_2),\ K_1, K_2 \subseteq \{1, \ldots, n\},\ K_1 \cap K_2 = \emptyset. \quad (1)$$

To do this, a monotone Boolean function $A(x_1, \ldots, x_n)$ is represented by an oriented tree D with n leaves containing Boolean variables $x_1, \ldots, x_n$. The edges of the tree D are directed from the leaves to the root of the tree. Each vertex of the D tree that is not a leaf has a conjunction or disjunction icon. Each such vertex has two edges, and one edge comes out of it. By mathematical induction, it is easy to prove that in a tree D the number of vertices that are not leaves is equal to $n-1$.

At each step of the algorithm for determining the Boolean function $A(x_1,\ldots,x_n)$, two sheets are allocated, from which the edges are directed directly to one vertex containing the conjunction or disjunction operation. After calculating this Boolean operation on the variables in the leaves, a new leaf is formed and as a result the total number of leaves is reduced by one. This algorithm consists of $n-1$ steps and continues until the tree has exactly one vertex. One step of calculating the Boolean function $A(x_1,\ldots,x_n)$ is shown in Fig. 1. By induction over n, we can prove that the definition of the Boolean function $A(x_1,\ldots,x_n)$ is correct, i.e. does not depend on the order of selection of pairs of sheets, the operation of conjunction or disjunction which leads to the formation of a new sheet (see Fig. 1, Fig. 2).

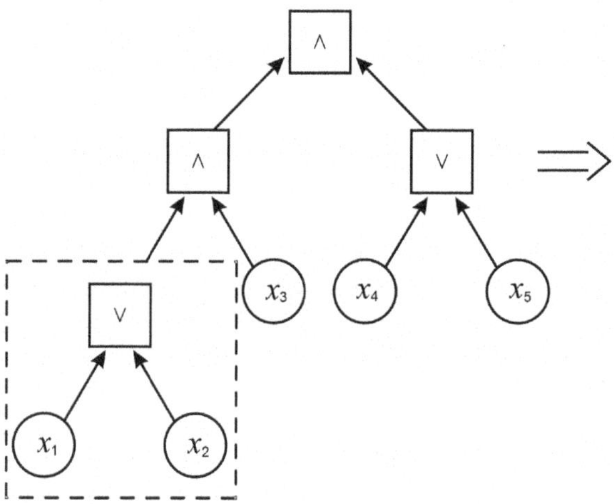

Fig. 1. Oriented tree that represents the construction of a function $A(x_1,\ldots,x_n)$.

It is worthy to remark, that oriented tree D, representing logical function $A(x_1,\ldots,x_n)$, may be compressed as follows (see Fig. 3) so that any path from a leave to a root includes a sequence of logical operations $\vee$, $\wedge$, which follow one after the other.

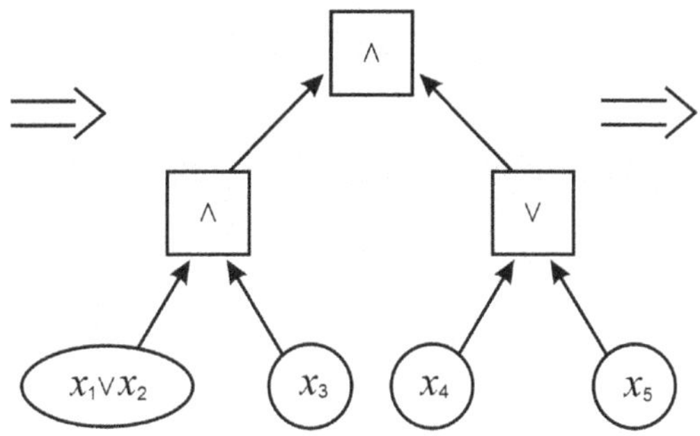

Fig. 2. A step in a calculation of the function $A(x_1, \dots, x_n)$.

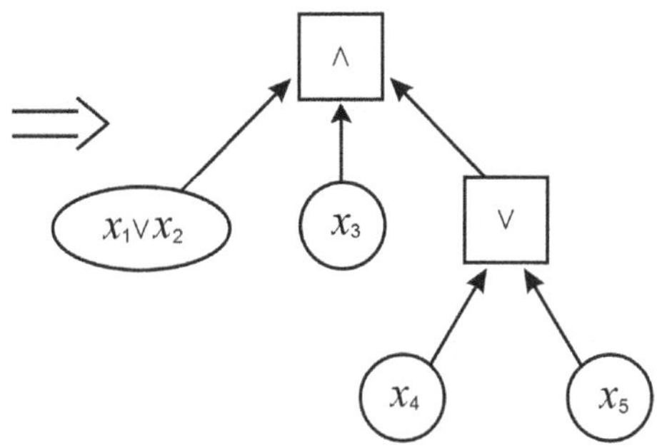

Fig. 3. A compression of the tree D, representing the function $A(x_1, \dots, x_n)$.

3 Results

Now let's go to the description of the algorithm for sequential probability calculation $P(A(x_1, \dots, x_n) = 1)$. For this purpose in leaves of a tree D we will replace Boolean variables $x_1, \dots, x_n$ with probabilities $p_1, \dots, p_n$. In turn, we will replace operation of conjunction $x_1 \wedge x_2$ with operation $p_1 \otimes p_2 = p_1 \cdot p_2$, and operation of disjunction $x_1 \vee x_2$ – with operation $p_1 \oplus p_2 = p_1 + p_2 - p_1 p_2$. As a result of this replacement, the algorithm for calculating the Boolean function $A(x_1, \dots, x_n)$ is converted to the algorithm for calculating the reliability function $P(A(x_1, \dots, x_n) = 1)$ (Figs. 4, 5, and 6).

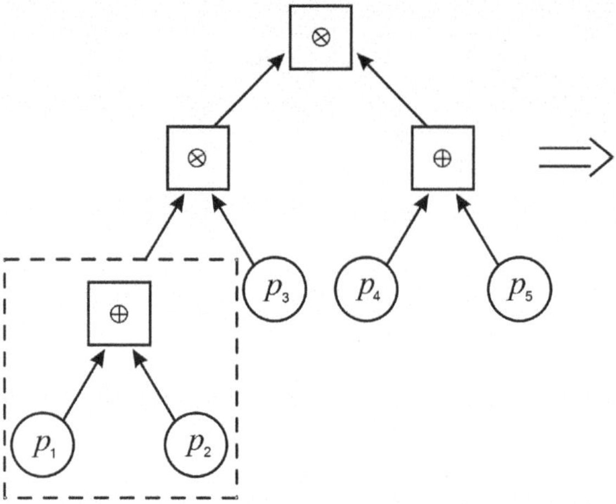

Fig. 4. Oriented trees that represent a sequence of reliability calculation operations $P(A(x_1, \ldots, x_n) = 1)$.

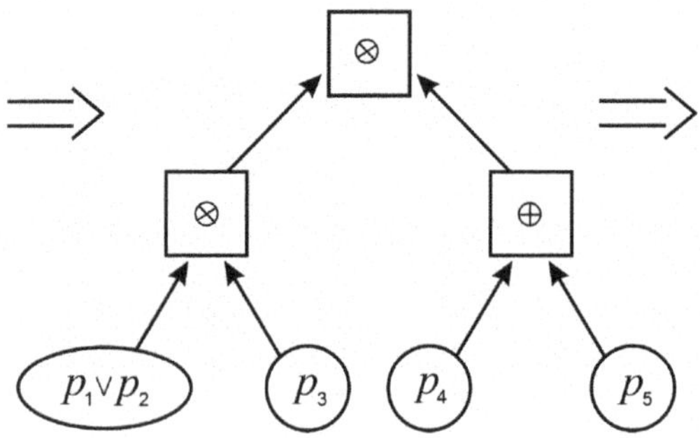

Fig. 5. Oriented trees that represent a sequence of reliability calculation operations $P(A(x_1, \ldots, x_n) = 1)$.

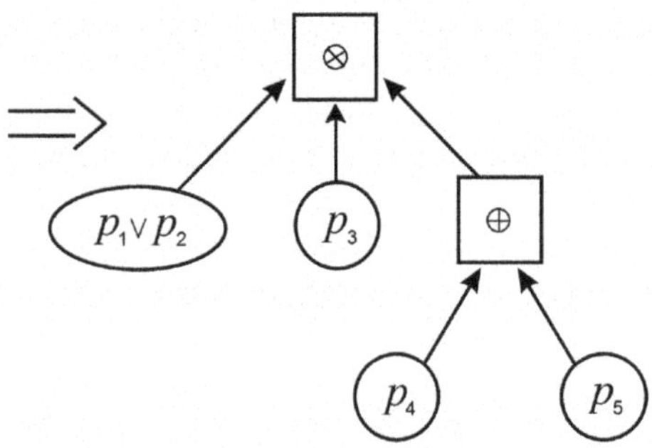

Fig. 6. Oriented trees that represent a sequence of reliability calculation operations $P(A(x_1, \ldots, x_n) = 1)$.

4 Discussions

It should be noted that the number of arithmetic operations N_1 for calculating the probability $P(A(x_1, \ldots, x_n) = 1)$ in the case when the system obeys the modular principle is determined by the equality $N_1 = 3n_{\oplus} + n_{\otimes} = O(n)$, where $n_{\oplus}$ – the number of vertices of the form $\oplus$, $n_{\otimes}$ – the number of vertices of the form $\otimes$ in the tree D, $n_{\oplus} + n_{\otimes} = n - 1$. For comparison, note that in general, when the Boolean function $A(x_1, \ldots, x_n)$ does not obey the modular principle, the probability $P(A(x_1, \ldots, x_n) = 1)$ is determined by the equality

$$P(A(x_1, \ldots, x_n) = 1) = \sum_{x_1, \ldots, x_n = 0}^{1} A(x_1, \ldots, x_n) \prod_{k=1}^{n} p_k^{x_k}. \qquad (2)$$

In the formula (2), the value p^x, $0 \leq p \leq 1$, $x = 0, 1$, is determined by the relations: if $x = 1$, then $p^x = p$, if $x = 0$, then $p^x = 1 - p$. In the right part of the formula (2) contains 2^n terms, so the number of N_2 arithmetic operations to calculate the probability of $P(A(x_1, \ldots, x_n) = 1)$ satisfies the inequality $N_2 \geq 2^n$.

The Boolean function $A(x_1, \ldots, x_n)$, satisfying the modular principle, can be represented as an indicator of the connection of the initial and final vertices of a bipolar with n edges. To do this, the leaves of the D tree are replaced with two-poles representing individual elements of the system.

Next, replace the two sheets connected in the tree D Boolean operation of conjunction or disjunction (see Fig. 1), for serial or parallel connection of two-poles (see Fig. 3) placed in place of these sheets. This procedure continues recursively until the construction of a two-pole corresponding to the logical function A (Fig. 7).

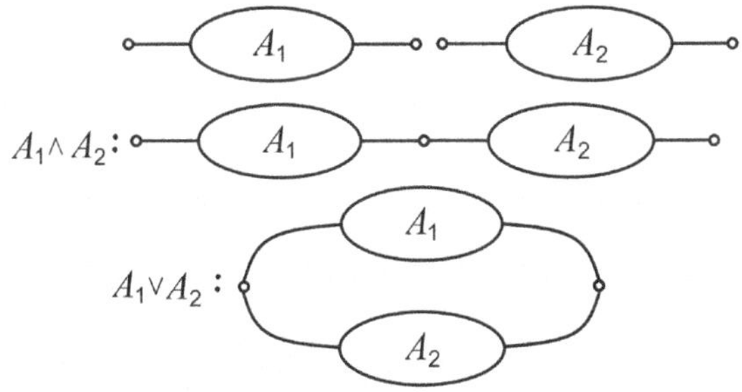

Fig. 7. Two-poles, representing Boolean functions A_1, A_2, $A_1 \wedge A_2$, $A_1 \vee A_2$.

Thus, a system constructed on a modular principle may be interpreted as a two-pole system built on a parallel-sequential principle. The class $\mathcal{K}$ of two-poles built on the parallel-sequential principle has the following properties. If $\Gamma_1, \ldots, \Gamma_r \in \mathcal{K}$, then the parallel (then the serial) connection of these two-poles also belong to the class $\mathcal{K}$. This property allows us to consider now not only oriented trees D, defining the logic function A, in which each vertex includes two incoming edges, but also trees of a more general type, in which each vertex can include $r > 1$ incoming edges, independently on icon of this vertex (disjunction or conjunction).

5 Conclusion

Thus, the modular principle of building a complex system begins with the definition of a logical function A, satisfying certain properties, goes to the construction of an oriented tree D, with n leaves-logical variables, and in the other vertices logical operations of conjunction and disjunction. Next, the tree D is transformed into a two-pole, built on a parallel-sequential principle, the reliability of which can be calculated using $O(n)$ arithmetic operations.

Therefore, if you know the probabilities p_k, $k = 1, \ldots, n$, characterizing health of individual elements, you can calculate the probability of health of the entire system using recursive formulas for the probability of serial and parallel connections of individual elements.

The proposed model of a complex system arranged on a modular principle allows you to control the reliability of a complex system with fairly simple computing tools.

References

1. Riabinin, I.A.: Logical-probabilistic calculus: a tool for studying the reliability and safety of structurally complex systems. Autom. Remote Control **64**(7), 1177–1185 (2003)
2. Barlow, R.E., Poschan, F.: Mathematical Theory of Reliability. Wiley, New York (1965)
3. Malinina, L.A., et al.: Basics of Computer Science. Phenics, Rostov-on-Don (2006). (In Russian)
4. Lokteva, S.E.: The Computer controlled Machine Tools and Industrial Robots. Engineering, Moscow (1986). (In Russian)
5. Kim, S.F., et al.: Modular principle of construction of mathematical models of apparatuses and flow charts of oil trade preparation. Oil Refin. Petrochem. **10**, 41–44 (2013). (In Russian)
6. Zhuravlev, Y., Flerov, Y.A., Fedko, O.S.: Discrete Analysis. Combinatorics. Algebra of Logic. Graph Theory. Textbook, MPhTI, Moscow (2012). (In Russian)

Teaching Decision Tree Using a Practical Example

Zdena Dobesova(✉)

Department of Geoinformatics, Faculty of Science, Palacky University, 17. listopadu 50, 779 00 Olomouc, Czech Republic
zdena.dobesova@upol.cz

Abstract. The positive experience with student gathering data in the Data Mining course is mentioned in this article. The inspiration for a practice lecture was a literature example of a decision tree for classification of sex from the weight and height of persons. Therefore, students fill anonymously simple questionnaire with personal weight, height, and sex. The data set was used as training data for the construction of a decision tree. Decision tree as supervised learning produces rules for classification of sex based on the input attributes. The final decision tree as a result of the training phase was used also for prediction of class (sex) on newly collected testing data. Both parts – construction of a decision tree and prediction was practically demonstrated. The data mining software Orange was selected for practical lectures. The Orange advantages are intuitiveness and easy design of workflow. The article shows result decision trees and results of prediction on real data. Teacher final finding is that the active collecting data make students more involved in the topic and assure a deep understanding of the lectured topics like decision trees.

Keywords: Data mining · Decision tree · Motivation · Engagement · Lecturing

1 Introduction

The author of the article guarantees the subject Data Mining in the first grade of master study program Geoinformatics at Palacký University in Olomouc, Czech Republic. The construction of a decision tree as a prediction model is one of the topics that belong to that course. The decision tree is a graphical expression of a set of rules that predict the final category. There are existing algorithms and their implementation in software for data mining and machine learning. The software automates the construction of classification trees. Students are familiar with two of the software for data mining tasks. The first is WEKA and the second is Orange [1, 2]. Both are non-commercial software for free use. The Orange is primarily aimed for educational purposes.

The practical examples help in the understanding of algorithms and software usage by university students. The author of the article has a positive experience with using a practical example to fix the knowledge of students in other courses. Contributive experience is mentioned in the article about the relational database design for the botanical garden plant database under the BotanGIS project [3].

R. Silhavy (Ed.): CSOC 2020, AISC 1226, pp. 247–256, 2020.
https://doi.org/10.1007/978-3-030-51974-2_23

2 Data and Methods

The important thing is the motivation of students to take part in some practical lessons. Acquiring knowledge is much stronger when students are engaged in gathering source data and when subsequently process this data as example data.

2.1 Decision Tree

The decision tree is a nonparametric algorithm of machine learning. The nonparametric method means that the learning process does not produce a form of function with learned coefficients like linear or logistic regression. Nonparametric algorithms are called machine learning algorithms. By not making assumptions, they are free to learn any functional form from the training data. Non-parametric methods are often more flexible, achieve better accuracy but require a lot more data and training time. Examples of include Support Vector Machines, Neural Networks and Decision Trees. Decision trees are an example of a low bias algorithm [4].

The graphical representation of the decision tree model is mostly a binary tree. Each node in a tree represents a single input variable and a split point on that variable. The leaf nodes of the tree contain an output variable (or target) which is used to make a prediction. Predictions are made by walking the splits of the tree until arriving at a leaf node and output the class value at that leaf node. Decision Trees are an important type of algorithm for predictive modeling machine learning [4]. Training data contains the output attribute – class value. Decision trees belong to supervised machine learning. They are also often accurate for a broad range of problems and do not require any special preparation for your data (numeric, categorical).

2.2 Literature Example as Inspiration

The literature contains some examples of decision trees. Most often is presented the dataset about playing tennis or golf (Yes/No) under various weather conditions (temperature, outlook, windy and humidity). The tennis-weather data is a small open data set with only 14 examples. In RapidMiner it is named Golf Dataset, whereas software WEKA has two data set: weather.nominal.arff and weather.numeric.arff [1, 5]. The resented final decision tree contains (Fig. 1) a first decision node (Outlook) with three branches (Sunny, Overcast and Rainy). Leaf node (Play) represents a classification or decision in the tree [6]. Figure 2 shows the decision tree based on training data set about playing tennis in WEKA software.

Book Master Machine Learning Algorithms [4] presents dataset with two input values of height in centimeters and weight in kilograms of some persons. The output value is sex as male or female. For demonstration purposes only, the binary classification decision tree is presented in that book (Fig. 3).

Outlook	Temp	Humidity	Windy	Play Golf
Rainy	Hot	High	False	No
Rainy	Hot	High	True	No
Overcast	Hot	High	False	Yes
Sunny	Mild	High	False	Yes
Sunny	Cool	Normal	False	Yes
Sunny	Cool	Normal	True	No
Overcast	Cool	Normal	True	Yes
Rainy	Mild	High	False	No
Rainy	Cool	Normal	False	Yes
Sunny	Mild	Normal	False	Yes
Rainy	Mild	Normal	True	Yes
Overcast	Mild	High	True	Yes
Overcast	Hot	Normal	False	Yes
Sunny	Mild	High	True	No

Fig. 1. Example of data and corresponding decision tree for play golf [6].

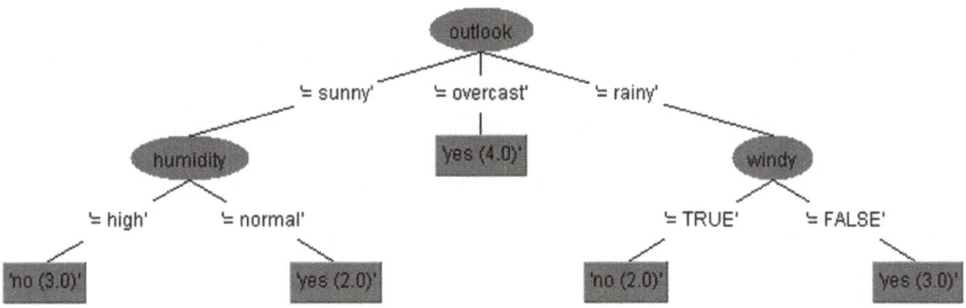

Fig. 2. Example of the decision tree for playing tennis in WEKA software [7].

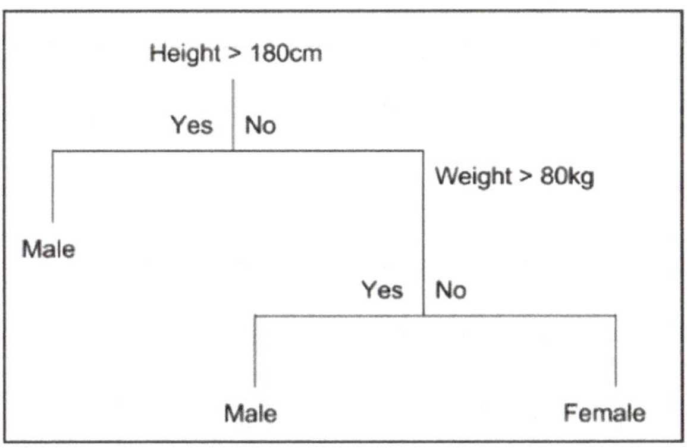

Fig. 3. Example decision tree of sex presented in the literature that is fictitious for demonstration purpose only [4].

The decision tree could be simply rewritten to the set of three rules [4]:

```
If  Height > 180 cm then Male
If  Height <= 180 cm  AND Weight > 80 kg then Male          (1)
If  Height <= 180 cm  AND Weight <= 80 kg then Female
```

This second literature example was an inspiration for practicing at university course lectured by me. The reason was that the example is understandable for students without any supplementary knowledge from a specific area. The next reason was that the gathering of training data was easy. I asked a group of students to fill the data anonymously to the questionnaire about their bodies.

3 Practical Example

The data set for practical example was collected for two years during lecturing course at university. The students were asked to fill the questionnaire. The questionnaire was prepared by Google Forms. It contains only three attributes: Weight, Height and Sex. No other information about the name or surname was collected. The group was relatively homogenous, and the age was from 20 to 26 years approximately. The collection of the data was assured at the beginning of the lecture about classification and regression decision trees. The link to the final collected data was handed on to the students. Students could freely download data and start to process them at the lecture. There was one interesting situation. One record was wrong due to mixing up weight and height by one respondent in the questionnaire. It was not hard to detect it and repair it. The correction of real data had also an educational effect. Real data very often contains mistakes, on the contrary, an official training data. The data are depicted in Table 1.

Table 1. Overview of gathered training data set for sex classification

Characteristic	Value
Number of records	58
Number of females	18
Number of males	40
Age	20–26 year
Weight: Min-Max	46–93 kg
Average weight	69.4 kg
Height: Min-Max	152–195 cm
Average height	176.3 cm

Firstly, students used the WEKA software. The WEKA implements J48 algorithm based on C4.5 [8] as an extension of the former ID3 algorithm [9, 10]. The constructed decision tree is in Fig. 4a. The condition for branching considers the value of 169 cm for person height. Totally six instances are incorrectly classified. The higher error is in the case of females where four females are classified as males. The WEKA also create a summary report with statistics (Fig. 4 b).

Secondly, students used the data mining software Orange. The algorithm for decision tree is designed in-house in Orange. There are three implementations

a)

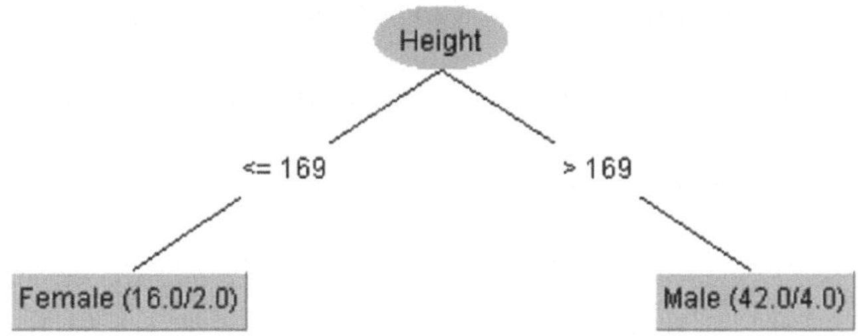

b)

```
=== Summary ===

Correctly Classified Instances          52               89.6552 %
Incorrectly Classified Instances         6               10.3448 %
Kappa statistic                          0.7507
Mean absolute error                      0.1851
Root mean squared error                  0.3043
Relative absolute error                 43.0097 %
Root relative squared error             65.7591 %
Total Number of Instances               58

=== Detailed Accuracy By Class ===

                 TP Rate  FP Rate  Precision  Recall   F-Measure  MCC      ROC Area  PRC Area  Class
                 0,778    0,050    0,875      0,778    0,824      0,753    0,864     0,750     Female
                 0,950    0,222    0,905      0,950    0,927      0,753    0,864     0,894     Male
Weighted Avg.    0,897    0,169    0,896      0,897    0,895      0,753    0,864     0,849

=== Confusion Matrix ===

  a  b   <-- classified as
 14  4 |  a = Female
  2 38 |  b = Male
```

Fig. 4. Decision tree in WEKA software (a) and text summary (b).

(TreeLearner, SklTreeLearner, SimpleTreeLearner) [11]. The processing of data is designed like a workflow in Orange software (Fig. 5). The design is easy by drag and drop nodes to the canvas. The first node File (Men_Women dataset) at the left connects the source data. The setting of the decision tree is arranged by node Tree (in the middle). The dialogue offers the parameters like "Minimum number of instances in leaves", "Limit the maximal tree depth", etc. (Fig. 5). The parameters influence the pruning of a tree [12]. The last node on the right Tree Viewer visualizes the tree in the workflow.

The decision tree shows which attribute splits the best dataset. In the case of sex classification, it is the height (Fig. 6). The attribute height is more important that the attribute weight of a person in case the sex classification of persons. Only four instances are classified incorrectly. Orange uses the hue of color to depict the homogeneity of set in each node. The light and dark blue are for Female class in the presented tree. The light red and tones of red color are for Male class. Moreover, small circular diagrams on the right edges of rectangle nodes depict the structure of the

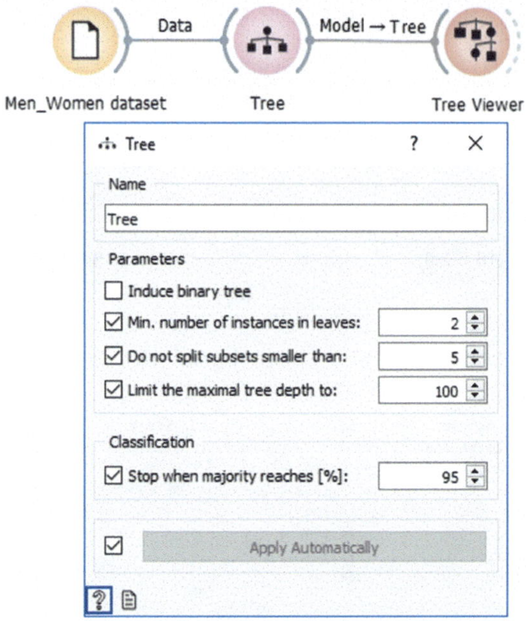

Fig. 5. Workflow for construction of a decision tree in Orange software with the dialogue of the node Tree.

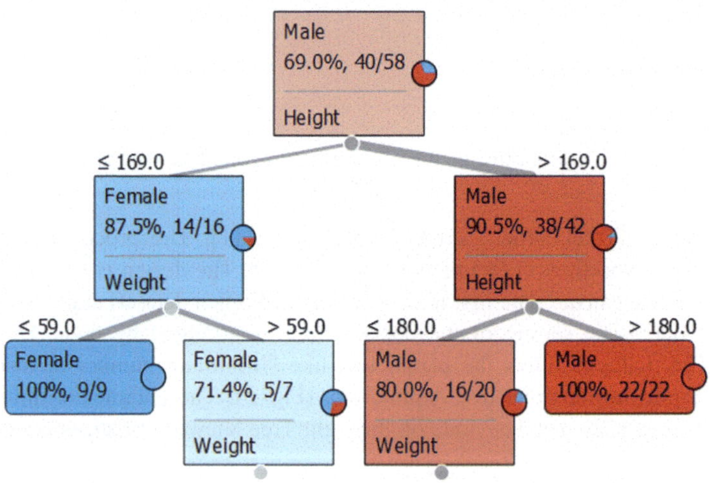

Fig. 6. Decision tree in Orange software for prediction of sex trained on 58 records.

dataset. The graphical expression of a tree is very illustrative and helps in the interpretation of a tree.

The final decision tree is possible to use for the prediction of new instances [13]. We used the newest dataset from the contemporary class of students as data for prediction. The new dataset contains 13 persons (5 women and 7 men). The constructed workflow is the only an extension of the previous workflow (Fig. 5). New node File and Predictions are added to the workflow. Figure 7 shows the workflow and output table of node Predictions. The table contains all instances with the comparison of predicted sex and original data. The tree predicts eight instances correctly and five predicts incorrectly. Four females are predicted as males and one male is predicted as female. All women have a height higher than 169 cm in incorrectly classified instances.

It is evident that the prediction depends strongly on the training data for decision tree construction. The following step is trying to construct the decision tree using all data (totally 71 instances). The decision tree contains a top node ones again the weight, but the limit is 176 cm (Fig. 8). This practical experiment with different input dataset

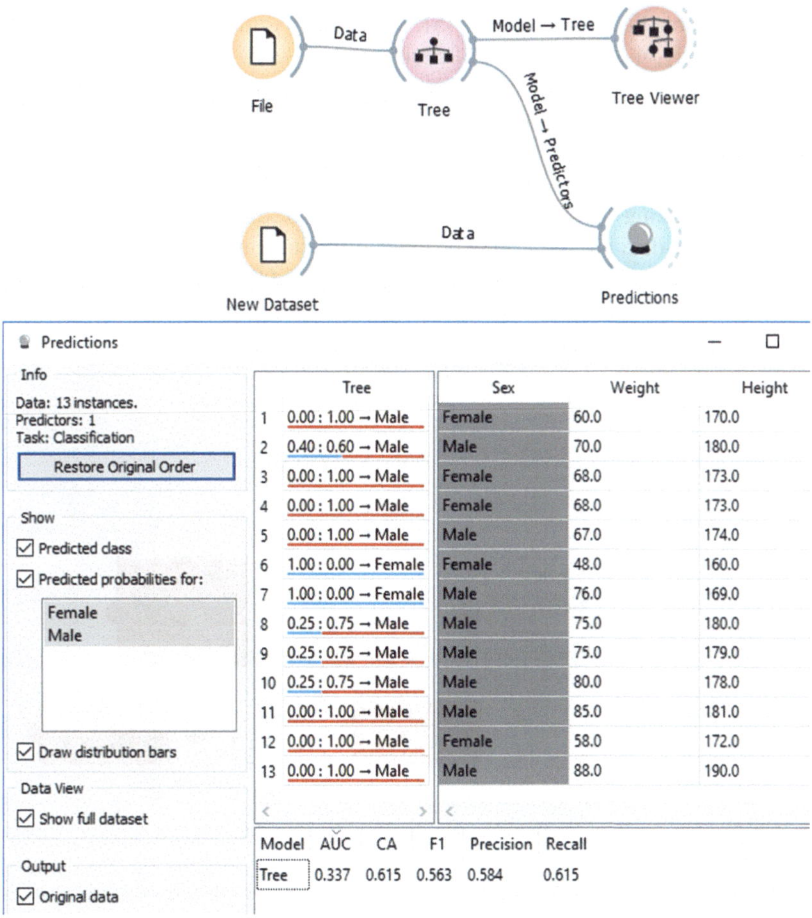

Fig. 7. Prediction of class sex in Orange software based on the pre-trained decision tree.

shows the influence of data and sensitivity on data values in practical lecture. Also, the situation of overfitting is preset.

There is space to compare and try the prediction with other methods like logistic regression [14]. The accuracy of the decision tree is 81% and logistics regression has 74%. In this case of all 71 instances, the logistic regression produces a worse result than the decision tree.

Beside this presented example also one more example is practised in lectures. The source data are also gathered locally at Palacky University. This data set is data about the dissemination of information about study branch Geoinformatics and geography between applicants for university study at secondary schools (the title is Dissemination of Study Information). The dissemination of information (leaflets, Geaudeamus fair-trade, Open Days, advertisements) is gathered by questionnaire for applicants [15]. The dataset has been systematically collected for four years, from 2016 to 2019.

The system of data collection and structure are presented at the one lecture of the Data Mining course like in the article [15]. Subsequently, the dataset is used two times in lecturing. Firstly, the construction of decision trees was used for practical prediction of the likelihood of high school students enrolling to study branch Geoinformatics and geography [16]. The second utilization is at the lecture about finding association rules [15] as another data mining method.

Both mentioned live data set (Sex Prediction and Dissemination of Study Information) are interesting for a practical demonstration of data processing in the course Data Mining. The data are more understandable for students than book examples and more attract them. On the other side, the interpretation of live data is more demandable because they may contain some outliers, errors or missing data.

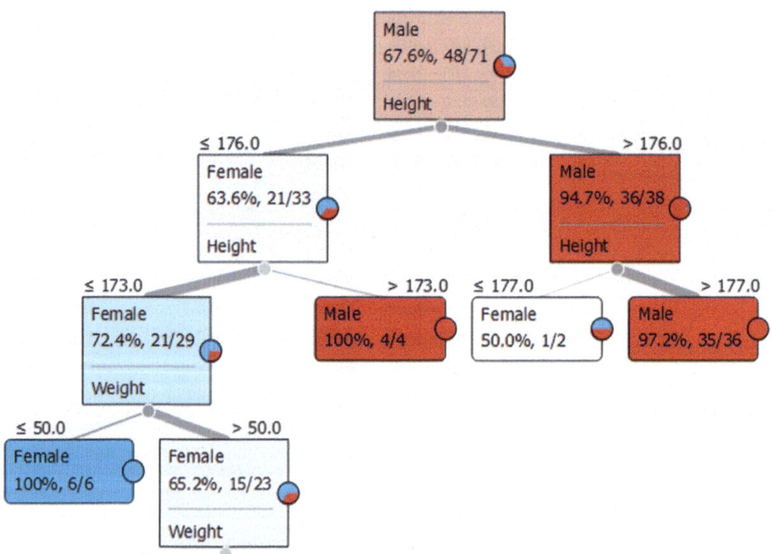

Fig. 8. Decision tree in orange for an extended data set of 71 instances.

4 Conclusion

The use of a practical example is promising. Students positively mentioned that example at written and oral exam while they finished the course. Therefore, this example will be used once again next year in the syllabus of subject Data Mining. Moreover, the data set could be easily extended by new records of new student personal measures. The model could be tested from the point of prediction and measuring model accuracy in case of new data records. The collected data also could be stored in the Moodle e-learning system for the course that runs at university. There is a potential to adapt them to one unified educational institution's datasets to serve for more courses and study branches like in other universities [17].

Moreover, there is a very good experience with Orange software for data mining tasks as educational software. It was the first experience with practicing Orange in the academic year 2019/20 at the Department of Geoinformatics as alternating for WEKA software. The use of Orange software is very intuitive and allows the quick design of workflow. The results of data mining are easily accessible for students in the process of acquiring new knowledge like using a decision tree method as one of the basic machine learning methods.

References

1. Machine Learning Group at the University of Waikato: WEKA, The workbench for machine learning. https://www.cs.waikato.ac.nz/ml/weka/. Accessed 15 Jan 2020
2. University of Ljubljana: Orange. https://orange.biolab.si/. Accessed 04 May 2019
3. Dobesova, Z.: Teaching database systems using a practical example. Earth Sci. Inf. **9**(2), 215–224 (2016)
4. Brownlee, J.: Master Machine Learning Algorithms: Discover How They Work and Implement Them From Scratch. Jason Brownlee (2016)
5. Nicôme - The Data Blog. https://gerardnico.com/data_mining/start. Accessed 10 Jan 2020
6. An Introduction to Data Science, Decision Tree - Classification. http://www.saedsayad.com/decision_tree.htm. Accessed 15 Jan 2020
7. Witten, I.H.: Data Mining. Morgan Kaufmann, Burlington (2009)
8. Quinlan, J.R.: C4.5: Programs for Machine Learning. Morgan Kaufmann Publishers Inc. (1993)
9. Quinlan, J.R.: Induction of decision trees. Mach. Learn. **1**, 81–106 (1986)
10. Pavel, P.: Metody Data Miningu, Part 2. University of Pardubice, Economic-administrative faculty, Pardubice (2014)
11. University of Ljubljana: Orange Data Mining Library, Classification https://docs.biolab.si//3/data-mining-library/reference/classification.html#classification-tree. Accessed 16 Jan 2020
12. Tan, P.N., Steinbach, M., Kumar, V.: Introduction to Data Mining, 1st edn. Addison-Wesley Longman Publishing Co., Inc., Boston (2005)
13. University of Ljubljana: Getting Started with Orange 06: Making Predictions. https://www.youtube.com/watch?v=D6zd7m2aYqU. Accessed 10 Dec 2019
14. University of Ljubljana: Getting Started with Orange 07: Model Evaluation and Scoring. https://www.youtube.com/watch?v=pYXOF0jziGM. Accessed 16 Jan 2020

15. Dobesova, Z.: Discovering association rules of information dissemination about geoinformatics university study. In: Silhavy, R. (ed.) Artificial Intelligence and Algorithms in Intelligent Systems, vol. 764, pp. 1–10. Springer, Cham (2019)
16. Dobesova, Z., Pinos, J.: Using decision trees to predict the likelihood of high school students enrolling for university studies. Advances in Intelligent Systems and Computing, vol. 859, pp. 111–119 (2019)
17. Machova, R., Komarkova, J., Lnenicka, M.: Processing of Big Educational Data in the Cloud Using Apache Hadoop. IEEE (2016)

Comparison of Discrete Autocorrelation Functions with Regards to Statistical Significance

Tomas Barot[1(✉)], Harald Burgsteiner[2], and Wolfgang Kolleritsch[2]

[1] Faculty of Education, Department of Mathematics with Didactics, University of Ostrava, Fr. Sramka 3, 709 00 Ostrava, Czech Republic
Tomas.Barot@osu.cz

[2] Institute for Digital Media Education, University College of Teacher Education Styria, Hasnerplatz 12, 8010 Graz, Austria
{Harald.Burgsteiner,Wolfgang.Kolleritsch}@phst.at

Abstract. For purposes of signal analysis, a wide spectrum of methods has appeared in the mathematical statistics. With regards to a random behavior of considered signals, the methods are based on the probability theory. In the signal processing theory, the autocorrelation functions can be considered as a suitable tool for an identification of a tightness of bindings within an analyzed signal. However, this type of analysis is based only on the descriptive approach. A comparison of autocorrelation functions based on principles of the testing hypotheses can be advantageous for an evaluation of a tightness of bindings across more than one analyzed signal. This type of comparison of signal properties has not been so widely presented yet. In favor of this aim, a proposal of a comparison of the estimated autocorrelation functions is presented in this paper. The real acoustic signals were recorded in the Campus studio of RadioIgel at University College of Teacher Education Styria. Then these signals were modified using particular types of sound effects. The proposed analysis is presented for a selected part of the acoustic signals before and after modifications.

Keywords: Signal processing · Discrete autocorrelation functions · Acoustic signals · Statistical significance · Paired tests · Hypothesis testing

1 Introduction

In the frame of the general signal processing [1, 2], a wide spectrum of methods for purposes of signals analyses are primarily based on the probability theory [3, 4]. These approaches are advantageous as well as for the processing of the random based signals [5, 6] influenced by a noise. A further utilization can be suitable for an identification of real systems, e.g. [7, 8], bounded with principles of various types of approximations [9, 10] and models in the time analyses [11–15].

In the signal processing, the important procedures are built on principles of the regression and the correlation [16, 17]. However, application of the testing hypotheses [18] has not been there so widely considered in case of analyses or comparisons of

R. Silhavy (Ed.): CSOC 2020, AISC 1226, pp. 257–266, 2020.
https://doi.org/10.1007/978-3-030-51974-2_24

properties of signals. The comparison of statistical properties with regards to the significance level [18] can be a particular form of a guarantee of formulated conclusions about signal properties, as can be seen in this paper.

Autocorrelation functions [19] have been applied as suitable tools for identification of a tightness of bindings within an analyzed signal. The continuous version of the autocorrelation function is primarily considered in the theoretical plane. The obtainment of the discrete version is bound with software computational methods. However, this type of the analysis is based only on the descriptive approach. A comparison of autocorrelation functions based on principles of the testing hypotheses can be advantageous for an evaluation of the tightness of the binding across more than one analyzed signal. This type of comparison of signal properties has not been so widely presented yet.

Obviously, research results can be simple processing with regards to the descriptive principles (e.g. correlations, autocorrelation functions). However, another utilization, concretely, the quantitative methods [18] have not been so often seen. Particularly in cases, if only descriptive descriptions are established.

In this sense, the established type of the signal analysis by autocorrelation functions is extended by a statistical complementary analysis in this paper. Concretely, statistical comparisons of pair of progresses of discrete versions of autocorrelation functions are proposed and described. A discussion of advantages of this proposal is considered for obtained results of an analysis of signals from a practice. At the Institute for Digital Media Education, signals were exported from sound records in the Campus studio of RadioIgel and IgelTV at the University College of Teacher Education Styria. Modified versions of the original signal were achieved after the application of the particular sound effects. The comparison of signals' discrete autocorrelation functions is realized in this paper with regards to the statistical significance.

2 Computation of Tightness of Binding Within Analyzed Signal

Signal processing methods [1, 2] can be classified as suitable approaches for their front advantage which is based on a stochastic character [3, 4] of analyzed variables.

With regards to time-serialized data and various types of models, e.g. [11–15], a wide spectrum of signal analyses can be provided. Their mathematical background is primarily based on correlations or regression approaches e.g. [16, 17].

Autocorrelation functions [19] can be considered as an appropriate tool for an identification of a tightness of the binding within an analyzed signal. As an estimation, this characteristic is built. The continuous version of the autocorrelation function considers a definition with the limit, as can be seen in Eq. (1) [19].

$$R_{x,x}(\tau) = \lim_{T_c \to \infty} \left(\frac{1}{2T_c} \int_{-T_c}^{T_c} x(t)x(t+\tau)dt \right) \tag{1}$$

The dynamics between signal's parts is considered on the observed time-interval T_c. Variable τ is a difference between concrete selected times in these comparisons and it is the independent variable. The important property for the progress of the autocorrelation function is the initial value equal to variance of signal values [19].

The discrete version of the autocorrelation function respects the sampling of the continuous signal with regards to the determination of a sampling period. A number of samples is N. An expression of its final computation can be seen in (2) and (3) [19].

$$R_{x,x}(k) = \frac{1}{N}\sum\nolimits_{i=1}^{N-k} x(i)x(i+k) \tag{2}$$

$$k = 0, 1, \ldots, m; m \approx \frac{1}{10}N \tag{3}$$

The final estimation for the discrete version of the autocorrelation function has m values, which can be plotted in the graphical form. It is recommended to set m as N divided by 10. Iteration parameter k then influents the final value of the discrete autocorrelation function [19].

3 Considered Techniques of Quantitative-Research Approach

The quantitative research [18] has been primarily applied in the social sciences as a main tool for an identification of effects of some applied phenomenon e.g. [20–22]. The feedback can be concluded in two types of results i.e. as descriptive and as results of the testing hypotheses. The utilization of the testing hypotheses has been also obviously used in the technical fields. Particular applications of the quantitative principles including the statistical significance can be seen e.g. in Model Predictive Control [23]. In all current available software solutions for the statistical processing, the quantitative research procedures can be suitably provided. According to a concrete field of research, the significance level is declared and software results (in the form of p-values) are then compared to this level.

Whether the tested data fulfill the normality [24], the parametrical tests are selected for the purpose of the testing the hypotheses. The parametrical tests are bound with the Gaussian probability distribution [18] of the source data. Algorithms of these methods are based on estimations of means or variances. In the opposite case, the core of the algorithms can not be based on parameters of the Gaussian probability distribution and the procedure of the resulting the p-value is depended on medians. The situation for a rejection of an assumption of the normality [24] causes the selection of the non-parametrical statistical tests for the purpose of the testing hypotheses [18].

In this paper, the main idea is based on the inclusion of principles of a paired comparison in the mathematical apparat into the signal processing application. The paired T-test or Wilcoxon test [18, 23] is then the appropriate option for the testing. In case of fulfilling the normality of data, the paired T-test should be selected. In the opposite case, the Wilcoxon non-parametrical test should be used. A definition of the k-

th zero (kH_0) and of the k-th alternative (kH_1) hypothesis has the following form (Table 1) [18]:

Table 1. Hypotheses determined for statistical paired tests

Hypothesis	Assumption
k-th H_0	There are not statistically significant differences between each pair of values in the analyzed data file on the significance level α
k-th H_1	There are statistically significant differences between each pair of values in the analyzed data file on the significance level α

Results obtained in testing hypothesis have a form of p-values. If p is greater or is equal to the significance level α, then there are not the statistical significant differences between pairs of measured values. In the opposite case, there are the statistical significant differences between pairs of values in the observed data file.

4 Proposal of Paired Comparison of Discrete Autocorrelation Functions with Regards to Statistical Significance

As a secondary analysis of time-serialized signals, quantitative methods [18] have not been so widely used. The secondary analysis can be considered as a complementary type of the description for obtained results. The statistical significance can suitably extend the autocorrelation-function analysis by a comparison. Especially, a pair of progresses of autocorrelation functions would be compared on the significance level.

As a reason for a utilization of the following proposed approach, an increase of the guarantee of the methods used can be considered by the statistical significance. Used methods based on possibilities of analyses of signals can be suitably improved using the procedures with principles of the statistical induction.

The improving the descriptive approaches can be realized by the testing hypotheses in the additional form. Considering the significance level 0.001, the guarantee of the expression value of obtained results of the signal-processing methods can be improved. On this significance level, scientific results e.g. in medical science e.g. [25, 26], in engineering practices have been often presented.

In the quantitative research, the paired tests [18] are aimed for testing the statistical significant differences across the analyzed samples of data. The necessary condition for the application of statistical paired tests is the same length of data file before and after some modification. For case of the fulfilling the normality of data, the Paired T-test is selected. In the opposite case, the Wilcoxon test is used [18].

In this paper, the hypotheses declared in Table 1, are tested on the significance level $\alpha = 0.001$. As a modified approach, the extended analysis of pairs of the discrete autocorrelation functions using the statistical paired tests is proposed. The measurement can be applied on pairs of progresses of the discrete autocorrelation functions, e.g. in case of the original and the modified signal. In Table 2, the structure of measured data is proposed in order to the comparison of both autocorrelation functions. Using symbol *, the modified version of signal is presented in the table.

Table 2. Proposal of structured of data for comparison of discrete autocorrelation functions by statistical paired tests

k	Progress of $R_{xx}(k)$ for Original Signal	Progress of $R^*_{xx}(k)$ for Modified Signal
1	$R_{x,x}(0) = \frac{1}{N}\sum_{i=1}^{N-1} x(i)x(i+0)$	$R^*_{x,x}(0) = \frac{1}{N}\sum_{i=1}^{N-1} x^*(i)x^*(i+0)$
2	$R_{x,x}(1) = \frac{1}{N}\sum_{i=1}^{N-2} x(i)x(i+1)$	$R^*_{x,x}(1) = \frac{1}{N}\sum_{i=1}^{N-2} x^*(i)x^*(i+1)$
⋮	⋮	⋮
m	$R_{x,x}(m) = \frac{1}{N}\sum_{i=1}^{N-m} x(i)x(i+m)$	$R^*_{x,x}(m) = \frac{1}{N}\sum_{i=1}^{N-m} x^*(i)x^*(i+m)$

5 Results

For purposes of the comparison of estimations of the discrete autocorrelation functions, some real acoustic signals were obtained and extracted from the radio records in the Campus studio of RadioIgel and IgelTV at Graz's Institute for Digital Media Education at the University College of Teacher Education Styria.

The application of the paired comparison of the discrete autocorrelation functions assumes the existence of an original and a modified version of a signal. The following changed signals of the original acoustic record are considered after application of the particular sound effects.

The audio record was prepared and then 4 effects were applied. For all cases a sampling frequency of 44100 Hz was applied. The original record of the radio signal had a discrete based sampled structure with 44100 values in each second window. In this paper, each 100th sample (from original and from all modifications) was considered by the sampling period $T = 0.00227$ s, i.e. with the consideration of 441 values in one second.

For the set of the first 100 initial discrete samples, a number of discrete samples N was equal to 100. Therefore, constraint m was estimated as 10. In the computation of the concrete discrete autocorrelation functions, the iteration parameter k belongs to interval 0, 1, …, 10. The original and modified 100 discrete samples are displayed with the time values in seconds in Fig. 1, 2, 3, and 4. On the vertical axis the values are dimensionless, being a sample value normalized to the range from −1 to 1.

In Table 3, this selection of discrete samples with the assignments of the concrete discrete autocorrelation functions can be seen.

The discrete autocorrelation functions (Table 3) [19] were computed using rules (2) and (3) and displayed in Fig. 5. Concrete obtained values are included in Table 4.

The proposed approach based on paired analysis of values of the discrete autocorrelation functions (Table 3) respected the structure given by Table 2. The results of

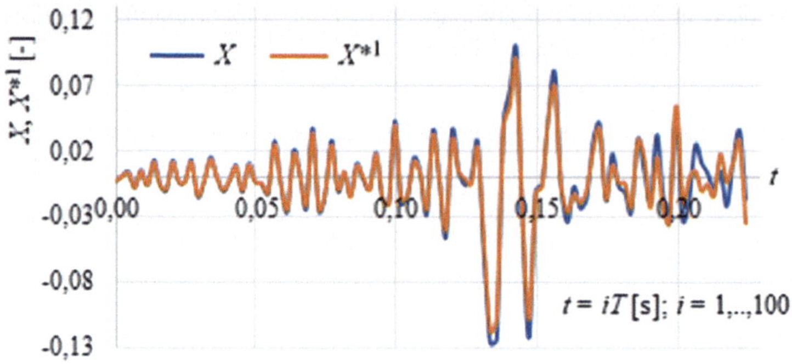

Fig. 1. Discrete samples of original measured signal X and of modified version X^{*1}

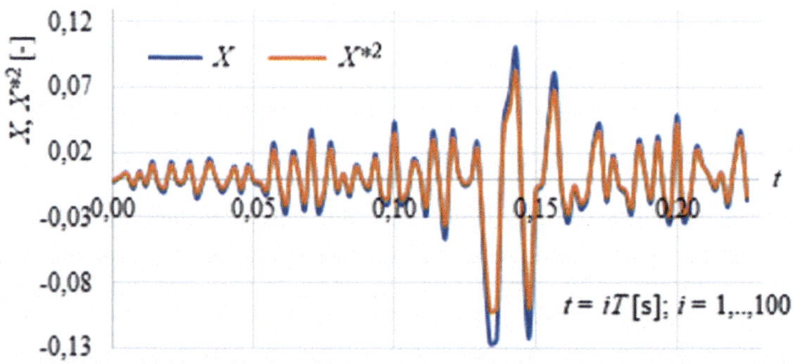

Fig. 2. Discrete samples of original measured signal X and of modified version X^{*2}

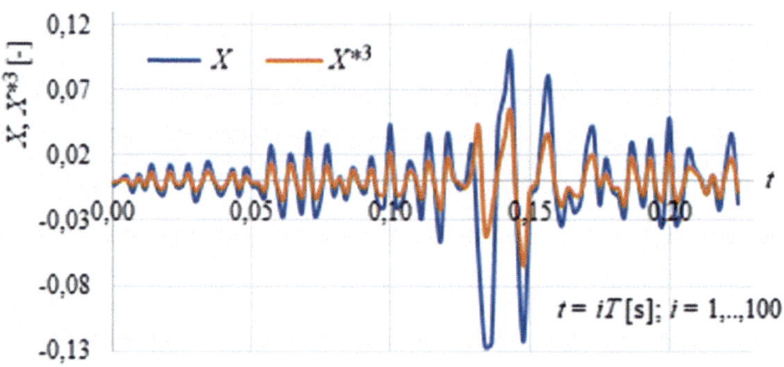

Fig. 3. Discrete samples of original measured signal X and of modified version X^{*3}

testing normality of data and providing the selected paired tests were achieved in IBM SPSS Statistics 26. The data source was like described in Table 4 with values of the discrete autocorrelation functions.

According to results in Table 5, the unfulfilling the normality of data occurred. Therefore, the non-parametrical Wilcoxon paired test was selected as an option for the

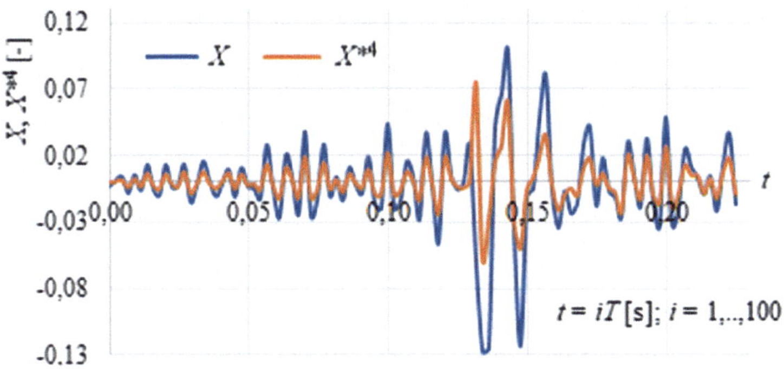

Fig. 4. Discrete samples of original measured signal X and of modified version X^{*4}

Table 3. Set of measured acoustic signals with description of their modifications

Variable	Discrete autocorrelation function	Description of signals
X	$R_{x,x}(k)$	Original record of radio broadcast
X^{*1}	$R^{*1}_{x,x}(k)$	Modified original track by echo effect
X^{*2}	$R^{*2}_{x,x}(k)$	Modified original track by compressor effect
X^{*3}	$R^{*3}_{x,x}(k)$	Modified original track by phaser effect
X^{*4}	$R^{*4}_{x,x}(k)$	Modified original track by wah-wah effect

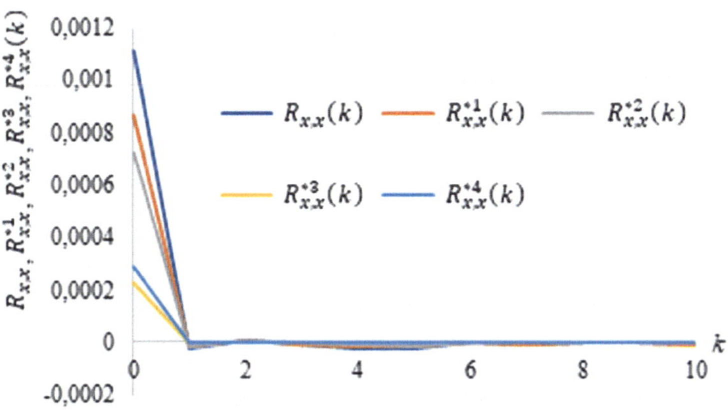

Fig. 5. Estimation of discrete autocorrelation functions of measured signal X and modified versions (after application of particular effects) X^{*1}, …, X^{*4}

testing hypotheses on existence of the statistical significant differences between progresses of the discrete autocorrelation functions. In this testing, non-existences of the statistically significant differences were identified. Advantageous of this claim is the guarantee of the statistical significance in comparison with classical approach (Table 4, Fig. 5).

Table 4. Achieved values of discrete autocorrelation functions

k	$R_{x,x}(k)$	$R^{*1}_{x,x}(k)$	$R^{*2}_{x,x}(k)$	$R^{*3}_{x,x}(k)$	$R^{*4}_{x,x}(k)$
0	$1{,}11 \times 10^{-3}$	$8{,}70 \times 10^{-4}$	$7{,}21 \times 10^{-4}$	$2{,}28 \times 10^{-4}$	$2{,}85 \times 10^{-4}$
1	$-2{,}62 \times 10^{-5}$	$-1{,}91 \times 10^{-5}$	$-1{,}35 \times 10^{-5}$	$9{,}12 \times 10^{-7}$	$2{,}62 \times 10^{-6}$
2	$6{,}69 \times 10^{-6}$	$5{,}21 \times 10^{-6}$	$3{,}54 \times 10^{-6}$	$2{,}60 \times 10^{-6}$	$2{,}35 \times 10^{-6}$
3	$-5{,}55 \times 10^{-6}$	$-5{,}70 \times 10^{-6}$	$-2{,}93 \times 10^{-6}$	$1{,}06 \times 10^{-7}$	$-6{,}34 \times 10^{-7}$
4	$-2{,}07 \times 10^{-5}$	$-1{,}44 \times 10^{-5}$	$-1{,}09 \times 10^{-5}$	$-1{,}55 \times 10^{-6}$	$-2{,}36 \times 10^{-6}$
5	$-2{,}42 \times 10^{-5}$	$-1{,}51 \times 10^{-5}$	$-1{,}27 \times 10^{-5}$	$4{,}61 \times 10^{-7}$	$2{,}45 \times 10^{-6}$
6	$-3{,}44 \times 10^{-6}$	$-2{,}35 \times 10^{-6}$	$-1{,}82 \times 10^{-6}$	$-2{,}67 \times 10^{-7}$	$-4{,}00 \times 10^{-7}$
7	$-5{,}83 \times 10^{-6}$	$-5{,}08 \times 10^{-6}$	$-2{,}85 \times 10^{-6}$	$1{,}73 \times 10^{-6}$	$1{,}02 \times 10^{-6}$
8	$-1{,}04 \times 10^{-6}$	$-5{,}95 \times 10^{7}$	$-4{,}86 \times 10^{-7}$	$6{,}20 \times 10^{-7}$	$1{,}01 \times 10^{-6}$
9	$-4{,}13 \times 10^{-6}$	$-3{,}31 \times 10^{-6}$	$-2{,}40 \times 10^{-6}$	$-3{,}92 \times 10^{-7}$	$-5{,}35 \times 10^{-7}$
10	$-7{,}35 \times 10^{-6}$	$-4{,}67 \times 10^{-6}$	$-3{,}90 \times 10^{-6}$	$-1{,}73 \times 10^{-7}$	$-3{,}56 \times 10^{-7}$

Table 5. Statistical paired comparison of discrete autocorrelation functions

Compared pairs	Testing normality of data Using Shapiro-Wilk test		Testing hypotheses on statistical sign. Differences		
Discrete Autocorrelation functions	p-value	Normality	Finally selected paired test	p-value	Conclusion of testing hypothesis kH
$R_{x,x}, R^{*1}_{x,x}$	$p = 0.000$ < 0.001	Unfulfilled	Wilcoxon test	$p = 0.182$ > 0.001	Fail to reject $1H_0$
$R_{x,x}, R^{*2}_{x,x}$	$p = 0.000$ < 0.001	Unfulfilled	Wilcoxon test	$p = 0.155$ > 0.001	Fail to reject $2H_0$
$R_{x,x}, R^{*3}_{x,x}$	$p = 0.000$ < 0.001	Unfulfilled	Wilcoxon test	$p = 0.110$ > 0.001	Fail to reject $3H_0$
$R_{x,x}, R^{*4}_{x,x}$	$p = 0.000$ < 0.001	Unfulfilled	Wilcoxon test	$p = 0.110$ > 0.001	Fail to reject $4H_0$

6 Conclusion

In this paper, pairs of the generally used descriptive based autocorrelation functions were practically compared using the Wilcoxon test with regards to the guarantee of the statistical significance. Particularly, the tightness of bounding between two signals were compared. As can be seen on obtained results, the utilization of the statistical methods of the mathematical induction can improve the guarantee of the analysis of properties of signals. This approach has not been widely seen in the routines in the signal processing. The proposed extension including the pair tests in the analysis of autocorrelation functions was utilized on real data of radio signals in the discrete variant. Signals were analyzed by the proposed procedure in the frame of the consideration of the statistical significance by the strictly defined significance level 0.001 generally used in the technical fields.

References

1. Lathi, B.P.: Signal Processing and Linear Systems. Oxford University Press, Oxford (2009)
2. Giron-Sierra, J.M.: Digital Signal Processing with Matlab Examples, Volume 1. SCT. Springer, Singapore (2017). https://doi.org/10.1007/978-981-10-2534-1
3. Tuckwell, H.C.: Elementary Applications of Probability Theory: With an Introduction to Stochastic Differential Equations. CRC Press, Boca Raton (2018)
4. Naghettini, M.: Elementary probability theory. In: Naghettini, M. (ed.) Fundamentals of Statistical Hydrology, pp. 57–97. Springer, Cham (2017). https://doi.org/10.1007/978-3-319-43561-9_3
5. Qian, F., Wu, Z., Zhang, M., et.al.: Youla parameterized adaptive vibration control against deterministic and band-limited random signals. Mech. Syst. Sig. Process. **134**(1) (2019). Elsevier. ISSN 0888-3270. https://doi.org/10.1016/j.ymssp.2019.106359
6. Shuiqing, X., Congmei, J., Yi, C., et al.: Nonuniform sampling theorems for random signals in the linear canonical transform domain. Int. J. Electron. **105**(6) (2018). Taylor and Francis. ISSN 0020-7217. https://doi.org/10.1080/00207217.2018.1426115
7. Marholt, J., Gazdos, F.: Modelling, identification and simulation of the inverted pendulum PS600. Acta Montanistica Slovaca **15**(1), 14–18 (2010). Acta Montanistica Slovaca. ISSN 1335-1788
8. Bobal, V., Kubalcik, M., Chalupa, P.: Use of MATLAB/simulink environment for identification of real system: case study. In: 33rd International ECMS Conference on Modelling and Simulation, ECMS 2019, pp. 138–144. European Council for Modelling and Simulation (2019). https://doi.org/10.7148/2019
9. Xue, X., Zhang, X., Feng, X., et al.: Robust subspace clustering based on non-convex low-rank approximation and adaptive kernel. Inf. Sci. **513**, 190–205 (2020). https://doi.org/10.1016/j.ins.2019.10.058. Elsevier. ISSN 0020-0255
10. Schmidt, W.M., Summerer, L.: Simultaneous approximation to two reals: bounds for the second successive minimum. Mathematika **63**(3), 1136–1151 (2017). https://doi.org/10.1112/s0025579317000274. Cambridge University Press. ISSN 0025-5793
11. Jurek, M., Wagnerova, R.: Mathematical model of real CNC machine. In: 20th International Carpathian Control Conference, ICCC 2019 (2019). Institute of Electrical and Electronics Engineers. https://doi.org/10.1109/carpathiancc.2019.8766023
12. Kral, E., Capek, P.: A comparison of the performance of a mathematic expression parser in heat load modelling. Int. J. Math. Comput. Simul. **9**, 9–13 (2015). North Atlantic University Union. ISSN 1998-0159
13. Ruzickova, M., Dzhalladova, I., Laitochova, J., et al.: Solution to a stochastic pursuit model using moment equations. Discrete Continuous Dyn. Syst. Ser. B **23**(1), 473–485 (2018). https://doi.org/10.3934/dcdsb.2018032. American Institute of Mathematical Sciences (2018). ISSN 1531-3492
14. Chramcov, B., Balate, J.: Time series analysis of heat demand. In: 22nd European Conference on Modelling and Simulation, ECMS 2008, pp. 213–217. European Council for Modelling and Simulation (2008). https://doi.org/10.7148/2008-0213

15. Kozel, D.: Time models in myth and music of the 20th century. Muzikoloski Zbornik. **55**(1), 177–194 (2018). https://doi.org/10.4312/mz.55.1.177-194. University of Ljubljana. ISSN 0580-373X
16. Koukouvinos, C., Jbilou, K., Mitrouli, M., Turek, O.: An eigenvalue approach for estimating the generalized cross validation function for correlated matrices. Electron. J. Linear Algebra **35**(1), 482–496 (2019). https://doi.org/10.13001/1081-3810.4090. International Linear Algebra Society. ISSN 1081-3810
17. Hanus, R.: Time delay estimation of random signals using cross-correlation with Hilbert Transform. Meas. J. Int. Meas. Confederation **146**, 792–799 (2019). https://doi.org/10.1016/j.measurement.2019.07.014. Elsevier. ISSN 0263-2241
18. Kitchenham, B., et al.: Robust statistical methods for empirical software engineering. Empir. Softw. Eng. **22**(2), 579–630 (2016). https://doi.org/10.1007/s10664-016-9437-5
19. Hunt, N.H.: Autocorrelation function, mutual information, and correlation dimension. In: Nonlinear Analysis for Human Movement Variability, pp. 301–342. CRC Press (2016)
20. Lackova, L.: Protective factors of university students. The New Educational Review. **38**(4), 273–286 (2014). Adam Marszalek Publishing House. ISSN 1732-6729
21. Dobesova, Z., Lackova, L., Malcik, M.: Perception and cognition of different types of graphic notations as a source of information in applied informatics. In: 10th Jubilee International Symposium on Applied Computational Intelligence and Informatics (SACI), pp. 209–214. IEEE (2015)
22. Korenova, L., Barot, T., Krpec, R., et al.: Analysis of applied mobile technologies in university math education. In: 11th Annual International Conference on Education and New Learning Technologies, EDULEARN19. IATED (2019). https://doi.org/10.21125/edulearn.2019.1619
23. Barot, T., Krpec, R., Kubalcik, M.: Applied quadratic programming with principles of statistical paired tests. In: Silhavy, R., Silhavy, P., Prokopova, Z. (eds.) CoMeSySo 2019 2019. AISC, vol. 1047, pp. 278–287. Springer, Cham (2019). https://doi.org/10.1007/978-3-030-31362-3_27
24. Vaclavik, M., Sikorova, Z., Cervenkova, I.: Analysis of independences of normality on sample size with regards to reliability. In: Silhavy, R., Silhavy, P., Prokopova, Z. (eds.) CoMeSySo 2019 2019. AISC, vol. 1047, pp. 304–311. Springer, Cham (2019). https://doi.org/10.1007/978-3-030-31362-3_29
25. Svoboda, Z., Honzikova, L., Janura, M., et al.: Kinematic gait analysis in children with valgus deformity of the hindfoot. Acta of Bioeng. Biomech. **16**(3), 89–93 (2014). Institute of Machine Design and Operation. ISSN 1509-409X

Static Compensator for Nonsquare Systems – Application Example

Daniel Honc(✉), František Dušek, and Jan Merta

University of Pardubice, Pardubice 532 10, Czech Republic
daniel.honc@upce.cz

Abstract. The paper deals with the decentralized control of multivariable systems with the number of manipulated variables greater than the number of controlled variables. Proposed static compensator ensures automatic creation of input/output pairs for the simple control loops. The compensator provides steady-state autonomy and unit gain of the controlled system. Steady-state gain matrix and vector of the offsets are enough information for the compensator design. Laboratory example is presented to demonstrate innovative compensator design and its application.

Keywords: Multivariable system · Decentralized control · Static compensator

1 Introduction

One set-point change and subsequent manipulated variable change acts as a disturbance to the other control loops in case of Multi-Input Multi-Output (MIMO) system control. Two approaches for controlling MIMO systems are possible - see Fig. 1.

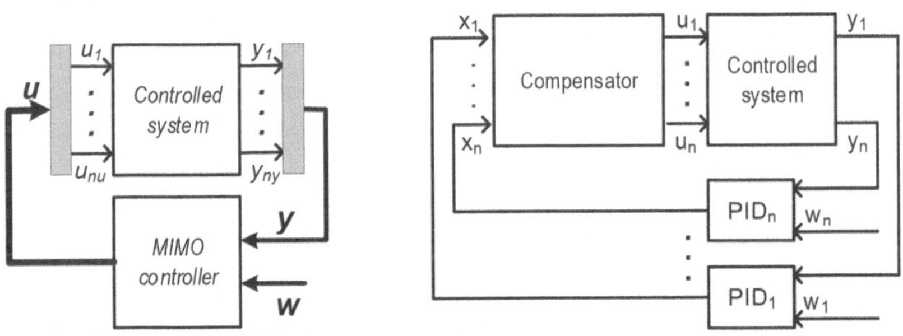

Fig. 1. Multivariable controller and decentralized control.

Multivariable controller calculates the values of all manipulated variables simultaneously from all controlled variables and set-points. Special block of the compensator can be placed in front of the controlled system to ensure the autonomy of the controlled system in case of the decentralized control. Dynamical autonomy of square systems means that changing in input u_i will only cause a change to the corresponding output y_i.

R. Silhavy (Ed.): CSOC 2020, AISC 1226, pp. 267–273, 2020.
https://doi.org/10.1007/978-3-030-51974-2_25

It is necessary to know linear dynamical model of the controlled system. Dynamic compensator ensuring autonomy in transient states can be complex and not physically feasible [1, 2]. Because of this only limited autonomy compensators are designed. For the practical applications of decentralized control even without the compensator the input/output pairing and control loop tuning can be challenging task.

The paper deals with the specific type of autonomous control of the systems with more inputs than outputs. Steady state autonomy is realized with innovative static compensator - see Fig. 2.

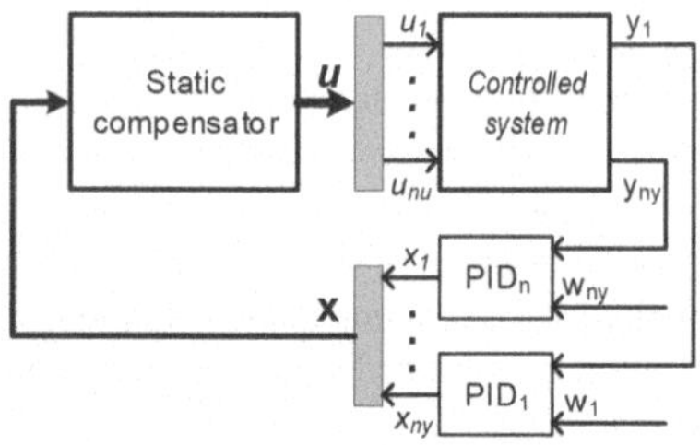

Fig. 2. Decentralized control with static compensator.

The introduction of new manipulated variables x_i assigned to the corresponding controlled variables y_i solves the problem of creating the control loops. The new control variables (vector **x**) are recalculated by the static compensator to the system inputs (vector **u**). The gain matrix between the new control variables and the system outputs (vector **y**) is unit matrix - in steady state **y** = **x**. Some disadvantage can be that the use of a static compensator may make worse the frequency properties of the closed loop control circuit [4–6]. The proposed static compensator design is a general procedure for the application of the multivariable Split Range method [7]. Controlled process gain matrix estimation and compensator design was published in [3].

In the paper gain matrix estimation in Excel and laboratory application to a process with four inputs and three outputs and compact control system AMiNi4DS is presented.

2 Gain Matrix Estimation

Experimental data - steady-state output values for different input combinations are needed to estimate the gain matrix of the controlled system. The number of the samples must be greater than the number of the inputs. The method is described in detail in [3].

Steady-state model of the controlled process with inputs **u** (**u**: $n_u \times 1$) and outputs **y** (**y**: $n_y \times 1$) has a form

$$\mathbf{y} = \mathbf{Z}.\mathbf{u} + \mathbf{z}_0 \tag{1}$$

where **Z** is the gain matrix (**Z**: $n_y \times n_u$) and

$\mathbf{z}_0$ is an offset vector ($\mathbf{z}_0$: $n_y \times 1$).

The gain matrix **Z** and the vector of the offsets $\mathbf{z}_0$ is estimated by least square matrix method from N samples of the steady state input/outputs pairs of the controlled system. We arrange the data into a matrix of output values **Y** (**Y**: $N \times n_y$) and extended matrix of input values $\mathbf{U}_r$ ($\mathbf{U}_r$: $N \times n_u + 1$) as follows

$$\mathbf{Y} = \begin{bmatrix} \mathbf{y}^T(1) \\ \vdots \\ \mathbf{y}^T(N) \end{bmatrix} = \begin{bmatrix} y_1(1) & \cdots & y_{ny}(1) \\ \vdots & \ddots & \vdots \\ y_1(N) & \cdots & y_{ny}(N) \end{bmatrix},$$
$$\mathbf{U}_r = \begin{bmatrix} \mathbf{u}^T(1) & 1 \\ \vdots & \vdots \\ \mathbf{u}^T(N) & 1 \end{bmatrix} = \begin{bmatrix} u_1(1) & \cdots & u_{nu}(1) & 1 \\ \vdots & \ddots & \vdots & \vdots \\ u_1(N) & \cdots & u_{nu}(N) & 1 \end{bmatrix}$$

Solution - the extended $\mathbf{Z}_r$ gain matrix is obtained as

$$\mathbf{Z}_r = \mathbf{Y}^T.\mathbf{U}_r.\left(\mathbf{U}_r^T.\mathbf{U}_r\right)^{-1} \tag{2}$$

The gain matrix **Z** is the first n_u columns and the vector of the offsets $\mathbf{z}_0$ is the last column of the extended matrix $\mathbf{Z}_r$.

3 Static Compensator Design

Compensator together with the original system create new system with the same number of new inputs **x** (**x**: $n_y \times n_y$) as the outputs **y**. The gain matrix of the new system is unit matrix **I** (**I**: $n_y \times n_y$)

$$\mathbf{y}_\infty = \mathbf{Z}.\mathbf{u} + \mathbf{z}_0, \quad \mathbf{u} = \mathbf{K}.\mathbf{x}, \quad \mathbf{y}_\infty = \mathbf{x} \tag{3}$$

In case of the systems, where the number of inputs is greater than the number of outputs, there are infinity combinations of inputs leading to the desired outputs. It is possible to introduce an additional requirement (preferred, desired vector of inputs $\mathbf{u}_w$). Compensator calculation is done by minimizing the deviation of the input vector **u** from the preferred input vector $\mathbf{u}_w$, with the constraint respecting the relationship between inputs and outputs $\mathbf{y} = \mathbf{Z}.\mathbf{u} + \mathbf{z}_0$ and the requirement of the unit gain $\mathbf{y}_\infty = \mathbf{I}.\mathbf{x}$

$$\begin{aligned} &\min_{\mathbf{u}} J(\mathbf{u}), \quad J(\mathbf{u}) = (\mathbf{u} - \mathbf{u}_w)^T.\mathbf{M}.(\mathbf{u} - \mathbf{u}_w) \\ &\text{subject to}: \ \mathbf{y}_\infty = \mathbf{Z}.\mathbf{u} + \mathbf{z}_0 = \mathbf{I}.\mathbf{x} \end{aligned} \tag{4}$$

where the optional matrix **M** (**M**: $n_u \times n_u$) includes possible weighting requirements for the individual inputs or their combinations (it can be chosen as an unit matrix). The matrix **M** and the vector of the preferred inputs $\mathbf{u}_w$ allows to include additional requirements to the control - in addition to the set-points following.

The minimization of the cost function (5) is described in detail in [3] and the solution is

$$\begin{bmatrix} \mathbf{u} \\ \boldsymbol{\lambda} \end{bmatrix} = \begin{bmatrix} \mathbf{M}^T & \frac{1}{2}\mathbf{Z}^T \\ \mathbf{Z} & 0 \end{bmatrix}^{-1} \cdot \begin{bmatrix} \mathbf{M}^T.\mathbf{u}_w \\ \hline \mathbf{y}_\infty - \mathbf{z}_0 \end{bmatrix} \tag{5}$$

Let us use the equality $\mathbf{y}_\infty = \mathbf{x}$, introduce the notation $\mathbf{S}$ ($\mathbf{S}$: $n_u + n_y \times n_u + n_y$) for the inverse matrix and denote two submatrices $\mathbf{R}$ ($\mathbf{R}$: $n_u \times n_u$) and $\mathbf{K}$ ($\mathbf{K}$: $n_u \times n_y$) as indicated in the equation

$$\begin{bmatrix} \mathbf{u} \\ \boldsymbol{\lambda} \end{bmatrix} = \mathbf{S}.\begin{bmatrix} \mathbf{M}^T.\mathbf{u}_w \\ \mathbf{x} - \mathbf{z}_0 \end{bmatrix}, \quad \mathbf{S} = \begin{bmatrix} \mathbf{R} & \mathrm{K} \\ \mathbf{X}_1 & \mathbf{X}_2 \end{bmatrix} \tag{6}$$

Since we are interested in the solution with respect to the vector $\mathbf{u}$, we will only use the submatrices $\mathbf{R}$ and $\mathbf{K}$. The static compensator can be written in the matrix form as

$$\mathbf{u} = \mathbf{K}.(\mathbf{x} - \mathbf{z}_0) + \mathbf{R}.\mathbf{M}^T.\mathbf{u}_w \tag{7}$$

and its block scheme is in Fig. 3.

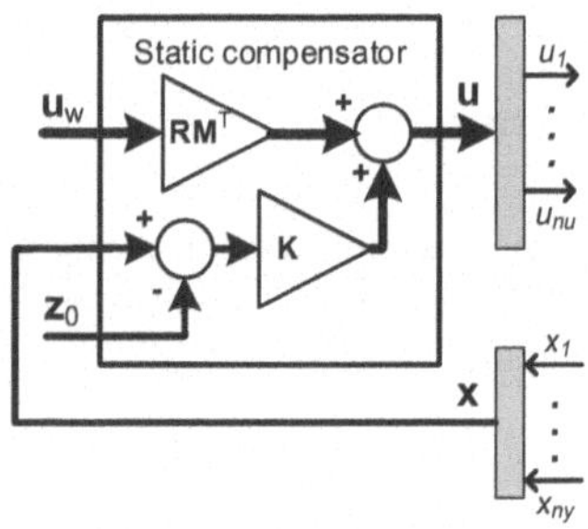

Fig. 3. Static compensator.

4 Laboratory Application

The calculation of the static compensator in Microsoft Excel and the program implementation with standard PID controllers in the compact control system AMiNi4DS is presented on example of a system with four inputs and three outputs. 15 steady-state input/output pairs were measured for the gain matrix estimation. A hardware-based simulator of the dynamic systems was used as a controlled system for the application.

4.1 Gain Matrix Estimation in Excel

Example of the static compensator calculation in Excel is shows in Fig. 4. The progress of the solution is indicated by arrows with the numbers of each calculation step. The measured values in the Measured data area are supplemented with a column of ones. Auxiliary matrices are calculated in the steps 1 to 3 and calculation of the extended gain

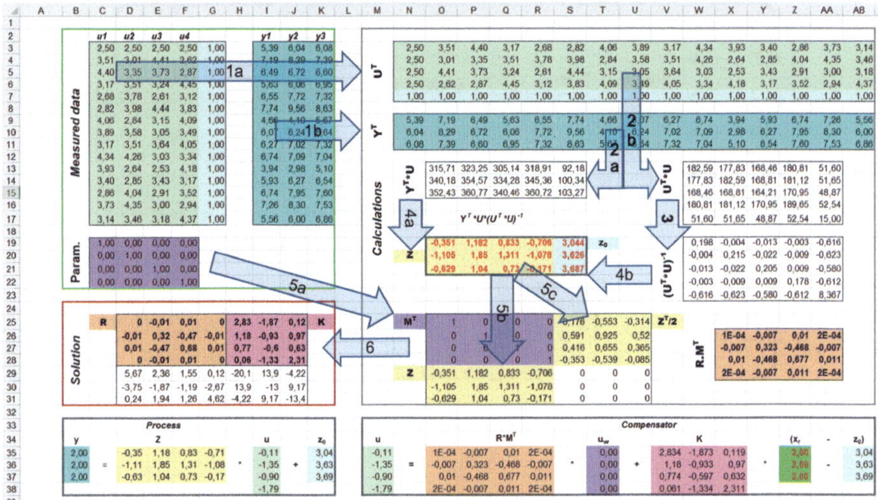

Fig. 4. Calculations in excel.

matrix $\mathbf{Z}_r$ according to the Eq. (2) is completed the in the step 4. This matrix contains the process gains **Z** and the offsets $\mathbf{z}_0$. The offsets $\mathbf{z}_0$ can be updated at any time (process outputs for zero inputs) without having to recalculate the compensator matrices which depends only on the gain matrix **Z**.

Auxiliary matrix **S** for the calculation of the compensator in the step 5 according to the Eq. (6) is based on the gain matrix **Z** of the process and the chosen weighting matrix **M** (area Param.). In the last step 6 the compensator matrices **K**, **R** and the $\mathbf{RM}^T$ matrix for the implementation of the compensator is calculated. The static compensator represents a matrix Eq. (7) for the recalculation of the vector **x** (3 outputs of the PID controllers) to vector **u** (4 inputs of the controlled system) respecting the system offset $\mathbf{z}_0$ and the vector $\mathbf{u}_w$ of the preferred input vector.

4.2 Application in the Control System AMiNi4DS

An application of the proposed compensator is presented in the compact control system AMiNi4DS. This control system supports matrix multiplication and provides also digital PID controllers so the control including a static compensator is easy to implement.

Static compensator and three PID controllers were realized programmatically using the design environment supplied with the control system according to the scheme in Fig. 5. The $\mathbf{R}_m$ and **K** matrices used by the compensator were manually filled with the values calculated in Excel. The offset vector $\mathbf{z}_0$ and the preferred input vector $\mathbf{u}_w$ can be changed from the control panel.

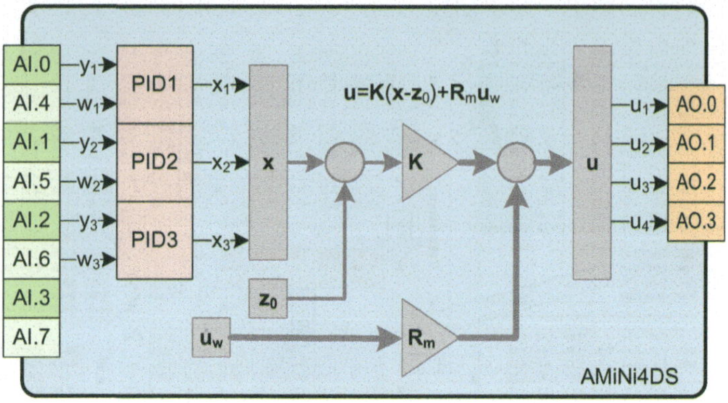

Fig. 5. Controllers and compensator in AMiNi4DS.

Controllers and the static compensators are calculated in one process. Part of the program code in the form of structured text is shown in Fig. 6. Controller variables of the controllers (standard PID module) are directly connected to the analog inputs AI.0-2 and the set-points to inputs AI.4-6. Units for all inputs and outputs variables of the controllers Volts. The outputs of the PID controllers are merged into the vector **x**. Vector **x** is recalculated according to the matrix equation of the static compensator to the vector **u** and split to the individual analog outputs AO.0-3.

```
// input vector actualization AI[0-7]
// reading AI0 y1
AnIn #AI00_0, AI[0,0], 10.000, 0.000, 10.000, 0.000, 10.000
…
// reading AI4 w1
AnIn #AI00_4, AI[4,0], 10.000, 0.000, 10.000, 0.000, 10.000
…
// 3 digital PID controllers
// PID no.1 w:AI.4,y:AI.0,u:x
PID AI[4,0], AI[0,0], x, Mod1, Par1
Let xr[0,0]=x
…
// static compensator
MtxSub x3, xr, z0          // x3=xr-z0
MtxMul u4b, K, x3          // u4b=K*x3=K*(xr-z0)
MtxMul u4a, Rm, uw         // u4a=Rm*uw
MtxAdd u, u4a, u4b         // u=u4a+u4b=Rm*uw+K*(xr-z0)
// output vector actualization and saturation
// writing u
AnOut #AO00_0, u[0,0], 10.000, 0.000, 10.000, 0.000, 10.000
…
```

Fig. 6. Code in AMiNi4DS.

5 Conclusions

The paper describes design of an innovative simple and practically usable multivariable decentralized control system for the processes with higher number of manipulated variables then the controlled variables. The method does not require knowledge of the dynamical mathematical model of the controlled system. All necessary information for the design can be obtained by simple evaluation of the steady-state experimental data. It is shown how to estimate the process gain matrix and how to calculate the matrices of the static compensator.

The use of the static compensator does not eliminate the dynamical coupling of the individual control loops, nor does it ensure control stability. The main benefit is a general solution for the pairing of the manipulated and controlled variables for individual control loops in the case of multivariable systems. In addition, the use of the static compensator provides unit gain of each pair, the invariance in steady state and suppression of the offsets. These features make tuning of individual PID controllers easier.

Acknowledgment. This research was supported by SGS grant at Faculty of Electrical Engineering and Informatics, University of Pardubice.

References

1. Nordfeldt, P., Hägglung, T.: Decoupler and PID controller design of TITO systems. J. Process Control **10**, 923–936 (2006)
2. Waller, M., Waller, J.B., Waller, K.V.: Decoupling revisited. Ind. Eng. Chem. Res. **42**, 4575–4577 (2003)
3. Dušek, F., Honc, D., Merta, J.: Static compensator for decentralized control of nonsquare systems. Adv. Intell. Syst. Comput. **1047**, 1–6 (2019)
4. Skogestad, S., Postlethwaite, I.: Multivariable Feedback Control: Analysis and Design, 2nd repr. ed., vol. xiv, p. 574. Wiley, Chichester (2008)
5. Lee, J., Kim, D.H., Edgar, T.F.: Static decouplers for control of multivariable processes. AIChE J. **51**(10), 2712–2720 (2005)
6. Skogestad, S., Morari, M.: Implications of large RGA elements on control elements. Ind. Eng. Chem. Fundam. **26**, 2323–2330 (1987)
7. Bequette, B.W.: Ratio, selective, and split-range control, Chap. 12. In: Process Control: Modeling, Design, and Simulation, vol. xxix, 769 p. Prentice Hall PTR, Upper Saddle River, N.J. (2003)

A Review of the Determinant Factors of Technology Adoption

Kayode Emmanuel Oyetade(✉), Tranos Zuva, and Anneke Harmse

Department of Information and Communication Technology, Vaal University of Technology, Vaal, South Africa
kayoyetade@gmail.com, {tranosz,anneke}@vut.ac.za

Abstract. Technology adoption has been researched extensively in the literature. New or modified models are emerging with different variables but not much re-search has been done to find out the variable that has been consistent in most model and which ones has been added to modify models. Eighty papers published in seventy-three journals and seven conference proceedings between the years 1992–2019 were reviewed. This study was conducted with three objectives in mind (1) to highlight the mostly used factors from the reviewed literature, (2) to investigate technology adoption factors that were found significant and non-significant from an analytic point of view, (3) to identify factors to be used as core factors of a generic adoption model. Results of the study revealed that the identified factors are mostly derived from TRA, TAM, TPB, MPCU, DOI, SCT, UTAUT and its extensions.

Keywords: Technology adoption · Content analysis · Factors · ICT

1 Introduction

The prevalence of Information and Communication Technology (ICT) in all aspects of human life is making a potential impact and changing the way businesses operate, communicate, access information and services [11, 42]. This is evident by advances in computing and new technologies with additional immersive experience content which include Robotics, Artificial intelligence, Internet of Things, Blockchain, Drones, 3-D printing, Augmented Reality, Virtual Reality, and Mixed Reality etc. These digital technologies are believed will become the computing platforms of the future [72] changing how we design and create and experience everything from retail to agriculture, health industries, engineering, 21st century education, entertainment which will affect and disrupt current relevant technologies [72, 76]..

Despite the possibilities of innovative ways and approaches to commit to novel technologies, a digital, positive attitude and an ability to cope with this technology are required by people using this technology resulting in a renewed emphasis on the process of technology adoption [94, 95]. Moreover, the real benefits of deciding on the usage or non-usage of technology are always subject to change based on the adoption of other similar technologies.

Technology adoption is gaining considerable attention in recent times as relevant stakeholders (researchers, organization, etc.) are exploring factors that influence an

R. Silhavy (Ed.): CSOC 2020, AISC 1226, pp. 274–286, 2020.
https://doi.org/10.1007/978-3-030-51974-2_26

individual's adoption of technology. Proposed theories and models include IDT, TAM, TRA, and UTAUT etc. which have been used to explain users' acceptance and rejection of technologies.

Technology adoption can be described as an individual's willingness to use an innovation while working on accomplishing their daily tasks. It is therefore important to investigate core factors that influence an individual's technology acceptance.

Reviewing the studies on "technology adoption" should reveal mostly used significant factors and provide us insight into explaining behaviors that predicts technology adoption. Moreover, revealing these factors will guide future re-searchers in creating models for innovation, diffusion and adoption when con-ducting their research studies. On this basis, 80 papers that derived their results from models and framework of Technology Adoption were analyzed with three major objectives: (1) to highlight the most used factors of technology adoption from reviewed literature; (2) to investigate Technology adoption factors that were found significant and non-significant in the reviewed literature from an analytic point of view; (3) to identify factors that will be used to develop and test a generic model.

2 Theoretical Framework

Technology adoption is the choice to acquire and fully utilize innovation as the best available action [71]. It examines the choices an individual makes to accept or reject a particular innovation and the extent to which that innovation is integrated into the appropriate context [78]. Theories and models proposed to explain technology acceptance or rejection include but not limited to TRA, TPB, TAM, TAM2, TAM3, TOE, DOI, Decomposed –TPB, and the UTAUT [3, 5, 30, 31, 60, 71, 78, 82, 91].

Innovation Diffusion Theory (IDT): Developed by [71] who theorize that an individual accepts or rejects an innovation temporarily or permanently based on his/her prior experience of that innovation [64, 65]. This individual's perception of adopting new technologies is influenced by the five attributes: relative ad-vantage (RA), complexity, compatibility, observability, trialability [73]. This innovation is communicated through certain channels over time among members of a social system known as diffusion [71]. According to [71], RA is an innovation's degree of superiority and attractiveness over existing idea it is replacing; compatibility is an innovation's degree of consistency with the existing potentials adopters, past experiences, and values; complexity is the degree of difficulty to understand and use an innovation; trialability is an innovation's degree to which it can be tested out on a limited basis; and observability is an innovation's degree of its visibility to others.

Theory of Reasoned Action (TRA): Formulated to predict a wide range of behaviors [36, 37]. TRA posits an individual's behavior is determined by his/her behavioral intention and beliefs for that behavior [59, 77, 78]. TRA helps to gain insight in explaining the relation between our intentions to perform and our attitudes and beliefs. This intention is determined by two factors, attitude towards behavior and subjective norm.

Theory of Planned Behavior (TPB): An extension of TRA helps us understand the potential user's behavioral intention to make use of new technologies. It posits that specific striking beliefs influence behavioral intentions and subsequent behavior [5]. This intention is influenced by attitude toward behavior, subjective norm and perceived behavioral control (PBC). PBC is an individual's perception of one's ability to act out a given behavior easily [5, 18, 60].

Technology Acceptance Model (TAM): The Technology Acceptance Model (TAM): Influenced by TRA and TPB [31], helps to understand and explain an individual's behavioral intention to adopt new technologies [30, 31, 64]. According to TAM, two factors determine the intention to adopt new technologies which are Perceived Usefulness (PU) and Perceived Ease of Use (PEOU). PU is an individual's perception of the usefulness of technology and PEOU is the degree of an individual's perception a given system is free of effort [34].

Technology Acceptance Model 2 (TAM 2): An extension of TAM which posits that PU and PEOU, facilitate the effects of external factors, such as system characteristics, training, development process, on intention to use the system. TAM2 includes subjective norms as additional determinants of intention to use technology [9].

The Unified Theory of Acceptance and Use of Technology (UTAUT): Used to explain the acceptance of innovations and develop a theoretical understanding behind the introduction, diffusion and acceptance of technology [91]. A combination of several theories and models to identify the behavioral intention (age, gender, voluntariness, and experience) among variables (effort expectancy, performance expectancy, facilitating conditions, and social influence assumed to influence adoption and use [91].

3 Method

This study reviewed eighty papers from different journals and conference proceedings to retrieve factors related to Technology Adoption. Content analysis by an inductive approach was used to analyze the identified factors making this study adds value to the family of related literature on technology adoption research.

3.1 Scope of the Study

This study reviewed journals and conference papers that founded their studies on different theoretical constructs of Technology Adoption. We used the keyword "Technology Adoption as a Review", 'Technology Adoption Related Studies' as the search criteria from the ACM Digital, Google Library, ResearchGate, Google Scholar, PsycINFO, and IEEE Xplore database. A list of 120 papers was found from the search. We then applied some selection criteria to refine the articles that apply to the scope of this study. Thereafter, a qualitative content analysis through an inductive approach was done to analyze further.

3.2 Article Selection

These articles were then reviewed to check and eliminate for duplicates and reviews that do not fit into the scope of this study. References of the included articles were checked for other articles eligible for this review (snowball method). We arrived at 80 selected articles listed according to their author name, publication date, and article name which were screened based on the inclusion criteria presented below.

From the 80 selected articles, data obtained from them was entered into a data extraction form (excel) using the variable name, author name, frequency of variable in the reviewed literature, level of significance and non-significance of the identified variable. Content analysis was applied to derive relations and concepts that will explain the collected data by bringing together similar data within a framework of certain factors. We then interpret these factors by arranging them in a way that readers can understand through the systematic classification process of coding and identifying themes or patterns [46, 98].

Once all the factors were identified, a table was created to list them according to their significance and non-significance and observed frequencies in the reviewed literature. Factors similar in definition or close to each other were grouped under a new name. We observed the frequencies for these new groups before ordering according to the level of significance to arrive at our list of identified factors. The numerical transformation was done on the new list of factors that were then listed based on the ratio of the significant variable to their frequency. Then, the most significant and widely used per weight average was determined to arrive at our choice of selection with a value of 0.8 and above. Finally, the obtained findings were interpreted.

3.3 Applied Selection Criteria

- The research aimed at investigating factors that influence the adoption of technology
- All variables that are significant and non-significant variables in the researched study in literatures
- Variables that are measured with the same instrument/questionnaire are grouped as one factor. For example self-efficacy, computer self-efficacy, online self-efficacy, and perceived self-efficacy were grouped as self-efficacy.
- Factors with an index of 0.0 or no significance irrespective of their frequency in literature were considered.
- Factors with only one significance irrespective of their frequency in literature were considered.
- Factors with 2–3 significance irrespective of their frequency in literature were considered.
- Those with the index of 0.8 and greater using the total frequency of variables that were been significant in literature over the total frequency of the variable in literature were used as our generic model of technology adoption.

4 Result

From the total number of 153 factors identified from the literature, Table 1 illustrate factors with no significance or has a variable effect of 0.0 irrespective of their frequency in the reviewed literature. The study grouped this factors with 0 significance and remove 82 factors. Furthermore, the study grouped and removed 44 variables with only one significance irrespective of its frequency in the study reviewed to arrive at 27 factors.

Table 1. Factors with no significance or variable effect of 0.0

Author	Factors
[4, 13, 14, 25–27, 29, 39, 47, 51, 52, 65, 70, 75, 80, 81, 83, 84, 87, 99–101, 103]	Flow experience, intention, intention to recommend, awareness of mobile banking services, support

Factors that suit the criteria of 2–3 significance in the researched study are presented in Table 2 irrespective of their frequency in the reviewed literature. This allowed the study to group and remove 15 factors.

Table 2. Factors with 2 and 3 significance

Authors	Factors
[1, 4, 6–8, 14, 16, 17, 21, 25–27, 29, 39, 40, 47, 51–54, 56, 63, 68–70, 75, 80, 81, 83–87, 89, 96, 99, 101, 103, 104]	Trust, cost, perceived behavioral control, perceived risk, hedonic motivation, prior experience, observability, market scope/e-readiness, voluntariness, organizational culture, perceived technology security, job relevance, complexity, security

Table 3 presents the remaining list of 12 factors after the study removed factors with 0–3 significance irrespective of their frequency in reviewed literature.

Table 4 presents our proposed generic technology factors which will be used to develop our generic technology adoption model. To arrive at these factors, the study considered only factors with frequency over the total significance effect greater than 0.8 from the list of factors presented in Table 3 above.

Table 3. List of factors from reviewed literature

Factors	Related studies	Frequency	Sig	Non-sig	Variable effect
Perceived ease of use	[1, 7, 8, 10, 20–24, 32, 33, 35, 38, 39, 41, 43, 44, 50, 52, 54, 56, 66, 74, 75, 79, 83, 88–90, 93, 102, 104]	33	22	5	0.8
Perceived usefulness	[1, 7, 8, 10, 20, 22–24, 32, 33, 35, 38, 41, 45, 50, 54–56, 66, 74, 75, 79, 83, 88, 89, 92, 93, 97, 102, 104]	27	17	2	0.9
Social influence	[2, 6, 12, 13, 15, 17, 19, 24, 28, 39, 40, 43, 48, 51, 54, 55, 57, 58, 61, 62, 66, 69, 75, 88, 93, 96, 97, 102, 104–106]	25	15	10	0.6
Behavioral intention to use	[1, 6, 7, 12, 16, 19, 20, 24, 32, 33, 35, 43, 45, 48, 49, 51, 53, 54, 56, 57, 62, 67–69, 79, 83, 86, 89, 92, 93, 97, 102, 106]	33	15	2	0.9
Effort expectancy	[2, 6, 12, 15–17, 19, 28, 40, 43, 48, 51, 57, 62, 68, 69, 96, 105, 106]	19	14	5	0.7
Performance expectancy	[2, 6, 15–17, 19, 40, 43, 48, 51, 57, 62, 68, 69, 96, 105, 106]	18	13	3	0.8
Facilitating conditions	[6, 15, 17, 19, 28, 35, 38, 43, 48, 51, 57, 62, 69, 96, 105, 106]	16	13	2	0.9
Attitude towards use	[1, 8, 17, 23, 24, 32, 35, 38, 49, 50, 53, 54, 56, 61, 67, 74, 79, 83, 93, 102]	20	12	1	0.9
Subjective norm	[1, 33, 38, 53, 54, 56, 67, 81, 83]	9	6	2	0.8
Relative advantage	[14, 25–27, 39, 47, 52, 70, 75, 80, 81, 83, 84, 87–89, 92, 99]	18	6	1	0.9
Self-efficacy	[1, 8, 35, 40, 44, 52, 56, 74, 81, 83]	9	6	1	0.9
Perceived enjoyment	[10, 24, 49, 97]	4	4	2	0.7

Table 4. Probable core factors of generic technology adoption models

Factors	Frequency	Significance	Non-significance	Variable effect
Perceived ease of use	33	22	5	0.8
Perceived usefulness	27	17	2	0.9
Performance expectancy	19	17	4	0.8
Behavioral intention to use	33	15	2	0.9
Facilitating conditions	16	13	2	0.9
Attitude towards use	20	12	1	0.9
Relative advantage	18	6	1	0.9
Self-efficacy	9	6	1	0.9
Subjective norm	9	6	2	0.8

5 Discussion and Conclusion

This study was limited to 80 articles sourced from renowned databases such as ACM Digital, Google Library, Google Scholar, ResearchGate, PsycINFO, and IEEE Xplore etc. using keywords "Technology Adoption" and "Technology Adoption as a review" as our search criteria. Findings from this study will con-tribute to existing literature from another viewpoint. An individual's technology adoption process cannot be explained by only one variable, this is due to the complex nature of man which requires several factors to add up to explaining technology adoption.

The findings of this study can be summarized to determine technology adoption despite various attempt to integrate and modify variables of technology adoption. As much as the criteria of 0.8 were used, the study found that most variables had a trajectory to be continuous and still recent which makes our factors relevant.

The variable "Perceived Ease of Use" was found to be the most used factor which was highly significant in the reviewed literature. Also, the factors "Perceived Usefulness" and "Performance Expectancy" were found significant and highly used in the study reviewed. The variable "Self Efficacy" and "Subjective Norm" was were significant corresponding to the frequency it was used in the literature reviewed by the researcher.

All these variables (Table 4) based on the screening, analysis and selection criteria contribute are preferred and contribute "Technology Adoption Models" developed by the researchers.

Future research will pay closer attention to the factors which were not significant by conducting an in-depth analysis to (i) find the nature of research and the environment the study was conducted (ii) code the non-significant variable to find the trend where they are not valid. Factor Analysis will be applied to shrink factors to a smaller data set. In conclusion, a sample size will be selected from a defined population that will be tested and validated quantitatively to find a generic model that everybody can fit in to test in another environment.

References

1. Abramson, J., Dawson, M., Stevens, J.: An examination of the prior use of e-learning within an extended technology acceptance model and the factors that influence the behavioral intention of users to use m-learning. SAGE Open **5**(4), 2158244015621114 (2015)
2. AbuShanab, E., Pearson, J.M.: Internet banking in Jordan: the unified theory of acceptance and use of technology (UTAUT) perspective. J. Syst. Inf. Technol. **9**(1), 78–97 (2007)
3. Agarwal, R., Prasad, J.: Are individual differences germane to the acceptance of new information technologies? Decis. Sci. **30**(2), 361–391 (1999)
4. Aghaunor, L., Fotoh, X.: Factors affecting ecommerce adoption in Nigerian banks (2006)
5. Ajzen, I.: The theory of planned behavior. Organ. Behav. Hum. Decis. Process. **50**(2), 179–211 (1991)
6. Alalwan, A.A., et al.: Examining factors influencing Jordanian customers' intentions and adoption of internet banking: extending UTAUT2 with risk. J. Retail. Consum. Serv. **40**, 125–138 (2018)
7. Alambaigi, A., Ahangari, I.: Technology Acceptance Model (TAM) as a predictor model for explaining agricultural experts behavior in acceptance of ICT. Int. J. Agric. Manage. Dev. **6**(2), 235–247 (2016)
8. Al-Azawei, A., Lundqvist, K.: Learner differences in perceived satisfaction of an online learning: an extension to the technology acceptance model in an Arabic sample. Electron. J. e-Learn. **13**(5), 408–426 (2015)
9. Alharbi, S., Drew, S.: Using the technology acceptance model in understanding academics' behavioural intention to use learning management systems. Int. J. Adv. Comput. Sci. Appl. **5**(1), 143–155 (2014)
10. Al-hawari, M.A., Mouakket, S.: The influence of technology acceptance model (TAM) factors on students'e-satisfaction and e-retention within the context of UAE e-learning. Educ. Bus. Soc. Contemp. Middle East. Issue **3**(4), 299–314 (2010)
11. Alkhalifah, A.: A research methodology to explore the adoption of e-government. Int. J. Comput. Eng. Inf. Technol. **9**(9), 216 (2017)
12. Al-Qeisi, K., Dennis, C., Abbad, M.: How viable is the UTAUT model in a non-Western context? Int. Bus. Res. **8**(2), 204–219 (2015)
13. Alrawabdeh, W.: Environmental factors affecting mobile commerce adoption-an exploratory study on the Telecommunication firms in Jordan. Int. J. Bus. Soc. Sci. **5**(8) (2014)
14. Alshamaila, Y., Papagiannidis, S., Li, F.: Cloud computing adoption by SMEs in the north east of England: a multi-perspective framework. J. Enterp. Inf. Manage. **26**(3), 250–275 (2013)
15. Alshehri, M., Drew, S., AlGhamdi, R.: Analysis of citizens acceptance for e-government services: applying the UTAUT model (2013). arXiv preprint arXiv:1304.3157
16. Alsheikh, L., Bojei, J.: Determinants affecting customer's intention to adopt mobile banking in Saudi Arabia. Int. Arab J. e-Technol. **3**(4), 210–219 (2014)
17. Amadin, F., Obienu, A.: Intention to use university e-mail system based on modified utaut model: perspectives of University of Benin Postgraduate Students. In: Proceedings of the 6th iSTEAMS Multidisciplinary Cross_Border Conference, University of Professional Studies, Accra Ghana (2016)
18. Armitage, C.J., Conner, M.: Efficacy of the theory of planned behaviour: a meta-analytic review. Br. J. Soc. Psychol. **40**(4), 471–499 (2001)
19. Attuquayefio, S., Addo, H.: Using the UTAUT model to analyze students' ICT adoption. Int. J. Educ. Develop. using Inf. Commun. Technol. **10**(3), 75–86 (2014)

20. Baharin, A.T., et al.: Evaluating effectiveness of IDEWL using technology acceptance model. Procedia Soc. Behav. Sci. **171**, 897–904 (2015)
21. Beldad, A.D., Hegner, S.M.: Expanding the technology acceptance model with the inclusion of trust, social influence, and health valuation to determine the predictors of German users' willingness to continue using a fitness app: a structural equation modeling approach. Int. J. Hum. Comput. Interact. **34**(9), 882–893 (2018)
22. Bresciani, S., Eppler, M.: Extending tam to information visualization: a framework for evaluation. Electron. J. Inf. Syst. Eval. **18**(1), 46–58 (2015)
23. Chang, C.-C., Yan, C.-F., Tseng, J.-S.: Perceived convenience in an extended technology acceptance model: mobile technology and English learning for college students. Australas. J. Educ. Technol. **28**(5), 809–826 (2012)
24. Chen, H., et al.: An extended technology acceptance model for mobile social gaming service popularity analysis. Mobile Inf. Syst. **2017**, 1–12 (2017)
25. Chiu, C.-Y., Chen, S., Chen, C.-L.: An integrated perspective of TOE framework and innovation diffusion in broadband mobile applications adoption by enterprises. Int. J. Manage. Econ. Soc. Sci. **6**(1), 14–39 (2017)
26. Chong, A.Y.-L., et al.: Factors affecting the adoption level of c-commerce: an empirical study. J. Comput. Inf. Syst. **50**(2), 13–22 (2009)
27. Chong, S., Bauer, C.: A model of factor influences on Electronic Commerce adoption and diffusion in small-and medium-sized enterprises. In: PACIS Proceedings, p. 23 (2000)
28. Dadayan, L., Ferro, E.: When technology meets the mind: a comparative study of the technology acceptance model. In: Wimmer, M.A., Traunmüller, R., Grönlund, Å., Andersen, K.V. (eds.) Electronic Government. Lecture Notes in Computer Science, vol. 3591. Springer, Heidelberg (2005). https://doi.org/10.1007/11545156_1
29. Daniel, E.M., Grimshaw, D.J.: An exploratory comparison of electronic commerce adoption in large and small enterprises. J. Inf. Technol. **17**(3), 133–147 (2002)
30. Davis, F.D.: Perceived usefulness, perceived ease of use, and user acceptance of information technology. MIS Q. **1989**, 319–340 (1989)
31. Davis, F.D., Bagozzi, R.P., Warshaw, P.R.: User acceptance of computer technology: a comparison of two theoretical models. Manage. Sci. **35**(8), 982–1003 (1989)
32. Diop, E.B., Zhao, S., Van Duy, T.: An extension of the technology acceptance model for understanding travelers' adoption of variable message signs. PLoS ONE **14**(4), e0216007 (2019)
33. Dumpit, D.Z., Fernandez, C.J.: Analysis of the use of social media in higher education institutions (heis) using the technology acceptance model. Int. J. Educ. Technol. High. Educ. **14**(1), 1–16 (2017). https://doi.org/10.1186/s41239-017-0045-2
34. El-Gohary, H.: Factors affecting E-Marketing adoption and implementation in tourism firms: an empirical investigation of Egyptian small tourism organisations. Int. J. Educ. Technol. High. Educ. **33**(5), 1256–1269 (2012)
35. Fathema, N., Shannon, D., Ross, M.: Expanding the Technology Acceptance Model (TAM) to examine faculty use of Learning Management Systems (LMSs) in higher education institutions. J. Online Learn. Teach. **11**(2), 210–232 (2015)
36. Fishbein, M.: A theory of reasoned action: some applications and implications. In: Nebraska Symposium on Motivation. Nebraska Symposium on Motivation (1980)
37. Fishbein, M., Ajzen, I.: Understanding attitudes and predicting social behavior (1980)
38. Gagnon, M.P., et al.: Using a modified technology acceptance model to evaluate healthcare professionals' adoption of a new telemonitoring system. Telemed. e-Health **18**(1), 54–59 (2012)

39. Ghane, F., et al.: The role of social influence and innovation characteristics in the adoption of Integrated Pest Management (IPM) practices by paddy farmers in Iran. In: International Conference on Social Science and Humanity-IPEDR, Singapore (2011)
40. Giannakos, M.N., Vlamos, P.: Educational webcasts' acceptance: empirical examination and the role of experience. Br. J. Edu. Technol. **44**(1), 125–143 (2013)
41. Godoe, P., Johansen, T.: Understanding adoption of new technologies: technology readiness and technology acceptance as an integrated concept. J. Eur. Psychol. Stud. **3**(1), 38–52 (2012)
42. Gunbatar, M.S.: Examination of undergraduate and associate degree students' computer programming attitude and self-efficacy according to thinking style, gender and experience. Contemp. Educ. Technol. **9**(4), 354–373 (2018)
43. Gupta, B., Dasgupta, S., Gupta, A.: Adoption of ICT in a government organization in a developing country: an empirical study. J. Strateg. Inf. Syst. **17**(2), 140–154 (2008)
44. He, Y., Chen, Q., Kitkuakul, S.: Regulatory focus and technology acceptance: perceived ease of use and usefulness as efficacy. Cogent Bus. Manage. **5**(1), 1459006 (2018)
45. Horton, R.P., et al.: Explaining intranet use with the technology acceptance model. J. Inf. Technol. **16**(4), 237–249 (2001)
46. Hsieh, H.-F., Shannon, S.E.: Three approaches to qualitative content analysis. Qual. Health Res. **15**(9), 1277–1288 (2005)
47. Hussin, H., Noor, R.M.: Innovating business through e-commerce: exploring the willingness of Malaysian SMEs. In: Proceedings of the Second International Conference on Innovation in IT. Citeseer (2005)
48. Ibrahim, A., Adu-Gyamfi, M., Kassim, B.A.: Factors affecting the adoption of ICT by administrators in the university for development studies tamale: empirical evidence from the UTAUT model. Int. J. Sustain. Manage. Inf. Technol. **4**(1), 1 (2018)
49. Isiyaku, D.D., Ayub, A.F.M., Abdulkadir, S.: Empirical modeling of information communication technology usage behaviour among business education teachers in tertiary colleges of a developing country. S. Afr. J. Educ. **35**(4), 1–14 (2015)
50. Julianto, I.P., Yasa, I.N.P.: The analysis of technology acceptance model (TAM) on the use of accounting information system. In: International Conference on Tourism, Economics, Accounting, Management, and Social Science, TEAMS 2018. Atlantis Press (2019)
51. Kijsanayotin, B., Pannarunothai, S., Speedie, S.M.: Factors influencing health information technology adoption in Thailand's community health centers: applying the UTAUT model. Int. J. Med. Informatics **78**(6), 404–416 (2009)
52. Lee, J., et al.: Adoption of Internet technologies in small businesses. In: PACIS Proceedings, p. 71 (2001)
53. Liao, S., et al.: The adoption of virtual banking: an empirical study. Int. J. Inf. Manage. **19** (1), 63–74 (1999)
54. Lu, C.-T., Huang, S.-Y., Lo, P.-Y.: An empirical study of on-line tax filing acceptance model: integrating TAM and TPB. Afr. J. Bus. Manage. **4**(5), 800–810 (2010)
55. Lu, J., Yao, J.E., Yu, C.-S.: Personal innovativeness, social influences and adoption of wireless Internet services via mobile technology. J. Strateg. Inf. Syst. **14**(3), 245–268 (2005)
56. Maduku, D.K.: Behavioral intention towards mobile banking usage by South African retail banking clients. Invest. Manage. Financ. Innov. **11**(3), 37–51 (2014)
57. Maduku, D.K.: Understanding behavioural intention towards e-books use: does gender really matter. In: Proceedings of 31st International Business Research Conference (2015)
58. Marques, A., et al.: Medical records system adoption in European hospitals. Electron. J. Inf. Syst. Eval. **14**(1), 89 (2011)
59. Masrom, M.: Technology acceptance model and e-learning. Technology **21**(24), 81 (2007)

60. Mathieson, K.: Predicting user intentions: comparing the technology acceptance model with the theory of planned behavior. Inf. Syst. Res. **2**(3), 173–191 (1991)
61. Phung, T.M., Thi, L., Pham, M., Do, N.H.: Effect of eWOW and social influence on product adoption intention. In: Proceedings of 11th International Days of Statistics and Economics (2017)
62. Moghavvemi, S., et al.: The entrepreneur's perception on information technology innovation adoption: an empirical analysis of the role of precipitating events on usage behavior. Innovation **14**(2), 231–246 (2012)
63. Nasri, W.: Factors influencing the adoption of internet banking in Tunisia. Int. J. Bus. Manage. **6**(8), 143–160 (2011)
64. Nchunge, D.M., Sakwa, M., Mwangi, W.: User's perception on ICT adoption for education support in schools: a survey of secondary school teacher's in Thika district Kenya. Int. J. Humanit. Soc. Sci. **2**(10), 17–29 (2012)
65. Oliveira, T., Martins, M.F.: Understanding e-business adoption across industries in European countries. Ind. Manage. Data Syst. **110**(9), 1337–1354 (2010)
66. Park, N., et al.: Factors influencing smartphone use and dependency in South Korea. Comput. Hum. Behav. **29**(4), 1763–1770 (2013)
67. Parker, D., et al.: Intention to commit driving violations: An application of the theory of planned behavior. J. Appl. Psychol. **77**(1), 94 (1992)
68. Rahi, S., Ghani, M.A.: The role of UTAUT, DOI, perceived technology security and game elements in internet banking adoption. World J. Sci. Technol. Sustain. Develop. **15**(4), 338–356 (2018)
69. Rahi, S., Ghani, M., Ngah, A.: A structural equation model for evaluating user's intention to adopt internet banking and intention to recommend technology. Accounting **4**(4), 139–152 (2018)
70. Ramdani, B., Kawalek, P., Oswaldo, L.: Knowledge management and enterprise systems adoption by SMEs. J. Enterp. Inf. Manage. **22**(1/2), 10–24 (2009)
71. Rogers, E.M.: Diffusion of preventive innovations. Addict. Behav. **27**(6), 989–993 (2002)
72. Sachs, G.: Profiles in Innovation: Virtual & Augmented Reality - Understanding the race for the next computing platform. Goldman Sachs Group Inc. (2016)
73. Sahin, I.: Detailed review of Rogers' diffusion of innovations theory and educational technology-related studies based on Rogers' theory. Turk. Online J. Educ. Technol. (TOJET) **5**(2), 14–23 (2006)
74. Sánchez, R.A., Hueros, A.D.: Motivational factors that influence the acceptance of Moodle using TAM. Comput. Hum. Behav. **26**(6), 1632–1640 (2010)
75. Shah Alam, S., Ali, M.Y., Jani, M.F.M.: An empirical study of factors affecting electronic commerce adoption among SMEs in Malaysia. J. Bus. Econ. Manage. **12**(2), 375–399 (2011)
76. Shahroom, A., Hussin, N.: Industrial revolution 4.0 and education. Int. J. Acad. Res. Bus. Soc. Sci. **8**(9), 314–319 (2018)
77. Southey, G.: The theories of reasoned action and planned behaviour applied to business decisions: a selective annotated bibliography. J. New Bus. Ideas Trends **9**(1), 43–50 (2011)
78. Straub, E.T.: Understanding technology adoption: theory and future directions for informal learning. Rev. Educ. Res. **79**(2), 625–649 (2009)
79. Šumak, B., et al.: Factors affecting acceptance and use of Moodle: an empirical study based on TAM. Informatica **35**(1), 91–100 (2011)
80. Tan, K.S., Eze, U.C.: An empirical study of internet-based ICT adoption among Malaysian SMEs. Commun. IBIMA **1**(1), 1–12 (2008)
81. Tan, M., Teo, T.S.: Factors influencing the adoption of Internet banking. J. Assoc. Inf. Syst. **1**(1), 5 (2000)

82. Taylor, S., Todd, P.A.: Understanding information technology usage: a test of competing models. Inf. Syst. Res. **6**(2), 144–176 (1995)
83. Teo, T.S., Pok, S.H.: Adoption of WAP-enabled mobile phones among Internet users. Omega **31**(6), 483–498 (2003)
84. Thong, J.Y.: An integrated model of information systems adoption in small businesses. J. Manage. Inf. Syst. **15**(4), 187–214 (1999)
85. Tu, M.: An exploratory study of Internet of Things (IoT) adoption intention in logistics and supply chain management: a mixed research approach. Int. J. Logist. Manage. **29**(1), 131–151 (2018)
86. Van den Berg, J., Van der Lingen, E.: An empirical study of the factors affecting the adoption of mobile enterprise applications. S. Afr. J. Ind. Eng. **30**(1), 124–146 (2019)
87. Van Slyke, C., Belanger, F., Comunale, C.L.: Factors influencing the adoption of web-based shopping: the impact of trust. ACM SIGMIS Database Database Adv. Inf. Syst. **35** (2), 32–49 (2004)
88. Van Slyke, C., Belanger, F., Hightower, R.: Understanding gender-based differences in consumer e-commerce adoption. In: Proceedings of the 2005 Southern Association of Information Systems Conference (2005)
89. Van Slyke, C., et al.: The influence of culture on consumer-oriented electronic commerce adoption. J. Electron. Commer. Res. **11**(1) (2010)
90. Venkatesh, V., Davis, F.D.: A theoretical extension of the technology acceptance model: four longitudinal field studies. Manage. Sci. **46**(2), 186–204 (2000)
91. Venkatesh, V., et al.: User acceptance of information technology: toward a unified view. MIS Q. **2003**, 425–478 (2003)
92. Wang, Y., Meister, D., Wang, Y.: Reexamining relative advantage and perceived usefulness: an empirical study. Int. J. Inf. Commun. Technol. Educ. **7**(1), 46–59 (2011)
93. Weng, F., et al.: A TAM-based study of the attitude towards use intention of multimedia among school teachers. Appl. Syst. Innov. **1**(3), 36 (2018)
94. Yucel, U., Gulbahar, Y.: Technology acceptance model: a review of the prior predictors. J. Fac. Educ. Sci. **46**(1), 89–109 (2013)
95. Yuen, A., Ma, W.: Exploring teacher acceptance of E-learning technology. Asia Pac. J. Teach. Educ. **36**(3), 229–243 (2008)
96. Zahir, M., Gharleghi, B.: Adoption of internet banking in maldives, the most important determinants. Asian Soc. Sci. **11**(2), 181–189 (2015)
97. Zhang, S., Zhao, J., Tan, W.: Extending TAM for online learning systems: an intrinsic motivation perspective. Tsinghua Sci. Technol. **13**(3), 312–317 (2008)
98. Zhang, Y., Wildemuth, B.M.: Qualitative analysis of content. In: Applications of Social Research Methods to Questions in Information Library Science, vol. 308, p. 319 (2009)
99. Zhu, K., et al.: Innovation diffusion in global contexts: determinants of post-adoption digital transformation of European companies. Eur. J. Inf. Syst. **15**(6), 601–616 (2006)
100. Zhu, K., Kraemer, K., Xu, S.: Electronic business adoption by European firms: a cross-country assessment of the facilitators and inhibitors. Eur. J. Inf. Syst. **12**(4), 251–268 (2003)
101. Zhu, K., Kraemer, K.L.: Post-adoption variations in usage and value of e-business by organizations: cross-country evidence from the retail industry. Inf. Syst. Res. **16**(1), 61–84 (2005)
102. Hsu, C.-L., Lu, H.-P.: Why do people play on-line games? An extended TAM with social influences and flow experience. Inf. Manage. Sci. **41**(7), 853–868 (2004)
103. Kuan, K.K., Chau, P.Y.: A perception-based model for EDI adoption in small businesses using a technology–organization–environment framework. Inf. Manage. Sci. **38**(8), 507–521 (2001)

104. Nassuora, A.B.: Understanding factors affecting the adoption of m-commerce by consumers. J. Appl. Sci. **13**(6), 913–918 (2013)
105. Muriithi, P., Horner, D., Pemberton, L.: Factors contributing to adoption and use of information and communication technologies within research collaborations in Kenya. Inf. Technol. Develop. **22**(Suppl. 1), 84–100 (2016)
106. Mosweu, O., Bwalya, K.J.: A multivariate analysis of the determinants for adoption and use of the Document Workflow Management System in Botswana's public sector. S. Afr. J. Libr. Inf. Sci. **84**(2), 27–38 (2018)

Vulnerability of Smart IoT-Based Automation and Control Devices to Cyber Attacks

Tibor Horák[1], Marek Šimon[2], Ladislav Huraj[2(✉)], and Roman Budjač[1]

[1] Institute of Applied Informatics, Automation and Mechatronics Faculty of Materials Science and Technology in Trnava, Slovak University of Technology in Bratislava, Trnava, Slovakia
{tibor.horak, roman.budjac}@stuba.sk

[2] Department of Applied Informatics, University of SS. Cyril and Methodius, Trnava, Slovakia
{marek.simon, ladislav.huraj}@ucm.sk

Abstract. In modern industrial enterprises, safety always comes first. Regardless of whether it is safety against weather conditions such as fires, floods, or it is securing the building from the intrusion of unwanted person. With these sensors, these services can also provide IoT smart security devices. In the connected world of Industry 4.0, there are way too many opportunities to take control of such devices, and so with a help of a cyber attack, computer attackers would be able, in two ways, to make the device impossible to operate. The first way is to manipulate the device, disable alarm sensors, and steal the data. The second way is to misuse the device for attacking another in a reflected way. The article illustrates the possibility of how the safety sensors can be disabled, and how this safety device can be used to attack another IoT device - the thermostat. Finally, the case study demonstrates inability of IoT-based automation and control device to send alarm notifications when a threat is detected by its sensors during the DDoS attacks.

Keywords: IoT security device · DDoS attack · Reflection attack · Industry 4.0

1 Introduction

Every modern, industrial enterprise Industry 4.0 is equipped with IoT devices. IoT introduces modern devices controlled remotely by various mobile applications over the Internet. These devices can use sensors to speed up and streamline the production processes. However, the rapid deployment of these devices has also caused lots of problems. One of these problems is the susceptibility to cyber attacks [1].

The article focuses on the IoT smart device that deals with security. This smart device is equipped with a control unit and sensors. These sensors are designed to protect industrial plant buildings from weather conditions such as unauthorized intrusion, floods, and fires. They can also detect an event in the building and then notify the building manager that something is happening, so the building manager can immediately call the security services. Because of the increasing growth of cyber

R. Silhavy (Ed.): CSOC 2020, AISC 1226, pp. 287–294, 2020.
https://doi.org/10.1007/978-3-030-51974-2_27

attacks, which is a chronic problem of the Internet of Things, it could disrupt the operation of the security device [2].

There are two main types of attacks. In the first attack, the attacker concentrates on disabling the functional device by overloading it with packets from multiple sources. In the case of a security device, this could result in the shutdown of sensors which monitor the object and protect against weather and intrusion by unauthorized people who could cause considerable damage. The attack is called DDoS (Distributed Denial of Service) [3]. In the second attack, the device is misused for attacking other devices by sending packets to a non-existent port, and the device will send a response containing an error message to another, unsuspecting device. In this case, the attack is called DRDoS (Distributed Reflection Denial of Service) [4]. According to Maire O'Neill, as early as 2014, an IoT lightbulb, which was plugged into the network, was attacked by the attackers who managed to retrieve passwords from the WI-FI network [5]. Although various defense techniques based on artificial intelligence, high performance computing, or intrusion prevention system are used to defend against such attacks, this is not possible with IoT devices with limited resource capacity [6, 7].

The goal of this article is to examine IoT security system Fibaro, in greater detail, from the perspective of cyber attacks DDoS and DRDoS performed on this device.

2 Background

The paradigm of Internet of Things applies to scenarios in which network connectivity and computing capacity are enhanced by built-in sensor. It is allowing these devices to create, exchange and consume data with a minimal human intervention. This type of paradigm is being implemented and facilitated by critical advances in computing power, electronics miniaturization, and networking. This also includes the security intelligent kit from Fibaro company, which is part of the wireless system. The miniaturized control unit is the brain of the Fibaro system on the Nginx 1.9.5 platform and the SSH server Dropbear 2015.67. This control unit listens and manages all wireless modules over the Z-Wave network and can also be a gateway to the Internet and WiFi network. Wireless modules serve as a motion detector, flood detector, smoke detector, wireless contact, and wireless socket. With the help of these wireless modules, the administrator is informed about all of the events in the object, and that way it is possible to manage all settings in one place. This kind of a wireless, efficient, and low-installation security device can save households and companies a lot of money, both from the point of view of the installation and also from timely information about the threat of danger.

Fibaro company, which started producing IoT devices, was established in 2010 when these devices were not widely used yet. Nowadays, the situation is completely different. The number of IoT devices will exceed nearly a billion in just a few years. The Fibaro plant building in Poland produces about a million different devices each year - from intelligent plugs, lamps, motion sensors and flood sensors to devices that have a direct or indirect impact on the safety of the buildings which are equipped with them. In addition, Fibaro sales in Russia in 2018 increased nearly 10 times compared to

2017. As the company clearly plays an important role in the IoT market, the study of the intelligent Fibaro intelligent device is truly up to date.

2.1 Attacks in IoT Environment

While the deployment of IoT devices in a variety of areas introduces a wealth of benefits, their specific attributes, along with networking, present new security challenges. For example, competitive costs, IoT equipment technical restrictions, and cost reductions are already forcing the manufacturers not to deal with their safety issues, which leads to security vulnerabilities. In addition to the potential security gaps, the increase in the number, type and nature of Internet of Things, devices will undoubtedly increase the usage. Vulnerable IoT devices can serve as entry points for attackers by allowing harmful reprogramming or malfunctioning of the device. Besides that, poorly designed devices can expose user data to a theft by leaving data flows inadequately protected. Scanning of the IoT Fibaro security device found that the port 80 is running Nginx version 1.9.5 web server, and the port 22 is running SSH Dropbear 2015.67 server. The surprise itself was that the device revealed the information of itself very easily. The Nginx version 1.9.5 web server is vulnerable to Buffer over-read attacks.

The system works with a current buffer that is used to store data temporarily, but when a program or system process receives more data than was originally allocated for storage, the data overflows. In this attack, the attacker's goal is to access data that reveals private information [8].

Even SSH Dropbear version 2015.67 server is not completely error-free. It is prone to a Format String attack where the submitted data is evaluated as a command through the application, what allows the attacker to run the code and read the stack, or cause segmentation errors in the running application. It may compromise the security and the stability of the system [9].

The greatest fear for IoT devices is cyber attacks in the form of DDoS attacks. The attack can be conducted from multiple sources simultaneously in the form of a large number of small requests to disrupt service delivery by overloading primary resources by overloading the CP and RAM memory. This results in overloading the service provider's network infrastructure [10]. A very similar attack is the DRDoS attack that differs from DDoS in such a way that it is a misuse of the device to attack another in a way that the attacker sends the packets to the wrong port to the specific device, and it reflects in the way that ICMP error packets are sent to another device [4].

The most known types of DDoS attacks are UDP flood and SYN flood. The first one is the UDP flood. The UDP attack is the most common form of attack and works on a very simple principle. An attacker sends a large number of UDP packets to random ports of the target server. The target server has to respond to packets. A huge amount of processed data will prevent the server from communicating. This type of an attack is popular probably because defense against it is very difficult [11]. The second one is the SYN flood attack. The SYN flood attack is the second most common attack that uses the key TCP Internet protocol and works by the whole process consisting of SYN messages, then SYN-ACK, and finally ACK. This is also called a three-way handshake. Although the target server receives a SYN message and responds to it with its SYN-ACK, it no longer deliberately receives the necessary ACK confirmation message. Therefore, the server does not know the status of the communication and is waiting to receive this message, which could be delayed, for example, due to a network traffic. The whole situation could result in that the server has to keep the connection half open for some time. The whole situation can lead to depletion of server resources, which leads to its malfunction or direct malfunction [12].

3 Experiment Design

A test network infrastructure was set up in order to test IoT Fibaro's security under cyber-attacks. It consists of a high-speed router ASUS-RT-AC66U, which can transfer data up to 1.75 Gbps. A computer with a Kali-Linux operating system is connected to the router, and it is equipped with a set of hping3 tools in order to perform demonstration tests of DDoS and DRDoS cyber attacks. As shown in Fig. 1, this computer is connected via an Ethernet interface at speed of 1000 Mbps. The Fibaro IoT security device is also connected via Ethernet, but only with a speed of 100 Mbps. Another connected device in the network is the IoT thermostat from Honeywell, which serves for intelligent temperature control at homes and in industrial buildings. This device is connected to a wireless Wi-Fi network with a speed of over 54 Mbps, and it will be used in this network infrastructure to test for the reflected DRDoS attack. Also, a simple, star-shaped topology was chosen to connect the devices.

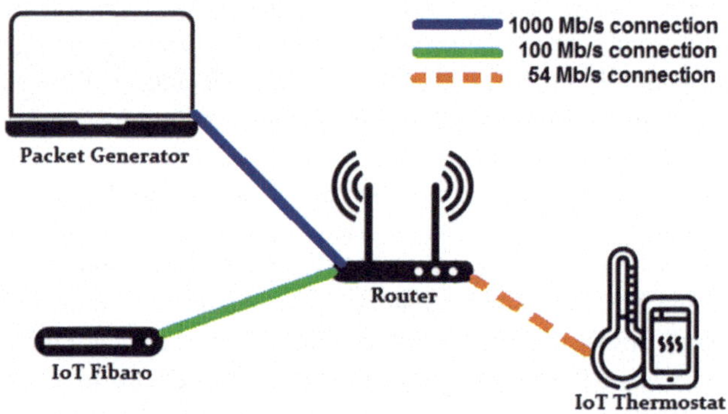

Fig. 1. The topology of the testing network

4 Performance Tests

DDoS and DRDoS attacks were performed on IoT devices. Specific types of DDoS attacks were directed straight to the IoT Fibaro security device. A reflected DRDoS attack was performed on the IoT intelligent thermostat in which IoT Fibaro misused the attack. This also ensured the real attacker's anonymity. The attacks were implemented by using the Kali Linux operating system, which is equipped with hping3 tools (http://hping.org). These tools support TCP, UDP, and ICMP protocols. These tools are used by security analysts in order to verify the security of managed network infrastructures.

The control and generation of packets were controlled by commands via the command line.

The first scenario of DDoS attacks was Syn flood attack. As can be seen in the Fig. 2, two attacks were performed on the Fibaro. They both took 60 s to perform. The packets were sent at the highest possible speed as allowed by the network infrastructure. During the first attack on an existing port 8000, a round-trip time measurement of port 80 was performed. The average response value of the round-trip time was 0.413 s. The Fibaro was still working during the attack, no shutdown occurred. The second Syn flood attack, was led on the port 80. It was not possible to measure the round-trip time during the attack, since the Fibaro device was flooded and unavailable, as can be seen from Fig. 2 the Fibaro device responded very rarely.

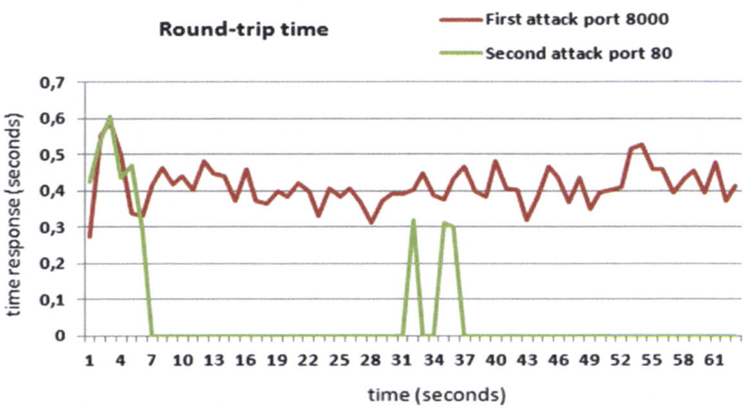

Fig. 2. Measurements of the round-trip time during Syn flood attacks

Another scenario of a DDoS attack is an HTTP Get flood attack. Within 60 s, the packets were sent to the Fibaro device at the highest possible speed from more than 9,000 IP addresses which corresponds to a medium-sized botnet attack. To verify the Fibaro availability the availability of default website of the Fibaro device was used. This website is automatically accessible once the device is installed. During this attack, a legitimate client tried to load the website every second after the attack began. The Fig. 3 shows the values from 0–10. If the speed of homepage load rate was 10, then the attempt of client loading the website was unsuccessful. Out of the 58 attempts to display the homepage to valid clients, 44 attempts were unsuccessful. In 14 attempts, the webpage was loaded with difficulties with an average time of 2.14 s. The webpage was successfully loaded only in less than one quarter of the requests, and the device was more than 76% unavailable.

The scenario of DRDoS attack on Fibaro device was also tested. The attack, shows possibility to misuse the Fibaro device for a reflective attack on the IoT Honeywell thermostat, as follows: an attacker uses an ICMP echo flood attack during 60 s time period. The attacker forwarded packets with error messages to IoT Fibaro on the non-existent port 600. Then, the Fibaro device forwarded the packets with an error message

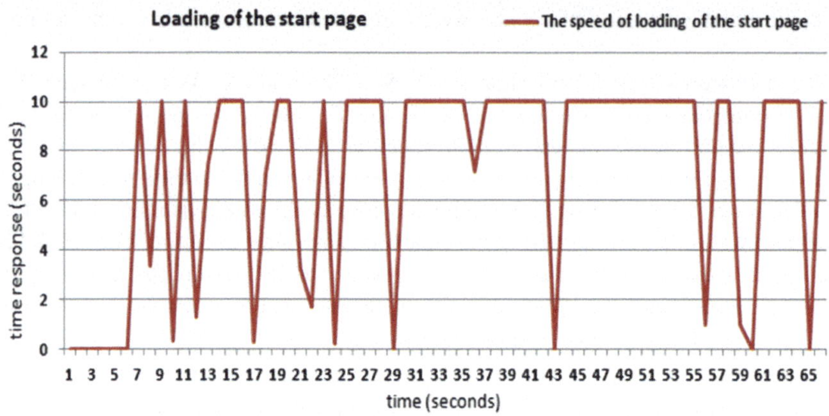

Fig. 3. The speed of loading of the web page during the attack

in reflected way on a spoof IP address used by the Honeywell IoT thermostat. Packets were sent at the highest possible speed. As can be seen in the Fig. 4, the attack was effective since the IoT thermostat stopped communicating with the unsuspecting IoT Fibaro only 13 s after commencing the attack.

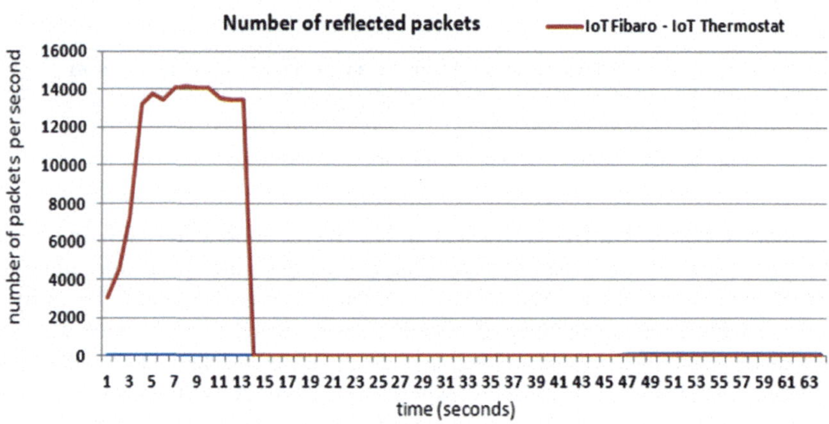

Fig. 4. Number of reflected packets per second transmitted to the network by the IoT devices

It should be underline that during HTTP Get flood cyber attack, as soon as the device was unavailable due to overload by flood packets, it was unable to send alarm notifications to the mobile application from the motion sensor when the sensor detected a movement. The alarm notification was only delivered to the mobile application approximately after 30 s after the attack has-finished.

The case study of the experimental cybernetic attacks on the specific IoT devices have clearly shown that they are very too-unsafe, both in the case of the attack and in the case of device misuse in order to attack other IoT devices. This case study also revealed a serious problem with the Fibaro security system which is unable to send alarm notifications when a threat is detected by its sensors during the DDoS attack.

5 Conclusions

During this study, it was found that the threats of cyber attacks on IoT devices are really justified. The threats are justified whether they are attacks on specific devices in the form of DDoS attacks or misuse of devices to attack DRDoS on others. The subject of this study was to test a specific IoT-based Fibaro automation and control device and to determine its vulnerability and risk from the view of both types of attack. During the study, it was demonstrated that the Fibaro device was inaccessible in most cases during the DDoS attack. During the attack, the device was unable to send alarm notifications to the mobile application and so failed to inform in time the administrator of the threats detected by the sensors during the attack. An advantage of a smart IoT-based automation and control device would be if a manufacturer is able to overcome these problems over time with firmware updates. Therefore, it is important to buy an-equipment with such support from the manufacturer.

Acknowledgements. The work was supported by the grant VEGA1/0272/18 *Holistic approach of knowledge discovery from production data in compliance with Industry 4.0 concept.*

References

1. Abomhara, M., Køien, G.: Cyber security and the internet of things: vulnerabilities, threats, intruders and attacks. J. Cyber Secur. Mobil. **4**, 65–88 (2015)
2. Ali, B., Awad, A.I.: Cyber and physical security vulnerability assessment for IoT-based smart homes. Sensors **18**(3), 817 (2018)
3. Zhang, C., Green, R.: Communication security in internet of thing: preventive measure and avoid DDoS attack over IoT network. In: Proceedings of the 18th Symposium on Communications & Networking, CNS '15, pp. 8–15 (2015)
4. Arukonda, S., Sinha, S.: The innocent perpetrators: reflectors and reflection attacks. ACSIJ Adv. Comput. Sci. Int. J. **4**, 94–98 (2015)
5. O'Neill, M.: Insecurity by design: today's IoT device security problem. In: Secure Digital Systems at the Center for Secure Information Technologies, vol. 2, pp. 48–49. Engineering (2016)
6. Vishwakarma, R., Jain, A.K.: A survey of DDoS attacking techniques and defence mechanisms in the IoT network. Telecommun. Syst. **73**, 3–25 (2020)
7. Dudáš, A., Voštinár, P., Škrinárová, J., Siláči, J.: Improved process of running tasks in the high performance computing system. In: 16th International Conference on Emerging eLearning Technologies and Applications (ICETA), pp. 133–140. IEEE (November 2018)
8. Wang, X., Qi, Y., Zhang, Ch., et al.: SecretSafe: a lightweight approach against Heap Buffer Over-Read Attack. In: IEEE 41st Annual Computer Software and Applications Conference, pp. 628–636 (2017)
9. Kilic, F., Kittel, T., Eckert, C.: Blind format string attacks. In: Tian, J., Jing, J., Srivatsa, M. (eds.) SecureComm 2014. LNICST, vol. 153, pp. 301–314. Springer, Cham (2015). https://doi.org/10.1007/978-3-319-23802-9_23

10. Ahmed, M.E., Ullah, S., Kim, H.: Statistical application fingerprinting for DDoS attack mitigation. IEEE Trans. Inf. Forensics Secur. **14**, 1471–1484 (2019)
11. Bijalwan, A., Wazid, M., et al.: Forensics of Random-UDP Flooding Attacks. J. Netw. **10**, 287–293 (2015)
12. Bogdanoski, M., Suminoski, T., Risteski, A.: Analysis of the SYN flood DoS attack. Int. J. Comput. Netw. Inf. Secur. **5**(8), 1–11 (2013)

Virtual Reality Technology Application to Increase Efficiency of Fire Investigators' Training

Irina Pozharkova(✉), Andrey Lagunov, Alexander Slepov, Maria Gaponenko, Eugeniy Troyak, and Alexander Bogdanov

FSBEE HE Siberian Fire and Rescue Academy, EMERCOM of Russia, RussiaZheleznogorsk 1 Severnaya Street, 662972
pozharkova@mail.ru

Abstract. The paper describes experience of virtual reality technology application to increase efficiency of fire investigators' training. It presents a virtual simulator that provides students with means for acquiring and improving professional skills of fire investigation. We reviewed efficiency evaluation outcomes relevant to formation of professional competencies related to fire investigation implemented in learning process of competency-oriented didactic assistance of professional training based on virtual reality technologies. #CSOC1120

Keywords: Virtual reality · Virtual simulator · Advanced educational technologies · Competency-based education · Fire investigation

1 Introduction

Nowadays, learning process is one of the most promising directions of developing and implementing advanced information technologies. Virtual reality is one of them. Being an efficient alternative to conventional learning methods, it represents new opportunities of theoretical learning and practical exercises.

According to several studies [1], training of fire engineering professionals is based on science-based competency-based educational program whose purposes are focused on achievement of declared qualification and a graduate's readiness for independent post-graduate professional practice. Along with this, as a rule, various full-scale experiments should be conducted to form special practical skills. It is often either not physically possible or requires considerable resources. Therefore, advanced didactic tools based on information technologies should be implemented in learning process related to training of fire engineering professionals at the moment. In particular, as a more efficient and cheaper alternative to conventional learning methods virtual reality technology represents new opportunities of theoretical learning and practical exercises.

Fire Investigation subject is a backbone of professional training of a fire investigator. Its purpose is to form knowledge and practical skills relevant to legal nature of professional activities of inquiry officers of EMERCOM of Russia during inspection activities, criminal and administrative investigations of fire and fire safety breach cases.

R. Silhavy (Ed.): CSOC 2020, AISC 1226, pp. 295–303, 2020.
https://doi.org/10.1007/978-3-030-51974-2_28

Practical trainings and laboratory practicals where students acquire tactical technical overhauling skills inter alia represent a backbone to form professional competencies that are of interest in the paper.

Mandatory review of photo and video records is a feature of Fire Investigation practical trainings. However, scattered segments of a fire point presented as individual photo or video records are not always a suitable means to ensure a comprehensive evaluation of a fire circumstances, detection of basic burning principles, prediction of a heat source presence and a potential fire cause [2]. Fire site simulation is a technically challenging activity because a full-scale training range and multiple combustion scenarios cannot be implemented because of restricted area of premises. Nevertheless, a virtual training range based on virtual reality technology is free from these disadvantages. For example, an actual fires data base can be developed on the basis of panoramic photos in software environment. In this case, students have full access to information that in real practice would be available for an investigator after a static inspection.

As some investigative activities are associated with photographic evidence, guidelines on digital photography methods and techniques related to fire investigation must be practiced during a practical training [3]. 360° panoramic images can be created by means of advanced devices and imaging technologies. In addition such images demonstrate high technical quality (e.g., an image is seamless, and no distortions are ob-served). Moreover, these panoramic images meet all photo recording standards, and they can be used as evidence. Also, they are ready to be implemented in a virtual training range.

Since virtual reality technologies can broaden functionality of laboratory practicals in regard to fire engineering technical expertise, a student can gain experience of a non-intrusive fire site investigation. In turn, this can reduce financial and time expenditures for training of these very professionals [3].

Another thing is that virtual reality enables implementation of a so-called "dynamic training range" dedicated to practical trainings whose advantages include:

- variable location and basic conditions of environment;
- multiple use to provide practical trainings;
- relatively low cost.

2 Methods

A virtual simulator intended to train fire engineering professionals was developed and implemented in learning process in the FSBEE HE Siberian Fire and Rescue Academy EMERCOM of Russia. In this context, software and hardware complying with the assigned task and providing further increase in its functionality were selected as a result of a comparative analysis. Moreover, simulator architecture and its functional handling models were built and implemented, software was set, and virtual training ranges simulating a real fire site were developed [4]. See technical and software architectures of the virtual simulator in Figs. 1 and 2.

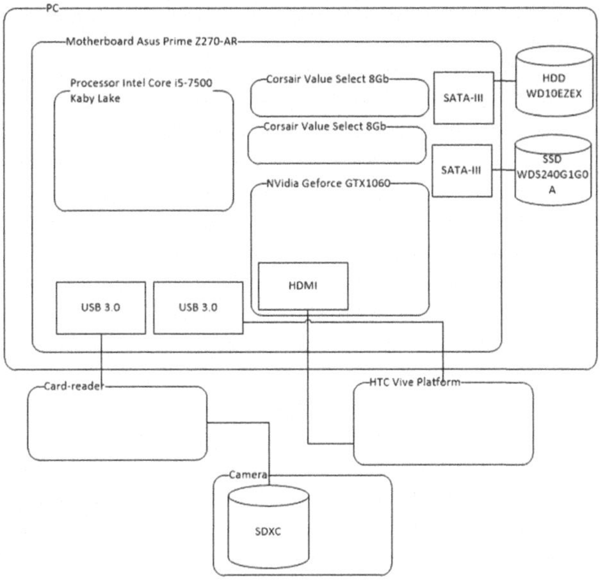

Fig. 1. Technical architecture of the virtual simulator.

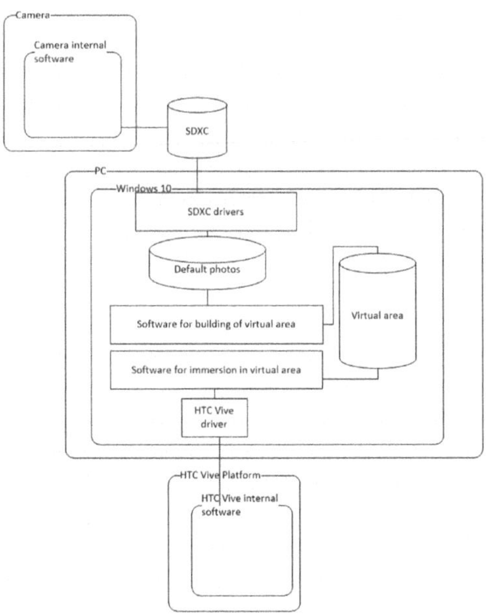

Fig. 2. Software architecture of the virtual simulator.

See decomposition of a functional model of a practical training using a virtual simulator in Fig. 3.

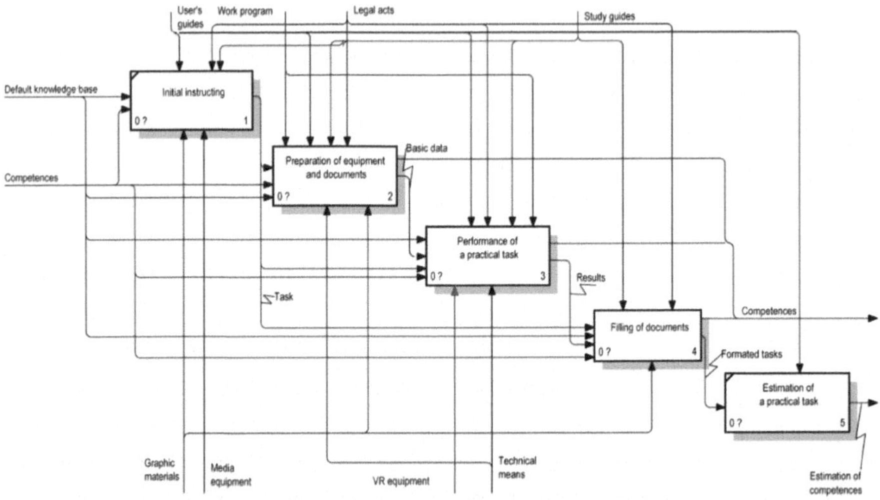

Fig. 3. Decomposition of Fire Case Investigative Activities lesson (virtual simulator, IDEF0 functional model).

Reviewed architecture of the virtual simulator (Figs. 1 and 2) and developed procedure of its application to a practical training (Fig. 3) provide its implementation in learning process without any significant modifications of educational program steering document. Major changes include a partial supersede of Fire Case Investigative Activities didactic graphic matter by a virtual dynamic training range. This decreases students' workload during training materials review and increases students' involvement in learning process.

3 Results

To review efficiency of professional fire investigation competencies formation while implementing competency-based didactic assistance of professional training based on virtual reality technologies in learning process several test runs were performed. During these tests we investigated a study group of Fire Investigation students using a VR simulator and a control group applying conventional educational methods (including a practical fire site inspection training based on examination of photo records). These groups included 89 and 100 subjects, respectively.

Parameters used to evaluate implementation quality of the learning process are mentioned below:

- students' involvement in the learning process;

- current academic performance, the subject interim assessment scores, retained knowledge;
- students' workload related to the subject.

Students' workload evaluation was based on student-reported time spent on performance of practical exercises included into the educational program steering document.

See outcomes of evaluated students' workload related to practical fire site inspection exercises in Fig. 4.

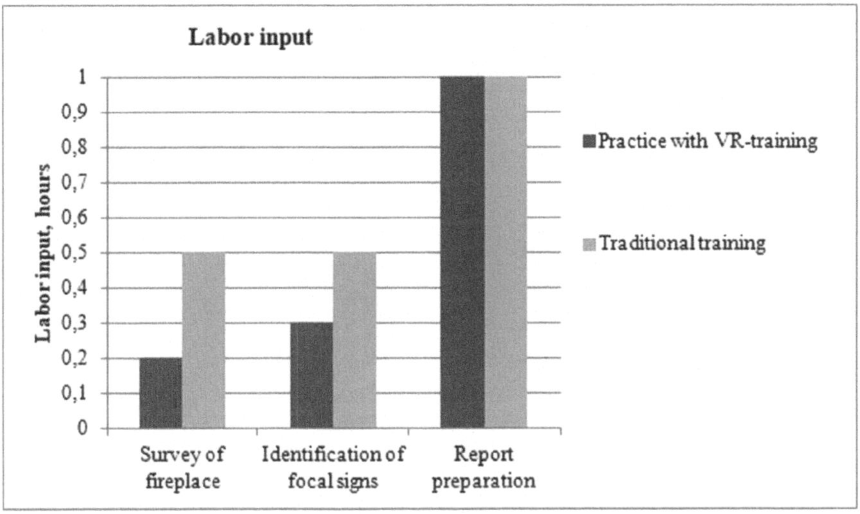

Fig. 4. Comparison of students' workload related to practical fire site inspection exercises.

Also, students' involvement in the learning process was evaluated in the light of timely performance of practical fire site inspection exercises (Fig. 5).

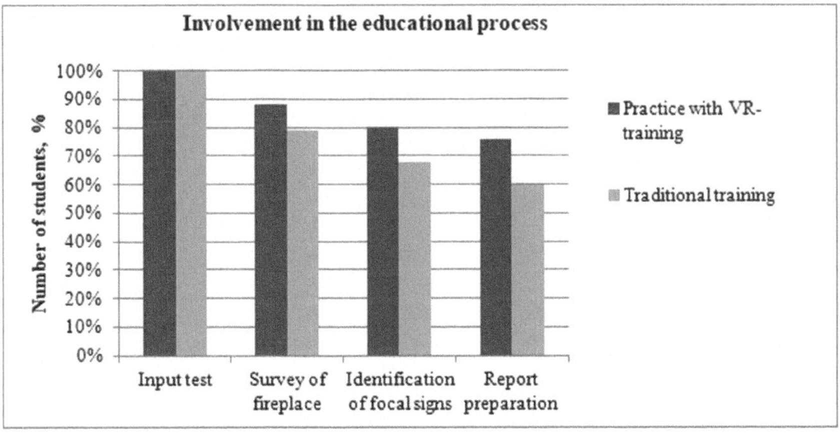

Fig. 5. Comparison of students' involvement in a practical fire site inspection training.

Outcome analysis showed that the virtual simulator is associated with higher students' involvement in the learning process and lower students' workload than conventional lessons. Along with this, a students' performance diagram (see Fig. 6) demonstrates increased learning efficiency associated with the virtual simulator.

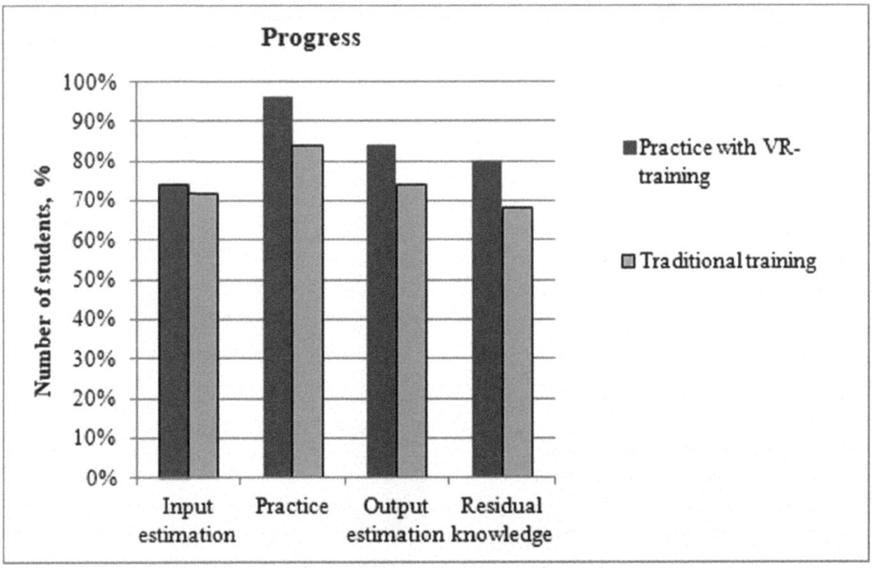

Fig. 6. Comparison between performance of Fire Investigation students.

χ^2 criterion was selected to compare scores of study and control groups demonstrating students' knowledge before and after the experiment. These scores are presented in a 3-point order scale (Figs. 7 and 8). The group dimension was a number of students who achieved the target score at incoming and final assessments. χ^2 statistic test was selected as one providing the most reliable match and difference of findings along with low L value of scale steps [5] because L = 3 stands for «3», «4» and «5» grades in our case.

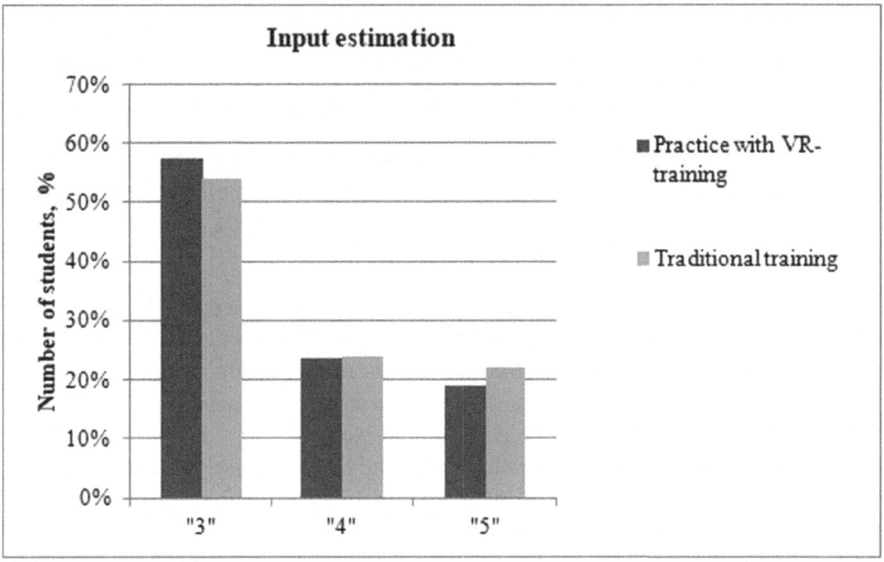

Fig. 7. Bar graphs: incoming assessment scores of students using the virtual simulator and conventional methods.

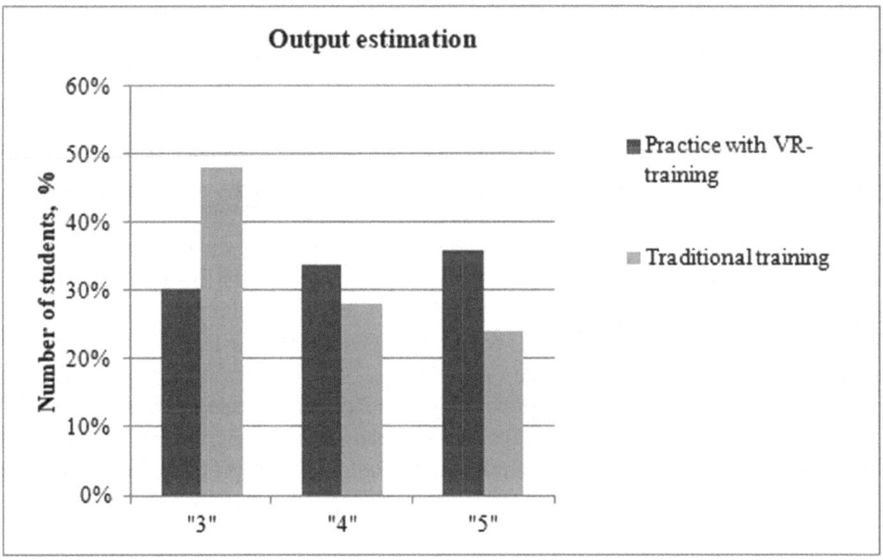

Fig. 8. Bar graphs: final assessment scores of students using the virtual simulator and conventional methods.

4 Discussion

The empirical value of χ^2 criterion resulted from the incoming assessment of the experimental and control groups (Fig. 7) is $\chi^2_{emp} = 0.29$. The value is less than the tabulated one ($\chi^2_{tab} = 5.99$ (L value of order scale steps is 3, significance level = 0.05). Thus, this indicates match of compared sample characteristics (significance level = 0.05).

Comparison of final assessment criterion values (Fig. 8) (i.e., empirical $\left[\chi^2_{emp} = 6.47\right]$ and tabulated $\left[\chi^2_{tab} = 5.99\right]$ ones; $\chi^2_{emp} > \chi^2_{tab}$), demonstrates 95% reliability of differences between characteristics of compared samples. Hence, we can conclude that the effect of these changes is associated with application of the method related to formation of professional competencies in the sphere of fire investigation implemented in the learning process of competence-based didactic assistance of professional training based on virtual reality technologies.

High efficiency of the virtual simulator used the learning process is associated with increased students' involvement, decreased students' efforts during practical material consolidation due to application of advanced representative technologies of immersion into a working area, as well as a great number of various training ranges improving learning outcomes.

The study outcomes confirmed that competency-based didactic assistance based on virtual reality technologies will make formation of professional students' competencies in the sphere of fire investigation more efficient.

5 Conclusion

Our experiments confirmed that competency-based didactic assistance based on virtual reality technologies will make formation of professional students' competencies in the sphere of fire investigation more efficient. Along with this high efficiency of the virtual simulator implementation into learning process is associated with increased students' involvement, decreased students' efforts during practical material consolidation due to application of advanced representative technologies of immersion into a working area, as well as a great number of various training ranges improving learning outcomes.

The study does not cover all the opportunities of virtual reality technologies considered as a base to develop competency-based didactic assistance of learning process, and it represents one of potential ways of its implementation. Further development of VR-technologies application for training of fire engineering professionals is still of immediate interest.

References

1. Subacheva, A.: Computer modelling-based didactic assistance of professional training of fire safety engineers. Ekaterinburg (2012)
2. Galishev, M., Sharapov, S., Popov, A.: Fire Investigation: Manual. University of Fire and Rescue Academy EMERCOM of Russia, Saint Petersburg (2013)

3. Cheshko, I., Yun, N., Plotnikov, V.: Fire Site Inspection: Guideline. Federal State Institution Scientific Research Institute of Fire Protection, Moscow (2004)
4. Pozharkova, I., Lagunov, A., Slepov, A., Gaponenko, M., Troyak, E., Bogdanov, A.: Increase in efficiency of fire investigator training by means of virtual reality technologies. Siberian Fire Rescue Bull. Anal. Res. J. **4**, 96–100 (2019)
5. Novikov, D.: Statistical Methods of An Educational Research (Typical Cases). MZ-Press, Moscow (2004)

Real-Time Data Compression System for Data-Intensive Scientific Applications Using FPGA Architecture

Mohammed Bawatna(✉), Oliver Knodel, and Rainer G. Spallek

Institut für Technische Informatik, Technische Universität Dresden, Dresden, Germany
m.bawatna@hzdr.de

Abstract. Particle accelerators are continually advancing and offer insights into the world of molecules, atoms, and particles on the ever shorter length and timescales. A variety of detectors, which are connected to different front-end electronics are installed in various kinds of Data Acquisition (DAQ) systems, to collect a huge amount of raw data. This goes along with a rapid and highly accurate transformation of analog quantities into discrete values for electronic storage and processing with exponentially increasing amounts of data. Therefore, data reduction or compression is an important feature for the DAQ systems to reduce the size of the data transmission path between the detectors and the computing units or storage devices. The flexibility of the Field Programmable Gate Arrays (FPGAs) allows the implementation of real-time data compression algorithms inside these DAQ systems. In this contribution, we will present our developed real-time data compression technique for continuous data recorded by high-speed imaging detectors at the terahertz source facility at ELBE particle accelerator. The hardware implementation of the algorithm proved its real-time suitability by compressing one hundred thousand consecutive input signals without introducing dead time.

Keywords: Lossless data compression · FPGA · Real-time · Data-intensive

1 Introduction

Data compression is a process in which the amount of digital data is compressed or reduced. This reduces the required storage space and decreases the transmission time of the data. The data compression methods attempt to remove redundant information. The reversal is called data decompression. There are two types of data compression methods: lossy and lossless methods. In the case of lossless compression, as in [1], all original data can be obtained again from the compressed data. However, in the case of lossy compression or irrelevance reduction, the original data cannot be recovered exactly from the compressed data, which means that part of the information is lost. In addition to a high data compression performance, the algorithms used for data compression should meet other requirements, as in [2], such as high adaptability, high robustness, low latency, and low complexity.

R. Silhavy (Ed.): CSOC 2020, AISC 1226, pp. 304–313, 2020.
https://doi.org/10.1007/978-3-030-51974-2_29

The adaptation to the signal statistics allows higher compression rates. The better the algorithm can adapt to the data, the fewer changes in the signal statistics can have a negative effect. The robustness of the data compression ensures that the vulnerability of an algorithm to transmission errors is neglected. It would, therefore, be necessary to be able to reconstruct large parts of the data despite the holes in the received code. The low latency times for remotely controllable systems is important for the bidirectional image and sound transmission such as in video conference, where very short delay times are necessary. The data compression method should also have a low level of complexity. This reduces the programming effort for FPGA processors and the costs for chip production Application-Specific Integrated Circuits (ASICs), and storage space for Digital Signal Processing (DSP) solutions can be saved.

The data compression requirements are correlated. High compression rates can be achieved by very good adaptation to the signal, but require additional operations and processing time. Choosing the right compression strategy as well as new developments, require knowledge of the practical area of application.

The use of FPGAs architecture is becoming increasingly important in data processing due to new hardware-based developments. Applications that are too slow on common PC systems can be accelerated with reconfigurable hardware. Many DAQ systems and frame grabbers are now equipped with an FPGA. In addition to the tasks of data acquisition, additional algorithms can be executed by using an FPGA. A typical field of application is the reduction of the amount of raw data in order to minimize transmission times. The current data compression methods and its implementations do not achieve the real-time processing speed at high frame rates. There is a need for new methods and efficient implementations that allow for sufficient reduction performance without distorting the recorded data. Another area of responsibility is the reduction of data for an application-specific case. In this case, the flexibility of FPGAs is advantageous because they are not fixed to a specific application. In literature, the current implementation built with discrete logic devices, as in [3–6], reaches only less than 200 frame rates per second and is insufficient for modern DAQ systems that install high-speed ADCs. An FPGA-based solution that can communicate directly with the image sensor can achieve the timing constraints requirements.

This paper organized as follows: first, a brief discussion about the current hardware accelerators such as ASICs and FPGAs. Second, we will explain the characteristics of the digitized detector data at the terahertz radiation source in the ELBE particle accelerator. Third, a comparison of different lossless methods will be discussed. Fourth, we will present our current development progress of the real-time data compression on the FPGA-based DAQ system and its prospects. After summarizing the results, foreseeable future development and upgrades are discussed.

2 Hardware Acceleration

Software solutions for data compression often cannot meet the time requirements; therefore, many data compression algorithms are already implemented in FPGA hardware as in [7] and [8]. The efficient implementations in hardware make the implementation performance and cost advantages more attractive.

As an alternative to microprocessors (CPUs), DSPs and ASICs can be used. DSPs allow the treatment of complex algorithms by sequential processing. This keeps them flexible and adaptable to the problem. They also include fast calculation units for floating-point arithmetic as well as analog functions such as ADC and DAC converters and interfaces for special device drivers and timers. However, the commands and data are executed by a von Neumann Architecture in the CPUs. This circumstance forms a bottleneck and limits the processing speed, which leads to a technical limit for real-time applications. This architecture indicates that the program and data are transferred to the same storage space and the same main memory via the data buses. In the ASICs architecture, the processing of algorithms takes place in parallel, or the data are pushed through processing chains (pipelines) to start loops. By simple case distinctions, multiplexers are controlled, which switches the data through to different processing units. When high parallel processing, high speeds, or low power consumption are required, application-specific architectures offer significant advantages.

FPGAs have long been used in the area of prototyping development to provide hardware algorithms e.g., for ASIC implementations. They have the same advantages over CPUs and DSPs as ASICs. Compared to ASICs, the advantage of FPGAs is lower cost and re-programmability. Design changes can, therefore, be made quickly and at a little additional cost (rapid prototyping). In the industrial sector, so far, ASIC solutions have dominated since the delay times of comparable FPGAs are higher by a factor of 10 to 100. Due to the strong residency of FPGA resourcing, new applications are possible. However, FPGAs have a complexity limit, the more complex the problem, the harder the circuit. Programming FPGAs requires expert knowledge. Compared to the ASIC design, however, less physical and electro-technical knowledge is required. Prefabricated design bells can start on a higher level; the testing of the circuit on the hardware platform offers significant benefits. The advancement of hardware description languages and new language concepts, such as hardware description languages VHDL and Verilog, also offer a speedy improvement. In the field of data compression, the use of FPGAs and ASICs can be more efficient in achieving solutions.

While ASICS are hard-coded, FPGAs allow them to adapt to changes in standards or use new, improved enhancements. Now that new standards are being designed with hardware in mind, they are less complex and do not require fewer hardware resources than the old standards. Frame grabbers use the FPGA to capture the data and allow the connection of different camera types. In addition to this task, there is often potential for further data compression

3 Characteristics of the Digitized Detector Data at Terahertz Radiation Source in Elbe

The radiation source ELBE (Electron Linear accelerator for beams with high Brilliance and low Emittance) as in [11] delivers multiple secondary beams, both electromagnetic radiation, and particles. The characteristics of these beams make ELBE an outstanding research instrument for external users as well as scientists of the Helmholtz Zentrum Dresden Rossendorf (HZDR) as in [9], and [10].

As part of the ELBE accelerator, one electron beamline has been modified to allow for the generation and acceleration of ultra-short, less than 150 femtoseconds, highly charged electron bunches. This upgrade enables the operation of high-field terahertz sources, as in [9], and [10], based on super-radiant terahertz emission at the ELBE accelerator and thereby opens up the opportunity to generate carrier-envelope phase-stable high-field THz pulses with extremely flexible parameters with respect to repetition rate (between a few tens of Hz to eventually 13 megahertz), pulse form, and polarization.

3.1 Diagnostics and Control Signals

The femtosecond level diagnostic and control of sub-picosecond electron bunches is an essential topic in modern accelerator research. Accurate timing of an accelerator to an external laser system can be accomplished in several ways. Several signal processing operations need to be done on the measured arrival time information to achieve an excellent time resolution, and performance accuracy of a few ten fs. For each THz pulse, a signal consists of 2048 pixels is recorded, for diagnostics purposes, by a linear array detector camera as in Fig. 1.

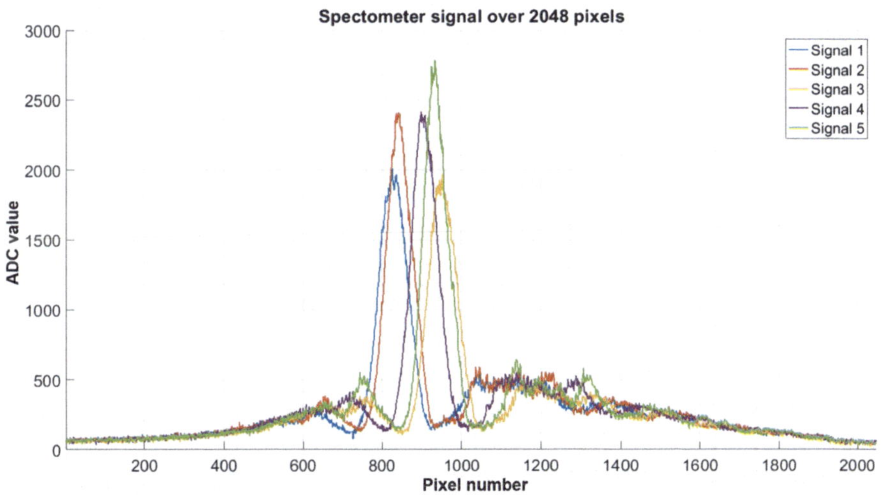

Fig. 1. The signal measured by the imaging detector for five consecutive signals of THz pulse at TELBE at repetition rate of 100 kHz.

The demand for high SNR and more accurate time resolution has lead in the recent years to enhancements of the sensor technology and ADC architecture. However, increasing the clock rate in the ADC, and thus the increase in the data acquisition rate, make the task of saving and processing the data more complex. In the current DAQ system at TELBE, the raw data is saved in offline storage devices. The current DAQ system at TELBE faces the challenge of allocating the required resources to save the longtime scale data of a few weeks at 100 kHz repetition rates of THz pulses. For each day, tens of loops are measured. The required disk capacity to store one loop of raw data is 89.7 Gigabytes.

3.2 Real-Time DAQ System at TELBE

The online DAQ system, as in [11], enables continuous data acquisition and analysis at the highest, which provides opportunities for new experiments. The arrival time information for each THz pulse is recorded as a vector of 256 pixels using a KALYPSO linear array detector [12]. The measure photon energy at each pixel of the size of 50 µm × 3 mm is converted into a digital value with a 12-bit resolution.

The architecture of the DAQ system as shown in Fig. 2 is divided into three main parts: KALYPSO front-end, a Xilinx Virtex-7 XC7VX330T FPGA board and a GUI used for controlling the parameters such as the image exposure time, noise threshold, as well as, storing the raw data and performing the online signal processing. The imaging detector front-end is connected to the FPGA board by an FPGA Mezzanine Card (FMC) connector. When the exposure or integration time is provided to the sensor, and the FPGA issues a frame request command, the image is stored in the pixel-matrix. The pixel values are digitized. These values are transferred using Low Voltage Differential Signal (LVDS) channels. Each LVDS channel is responsible for a group of adjacent columns of the pixel matrix. A standard PCI Express (PCIe) connection is used to transfer the data from the camera directly either to NVIDIA GPU card or to the main computer memory to avoid the bottleneck between data acquisition and the limitation of the camera's internal buffers.

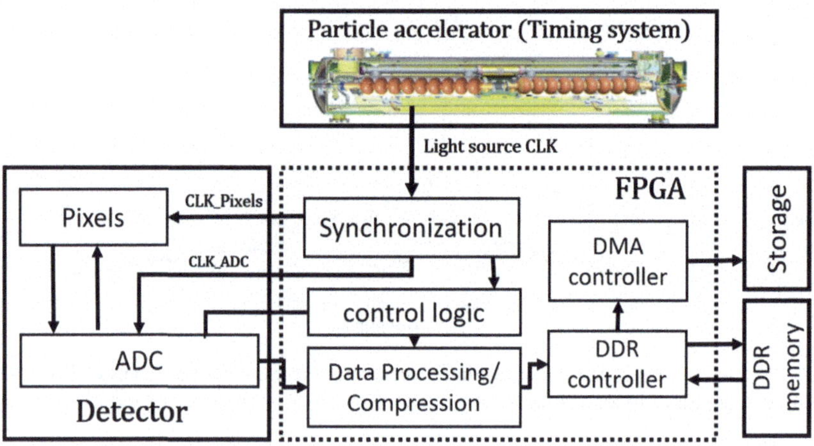

Fig. 2. The block diagram of the developed DAQ system at TELBE facility.

4 Comparison of Different Lossless Methods

Considering the importance of not distorting the recorded data by the various detectors at the particle accelerators, lossless compression methods are preferred instead of lossy methods. Most lossless compression programs perform two steps in sequence: the first step generates a statistical model for the input data, and the second step uses this model to map input data to bit sequences. The three primary lossless algorithms, as in [1, 2, 13], and [14] are Run-length, Huffman coding, and arithmetic coding.

In order to select the best lossless compression method to implement in the FPGA installed in our DAQ system to achieve the real-time data compression, we performed a comparison between the three primary lossless methods as shown in Table 1. The sample file with TIFF format was used in this comparison with the size of 40960.261 KB. The file is then compressed with Run-length, Huffman, and Arithmetic.

Table 1. Performance comparison of lossless compression algorithms.

Parameter	Lossless algorithm		
	Run-length	Arithmetic	Huffman
Compression size (KB)	22505.64	5657.49	11669.59
Compression ratio	1.82	7.24	3.51
Compression speed (bits/sec)	13.8	15.6	28.1
Decompression speed (bits/sec)	13.75	19.2	29.5

The performance parameters, as shown in Table 1, was calculated using Eqs. 1, and 2 as follows:

$$\text{Compression Ratio} = \frac{\text{Uncompressed Size}}{\text{Compressed Size}} \quad (1)$$

$$\text{(De)compression Speed} = \frac{\text{Compressed bits}}{\text{Seconds to (de)compress}} \quad (2)$$

These results indicate that the Arithmetic has an advantage in compression size over the Run-length and Huffman algorithms. On the other hand, the Huffman has an advantage in speed to compression and decompression over the Run-length and Arithmetic, which makes it more convenient for real-time data compression. These data are expected to be a reference for next developing in the FPGA architecture.

5 Implementation of the Algorithm in an FPGA

As discussed in section four, it is important not to distort detector data, as well as the fastest method in both compression and decompression. In this section, we will present our real-time data compression implementation on the FPGA architecture, shown in Fig. 2, using the Huffman method.

5.1 Algorithm

The Huffman encoder determines the character frequency of all characters in the file as in [13]. A binary tree is then created on the basis of this data, whereby the two nodes with the lowest number of characters are always attached to a common parent node, which bears the sum of the number of characters of its daughter nodes. The procedure is analogous for all nodes until all nodes are in a tree with a root, the number of characters of which is 100%. The resulting tree can only be used to encode the file. The left branches of the tree are labeled 0, the right branches 1, which means that all characters are encoded using a specific sequence of 0 and 1.

The generated tree must be saved for subsequent decoding. There is also the possibility that only the frequency distribution is saved with the encoded file, although it must be ensured that the encoder and decoder generate the same tree from this frequency distribution. The Huffman coding is a simple and effective compression method, which, however, depends on a favorable frequency distribution of the characters. The optimal compression would only be achieved if the number of characters was powers of two, and each character appeared at different times.

5.2 Hardware Implementation

The Huffman algorithm, as mentioned above, can be stored in a table together with an unspecified statement. The Huffman table can be implemented in a lookup table (LUT), which is addressed above the characters to be coded. The consolidation of the individual code-words into codes of fixed length is complex and was implemented as in [15]. The design is shown in Fig. 3.

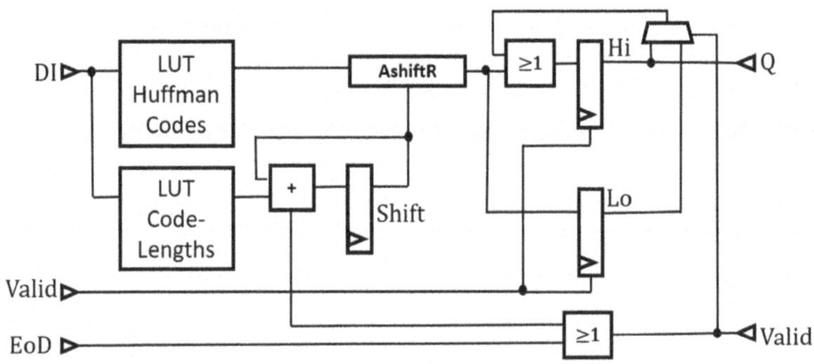

Fig. 3. Construction of the LUT-based Huffman encoding.

The LUTs are codes of the array shifters (AshiftR), which shift each bit array n bits right one clock at a time. The code lengths are added up and stored in the register Shift, which controls the array shifter. If the adder signals a carry-out, the Hi register is loaded off the shifted code-word and the Lo register. The End-Of-Data (EoD) input resets the shift register and generates a valid signal for the remainder.

In the FPGA, it can happen that the LUT becomes too big. In addition, many bits are unused in the LUT for the code-words due to the variable length of the codes. The width of the table must be based on the longest code-word. For a Xilinx XC7VX330T series, FPGA capable of storing thirty-two bits in a CLB, which requires a total of 148 CLBs for the table. Additional logic, such as the multiplexer, is necessary to drive the LUTs. It can be assumed that approximately one-third of the bits remain unused due to the variable code length. A single LUT needs forty bits, leaving thirteen bits unused. Breaking into bit-oriented LUTs yields three LUTs, each with eight bits, one LUT with four, and one with two bits. The selection of the two small LUTs is made via simple logic links, and the number of redundant bits is reduced to three bits. A disadvantage of the current implementation is the necessary additional control logic, which may necessitate the incorporation of further register stages and may lead to slightly increased latencies.

5.3 Speed and Resources

For 8-bit pixels and a maximum code string of 13 bits, the two tables with control logic occupy approximately 210 CLBs. The rest of the logic occupies about ninety more CLBs, and the array shifter must be able to accommodate two maximum code-words. The two output registers must share three code-words together in the worst case. This gives 32 bits for the array shifter and 24 bits each for the output registers. Due to the LUTs and the array shifter, the design runs at a maximum of 100 MHz and processes one pixel per clock. The reduction performance of this lossless process depends on the Huffman code used. A code adapted to the signal entropy achieves a very good compression, which does not differ from a software-based solution.

5.4 Hardware Results

In order to evaluate the implemented data compression algorithm in the FPGA which is installed in our DAQ system, we performed the similar test as in Sect. 4 of this paper. The sample file with TIFF format was used in this evaluation with the size of 40960.261 KB. This TIFF image consists of one hundred thousand consecutive signals, each signal is stored as one row in this TIFF image (2048 pixels, 16 bits resolution as shown in Fig. 1). The file is then compressed with the implemented Huffman algorithm. The hardware implementation of the algorithm proved its real-time suitability without introducing dead time. Only an average latency that varies between 10 to 45 clock cycles. The designed data compression intellectual property block is available for an implementation in current and future Front End Electronics (FEE) ultra-fast detectors inside FPGAs.

6 Summary and Future Work

The developed lossless compression method on FPGA architecture in this contribution is based on a Huffman coding. This method has been selected due to its high compression efficiency and speed in comparison with other methods and implemented on FPGA architecture to achieve a real-time data compression for the high-speed imaging detector installed in our DAQ system at TELBE user facility.

The suitability of FPGAs as a hardware architecture for data compression for modern high speed detectors in DAQ systems was discussed in this contribution. The flexibility of FPGAs is important in application-oriented data compression, and allow us to achieve high processing speed. The new FPGAs generations decrease the limitations due to complexity and resource boundaries. A software-based solution on a conventional CPU could be flexible and also benefits from the advancement in processor development. However, it still lags in speed due to the von Neumann architecture. ASICs are faster than FPGA-based solutions due to the SRAM and multiplexer based control used in FPGAs, which is eliminated in ASICs. However, ASICs are not reprogrammable and can only be used for the specified problem. The speed advantage of ASICs over FPGAs, as well as the favorable cost factor for large chip counts, becomes less important in the data compression field as far as the requirement of real-time image processing is achieved. The real-time limit was fixed at about one thousand frames per second. However, the installed imaging detector in our DAQ system allows for a frame rate of up to 2.7 MHz. In order to perform the data compression above this limit, which is in our near future plan, either a new FPGA hardware platform with a higher clock rate must be used, or a GPU card should be run in parallel with our DAQ system.

References

1. Sharma, K., Gupta, K.: Lossless data compression techniques and their performance. In: 2017 International Conference on Computing, Communication and Automation (ICCCA) (2017). https://doi.org/10.1109/ccaa.2017.8229810
2. Uthayakumar, J., Vengattaraman, T., Dhavachelvan, P.: A survey on data compression techniques: from the perspective of data quality, coding schemes, data type and applications. J. King Saud Univ. - Comput. Inf. Sci. (2018). https://doi.org/10.1016/j.jksuci.2018.05.006
3. Nishikawa, Y., Kawahito, S., Inoue, T.: Parallel image compression circuit for high-speed cameras. Real-Time Imaging IX **10**(1117/12), 588030 (2005)
4. Lien, J.-M., Kurillo, G., Bajcsy, R.: Multi-camera tele-immersion system with real-time model driven data compression. Vis. Comput. **26**, 3–15 (2009)
5. Tawel, R.: Real-time focal-plane image compression. In: [Proceedings] DCC 1993: Data Compression Conference. https://doi.org/10.1109/dcc.1993.253109
6. Patauner, C., Marchioro, A., Bonacini, S., Rehman, A.U., Pribyl, W.: A lossless data compression system for a real-time application in HEP data acquisition. In: 2010 17th IEEE-NPSS Real Time Conference (2010). https://doi.org/10.1109/rtc.2010.5750389
7. Fajardo, C.A., Angulo, C.A., Mantilla, J.G., Obregon, I.F., Castillo, J., Pedraza, C., Reyes, O.M.: Computational architecture for fast seismic data transmission between CPU and

FPGA by using data compression. In: 2016 Data Compression Conference (DCC) (2016). https://doi.org/10.1109/dcc.2016.76
8. Rigler, S., Bishop, W., Kennings, A.: FPGA-based lossless data compression using Huffman and LZ77 algorithms. In: 2007 Canadian Conference on Electrical and Computer Engineering (2007). https://doi.org/10.1109/ccece.2007.315
9. Bawatna, M., Green, B., Deinert, J.-C., Kovalev, S., Knodel, O., Spallek, R., Cowan, T.: Pulse-resolved data acquisition system for THz pump laser probe experiments at TELBE using super-radiant Terahertz sources. In: 2019 IEEE MTT-S International Microwave Workshop Series on Advanced Materials and Processes for RF and THz Applications (IMWS-AMP) (2019). https://doi.org/10.1109/imws-amp.2019.8880116
10. Kovalev, S., Green, B., Golz, T., Maehrlein, S., Stojanovic, N., Fisher, A.S., Kampfrath, T., Gensch, M.: Probing ultra-fast processes with high dynamic range at 4th-generation light sources: arrival time and intensity binning at unprecedented repetition rates. Struct. Dyn. **4**, 024301 (2017)
11. Bawatna, M., Green, B., Kovalev, S., Deinert, J.-C., Knodel, O., Spallek, R.G.: Research and implementation of efficient parallel processing of big data at TELBE user facility. In: 2019 International Symposium on Performance Evaluation of Computer and Telecommunication Systems (SPECTS) (2019). https://doi.org/10.23919/spects.2019.8823486
12. Lorenze, R., et al.: KALYPSO: a Mfps linear array detector for visible to NIR radiation. In: IBIC2016, 14–16 September, Barcelona, Spain, pp. 740–743 (2017). https://doi.org/10.18429/JACoW-IBIC2016-WEPG46
13. Huffman Coding. Springer Reference. https://doi.org/10.1007/springerreference_73181
14. Pu, I.M.: Run-length algorithms. In: Fundamental Data Compression, pp. 49–65 (2006)
15. Bhaskaran, V., Konstantinides, K.: Image and Video Compression Standards: Algorithms and Architectures. Kluwer Academic Publishers, Boston (1995)

Method for Evaluating Statistical Characteristics of Fluctuations in the Total Electronic Content of the Ionosphere Based on the Results of its GPS-Sensing

M. V. Peskov, V. P. Pashintsev, A. F. Chipiga(✉), M. A. Senokosov, and I. V. Anzin

North Caucasus Federal University, Stavropol, Russia
mvpeskov@hotmail.com, afchipiga@ncfu.ru

Abstract. A method that allows to determine the change in time of the average value of the total electronic content of the ionosphere and the standard deviation of its small-scale fluctuations based on data obtained with using a two-frequency receiver of GLONASS/GPS satellite radio navigation systems has been developed. The 6th-order Butterworth digital filter is used to determine the average value of the total electronic content and its small-scale fluctuations. #CSOC1120.

Keywords: Ionosphere · Full electronic content · GPS monitoring

1 Introduction

It is known [1–5] that the occurrence of intense fluctuations in the phase and the amplitude of the received signals (flickering, fading), which cause a significant decrease of the quality indicators of satellite radio systems, is due to the diffraction of radio waves on small-scale inhomogeneities of the electronic concentration of the ionosphere.

The study of the thin structure of the ionosphere and the influence of small-scale inhomogeneities of its electronic concentration on the quality indicators of space radio systems is traditionally based on the analysis of statistical parameters of received radio signals (the average signal level, the flicker index, the standard deviation (SD) of fluctuations in the phase, correlation functions, etc.) [1–4].

A thorough analysis of the influence of conditions transionospheric radio wave propagation of the quality indicators of satellite radio systems, which had allowed to establish their relationship with the statistical characteristics of fluctuations ΔN_T of the total electronic content (TEC) of the ionosphere N_T (its average value $\bar{N}_T$ and SD $\sigma_{\Delta N_T}$), was carried out [5–7]. However, methods for evaluating the quality indicators of space radio systems, which are developed in [5–7], are based on averaged data on ionospheric TEC obtained during numerous experiments on creating artificial ionospheric disturbances.

R. Silhavy (Ed.): CSOC 2020, AISC 1226, pp. 314–320, 2020.
https://doi.org/10.1007/978-3-030-51974-2_30

At the same time, it is possible to determine the ionosphere's TEC in real time using a specialized two-frequency receiver of satellite radio navigation systems (SRNS) GLONASS/GPS [8–10]. A method that allows to select from the time series of changes in the ionosphere's ($N_T(t)$) TEC such components that describe its small-scale fluctuations with using a simple digital moving average filter was developed in [9].

The need to refine the method [9] is due to the lack of a description of the process of determining the statistical characteristics of small-scale fluctuations of TEC and the disadvantages of the applied smoothing procedure using a simple (arithmetic) moving average: reducing the amplitude of the smoothed time series and the dependence of the digital filter structure on the smoothing parameters.

The purpose of the article is to develop a method for determining the statistical characteristics of small-scale fluctuations in the total electronic content of the ionosphere, which allows to eliminate the shortcomings of the known method [9].

2 Method

According to [9], the accumulation of discrete values of the ionosphere's TEC coming from the output of a specialized two-frequency receiver of GLONASS/GPS SRNS allows you to form a time series

$$N_T(t) = \bar{N}_{T0}(t) + \sum_{l=l_{\min}}^{l_{\max}} \Delta N_T(t, l), \qquad (1)$$

that reflects the change in time of TEC's fluctuations $\Delta N_T(t, l)$ relative to the average (background) value $\bar{N}_{T0}(t)$, caused by the presence of inhomogeneities of electronic concentration with scales on the propagation path $l_{\min} \le l \le l_{\max}$. The first term of the series (1) characterizes a relatively slow (about tens of minutes and units of hours) change in the TEC due to the movement of the navigation spacecraft (NSC), which leads to a change in the path length of the wave in the ionosphere with an inclined propagation [4, 8, 9].

It is obvious that the change in time $\Delta N_T(t, l)$ of each of the components of the second term of the series (1) will be determined by the movement of a set of inhomogeneities distributed along the propagation path of the electronic concentration $\Delta N(l)$ within the zone (with radius l_F) essential for the propagation of radio waves (the first Fresnel zone), with some effective scanning speed v_e [1–3, 9].

Thus, it is possible to make a transition from the spatial characteristics of inhomogeneities of the electronic concentration of the ionosphere to the temporal (frequency) characteristics of fluctuations of its TEC, since the components of the second term of the series (1) will no longer be characterized with linear sizes l of inhomogeneities, but with a period $\tau_f = l/v_e$ (frequency is $f_F = 1/\tau_F = v_e/l$) of fluctuations of the TEC.

It is known [1, 4, 5] that fading of radio signals is caused by scattering of radio waves on small-scale inhomogeneities of the electronic concentration of the ionosphere, whose linear size $l \le l_F$ does not exceed the radius l_F of the first Fresnel zone.

Therefore, from the point of view of the manifestation of the diffraction properties of the ionosphere, inhomogeneities (with sizes $l_{\min} \leq l \leq l_F$) that cause fluctuations in the TEC $\Delta N_T(t)$ with periods $l_{\min}/v_e \leq \tau_F \leq l_F/v_e$ and corresponding frequencies $v_e/l_F \leq f_F \leq v_e/l_{\min}$ should be considered as small-scale ones.

It is assumed that the inhomogeneities (with sizes $l > l_F$) of the electronic concentration correspond to fluctuations of the TEC with periods $\tau_F > l_F/v_e$ (frequencies are $f_F < v_e/l_F$), which together with the first term of the series (1) cause the manifestation of only the dispersion properties of the ionosphere. Therefore, in the future, the average value of the ionosphere's TEC $\bar{N}_T(t)$ will be understood as

$$\bar{N}_T(t) = \bar{N}_{T0}(t) + \sum_{l=l_F}^{l_{\max}} \Delta N_T(t,l). \tag{2}$$

In this case, taking into account (2), the expression (1) will take the form

$$N_T(t) = \bar{N}_{T0}(t) + \sum_{l=l_F}^{l_{\max}} \Delta N_T(t,l) + \sum_{l=l_{\min}}^{l_F} \Delta N_T(t,l) = \bar{N}_T(t) + \Delta N_T(t). \tag{3}$$

Random fluctuations of the TEC $\Delta N_T(t)$, which are contributed by a set of electronic concentration inhomogeneities distributed along the propagation path, are characterized with the SD

$$\sigma_{\Delta N_T}(\tau_{\text{SD}}) = \sqrt{\frac{1}{n}\sum_{i=1}^{n}[\Delta N_T(t_i) - \langle \Delta N_T(\tau_{\text{SD}})\rangle]^2} = \sqrt{\frac{1}{n}\sum_{i=1}^{n}\Delta N_T^2(t_i)} \tag{4}$$

and zero mathematical expectation $\langle \Delta N_T(\tau_{SD})\rangle = \frac{1}{n}\sum_{i=1}^{n}\Delta N_T(t_i) = 0$ on the interval $\tau_{\text{SD}} = t_n - t_1 = l_F/v_e$ [5].

According to [9], to select individual components of a time series (1), a simple (arithmetic) moving average smoothing procedure is applied to it sequentially (with window widths $\tau_{\text{smth}\,1}$ и $\tau_{\text{smth}\,2}$). The parameter $\tau_{\text{smth}\,1}$ is chosen so to eliminate in the original time series (1) fluctuations with the periods $\tau_F > l_F/v_e$, and the parameter $\tau_{\text{smth}\,2}$ provides definition of small-scale fluctuations with minimal period $\tau_F \geq 5\tau_s$, where τ_s is the sampling interval of dual frequency receiver of SRNS.

Analysis of [9] allows us to conclude that the procedure for determining the average value of the ionosphere's TEC $\bar{N}_T(t)$ is equivalent to passing the initial series (1) through a low-pass filter (LPF) with a cutoff frequency $f_{\text{co}1} = 1/\tau_{\text{smth}\,1} = v_e/l_F$, and the procedure for determining small-scale fluctuations of the ionosphere's TEC is equivalent to passing a series (1) through a pass band filter (PBF) with cutoff frequencies $f_{\text{co}\,1}$ and $f_{\text{co}2} = 1/\tau_{\text{smth}\,2} = 1/5\tau_s = f_s/5$. It is worth mentioning that after filtering with the simple moving average method, the amplitude of TEC's fluctuations $\Delta N_T(t)$ decreases (the filter gain is $G_f < 1$ in the bandwidth), which is a significant disadvantage of using this method when studying the thin structure of the ionosphere [10].

Each k-th output (smoothed) sample $y(k)$ of such a digital filter is determined with the expression

$$y(k) = \frac{1}{n}\sum_{m=1}^{n} x\left(k - \frac{n}{2} + m\right), \tag{5}$$

where $n = \tau_{\rm smth}/\tau_{\rm s}$ is the number of input samples x with a sampling interval $\tau_{\rm s}$ participating in the formation of the output sample $y(k)$.

Since the small-scale fluctuations of the TEC $\Delta N_T(t) \ll \bar{N}_T(t)$ are generally much smaller than its average value $\bar{N}_T(t)$ and their frequencies are almost evenly distributed in the range $f_{\rm F} = v_{\rm e}/l_F \ldots f_{\rm s}/5$, the amplitude-frequency response of the digital filter used to determine them should be as smooth as possible in the pass band and suppression band and also be quite steep in the transition band. In addition, the digital filter should not lead to a decrease in the amplitude of small scale fluctuations of the TEC ($G_{\rm f} = 1$). These conditions are met by the Butterworth digital filter, which is successfully used for smoothing the parameters of received radio signals (amplitude and phase) in ionospheric flickering studies and implemented in hardware in modern specialized two-frequency receivers [4, 11–13].

Each k-th output sample $y(k)$ of such a digital filter is determined with the expression

$$y(k) = \frac{1}{a_0}\left[\sum_{m=0}^{k} b_m x(k-m) - \sum_{m=1}^{k} a_m y(k-m)\right], \tag{6}$$

where coefficients a_m and b_m are defined with cutoff frequency (frequencies) $f_{\rm co}$ of the digital filter and data sampling frequency $f_{\rm s}$, and m defines the order of the filter [4, 13]. To smooth out the results of measuring the amplitude and phase of received radio signals in the study of ionospheric flickering, the 6th-order Butterworth digital LPF with a 3 dB attenuation at the cutoff frequency $f_{\rm co}$ is used [4, 11]. To facilitate hardware implementation, the 6th-order filter is often presented as three sequentially connected 2nd-order sections [4].

The advantage of using a digital Butterworth filter to determine the average value of the ionosphere's TEC $\bar{N}_T(t)$ and its small-scale fluctuations $\Delta N_T(t)$ is also the immutability of its structure when the cutoff frequency (frequencies) $f_{\rm co}$ and data sampling frequency $f_{\rm s}$ change. This means that only the coefficient values a_m and b_m will change in expression (6), not their number and the number of input samples, in contrast to the moving average filter (5), in which the number of smoothed input samples $n = \tau_{\rm smth}/\tau_{\rm s}$ can change depending on the width of the smoothing window $\tau_{\rm smth}$ and the data sampling interval $\tau_{\rm s}$.

Thus, to determine the average value of the ionosphere's TEC $\bar{N}_T(t)$ and its small-scale fluctuations $\Delta N_T(t)$ based on TEC $N_T(t)$ measurements with a sampling frequency $f_{\rm s}$, it is proposed to use the 6th-order digital Butterworth LPF with a 3 dB attenuation at the cutoff frequency $f_{\rm co\,1} = v_{\rm e}/l_F$ and the 6th-order digital Butterworth BPF with a 3 dB attenuation at the cutoff frequencies $f_{\rm co\,1}$ and $f_{\rm co\,2} = f_{\rm s}/5$.

The block diagram of the algorithm that implements the developed method is shown in Fig. 1.

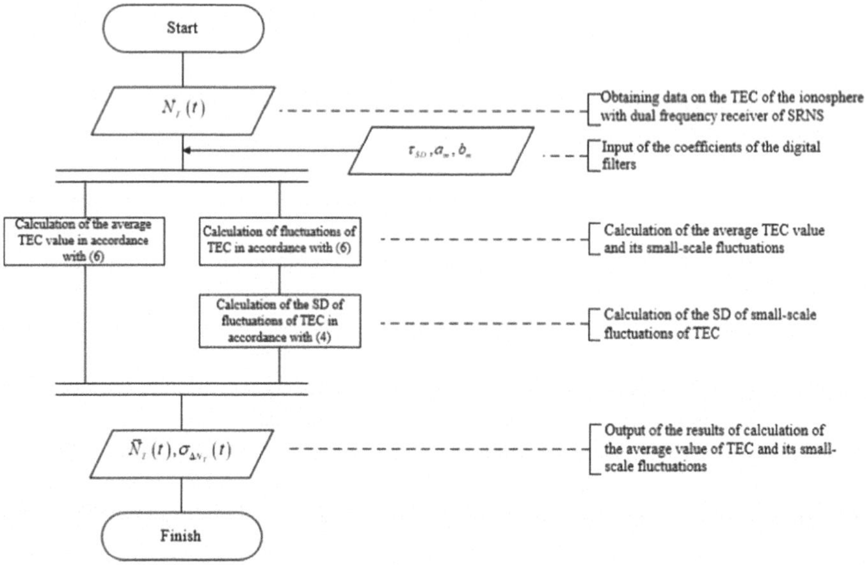

Fig. 1. Block diagram of the algorithm for determining the change in time of the average value and the standard deviation of the ionosphere's TEC based on data from the dual-frequency receiver of GLONASS/GPS

It follows that the method for determining the change in time of the average value of the ionosphere's TEC $\bar{N}_T(t)$ and SD $\sigma_{\Delta N_T}(t)$ of its fluctuations on the basis of data obtained using the two-frequency receiver of GLONASS/GPS SRNS should include three stages.

1) transmission of the initial time series (1) through the 6th-order Butterworth digital LPF with 3 dB attenuation at the cutoff frequency $f_{\text{co}\,1} = v_e/l_F$ to determine the average value of the ionosphere's TEC $\bar{N}_T(t)$;
2) transmission of the initial time series (1) through the 6th-order Butterworth digital BPF with 3 dB attenuation at the cutoff frequencies $f_{\text{co}\,1}$ и $f_{\text{co}\,2} = f_s/5$ to determine small-scale fluctuations $\Delta N_T(t)$ of the ionosphere's TEC;
3) calculation of SD $\sigma_{\Delta N_T}(t)$ of small-scale fluctuations $\Delta N_T(t)$ of the ionosphere's TEC obtained in the previous step, according to expression (4) on the interval $\tau_{\text{SD}} = l_F/v_e$.

3 Discussion

The developed method of determining the statistical characteristics of small-scale fluctuations of the total electronic content of the ionosphere allows to eliminate the disadvantages of the known method [9], as the 6th-order digital Butterworth filter used for determining the average TEC of the ionosphere $\bar{N}_T(t)$ and its small-scale fluctuations $\Delta N_T(t)$, has a very smooth frequency response in the pass bands and suppression bands and does not reduce the amplitude of the output signal compared to the input.

The results of determining the statistical characteristics of small-scale fluctuations in the total electronic content of the ionosphere can be used to predict the quality indicators of satellite radio systems.

The method also can be used in the research of fluctuations of the ionosphere's TEC caused by inhomogeneities of its electronic concentration with scales $l \gg l_F$ with calculating the coefficients a_m and b_m corresponding to the low-pass filter and pass band filter without changing their structure, which is especially important in their hardware implementation in a specialized two-frequency receiver of GLONASS/GPS SRNS.

4 Conclusion

The developed method allows to determine the average value of the total electronic content of the ionosphere and the standard deviation of its small-scale fluctuations based on the results of two-frequency measurements of radio navigation parameters using signals from the receiver GPStation-6satellite of radio navigation systems GLONASS/GPS and processing these data using digital 6-th order Butterworth filters (low-frequency and band-pass) with reasonable values of cutoff frequencies.

Acknowledgements. The work was supported by the Russian Foundation for Basic Research, project No. 18-07-01020.

References

1. Yeh, K.C., Liu, C.H.: Radio wave scintillations in the ionosphere. Proc. IEEE **4**(70), 324–360 (1982)
2. Crane, R.K.: Ionospheric scintillation. Proc. IEEE **2**(65), 180–204 (1977)
3. Aarons, J.: Global morphology of ionospheric scintillations. Proc. IEEE **4**(70), 360–378 (1982)
4. Fremouw, E.S., Leadabrand, R.L., Livingston, R.C., Cousins, M.D., Rino, C.L., Fair, B.C., Long, R.A.: Early results from the DNA Wideband satellite experiment – complex-signal scintillation. Radio Sci. **1**(13), 167–187 (1978)
5. Maslov, O.N., Pashintsev, V.P.: Models of transionospheric radio channels and noise immunity of space communication systems. Supplement to the journal «Infocommunication technologies» Issue 4. PSATI, Samara (2006)

6. Pashintsev, V.P., Solchatov, M.E., Gakhov, R.P.: Influence of the ionosphere on the characteristics of space information transmission systems: monography. Fizmatlit, Moscow (2006)
7. Pashintsev, V.P., Katkov, K.A., Gakhov, R.P., Malofey, O.P., Shevchenko, V.A.: Satellite navigation during ionospheric disturbances. NCFU, Stavropol (2012)
8. Afraimovich, E.L., Perevalova, N.P.: GPS monitoring of The Earth's upper atmosphere. State institution «Scientific center for reconstructive and regenerative surgery of the East Siberian scientific center of the Siberian branch of the Russian Academy of medical sciences», Irkutsk (2006)
9. Pashintsev, V.P., Peskov, M.V., Smirnov, V.M., Smirnova, E.V., Tynyankin, S.I.: Procedure for extraction of small-scale variations in the total electron content of the ionosphere with the use of transionospheric sounding data. J. Commun. Technol. Electron. **12**(62), 1336–1342 (2017)
10. Yasyukevich, Y.V., Perevalova, N.P., Edemskiiy, I.K., Polyakova, A.S.: Response of the ionosphere to Helio-and geophysical perturbing factors according to GPS data: monograph. ISU, Irkutsk (2013)
11. GPStation-6. GNSS Ionospheric Scintillation and TEC Monitor (GISTM) Receiver User Manual (OM-20000132), Rev. 2 (2012)
12. Shanmugam, S., Jones, J., MacAulay, A., Van Dierendonck, A.J.: Evolution to modernized GNSS ionoshperic scintillation and TEC monitoring. In: Proceedings of IEEE/ION PLANS 2012, Myrtle Beach, South Carolina, pp. 265–273 (2012)
13. Introduction to digital filtering. Under the editorship of Bogner R. and Constantinidis, A. Mir, Moscow (1976)

Communication System for Strain Analysis over Metals on the Base of Tensoresistor Transducers

Michail Malamatoudis[1], Panagiotis Kogias[1(✉)], Dionisia Daskalaki[2], and Stanimir Sadinov[2]

[1] Department of Physics, International Hellenic University, University Campus II Ag. Loukas, 65404 Kavala, Greece
malamatoudismichail@yahoo.gr, kogias@teikav.edu.gr
[2] Department of Communications Equipment and Technologies, Technical University of Gabrovo, H. Dimitar Street 4, 5300 Gabrovo, Bulgaria
d_daskalaki@yahoo.gr, murry@tugab.bg

Abstract. In this paper an architecture of a system for strain investigation and analysis over metals with remote access is proposed. The deformations of the test turners are measured by two strain gauges connected to adjacent arms of a Winston bridge. The models are examined with a high coefficient of determination R2 above level 0.98. The results according to synthesizing artificial neural networks in MATLAB environment about determination the amount of measuring transducers in detection the loads of experimental cantilever beam are presented. Two neural models with 9 and 6 hidden neurons about variables "Uout" and combination "F and Uout" with correct classification of test data were selected. Levels of the mean square error related to the synthesized neural network in two 9.9631e−04 compared to the network in one input parameter 0.0832 are observed, respectively.

Keywords: Tensoresistor transducers · Metals · Strain measurement system · Regression analysis · Artificial neural network

1 Introduction

The impact of forces of various magnitude on parts and structures can be recorded by applying a wide range of transducers. These include resistive, inductive, capacitive, piezoelectric, electromagnetic, magnetoelastic, galvanic, vibrational, acoustic, gyroscopic and others [1]. One of the most commonly used sensor types are strain gauges. The strain gauges are divided into metal and semiconductor ones. They are compared against different criteria such as measuring range, sensitivity, resistance, resistance tolerance and dimensions. They are connected mainly in DC bridge circuits in different configurations with one working transducer, two working transducers in adjacent circuits, two working transducers in the opposite circuits and four working transducers [2].

In scientific research different systemic solutions for measurement of forces are examined. Automated wireless network monitoring systems related to the measurement of intense elastic surface deformations in serviced plastic-coated metal pipelines are

R. Silhavy (Ed.): CSOC 2020, AISC 1226, pp. 321–328, 2020.
https://doi.org/10.1007/978-3-030-51974-2_31

widespread [3, 4]. As described in [5], a multi-channel measurement system has been designed to evaluate sensor elements for deformation using a USB communication interface. The test modules were equipped with an 8-bit microcontroller, an adjustable gain amplifier, a Bessel low frequency filter and an analog-to-digital converter controlled by SPI and other control signals.

Regression analysis apparatus and artificial neural networks are used in the field of communications for deriving models for predicting teletraphic flows in voice services and Markov circuits, modeling of digital recursive filter units and others [6–8]. This paper proposes the architecture of a system for measuring and regressive prediction in the change of forces applied on metals by one or two strain gauges, as well as the determination of the number of detecting transducers using artificial intelligence, based on information gathered via serial communication devices controlled by WEB integrated LabVIEW applications. Options for statistical analysis and storage of results on server-based databases are offered, providing a higher level of information security.

2 Exhibition

A system for examining and analyzing the degree of straining in parts due to forces of different magnitude is designed, the architecture of which is shown in Fig. 1. The tested object of the study is a metal cantilever beam, to which forces are applied perpendicularly. Two strain gauge transducers are glued to the surface, arranged as to provide the required temperature compensation without the need for an additional sensor element. The strain gauge transducers are connected in the adjacent branches of a standard DC bridge. As a result of the beam straining, the electrical and mechanical parameters of the sensors change, as well as the voltage in the measuring diagonal of the bridge, the value of which is measured with a National Instruments' module NI 6002.

Through LabVIEW virtual applications, the bridge output voltage is monitored in real time using a configured serial communication channel between the NI 6002 and a PC. Statistical analysis of the recorded data is performed with respect to various indicators such as minimum, average and maximum values, standard deviation, maximum and minimum times, etc. On the basis of previously obtained STATISTICA regression models, set in LabVIEW using a specified sub-virtual instrument, the impact force is calculated in the case of one or two strain gauges. Using MATLAB SCRIPTS, the Parameters of Trained Artificial Neural Networks (ANN) are loaded in LabVIEW to determine the amount of sensor elements necessary to measure the input non-electrical value. The availability of a WEB server allows for remote access to the tools for setting up the connection to an USB module, processing and visualization of measuring, statistical and forecast results within the Internet environment.

As you can see in Fig. 1, there are a Process of measurement, USB cable, WEB applications, Measuring and statistical results visualization and Software Processing. More analytically:

- STATISTICA Regression models for predicting F.
- MATLAB Neural networks for quantitative identification of strain gauges.
- MSSQL Data bases.

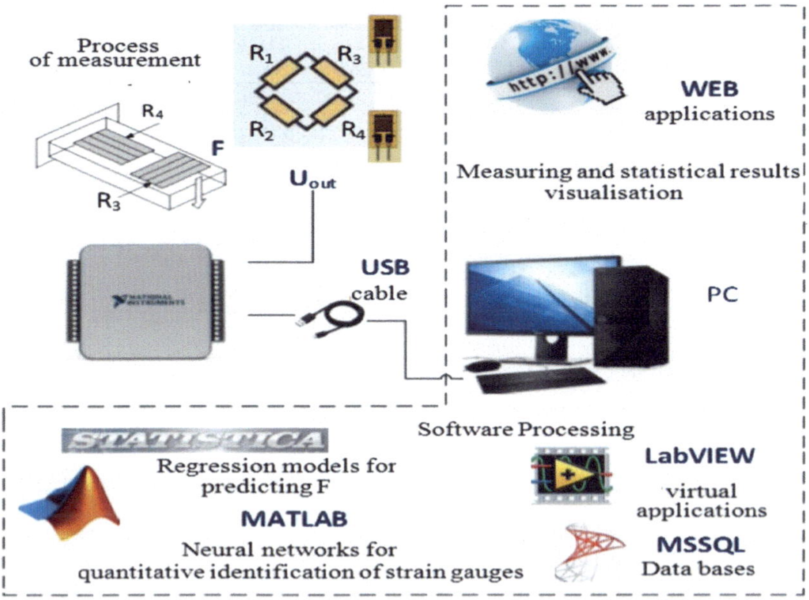

Fig. 1. Architecture of a force analysis system on metals inside strain gauges

When testing the experimental setup, experimental data were obtained using one and two working transducers, applying identical loading forces on the studied object. Information categories have been formed, including 26 entries for each "F (regression methods for predicting F-matrix entries) and U_{out1}" and "F (regression methods for predicting F-matrix entries) and U_{out2}" groups. Based on the experimental data and using regression analysis, the results of which are given in Fig. 2, mathematical linear models for the prediction of forceful parametric changes "y" depending on the output voltage of the bridge "x" are derived for both using one and two sensitive sensor elements cases. When evaluating the data regarding models (1) for one and (2) for two transducers with respect to the significance of the experimental regression coefficients b_i, no unignificant coefficients were found at the baseline significance level $\alpha = 0.05$. With respect to the coefficients of certainty of R^2 relative to the obtained forecast models, close high levels were found, respectively $R^2 = 0.99897588$ for one and $R^2 = 0.99887101$ for two detecting strain gauges. Models (1) and (2) can be defined as adequate and completely describing the experimental data obtained during the diagnosis.

$$y = -0.4749 + 394.7272x \tag{1}$$

$$y = -0.4334 + 788.5744x \tag{2}$$

In order to confirm the correctness of the analysis performed, normal probability residual diagrams are presented, based on the derived predictive models of the load force on the workpiece (Fig. 3). In view of the dependencies shown, a good

	Regression Summary for Dependent Variable: y (Spreadsheet2) R= ,99948781 R?= ,99897588 Adjusted R?= ,99893320 F(1,24)=23411, p<0,0000 Std.Error of estimate: ,29978					
N=26	b*	Std.Err. of b*	b	Std.Err. of b	t(24)	p-value
Intercept			-0,4749	0,122680	-3,8707	0,000730
x	0,999488	0,006532	394,7272	2,579826	153,0054	0,000000

a)

	Regression Summary for Dependent Variable: y (Spreadsheet1) R= ,99943534 R?= ,99887101 Adjusted R?= ,99882396 F(1,24)=21234, p<0,0000 Std.Error of estimate: ,31475					
N=26	b*	Std.Err. of b*	b	Std.Err. of b	t(24)	p-value
Intercept			-0,4334	0,128564	-3,3714	0,002529
x	0,999435	0,006859	788,5744	5,411632	145,7184	0,000000

b)

Fig. 2. Regression results on models for a) one and b) two working transducers

arrangement of residues is observed near the 45° line. This gives grounds for confirming the random nature of the distribution of the differences between the theoretically expected experimental and predictive results and the correctness of the apparatus.

Tables 1 and 2 present data on the criteria for accuracy and root mean square error in synthesizing artificial neural networks with one (F) and two (F and Uout) input variables in order to predict the behavior of the operating strain gauge transducers. Architectures with one intermediate layer were studied for a Tanges-sigmoidal activation function with variations in neurons from 5 to 15.

The output groups with, respectively, "one working transducer" and "two working transducers", were defined by separate output neurons of linear activation type. In the case of neural training, sets containing 52 information patterns (26 for each test class) were used.

As a result of the one-variable network study, a relatively larger range of accuracy changes was observed from 50.0% with 15 to 100.0% with 9 neurons. A similar conclusion can be drawn with respect to the second criterion, varying from 0.0832 with 9 to 0.4670 with 12 hidden neurons. In relation to the neural models with two input variables, a maximum accuracy of 100.0% was found with 6, 8–12, and 13 neurons. The minimum root mean square error equals 9.9631e−04 with 6, while its highest levels reach 0.1139 with 13 neural units. The selected networks with the best indicators with 9 and 6 neurons in the hidden layers in the cases with one and two input variables are shown in Fig. 4.

In relation with the selected neural models, linear regression dependencies were generated in reference to their outputs (Fig. 5 and Fig. 6). There is a significant difference in the levels of R correlation coefficients, varying during training with "F" from 0.54 to 0.57 and about 0.98 with "F and U_{out}" data types. In the model on Fig. 4.b) a significantly better linear relationship between targeted and obtained network results is observed (Fig. 7).

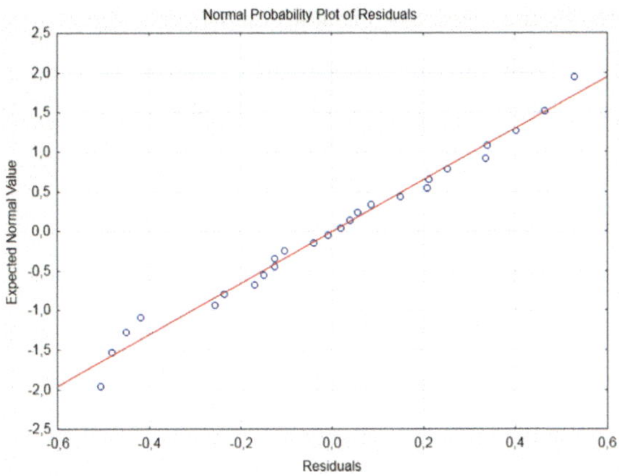

a)

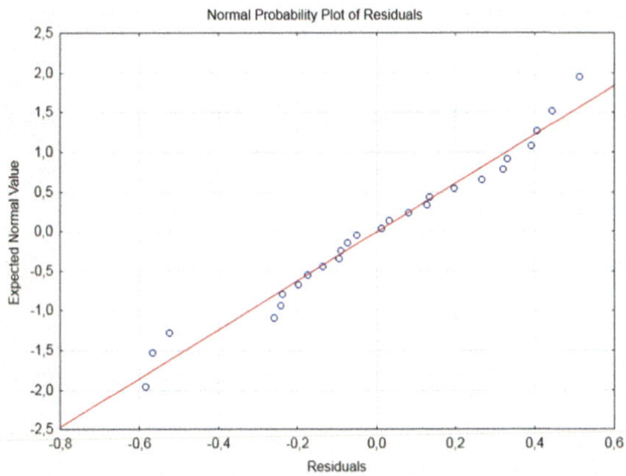

b)

Fig. 3. Normal probability graphs for models for a) one and b) two working transducers

Table 1. Results from testing artificial neural networks at one input variable

Hidden neurons	Accuracy, %	RMS error
5	87.5	0.0973
6	62.5	0.1861
7	62.5	0.2590
8	87.5	0.1447
9	**100.0**	**0.0832**
10	50.0	0.3517
11	75.0	0.2813
12	62.5	0.4670
13	87.5	0.1331
14	75.0	0.1585
15	50.0	0.3843

Table 2. Results from testing artificial neural networks at two input variables

Hidden neurons	Accuracy, %	RMS error
5	87.5	0.0349
6	**100.0**	**9.9631e−04**
7	87.5	0.0694
8	100.0	0.0065
9	100.0	0.0176
10	100.00	0.0029
11	100.0	0.0044
12	100.0	0.0349
13	87.5	0.1139
14	100.00	0.0124
15	87.5	0.0818

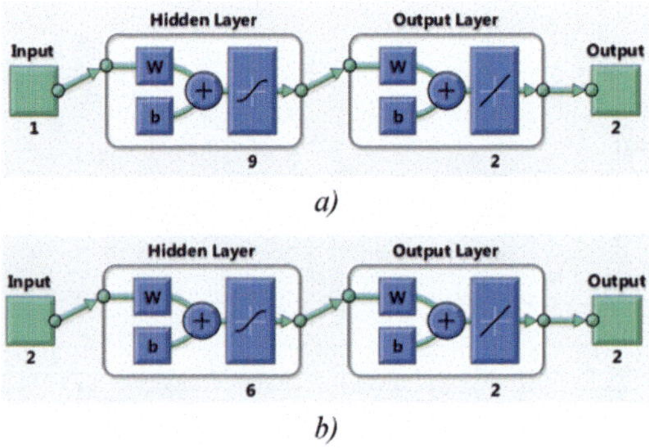

Fig. 4. ANN synthesized for quantitative prediction of strain gauges for a) one and b) two input variables

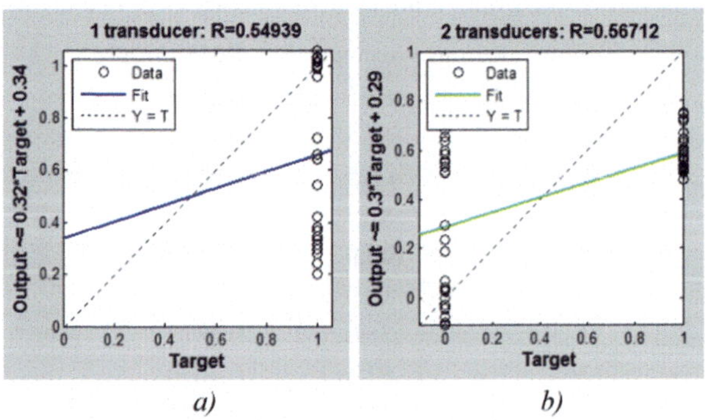

Fig. 5. Linear regression dependencies for network outputs - a) one and b) two transducers of the selected ANN with one input variable

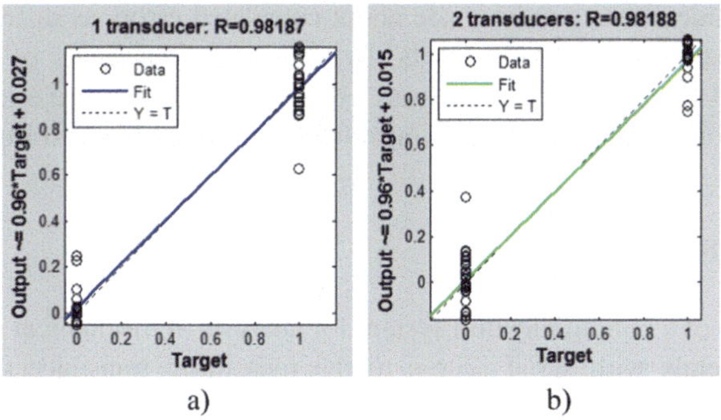

Fig. 6. Linear regression dependencies for network outputs - a) one and b) two transducers of the selected ANN for two input variables

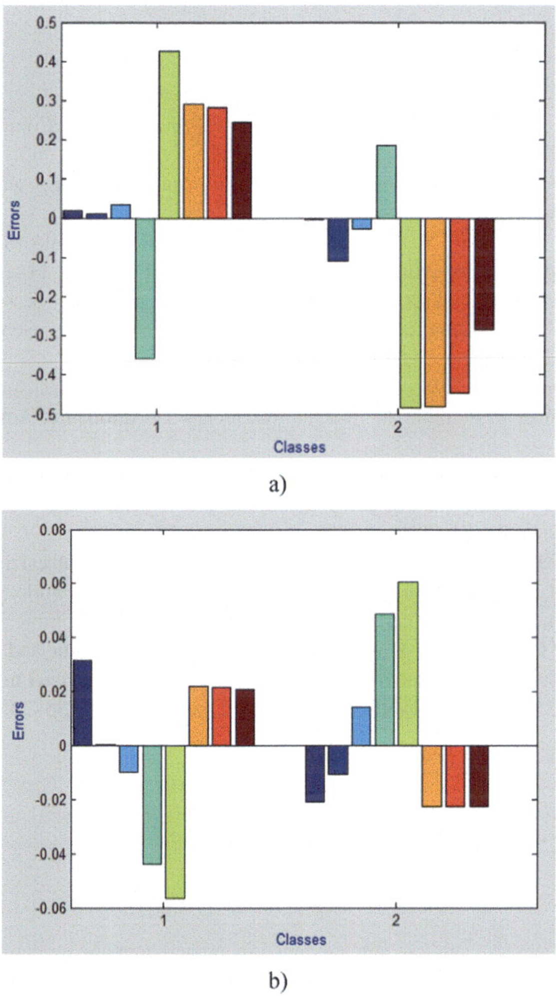

Fig. 7. Network errors related to synthesized ANNs for a) one and b) two input variables

Similar considerations can be made about calculated errors as differences between expected and estimated network results, by comparing the limits of their variation. Ranges from −0.4852 to 0.4263 and from −0.0563 to 0.0607 were established for the experimentally obtained estimated neural models with one and two input variables.

3 Conclusion

The architecture of a communication system for analysis of mechanical deformations in the study of metals with serial access to the measuring transducers, controlled by applications accessible within the Internet environment, is proposed. The regression apparatus and the neural toolkit have been successfully applied for the derivation of mathematical models and synthesis of networks for the estimation of nonelectric effects and the number of working sensory elements according to the defined quality indicators.

References

1. Stefanesko, D.: Handbook of Force Transducers: Principles and Components. Springer, Heidelberg (2011)
2. New Jersey – Department of Transportation. Design manual for bridges and structures, Sixth edition. The State of New Jersey (2016)
3. Druzhynin, A., Khoverko, Y., Ostrovkyi, I., Koretskyi, R., Nichkalo, S.: Remote control measuring based on strain sensors. Comput. Prob. Electr. Eng. **2**(1), 11–14 (2012)
4. Hongell, T., Kivela, I., Hakala, I.: Wireless strain gauge network - best-hall measurement case. In: IEEE Ninth International Conference on Intelligent Sensors, Sensor Networks and Information Processing, pp. 1–6 (2014)
5. Dostalek, P., Dolinay, J., Vasek, V.: Design of the multichannel measurement system for strain gauge sensor evaluation. In: Recent Researches in Automatic Control, pp. 245–248 (2014)
6. Balabanova, I., Georgiev, G., Kogias, P., Sadinov, S.: Selection of plan of experiment by statistical analysis of the parameters of teletraffic model with voice services. J. Eng. Sci. Technol. Rev. **9**(6), 76–81 (2016)
7. Balabanova, I., Georgiev, G., Sadinov, S., Kostadinova, S.: Synthesizing of models for identification of teletraffic Markov chains by artificial neural networks and decision tree method. J. Electr. Eng. (Slovakia) **69**(5), 379–384 (2018)
8. Balabanova, I., Georgiev, G., Kostadinova, S.: Computer modeling and investigation into web-based application of digital IIR filters with LabVIEW and artificial neural networks. In: Booklet of the 55-th Science Conference of Ruse University, pp. 235–245 (2016)

Information Technology in Mathematics Training

V. I. Temnyh[1], T. P. Pushkaryeva[1(✉)], V. V. Kalitina[2], and T. A. Stepanova[3]

[1] Siberian Federal University, 79, Svobodny Prospect, Krasnoyarsk 660041, Russia
a_tatianka@mail.ru

[2] Krasnoyarsk State Agrarian University, 90, Prospect Mira, Krasnoyarsk 660049, Russia

[3] Krasnoyarsk State Pedagogical University named after V.P. Astafiev, 89, Lebedevoy, Krasnoyarsk 660049, Russia

Abstract. There are three reasons for the need to apply information technology in mathematics training and for an innovative digital education environment: the digital economy and the new human resources requirements it creates; digital generation (new generation of students with special socio-psychological characteristics); new digital technologies shaping and evolving the digital environment. The main idea of the work is to identify the peculiarities of mathematics training in the conditions of the economy and education digitalization and to build from these positions the technique of mathematics training. The analysis of literature has led to the conclusion that in order to achieve success in professional activity SoftSkills and DigitalSkills are required first of all. The main socio-psychological characteristics of the modern student are the increase in visual perception; clip thinking; the desire to work together in the form of "Wiki"; information congestion. The corresponding technique of mathematics training with continuous application of information technologies is described. For the construction of the electronic course the LMS Moodle system is chosen. #CSOC1120.

Keywords: Information technology · Mathematics training

1 Introduction

In the structure of most professional education, mathematics is identified as one of the important disciplines. In conditions of global informatization, the main tasks of mathematical training in addition to mastering methods of mathematical activity include the formation of skills of independent search and processing of information, ability to model processes and solve problems with the help of information technologies. The rapid development of digital technologies and their penetration into the professional, educational and social life of people has led to the emergence of a new entity – a representative of the digital generation, having its own, distinctive, ability to perceive information and process it. All of it generates the necessity of replacement of

R. Silhavy (Ed.): CSOC 2020, AISC 1226, pp. 329–339, 2020.
https://doi.org/10.1007/978-3-030-51974-2_32

the existent approaches and models of mathematics training on new that correspond to the modern stage of society development and contribute to the formation of a competitive specialist.

Presently the insufficient structured of training material, the weak demand for methods and means of developing mechanisms of perception and processing of mathematical information and the low level of application of information technology (IT) tools are watching in on-line tutorials and methodologies of mathematics training.

The main purpose of this work is to identify competences, the formation of which will ensure the necessary level of mathematics training of the future engineer, to identify the psycho-physiological peculiarities of modern students and to determine the methods and tutorials of mathematics on their basis.

The formation and development of the digital economy significantly change the order to education: the main task becomes the formation of new digital competences, regardless of the chosen direction of training. Conducted literature analysis showed that until recently two groups of competences were allocated – professional skills (HardSkills) and skills closely related to personal qualities of the employee (SoftSkills), such as work on a team, finding compromises, ability to have a friend in communication [1–6]. Professional skills were thought to be fundamental in hiring not.

Today, due to the general computerization and digitalization of all areas of activity, a third group – digital skills (DigitalSkills) – has been identified. Scientists at Harvard University, after conducting a study, concluded that 15%–25% of HardSkills and 75%–85% of SoftSkills and DigitalSkills are required to achieve professional success.

The World Economic Forum Expert Group in Davos highlighted ten skills that will be in demand in 2020, among them:

- complete and multilevel solution of problems;
- creativity;
- interaction with people;
- flexibility of mind.

Based on the above, it is possible to highlight the main directions for training specialists of the future [2]:

- development of adaptability, critical and system thinking, ability to self-training according to the concept of "lifelong learning";
- formation of skills of search, processing and analysis of information, media literacy;
- development of high communication abilities (ability to work in team, cooperation, skills of the self-presentation, skills of business negotiations).

The rapid development of digital technologies and their penetration into the professional, educational and social life of people has created a new person, which is commonly called a representative of the digital generation [7].

Their main characteristics are [7, 8]:

- the increased level of visual perception;
- clip thinking;
- aspiration to joint activities for the vikideystviye type;
- information congestion.

Therefore, it is almost impossible to integrate the representatives of the digital generation into the traditional educational process. New methods and means of teaching mathematics are needed.

This article describes the technique of mathematics training for the students of technical university of artistic direction, based on active application of information technologies, contributing to formation of skills necessary today, and taking into account peculiarities of thinking of modern students.

2 Methods

The basis for construction of the mathematics training is chosen personally-centered approach, the essence of which is that the teaching process is built around the person. With this approach, the students choose not only what to learn, but also can choose the appropriate place, time, methods and tutorials of study.

In this regard, the course structure is designed according to a modular scheme that ensures non-linearity of the educational process. Modular organization of educational content refers to the presentation of discipline content in the form of information blocks, which are learned in accordance with didactic purpose. The respective objectives and expected training results are specified for each module. The didactic objective of each module contains not only an indication of the amount of content studied, but also the level of its absorption. This allows the student to build their individual trajectory of training in mathematics.

Digitalization of the economy leads to the fact that the production process from the long, distributed on the technological stages, becomes compact, autonomous, completed project, the implementation of which is carried out not by a separate specialist, but by a team. This makes it necessary to introduce the project method into the educational process and to build skills to work in the team.

The clip thinking of a modern student requires new forms of presentation of learning mathematical information, compact and convenient for rapid perception and application. In addition to the traditional method of presentation, forms of data submission such as mental maps, infographics should be used.

The peculiarities of the application of information technologies in mathematical training should be determined by the peculiarities of students training in mathematics. In this case, the feature is that as future engineers these students training in mathematics according to the program corresponding to the technical university, but, studying on the artistic direction, most of them, according to testing, have a humanitarian warehouse of mind.

This requires consideration of personal psycho-physiological features of perception, memorization and extraction (thinking) of abstract mathematical information.

The construction of the information model of thinking made it possible to conclude that to increase the level of perception of abstract mathematical concepts its visualization is necessary, and to increase the level of memorization – dynamic visualization, which allow IT [9]. Such a presentation of tutorial material is quite consistent with one of the characteristic properties of members of the digital generation – an increased level of visual perception.

Mathematics training based on a person-centered approach implies creating a set of conditions and opportunities that allow students to independently build an individual educational route, choose the time and place of study, methods and tutorials. The creation of a digital information and education environment will ensure compliance with these requirements. In such an environment, the teacher acts as a consultant, a project specialist, and a developer of students individual educational paths. Its main task is to train the student to independently set goals and tasks, to choose suitable methods and means of training for him, to find and analyze information, to evaluate the results of his work, to work in a team.

3 Results

3.1 The Training Environment LMS MOODLE

Based on the above-mentioned conclusions an electronic course on mathematics for students of the profile "Technologies of artistic processing of materials" is implementing at the Siberian Federal University [10].

To implement the course one of the most common educational technologies – Moodle (Modular Object-Oriented Dynamic Learning Environment) – is chosen. The Moodle system is focused on organizing teacher-student interaction, and is also suitable to support face-to-face and blended learning.

Learning Management System (LMS) Moodle specifically designed to create online courses by educators. This free of charge expandable program complex with the functional possibilities, mastering simplicity and use comfort satisfies to most requirements produced by users to the electronic departmental teaching. Currently, the Moodle system is used for education by the largest universities in the world.

The electronic course at Moodle includes:

- Organizational materials:
  - ✓ Workflow for the organization of the training process.
- Materials for theoretical studying:
  - ✓ Lectures (.pdf files), infographics, mind maps, presentations, tutorial, additional resources.
- Tasks for practical work and methodical materials on their performance:
  - ✓ Methodological materials on performance for the laboratory and practical works.
  - ✓ Wiki.
  - ✓ Glossary.
  - ✓ Tests.

3.2 The Training Model

We have chosen the blended learning as the learning model. The blended learning is the integration of a traditional (face-to-face learning system) and distributed learning system centered on technology. When implementing the blended training, two systems are merged: traditional, characterized by synchronous interaction of individuals, and distributed, which is characterized by asynchronous actions – interaction of individuals, regardless of time and place [11, 12].

In the context of digitalization of education, it is possible to optimally combine the "strengths" of traditional learning with the advantages of remote technologies. The model of the blended learning allows face-to-face classes to be made more saturated and effectively organized, as a large part of the material students learn independently in an information and educational environment (IOS) with the help of an electronic course. The created IOS provides communication, accessibility of educational information, allows to realize to the full potential abilities of each student.

Figure 1 shows a diagram of a mixed technology learning process.

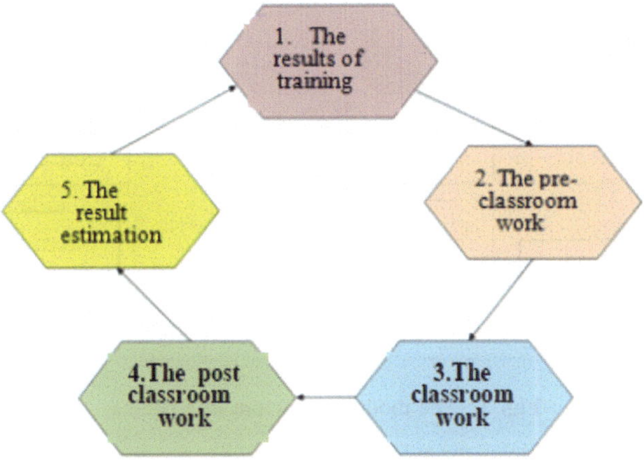

Fig. 1. The blended model learning diagram

3.3 The Structure of a Course

In accordance with the curriculum and working program of the discipline, the contents of the course are divided into 6 modules:

- linear algebra (M1);
- vector algebra and analytical geometry (M2);
- differential calculus (M3);
- integral calculus (M4);
- ordinary differential equations (M5);
- probability theory and mathematical statistics (M6).

Study objectives and expected results are defined for each module (Fig. 2).

The main objectives of the study are to develop students understanding of basic concepts and methods of calculating these sections of mathematics and to develop skills of their practical use in solving educational and professional tasks with the help of IT; development of logical and algorithms thinking.

The planned results are the ability to quickly identify and apply mathematical methods and IT to solve a specific task in professional activity.

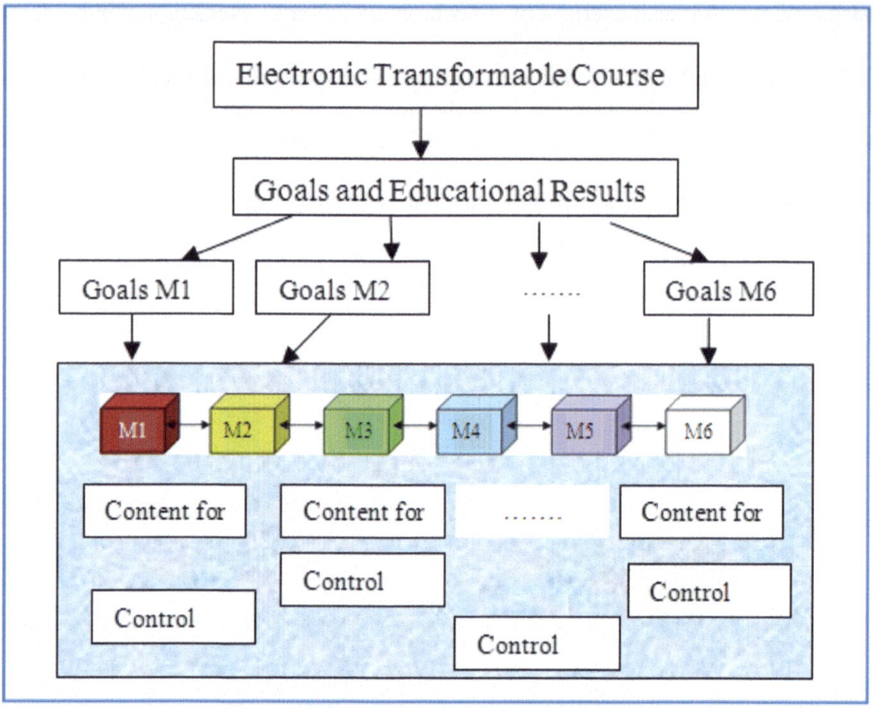

Fig. 2. The modular structure of a course

In accordance with the curriculum, the process of mathematics training for the future engineers involves lectures, seminars, laboratory works and out-of-school independent work. The entire learning process is built on continuous use of IT tools.

3.4 The Training Forms

Lectures. Attention, interest and compactness play the main role in ensuring the necessary level of perception, imagination and memorization of tutorial material. What is emotionally saturated, fascinating and unduly causes intense concentration.

To ensure the marked characteristics of attention and visualization of the presented material, all lectures in the electronic course are in the form of presentations, as well as mental maps and infographics (Fig. 3, 4). Videos created in Macromedia Flash and attached as hyperlinks to concepts on slides are actively used in lectures to help create an image of abstract mathematical concepts.

At the first lecture it is proposed to discuss finding a solution to one of the applied orientation problem in order to explain the importance and necessity of mathematical methods, including the method of mathematical modeling, in solving such problems. Then a mind map is presented, displaying in stages mathematical sections, knowledge

a) *b)*

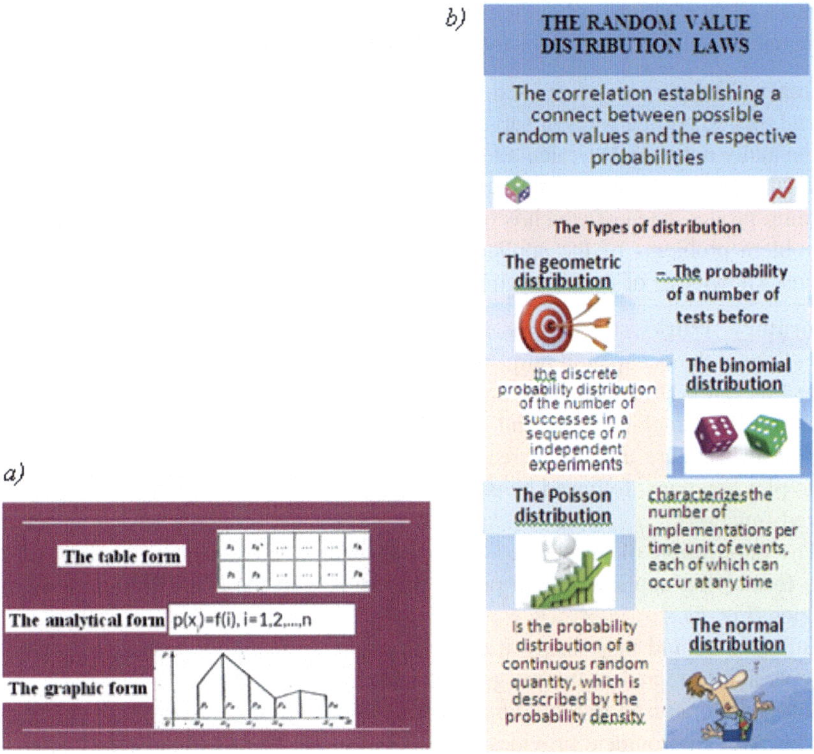

Fig. 3. *a*) The slide of lecture - presentation on the probability theory *b*) The infographics slide

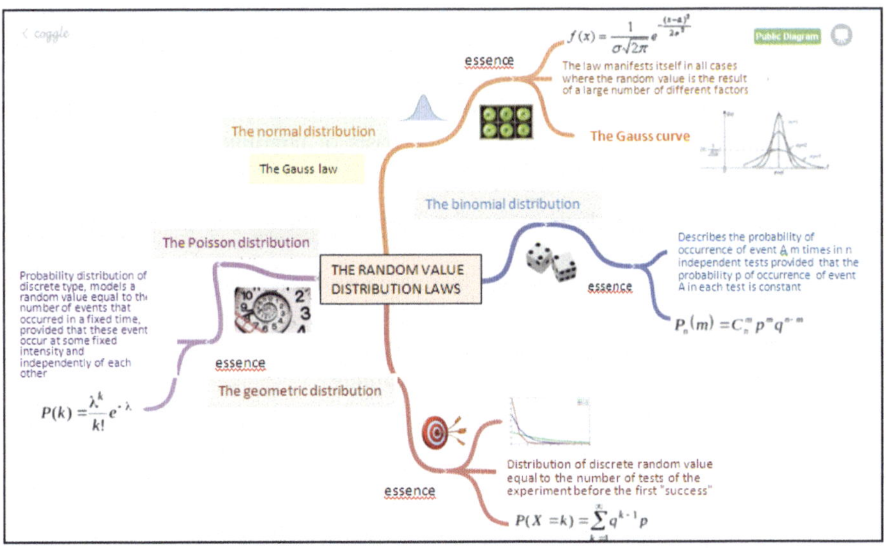

Fig. 4. The part of the mind map on the probability theory

of which is necessary to solve the problem. This provides motivation for the study of mathematics and the integrity of the perception of the mathematical course.

Seminars. The purpose of the seminar sessions is to identify the essence of the presented tasks, to outline ways of solving, to discuss possible methods of solving, to carry out manually not complex calculations. The main form of carrying out seminar classes is work in groups, namely individual-group work. For seminars, students are divided into mini groups, each of which is offered a list of tasks to solve. In this regard, a base of problem problems of the applied direction has been built, the solution of which requires knowledge of certain sections of mathematics.

Laboratory Works. Laboratory works are carried out with use of various IT depending on the occupation purposes (spreadsheets of Excel, a mathematical package of the MATHCAD program, the program for creation of mind maps, the program for creation of infographics, Wiki and etc.).

Spreadsheets are used to increase the level of memory of mathematical formulas and to plot functions, followed by comparison with a graph obtained analytically using derived functions.

The application of specialized mathematical packages ensures the formation of skills for solving application and profile problems with the help of information technologies (Fig. 5).

Building a mind map makes it easier to create an image of abstract mathematical concepts and to represent the integrity of mathematical sections.

Infographic is used to draw up a short lecture project.

The Wiki environment provides the skills of group work and the ability to analyze and comment on the work of other students.

Out-of-school independent work of students is organized on the basis of project method and application of IT, including cloud technologies. The students carry out research and training projects, which are presented in the form of presentations, booklets and websites at a planned seminar conference. Cloud technologies, in particular Google disk programs, are used to work on the project.

The carried out pedagogical experiment confirmed the effectiveness of the proposed methodology of teaching mathematics on the basis of continuous application of IT, which provides:

- the increasing of the students motivation level to training and the level of understanding of theoretical material;
- the increasing of the level of their activity when studying mathematics in the information environment and formation of communication competences;
- the development by students of new types of activity;
- the flexibility, mobility and availability of training.

All this as a result contributes to the formation of competences required for successful professional activity by modern employers.

9

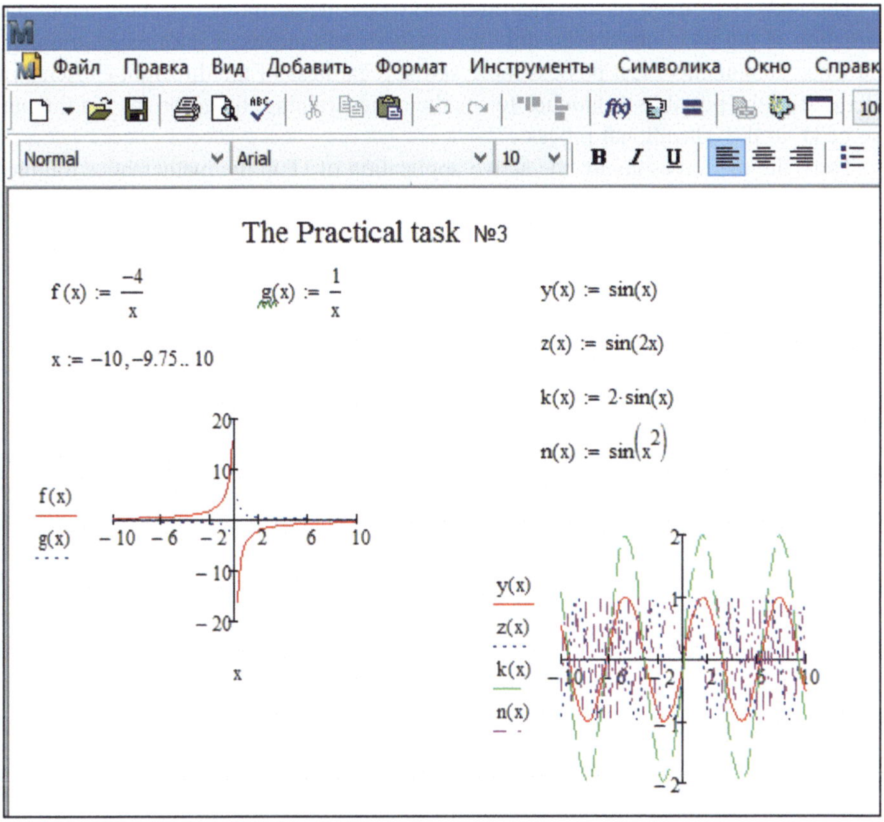

Fig. 5. The functions calculate and their plots in MathCad

4 Discussion

The development of the digital economy inevitably leads to the emergence of new conditions for the development of industry, increased competition, redistribution of labour, which entails new requirements for specialists, bringing the personal qualities of the employee and his digital competences to the first place, and sidelining professional skills. As a result, the culture of behavior and communication changes, the ways of perception of information and thinking, the processes of identification and socialization of the person change, and new requirements to creativity are made.

According to the program «Digital Economy of the Russian Federation» [13] by 2025, Russian universities must produce at least 120,000 IT-related specialists, and 40% of the total population of the Russian Federation will have digital skills.

The implementation of the requirements is possible only if the entire education system is improved, the necessary conditions are created for the formation of digital skills, the methodological basis for the development of competences in the field of digital technologies is built.

The authors of most of the scientific works considered by us, speaking about the formation of digital competences, describe methods and means of informatics training. But since all students must possess these skills, regardless of the direction chosen, we offer one of the possible techniques for mathematics training, based on the continuous use of IT in the educational process.

Thus, the main reasons for the active application of IT in the mathematics training of future engineers are: [3]:

- digital economy;
- digital generation;
- new digital technologies.

The development of the digital economy puts new demands on specialists, bringing to the first place the personal qualities of the employee and his digital competence, and putting the professional skills on the second plan.

Psychological and pedagogical features of representatives of digital generation ("generation Z"), such as clip thinking, inability to long keep attention, ability to parallel processing of information cause need of search of new pedagogical technologies.

New digital technologies allow to build a digital information and educational environment, providing the formation of new, necessary today, competences, and to create a complex of conditions for independent development of students.

5 Conclusion

The IT used in mathematics training described in this article certainly does not exhaust all existing ones today. As observations and pedagogical experience show, the modern students have a mobile device most highly sought. Even at presence of the printed textbook and computer for viewing of educational information they use smartphone mostly. The relatively recently introduced specialized applications allow to use mobile devices in education also, providing remote access to educational resources, choice of place and time of study of the material. Our further research is related to this direction.

References

1. Priority project in the field of Education "Modern digital educational environment in the Russian Federation" (approved by the Presidium of the Council under the President of the Russian Federation on Strategic Development and Priority Projects, Protocol dated 25.10.2016 № 9)
2. Butenko, V., Polunin, K., Kotov, I.: Russia 2025: From Personnel to Talent 2017, October, The Boston Consulting Group. http://image-src.bcg.com/Images/Russia-Skills_Outline_v.1.8_preview_tcm27–177753.pdf. Accessed 12 Jan 2018

3. Blinov, V.I., Dulinov, M.V., Esenina, E.U., Sergeev, I.S.: The project of didactic concept of digital professional education and training. Pero, Moscow, Russia (2019)
4. Quarles, A.M.: Integrating digital/mobile learning strategies with students in the classroom at the historical black college/university. In: Proceedings–2017 (HBCU) Handbook of Research on Digital Content, Mobile Learning, and Technology Integration Models in Teacher Education, 13 July, pp. 390–408 (2017)
5. Maxwell, A.: Mobile learning for undergraduate course through interactive apps and a novel mobile remote shake table laboratory. In: Proceedings – 2017 Annual Conference and Exposition, Conference Proceedings, June (124th ASEE Annual Conference and Exposition; 25 June 2017 to 28 June 2017; Code 129594) Columbus; United States, vol. 24 (2017)
6. Instefjord, E., Munthe, E.: Preparing pre-service teachers to integrate technology: an analysis of the emphasis on digital competence in teacher education curricula. Eur. J. Teach. Educ. **39** (1), 77–93 (2016)
7. Maksimova, O.A.: Digital generation: lifestyle and identity design in virtual space. J. Chelyabinsk State Univ. **22**(313), 6–10 (2013)
8. Osipov, M.V.: Identification of student - representative of digital generation. In: Youth Science Forum: Humanities. Electronic Collection of Articles on the Materials of the XV Student International Correspondence Scientific and Practical Conference, ICNO, vol. 8 (15) (2014). http://www.nouchforum.ru/archive/MIVF_humanities/8(15).pdf. Accessed 21 Dec 2019
9. Park, N.I., Pushkaryeva, T.P.: Principles of mathematical training of students from the point of view of the information model of thinking. Open Educ. **5**(94), 4–11 (2012)
10. Pushkaryeva, T.P., Kalitina, V.V.: On formation of mathematical competence of students on the basis of design-target approach. Mod. Prob. Sci. Educ. **4** (2019). http://www.science-education.ru/article/view?id=29081. Accessed 06 Aug 2019
11. Mijares, I.: Blended learning: are we getting the best from both worlds? LiteratureReview-forEDST. http://elk.library.ubc.ca/bitstream/handle/2429/44087/EDST561-LRfinal-1.doc.docx?sequence=1. Accessed 03 Apr 2015
12. Griff, R.: Athabasca University. Learning Analitics: On the Way to Smart Education. http://distant.ioso.ru/seminar_2012/conf.htm. Accessed 10 Apr 2015
13. The State Program “Digital Economy of the Russian Federation”. http://static.government.ru/media/files/9gFM4FHj4PsB79I5v7yLVuPgu4bvR7M0.pdf

Constructing the Functional Voxel Model for Terrain on the Basis of Bilinear Interpolation of Triangulated Network

A. V. Tolok[1(✉)] and N. B. Tolok[2]

[1] Moscow State Technology University "Stankin", 3 A, Vadkovskiy per., Moscow 127994, Russia
a.tolok@stankin.ru

[2] Trapeznikov Institute of Control Sciences, Russian Academy of Sciences, Moscow, 65, Profsoyusnaya st., Moscow 117997, Russia

Abstract. One of the approaches to Constructing the functional voxel model (FV-Model) of the topography is considered. This approach allows to get the local functional dependence for each point of the complex shaped terrain on the extended land area. Bilinear interpolation of the values of components of the average normal in the triangulated topographical network vertices is the smoothing functional principle. #CSOC1120.

Keywords: Topographical network · Triangulated surface · Topography · Terrain · The functional voxel method · Bilinear interpolation

1 Introduction

The analytical representation of the complex shaped and extended terrain is a very time-consuming and arduous task that requires considerable resources for storing and processing this data during the computer calculation of various geometric characteristics.

As an example, we can review different approaches devoted to designing means of bilinear and cubic spline interpolation of regular and irregular polygon meshes [1–4]. In general, such mathematical approaches can solve the problem but as the result of these methods of modeling we'll get the piecewise-analytical representation of a model that will cause the further complexities in its general perception while using in computer calculations and processing.

One of the main problems is to provide the cubic interpolation over the unstructured (irregular) control grid though topography data imply description of the terrain exactly in this way. Adduction to structured description of the control grid will cause the loss of accuracy for the obtained computer model. Bilinear interpolation of the values in the vertices if the geodesic network does not allow to reach the expected smoothing effect. At the same time each of these approaches imply the storage of the interpolation matrixes and control grids. So, obtaining the height measure at the point will require the implementation of multiple calculations.

R. Silhavy (Ed.): CSOC 2020, AISC 1226, pp. 340–347, 2020.
https://doi.org/10.1007/978-3-030-51974-2_33

The researched objective is the application of the functional voxel computer modeling method (FVM) to constructing the complete analytical representation of complex terrain where each point is calculated by a homogeneous local function. Such approach allows to consider the functional voxel model (FV-Model) as the analytical for each point of discrete space.

The detailed functional-voxel method description was given in [5–11]. Using linear approximation in the neighborhood of each of the R^n space points, FV-method allows for any domain of the analytical function defined in R^n-space the further transformation in homogeneous equations of the tangents $a_1x_1 + a_2x_2 + \ldots + a_nx_n + a_{n+1}x_{n+1} = 0$ (local function). While making normalization of the coefficients of the equation by the norm $N = \sqrt{a_1^2 + \ldots + a_{n+1}^2}$ as a result in the space R^{n+1} we'll obtain the normal vector containing the local function characteristics determined on the interval [−1, 1].

For the purpose of providing compactness and visibility of computer representation let us establish some correspondence between the local geometric characteristics with the intensity gradation of the tone of the monochrome palette, for example, [0…255]. As a result, we obtained *n + 1* voxel images where *n* indicates the dimensionality. In this case, the term voxel is understood as a color characteristic in a point of space with given dimensionality, which serves as a means of displaying a local geometric characteristic.

Let us represent voxel geometric model of a function $z = 5(y\sin\pi x + x^2\cos\pi y)$ expressed by four local geometric characteristics (a_1, a_2, a_3, a_4) for the local function $z = \frac{a_4}{a_3} - \frac{a_1}{a_3}x - \frac{a_2}{a_3}y$ (Fig. 1):

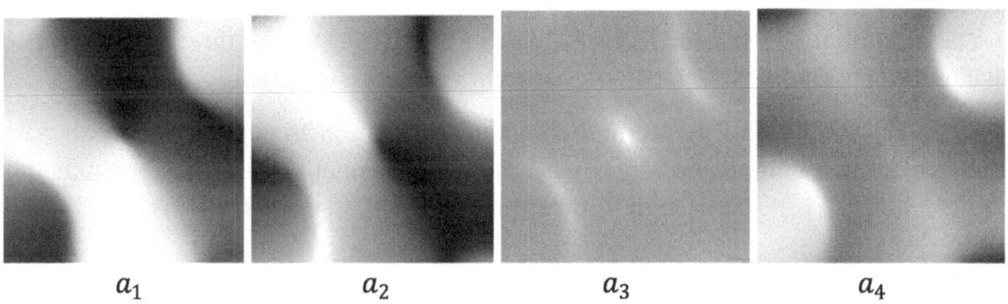

Fig. 1. Voxel images of four local geometric characteristics (a_1, a_2, a_3, a_4) with dimensionality n = 2.

The proposed method of Bilinear interpolation allows to define intermediate values for components of the normal at each point of the given domain on the basis the values of components of the average normal in the triangulated topographical network vertices. Thus, the proposed method can be easily adaptable to the principles of The voxel-functional modeling method (VFM).

2 Method of Bilinear Interpolation of the Normal Vectors' Components in the Triangulated Network Vertices (Phong Interpolation Method)

Method of Bilinear interpolation of the normal components is better known as The Phong Interpolation Method in the computer graphics modeling applications [12]. This method allows smooth distribution of normal components' values inside the given domain on the basis of components of average normal in the polygon network vertices and further computation of the reflected light intensity model in the rendering problems.

The fundamental difference between the proposed approach and well-known methods is that it is sufficient for rendering problems to consider 3-component model of the normal (n_x, n_y, n_z) and the distance from a current point of an object to an observer ρ. Three components of the normal are not sufficient to generate the normal form of equation of a plane $n_x x + n_y y + n_z z - p = 0$ in the considered point.

The coefficient ρ specifies the location of the current point with reference to the origin (distance). Here the coefficient ρ should be added as the fourth component of the normal vector (n_1, n_2, n_3, n_4) in order to achieve unambiguity while representation of the local geometry of an object. It is sufficient to consider the triangle of the control grid to describe a plane, passing through three points, by the equation $ax + by + cz + d = 0$. Then we should normalize the coefficients a, b, c, d by the norm $N_4 = \sqrt{a^2 + b^2 + c^2 + d^2}$. As the result we will obtain the four components of the normal that describe local equation $n_1 x + n_2 y + n_3 z = n_4$ at each point of the terrain surface.

The algorithm of Bilinear interpolation can be represented by three basic stages of calculation.

Stage 1. Definition of the normal components for each triangle of the mesh.
To calculate four components (n_1, n_2, n_3, n_4) of the normal to the plane determined by points (A_i, B_i, C_i) of the triangle ***i***, determinant of a 4 × 4 matrix is used:

$$\begin{vmatrix} x & y & z & 1 \\ x_A & y_A & z_A & 1 \\ x_B & y_B & z_B & 1 \\ x_C & y_C & z_C & 1 \end{vmatrix} = 0$$

Stage 2. The four components of the average normal in vertex k of the mesh are defined as the summation of the corresponding components of the normals to the planes of the triangles containing vertex k.

Stage 3. Double linear interpolation:

1. Linear interpolation of components of the normal between vertices of line segments [AB] and [AC]:
   a. $P_i^1(u) = (1-u)a_j^{A_i} + ua_j^{B_i} \qquad n = \frac{|AP^1|}{|AB|}, \quad j = 1\ldots4,$
   b. $P_i^2(v) = (1-v)a_j^{A_i} + ua_j^{C_i} \qquad n = \frac{|AP^2|}{|AC|}, \quad j = 1\ldots4.$

2. Linear interpolation of components of the normal between intermediate points of the line segment $[P^1P^2]$:

$$a_j^i(t) = (1-t)P_i^1 + tP_i^2 \qquad t = \frac{|P^1a_j|}{|P^1P^2|}, \qquad j = 1\ldots4,$$

where $\boldsymbol{i}$ – the number of a triangle.

As the result of Bilinear interpolation at each point of the space of the discretely defined function we obtain the coefficients (a_1, a_2, a_3, a_4) of the equation of a plane using triangulated surface.

3 The Voxel-Functional Representation of the Model

Having obtained the local geometric characteristics (a_1, a_2, a_3, a_4) for each point of filling domain of triangulated topographical network, after normalization $n_j = a_j/\sqrt{a_1^2 + a_2^2 + a_3^2 + a_4^2}$ we receive components of the normal: $\vec{n}(n_1, n_2, n_3, n_4)$.

Let us establish some correspondence between every value of n_j and the value of intensity gradation of the tone of the monochrome palette C_j:

$$C_j = \frac{(n_j + 1)P}{2},$$

where P – is the upper value of the color intensity of the palette, $C_j \to [0, P]$, for example $P = 256$.

As the result of calculations, we'll obtain four images of local geometrical characteristics (C_1, C_2, C_3, C_4) illustrated on Fig. 1.

Let's apply the proposed mathematical apparatus of bilinear interpolation to preparation of image data for each point of the complex shaped terrain.

Let's consider the real example prepared especially for studying the features of the terrain overlaying the regular mesh with 50 × 50 vertices over its heightmap (Fig. 2).

The results of the algorithm of bilinear interpolation for the area of the complex shaped terrain represented by the regular mesh with 2500 vertices are shown at Fig. 3.

The main advantage of such approach to representation of the terrain model is obtaining images of the local geometrical characteristics (Fig. 4) that provides expeditious representation of the local function for each of the points of the considered area

$$z = \frac{a_4}{a_3} - \frac{a_1}{a_3}x - \frac{a_2}{a_3}y, \quad \text{where } n_j = \frac{2C - P}{P}, (j = 1\ldots4).$$

Another advantage of the FV-model is that represented components of the normal (n_1, n_2, n_3, n_4) can be easily transitioned into the lower-dimensional projections. For example, it is sufficient to carry out the following transformation:

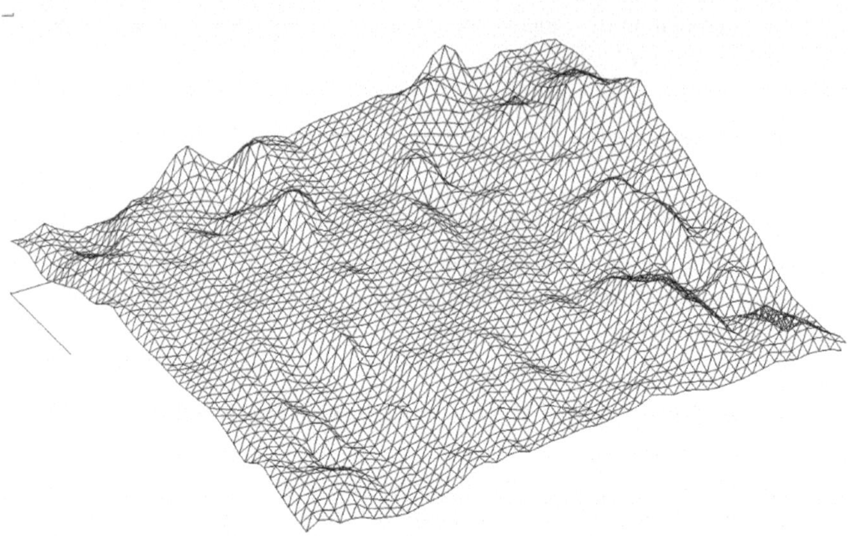

Fig. 2. The discretely defined function of the terrain

Fig. 3. An example of the halftone gradation for the height datum z of the area obtained by the bilinear interpolation of the given heightmap data in the overlaying regular mesh vertices.

$$n_j = \frac{2C - P}{P}, (j = 1\ldots4), n_i = \frac{n_i}{\sqrt{n_1^2 + n_2^2 + n_3^2}}, (i = 1\ldots3).$$

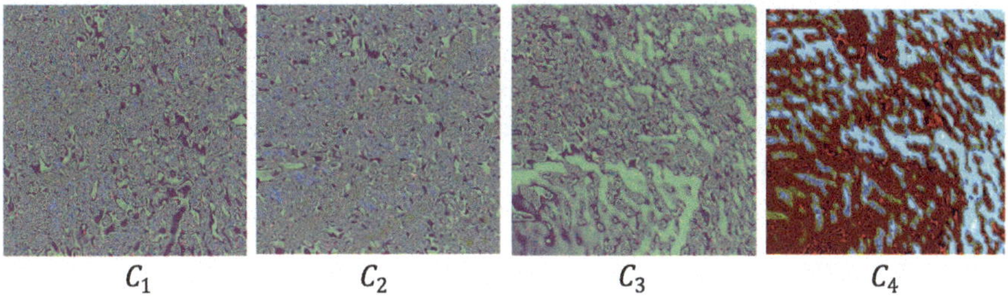

Fig. 4. Voxel Images of the local geometrical characteristics obtained on the basis of bilinear interpolation using RGB Palette P = 256 × 256 × 256 = 16777216 gradation of colour.

to get the three-component model (n_x, n_y, n_z) on the basis of the four-component model (n_1, n_2, n_3, n_4).

Thus, we reduce the amount of the normalized components and the components of the norm. In this case the model based on the three-component normal (n_x, n_y, n_z) becomes suitable for rendering.

Obtaining two images C_x, C_y (Fig. 5) of the 2-component normal vector (n_x, n_y) allows application of The gradient descent algorithm contained in the FV-method Tools and described in [8]

$$n_x = \frac{n_1}{\sqrt{n_1^2 + n_2^2}}, n_y = \frac{n_2}{\sqrt{n_1^2 + n_2^2}}, C_x = \frac{(n_x + 1)P}{2}, C_y = \frac{(n_y + 1)P}{2}, P = 256.$$

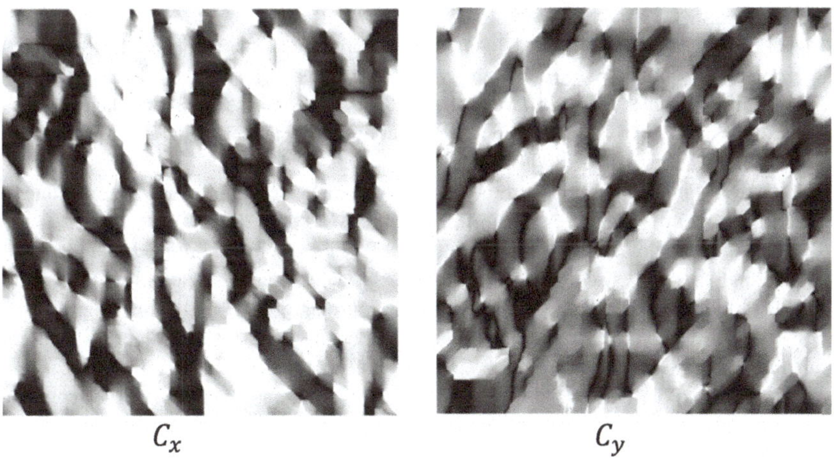

Fig. 5. Voxel images for the gradient descent algorithm from FV-method Tools

Testing calculation accuracy of the received model showed good results. In the process of automatic defining of calculation maximum of the local function z over the specified domain of the functional voxel model compared with maximum of height datum z in the given mesh showed very low discrepancy between the results which equals few thousandth.

4 Conclusions

The proposed approach to the computer representation of geodesic information creates the local function in the points of discrete space. This approach allows to apply obtained description in various analytical calculations having possibility of simple calculation of differential characteristics which reflect the local properties of the surface shape.

References

1. Kolesnikov, Yu.N.: Geometricheskoe modelirevanie v grapgicheskih sistemah realnogo vremeni (Geometric Modeling in Graphical Real-time Systems), p. 218. PSU, Penza (2006)
2. Lee, V.G.: Geometricheskiy instrumentariy sinteza sredy virtualnoy realnosti primenitelno k trenazhoram (Geometrical instruments for the synthesis of virtual reality environment applied to simulators). Diss. Doc. Tech. sciences: 05.01.01, Kiev, p. 320 (2000)
3. Hoang, T.H., Kosnikov, Yu.N., Zimin, A.P., Aleksandrova, N.V.: Smeshivaushie funkcii v geomertichescom modelirovanii I visualizacii poverhnostey cvobodnih form (Blending Functions in Geometrical Modeling and Visualization of Freeform Surfaces), XXI century: Resumes of the Past and Challenges of the Present plus: Scientific Periodical, Series: Engineering Sciences, Information technologies. Penza, PenzSTU, № 03 (issue 25), pp. 51–60 (2015). ISSN 2221-951X
4. Kosnikov, Yu.N., Kuzmin, A.V., Hoang, T.H.: Morphing of spatial objects in real time with interpolation by functions of radial and orthogonal basis. J. Phys. Conf. Ser. **1015**, 032066 (2018). http://iopscience.iop.org/article/10.1088/1742-6596/1015/3/032066/meta
5. Tolok, A.V.: Using voxel models in automation of mathematical modeling. Autom. Remote Control **70**(6), 1067–1079 (2009)
6. Tolok, A.V.: Funktsional'no-voksel'nyi metod v komp'yuternom modelirovanii (Functional Voxel Method in Computer Modeling). Fizmatlit, Moscow (2016)
7. Tolok, A.V., Tolok, N.B., Funkcionalno-vokselniy metod komputernikh vichisleniu (Functional Voxel Method of Computer Calculations). Scientific Visualization. M. MEPhI. T.2, Issue 9, 1–12 (2017). http://sv-journal.org/2017-2/01.php?lang=ru
8. Grigoryev, S.N., Tolok, N.B., Tolok, A.V.: Local search gradient algorithm based on functional voxel modeling. Program. Comput. Softw. **43**(5), 300–306 (2017)
9. Tolok, A.V., Tolok, N.B., Loktev, M.A.: Modeling function domain for curves constructed based on a linear combination of basis bernstein polynomials. Program. Comput. Softw. **44** (6), 526–532 (2018)

10. Tolok, A.V., Tolok, N.B., Loktev, M.A.: Voxel representation of local geometry. In: International Conferences on Interfaces and Human Computer Interaction 2018 Game and Entertainment Technologies 2018 and Computer Graphics, Visualization, Computer Vision and Image Processing 2018. Part of the Multi Conference on Computer Science and Information Systems, pp. 435–438 (2018)
11. Tolok, A.V., Tolok, N.B.: Mathematical programming problems solving by functional voxel method. Autom. Remote Control **79**(9), 1703–1712 (2018)
12. Rogers, D.F., Adams, J.: Mathematical Elements for Computer Graphics. McGraw-Hill, New-York (1990)

Development of the Information Support System Components for Personnel Security Management of the Region

V. V. Bystrov[1], D. N. Khaliullina[1](✉), and S. N. Malygina[1,2]

[1] Institute for Informatics and Mathematical Modeling Kola Science Centre of the Russian Academy of Sciences, Apatity, Murmansk Region, Russia
{bystrov, khaliullina, malygina}@iimm.ru

[2] Apatity Branch of Murmansk Arctic State University, Apatity, Murmansk Region, Russia

Abstract. The article is devoted to the development of methods and means of network-centric management of regional socio-economic systems. For the organization of information and analytical support of decision-making in the field of personnel security of the region, the authors propose to use a combination of three approaches to management: functional-target, process and network-centric. The article describes the features of the mutual application of these approaches to solving issues of staffing. A model representation of the architecture of a multi-agent support system for network-centric management of regional personnel security is presented. The authors describe in detail the General algorithm of functioning of the developed software system, based on the PDCA cycle of project management.

Keywords: Functional-targeted approach · Process approach · Network-centric management · Multi-agent system · Personnel security · Socio-economic system

1 Introduction

The creation of methods and tools for management of regional socio-economic systems is a topical issue of scientific and practical research long time. Under this subject the development of information and analytical support for management is one of the promising areas. The development of such decision support tools is primarily related to the modeling the subject area as a way of representing the subject of the study in a formal form.

Modeling of socio-economic systems is a rather time-consuming process associated with the presence of a large number of heterogeneous elements and relationships between them. Using this type of systems is limited by the mental capabilities of a person. One possible way to solve this problem is to represent complex systems as hierarchical tree structures.

The method of constructing hierarchical tree structures is called the functional-target approach. The approach is being developed at the Institute of Informatics and

R. Silhavy (Ed.): CSOC 2020, AISC 1226, pp. 348–361, 2020.
https://doi.org/10.1007/978-3-030-51974-2_34

mathematical modeling of the Russian Academy of Sciences by researchers of the scientific school under the leadership of V. A. Putilov [8]. The main idea of this approach is to form a hierarchical system of goals, then to construct such tree of functions, the performance of the functions leads to the achievement of the corresponding goals.

This article presents the interim results of the implementation of research projects such as "Models and methods of configuration of adaptive multi-level network-centric systems of regional security management in the Arctic zone of the Russian Federation" and "Methods and means of information support of personnel security management of the regional mining and chemical cluster". The common task for these projects is to develop a model for network-centric management of heterogeneous elements of regional socio-economic systems. To solve the task we propose to use a combination of known approaches to managing complex systems such as functional-target approach, process approach and network-centric approach. This article discusses in more detail the features of the joint application of these approaches. As a result of complex application of the approaches a multi-level recurrent model of hierarchical management of regional personnel security is formed. According to the current study the subject area is the personnel security of the region. Regional personnel security is provided by interrelated socio-economic systems of the region, such as the regional labor market, the regional training and retraining system, the regional labor consumption system and others. The choice of this subject area is not accidental. In recent years the problem of meeting the personnel needs of the economy of the northern and arctic regions of the Russian Federation has become more acute. The solution to this problem forces us to develop new approaches to personnel policy management. In particular it forces us to create new information technologies to support personnel management in the regional economy (including the network-centric paradigm for managing complex systems).

2 Background

We reviewed open sources of information on the development of decision support tools for human resources management. We can conclude that most of the work is applied in nature and is aimed at specific practical tasks in this area. A relatively small part of scientific publications is devoted to the development of methodological and software tools for personnel support in the whole region [1–7].

The work [2] analyses the relationship between human resources and the regional economy and identifies the main problems of regional management of human resources. The authors propose some human resource development measures to improve the regional economy.

Article [5] defines the concept of Regional Strategic Human Resources Management (RSHMR). At the same time RSHRM is combined with the regional industrial. To support the sustainable development in Dalian the demand for categories of human resource is predicted.

Work [3] that raises the issue of the human resources development as the key factor affecting regional economic development is close enough to the subject of our study. The authors of article [3] propose to build regional human resources management system to develop the regional economic. This approach is intersected with the research we are conducting.

The works [6, 7] affect the regional level, but the use of decision support systems for personnel management, in the end, is reduced to the analysis of personnel movements in the enterprise. For example, paper [6] investigates the effects of reduced workforce levels on the production outputs of industry sectors in the aftermath of a hurricane. This work develops a workforce recovery model to assess workforce disruption scenarios in the aftermath of a hurricane. This is accomplished by accounting regional data and historical scenarios associated with the formulation of the workforce disruption model.

From the point of view of software development, the closest work is a study aimed at developing a regional human resource management decision support system which is based on the technology of data warehouse [4]. The data cube model is employed to construct the data warehouse in this system.

Unfortunately, during the analysis of publications, direct analogues of our research related to the development of complex software decision-making tools in the field of personnel security management were not identified.

3 Methods

3.1 Functional-Target Approach

Applying of functional-target approach to any subject area is inextricably linked with such concepts as goal, task, function. In the current work the goal of the system is considered as the desired state of the system or the desired result of its behavior. The process of achieving the goal by the system is inextricably linked to the notion of "task". In this case the task is semantically interpreted as a problem situation that has a clearly defined goal to be achieved. "Function" is another term actively used in the functional-target approach. Here the function is perceived as an action of the system aimed at solving the corresponding task. A basic operation "covering the goal with action" is used to establish the correspondence of a specific function to a specific goal [8, 9].

Finally using the goal-setting principle and operation of covering, two tree-like hierarchical structures are formed. The structures reflect the decomposition of the goals and the actions (functions) covering them. At that the initial vertex of the goal tree is the global goal.

When building a tree of goals of any complex system, you can set different global goals. In accordance with this statement, we define the set of global goals as: $\{G_i^0\}, i = \overline{1, n}$. According to the functional-target approach, a goal decomposition procedure is carried out to construct the goal tree of the subject area. Decomposition consists in recurrent division of the current goal into multiple sub-goals. The joint achievement of the sub-goals ensures the achievement of the current goal. This process is carried out until you get a set of indivisible goals (primitives). The condition of the indivisibility of the goal is the possibility to achieve this goal by an elementary action (function). The decomposition results in a goal tree that can be represented by a graph:

$Tr^G = \langle V^G, L^G \rangle$ – goal tree.
Here, $V^G = \{V^{G_k}\}$ – the set of vertices of the tree representing the goals, where k – the level number of the hierarchy, $k = 0, \ldots, N$, N – number of hierarchy levels.
$V^{G_k} = \{v_i^{G_k}\}$ – the set of vertices of the k-th level of the hierarchy, where i – goal number, $i = 1, \ldots, m_k$,
$L^G = \{l_{ij}^G\}$ – the set of relationships between the goals, $i, j = 0, \ldots, m_k$,
$l_{ij}^G = \langle v_i^{G_k}, v_j^{G_{k+1}} \rangle$ – relationship between k-level i-th goal and j-level $(k + 1)$-th goal.

It is worth noting that for each global goal its own goal-tree is built.

To represent the goals decomposition into the sub-goal, we will introduce the concept of recursion on the goal tree:

$$V_i^{G_{k+1}} = F_r\left(v_i^{G_k}\right),$$

$$F_r : V^{G_k} \rightarrow V^{G_{k+1}},$$

$\forall v_i^{G_k} \in V^{G_k} \exists V_i^{G_{k+1}} \subset V^{G_{k+1}} : \forall v_j^{G_{k+1}} \in V_i^{G_{k+1}} \exists l^G = \langle v_i^{G_k}, v_j^{G_{k+1}} \rangle$, at that, there is a case when: $V_i^{G_{k+1}} = \emptyset$.

According to the concept of functional-target approach, the tree of functions (actions) is maped on the tree of goals using the procedure "covering the goal with action" [8]. Applying this we obtain the corresponding function tree from the goal tree (Fig. 1).

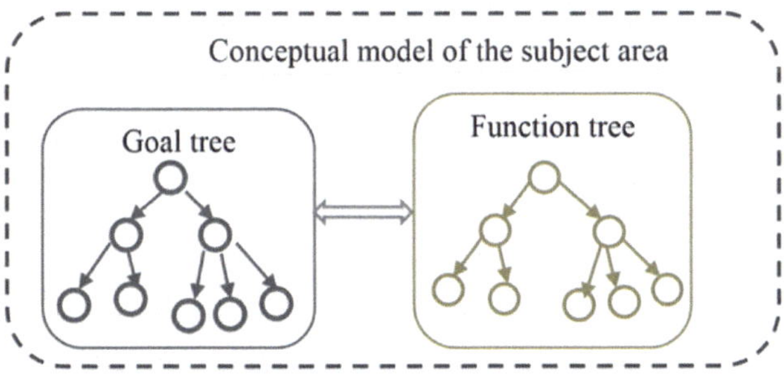

Fig. 1. Conceptual model of the subject area according to the functional-target approach

This action tree can be represented by the following formal description:

$Tr^A = \langle V^A, L^A \rangle$ – action (function) tree.
Here, $V^A = \{V^{A_k}\}$ – the set of vertices of the tree representing the actions, where k – the level number of the hierarchy, $k = 0, \ldots, N$, N – number of hierarchy levels.
$V^{A_k} = \{v_i^{A_k}\}$ – the set of vertices of the k-th level of the hierarchy, where i – action number, $i = 1, \ldots, m_k$, at that, V^{A_k} covers corresponding level V^{G_k} of the goal tree.
$L^A = \{l_{ij}^A\}$ – the set of relationships between the actions, $i, j = 0, \ldots, m_k$,

$l_{ij}^{A} = \langle v_i^{A_k}, v_j^{A_{k+1}} \rangle$ – relationship between k-level i-th action and j-level $(k + 1)$-th action.

In contrast the goal tree, the function tree is built in reverse order from bottom to top.

3.2 Process Approach

In the proposed management model, each function tree primitive (F_i^P) can be described by the corresponding process or chain of processes. According to the international standard ISO 9001:2000, the process is a set of interconnected or interacting activities, converting inputs to outputs [10]. At the same time, there are executors $\{Ex_l\}$ of the process, and various regulators act on the process. When describing the primitives of the function tree, it is necessary to take into account that processes can occur both sequentially and in parallel (Fig. 2).

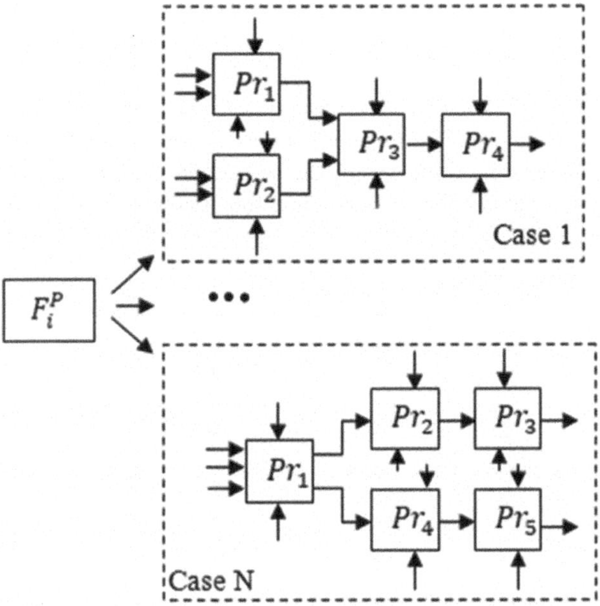

Fig. 2. Representation of function tree primitive (F_i^P) as alternative process chains

For example, consider a function tree primitive such as "To organize retraining that occurs on enterprises and specialized retraining centers". We can present it as different process chains:

1. "Determining the number of persons for retraining" → "Formation of the Educational Program of Retraining" & "Formation of the teaching staff" & "Formation of Material and Technical Resources" → "Formation of a group of students" → "Implementation of a training program" & "Maintenance of accompanying reporting documentation" → "Carrying out a qualification test" → "Registration of qualification documents".
2. "Determining the number of persons for retraining" → "Formation of the Educational Program of Retraining" → "Formation of the teaching staff" → "Formation of Material and Technical Resources" → "Formation of a group of students" → "Implementation of a training program" → "Carrying out a qualification test" → "Assignment of qualifications".
3. Other combinations of processes.

In this example, the symbol "→" should be interpreted as a sequential execution of the actions, and the symbol "&" as simultaneous/parallel processes.

To form the vertices of the tree of functions/actions of a higher hierarchy level it is necessary to choose an alternative chain of processes. At that the task of indicator evaluation of alternatives is set. To obtain an objective assessment of the process chain it is proposed to apply a system of indicators that includes several categories of process indicators (finance, time, economic effect, etc.).

3.3 Network-Centric Approach

In recent years, the theory of network-centric management of various socio-economic systems has been actively developing. The employees from the Institute for Informatics and Mathematical Modelling of Technological Processes of the Kola Science Center Russian Academy of Sciences Masloboyev A.V. and Putilov V.A. offered a number of conceptual and methodological developments in this scientific area. In particular, they developed a multi-level recurrent model of hierarchical management of complex security of regional socio-economic systems. According to these authors, "the use of the recurrent model makes it possible to form mathematical models of security management of regional socio-economic systems operating in conditions of uncertainty" [11].

This article continues to develop the concept of network-centric management systems by integrating functional-target and hierarchical approaches to the management of regional socio-economic systems.

At the current stage of the study, the main assumption is that all participants in the processes of ensuring personnel security have a sufficient level of interest and motivation to solve the tasks assigned to them. Otherwise, it makes no sense to talk about creating an effective management system.

Figure 3 shows the conceptual scheme of the regional personnel security management model. It shows the main levels of the management model at that the individual stages of decision-making are implemented. The model includes three main levels: conceptual level, virtual level and organizational level.

The organizational level is a set of organizational structures. The structures are directly or indirectly involved in the planning, implementation and monitoring of

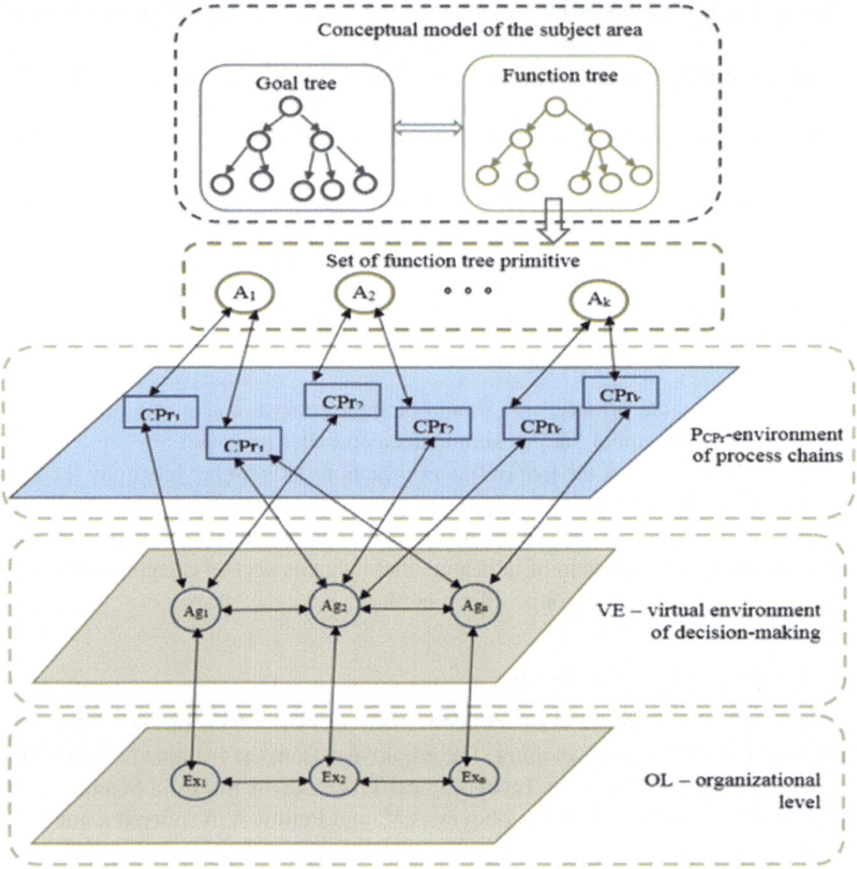

Fig. 3. Conceptual scheme of personnel security management model

actions to achieve the global goal. The atomic elements of the organizational level are representatives of organizations and enterprises participating in the operation of the developed software of supporting network-centric management of personnel security in the region. In the proposed structure of the management model interaction is carried out between elements of the organizational and virtual levels. This interaction is manifested in the form of interaction between the representative of the organizational structure and the user interface of the corresponding software agent.

The virtual environment is a network-centric structure that has the following characteristics:

1. a large number of heterogeneous agents and sources of information;
2. network architecture with three types of software agents;
3. organization of distributed information storage;
4. operating large volumes of information;

5. the decision center is decentralized and can change (migrate over the network) depending on the task being solved;
6. availability of mechanisms for coordinating goals and for coordinating actions and/or management between nodes of a multi-agent network.

The virtual level of such personnel security management model interacts with both organizational and conceptual level objects.

The conceptual level of the model is formed by following elements:

1. formalized conceptual model descriptions that include a graph representation of the trees of goals and functions (in accordance with the functional-target approach),
2. previously developed conceptual models of personnel logistics in the region [12] and models of regional security management based on project management [13].

In practice formalized conceptual descriptions will be presented in the form of applied ontologies of OWL format.

When realizing the proposed model into a software product, the formalized conceptual descriptions will be represented by applied ontologies of the OWL format.

The interaction between the conceptual and virtual levels represents the appointment of software agents as executors of certain processes from the formed space of action chains. At that various entities can act as executors of the process. They can be entities responsible for the implementation of the process; stakeholders; directly the executors of the process, etc. As a result, each executor of the process is assigned an agent from the virtual decision-making environment.

4 Results

We offer to implement the above-described approaches to network-centric management of personnel security in the region in the form of a prototype decision support system (DSS). The general algorithm of functioning of this prototype consists of fourteen key stages. The main purpose of the offered tools is to organize information and analytical support for the process of solving the problem in the field of personnel security. It is proposed to consider the problem from the point of view of the project approach. According this, the task to be solved presents in the form of a global project [13]. Given this, it was decided to build a general algorithm for the functioning of the prototype of the decision support system based on the Deming Cycle [14] used in project management.

All stages of the prototype functioning were divided into groups according to the Deming Cycle. The first eight stages of the proposed algorithm relate to the global project planning. The next two stages relate to the implementation of the project. The eleventh stage relates to the verification of the project, and the rest of them – to corrective action generation. Consider these stages in more detail.

The first stage is the initiation of the problem to be solved. At this stage, one of the participants in the organizational environment raises the problem of the socio-economic development of the region by referring to the virtual environment through its agent-representative. There may be situations where the problem is raised by a coalition of agents. For example, the problem to provide labour resources for the economy of the region is submitted by one participant of the organizational environment, in particular, by the Ministry of economic development of the Murmansk region. For example, the problem of the need to provide labour resources for the economy of the region is submitted to one participant of the institutional environment, in particular, the Ministry of Economic Development of the Murmansk region. But the problem of training a certain number of qualified workers of a given profile can be raised by a coalition of mining and chemical cluster enterprises.

The second stage is the formulation of the global goal of the initiated problem. It is assumed that the result of this stage is a formulated global goal. The agent-initiator and other agents of the virtual environment that form this goal have the certain area of responsibility to solve issues related to the problem under consideration.

There may be cases when the initiated problem corresponds to several formulations of global goals that have different semantic. For example, there is the problem "The presence of personnel shortages in the Murmansk region". For this problem it is possible to formulate following global goals:

1. "to balance the supply and demand of the regional labor market";
2. "to organize attracting personnel to work by a shift method";
3. etc.

The third stage is the formation of a goal tree corresponding to the selected global goal. Here, the global goal decomposition procedure is carried out under the control of the virtual environment agent that initiated the problem. This agent becomes the coordinator of the process of dividing the global goal into sub-goals. The dividing is carried out through the interaction of the agent coalition (CoAg) that is a subset of the agent set {Ag} in the virtual environment.

$$CoAg = \{Ag_j\}, j = \overline{1, ..H}$$

$$CoAg \subseteq Ag, Ag = \{Ag_i\}, i = \overline{1, ..N}, H \leq N,$$

where H, N - the number of agents in the coalition and virtual environment respectively.

In the coalition there are two roles of agents: agent-coordinator and agent-expert. Expert agents are ranked by degree of awareness (competence) of the subject area. As a result of the joint work of the agent-coordinator with the agents-experts, the goals tree is formed at the exit of this stage.

The fourth stage is the formation of the function tree. Unlike the previous stage, the process of building the function tree is carried out in the opposite direction ("bottom-up"). Here, the top-level vertex of the action tree is obtained by co-executing the functions of the lower level of the hierarchy. This procedure ends when the vertex corresponding to the global goal is reached. In the general case, the set of expert agents involved at this stage may differ from the set of experts involved in the implementation of the previous stage. The described stages result in two tree structures such as the goal tree and the corresponding function tree.

The fifth stage is the formation of process chains. The coalition of agents for each primitive of the function tree generates either one chain of processes or a set of alternative chains (see Fig. 2).

The sixth stage is the selection of the processes set. The main objective of this stage is to limit the set of processes considering in the solved problem. The main objective of this stage is to limit the set of processes considering in the solved problem. The limitation is carried out by cutting off alternative process chains in accordance with the given criteria. Selection criteria are set by the coordinating agent and may include financial, time, material and other types of indicators. A weighted sum of indicators for each indicator category can be used as an integral assessment of process chain alternatives.

$$Ind_{CP} = \sum\nolimits_{i=1}^{N} \sum\nolimits_{j=1}^{N} (w_{i,j} \cdot Ind_{i,j}),$$

where i – number of indicator category, j – number of indicator in the i-th indicator category, $w_{i,j}$ – weight coefficient of a specific indicator, $Ind_{i,j}$ – value of a specific indicator.

The sixth stage results a set of necessary process chains (CPr) each of which uniquely corresponds to an element from the set of primitives of the function tree.

$$CPr = \{CPr_i\} \rightarrow \{F_i^P\}, i = \overline{1,..N},$$

where N – number of primitives.

The seventh stage is the synthesis of the network of processes of the problem being solved. The main objectives of this stage are:

1. To define relationships between processes and process chains according to the function tree.
2. To define the order (sequence) of execution of action chains.
3. To coordinate temporal characteristics of the processes.
4. To evaluate the results of planning of the network of current processes.

This stage verifies if the process chains cover all vertices for each level of the function tree hierarchy. If conflict situations are identified when relationships between processes or between their temporal characteristics are being established, it is necessary to make changes to previously adopted decisions. The correction can be made at almost all of the above stages, they are refining (detailing) the global goal, rebuilding the goal tree, modifying the function tree (in particular, modifying a set of the primitives), forming new process chains or regrouping of already formed process chains, etc. Depending on the specific case, it is possible to return to one or another stage of the General scheme of work of the developed tools for network-centric management of personnel security. At that, each corrective action has limited number of allowed iterations that must be performed to obtain a satisfactory result. If the number of iterations for corrections is over and result of the correction is negative, a transition to an earlier stage of the general scheme of work is carried out. For example, if you try to regroup already formed process chains and the problem does not disappear, the

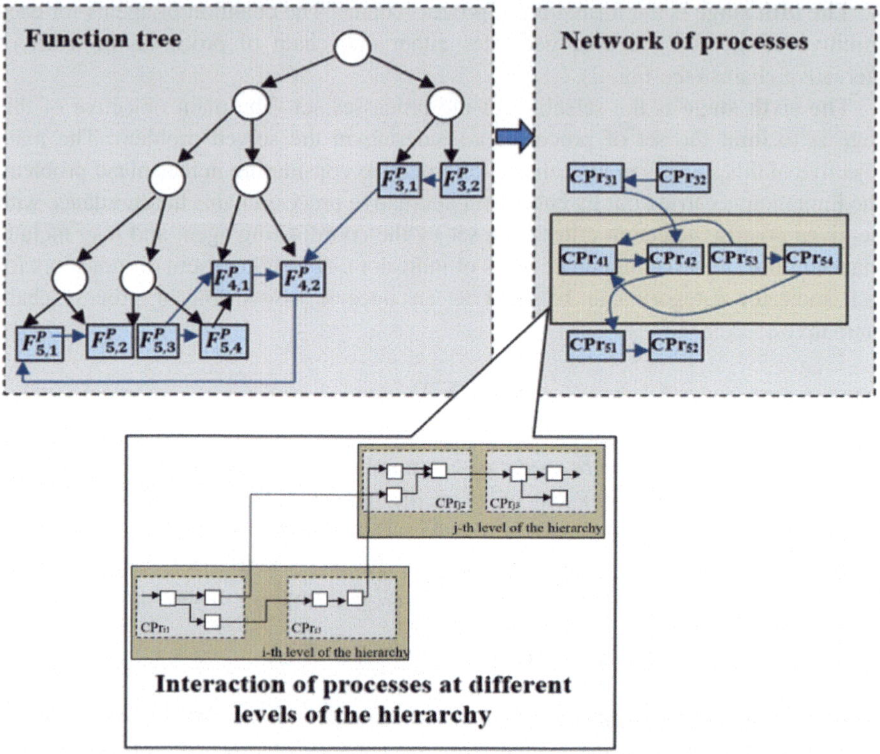

Fig. 4. Formation of a network of actual processes

transition to the stage of formation of alternative process chains (the fifth stage) is carried out. At the output of this stage, a coordinated network of actual processes is formed (Fig. 4).

In the stages from 5-th to 7-th coalitions of agents involved in this task may change. This may be due to the appearance of new experts with the necessary competencies, the exclusion of members from the expert group with insufficient knowledge, skills and abilities.

The eighth stage is selection of executors. For each function tree primitive, which at this stage is already a chain of processes, a separate working group of virtual environment agents is formed. One of the agents of each working group is given the authority of coordinator. The mechanisms for determining the coordinator may be the following:

1. the self-nomination of an agent;
2. the delegation of authority by a higher managing agent,
3. the nomination of agent by an agreed decision of the entire working group.

It is possible that the initiator of the problem himself becomes the coordinator of a separate working group.

The following set of actions describes the algorithm for assigning the executor of a separate process:

1. A coordinated selection of executors from a variety of agents who have shown the initiative to be the executor of the process. Self-nomination by the participants of the working group corresponds to the principles of network-centric management.
2. Joint decision of the working group to nominate agents as executors of processes if there are processes without executors in the process network.
3. If there is no full coverage of the set of processes by the executors after the first two actions, the executors of the processes are assigned by the coordinator of this working group.

The key parameter in appointing the executors of a particular process is its competence area that is a set of its knowledge, skills and abilities to solve problems from a specific subject area. In each of the considered actions of the algorithm for assigning the executor, preference is given to a more competent agent. The determination of the area of competence is a separate scientific task that can be reduced to calculating integral assessment of a certain set of agent parameters such as experience in implementing projects, subject area, area of responsibility, etc.

When determining the executors of a particular process, there may be a situation when the number of executors of a given process is more than one. In this case, it is rational to carry out the detailing of the process into many subprocesses so that each subprocess has one executor. At the current stage of the study, we introduce a restriction that each executor of the process is responsible for the results of the implementation of the process assigned to him. This restriction is not required in the theory of project management.

The ninth stage is the initialization of the global project. It consists in pointing out specific executors, project deadlines, descriptions of financial aspects, etc. All of these parts of project are documented and accepted by all participants in the global project.

The tenth stage is the launch of a global project. All participants in the global project begin to perform their duties in accordance with their roles, areas of responsibility and the schedules set out in the project documentation.

The eleventh stage is to control the implementation of the global project. The coordinators of each working group collect information on indicators characterizing the progress of the processes that are in their area of responsibility.

The twelfth stage is to develop corrective actions. The proposed management model assumes active use of forecasting to analyze possible ways to implement the planned global project. Forecasting will be implemented using simulation modeling. It allows you to quickly obtain forecasts in the set scenario conditions both about individual processes of the global project and about the project as a whole. The coalition of agents that constitutes the working group on the implementation of the process chain makes a collective decision to change the implementation plan of the global project taking into account the forecast information and monitoring data.

The thirteenth stage is to bring corrective actions to the executors. This stage involves notifying all interested executors by virtual environment tools whose activities affect the developed corrective action.

It should be noted that from the ninth to the thirteenth stages the realization of the formed plan for the implementation of the global project is carried out at the organizational level of the regional personnel security management model. That is representatives of organizations by interacting with software agents of the virtual environment transmit information about intermediate and final results and progress of the processes assigned to them and also take part in the development of corrective actions.

The fourteenth stage is to analyze the degree of achievement of the global goal based on the results of individual stages of the project. When analyzing the interim results of a global project there may be a situation when corrective actions do not lead to the achievement of the original global goal. In this case, the question is raised to continue the global project but with worse results or to terminate the global project and return and to the first stage of the general algorithm.

The general algorithm of functioning of personnel security decision support system operation given in this section reflects the concept of combined use of functional-target, process, project and network-centric approaches to management of social and economic systems.

5 Conclusion

To date, the creation of methodological and software tools of information and analytical support for the management of socio-economic systems is a popular scientific and technical direction. As part of the development of this direction, it is proposed to use a combination of known methods and approaches for managing personnel security in the region. We apologies the integration of different approaches to the management of complex systems can have a new synergistic effect affecting the issues of efficiency, quality and validity of decision-making in the field of personnel policy.

At the same time, it is worth noting that the article describes the conceptual aspects of the proposed solutions in the field of personnel security management. A number of issues raised remain open to further scientific research. In particular, the development of mechanisms for coordinating the goals of different project participants and resolving conflict situations during its implementation remain open to further scientific research.

Acknowledgements. The work was supported by the Ministry of science and higher education of the Russian Federation and partially by The Russian Foundation for basic research-project № 19-07-01193 A.

References

1. Wang, C.: Comprehensive evaluation model of regional human resources management - an empirical study from the employees' perspective. In: Proceedings of the Second China Private Economy Innovation International Forum, pp. 527–532 (2009)

2. Jiao, Q., Qiu, L.: Strategy of human resource development for regional economy. In: Proceedings of 2008 Conference on Regional Economy and Sustainable Development, pp. 871+ (2008)
3. Wang, L., Xie, M.: Analysis on the management of regional human resources development. In: Human Resources Management in the Knowledge Economy Era, vols. I and II, pp. 672+ (2009)
4. Xiao, H.R., Zhang, H.Y., Li, M.Q.: Regional human resource management decision support system based on data warehouse. In: Proceedings of the 4th World Congress on Intelligent Control and Automation, vols. 1–4, pp. 2118–2121 (2002). https://doi.org/10.1109/wcica.2002.1021459
5. He, M., Liu, D., Yang, D.: Regional strategic human resource management exploration. In: Proceedings of the 2005 Conference of System Dynamics and Management Science, Sustainable Development of Asia Pacific, vol. 1, pp. 331–339 (2005)
6. Akhtar, R., Santos, J.R.: Risk analysis of hurricane disruptions on workforce and interdependent regional sectors. In: IEEE Systems and Information Engineering Design Symposium (SIEDS), pp. 41–46 (2013)
7. Chenthamarakshan, V., et al.: Effective decision support systems for workforce deployment. IBM J. Res. Dev. **54**(6), № 5, 5:1–5:15 (2010)
8. Kuzmin, I.A., Putilov, V.A., Filchakov, V.V.: Distributed information processing in scientific research. Science (1991). (in Russia)
9. Putilov, V.A., Gorokhov, A.V.: System dynamics of regional development. RC «Pazori», Murmansk (2002). (in Russia)
10. International Standard ISO 9001:2000. http://niits.ru/public/2003/069.pdf. Accessed 22 Nov 2019
11. Masloboev, A.V., Putilov, V.A., Syutin, A.V.: Multilevel recurrent model for hierarchical control of complex regional security. Sci Tech. J. Inf. Technol. Mech. Opt. **6**(94), 163–170 (2014). (in Russia)
12. Malygina, S.N., Bystrov, V.V., Khaliullina, D.N.: Logistics of personnel provision in the region: formalization and structure of the poly-model complex. Transactions of the Kola science centre. Inf. Technol. (9), 36–47 (2018). Apatity: KSC RAS (2018). https://doi.org/10.25702/ksc.2307-5252.2018.10.36-47. (in Russia)
13. Bystrov, V.V., Masloboev, A.V., Putilov, V.A.: Application of project management in regional security management tasks: approach and formal apparatus. Reliab. Qual. Complex Syst. **4**, 73–84 (2017). (in Russia)
14. Repin, V.V., Eliferov, V.G.: Process approach to management. Business process modeling. RIA «Standards and quality», Moscow (2008). (in Russia)

Automated Detection of Anthropogenic Changes in Municipal Infrastructure with Satellite Sub-meter Resolution Imagery

D. K. Mozgovoy[1], D. V. Kapulin[2], D. N. Svinarenko[1], A. I. Sablinskii[3], T. N. Yamskikh[2], and R. Yu. Tsarev[2,4(✉)]

[1] Oles Honchar Dnipro National University, 72, Gagarin Prospect, Dnipro 49000, Ukraine

[2] Siberian Federal University, 79, Svobodny Prospect, Krasnoyarsk 660041, Russia
tsarev.sfu@mail.ru

[3] Kemerovo State University, 73, Sovetsky Prospect, Kemerovo 650043, Russia

[4] Krasnoyarsk State Agrarian University, 90, Procpect Mira, Krasnoyarsk 660049, Russia

Abstract. This paper discusses an efficient method for the automated detection of anthropogenic changes in urban development with multispectral satellite producing sub-meter resolution imagery. The proposed method efficiency is confirmed with vector layers of detected changes in urban development obtained as a result of processing infrared and visible multidate images. The developed technique significantly reduces complexity and thus increases operational efficiency of map updating due to automation of satellite imagery processing.

Keywords: Satellite monitoring · Urban development · Multidate images · Image processing · Map updating

1 Introduction

The intensive development of new construction technologies in recent decades has been one of the main reasons for the rapid dynamics of urban development worldwide. This trend is observed both in the high-rise and low-rise housing and in the construction of industrial and commercial facilities. Therefore, in the field of information technologies there is an urgent task to develop highly efficient methods for automated detection of changes in urban development. This can be judged from the numerous scientific publications of recent years in the field of applied use of aerospace images [1–5].

Ground-based measurements and aerial imagery have traditionally been used to regularly update urban spatial databases. At the same time, the periodicity of updating urban maps was extremely low (usually once every few years) due to high complexity of data collection and processing. In recent years significant progress in the development of optoelectronic multispectral scanners (Table 1) has fostered the use of satellite imagery for assessing the evolution of land-based objects and updating urban maps [6–8]. Their use is especially effective for the automated detection of changes in various anthropogenic objects, such as buildings, roads, engineering structures, etc. [9–13].

R. Silhavy (Ed.): CSOC 2020, AISC 1226, pp. 362–370, 2020.
https://doi.org/10.1007/978-3-030-51974-2_35

Table 1. Satellites with multispectral scanners of sub-meter resolution.

Satellite	Year of Launch	Operator (country)	Resolution			Swath, km	Accuracy of geolocation, m
			PAN/MS, m	Radiometric, bit	No. of bands		
EROS-B	2006	Israel	0,7	10	1	7	N/A
Cartosat-2	2007	India	0,8	10	1	9,6	N/A
WorldView-1	2007	USA	0,5	11	1	17,6	5
GeoEye-1	2008	USA	0,4/1,6	11	4	15	2…3
WorldView-2	2009	USA	0,46/1,84	11	8	16	5
Cartosat-2B	2010	India	0,8	10	1	9,6	N/A
Pleiades-1A	2011	France	0,5/2	12	4	20	4.5
Kompsat-3	2012	Korea	0,5/2	14	4	17	13
Pleiades-1B	2012	France	0,5/2	12	4	20	4.5
SkySat-1	2013	USA	0,8/2,0	11	4	8	N/A
Gaofen-2	2014	China	0,8/3,24	10	4	45	50
ASNARO-1	2014	Japan	0,5/2,0	11	6	10	10
WorldView-3	2014	USA	0,3/1,2	11/14	8	13	3.5
SkySat-2	2014	USA	0,8/2,0	11	4	8	N/A
Kompsat-3A	2015	Korea	0,4/1,6	14	4	12	13
PeruSat-1	2016	Peru	0,7/2,0	12	4	10	N/A
SkySat-3…7	2016	USA	0,8/2,0	11	4	8	N/A
Gokturk-1A	2016	Turkey	0,7/2,8	12	4	20	10
SuperView-1A/B	2016	China	0,5/2	11	4	12	20
WorldView-4	2016	USA	0,3/1,2	11	4	13	3
Mohammed-VIA	2017	Morocco	0,7/2,8	12	4	20	N/A
SkySat-8…13	2017	USA	0,8/2,0	11	4	8	N/A
SuperView-1C/D	2018	China	0,5/2	11	4	12	20

*N/A - no information available

In order to detect changes in multidate satellite images of submeter spatial resolution, differential images are used wherein detecting changes by difference in values between adjacent pixels having the same geographical coordinates on two images reduced to the same spatial resolution. As a rule, panchromatic band data providing the highest granularity level and sensitivity are used. However, such a method has a significant disadvantage in its simplicity: it does not allow to detect changes in cases when compared objects have different level of brightness in multispectral bands, but similar in panchromatic band.

The research aims at improving the method for the automated detection of anthropogenic changes in urban development with multispectral satellite producing sub-meter resolution imagery in the visible and infrared bands by combining the data of all spectral bands in order to increase the efficiency and reliability of urban map updating.

2 Research Methodology

The territory of Healdsburg in Sonoma County, California, USA, was chosen as a test site to improve the method for the automated detection of anthropogenic changes in urban development as there a rather high dynamics of urban infrastructure development, including changes in housing and commercial facilities. Besides, the largest oil and gas pipelines are located in or near Healdsburg, which requires regular monitoring of changes in development in accordance with the existing standards. Multidate satellite images in visible and infrared bands acquired by Pleia-des-1A Satellite Sensor on May 31, 2012 and June 18, 2014 were selected for the analysis (Fig. 1).

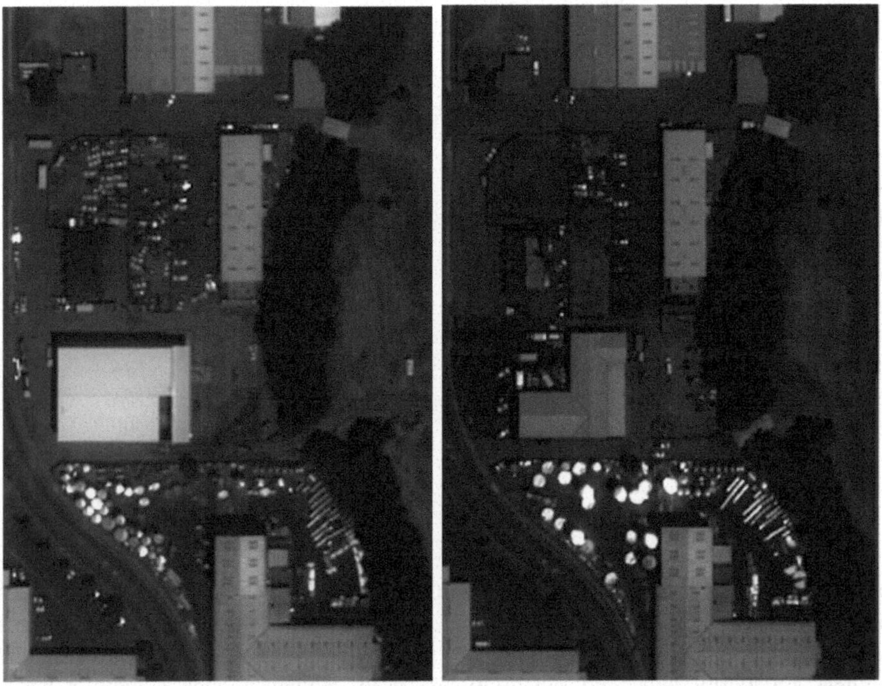

Fig. 1. Fragments of initial satellite images acquired on May 31, 2012 (left) and June 18, 2014 (right).

The research comprised the following stages of image processing and analysis [14]:

- Pre-processing (normalization) of multidate satellite images, the same for the initial and the next images, including purification, augmentation of spatial resolution and correlative alignment to the reference image (Fig. 2);

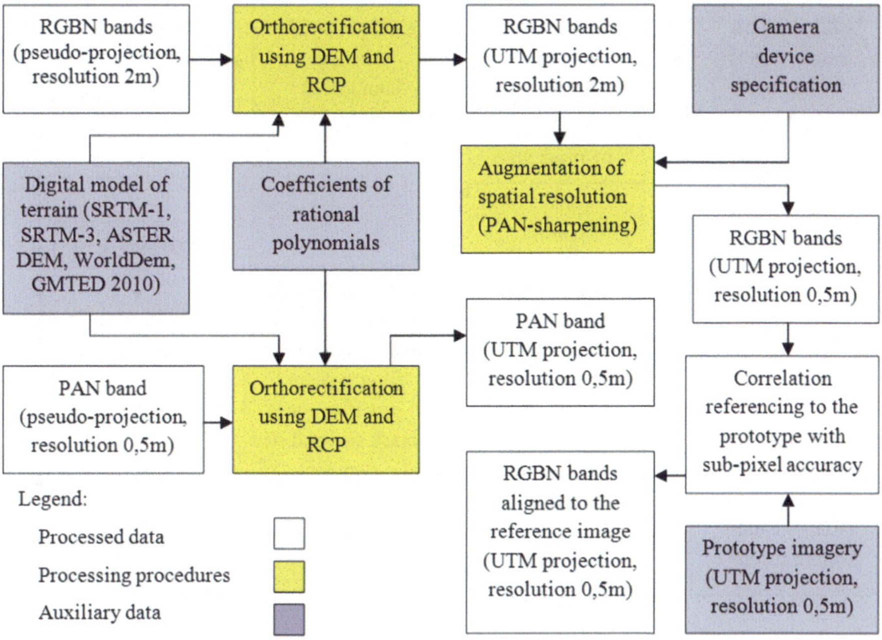

Fig. 2. The main stages of preliminary satellite images processing.

– Thematic processing of normalized satellite images, including filtering of shadows and small mobile objects (cars, itinerant trade locations, etc.), calculation of the 1st principle component, two-threshold binarization, morphological and linear filtering and vectorization of identified changes (Fig. 3).

The most widely used methods for estimating the accuracy of satellite image change detection are [15]:

– comparing the results of automated remote sensing data classification with synchronous ground based observations and measurements performed simultaneously with satellite imagery (or with a small time interval);
– comparing the results of automated remote sensing data classification with the results obtained with certified satellite image processing software packages (but it is difficult or impossible to assess the accuracy of the standard measure);
– comparing the results obtained with manual classification of changes using remote sensing data with the operators' results evaluated by the expert group (this method is used for relatively small amount of data or for a limited set of test areas to be evenly spread across the investigated territory).

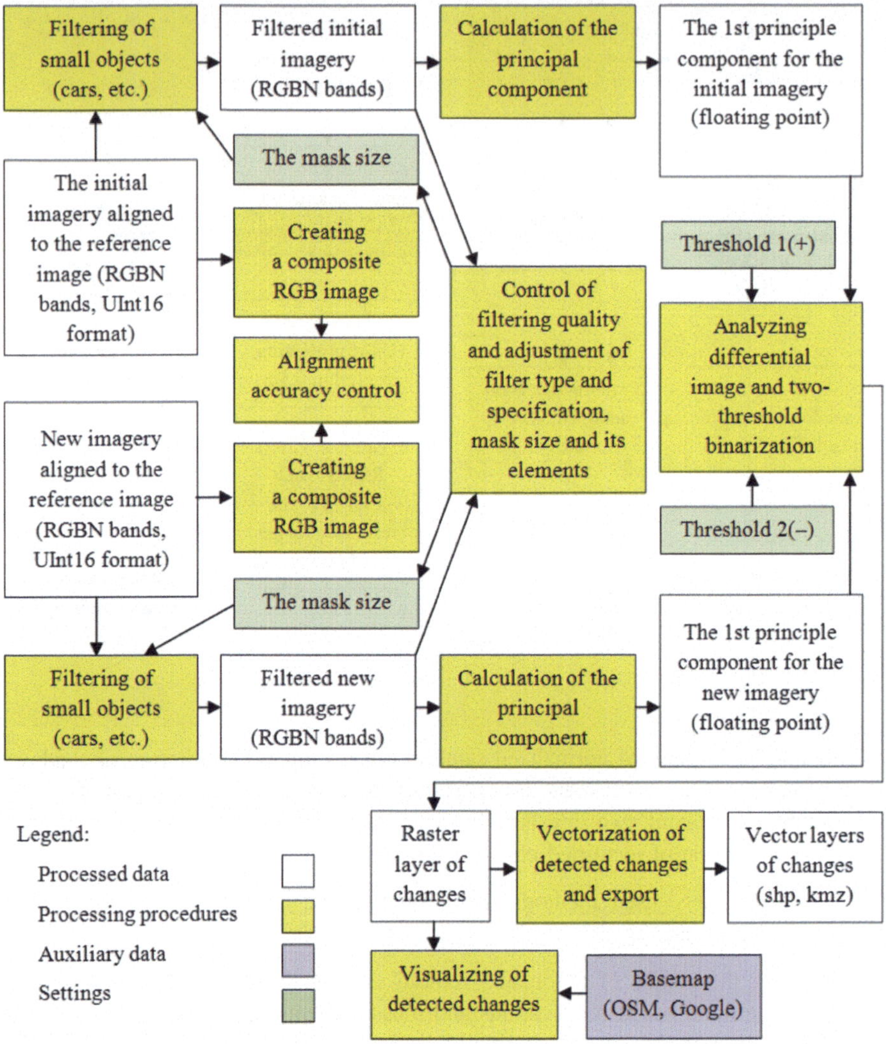

Fig. 3. The main stages of thematic satellite images processing.

Due to the lack of ground-based measurement data for the area under study, the last of the above mentioned methods was used to assess the accuracy of anthropogenic changes automated detection. Which means that comparison of the results obtained with manual classification of changes was performed for separate fragments of RGB image in interactive mode within the QGIS software environment. The expert assessment of the standard measure accuracy thus obtained averaged 5… 9%.

3 Results and Discussion

The file sizes of high-spatial resolution multi-spectrum images are usually large enough (for example, a scene imaged in the visible and IR range can take several gigabytes). Therefore, it is desirable to use modern computers with Intel Core i7 processors or higher and at least 64 GB of RAM to quickly process real-time images taken by the SuperView 1A. The software can be either commercial (ERDAS, ENVI, ArcGIS, etc.) or free (SNAP, SAGA, GRAAS, QGIS, etc.), run in both MS Windows and Linux environments. To provide more efficient automation of processing procedures it is possible to use tools (e.g. Imagine Model Maker in ERDAS package, Graph Builder in SNAP package) or programming languages and specialized utilities (e.g. IDL in ENVI package, Python GDAL in QGIS system).

A differential image was obtained after normalization and filtering of multispectration satellite images in visible and IR ranges acquired by the Pleiades-1A sensor on May 31, 2012 and June 18, 2014. It was further used to determine changes in urban development (Fig. 4).

Detected changes   Vector + initial image   Vector + next image

Fig. 4. Detected changes and vector layers overlay on a raster.

The boundaries of the detected building changes were sufficiently accurately identified on various test areas. The high level of the proposed method stability was confirmed without using additional vegetation and water masks.

The main advantages of the proposed method in comparison with those using only panchromatic band data are the ability to detect changes in cases when initial and next images of the objects being analyzed have the same albedo values in the panchromatic band. This contributes to significant improvement of change detection accuracy. Besides, due to the high degree of satellite image processing automation, the methodology can significantly reduce complexity and thus increase the operational efficiency of map updating. The software implementation of this methodology in the form of a geo-information web service, compared to traditional software and hardware, has significant organizational, technical and economic advantages, such as:

- it works directly in the browser, which does not require installation of additional software;
- software and hardware independence, which ensures using this web service on mobile devices;
- the results of image processing are stored on the server, which allows all customers to use the web service regardless of their location;
- high economic efficiency (does not require purchase of powerful graphic stations and expensive software);
- minimum requirements for the end-user training (there is no need to spend time studying large and complex software packages).

The practical application of the proposed method is quite extensive as it provides automated detection of changes not only in urban development, but also in any other natural and man-made objects (road network, plant and water objects, etc.). Therefore it could be used for the benefit of various State services, private companies and commercial entities. With software implementation of this methodology in the form of a geo-information web service, it can be used for informing the mass user - population.

The results of this research are included in the educational materials for lectures and laboratory classes being a part of the module "Ultra-high spatial resolution satellite images processing" for senior students of the Dnepropetrovsk National University named after Oles' Honchar within the framework of the course "Remote sensing systems" and also used to prepare course and diploma papers. Students test the proposed methodology experimentally using multispectral images of various Earth areas acquired by the existing RSS satellites.

4 Conclusion

An efficient method for the automated detection of anthropogenic changes in urban development with multispectral satellite producing sub-meter resolution imagery in visible and IR-bands is proposed. It significantly reduces complexity and thus increases operational efficiency and accuracy of map updating due to automation of satellite imagery processing and can be used to solve various applied tasks that are associated with construction, communication lines, engineering communications, etc.

Directions of further research. Currently, the proposed methodology is being developed using multispectral images of different Earth areas acquired by the existing submeter resolution satellites in order to determine the optimal operation parameters for the main types of modern on-board sensors taking into account the environmental and imaging conditions. In addition, this method is being improved with regard to the use of additional vegetation and water masks in order to enhance stability and quality of detection.

Acknowledgments. This work was supported by the Ministry of Education and Science of the Russian Federation in the framework of the Federal target program «Research and development of priority directions of development of the scientific-technological complex of Russia for 2014-2020» (unique ID project RFMEFI60519X0185).

References

1. Lu, D., Li, G., Moran, E.: Current situation and needs of change detection techniques. Int. J. Image Data Fusion **5**, 13–38 (2014)
2. Tewkesbury, A., Comber, A., Tate, N., Lamb, A., Fisher, P.: A critical synthesis of remotely sensed optical image change detection techniques. Remote Sens. Environ. **160**, 1–14 (2015)
3. Aleksandrowicz, S., Turlej, K., Lewiński, S., Bochenek, Z.: Change detection algorithm for the production of land cover change maps over the European union countries. Remote Sens. **6**, 5976–5994 (2014)
4. Tian, J., Chaabouni-Chouayakh, H., Reinartz, P., Krauss, T., D'Angelo, P.: Automatic 3D change detection based on optical satellite stereo imagery. In: ISPRS Technical Commission VII Symposium on Advancing Remote Sensing Science, vol. 38, pp. 586–591. ISPRS, Vienna (2010)
5. Rottensteiner, F.: Automated updating of building databases from digital surface models and multi-spectral images. In: The International Archives of the Photogrammetry, Remote Sensing and Spatial Information Sciences, Beijing, China, vol. XXXVII, part B3a, pp. 265–270 (2008)
6. Deng, J., Wang, K., Deng, Y., Qi, J.: PCA-based land-use change detection and analysis using multitemporal and multisensor satellite data. Int. J. Remote Sens. **29**, 4823–4838 (2018)
7. Champion, N., Boldo, D., Pierrot-Deseilligny, M., Stamon, G.: 2D building change detection from high resolution satellite imagery: A two-step hierarchical method based on 3D invariant primitives. Pattern Recogn. Lett. **31**, 1138–1147 (2010)
8. Le Bris, A., Chehata, N.: Change detection in a topographic building database using submetric satellite images. In: International Archives of Photogrammetry, Remote Sensing and Spatial Information Sciences, vol. 38(3/W22), pp. 25–30. ISPRS, Munich (2011)
9. Malpica, J., Alonso, M.: Urban changes with satellite imagery and LiDAR data. In: International Archives of the Photogrammetry, Remote Sensing and Spatial Information Sciences - ISPRS Archives, vol. 38, pp. 853–858. ISPRS, Kyoto (2010)
10. Awrangjeb, M., Zhang, C., Fraser, C.S.: Improved building detection using texture information. In: International Archives of the Photogrammetry, Remote Sensing and Spatial Information Sciences - ISPRS Archives, vol. 38 (3W22), pp. 143–147. ISPRS, Munich (2011)
11. Bouziani, M., Goïta, K., He, D.-C.: Automatic change detection of buildings in urban environment from very high spatial resolution images using existing geodatabase and prior knowledge. ISPRS J. Photogramm. Remote Sens. **65**, 143–153 (2010)
12. Matikainen, L., Hyyppa, J., Ahokas, E., Markelin, L., Kaartinen, H.: Automatic detection of buildings and changes in buildings for updating of maps. Remote Sens. **2**(5), 1217–1248 (2010)
13. Dini, G.R., Jacobsen, K., Rottensteiner, F., Al Rajhi, M., Heipke, C.: 3D building change detection using high resolution stereo images and a GIS database. In: International Archives of the Photogrammetry, Remote Sensing and Spatial Information Sciences - ISPRS Archives, vol. 39, pp. 299–304. ISPRS, Melbourne (2012)

14. Mozgovoy, D.K., Hnatushenko, V.V., Vasyliev, V.V.: Automated recognition of vegetation and water bodies on the territory of megacities in satellite images of visible and IR bands. In: ISPRS Annals of the Photogrammetry, Remote Sensing and Spatial Information Sciences, Beijing, China, vol. 4(3), pp. 167–172 (2018)
15. Mozgovoy, D., Hnatushenko, V., Vasyliev, V.: Accuracy evaluation of automated object recognition using multispectral aerial images and neural network. In: Proceedings of SPIE - The International Society for Optical Engineering, vol. 10806, art. 108060H. SPIE, Shanghai (2018)

Geometry-Based Automated Recognition of Objects on Satellite Images of Sub-meter Resolution

D. K. Mozgovoy[1], D. V. Kapulin[2], D. N. Svinarenko[1], T. N. Yamskikh[2], A. A. Chikizov[2], and R. Yu. Tsarev[2,3](✉)

[1] Oles Honchar Dnipro National University, 72, Gagarin Prospect, Dnipro 49000, Ukraine

[2] Siberian Federal University, 79, Svobodny Prospect, Krasnoyarsk 660041, Russia
tsarev.sfu@mail.ru

[3] Krasnoyarsk State Agrarian University, 90, Procpect Mira, Krasnoyarsk 660049, Russia

Abstract. The paper considers an algorithm for automated classification of mobile small size objects on multispectral satellite images of submeter spatial resolution using radiometric and geometric features. It ensures recognizing the desired classes of objects with high accuracy regardless of their orientation in the image. The geometric features of the objects classified in the binary image included the area of the object, the lengths of the principal and auxiliary axes of inertia, the eccentricity of the ellipse with the main moments of inertia, the area of a convex polygon described near the object, the equivalent diameter of a circle with the same area, and the convexity coefficient.

Keywords: Satellite monitoring · Sub-meter resolution · Image processing

1 Introduction

The most difficult task in automated processing of multispectral satellite images of sub-meter spatial resolution is searching and recognizing small objects (structures, vehicles, etc.) using radiometric, spectral, textural, statistical, geometric, and other decoding features [1–3]. The values of the upper and lower binarization threshold (so-called cut-off threshold) are usually used as radiometric decoding features for each spectral channel of the image, which are selected manually or read from the decryption feature library file. Geometric characteristics include area, perimeter, shape coefficient, describing the roundness and convexity of the object, eccentricity, center of gravity, coordinates of the describing rectangle, etc. [4–6].

Most of the existing modern software packages for processing satellite images ensure per-pixel classification of images or areas of interest in general, as a rule, taking into account only their spectral characteristics, without analyzing the geometric properties of separate objects [7–9]. Skeleton-based methods to determine geometric parameters of objects in binary images, (e.g., used in character recognition), are not invariant with respect to rotation and therefore cannot be used to recognize moving

R. Silhavy (Ed.): CSOC 2020, AISC 1226, pp. 371–379, 2020.
https://doi.org/10.1007/978-3-030-51974-2_36

objects with arbitrary orientation [10–12]. Recognition methods using neural networks developed in recent years require long term training, as well as high requirements for computing resources [13–15].

The research was aimed at developing and subsequent testing the methods and algorithms for automated classification of small objects on multispectral satellite images of sub-meter spatial resolution using geometric features invariant to rotation.

2 Mathematical Models and Methods

The simplest and most common geometric features that are invariant to rotation are the area and perimeter of the object.

The pixel area of a binary object is equal to the number of nonzero image elements belonging to the object. Moreover, the set of single readings (*x*, *y*) with coordinates (*x*, *y*) belonging to the region *A* is given as follows

$$g(x,y) = \begin{cases} 1, & (x,y) \in A, \\ 0, & \text{otherwise.} \end{cases}$$

If the coordinates of the upper-left and lower-right corners of the rectangle describing the region are ($X_{\min}$, $Y_{\min}$) and ($X_{\max}$, $Y_{\max}$), respectively, then the area is

$$S = \sum_{y=Y_{min}}^{Y_{max}} \sum_{x=X_{min}}^{X_{max}} g(x,y).$$

The center of gravity of the region is given by the coordinates (X_c, Y_c), defined as the mean value (*x*, *y*) of the coordinates belonging to the region according to the equation

$$X_c = \frac{1}{S} \sum_{(x,y) \in A} x,$$

$$Y_c = \frac{1}{S} \sum_{(x,y) \in A} y.$$

Determining the coordinates of the center of gravity of the object allows one to normalize the position of the object by determining the position of the origin in the image plane. An object is given a central position.

If the number of end readings of a region is *N*, then the perimeter length *P* is the sum of the distances between the adjacent boundary points

$$P = \sum_{i=1}^{N} r_i,$$

$$r_i = \sqrt{(x_{i+1} - x_i)^2 + (y_{i+1} - y_i)^2}.$$

A reading is end if at least one of the nearest neighboring readings does not belong to the region A.

To assess the compactness of the object, a form factor is used, defined as the ratio of the perimeter square to the area

$$K = \frac{P^2}{S}.$$

To estimate the roundness of the region the following coefficient is used

$$C = \frac{m_A}{\sigma_A}.$$

where m_A – is the average value of the distances from the center of gravity of the region to the end readings, determined by the formula

$$m_A = \frac{1}{N}\sum_{i=1}^{N} r_{ic},$$

where σ_A – is spectral reflectance coefficient of these distances, determined by the formula

$$\sigma_A = \sqrt{\frac{1}{N-1}\sum_{i=1}^{N}(r_{ic} - m_A)^2},$$

where r_{ic} – is the distance from i end reading to the center of gravity of the region.

The radiuses of the inscribed and circumscribed circles are also used to assess the compactness of the object. Moreover, statistical moments of the region are often used for recognition.

The discrete central moment m_{ij} of a region is defined as follows

$$m_{ij} = \sum_{(x,y)\in \mathrm{Reg}} (x - \tilde{x})^i (y - \tilde{y})^j$$

$$\tilde{x} = \frac{1}{n}\sum_{(x,y)\in \mathrm{Reg}} x,$$

$$\tilde{y} = \frac{1}{n}\sum_{(x,y)\in \mathrm{Reg}} y,$$

where n – is the total number of pixels in the region.

Also, the scale, translation and rotation invariant features of the region are used, for recognition.

For example, eccentricity, characterizing elongation and eccentricity of an object

$$elongation = \frac{m_{20} + m_{02} + \sqrt{(m_{20} - m_{02})^2 + 4m_{11}^2}}{m_{20} + m_{02} - \sqrt{(m_{20} - m_{02})^2 + 4m_{11}^2}}.$$

A statistical approach is used to normalize the orientation of an object in analyzing binary images. The object is described by some scattering ellipse.

The direction of the eigenvector x of the covariance matrix B of the coordinates of non-zero brightness readings, that is, belonging to the region A is chosen for orientation.

The eigenvector must correspond to the maximum eigenvalue λ of covariance matrix

$$B = \begin{pmatrix} B_{11} & B_{12} \\ B_{22} & B_{21} \end{pmatrix},$$

where B_{ij} are central moments of the second order;

B_{11} is the dispersion of x- coordinates of non-zero brightness readings;
B_{22} is the dispersion of y- coordinates of non-zero brightness readings;
B_{12} is the covariance (x, y) of non-zero brightness readings coordinates.

The eigenvalues λ are found using the equation

$$(B - \lambda \mathrm{E})x_{\lambda} = 0,$$

where E is the unity matrix,

x_{λ} is the eigenvector corresponding to the number λ.

The eigenvalue λ of a covariance matrix is found using the equation

$$B - \lambda \mathrm{E} = 0.$$

The parameters of the approximating ellipse estimated from the binary image are: the small half-axis a, the large half-axis b and the tilt angle of the ellipse major axis in accordance with the described statistical approach to the object binary image normalization.

The dimensions of the ellipse half-axes are defined as follows. The ratio of the covariance matrix eigenvalues (ellipse half-axes) is determined:

$$k = \sqrt{\frac{\mathrm{abs}(\lambda_2)}{\lambda_1}},$$

where λ_1 is the largest eigenvalue, λ_2 is the smallest eigenvalue.

If the ratio of the minor and major semi axes of the ellipse is *a/b* = *k,* then the area of the ellipse *Square* = *pab* = pkb^2.

The major half-axis of the ellipse is:

$$b = \sqrt{\frac{Square}{\pi k}}$$

where *Square* is the area of the binary image (the number of readings with non-zero brightness).

The minor half-axis of the ellipse is determined using the equation *a* = *kb.*

Satellite image classification technique.

The procedures for processing and analyzing satellite images include the following stages:

1) preprocessing (normalization) of satellite images, including orthorectification, augmentation of spatial resolution, radiometric enhancement, contour allocation;
2) thematic processing of normalized satellite images, including:

   - converting to grayscale;
   - thresholding binarization of the selected area;
   - morphological filtering of binarized objects;
   - segmentation of filtered binary objects;
   - filtering of segmented objects by area;
   - identification of geometric parameters;
   - calculation of cross-correlation coefficient between geometric parameters;
   - objects classification by geometric parameters taking into account a given threshold (nearness criterion);
   - optimization of classification parameters in order to obtain the required number of classes;
   - vectorization of recognized objects and exporting the results to a standard format.

The geometric features of the objects being classified in the binary image include:

- the area of the object (number of the object pixels);
- lengths of principle and auxiliary axes of inertia (the lengths of the axes, which represent the directions in the object corresponding to the semi-axes of inertia ellipsoid);
- the eccentricity of the object (eccentricity of an ellipse with principal moments of inertia equal to the principal moment of the object's inertia);
- the area of a polygon (the area of a convex polygon described near the object);
- equivalent diameter (diameter of a circle having the same area as the object; calculated using the formula *sqrt (4S/p)*, where *S* is the area of the object);
- object convexity coefficient (a coefficient that is equal to the ratio *S*/S*, where *S* is the area of the object; *S** is the area of the polygon).

Geometry-based classification of small-size objects assumes comparing all objects of the binary image following the each with each alignment algorithm by calculating the cross-correlation coefficient r between all values of the objects geometric features:

$$r = \frac{\frac{1}{n}\sum_{k=1}^{n}(x_{1k} - \bar{x}_1)(x_{2k} - \bar{x}_2)}{\sqrt{D_1 \cdot D_2}},$$

where x_1 and x_2 are mean values for two samples;

x_{1k} and x_{2k} – are the current sample values;
n – is the number of sample elements;
D_1 and D_2 – are the average variance for two samples.

Next, we set the threshold value of cross-correlation coefficient and select the pairs of objects which value is higher. We evaluate the obtained pairs of objects in order to combine them into classes on the basis of calculations results in a table form.

The threshold value of correlation coefficient depends on how much the objects combined into one class and the classes themselves differ from each other. Thus, the threshold value can be rough or insensitive to changes. Then in the limit case all objects are combined into one class. Or it can be more sensitive to the slightest differences between the objects. Then in the limit case several classes are created and only one object is added into each class. Therefore, to get the required number of classes it is very important to choose the optimal threshold value.

3 Results and Discussion

To research the automated object classification algorithm we used a fragment of a multispectral satellite sub-meter resolution image containing small moving objects. Preprocessing procedures included converting the original RGB image to grayscale and contour allocation (Fig. 1 and 2). This was followed by binarization, morphological filtering, segmentation and filtration by area (Fig. 3 and 4). Finally, we could obtain the image containing objects that were of interest for their subsequent classification.

Fig. 1. Original RGB image.

Fig. 2. Grayscale image.

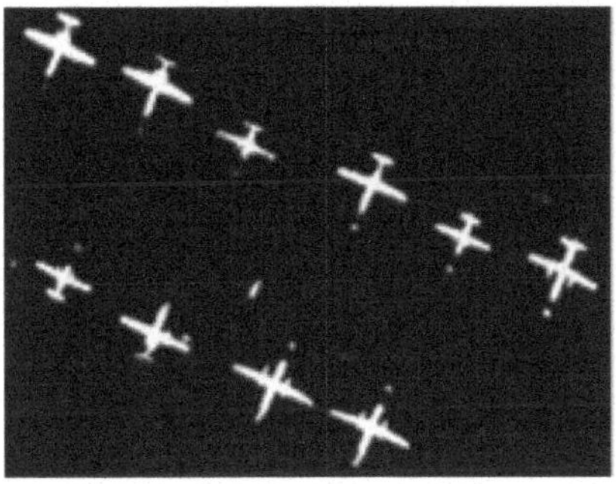

Fig. 3. The result of binarization.

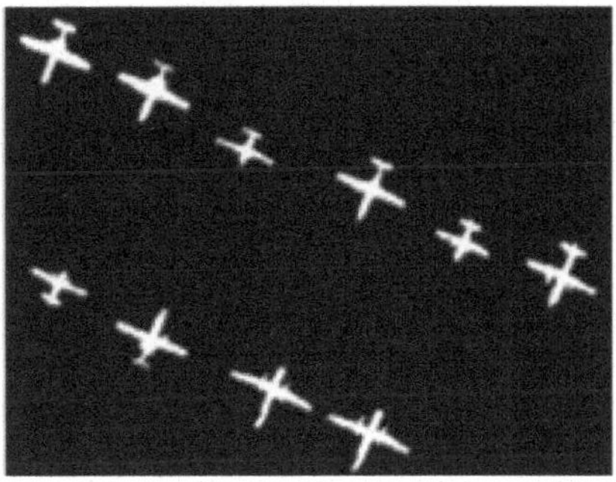

Fig. 4. The result of filtration by area.

Geometric features of all binary image objects were counted and normalized to one (Table 1).

Table 1. Geometric features of binary image objects.

Parameter	Object number									
	1	2	3	4	5	6	7	8	9	10
The area of the object	0.25	0.13	0.20	0.22	0.11	0.22	0.22	0.20	0.14	0.24
Length of the principle axis of inertia	0.84	0.67	0.83	0.86	0.67	0.95	0.94	0.81	0.65	0.78
Length of the auxiliary axis of inertia	0.69	0.46	0.59	0.60	0.43	0.63	0.63	0.68	0.50	0.76
The eccentricity of the object	0.56	0.72	0.70	0.71	0.76	0.74	0.73	0.54	0.63	0.23
The area of a polygon	0.52	0.25	0.46	0.48	0.26	0.55	0.55	0.49	0.28	0.53
Equivalent diameter	0.56	0.41	0.50	0.53	0.38	0.53	0.53	0.51	0.43	0.55
The coefficient of convexity	0.48	0.52	0.43	0.45	0.44	0.40	0.40	0.41	0.52	0.45

Based on the results of calculating cross-correlation coefficient between all values of the objects geometric parameters, it is easy to determine which objects belong to the same class with the selected threshold value of the correlation coefficient K, which determines each class boundaries.

If $K = 0.8$, then three classes are formed, i.e. the two closest in geometric characteristics classes of four classes are combined into one when $K = 0.95$. An object that has been allocated to a separate class when $K = 0.95$ moves to the first class when $K = 0.8$. When $K = 0.99$, the objects are combined into seven classes, as the second and third classes formed when $K = 0.95$ fall into five classes, and the first and fourth classes remain.

The high level of processing procedures automation and relative simplicity of the proposed technique ensures its implementation in the form of a geoinformation web service [17], which, compared to traditional software and hardware, has significant organizational, technical and economic advantages, such as:

- it works directly in the browser, which does not require installation of additional software;
- software and hardware independence, which ensures using this web service on mobile devices;
- the results of image processing are stored on the server, which allows all customers to use the web service regardless of their location;
- high economic efficiency (does not require purchase of powerful graphic stations and expensive software);
- minimum requirements for the end-user training (there is no need to spend time studying large and complex software packages).

The results of this research are included in the educational materials for lectures and laboratory classes being a part of the module "Ultra-high spatial resolution satellite images processing" for senior students of the Oles Honchar Dnipropetrovsk National University within the framework of the course "Remote sensing systems" and also used to prepare course and diploma papers. Students test the proposed technique experimentally using multispectral images of various Earth areas acquired by the existing RSS satellites.

4 Conclusion

A method for automated classification of moving objects using geometric features invariant to rotation was developed and tested. To classify small size objects by geometric features, all objects of the binary image were compared following the each with each alignment algorithm by calculating the cross-correlation coefficient between all values of the objects geometric features.

Special software was developed for experimental testing of the proposed technique. It ensures obtaining noise-free binary images and classifying filtered objects by geometric features in arbitrary orientation of the satellite image or separate objects, and constructing properties graphs for each object and each class at different threshold values of each class properties.

Experimental studies carried out on various satellite images confirmed that neither object orientation nor image rotation by an arbitrary angle considerably affect the result of classification (the probability of correct class recognition was in the range 0.97… 0.99).

Directions for further research. Currently, the proposed technique is being tested using multispectral images of various parts of the Earth obtained from various ultra-high resolution optical-electronic satellites in order to determine the optimal processing parameters for the main types of modern onboard scanners, taking into account the region and shooting conditions. Moreover, a simplified version of this technique (without preprocessing procedures) is successfully tested using multispectral aerial images acquired in the visible and infrared spectral ranges.

Acknowledgments. This work was supported by the Ministry of Education and Science of the Russian Federation in the framework of the Federal target program «Research and development of priority directions of development of the scientific-technological complex of Russia for 2014–2020» (unique ID project RFMEFI60819X0274).

References

1. Chen, G., Hay, G.J., Carvalho, L.M.T., Wulder, M.A.: Object-based change detection. Int. J. Remote Sens. **33**, 4434–4457 (2012)
2. Blaschke, T.: Object based image analysis for remote sensing. ISPRS J. Photogramm. Remote Sens. **65**, 2–16 (2010)
3. Hussain, M., Chen, D., Cheng, A., Wei, H., Stanley, D.: Change detection from remotely sensed images: from pixel-based to object-based approaches. ISPRS J. Photogramm. Remote Sens. **80**, 91–106 (2013)
4. Cheng, G., Han, J.: A survey on object detection in optical remote sensing images. ISPRS J. Photogramm. Remote Sens. **117**, 11–28 (2016)
5. Van Etten, A.: object detection in satellite imagery, a low overhead approach, Part I. 31 August 2016. https://medium.com/the-downlinq/object-detection-in-satellite-imagery-a-low-overhead-approach-part-i-cbd96154a1b7. Accessed 19 Jan 2020
6. An, Z., Shi, Z., Teng, X., Yu, X., Tang, W.: An automated airplane detection system for large panchromatic image with high spatial resolution. Opt. Int. J. Light Electron Opt. **125**, 2768–2775 (2014)
7. Yuan, J., Yang, H.H.L., Omitaomu, O.A., Bhaduri, B.L.: Large-scale solar panel mapping from aerial images using deep convolutional networks. In: Proceedings - 2016 IEEE International Conference on Big Data, pp. 2703–2708. IEEE, Washington (2016)
8. Malof, J.M., Bradbury, K., Collins, L.M., Newell, R.G.: Automatic detection of solar photovoltaic arrays in high resolution aerial imagery. Appl. Energy **183**, 229–240 (2016)
9. Saito, S., Aoki, Y.: Building and road detection from large aerial imagery. In: Proceedings of Society of Photographic Instrumentation Engineers (SPIE) – The International Society of Optical Engineering, vol. 9405, art. 94050K. SPIE, San Francisco (2015)
10. Chen, X., Xiang, S., Liu, C.L., Pan, C.H.: Vehicle detection in satellite images by hybrid deep convolutional neural networks. IEEE Geosci. Remote Sens. Lett. **11**(10), 1797–1801 (2014)
11. Audebert, N., Saux, B.L., Lefèvre, S.: Segment-before detect: vehicle detection and classification through semantic segmentation of aerial images. Remote Sens. **9**(4) (2017). Article no. 368
12. Razakarivony, S., Jurie, F.: Vehicle detection in aerial imagery: A small target detection benchmark. J. Vis. Commun. Image Represent. **34**, 187–203 (2016)
13. Yuan, J.: Automatic building extraction in aerial scenes using convolutional networks. arXiv:1602.06564 (2016).
14. Shu, Y.: Deep convolutional neural networks for object extraction from high spatial resolution remotely sensed imagery. Ph.D. thesis, University of Waterloo (2014)
15. Saito, S., Yamashita, T., Aoki, Y.: Multiple object extraction from aerial imagery with convolutional neural networks. J. Imaging Sci. Technol. **60**(1), 010402 (2016)
16. Mozgovoj, D.K., Kravets, O.V.: A classification of the low-sized objects on high resolution satellite images. Ecol. Noospherology **20**(1–2), 26–31 (2009)
17. Mozgovoy, D.K., Tsarev, R.Y., Almabekova, O.A., Pupkov, A.N.: Satellite monitoring of the drought consequences via medium and high resolution multispectral images. In: 18th International Multidisciplinary Scientific Geoconference. SGEM 2018, Albena, Bulgaria, vol. 18 (4.2), pp. 579–590 (2018)

Fog Computing and IoT for Remote Blood Monitoring

M. V. Orda-Zhigulina(✉) and D. V. Orda-Zhigulina

Southern Scientific Center of the Russian Academy of Sciences, St. Chehova, 41, Rostov-on-Don 344006, Russia
jigulina@mail.ru

Abstract. It is important to develop modern medical digital technologies for processing, transmitting and storing the data of biomedical research. This is necessary to integrate both existing medical research methods and those first developed in mobile healthcare (mHealth). In mHealth it is possible to carry out remote diagnostics, monitoring and treatment of diseases, population prevention of diseases, rendering assistance through telemedicine to people, etc. At the same time mobile computing devices allows to organize contact and exchange of information between doctor and patient all over the world anytime. Big data base needs a lot of calculations and huge amount of computers. So «fog computing» could optimize this big data. In this paper technology and principles of the mobile application mHealth, are discussed. It is presented the system of non-invasive blood test which is based on optoacoustic flow cytometry and "fog computing." Also the smart sensor which is consisted of infrared laser and piezoceramic sensor is described.

Keywords: Data collection · Processing and storage systems · Remote blood monitoring systems · Mobile healthcare · Fog computing · Internet of Things · Medical system

1 Introduction

Please note that the first paragraph of a section or subsection is not indented. The first paragraphs that follows a table, figure, equation etc. does not have an indent, either.

Subsequent paragraphs, however, are indented. At present time urgent scientific problem is the development of modern medical digital technologies for integration into existing methods of biomedical research based on new information technologies. One of this technology is fog computing. It becomes relevant to use such a new information technology as fog computing in medical systems [1, 2] because of it is reported in published data [1–4]. The term "fog computing" was coined by industry as a metaphor for explaining the architecture of a computing system: "fog" is software and hardware located between the "cloud" (data center) and the "earth" (end-user devices) [3]. Over and above mobile health (mHealth) is being developed widely at present time [5–9]. It allows remote diagnostics, monitoring and treatment of diseases, population-based disease prevention, and telemedicine assistance to people in remote areas, etc. [5, 8]. Mobile networks cover 85% of the world's population, and about seven billion people

R. Silhavy (Ed.): CSOC 2020, AISC 1226, pp. 380–386, 2020.
https://doi.org/10.1007/978-3-030-51974-2_37

have mobile phones [10]. Therefore, using mobile devices allows to organize contact and exchange of information between the doctor and the patient at any time and any geo-point almost. The large amount data is produced as the result of this interaction. It turns out a problem how to structured and processed data correctly. The tasks of creating and improving technologies, methods and data processing algorithms that arise in the application of existing biomedical methods for studying human health become relevant based on the above. An integral part of mHealth are data processing systems that using mobile devices and personal computing devices; including systems for collecting and processing medical data, complex functional tools and applications, channels and technologies for data transfer [6]. Is the ability using existing communication systems and any mobile computing devices for processing medical data and monitoring patient health is a significant advantage of mobile healthcare systems. This solution reduces the cost of the data processing system, as it allows saving on the infrastructure for creating new systems. It is reported in published data [6, 11] the economic advantage of mobile healthcare over traditional methods of biomedical research is: reduction in the cost of medical and preventive measures without loss of quality and effectiveness of health-saving measures [6, 8, 11].

The goal of research is developing the technology of photoacoustic (hereinafter - PA) noninvasive blood analysis and its subsequent integration into mHealth. This technology can be implemented as a mobile application for a non-invasive blood analysis which will be available in the usual rhythm of a person's life, without queues in hospitals and personal visit to a doctor even. It is reported in published data the PA method noninvasive blood test at present time is just in developing areas of medical diagnostics. This method has not clinical application yet, laboratory studies are being conducted for it [12–19], and data processing algorithms and systems have not yet been developed that would allow receiving, storing and processing data on the patient's blood parameters in real time and without visiting a medical institution. In addition the use of modern information and communication technologies in terms of operational data transfer and information processing requires the development of low-cost algorithms and principles for constructing a system, since the sensors used in flow cytometry PA measure large volumes of data on the amplitude and spectrum of acoustic waves which are not necessary only transmit but also analyze. It is advisable to process the obtained large amounts of data using the technology by fog computing [1–3] which uses resources "on the ground" - mobile phones, tablets and other digital devices rather than central network nodes as in "cloud computing".

Thus, data processing is carried out on site and data channels become less busy. In more detail the advantages of the fog computing concept are considered in [2], which describes the creation of a medical system with which it will be possible to observe a specific human organ in order to prevent its critical state. In addition, the use of mobile devices and communication systems can significantly reduce the cost of medical institutions to introduce an electronic system for processing medical data through the use of existing infrastructure when creating new high-tech jobs in hospitals. With this background advantages by using fog computing the authors proposed the mHealth technology non-invasive blood test system based on PA method, the structure of which is shown in Fig. 1.

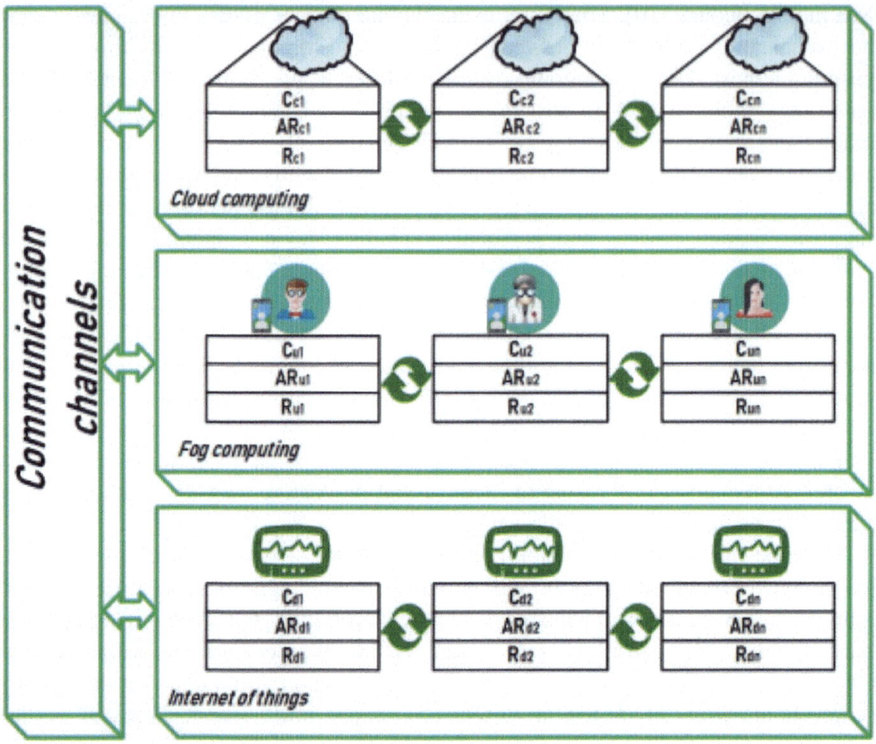

Fig. 1. The approach non-invasive blood test technology based on the PA method using fog computing.

Figure 1 introduces the following notation: 1 - diagnostic module (hereinafter - DM),

2 - a mobile device of the patient, on which the processing of the received information through a specialized mobile application,

3 - data transfer to the server on which a copy of the data is stored,

4 - data storage server,

5 - transfer of processed information to a doctor,

6 - receiving and analyzing the information received on the doctor's mobile device and making decisions,

7 - sending recommendations and comments from the doctor to the server where the data copy is stored,

8 - data transfer to a patient's mobile device.

Absolutely safely a patient at home can make non-invasive blood test using the diagnostic module (1); the data is transmitted from DM to mobile phone or any digital device (2) supports wireless technology and has an Internet connection. When data is transmitting (3, 7 and 8) to the server (4), the network is unloaded because of transmits not the entire volume of the received data but the already processed part (5), containing useful information about test for the doctor (6). The system with this architecture can be confidentiality ensured and the information is stored on the server can serve as evidence in the event of a dispute between the doctor and the patient. Using the proposed principles, it is possible to implement a monitoring system for the patient's blood parameters in the usual rhythm of life, without queues in hospitals and a personal visit to a doctor even.

2 Simulation Modeling

To test the theory a diagnostic module (1) was developed. Measured signal by a piezoceramic transducer is converted using software and hardware (hereinafter - SH) DM into data packets for the SH of the first mobile device (hereinafter – MD1).

Further on the CAN line control commands from the MD1 are sent back to the SH on the DM to control the infrared laser module and the piezoceramic transducer. In addition, SH on the MD1 form interactive graphic and text files specially for the doctor. The doctor contains information about the patient's blood. This information is transmitted by the Wi-Fi or GPRS between MD1 MD2 as well as to the medical center's by hospital's data center (hereinafter – HDC) where the information is archived and stored. In case of disputes the data from HDC can be sent to mobile devices of both the doctor and the patient and the administration of a medical institution. The interface between the patient and the diagnostic module is the original mobile application through which the patient controls the launch of the diagnostic module and through it receives recommendations from the doctor. The history of studies and the doctor's comments on them are also available through this mobile application in personal account. Information and control signals from the patient's MD1 are transmitted to the input of the transceiver module, which supports Wi-Fi and Bluetooth data transmission interfaces and protocols. Further signals via CAN (Controller Area Network) are transmitted to the control unit of the piezoelectric transducer and laser module which begins to generate a low-power optical signal directed to the patient's skin. The piezoceramic transducer receives an acoustic signal which is then passed through a filter and an amplifier. The spectrum of the received acoustic signal which carries information about the parameters of the patient's blood is shown in Fig. 2.

Next the received spectrum is sent to a device for recording and digital processing of analog signals which can generate a file with initially processed data that contains information about the level of the measured acoustic signal and these processed data are sent to the mobile computing device of each patient MD1. In this case the amount of computational load attributable to each patient's mobile device can be estimated according to the method described in [20]. These studies are planned to be carried out in the future to verify experimental studies of the developed technology of non-invasive blood test by the PA method in order to evaluate the reliability indicators and the computational load depending on the amount of data sent in the developed system.

The proposed approach non-invasive blood test technology based on the PA method is taken into account the experience of developing and implementing existing applications for mobile phones which allow, for example, monitoring blood sugar, keeping an electronic diary and receiving doctor's recommendations in real time [21–23]. There is no verification of the proposed combined approach, which should have

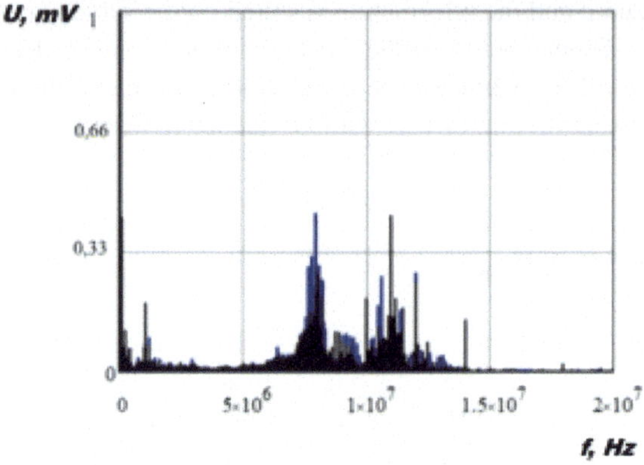

Fig. 2. Temporary implementation of the received photoacoustic signal.

given more weight to the paper, so it the system is still in developing and it will be in future to improve the quality and significance of research.

The developed approach provides the solution to the following main tasks:

- maintaining and updating a database containing information on the date and parameters of the patient's blood tests;
- display of graphical information about the parameters of the blood test of the patient in a form agreed with the doctor and approved by the administration of the medical institution;
- display of numerical and textual (message) diagnostic information in a form agreed with the doctor and approved by the administration of the medical institution,
- displays warnings and error messages about the operation of the system and the diagnostic module;
- displays prompts to the user;
- provides protection against unauthorized access and delimitation of the rights of users (patient, doctor and administration of a medical institution).

One of the advantages of the non-invasive blood test technology is the reduction of latency since the processing of medical data is sensitive to time delays. The reduction in processing time is directly proportional to the increase in the survival rate among patients [2].

Another important feature of the proposed approach is the ability to use the already existing computing resources of the patient and the medical institution (doctor)because of processing the data by using fog computing. At the same time the costs of the medical institution are minimized, and the patient's costs are to purchase a diagnostic system and a mobile application.

3 Conclusion

Implementation proposed by the authors of the non-invasive blood test technology based on the PA method using fog computing can save the patient's time and resources and reduce the waiting time for a doctor's consultation. On the part of the medical institution - to save costs on the maintenance of stationary workstations, through the use of the existing infrastructure of the medical institution and existing computing resources (mobile devices of the doctor and patients), to simplify the storage of information and increase the level of security while maintaining confidentiality and medical care.

Acknowledgement. The publication was prepared as part of GZ SSC RAS N GR project №AAAA-A19-119011190173-6 and RFBR project 18-05-80092.

References

1. Gia, T.N., Jiang, M., Rahmani, A.-M., Westerlund, T., Liljeberg, P., Tenhunen, H.: Fog computing in healthcare internet of things: a case study on ECG feature extraction. In: 2015 IEEE International Conference on Computer and Information Technology; Ubiquitous Computing and Communications; Dependable, Autonomic and Secure Computing; Pervasive Intelligence and Computing, pp. 356–363 (2015)
2. Kraemer, F.A., Braten, A.E., Tamkittikhun, N., Palma, D.: Fog computing in healthcare - a review and discussion. IEEE Access **5**, 9206–9222 (2017)
3. Aazam, M., Huh, E.N.: E-HAMC: leveraging fog computing for emergency alert service. In: 2015 IEEE International Conference on Pervasive Computing and Communication Workshops (PerCom Workshops), pp. 518–523 (2015)
4. Kaliaev, I.A., Mel'nik, E.V.: Metod mul'tiagentnogo raspredelenija resursov v intellektual'nyh mnogoprocessornyh vychislitel'nyh sistemah. Nauka Juga Rossii. Federal'noe gosudarstvennoe bjudzhetnoe uchrezhdenie nauki Juzhnyj nauchnyj centr Rossijskoj akademii nauk **3**(4), 37–46 (2007)
5. Agadzhanov, M.: mHealth — «mobil'noe» zdravoohranenie v sovremennom mire. [Electronic resource]. Electronic resource fora IT- specialists. [site] (2014). https://habr.com/company/medgadgets/blog/227159/. Accessed 20 Dec 2019
6. Mobil'noe zdravoohranenie. Novye gorizonty zdravoohranenija cherez tehnologii mobil'noj svjazi. Doklad o rezul'tatah vtorogo global'nogo obsledovanija v oblasti jelektronnogo zdravoohranenija Serija «Global'naja observatorija po jelektronnomu zdravoohraneniju», vol 3. Vsemirnaja organizacija zdravoohranenija (2013). 112 p.
7. Kolosov, A.S., Proshin, A.V.: Primenenie medicinskih mobil'nyh prilozhenij v praktike ambulatorno-poliklinicheskogo zvena. Mezhdunarodnyj nauchno-issledovatel'skij zhurnal **1** (67), 55–57 (2018)
8. Nikitin, P.V., Muradjanc, A.A., Shostak, N.A.: Mobil'noe zdravoohranenie: vozmozhnosti, problemy, perspektivy. Klinicist (4), 13–20 (2015)
9. Chestnov, O.P., Bojcov, S.A., Kulikov, A.A., Baturin, D.I.: Mobil'nye tehnologii na sluzhbe ohrany zdorov'ja. Zhurnal Medicinskie tehnologii (2), 6–10 (2015)
10. Pew Internet & American Life Project. Internet, Broadband, and Cell Phone Statistics. [Electronic resource] Pew Reseach Center: [сайт] (2010). http://www.pewinternet.org/Reports/2010/Internetbroadband-and-cell-phone-statistics.aspx?r=1. Accessed 20 Dec 2019

11. Proekt: Edinaja gosudarstvennaja informacionnaja sistema zdravoohranenija (EGISZ). TAdviser.ru – ploshhadka dlja top-zakazchikov i postavshhikov IT v Rossii. [Jelektronnyj resurs]. http://www.tadviser.ru/index.php/Proekt:Edinaja_gosudarstvennaja_informacionnaja_sistema_zdravoohranenija_(EGISZ). Accessed 20 Dec 2019
12. Nedosekin, D.A., Nolan, J., Cai, C., Bourdo, S.E., Nima, Z., Biris, A.S., Zharov, V.P.: In vivo noninvasive analysis of graphene nanomaterial pharmacokinetics using photoacoustic flow cytometry. J. Appl. Toxicol. **37**(11), 1297–1304 (2017)
13. Wang, Q., Zhou, Q., Yang, P., Wang, X., Niu, Z., Suo, Y., He, H., Gao, W., Tang, S., Wei, X.: Optimized signal detection and analysis methods for in vivo photoacoustic flow cytometry. In: Biophotonics and Immune Responses XII, vol. 10065 (2017)
14. Gnyawali, V., Strohm, E.M., Tsai, S.S.H., Kolios, M.C.: Simultaneous ultrasound and photoacoustics based flow cytometry. In: Photons Plus Ultrasound: Imaging and Sensing, vol. 10494 (2018)
15. Zhou, Q., Yang, P., Wang, Q., Pang, K., Zhou, H., He, H., Wei, X.: Labelfree counting of circulating cells by in vivo photoacoustic flow cytometry. In: Biophotonics and Immune Responses XIII, vol. 10495 (2018)
16. Kravchuk, D.A., Orda-Zhigulina, D.V., Sliva, G.Ju.: Jeksperimental'nye issledovanija optoakusticheskogo jeffekta v dvizhushhejsja zhidkosti. Izvestija JuFU. Tehnicheskie nauki **4**(189), 246–254 (2017)
17. Kravchuk, D.A., Starchenko, I.B.: Matematicheskoe modelirovanie optikoakusticheskogo signala ot jeritrocitov. Vestnik novyh medicinskih tehnologij **1**, 96–101 (2018)
18. Kravchuk, D.A.: Osobennosti formirovanija optoakusticheskih voln v biologicheskih tkanjah. Jelektronnyj zhurnal «Inzhenernyj vestnik Dona» **1**(48), 23 (2007)
19. Starchenko, I.B., Kravchuk, D.A., Kirichenko, I.A.: An optoacoustic laser cytometer prototype. Biomed. Eng. **51**(5), 308–312 (2018)
20. Mel'nik, E.V., Klimenko, A.B., Ivanov, D.Ja.: Model' zadachi raspredelenija vychislitel'noj nagruzki dlja informacionno-upravljajushhih sistem na baze koncepcii tumannyh vychislenij. Izvestija Tul'skogo gosudarstvennogo universiteta. Tehnicheskie nauki. Federal'noe gosudarstvennoe bjudzhetnoe obrazovatel'noe uchrezhdenie vysshego professional'nogo obrazovanija **2**, 174–187 (2018)
21. Bol'nica na domu s udalennym kontrolem vrachej. [Electronic resource]. mHealth. Mobil'naja medicina v Rossii i mire: [site] (2018). https://mhealthrussian.wordpress.com/2018/04/03/bol'nica-na-domu-s-udalennym-kontrolem/. Accessed 20 Dec 2019
22. Abdullaev, V.G., Chuba, I.V., T.K.A.: Mobil'nye prilozhenija dlja zdorov'ja. Medicinskaja informatika i inzhenerija, vol. 3, pp. 89–92 (2014)
23. Top-10 Mobil'nyh prilozhenij dlja sistemy zdravoohranenija. [Jelektronnyj resurs]. TELEMEDICINA.RU. Pervoe profil'noe SMI: [site] (2018). https://telemedicina.ru/news/world/top-10-mobilnyh-prilozheniy-dlya-sistemy-zdravoohraneniya. Accessed 20 Dec 2019

Application of Correcting Polynomial Modular Codes in Infotelecommunication Systems

Igor Kalmykov(✉), Nikita Chistousov, Andrey Aleksandrov, and Igor Provornov

North Caucasus Federal University, Stavropol, Russia
kia762@yandex.ru

Abstract. The purpose of the article is to develop algorithms for searching and correcting errors using polynomial modular code (PMC), which allows correcting an error cluster inside the code remainder. These codes have found its application in infotelecommunication systems, particularly, in OFDM technologies and digital signal processing (DSP). This is due to the fact that they perform parallelization of the computational process at the level of arithmetic operations. This allows increasing the speed of information processing. In this case, independent processing of input data, represented as remainders in PMC modules, is the basis for searching and correcting errors that arise during the computation process. Therefore, the development of algorithms for detecting and correcting errors that allow expanding the corrective capabilities of PMC when fixing a cluster of errors, and will increase the fault tolerance of OFDM's special processors (SP), is an urgent task. #CSOC1120.

Keywords: Digital signal processing · OFDM special processor · Polynomial modular code · Error correction · Position characteristic

1 Introduction

Modern infotelecommunication systems widely use special processors (SP) implementing orthogonal frequency-division multiplexing (OFDM). This is due to the fact that OFDM systems are highly stable under multipath radio signal propagation. Increase in the productivity of such OFDM SP is possible through the use of parallel computational algorithms. Though this entails an increase in circuit efforts, which negatively affect the reliability of the SP operation. This disadvantage can be eliminated by giving the OFDM SP a fault tolerance property by the use of correcting polynomial modular codes [1, 2]. Therefore, the expansion of the corrective properties of modular codes that allow correcting error clusters within a single PMC base is urgent.

R. Silhavy (Ed.): CSOC 2020, AISC 1226, pp. 387–398, 2020.
https://doi.org/10.1007/978-3-030-51974-2_38

2 Main Part. Methods

Currently, modular codes are widely used in many areas. So, the works [3–6] shows the appropriateness of using residual-class system codes when performing FFT and digital filters. The use of low-bit remainders and parallel processing of data allowed increasing the speed of FFT execution. The work [7] proposes to use modular codes for performing wavelet transforms. In works [8–13], the methods and algorithms for increasing the fault tolerance of the SP residual classes are presented.

It should be noted that the use of polynomial modular code allows increasing the speed of execution of orthogonal signal transformations. The base of such an implementation is the isomorphism generated by the Chinese remainder theorem (CRT). This allows us to move from one-dimensional to multidimensional signal processing moduli irreducible polynomials $p_i(z)$, where i = 1, …, k, according to

$$X_i(s) = \sum_{j=0}^{d-1} x_i(j) b_i^{jl} \bmod p_i(z) \tag{1}$$

where $x_i(j) \equiv x(j) \bmod p_i(z)$; $b_i^{\pm jl} \equiv b^{\pm jl} \bmod p_i(z)$; $X_i(s) \equiv X(s) \bmod p_i(z)$; b – a primitive root; $x(j)$ - an input signal sequence; $X(s)$ - spectral components of the input signal; $d = 2^v - 1$ - the dimension of the input vector.

The analysis of expression (1) shows that in order to perform such an orthogonal signal conversion, PMC can be used, in which the generating polynomials $p_i(z)$, where i = 1, …, k, are used as bases. Then the PMC code combination has the form

$$A(z) = (a_1(z),\ a_2(z), \ldots, a_k(z)) \tag{2}$$

where $A(z) \equiv a_i\ (z) \bmod p_i(z)$; i = 1,…, k.

The use of PMC allows us to increase the speed of addition, subtraction and multiplication operations by modulo, due to performing these operations on residues. However, the independent and parallel execution of these operations, as well as the lack of data exchange between the computational paths, can be used to construct corrective codes. It was proved in works [11, 12] that for two control bases $p_{k+1}(z)$ and $p_{k+2}(z)$, which satisfy

$$\deg p_{k+1}(z) + \deg p_{k+2}(z) \geq \deg p_{k-1}(z) + \deg p_k(z) \tag{3}$$

where deg $p_i(z)$ is the degree of the polynomial $p_i(z)$, the polynomial modular code allows correcting a single error, which is understood as the distortion of one digit of the code combination. The introduction of two reference bases extends the range to the full-range value

$$P(z) = \prod_{i=1}^{k+2} p_i(z) = P_1(z) \prod_{i=k+1}^{k+2} p_i(z) = P_1(z) P_2(z) \tag{4}$$

where $P_1(z) = \prod_{i=1}^{k} p_i(z)$ is the range of allowed combinations of PMC.

In this case, the allowed combination A(z) must satisfy

$$\deg A(z) < \deg P_1(z) \tag{5}$$

If there is a single error $\Delta a_i(z) = z^n$ in the PMC code, where $n = 0, \ldots, \deg p_i(z) - 1$ the allowed combination of A(z) turns into

$$A^*(z) = (a_1(z), \ldots a_i^*(z), \ldots, a_{k+2}(z)) = \left(a_1(z), \ldots, a_i(z) + \Delta a_j(z), \ldots, a_{k+2}(z)\right)$$

This leads to the overrunning of the polynomial $A^*(z)$ beyond $P_1(z)$, i.e.

$$\deg A^*(z) > \deg P_1(z) \tag{6}$$

Since PMCs are non-position codes, then the positional characteristics (PC) are used to search for and correct errors, which allow evaluating the execution of equality (5). In the work [2], an algorithm for determining the PC, which is called *the interval number*, is presented and is defined as

$$L(z) = [A(z)/P_1(z)] \tag{7}$$

If (5) is valid for the PMC code, then L(z) = 0. This means that it does not contain an error. Distortion of one bit of the code combination of PMC violates condition (5), and then $L(z) \neq 0$. In this case, every single error moves the allowed combination of A(z) into a certain interval, the value of which can uniquely correct the code combination $A^*(z)$.

Consider the situation when the error cluster occurred on one base $p_i(z)$, that is, several bits were distorted in the remainder. Let us prove that the use of two control bases satisfying condition (3) will allow correcting such error.

Theorem If two control bases that satisfy condition (3) are used in the ordered PMC code, i.e

$$\deg p_{k+1}(z) + \deg p_{k+2}(z) \geq \deg p_k(z) + \deg p_{k-1}(z)$$

then this code is able to correct an error cluster that distorts one remainder of the modular code.

Proof

It is known that if the PMC code does not contain an error, then the condition (5) is valid. Let the error in the PMC code occur on the i-th basis. Then the code has the form

$$A^*(z) = (\alpha_1(z),\ \ldots,\ \alpha_i^*(z), \ldots, \alpha_{k+2}(z)) \tag{8}$$

where $\alpha_i^*(z) = \alpha_i(z) + \Delta\alpha_i(z)$; $\Delta\alpha_i(z)$ - the depth of error

If the j-th base occurs, the PMC code has the form

$$A^{**}(z) = (\alpha_1(z),\ \ldots,\ \alpha_j^{**}(z), \ldots, \alpha_{k+2}(z)) \tag{9}$$

where $\alpha_j^{**}(z) = \alpha_j(z) + \Delta\alpha_j(z)$; $\Delta\alpha_j(z)$ - the depth of error; $j \neq i$.

Let us define the intervals including the erroneous code combinations of PMC

$$L^*(z) = \left[\frac{A^*(z)}{P_1(z)}\right] = \left[\frac{(A(z) + \Delta\alpha_i^*(z)B_i(z)) \bmod P(z)}{P_1(z)}\right] \tag{10}$$

$$L^{**}(z) = \left[\frac{A^{**}(z)}{P_1(z)}\right] = \left[\frac{\left(A(z) + \Delta\alpha_j^{**}(z)B_j(z)\right) \bmod P(z)}{P_1(z)}\right] \tag{11}$$

where $P(z) = \prod_{i=1}^{k+2} p_i(z)$ - a full range of PMC code

If the code combinations do not fall within the same interval, then

$$L^*(z) + L^{**}(z) \geq 1 \tag{12}$$

Let the combination A(z) = 0. Then the degenerations (10) and (11) can be represented in the form

$$L^*(z) = \left[\frac{(\Delta\alpha_i^*(z)B_i(z)) \bmod P(z)}{P_1(z)}\right] \tag{13}$$

$$L^{**}(z) = \left[\frac{\left(\Delta\alpha_j^{**}(z)B_j(z)\right) \bmod P(z)}{P_1(z)}\right] \tag{14}$$

It is known that the orthogonal bases of the PMC code are defined as

$$B_i(z) = m_i(z)\frac{P(z)}{p_i(z)} = m_i(z)\frac{P_1(z)p_{k+1}(z)p_{k+2}(z)}{p_i(z)} \tag{15}$$

$$B_j(z) = m_j(z)\frac{P(z)}{p_j(z)} = m_j(z)\frac{P_1(z)p_{k+1}(z)p_{k+2}(z)}{p_j(z)} \tag{16}$$

Substitute the equalities (15) and (16) into the expressions (8) and (9), respectively. Then we obtain

$$L^*(z) = \left[\frac{(\Delta\alpha_i^*(z)m_i(z)p_{k+1}(z)p_{k+2}(z)) \bmod P(z)}{p_i(z)}\right] \tag{17}$$

$$L^{**}(z) = \left[\frac{\left(\Delta\alpha_j^{**}(z)m_j(z)p_{k+1}(z)p_{k+2}(z)\right) \bmod P(z)}{p_j(z)}\right] \tag{18}$$

It is known that the interval number runs through all values modulo $P_2(z) = \prod_{i=k+1}^{k+2} p_i(z)$. Then the expressions (17) and (18) can be represented in the following form

$$L^*(z) = \left(\frac{\Delta\alpha_i^*(z)m_i(z)}{p_i(z)}\right) \bmod P_2(z) \tag{19}$$

$$L^{**}(z) = \left(\frac{\Delta\alpha_j^{**}(z)m_j(z)}{p_j(z)}\right) \bmod P_2(z) \tag{20}$$

Let us use the isomorphism generated by the Chinese remainder theorem in polynomials, and proceed to the multidimensional representation of intervals in the form of a modular code with two bases.

$$L^*(z) = \left(L_{k+1}^*(z),\, L_{k+2}^*(z)\right) = \left(\left|\frac{\Delta\alpha_i^*(z)m_i(z)}{p_i(z)}\right|_{p_{k+1}(z)}^+,\, \left|\frac{\Delta\alpha_i^*(z)m_i(z)}{p_i(z)}\right|_{p_{k+2}(z)}^+\right) \tag{21}$$

$$L^{**}(z) = \left(L_{k+1}^{**}(z),\, L_{k+2}^{**}(z)\right) = \left(\left|\frac{\Delta\alpha_j^{**}(z)m_j(z)}{p_j(z)}\right|_{p_{k+1}(z)}^+,\, \left|\frac{\Delta\alpha_j^{**}(z)m_j(z)}{p_j(z)}\right|_{p_{k+2}(z)}^+\right) \tag{22}$$

Let us suppose that when errors occur on the i-th and j-th bases of the PMC code, where $j \neq i$, the coincidence of intervals has occurred. Then the expression (12) takes the form

$$L^*(z) + L^{**}(z) = 0. \tag{23}$$

When using the PMC, the expression (23) can be written in the form

$$\left(L_{k+1}^*(z),\, L_{k+2}^*(z)\right) = \left(L_{k+1}^{**}(z),\, L_{k+2}^{**}(z)\right) \tag{24}$$

Denote

$$m_i(z)p_i^{-1}(z) = C_i(z),\; m_j(z)p_j^{-1}(z) = C_j(z) \tag{25}$$

Then the expression (24) can be written in the form

$$\left(\left|\Delta\alpha_i^*(z)C_i(z)\right|^+_{p_{k+1}(z)}, \left|\Delta\alpha_i^*(z)C_i(z)\right|^+_{p_{k+2}(z)}\right) = \left(\left|\Delta\alpha_j^{**}(z)C_j(z)\right|^+_{p_{k+1}(z)}, \left|\Delta\alpha_j^{**}(z)C_j(z)\right|^+_{p_{k+2}(z)}\right) \tag{26}$$

It means that

$$\left|\Delta\alpha_i^*(z)C_i(z)\right|^+_{p_{k+1}(z)} = \left|\Delta\alpha_j^{**}(z)C_j(z)\right|^+_{p_{k+1}(z)} \tag{27}$$

$$\left|\Delta\alpha_i^*(z)C_i(z)\right|^+_{p_{k+2}(z)} = \left|\Delta\alpha_j^{**}(z)C_j(z)\right|^+_{p_{k+2}(z)} \tag{28}$$

It is known that the degree of error depth does not exceed the degree of control bases

$$\begin{aligned} \deg \Delta\alpha_i^*(z) &< \deg p_{k+1}(z) \\ \deg \Delta\alpha_i^*(z) &< \deg p_{k+2}(z) \end{aligned} \tag{29}$$

$$\begin{aligned} \deg \Delta\alpha_j^{**}(z) &< \deg p_{k+1}(z) \\ \deg \Delta\alpha_j^{**}(z) &< \deg p_{k+2}(z) \end{aligned} \tag{30}$$

Hence, for the simultaneous fulfillment of equalities (27) and (28), it is necessary that condition fulfilled:

$$\left|C_i(z)\right|^+_{p_{k+1}(z)} = \left|C_i(z)\right|^+_{p_{k+2}(z)} \tag{31}$$

$$\left|C_j(z)\right|^+_{p_{k+1}(z)} = \left|C_j(z)\right|^+_{p_{k+2}(z)} \tag{32}$$

Though the irreducible polynomials $p_{k+1}(z)$ and $p_{k+2}(z)$ were chosen as the control bases. Hence, the equalities (31) and (32) cannot be satisfied. Consequently, the proposal made on the possibility of coincidence of intervals for the occurrence of single errors on different bases of PMC using two control bases satisfying the condition (3) is incorrect.

The proof is complete.

It is known that PMCs do not support the operation of division. Hence, it is necessary to represent (7) in the form of modular operations. To solve this problem, an algorithm based on the CRT was developed, according to which

$$A(z) = \sum_{i=1}^{k+2} \alpha_i(z)B_i(z) \bmod P(z) \tag{33}$$

where $B_i(z) = m_i(z)P(z)/p_i(z)$ – the orthogonal basis of the i-th base; $m_i(z)$ is the weight of the i-th orthogonal basis providing the condition $B_i(z) \equiv 1 \bmod p_i(z)$.

Then

$$L(z) = \left[\frac{A(z)}{P_1(z)}\right] = \left[\sum_{i=1}^{k+2} \alpha_i(z) B_i(z) \bmod P(z) \Big/ P_1(z)\right] \tag{34}$$

Let us use the similarity of the orthogonal bases $B_i(z)$ for excess PMC with bases $p_1(z), \ldots, p_{k+2}(z)$ and $B_i'(z)$ for PMC given by information bases $p_1(z), \ldots, p_k(z)$, that is

$$B_i'(z) \equiv B_i(z) \bmod P_1(z) \tag{35}$$

Then the following equality holds true

$$B_i(z) = R_i(z) P_1(z) + B_i'(z) \tag{36}$$

where $R_i(z) = [B_i(z)/P_1(z)]$; $i = 1, \ldots, k + 2$.

After substituting the equality (36) into the (34), and after simplifying, we obtain

$$L(z) = \sum_{i=1}^{k+2} \alpha_i(z) R_i(z) + \left[\sum_{j=1}^{k} \alpha_j(z) B_j'(z) \Big/ P_1(z)\right] + K(z) P(z))/P_1(z) \tag{37}$$

where K(z) – the rank of the complete base system of PMC.

Since the set of values of the interval number L(z) is a ring modulo $P_2(z)$, then the expression (37) can be represented

$$L(z) = \left|\sum_{i=1}^{k+2} \alpha_i(z) R_i(z) + K'(z)\right|^{+}_{P_2(z)} \tag{38}$$

where $K'(z) = \left[\sum_{j=1}^{k} \alpha_j(z) B_j'(z) \Big/ P_1(z)\right]$ - the rank of the non-redundant PMC system.

However, there are applications used in infotelecommunication systems that do not allow the use of PMC with two control bases $p_{k+1}(z)$ and $p_{k+1}(z)$. Therefore, an error correction algorithm was developed using a single reference base $p_{k+1}(z)$ satisfying the condition

$$\deg p_{k+1}(z) \geq \deg p_k(z) \tag{39}$$

Therefore, it is necessary to calculate two remainders

$$\alpha_{k+1}(z) = \sum_{i+1}^{k} \alpha_i(z). \tag{40}$$

$$\alpha_{k+2}(z) = \sum_{i+1}^{k} (i(z)\alpha_i(z)) \bmod p_{k+1}(z). \tag{41}$$

where $i(z)$ is the polynomial form of the i-th number, $\sum$ is addition modulo two.

To correct the error using information remainders $(\alpha_1(z), \ldots, \alpha_k(z))$, the calculations of excess remainders are made

$$\alpha'_{k+1}(z) = \sum_{i+1}^{k} \alpha_i(z) \tag{42}$$

$$\alpha'_{k+2}(z) = \sum_{i+1}^{k} (i(z)\alpha_i(z)) \bmod p_{k+1}(z) \tag{43}$$

The values $\alpha'_{k+1}(z)$, $\alpha'_{k+2}(x)$ allow calculating the error syndrome

$$\delta_1(z) = \alpha_{k+1}(z) + \alpha'_{k+1}(z) \tag{44}$$

$$\delta_2(z) = \alpha_{k+2}(z) + \alpha'_{k+2}(z) \tag{45}$$

If the error syndrome $\delta_1(z) = 0$ and $\delta_2(z) = 0$, then the combination does not contain an error. Otherwise, the values $\delta_1(z)$ and $\delta_2(z)$ can be used to correct a single error.

3 Research Results and Discussion

Let us use the developed algorithm (38) to study the correcting capabilities of the PMC code, with information bases $p_1(z) = z + 1$, $p_2(z) = z^2 + z + 1$, $p_3(z) = z^4 + z^3 + z^2 + z + 1$, and control modules $p_4(z) = z^4 + z^3 + 1$ and $p_5(z) = z^4 + z + 1$. Then the range of allowed combinations $P_1(z) = \prod_{i=1}^{3} p_i(z) = z^7 + z^6 + z^5 + z^2 + z + 1$.

Let $A(z) = z^4 + z^3 + z^2 + z = (0, z, 1, z^2 + z + z + 1, z^3 + z^2 + 1)$. For this PMC there are the following orthogonal bases

$$\begin{aligned}
B_1(z) &= R_1(z)P_1(z) + B'_1(z) = (z^7 + z^4 + z^2 + z)P_1(z) + z^6 + z^4 + z^3 + z^2 + 1, \\
B_2(z) &= R_2(z)P_1(z) + B'_2(z) = (z^7 + z^5 + z^2 + z + 1)P_1(z) + z^6 + z^4 + z + 1, \\
B_3(z) &= R_3(z)P_1(z) + B'_3(z) = (z^7 + z^4 + z^3 + z + 1)P_1(z) + z^5 + z^4 + z^3 + z^2 + z + 1, \\
B_4(z) &= R_4(z)P_1(z) = (z^7 + z^4 + z^3)P_1(z), \\
B_5(z) &= R_5(z)P_1(z) = (z^5 + z^4 + z)P_1(z).
\end{aligned}$$

Calculate the value of the rank of the non-redundant PMC

$$K'(z) = \left[\frac{(0 \cdot B_1'(z) + z \cdot B_2'(z) + 1 \cdot B_3'(z))}{z^7 + z^6 + z^5 + z^2 + z + 1}\right] = 1.$$

Then, using the algorithm (38), we obtain

$$L(z) = (z(z^7 + z^5 + z^2 + z + 1) + (z^7 + z^4 + z^3 + z + 1) + (z^7 + z^4 + z^3)(z^2 + z + 1) \\ + (z^3 + z^2 + 1)(z^5 + z^4 + z) + 1) \bmod (z^8 + z^7 + z^5 + z^4 + z^3 + z + 1) = 0.$$

Since $L(z) = 0$, then the PMC code does not contain an error.

Suppose that the error occurred at the base of $p_3(z)$ and its depth $\Delta\alpha_3(z) = 1$. Then $A^*(z) = (0,\ z,\ 0,\ z^2 + z + z + 1,\ z^3 + z^2 + 1)$. Let us calculate the rank of a PMC

$$K'(z) = \left[\frac{(0 \cdot B_1'(z) + z \cdot B_2'(z) + 0 \cdot B_3'(z))}{z^7 + z^6 + z^5 + z^2 + z + 1}\right] = 1.$$

Then, using the algorithm (38), we obtain

$$L(z) = (z(z^7 + z^5 + z^2 + z + 1) + (z^7 + z^4 + z^3)(z^2 + z + 1) + (z^3 + z^2 + 1) \\ \times (z^5 + z^4 + z) + 1) \bmod (z^8 + z^7 + z^5 + z^4 + z^3 + z + 1) = z^7 + z^4 + z^3 + z + 1.$$

The value $L(z) = z^7 + z^4 + z^3 + z + 1$ was determined by the error vector $\bar{e}(z) = (0,\ 0,\ 1,\ 0,\ 0)$, which allows us to correct the error

$$A(z) = A^*(z) + \bar{e}(z) = (0,\ z,\ 1,\ z^2 + z + z + 1,\ z^3 + z^2 + 1).$$

The carried out research has confirmed the statement of the theorem. So, the code shown in the example can solve 49 errors, compared to 15 errors, which was shown in the work [10]. Hence, the use of two control bases satisfying (3) allows us to correct the error clusters arising in one remainder.

Consider the application of the developed error correction algorithm, which uses a single reference base, in the SPN cipher Kuznyechik. This cipher uses a polynomial $p(z) = z^8 + z^7 + z^6 + z + 1$. One of the round operations is a non-linear transformation that implements the substitution operation. At the input of this block S, sized 256×8 bit, the byte X arrives, and the byte Y is removed from the output. Using PMC allows representing the input X and output Y bytes in the form of two remainders moduli $p_1(z) = z^4 + z + 1$ and $p_2(z) = z^4 + z^3 + 1$. As a result, two tables sized 256×4 bit are used. The substitution rules are shown in Table 1.

Table 1. Rules for the substitution of the S-table in modular code

X_{10}	0	1	2	…	253	254	255
$x_1(z) = \lvert X(z)\rvert^+_{p_1(z)}$	0	1	z	…	$z^3 + z^2 + z + 1$	$z^3 + z^2$	$z^3 + z^2 + 1$
$x_2(z) = \lvert X(z)\rvert^+_{p_2(z)}$	0	1	z	…	$z^2 + z + 1$	z^2	$z^2 + 1$
Y_{10}	252	238	221	…	75	99	182
$y_1(z) = \lvert Y(z)\rvert^+_{p_1(z)}$	$z^3 + z^2 + z$	$z^3 + z^2 + z + 1$	$z^3 + 1$	…	$z^2 + z + 1$	$z^3 + 1$	z^3
$y_2(z) = \lvert Y(z)\rvert^+_{p_2(z)}$	$z^2 + z$	$z^3 + z^2 + 1$	$z^3 + z^2$	…	z^2	$z^2 + z + 1$	$z + 1$

Let the byte X enter the input. Represent it in binary and polynomial form $X = 254_{10} = 11111110_2 = z^7 + z^6 + z^5 + z^4 + z^3 + z^2 + z$. Then in PMC it has the form $x_1(z) = |X(z)|^+_{p_1(z)} = z^3 + z^2$, $x_2(z) = |X(z)|^+_{p_2(z)} = z^2$. Then $X = 254_{10} = (z^3 + z^2,\ z^2 + z)$. In this case, the byte is removed from the S-block output $Y = 99_{10} = 01100011_2 = z^6 + z^5 + z + 1$. In the modular code we receive $y_1(z) = |Y(z)|^+_{p_1(z)} = z^3 + 1$ and $y_2(z) = |Y(z)|^+_{p_2(z)} = z^2 + z + 1$ Using these remainders, we calculate the control ones according to (40) and (41). Then

$$y_3(z) = \sum_{i=1}^{2} y_i(z) = (z^3 + 1) + (z^2 + z + 1) = z^3 + z^2 + z,$$

$$y_4(z) = \sum_{i=1}^{2} i(z)y_i(z) = ((z^3 + 1) + z(z^2 + z + 1)) = z^2 + z + 1.$$

We received excess PMC $Y(z) = (z^3 + 1,\ z^2 + z + 1,\ z^3 + z^2 + z,\ z^2 + z + 1)$.

Let the error occur on the first base and its depth is equal $\Delta y_1(z) = z^3$. Then we have $Y^*(z) = (1,\ z^2 + z + 1,\ z^3 + z^2 + z,\ z^2 + z + 1)$. Calculate the values of the control remainders using (42) and (43)

$$y_3'(z) = \sum_{i=1}^{2} y_i(z) = 1 + (z^2 + z + 1) = z^2 + z,$$

$$y_4'(z) = \sum_{i=1}^{2} i(z)y_i(z) = (1 + z(z^2 + z + 1)) = z^3 + z^2 + z + 1.$$

Let us calculate the error syndrome on the basis of (44) and (45)

$$\delta_1(z) = y_3(z) + y_3'(z) = (z^3 + z^2 + z) + (z^2 + z) = z^3,$$
$$\delta_2(z) = y_4(z) + y_4'(z) = (z^2 + z + 1) + (z^3 + z^2 + z + 1) = z^3.$$

Since the syndrome does not equal zero, the PMC contains an error. The application of the theorem allowed us to determine all the error vectors for single errors in PMC. For the error syndrome $\delta_1(z) = z^3$ and $\delta_2(z) = z^3$ the error vector is $\bar{e}(z) = (z^3, 0, 0, 0)$. Let us correct the error combination

$$Y(z) = Y^*(z) + \bar{e}(z) = (z^3 + 1, z^2 + z + 1, z^3 + z^2 + z, z^2 + z + 1).$$

4 Conclusion

The theorem presented in this paper, as well as the developed algorithm for searching and correcting errors, allows us to correct not only single errors, but also errors clusters occurring within one remainder. Expansion of the correcting ability of PMC codes allowed increasing by 3.27 times the number of corrected errors when using two control bases $p_4(z) = z^4 + z^3 + 1$ and $p_5(z) = z^4 + z + 1$ in the PMC code with information grounds $p_1(z) = z + 1$, $p_2(z) = z^2 + z + 1$, $p_3(z) = z^4 + z^3 + z^2 + z + 1$ and control modules. In order to reduce the introduced redundancy, an algorithm was developed that makes it possible to correct the error clusters when using a single control base. Theoretical calculations were confirmed by numerical examples.

Acknowledgments. This work was supported by the Russian Foundation for Basic Research, project No. 18–07-01020

Referenses

1. Beckmann, P.E., Musicus, B.R.: Fast fault-tolerant digital convolution using a polynomial residue number system. IEEE Trans. Signal Process. **41**(7), 2300–2313 (1993)
2. Makarova, A.V., Stepanova, E.P., Toporkova, E.V.: The use of redundant modular codes for improving the fault tolerance of special processors for digital signal processing. In: CEUR Workshop Proceedings, vol. 1837, pp. 115–122 (2017)
3. Chervyakov, N.I., Veligosha, A.V., Ivanov, P.E.: Digital filters in a system of residual classes // Izvestiya Vysshikh Uchebnykh Zavedenij. Radioelektronika **38**(8), 11–20 (1995)
4. Veligosha, A.V., Kaplun, D.I., Bogaevskiy, D.V., Voznesenskiy, A.S. Adjustment of adaptive digital filter coefficients in modular codes. In: Proceedings of the 2018 IEEE Conference of Russian Young Researchers in Electrical and Electronic Engineering, ElConRus 2018, pp. 1167–1170 (2018)
5. Omondi, A., Premkumar, B.: Residue Number Systems: Theory and Implementation. Imperial College Press, UK (2007)
6. Mohan, P.V.: Residue Number Systems. Algorithms and Architectures. Springer, Heidelberg (2002)
7. Kalmykov, I.A.E., Katkov, K.A., Timoshenko, L.I., Dunin, A.V.E., Gish, T.A.: Application of modular technologies in the large-scale analysis of signals. J. Theor. Appl. Inf. Technol. **80**(3), 391–400 (2015)

8. Veligosha, A.V., Kaplun, D.I., Klionskiy, D.M., Gulvanskiy, V.V.: Parallel-pipeline implementation of digital signal processing techniques based on modular codes. In: Proceedings of the 19th International Conference on Soft Computing and Measurements, SCM 2016, 7519731, pp. 213–214 (2016)
9. Kaplun, D.I., Klionskiy, D.M., Bogaevskiy, D.V., Veligosha, A.V.: Error correction of digital signal processing devices using non-positional modular codesError correction of digital signal processing devices using non-positional modular codes. Autom. Control Comput. Sci. **51**(3), 167–173 (2017)
10. Stepanova, E.P., Toporkova, E.V., Kalmykov, M.I., Katkov, R.A., Rezenkov, D.N.: Application of the codes of a polynomial residue number system, aimed at reducing the effects of failures in the AES cipher. J. Digital Inf. Manage. **14**(2), 114–123 (2016)
11. Chu, J., Benaissa, M.: Polynomial residue number system GF(2^{m}) multiplier using trinomials. In: 17th European Signal Processing Conference, Glasgow, Scotland, 24–28 August 2009
12. Chu, J., Benaissa, M.: Error detecting AES using polynomial residue number system. Microprocess. Microsyst. **37**(2), 228–234 (2013)
13. Yatskiv, V., Yatskiv, N., Jun, S., Sachenko, A., Zhengbing, H.: The use of a modified correction code based on a reside number system in WSN. In Proc. 7-th IEEE International Conference Intelligent Data Acquisition and Advanced Computing Systems, (IDAACS 2013), Berlin, Germany , vol. 1, pp. 513–516 (2013)

Opportunities for Application of Blockchain in the Scenario of Intelligence and Investigation Units

Gleidson Sobreira Leite(✉) and Adriano Bessa Albuquerque

Universidade de Fortaleza, Fortaleza, CE, Brazil
gleidson.sleite@gmail.com, adrianoba@unifor.br

Abstract. In the context of combating crime, government institutions in several countries have instituted units or sectors specialized in investigation and intelligence activities to act in different areas and expertise's. Due to the considerable concern with security and confidentiality, as well as the characteristics inherent to the context of these units, an opportunity arises to adopt alternative solutions aimed at the management, storage and sharing of digital assets. Due to the characteristics of decentralization, enhanced security, tamper-proof, traceability and transparency, in recent years, blockchain technology has become very trendy and penetrated different domains. However, blockchain adoption and use in the context of in investigation and intelligence units is rather unexplored in academic literature. Exploring the main characteristics of blockchain technology, this paper presents an overview of different application trends of blockchain technology, and propose the use of blockchain as a support mechanism in the management, storage and sharing of generated digital assets in the context of these specialized units.

Keywords: Blockchain · Information security · Intelligence units · Investigation units

1 Introduction

There are several variations in crime rates presented in statistical studies at different times and places in the world where, being an emerging and very important topic mainly due to the negative economic and social impacts, crime prevention is a worldwide concern.

Many of the benefits of globalization and the rise of technology in society, such as, for example, agility and ease of communication, financial movement and mobility, also created opportunities for the flourishing, diversification, expansion and organization of criminal groups [1].

A survey conducted in March 2018 by [2], attended by 2,373 senior managers from large global organizations across 19 countries, found about $ 1.45 trillion of total lost turnover estimated as a result of financial crimes, and statistical studies by [3] pointed out that in 2018 there were an estimated 1,206,836 violent crimes in the United States alone.

R. Silhavy (Ed.): CSOC 2020, AISC 1226, pp. 399–414, 2020.
https://doi.org/10.1007/978-3-030-51974-2_39

These and several other studies point to concerns about the evolution and expansion of crime, where, due to the diversity and volume of existing crime practices, preventive, repressive (impediment of continuity), control or punitive action by government institutions is essential in order to minimize the damage caused to society.

To this end, several government institutions in various countries, which work in the fight against crime, have established units or sectors specialized in investigation and intelligence activities to act in different areas and expertise, such as, for example, FBI or EUROPOL, which have units specialized in combating crimes such as corruption, criminal organizations, violent crimes, white-collar crime, financial crimes, among others [4, 5].

However, in the case of intelligence and investigation units, there is the question that they work with classified information and restricted activities, very specific and with considerable complexity [6]. These issues generate a high concern with security and confidentiality [7] during all internal operational processes including the generation, storage and sharing of information or assets between members of the same unit or even between different units or institutions.

For these situations, in the event of exposures or leaks of information or even assets like standards, methodologies, techniques, artifacts or strategies adopted, it can harm the performance not only of a specialized sector, but also of the institution as a whole and others involved [8].

Although current approaches related to digital asset management (DAM) has increasing benefits in booming global Internet economy, it is still a great challenge for providing an effective, secure, verifiable and traceable way to manage, store, share and retrieve digital assets. In some cases, due to the involvement of trusted third party, many approaches lack trust, transparency, security, and immutability [9].

Motivated by this scenario, and due to the considerable concern with security and confidentiality, as well as the characteristics inherent to the context of these units, an opportunity arises to adopt alternative solutions aimed at the management, storage and sharing of digital assets.

Based on this context and exploring the main characteristics of blockchain technology, this paper presents an overview of different application trends of blockchain technology, and propose the use of blockchain as a support mechanism in the management, storage and/or sharing of generated digital assets in the context of these specialized units.

This work also intends to be a contribution to the body of knowledge related to studies regarding the use of information technology in the fight against crime, and can be adopted to help both practitioners and researchers in order to identify possible applications trends.

This paper is organized as follow: the methodology and procedures performed in the work (Sect. 2); background and related work (Sect. 3); blockchain applications trends (Sect. 4); proposed application and discussions (Sect. 5), followed by the final considerations (Sect. 6).

2 Research Methodology

To accomplish the objective of this work, four specific actions were carried out. They were:

1) Research and exploration of the main characteristics of blockchain technology presented in academic studies and application trends in different domains.
2) Bibliographic research and selection of approaches proposed in academic research and related to applications of blockchain in the context of data/digital assets management (e.g. registry and inventory), storage and/or sharing.
3) Propose an approach with the use of blockchain as a support mechanism in the management, storage and/or sharing of generated digital assets in the context of in investigation and intelligence units.
4) Finally, discussions about how the application of blockchain technology and cryptography as support mechanisms may impact as a contribution with security, verifiability and traceability in management, store, share and retrievement of digital assets. Issues related to auditability, versioning identification, authenticity, non-repudiation, integrity verification, confidentiality, security, decentralization and transparency are considered and discussed.

3 Background and Related Work

Being first introduced as the core technology behind the Bitcoin cryptocurrency proposed by Satoshi Nakamoto in a pseudonymous paper in 2008 [10], blockchain has been attracting more and more attention from both the industry and academia.

With the growing interest in information and communication technologies globally, one can foresee the future of blockchain as one of the progressing technologies of current era. In December 2018, for example, Statista conducted a statistical survey where from 2019 to 2023, a huge increase has been observed in the size of the blockchain technology market worldwide (Fig. 1).

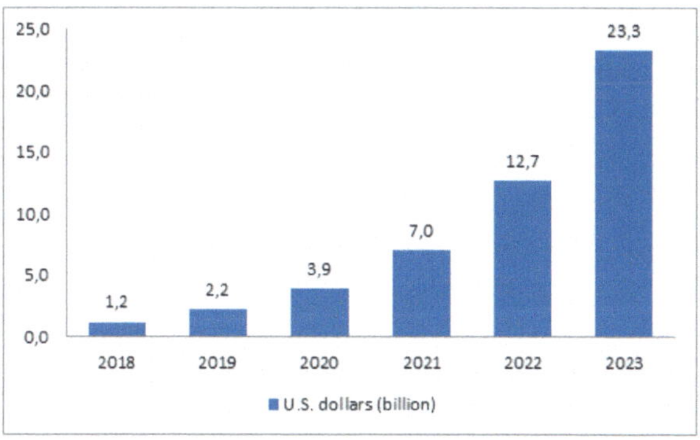

Fig. 1. Size of the blockchain technology market worldwide from 2018 to 2023 (in billion U.S. dollars). Survey period: 2018. Publication date: December 2018. Source: statista.com

According to [11], blockchain can be defined as a distributed and decentralized peer-to-peer network where transactions performed by participants are stored in strict order in an immutable, transparent append-only ledger, that is composed of blocks connected to each other via cryptographic hashes [12–15].

The basic idea behind the blockchain technology is that it allows actors in a system (called nodes) to transact data or digital assets using a P2P network that stores these transactions in a distributed way across the network. The owners of the assets, and the transactions involving change of ownership, are registered on the ledger by the use of public key cryptography and digital signatures. Every transaction is validated by the nodes in the network by employing some kind of a 'consensus mechanism' (a consensus protocol) [16, 17].

Whenever a transaction is entered into the P2P network, the nodes first validate the transaction. If the nodes agree on its legitimacy, they confirm the transaction and this decision is laid down in a block. This new block is added to the previous chain of blocks and, just like the previous ones, locked in the chain. In this way, the latest block maintains a shared, agreed-upon view of the current state of the blockchain [18].

Regarding the content of the blocks, they are hashed, forming a unique block identifier stored in the current and subsequent block. Each block contains time-stamped data transactions, whose integrity and authenticity are guaranteed thanks to hashing and public-key cryptographic algorithms. From the deterministic and irreversible result of the hash function, it is possible to verify if the content of the block has been modified. Each block references the hash of the block that came before it, establishing a link between them, and thus, creating the blockchain. Following hashes from the current block ends with the first created block on a specific blockchain (called the genesis block) [19–21].

An illustration example of the chain data structure of blockchain is presented in Fig. 2.

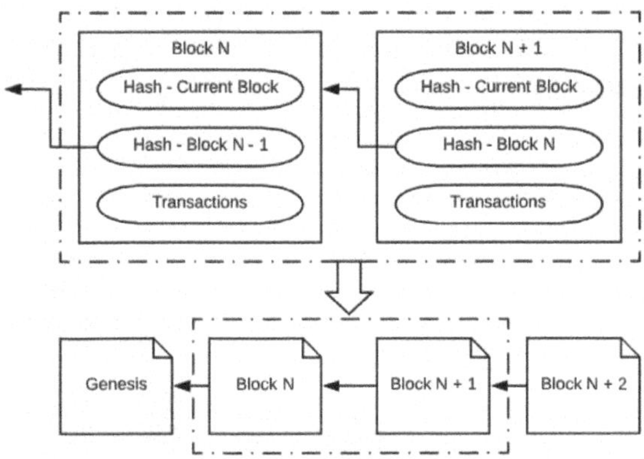

Fig. 2. Chain data structure of blockchain

The data (transactions) are digital signed, broadcast by the participants, and grouped in to blocks chronologically ordered. Once a new block is verified and written to the ledger, transactions cannot be included in the ledger without the consensus of the network majority [22].

To be able to maintain the data consistency in a distributed ledger, consensus algorithms are used to ensure the integrity of the data of the existing replicas in each peer (node) of the network. Multiple nodes must run a consensus algorithm to reach an agreement at the commit order of transactions, and thus, reach an agreement on a suitable decision among nodes [22–24].

In order to digitally formalize and secure relationships over a network, a smart contract may be used to automate agreement process between parties of the decentralized network.

First introduced by [25], a smart contract is an application (like a digital protocol) that verifies, executes and enforces contracts' terms that have been agreed between parties, helping transactions to be executed automatically in a transparent, conflict-free, undeniable, faster and more secure way, without having to rely on third parties. A smart contract adds functionality to blockchain and allows the addition of constraints, validations, and business logic to transactions [12, 23, 26].

With the continuous development of blockchain technology, the application scope of blockchain is very wide and can be roughly divided into three types: public blockchain, consortium blockchain, and private blockchain.

Table 1 presents a general comparison on types of blockchain [20, 21, 24, 26–28].

Table 1. Types of blockchain (public, private and consortium)

	Public	Private	Consortium
Definition	All participant can read, write, monitor, transact and engage in the consensus process	Used in a single organization. Only enables chosen nodes to join the network with restricted participations	Operates under the leadership of a group of organizations. Only enables chosen nodes to join the network with restricted participations
Access	Read: Open Write: Open Consensus: Open	Read: Open/Permissioned Write: Permissioned Consensus: Permissioned	Read: Open/Permissioned Write: Permissioned Consensus: Permissioned
Security	Proof-of-work [10], proof-of-stake [29], other consensus protocols	Practical Byzantine Fault-Tolerance [30], Raft [31], legal contracts, proof of authority	Practical Byzantine Fault-Tolerance [30], Raft [31], legal contracts, proof of authority
Privacy	Medium	High	High
Environment	Untrusted	Trusted	Trusted
Speed	Slow	Fast	Fast
Cost	High	Medium	Low
Architecture	Decentralized	Partially decentralized	Partially decentralized or centralized
Transaction Rate	Slower	Faster	Fastest
Membership	Anonymous/Pseudonymous	Known identity	Known identity
Typical Platforms	Bitcoin [10], Ethereum [18], Litecoin [32]	HyperLedger [23, 33], R3 Corda [34]	HyperLedger [23, 33], R3 Corda [34]

According to [11], the fundamental characteristics of blockchain technology enables implementation in a wide range of processes for asset registry, inventory, and information exchange, both hard assets like physical property, and intangible assets like votes, patents, ideas, reputation, intention, health data, information, etc. The essence of a blockchain is that organizations can keep track of a 'ledger' and that organizations jointly create, evolve and keep track of one immutable history of transactions and determine successive events.

Related to applications of blockchain in the context of data and digital assets management, storage and/or sharing, different approaches were proposed in academic researches.

[35] presented general discussions on how a blockchain based system can be modified to provide a solution for dynamic asset sharing amongst coalition members, enabling the creation of a logically centralized asset management system by a seamless policy-compliant federation of different coalition systems.

A blockchain based secure data sharing and delivery of digital assets framework was proposed by [36]. The main aim of the proposed scenario is to provide data authenticity and quality of data to buying customers as well as a stable business platform for owner. In this scenario, decentralized storage, Ethereum blockchain, encryption, benefits of interplanetary file system (IPFS), smart contract and incentive mechanism were combined.

Regarding the healthcare domain, [19] propose a novel drug supply chain management using Hyperledger Fabric based on blockchain technology to handle secure drug supply chain records. The proposed system conducted drug record transactions on a blockchain to create a smart healthcare ecosystem with a drug supply chain. A smart contract was launched to give time-limited access to electronic drug records and also patient electronic health records.

In concern of disseminating medical data beyond the protected cloud of institutions, [37] proposed a blockchain based data sharing framework (permissioned blockchain) that sufficiently addresses the access control challenges associated with sensitive data stored in the cloud using immutability and built-in autonomy properties of the blockchain.

Also focused in data-sharing-transaction application, [38] proposed a layered authorization transaction multilayer blockchain model based on mass data support focused in the current Internet of Things environment. The paper proposed a double-chain model in which the main alliance chain is responsible for processing transactions, and the private chain is responsible for storing transaction data. The alliance-chain network is deployed between each subject (e.g. organizations), and the private-chain network is deployed within each subject. The blockchain platform uses chain storage to store real data in the IPFS cluster server built by the alliance.

4 Blockchain Applications Trends

In concern of blockchain applications trends, companies and institutions from several sectors and industries of the global economy have been exploring the potential of blockchain technology in recent years.

Based on Cambridge Centre for Alternative Finance (CCAF) dataset of 67 live enterprise blockchain networks from 25 countries globally that have been deployed in production and are currently live, [39] pointed that 43% of the use cases list are applicable to finance and insurance sector. The accommodation and food services, as well as the healthcare and social assistance sectors came at a distant second place with 6% of all networks each.

In academics researches, there were several good survey papers on blockchain with different focus like on the specific aspects of blockchain such as security [40], privacy [41], architectures [42], consensus protocols [24, 28], smart contracts [43], applications, and Internet of Thing (IoT) and security related applications [20, 44, 45].

Other papers focused on surveying the security and privacy issues of specific blockchain platforms, such as Bitcoin [10] and Ethereum [18]. There was also a systematic survey on 41 highly-selected blockchain related papers [46], which the goal was to analyze the current research trend in blockchain area. The study identified some prototype applications developed and suggested for using Blockchain in other environments, such as IoT, smart contracts, smart property, digital content distribution, Botnet, and P2P broadcast protocols.

Regarding to government applications, [47] performed a systematic review research to understand the current research topics, challenges and future directions regarding blockchain adoption for e-Government.

From the 21 scientific articles published proposing blockchain integration within e-Government, most of the research (7 articles) discuss the application of blockchain for e-Government in general, discuss the idea, potential benefits, current issues, potential use, approach and evaluation of blockchain adoption. Blockchain applications in public healthcare received the highest attention with four articles. Meanwhile, three articles examined the use of blockchain in educational services, and also three articles in the context of smart cities. Two articles look into in the context of government to business supply chains, and single articles were dedicated to digital identity, e-voting, and the tax system. The results showed that the adoption of blockchain based applications in e-Government is still very limited.

The potential benefits in terms of strategic, organizational, economical, informational and technological aspects in government were identified in [48], and according to [49], in 2018 more than 100 blockchain projects created to transform government systems were being conducted in more than 40 countries around the world.

[50] also pointed that governments like in UK, Europe, China, United States of America (USA), among others, have released white papers and technical reports on blockchain to show their positive attitude toward the development of blockchain technology.

However, with regard to applications within the domain of intelligence and investigation units, there was a lack of approaches in both the industrial and academic aspects.

Motivated from [19, 36–38], in the next section the use of blockchain is proposed as a support mechanism in the management, storage and/or sharing of generated digital assets in the context of in investigation and intelligence units.

5 Proposed Approach

5.1 Scenario and Architecture Design

Figure 3 presents the scenario regarding the management, storage and/or sharing of digital assets of one or more specialized units in the context of intelligence and investigation activities.

In this scenario, during the execution of various activities inherent to these specialized units, different assets are generated and there may be a need to manage their use, storage and/or sharing within a unit or between different units (units may be located in one or more organizations).

In this case, specialists responsible for these assets need to not only perform CRUD operations (create, read, update, and delete) on the generated assets, but also have available means to perform access control, auditability, versioning identification, authenticity, non-repudiation, integrity verification, confidentiality and security of the generated digital assets.

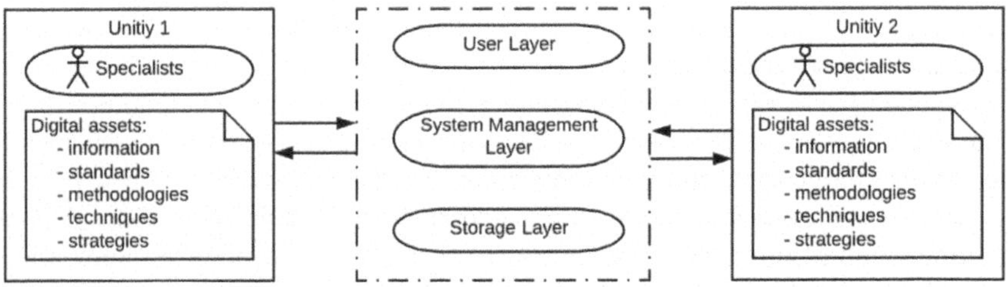

Fig. 3. Scenario: iteration of specialized units in the management, storage and/or sharing of digital assets

In the proposed design, 3 layers were adopted. The user layer is composed of one or more front-end applications to enable users to access the system management layer, perform operations with digital assets and other management activities. Applications can be developed in different programming languages or not, depending on the consensus among the participants.

Through this layer, users send transactions proposals to call backend services (e.g. CRUD operations, check log entries or sharing digital assets) provided by the blockchain network that transforms the ongoing data between the nodes.

The system management layer is where the blockchain network is implemented, and is composed of connected entities responsible for the secure establishment e efficient running of the scheme. Information about all operations is recorded in a distributed ledger where each block comprises a number of transactions that are hashed and encrypted.

In the scenario of specialized units, permissioned blockchains (private or consortium) are more appropriate because of the need to limit access to more reliable and restricted environments, as well as to identify participants and their operations.

Together with the use of smart contracts, immutable ledger, encryption, digital sign and access control policies, it is expected to restrict operations to authorized participants, as well as enable auditability, versioning identification, authenticity, non-repudiation, integrity verification, confidentiality and security of the generated digital assets.

In the system management layer, smarts contracts are used to provide controlled access to the ledger and to allow participants to execute certain aspects to transactions automatically. Invoked by the applications, the smart contract take the transactions and execute several kinds of queries and updates the ledger state by appending the transaction in blocks and returning the updated result to the application as a response.

Whereas a ledger holds facts about the current (world state) and historical (transaction logs) state of a set of business objects, a smart contract defines the executable logic that generates new facts that are added to the ledger. Taken together, these contracts lay out the business model that governs all of the interactions between transacting parties. Figure 4 illustrates the ledger operations using smart contract.

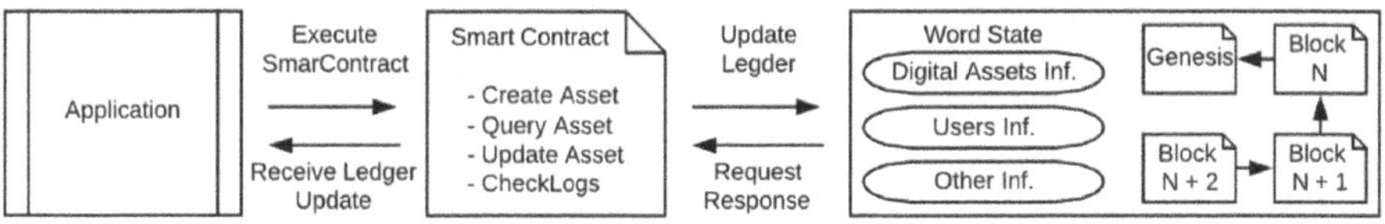

Fig. 4. Operations in the ledger using smart contracts

Finally, the third layer composes the infrastructure responsible for the storage of the actual digital assets (off-chain storage). Regarding storage, distributed (e.g. Interplanetary File System) or centralized (e.g. cloud storage or traditional databases) approaches can be adopted depending on the needs of the participants.

5.2 Transaction Process

Regarding the transaction flow, Fig. 5 illustrates the overall transaction process and role of the individual components of the network.

To allow higher parallelism and concurrency for the network, the transaction management is split between peers and orderers. Every transaction is executed in the peer using the world state, and if the transaction succeeds, it is signed with a peer's certificate.

Executing transactions prior to ordering allows each node to process multiple transactions at the same time. To enable the peers to trust all orderers and vice versa, the orderers do not re-execute the transaction, they will just order them and will not maintain the ledger (so peer and orderers can run independently).

The transaction process starts when the participant (user) application sends a transaction proposal to nodes. The user manager is responsible to issue the credentials to user application in order to authorize for sending transaction proposals.

The node which participates in the network can be either committers or endorsers. The endorser node is responsible to sign and authorize the proposal for transactions, while the committer node validates the results in responses to the transaction and writes a transaction block to the ledger. The endorser node also acts as a committing node and is used for receiving and executing a proposal for the transaction by invoking a smart contract without updating the ledger.

During the transaction, the endorser node collects and reads the RW (read–write) sets from the present world state, sign them, and returns the response to the user application for further processing. The user application then compiles the signed transaction in the form of a package and submits it to the consensus manager along with the RW sets.

The consensus occurs in parallel throughout the transaction process with the RW sets and signed transactions. After the consensus is sent to committer nodes in the form of blocks, each transaction is validated by comparing its RW set with the present world state.

The validated transaction is then written into the ledger, and the endorser will also update the current world state from the RW sets. Finally, whether it is successful or not, the committing node generates an asynchronous alert for the submitted transaction. The committer node also gets notified for every event by registering events through the user application.

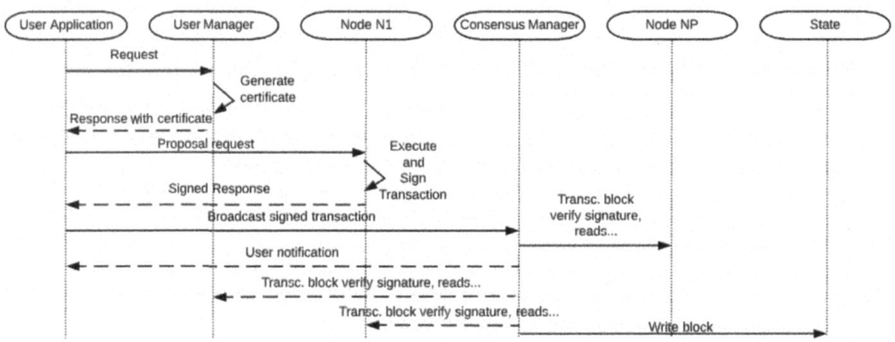

Fig. 5. Transaction flow

5.3 Network Topology Example

A sample example of the system management layer blockchain network topology is presented in the Fig. 6. The example consists of 4 specialized units (U1, U2, U3 and U4), where U1 and U2 belong to organization Org1, and U3 and U4 belong to organization Org2. All unities can communicate (access digital assets) with each other, but U1 and U2 also have needs for private communications within the Org1 network, as do U3 and U4 in Org2.

Hyperledger Fabric was adopted in the development of the proposed design for being a comprehensive, yet customizable, open-source framework and toolkit for developing enterprise blockchain applications. Another reason for the choice was because it is permissioned and offers the ability to create channels, work with smart contracts, as well as allow identity management, has efficient processing and chaincode functionality [19, 23, 33].

[33] also pointed out that Hyperledger is used in more than 400 prototypes, proofs-of-concept, and in production distributed ledger systems across different industries and use cases, and is, according to [39], the most used enterprise permissioned blockchain project in production.

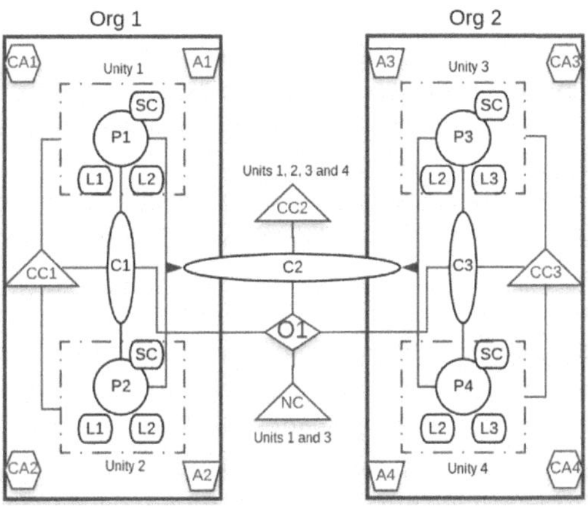

Fig. 6. Network Topology based on Hyperledger Fabric

To compose the blockchain network, each unity has pears nodes that maintain the state of the network, a copy of the ledger (L1, L2 and L3) associated with the channels used (C1, C2 and C3) in the communications, and smart contracts (S).

The network, under the control of unities U1 and U3, is governed according to policy rules specified in network configuration (NC). Each of the four organizations has a preferred certificate authority where are used for identification and access control.

User applications (A1, A2, A3 and A4) connect with other entities with the help of channels. A1 and A2 connect with C1 and C2, and A3 and A4 with C2 and C3.

A channel is a secure blockchain which is used to provide confidentiality and data isolation. Channel C2 acts as a central channel communication and is under control (channel configuration policy CC2) of all participants. C1 and C3 are exclusive channels used in cases of private and secure transactions to share confidential data directly without expose them to all units.

Channel C1 is governed according to the policy rules specified in channel configuration CC1 (under control of U1 and U2), and C3 according to CC3 (under control of U3 and U4).

The ordering service O1 is the node responsible for ordering the transaction into a block. It accepts endorsed transactions, orders them into a block, and delivers the blocks to the committing peers. Although the illustration shows only one ordering service, more than one can be used.

5.4 Discussions

The characteristics' inherent of blockchain technology, as well as the adoption of permissioned blockchains, cryptography and the use of Hyperledger Fabric in the proposed approach, may contribute in activities related to auditability, versioning identification, authenticity, non-repudiation, integrity verification, confidentiality, security, decentralization and transparency.

Through the tamper resistance characteristic of the blockchain, the integrity of the user data is reachable. All valid blocks and transactions recorded in the ledger are virtually immutable due to the need for validation by other nodes. With the immutable log of transactions traceability of changes, auditability and versioning identification are possible to accomplish with greater reliability.

Since all transactions are signed and registered with a generated hash value in the blocks, also adopting cryptography to allow more security and confidentiality when needed, is also possible provide integrity verification, non-repudiation analysis and authenticity. The use of private channels and/or cryptography increase the protection from data leaks, since only specific and authorized participants access the channel.

With the adoption of a permissioned blockchain, smart contracts and certificate authorities not only all transactions are identified and verified, but also allows that only permissioned participants can access and transact in the network.

Furthermore, the entire global ledger is synchronized between blockchain nodes according to a consensus mechanism, giving users greater confidence in the authenticity and accuracy of the data in the blockchain.

6 Conclusion

Due to the evolution and expansion of crime, as well as the diversity and volume of existing crime practices, researches aimed at helping or strengthening the performance of institutions that focus on fighting crime are essential.

Motivated by this scenario, and exploring the main characteristics of blockchain technology, this paper presented an overview of different application trends of blockchain technology, and propose the use of blockchain as a support mechanism in the management, storage and/or sharing of generated digital assets in the context of intelligence and investigation units.

Due to the characteristics' inherent of blockchain technology, as well as the adoption of permissioned blockchains, criptografy and the use of Hyperledger Fabric in the proposed approach, an alternative solution to contribute in activities related to auditability, versioning identification, authenticity, non-repudiation, integrity verification, confidentiality, security, decentralization and transparency of the generated digital assets was presented.

Future work recommends expanding the scope of this work by performing a case study of the approach in a real scenario of intelligence and investigation units.

References

1. UNODC: Organized Crime. United Nations Office on Drugs and Crime. https://www.unodc.org/unodc/en/organized-crime/intro.html. Accessed 18 Jan 2020
2. REFINITIV: Revealing the true cost of financial crime - 2018 Survey Report. https://www.refinitiv.com/content/dam/marketing/en_us/documents/reports/true-cost-of-financial-crime-global-focus.pdf. Accessed 20 Jan 2020
3. FBI: FBI Releases 2018 Crime Statistics. Federal Bureau of Investigation National Press Office, Department of Justice, United States, 30 September 2019. https://www.fbi.gov/news/pressrel/press-releases/fbi-releases-2018-crime-statistics. Accessed 16 Jan 2020
4. FBI: What We Investigate. Federal Bureau of Investigation Department of Justice, United States. https://www.fbi.gov/investigate. Accessed 13 Oct 2019
5. EUROPOL: Crime Areas – Fighting Crime on a Number of Fronts. European Union Agency for Law Enforcement Cooperation. https://www.europol.europa.eu/crime-areas-and-trends/crime-areas. Accessed 11 Jan 2020
6. Drezewski, R., Sepielak, J., Filipkowski, W.: The application of social network analysis algorithms in a system supporting money laundering detection. Inf. Sci. **95**, 18–32 (2015). https://doi.org/10.1016/j.ins.2014.10.015
7. UNDP: Investigation Guidelines. United Nations Development Programme. https://www.undp.org/content/dam/undp/library/corporate/Transparency/Investigation_Guidelines_ENG_August_2019.pdf. Accessed 2 Jan 2020
8. UNODC: Criminal Intelligence – Manual for Analysts. United Nations Office on Drugs and Crime, Vienna, United Nations (2011). https://www.unodc.org/documents/organized-crime/Law-Enforcement/Criminal_Intelligence_for_Analysts.pdf. Accessed 09 Jan 2020
9. Zhu, Y., Qin, Y., Zhou, Z., Song, X., Liu, G., Chu, W.C.-C.: Digital asset management with distributed permission over blockchain and attribute-based access control. In: 2018 IEEE International Conference on Services Computing (SCC) (2018). https://doi.org/10.1109/scc.2018.00032
10. Nakamoto, S.: Bitcoin: a peer-to-peer electronic cash system. https://bitcoin.org/bitcoin.pdf. Accessed 18 Jan 2020
11. Swan, M.: Blockchain: Blueprint for a New Economy. O'Reilly Media Inc., Sebastopol (2015)
12. Zheng, Z., Xie, S., Dai, H., Chen, X., Wang, H.: An overview of blockchain technology: architecture, consensus, and future trends. In: Proceedings of the 2017 IEEE International Congress on Big Data (BigData Congress), Boston, MA, USA, 11–14 December 2017, pp. 557–564 (2017)
13. Aste, T., Tasca, P., Di Matteo, T.: Blockchain technologies: the foreseeable impact on society and industry. Computer **50**, 18–28 (2017). https://doi.org/10.1109/MC.2017.3571064
14. Roehrs, A., da Costa, C.A., da Rosa Righi, R., Alex, R., Costa, C.A., Righi, R.R.: OmniPHR: a distributed architecture model to integrate personal health records. J. Biomed. Inform. **71**, 70–81 (2017)
15. Ren, Y., Leng, Y., Zhu, F., Wang, J., Kim, H.-J.: Data storage mechanism based on blockchain with privacy protection in wireless body area network. Sensors **19**(10), 2395 (2019). https://doi.org/10.3390/s19102395

16. Back, A., Corallo, M., Dashjr, L., Friedenbach, M., Maxwell, G., Miller, A., Wuille, P.: Enabling blockchain innovations with pegged sidechains. OPEN science review (2014). https://www.blockstream.ca/sidechains.pdf. Accessed 18 Jan 2020
17. Warburg, B.: How the blockchain will radically transform the economy. 2016. TEDSummit TED Talk, June 2016. https://www.ted.com/talks/bettina_warburg_how_the_blockchain_will_radically_transform_the_economy?language=en. Accessed 18 Jan 2020
18. Buterin, V.: Ethereum White Paper: A next-generation smart contract and decentralized application platform. Ethereum white paper (2014). https://www.weusecoins.com/assets/pdf/library/Ethereum_white_paper-a_next_generation_smart_contract_and_decentralized_application_platform-vitalik-buterin.pdf. Accessed 18 Jan 2020
19. Jamil, F., Hang, L., Kim, K., Kim, D.: A novel medical blockchain model for drug supply chain integrity management in a smart hospital. Electronics **8**(5), 505 (2019). https://doi.org/10.3390/electronics8050505
20. Alkurdi, F., Elgendi, I., Munasinghe, K.S., Sharma, D., Jamalipour, A: Blockchain in IoT security: a survey. In Proceedings of the 2018 28th International Telecommunication Networks and Applications Conference (ITNAC), Sydney, NSW, Australia, 21–23 November 2018, pp. 1–4 (2018). https://doi.org/10.1109/atnac.2018.8615409
21. Zheng, X., Zhu, Y., Si, X.: A survey on challenges and progresses in blockchain technologies: a performance and security perspective. Appl. Sci. **9**, 4731 (2019)
22. Mercenne, L., Brousmiche, K.-L., Hamida, E.B.: Blockchain studio: a role-based business workflows management system. In: 2018 IEEE 9th Annual Information Technology, Electronics and Mobile Communication Conference (IEMCON) (2018). https://doi.org/10.1109/iemcon.2018.8614879
23. Cachin, C.: Architecture of the hyperledger blockchain fabric. In: Proceedings of the Workshop on Distributed Cryptocurrencies and Consensus Ledgers, Chicago, IL, USA, 25 July 2016, vol. 310 (2016)
24. Yu, Z., Liu, X., Wang, G.: A survey of consensus and incentive mechanism in blockchain derived from P2P. In: Proceedings of the 2018 IEEE 24th International Conference on Parallel and Distributed Systems (ICPADS), Singapore, 11–13 December 2018, pp. 1010–1015 (2018). https://doi.org/10.1109/padsw.2018.8645047
25. Szabo, N.: Smart Contracts (1994). http://szabo.best.vwh.net/smart.contracts.html. Accessed 12 Jan 2020
26. Feng, Q., He, D.B., Zeadally, S.K., Muhammad, K.K.: A survey on privacy protection in blockchain system. J. Netw. Comput. Appl. **126**, 45–58 (2019)
27. Yang, J., Onik, M., Lee, N.-Y., Ahmed, M., Kim, C.-S.: Proof-of-familiarity: a privacy-preserved blockchain scheme for collaborative medical decision-making. Appl. Sci. **9**(7), 1370 (2019). https://doi.org/10.3390/app9071370
28. Cachin, C., Vukolic, M.: Blockchain consensus protocols in the wild. In: Leibniz International Proceedings in Informatics (LIPIcs). Schloss Dagstuhl–Leibniz-Zentrum fuer Informatik, vol. 91, pp. 1–16 (2017). https://doi.org/10.4230/lipics.disc.2017.1
29. Vasin, P.: Blackcoin's Proof-of-Stake Protocol v2 (2014). https://blackcoin.co/blackcoin-posprotocolv2-whitepaper.pdf. Accessed 10 Jan 2020
30. Castro, M., Liskov, B.: Practical Byzantine fault tolerance and proactive recovery. ACM Trans. Comput. Syst. **20**(4), 398–461 (2002)
31. Ongaro, D., Ousterhout, J.K.: In search of an understandable consensus algorithm. In: Proceedings of the USENIX Annual Technical Conference, pp. 305–319 (2014)
32. Litecoin - open source p 2p digital currency. https://litecoin.org/. Accessed 10 Jan 2020

33. Androulaki E., Barger, A., Bortnikov, V., Cachin, C., Christidis, K., De Caro, A., Enyeart, D., Ferris, C., Laventman, G., Manevich, Y., Muralidharan, S., Murthy, C., Nguyen, B., Sethi, M., Singh, G., Smith, K., Sorniotti, A., Stathakopoulou, C., Vukolić, M., Cocco, S.W., Yellick, J.: Hyperledger fabric: a distributed operating system for permissioned blockchains. In: Proceedings of the Thirteenth EuroSys Conference, ser. EuroSys 2018, pp. 30:1–30:15. ACM, New York (2018). https://doi.org/10.1145/3190508.3190538
34. Hearn, M.: Corda: A Distributed Ledger (2016). https://docs.corda.net/_static/corda-technical-whitepaper.pdf. Accessed 05 Jan 2020
35. Verma, D., Desai, N., Preece, A., Taylor, I.: A block chain based architecture for asset management in coalition operations. In: Ground/Air Multisensor Interoperability, Integration, and Networking for Persistent ISR VIII. https://doi.org/10.1117/12.2264911
36. Naz, M., Al-zahrani, F.A., Khalid, R., Javaid, N., Qamar, A.M., Afzal, M.K., Shafiq, M.: A secure data sharing platform using blockchain and interplanetary file system. Sustainability **11**, 7054 (2019)
37. Xia, Q., Sifah, E., Smahi, A., Amofa, S., Zhang, X.: BBDS: blockchain-based data sharing for electronic medical records in cloud environments. Information **8**(2), 44 (2017). https://doi.org/10.3390/info8020044
38. Xuan, S., Zhang, Y., Tang, H., Chung, I., Wang, W., Yang, W.: Hierarchically authorized transactions for massive internet-of-things data sharing based on multilayer blockchain. Appl. Sci. **9**, 5159 (2019)
39. Rauchs, M., Blandin, A., Bear, K., McKeon, S.: 2nd Global Enterprise Blockchain Benchmarking Study. Cambridge Centre for Alternative Finance. University of Cambridge Judge Busines School (2019). https://www.jbs.cam.ac.uk/fileadmin/user_upload/research/centres/alternative-finance/downloads/2019-ccaf-second-global-enterprise-blockchain-report.pdf. Accessed 03 Jan 2020
40. Kareem, A., Sulaiman, R.B., Farooq, M.U.: Algorithms and Security Concern in Blockchain Technology: A Brief Review (2018). https://papers.ssrn.com/sol3/papers.cfm?abstract_id=3234933. Accessed 03 Jan 2020
41. Henry, R., Herzberg, A., Kate, A.: Blockchain access privacy: challenges and directions. IEEE Secur. Priv. **16**, 38–45 (2018)
42. Mohan, C.: Blockchains and databases: a new era in distributed computing. In Proceedings of the 2018 IEEE 34th International Conference on Data Engineering (ICDE), Paris, France, 16–19 April 2018, pp. 1739–1740 (2018)
43. Wang, S., Ouyang, L., Yuan, Y., Ni, X., Han, X., Wang, F.: Blockchain-enabled smart contracts: architecture, applications, and future trends. IEEE Trans. Syst. Man Cybern. Syst. **49**, 2266–2277 (2019)
44. Al-Jaroodi, J., Mohamed, N.: Blockchain in industries: a survey. IEEE Access **7**, 36500–36515 (2019)
45. Roy, S., Ashaduzzaman, M., Hassan, M., Chowdhury, A.R.: BlockChain for IoT security and management: current prospects, challenges and future directions. In: Proceedings of the 2018 5th International Conference on Networking, Systems and Security (NSysS), Dhaka, Bangladesh, 18–20 December 2018, pp. 1–9 (2018)
46. Yli-Huumo, J., Ko, D., Choi, S., Park, S., Smolander, K.: Where is current research on blockchain technology? A systematic review. PLoS One **11**, e0163477 (2016)
47. Batubara, F.R., Ubacht, J., Janssen, M.: Challenges of blockchain technology adoption for e-government. In: Proceedings of the 19th Annual International Conference on Digital Government Research Governance in the Data Age – Dgo 2018, 30 May 2018–01 Jun 2018. https://doi.org/10.1145/3209281.3209317

48. Ølnes, S., Ubacht, J., Janssen, M.: Blockchain in government: benefits and implications of distributed ledger technology for information sharing. Gov. Inf. Q. **34**(3), 355–364 (2017). https://doi.org/10.1016/j.giq.2017.09.007
49. Jun, M.: Blockchain government - a next form of infrastructure for the twenty-first century. J. Open Innov.: Technol. Market Complex. **4**(1) (2018). https://doi.org/10.1186/s40852-018-0086-3
50. Zhang, R., Xue, R., Liu, L.: Security and privacy on blockchain. ACM Comput. Surv. **52**(3), 34 (2019). https://doi.org/10.1145/3316481. Article 51

Strategic Decision-Making and Risk-Management in Complex Systems

Vitaliy Nikolaevich Tsygichko(✉)

The Federal Research Center "Computer Science and Control" of the Russian Academy of Sciences, Moscow, Russia
vtsygichko@inbox.ru

Abstract. The present article sets the problem of assessing the negative consequences of strategic decisions in organization systems (OS). Research vocabulary for the subject matter "assessing the negative consequences of strategic decisions in organization systems" is presented in the article together with methodological concept of risk management in OS. A conceptual model of OS functioning has been constructed. The tasks of rational strategic making in the OS by choice from a limited multitude of alternatives have been formulated. The method for management of risk of negative consequences of the strategic decisions made in the OS is laid out. As a tool for assessment of the possible OS evolution trajectories influenced by the strategic decisions, we suggest a scenario method to make up the forecast area of the OS states in the changed living conditions.

Keywords: Organization System (OS) · Strategic decision · Negative consequences of decisions · OS conceptual model · OS condition · External environment · Emergency situation · Risk · Risk situation

1 Introduction

Assessment of risks in the process of management of the political, economic, social and other organization systems, where decision-making is done in the situation of high uncertainty of the condition of the managed system and the consequences the made decisions may yield, still remains a fundamental and outstanding problem of theory and practice of decision-making in an OS.

Most of the present time scientific research on this topic are focused on specific cases of individual application fields [1–7, 10]. But there is no general methodology concept that would define fundamental principles, conditions and possible ways of problem solution in organizational systems.

Making a choice between different alternatives of solving the problems that may occur in the complex OS, one should foresee not only the immediate effect of its implementation, but also the risks of the middle-term and long-term negative consequences of this decision for the elements of the system, its subsystems and the system as a whole. In many situations, the damage to the system and its structural components caused by the negative consequences of the made decisions may exceed the profit acquired from them. For example, severe deficit of the regional budget may be

R. Silhavy (Ed.): CSOC 2020, AISC 1226, pp. 415–432, 2020.
https://doi.org/10.1007/978-3-030-51974-2_40

compensated with an abrupt increase of the tax rate. It may solve the present problems of the region, but in the long run it may undermine its economy due to the bankruptcy of the companies who cannot bear the new tax load. The consequences of such a decision may cause significant shortages of the tax base and therefore make an impact on all spheres of life in the region.

Attempts of assessing the consequences of important strategic decisions have been made since ancient times. Unable to do a proper assessment, people would turn to oracles, sibyls and fortune-tellers who were very respected in the ancient civilizations. However, even today, despite the enormous scientific and technical achievements of our century, the problem remains outstanding. It is easy to give lots of examples of mistaken political, economic, and military decisions made without a proper research and forecast, which caused truly catastrophic consequences for countries, regions and the world as a whole.

Strategic decisions are intended to respond the challenges and threats of the socioeconomic development scenarios. They are called strategic for a reason: they are supposed to foresee the consequences of the developed and implemented strategies and doctrines, programs and projects, the consequences that cannot be predicted straightforwardly and correctly. How is it possible to assess such consequences in a responsible and grounded manner, in the situation of permanent uncertainty of our knowledge about the future, exposed to the various initially unknown factors and when the decisions are made, as a rule, with a deficit of information, time, and resources? This question is fundamental for scientific forecasting of, first of all, unexpected negative consequences, when the damage depreciates the expected results. The present article presents a methodological concept of this problem solution.

2 General Statements

Let us first introduce the main terms of the "assessment of the negative consequences of implementation of the strategic decisions in organization systems" domain.

An *organization system (OS)* is any level and element of the socioeconomic organization, from a community to a state in general to an individual socioeconomic unit, such as a separate enterprise (or a part of it).

Any OS is incorporated into a wider system, which makes up an *external environment* in which the organization exists.

Structural components of the OS interact with the external environment objects and influence each other in the process.

One of the basic terms of the studied domain is the term of a "condition". A condition of the OS, its subsystems and elements is an aggregation of the parameters that characterize it, fixed at a moment of time.

For every OS and its structural components, there are some acceptable condition parameters' change diapasons. When such diapasons are exceeded, an *emergency situation* occurs.

Every OS and its structural components exist in the given political, legal, economic, natural and other *living conditions,* in which it operates without changing its qualitative identity.

The object of this research is the *strategic decisions*, which are the decisions that change the state and the living conditions of the OS, its subsystems, elements and external environment.

It should be remark that the negative consequences occur, as a rule, in case of implementation of some *conflict strategic decisions,* i.e. the decisions that affect the economic, political and other interests of both the elements and subsystems of the OS itself and the external environment. A typical example of a conflict strategic decision may be the UK's decision to leave the European Union, which affects the interests of all EU states. Many possible negative consequences of this decision can be seen now. The government of the UK and the EU countries are now negotiating on the possible ways on the mitigation of the negative consequences.

The risk management-based procedure of choosing an acceptable option of a strategic decision in an OS based may be presented as the following sequence:

- evaluation of the problem situation and formulation of the strategic objectives of the OS in the given conditions;
- determination of the multitude of possible options of achieving the set goals;
- for every option, determination of a multitude of subsystems and their elements, which operate in a relation with its process of implementation; in other words, it requires building the cause-and-effect sequences to reflect the dynamics of the state of the OS and its elements in the process of implementation of every alternative decision;
- forecast of the further events that will take place in the OS and the external environment by making up the scenarios of possible changes of the living conditions and the state of the OS, its subsystems, and elements, resulting from the implementation of an alternative decision;
- analysis of the built scenarios in order to reveal the possible negative consequences of the implementation of the alternative decision for the OS, its structural components and the external environment;
- assessment of the risks of the negative consequences of such alternative decision;
- assessment of the possibility of compensating the negative consequences of the decision and the resources it might require;
- selection of an acceptable decision option based on the minimization of the risk of any negative consequences.

3 Strategic Decision-Making in OS

3.1 Conceptual Model of OS Functioning

The system research method requires a construction of a unified conceptual model of the management object. In other words, there should be an abstract image of the managed organization that reflects the main properties of the actual object, its structure, functioning mechanism, external and internal living conditions. The results of the conceptual model-based research may be used for the settlement of actual problems at all levels of management of socioeconomic processes. The conceptual mode terms may be used to describe any real management situation and a system analysis of the situation may be performed.

Let us take a hypothetic OS with a three-level hierarchic management system, with two A and B subsystems and two executive elements in every subsystem as an abstract object of our research (see Fig. 1). To create a formulated image of our object, let us turn to the approach developed in [8].

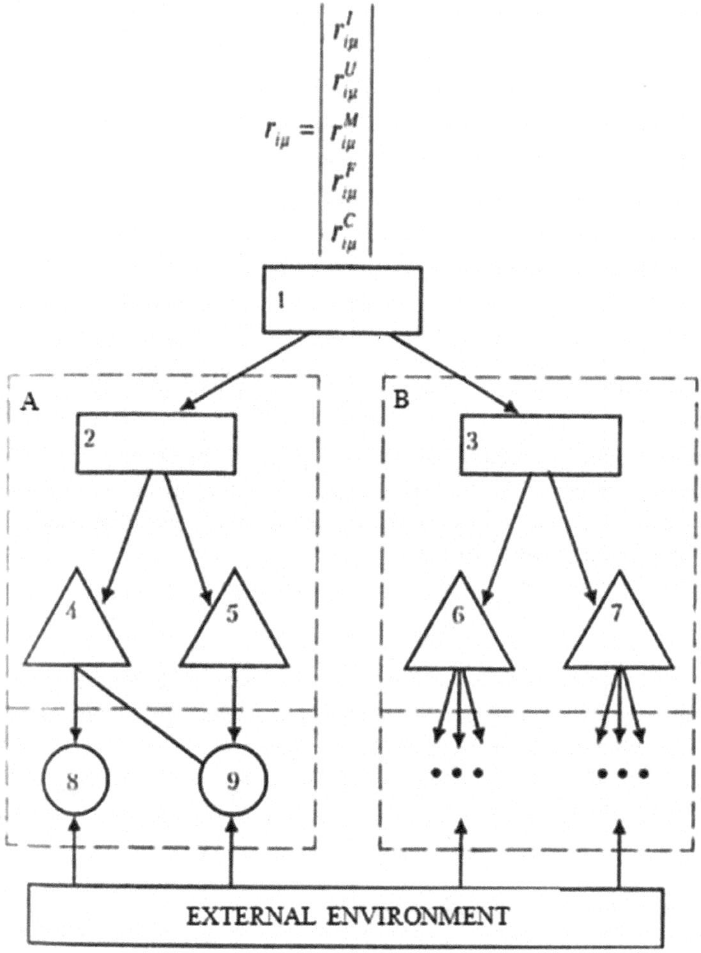

Fig. 1. Hypothetic OS structure.

Symbol $i = 1 - I$ is used to designate the numbers of the organization elements, where I is the number of elements in the OS structure. Rectangles designate the coordinating management elements, triangles stand for the control elements, and circles present the executive elements.

The relations between any two elements in the OS are described with a binary relations vector $r_{1\mu}$, where $\mu = \overline{1, I}$ and $\mu \neq 1$. Vector $r_{1\mu}$, a zero-unit column vector, indicates the presence or absence of connection between every i and all $\mu \in I$.

In our hypothetic OS, all the diversity of relations typical for the actual socioeconomic objects, will be presented as information relations $r^{I}_{i\mu}$, management relations $r^{U}_{i\mu}$, material exchange $r^{M}_{i\mu}$, financial exchange $r^{F}_{i\mu}$ and socio-political influence $r^{C}_{i\mu}$ (see Fig. 1). If more detailed analysis is necessary, every component of the vector $r_{i\mu}$ may, in its turn, be presented with a corresponding vector reflecting the structure of the information, management and other relations. The amount of the elements and relations is determined with the level of the system analysis.

The relations of the element μ with the element i are determined with the vector $r_{\mu i}$, which may be interpreted as an inverse relation vector to $r_{i\mu}$. For example, translation of information by the superior command level to the inferior ones is understood as direct relation, and translation of information up the hierarchy ladder is inverse relation. Multitudes $\{r_{i\mu}\}$ and $\{r_{\mu i}\}$ for all the elements of the studied system determine its structural state at the given stage of analysis. These multitudes make up a multidimensional mass of data, the size of which is determined with the number of the elements and the components of vector $r_{i\mu}$. Let us refer to this multidimensional mass of data as structural state matrix S.

$$S = \begin{vmatrix} 0 & r_{12} & \cdots & r_{1\mu} & \cdots & r_{1I} \\ \vdots & 0 & & & & \\ r_{i1} & & 0 & & & \\ \vdots & & & 0 & & \\ r_{\mu 1} & & & & 0 & \\ \cdots & \cdots & r_{1\mu} & \cdots & \cdots & 0 \end{vmatrix} \tag{1}$$

Here the line indicates the presence, if $r_{i\mu} = 1$, or the absence, if $r_{i\mu} = 0$, of any direct relations of the ith element with the other elements of the system, while the column stands for the presence, if $r_{i\mu} = 1$, or absence, if $r_{i\mu} = 0$, of the inverse (with regard to the ith element) relations of the system elements with the element i.

Matrix S describes the aggregate structures of the studied system. Any structure may be recovered from S by setting the required binary relations' components. For example, a management system is determined by binary relations' submatrix $r^{U}_{i\mu}$, and the financial system is determined by submatrix $r^{F}_{i\mu}$ etc. For hierarchic structures such as management structure, submatrix $r^{U}_{i\mu}$ reflects the subordination hierarchy of the system elements.

For many OS, the spatial location of its elements and their mutual positioning against each other, i.e. the system topology, is an important characteristic. Every element in the real system and the system itself are located on a limited area, the size and configuration of which may make a critical impact on the state of the system. In other situations, the relation may be inverse, i.e. the system topology is determined by its properties and state.

Let q_i be the border of the area occupied by an element, determined with the geographic coordinates φ, λ, then $q_i = q_i(\varphi_i, \lambda_i)$. In an special case, the element may be set as point $q_i = \{\varphi_i, \lambda_i\}$. Parameter q_i determines the geographic location and the configuration of the ith element in space. At the internal element structure study level, the spatial location of its constituent parts may be presented in the same way. Mutual spatial location of elements i and μ is determined with vector $q_{i\mu}$, the components of which are the distance between the elements and the orientation of element i against element μ:

$$q_{i\mu} = q_{i\mu}(q_i, q_\mu)$$

The aggregate spatial location of the elements q_i and their mutual positioning $q_{i\mu}$ totally determine the system topology. It can be presented as a matrix of the spatial state of the system – Q

$$Q = \begin{matrix} q_1 & \cdots & q_{1I} \\ \vdots & q_i & \\ q_{I1} & \cdots & q_i \end{matrix} \tag{2}$$

Another critical feature of the system is the internal condition of its elements. The international condition of the ith element is vector $P_i = \{P_i^J\}$, determined within the area of the possible values of the parameters P_i^J that characterize this element at the selected level of the analysis, where $j = 1 - J$ is the parameter number and J – is the amount of parameters describing the internal condition of the element.

The matrix of the internal condition of the system elements is determined with the column vector

$$P = \begin{Vmatrix} P_1 \\ \vdots \\ P_i \\ \vdots \\ P_I \end{Vmatrix} \tag{3}$$

The internal element condition matrix P brings out the lowest, the most detailed level of the OS description.

At the second level of OS presentation - P^2 is described as a structure consisting of subsystems, characterized with their integral features, P_k determined by the aggregate values of the parameters of the internal conditions of the elements belonging to each subsystem $\{P_k^l\}$

$$P^2 = \begin{vmatrix} P_1 \\ \vdots \\ P_k \\ \vdots \\ P_K \end{vmatrix} \tag{4}$$

$$P^2 = \{P_k^l\} = \overline{P_k}\{P_{ki}\}$$

where $k = 1 - K$ is the subsystem number, K is the amount of subsystems, $l = 1 - L$- is the number of the parameter and L – is the amount of parameters describing the internal condition of the subsystem;

$\overline{P_k}$ - operator of forming the internal condition indicators of subsystem k. Examples of such operators are standardized international and domestic methods of calculation of the current state of some branches of economy or the economy of the state as a whole.

At the highest generalization level, the internal condition of an OS as an integrated whole is presented with the macro indicators $m = 1 - M$ which generalize the internal condition of its subsystems with the operator $\overline{P}$.

$$P^3 = \begin{vmatrix} P_1 \\ \vdots \\ P_m \\ \vdots \\ P_M \end{vmatrix} \tag{5}$$

$$P^2 = \{P_m\} = \overline{P}\{P_k\}$$

One of the OS state parameters is the characteristic of the legal, economic, political, social and natural environment where it exists and operates. The vector of the environment of the OS existence may be presented as the following matrix

$$W = \begin{vmatrix} W_1 \\ \vdots \\ W_i \\ \vdots \\ W_I \end{vmatrix} \tag{6}$$

where W_i is the environment of existence of the element I.

It should be noticed that the environment of existence of the element W_i encompasses the parameters of the environment of existence of the OS as an integrated whole, the environment of existence of the subsystem, the element and the environment of the element existence in the subsystem belong to.

As any OS is a part of another wider system, the latter is making a continuous influence (deliberate or non-deliberate) on the studied system. This influence may be taken into account in several ways. First of all, the studied OS may be presented as a subsystem of a wider system. Then, using the structural state matrix S, it is possible to explicitly formulate all the relevant relations, i.e. the relations between the elements of the wider system and the studied system. This may also reflect the inverse impact of the system on the external environment.

Secondly, for every i^{th} element of the system it is possible to establish vector Γ_i, with the components γ_i that characterize various impacts made on this element by the external environment. The aggregation of the vectors for all the OS elements may be presented as matrix Γ of the external environment condition:

$$\Gamma = \begin{vmatrix} \Gamma_1 \\ \vdots \\ \Gamma_i \\ \vdots \\ \Gamma_I \end{vmatrix} \tag{7}$$

Variables S, Q, P, W and Γ make up the phased state space of the system. Let the point in the space $H = \{S, Q, P, W, \Gamma\}$ be the OS state

$$H = \begin{vmatrix} h_1 \\ \vdots \\ h_i \\ \vdots \\ h_I \end{vmatrix} \tag{8}$$

where h_i is the state of the element i.

Components of the vector h_i are the vector of communication of the element r_i with other elements of the system and the external system, the spatial position vector of the element Q_i, the vector of its internal state P_i, vector of the state of the environment where the element W_i exists and the external environment impact vector Γ_i:

$$h_i = \{r_i, Q_i, W_i, P_i, \Gamma_i\} \tag{9}$$

where the r_i- i^{th} line (direct relations) and the ith column (inverse relations) of the matrix S (1).

If the ith element has no direct management relations with the other elements of the system, i.e. $r^U_{i\mu}$, this element is an executive one.

Depending on the level of analysis, the executive elements of the system may be the lowest elements of the organization, with the internal structure being irrelevant at the selected level of analysis. This may be a workshop, a plant, an association or an economy branch.

Expression (9) is a formal way to describe the state of the system at the selected moment of time for any level of the given description hierarchy. The "state" characterizes the OS through its functional and organization structure, its location, internal state of its element, the condition of the environment where the system operates and the type of impact made on it by the external environment. If needed, additional coordinates may be introduced in the phase state space of the system. However, in our research the introduced indicators are enough for a formal interpretation of the OS state.

The process of OS functioning may be presented through the evolution of its state $H = \{S, Q, P, W, \Gamma\}$ in time. As a mathematic scheme for the description of the system dynamics, let us use the discrete method of presentation of the socioeconomic processes. Presentation of OS functioning as a finite discreet process, first of all, often corresponds to the periodicity of observations, and, secondly, simplifies the mathematic description to a great extent, as it relieves us of the need to study the delicate problems of existence and singularity of the solutions and the consistent implementation of functional analysis. The formal and theoretical description of the OS processes using a discreet mathematic scheme is a certain temporal sequence of the states of the system, determined with the position of point H in the phase space state $H(0)$.

$$\begin{aligned}&H(0), \Phi_1 : \{H(0)\} \to H(1), \Phi_2 : \{H(1), H(0)\} \to H(2), \ldots,\\&\Phi_n : \{H(n-1), H(n-2), \ldots, H(n-k)\} \to H(n), \ldots,\\&\Phi_N : \{H(N-1), H(N-2), \ldots, H(N-k)\} \to H(N)\end{aligned} \tag{10}$$

where n is the number of the process step; N is the number of the steps determining the duration of the system functioning under study; k is the number of steps in which the consequences of the previous activity of the system manifest themselves.

For the systems, the state of which at every moment depends only on their state at the previous functioning step, the equation of the trajectory of the point H in time looks as follows:

$$\begin{aligned}&H(0), \Phi_1 : \{H(0)\} \to H(1), \Phi_2 : \{H(1)\} \to H(2), \ldots,\\&\Phi_n : \{H(n-1)\} \to H(n), \ldots, \Phi_N : \{H(N-1)\} \to H(N)\end{aligned}$$

In the general case, transformation operators Φ_n depend on step number Φ of the OS functioning process, but for a large class of actual systems this dependence may be neglected. To keep it simple, let us assume that the operator Φ does not depend on time.

Let us refer to $\Phi : \{H(n)\}$ as an OS functioning process for all $n \in N$. It is assumed that the length of the time step of the discrete scheme of the OS dynamics description is invariably shorter than the shortest process in the system, and allows describing the functioning of any of its elements with the required degree of detail.

For description of the large-scale discreetly observed socioeconomic objects, the process step is usually the minimum period of reception of any statistic information.

OS is a purposeful system. In any way, the purpose of OS - Z may be presented as an aspiration to achieve the desired internal state of the multitude of its elements $\{P_i^z\}$, subsystems $\{P_k^z\}$, and the system as a whole P in the future.

$$Z = \{P_i^z\}, \{P_k^z\}, P \tag{11}$$

We suppose that the purposes of the OS remain unchanged within the studied time interval of the OS functioning, i.e. they do not depend on the step number n and the state variables H.

Purposeful behaviour of the system is manifested by the management system, the structure of which is determined with the binary control relations, $\left\| r_{i\mu}^U \right\|$ being the submatrices of the structural state matrix of the system S. Formalization of the control processes will be elaborated upon later. Here, let us remark that the state trajectory of the system $H(n)$ is the control function $H(n) = H\{U(n)\}$. At any n^{th} step of the process, the purposeful behaviour of the system is determined by the control vector:

$$U = |U_i|,$$

where U_i is the value of the control parameter of the ith element of the system.

The dynamics of the system are determined with the functioning of its elements in the existence environment, their interaction in such functioning process under the control effects formed at all levels of the OS management. As a result of the functioning and mutual influence of the system elements, they change their internal state, spatial location, structural relations and the influence made on them by the external environment.

As the system state and control parameters S, Q, P, W, Γ and U are interdependent functions, then the discrete description of the change in the state of the system $H(n)$ requires to determine the sequence of the change of such components at every n^{th} step of the process. In accordance with the contents of the OS functioning processes, the sequence of change of the state components of the system $H(n)$, i.e. the structure of operator Φ, may be presented as follows.

At every $(n+1)^{th}$ step of the process and in accordance with the state of the system developed at the previous step, $H(n)$, and the control vector $U(n)$ the system elements make an impact on each other as per their functional purposes, changing their internal state, i.e. forming the internal system state vector $P(n+1)$ at the $(n+1)^{th}$ step of the process.

The OS living conditions $W(n+1)$ may be modified as a result of a strategic decision made by the top management or under the effect of the external environment $\Gamma(n)$. These changes $W(n+1)$ may make a significant impact on the development of the vector of the internal state of the system $P(n+1)$.

If the dynamics of the spatial positions of the elements is critical for the process under study, then the new internal state of the system $P(n+1)$ together with the parameters $Q(n), S(n), W(n), \Gamma(n)$ determines the new spatial state of the system $Q(n+1)$.

The internal state of the elements, their positioning towards each other and he living conditions determine the new state of the relations of the system $S(n+1)$.

Evolution of the internal and spatial states of the system, the state of the environment and the structural state determine the new type of relations between the OS and a higher order system $\Gamma(n+1)$.

This is the sequence of formation of the system state at the $(n+1)^{th}$ step of the process $H(n+1)$, with regard to which the control over system $U(n+1)$ is found, and determining the direction of the OS evolution in accordance with its objectives at the next $(n+2)^{th}$ step of functioning.

Let us present the process $\Phi : \{H(N)\} \to H(n+1)$ of changing state and control of the system at the $(n+1)^{th}$ step as a series of consistent equations:

$$\begin{aligned}
&\overline{W} : \{P(n+1), S(n), Q(n+1), W(n), \Gamma(n), U(n)\} \to W(n+1)\\
&\overline{P} : \{P(n), S(n), Q(n), W(n), \Gamma(n), U(n)\} \to P(n+1)\\
&\overline{Q} : \{P(n+1), S(n), Q(n), W(n), \Gamma(n), U(n)\} \to Q(n+1)\\
&\overline{S} : \{P(n+1), S(n), Q(n+1), W(n+1), \Gamma(n), U(n)\} \to S(n+1)\\
&\overline{\Gamma} : \{P(n+1), S(n+1), Q(n+1), W(n+1), \Gamma(n), U(n)\} \to \Gamma(n+1)\\
&\overline{U} : \{P(n+1), S(n+1), Q(n+1), W(n+1), \Gamma(n+1), U(n), Z\} \to U(n+1)
\end{aligned} \tag{12}$$

i.e. operator Φ is a sequence of operators $\overline{W}, \overline{P}, \overline{Q}, \overline{S}, \overline{\Gamma}, \overline{U}$. Expression (12) describes the dynamics of the system at the $(n+1)^{th}$ step with no regard to the aftereffect, which may be taken into account by adding the parameters of the system states developed at the previous steps but still influencing the current state of the OS into the expression.

Progressive-cyclic type of functioning, assuming the progressive interchange of a number of states that constitute a cycle and then getting to the new initial position at the end of the cycle, i.e. spiral-like type of development is typical for the OS. Examples of this may be the economic cycle, technological cycle, cyclic work of the political election bodies etc. The state of the system is, to a great extent, determined with the phase of the cycle the system is in together with its subsystems or elements. The phase of the cycle is determined by the nature of the processes that occur in it, the state of the elements and the types of control. The temporal structure of the system functioning may be presented as a sequence in accordance with the selected step of the discrete description of the cyclic state matrix system:

$$\Delta = \begin{vmatrix} \delta_1 \\ \vdots \\ \delta_i \\ \vdots \\ \delta_I \end{vmatrix} \tag{13}$$

where δ_i is the phase of the cycle of the ith element.

Cycle forming operator Δ may be recorded as:

$$\overline{\Delta}:\{P(n+1),S(n+1),Q(n+1),W(n+1),\Gamma(n+1),U(n),\Delta(n)\}\rightarrow\Delta(n+1) \quad (13)$$

headings. Introduction of the cyclicity parameter makes it possible to include the temporal patterns of functioning of the system elements depending on the state of the system and its controls into the formal description.

The display system (12) supplemented with the Eq. (13) presents a *conceptual model of functioning of the OS.*

The components P_i of the operator P, implementing the functioning and interaction processes in every i^{th} element of the system at the $(n+1)^{th}$ step, may be presented as an equation:

$$P_i(n+1)=P_i\left\{\begin{array}{l} P_i(n), f_{\mu i}\left[P_i(n),P_\mu(n),q_{\mu i},r_{\mu i},(n),W_\mu(n),W_i(n),\Gamma_\mu(n),U_{\mu i}(n)\right], \\ f_{\mu i}\left[P_i(n),P_\mu(n),q_{i\mu},r_{i\mu},(n),W_\mu(n),W_i(n),\Gamma_{\mu i}(n),U_{i\mu}(n)\right], \\ W_i(n),\Gamma_i(n),\delta_i(n),U_i(n) \end{array}\right\} \quad (14)$$

where $U_{\mu i}(n)\subset U_\mu(n)$ and $U_{i\mu}(n)\subset U_i(n)$ for all $\mu\in I,\mu\neq i$.

Here the internal condition of the ith element at the $(n+1)^{th}$ step of the process is determined by its internal condition at the previous step $P_i(n)$, cyclicity parameter $\delta_i(n)$, condition of the environment of its existence $W_i(n)$, impact of the environment $\Gamma_i(n)$, control $U_i(n)$, as well as the results of interaction with other elements of the system expressed by functions $f_{\mu i}$ and $f_{i\mu}$, where $f_{\mu i}$ is the function of the impact of the element μ on i, and $f_{i\mu}$ is the function of the impact of the element i on μ. In the general situation, interaction is regarded as a phenomenon, constituting a process, separate from the internal functioning of its elements, such as competitive struggle in the global market etc. This process depends on the state of the interacting elements, their mutual positioning in space, the condition of the relations between them, the condition of the environment where they operate, the external effects and controls than may guide their activity. Corresponding dependencies may be made up for the other parameters describing the condition of the elements of the system $q_i, S_i, W_i, \Gamma_i, \delta_i$ and U_i.

The presented conceptual model of OS makes it possible to interpret and to analyse any processes happening in the system in its terms, and to formulate the general formal problem for the present research.

3.2 Formal Problem Statement for Strategic Decision-Making in OS

In the most general sense, the problem may be formulated as follows.

Let the parameters determining the current condition of the OS at the beginning of the strategic decision-making process $H_0=\{S_0,Q_0,P_0,W_0,\Gamma_0\}$ be known.

For the parameters of the internal condition of the OS and its subsystems and its elements at all the levels of presentation, their acceptable values, $P_{\partial on} = \{P_{i\partial\partial o}\}$, $P^2_{\partial on} = \{P_{k\partial\partial o}\}$, $P^3_{\partial on} = \{P_{l\partial\partial o}\}$ have been determined; the excess of these values is an emergency situation, which requires immediate reaction of the management of the corresponding level. The acceptable values of the parameters make up the multidimensional space of the acceptable internal states of the OS, its subsystems and elements.

Let us suppose that the OS has faced a problematic situation which requires a strategic decision to be made by its top management.

We know the limited multitude of the alternative solutions applicable in the situation $U^{cm} = \{U^{cm}_n\}$, where $n = 1 - N$ is the alternative decision number and N – is the number of alternatives.

It is supposed that these strategic decisions help achieving the set goal within the cycle Δ of the OS functioning with the length of ΔT with different grade of efficiency.

The objective of every strategic decision $U^{cm}_n \in U^{cm}$ is getting the "expected profit" C^n_z, which may be either tangible or intangible. The amount of the profit is the non-decreasing function of costs C^n for implementation of the decision $C^n_z(C^n)$.

Every decision brings its own profit. Let us suppose that any profit for the top management of our abstract OS may be presented in terms of value.

The implementation of the alternative decision $U^{cm}_n \in U^{cm}$ creates a new state of the external environment $W^n = \{W^n_i\}$, which, in its turn, in accordance with the OS conceptual model (12), changes all the components of the OS condition H and may, at the same time, make an adverse impact Γ on them, which may cause negative consequences for some elements and subsystems.

The negative consequences of a strategic decision may be the possible excess of the allowed value by at least one parameter of the internal state of the OS and its components, as well as tangible losses caused by the situation, i.e. the situation in which

$$P^n \notin P_{\partial on}, P^{2n} \notin P^2_{\partial on}, P^{3n} \notin P^3_{\partial on} \tag{15}$$

Let us suppose that for every decision $U^{cm}_n \in U^{cm}$ the mechanisms of the effect of the external environment on the condition of the OS and its components is known, i.e. there is a known multitude of functions $F = \{f^n_W(P), f^{2n}_W(P^2), f^{3n}_W(P^3)\}$ and a multitude of functions of interaction between the elements $F_{\text{вз}} = \{f_{\mu i}, f_{i\mu}\}$ (14), which allows forecasting the values of the OS internal condition parameters after implementation of every alternative decision

$$S^\Gamma = \{r^\Gamma_{i\mu}\} \subseteq S.$$

Let us assume that the functions F and $F_{\text{вз}}$ do not change in the decision implementation process, i.e. in the time interval ΔT.

For every decision, using the functions F and $F_{вз}$, from the structural state matrix S (1) the element submatrixes are recovered

$$S^{\Gamma} = \{r^{\Gamma}_{i\mu}\} \subseteq S,$$

and the submatrixes may be exposed to the direct or indirect (through the other related elements) negative effect Γ, determined by the changes happening in the external environment $W^n = \{W^n_i\}$ due to the implementation of the strategic decision $U^{cm}_n \in U^{cm}$.

Let us also consider the costs C^n of implementation of every decision $U^{cm}_n \in U^{cm}$ and the fiscal damage C^n_i of the possible negative consequences of their implementation for the elements, subsystems and the OS as a whole $C^n_I(H)$ to be known as well.

Let us suppose that the damage $C^n_I(H)$ is function F^n_C of the condition of the OS elements P^n.

$$C^n_I(H) = F^n_C(P^n) \tag{16}$$

Within the framework of the assumptions formulated above, evolution of the OS state influenced by a strategic decision is presented only by one definite trajectory $H^n(t)$ in the time interval ΔT which makes it possible to compare the damage C^n_I (16) of its negative consequences for every decision $U^{cm}_n \in U^{cm}$.

In this situation, when the decision implementation means are limited $C^n \leq C_{доп}$, the choice will be clear, a and the only decision $optU^{cm}_n \in U^{cm}$ would be the one that yields the maximum net "profit" $\max\Delta C^n_z$,

$$optU^{cm}_n <<\Leftrightarrow>> \max\Delta C^n_z = \max[C^n_z(C_{доп}) - C_{доп} - C^n_I] \tag{17}$$

The presented problem structures the process of strategic decision-making in the OS and determines the information required for the choice. However, the determined problem does not match the reality faced by the top management of the OS during strategic decision-making. The problem is that the credible and complete information, required for a rational decision-making, simply does not exist, as the implementation of every strategic decision generates a multitude of possible OS evolution trajectories, each of them yielding their own kind of damage in case of any negative consequences.

In this situation, for each decision the OS management needs to determine the $U^{cm}_n \in U^{cm}$ most dangerous OS evolution trajectory H^n_{eo}, i.e. the most adverse scenario, to assess the risks of such scenarios and to select the compromise between the expected profit and the risk of the negative consequences the decision may yield. This procedure of decision-making in the situation of uncertainty is usually referred to as *risk management*.

3.3 Management of the Risks of the Negative Consequences of the Strategic Decisions

According to the modern world outlook, OS management should be based on the assessment of risks of the adverse scenarios of these or those strategic management decisions.

General risk theory [5, 9, 10] defines risk as an activity related to overcoming uncertainty in the situation of inevitable choice, in the process of which there may be an opportunity of qualitative and quantitative assessment of the presupposed result, the failure and the deviation from the goal.

Risk management is a process related to identification, analysis of risks and decision-making intended to foresee the minimization of the negative consequences of the risk events. Examples of risk management procedures are the processes of choosing an activity strategy in the security market; selecting a risky but profitable investment project or selection of security measures for a critically important object etc.

Applicable to the object of the present study, risk is understood as likeliness R^n of the most adverse (dangerous) scenario to occur at the OS due to the made strategic decision

$$U_n^{cm} \in U^{cm}.$$

There is a wide range of published researches dedicated to the risk management, risk assessment and analysis methods in different domains of activity [1–7, 9, 10]. However, the problem of common theory and establishment of reliable methods for the quantitative assessment of the risk of negative consequences of strategic decisions made at the OS still remains unsolved. The root of the problem is, first of all, the complexity of the research subject as such, and the multidimensionality of the concept of risk itself.

Risk is a forecasting category that characterizes one of the important aspects of activity of the human society in the natural uncertainty of the condition of the natural, technology-related and social environments. Any purposeful activity of the human is associated with a possibility of emergence of any contingencies and coincidences that may cause some adverse results. This is the reason why any person assesses the chance for success and the risk of failure before starting any action.

In the widest sense, the risk may be defined as one's assessment of the consequences of the implementation of certain decisions for the achievement of the set goal in the conditions of uncertainty.

The concept of risk and its nature are closely associated to certain domains of human activity. This circumstance has generated a great number of possible risk situation forms, a number of definitions of risk and classifications based on different features, but there is one thing they have in common.

All possible kinds of risk situations are characterized with two elements: the likeliness of an adverse event and the consequences it may bring.

At the present moment, a common measure of risk is a combination of two indicators: the likeliness of an adverse event R^n and the degree of damage it may cause C_I^n.

The generalized damage indicator is the product of the likelihood of the risk event R^n and the amount of damage it may cause C_I^n.

$$Q^n = R^n C_i^n \tag{18}$$

This indicator can be used to characterize the degree of danger of different scenarios in case of implementation of certain strategic decisions, and to compare the decisions $U_n^{cm} \in U^{cm}$ be the severity of the possible negative consequences they may yield $\max Q^n$.

The resolution of any risky situation, i.e. the choice of the "best" strategic decision $optU_n^{cm} \in U^{cm}$ is a search of an acceptable compromise between the desire to acquire the maximum expected profit ΔC_z^n, the expenses C^n for implementation of the decision and its possible negative consequences. i.e. the amount of damage $C_I^n R^n$. The OS management should choose the most acceptable combination of the risk situation variable values for themselves.

$$\Delta C_z^n = C_z^n(C^n) - C^n - C_I^n R^n \tag{19}$$

Depending on the condition of the OS, the purposes and preferences of its management and the resources for the decision implementation that have developed to the present moment, the criterion for resolution of a risk situation may be any parameter, which can be used to formulate the three main decision-making problems $optU_n^{cm} \in U^{cm}$.

3.4 Problem One

Make the decision $optU_n^{cm} \in U^{cm}$ that provides the maximum net profit $\max \Delta C_z^n$ at the set implementation costs $C_{\partial on}^n$ and the damage of negative consequences not exceeding the acceptable threshold $C_I^n R^n \leq C_{I\partial}^n$.

$$optU_n^{cm} <<\Leftrightarrow>> \max \Delta C_z^n = C_z^n(C_{\partial on}) - C_{\partial on}^n - \min C_I^n R^n \tag{20}$$

3.5 Problem Two

Make the decision $optU_n^{cm} \in U^{cm}$ that provides the net profit that would not be lower than the established value $\Delta C_z^n \geq \Delta C_{z3}^n$ with the minimum implementation costs $\min C^n$ and the damage of negative consequences not exceeding the acceptable threshold $C_I^n R^n \leq C_{I\partial}^n$.

$$optU_n^{cm} <<\Leftrightarrow>> \Delta C_{z3}^n = C_z^n(\min C^n) - \min C^n - C_{I\partial}^{n\ n} \tag{21}$$

3.6 Problem Three

Select the option $optU_n^{cm} \in U^{cm}$ that ensures the minimum damage of the possible negative consequences of its implementation $\min C_l^n R^n$

$$optU_n^{cm} << \Leftrightarrow >> \min C_l^n R^n \tag{22}$$

The problems set above can be solved only provided that there is sufficient objective information on the expected damage C_l^n and the likeliness R^n of the most adverse scenarios for every $U_n^{cm} \in U^{cm}$. Determination of these parameters of a risk situation which may occur at the strategic decision-making in the uncertainty conditions constitutes the main task of the present research.

4 Scenario Method of Forecasting the Adverse Consequences of the Strategic Decisions Made at the OS

The main specificity of the strategic decisions made at the OS is their uniqueness, as they are always made in the conditions that have never occurred in the past. A coincidence of situations in the political, social, or economic spheres is an unlikely event. This circumstance makes it impossible to apply any regular forecasting methods to forecasting of the negative consequences that may occur due to the implementation of the strategic decisions, as there is no statistic or factual information of the current situation.

The only solution for this situation known today is the scenario forecasting method [8].

A tool for assessment of the possible ways of change of the OS evolution trajectory under the influence of the made strategic decisions may be the method of scenarios for the forecast area of OS states in the changed living conditions.

Outlining the forecast area means making hypotheses on the reaction of the OS, its subsystems and elements on the strategic decision, critical analysis of these hypotheses for logical consistency and correspondence to the OS and environment evolution tendencies developed by the moment of forecast. Upon the results of the criticism, the hypothesis is verified and transformed into a theory used to justify possible trajectories of the OS evolution in different situation that may occur in the future should the strategic decision be implemented. After that, some calculations and a logical analysis are carried out to study the acceptability of various trajectories of the OS evolution, and the borders of its possible states in the future are recovered.

The scenarios are developed by a team of experts; the assessments they make are used to calculate the damage and the likeliness of these scenarios. In the process of scenario-making, the experts perform the functions of the operators of the conceptual model of the OS functioning $\overline{W}, \overline{P}, \overline{Q}, \overline{S}, \overline{\Gamma}, \overline{U}$ (12) as the mechanisms of information transformation by such operators normally cannot be presented in an explicit way.

The possible scenarios of the strategic decision implementation may be interpreted as the results of solving the tasks of determining the possible OS evolution trajectories under certain conditions based on the conceptual OS model (12), as the conceptual model clearly determines the sequence and the contents of the scenario-making procedures.

5 Conclusion

The methodological concept of risk management in OS, conceptual model of OS functioning, the formulations of the problems of decision-making in risk situations presented in the article, make up a theoretical foundation for the development of an applied approaches to manage risks of the negative consequences of the decisions made in OS, and for the formulation of practical recommendations for the creation of a strategic decision-making support system based on the risk management procedure.

Acknowledgement. The research was supported by Russian Foundation for Basic Research (grant № 2819T-07-00572).

References

1. Howard, R., Abbas, A.: Foundations of Decision Analysis. Pearson Edition, London (2015)
2. Dorfman, M.S.: Introduction to Risk Management and Insurance, 9th edn. Pearson Edition, London (2007)
3. Kaneman, D., Slovic, P., Tversky, A.: Judgment Under Uncertainty: Heuristics and Biases. Cambridge University Press, Cambridge (1982)
4. Petrov, A.N., Suleimankadieva, A.E., Petrov, M.A.: Management of innovation risks in a corporation in the conditions of the cognitive economy. Innov. Econ. Issues **4**(9), 1543–1556 (2019)
5. Rygalovskii, D.M.: Risk management at enterprises: methodological and organizational issues. Mod. Manag. Technol. **12**(72), 20–30 (2016)
6. Simon, P., Hillson, D.: Practical Project Risk Management: The ATOM Methodology. Management Concepts Inc., Vienna (2012)
7. The Stanford Strategic Decision and Risk Management Program. http://scpd.stanford.edu/landing/sdrm_health.jsp. Accessed 10 Feb 2020
8. Tcygichko, V.N.: Socio-Economic Processes Prediction, 3rd edn. LIBERCOM Book House, Moscow (2009)
9. Sholomitckii, A.G.: Risk Theory. Choise in the Condition of Uncertainty and Risk Modeling. GU VSHE Publishing House, Moscow (2005)
10. Shahov, V.V., Millerman, A.S., Medvedev, V.G.: Theory and Management of Risks in Insurance. Finansy i statistika, Moscow (2014)

Dynamic Consensus: Increasing Blockchain Adaptability to Enterprise Applications

Alex Butean[1(✉)], Evangelos Pournaras[2], Andrei Tara[1], Hjalmar Turesson[3], and Kirill Ivkushkin[4]

[1] Lucian Blaga University of Sibiu, 10 Victoriei, Sibiu, Romania
alex@butean.com, research@andreitara.com
[2] University of Leeds, Leeds LS2 9JT, UK
e.pournaras@leeds.ac.uk
[3] York University, Keele Street, Toronto 4700, Canada
hturesson@schulich.yorku.ca
[4] Insolar Technologies GmbH, Hinterbergstrasse 49, Steinhausen, Canton of Zug, Switzerland
kirill.ivkushkin@insolar.io

Abstract. Decentralization powered by blockchain is validated for its capability to build trust like no other computational system before. The evolution of blockchain models has opened new use-cases that are becoming operational in many industry fields such as: energy, healthcare, banking, cross-border trade, aerospace, supply chain, and others. The core component of a decentralized architecture is the consensus algorithm - the set of rules that ensures an automated and fair agreement between the actors in the same network. Classic consensus algorithms are tailored to solve specific problems, but in an open ecosystem, each business case is unique and needs a certain level of customization. This paper introduces a new meta-consensus model called Dynamic Consensus, an architecture extension that allows multiple, complementary, consensus algorithms to run on the same platform. While classic consensus mechanisms are more appropriate for public or private systems (narrow set of rules), a dynamic approach would fit better for federated business consortiums (more rules and higher need for adaptability). The model is illustrated and analyzed as an ongoing experimental feature that can be added to enterprise blockchains designed to operate in cross-domain environments.

Keywords: Decentralized system · Blockchain · Consensus · Enterprise

1 Introduction

1.1 Classic Consensus Models

Proof of Work (PoW) was the first consensus model that applied the Nakamoto consensus [31] for blockchains. The protocol requires each block to contain a solution to a proof of work puzzle (i.e. the PoW) and to point to the previous valid block with the best PoW. Despite its energy inefficiency, variations of this algorithm are used by many platforms (Bitcoin, Ethereum, etc,) that are implementing fairly simple transaction

R. Silhavy (Ed.): CSOC 2020, AISC 1226, pp. 433–442, 2020.
https://doi.org/10.1007/978-3-030-51974-2_41

systems [14]. Even if the algorithm is computationally intensive and slow, it provides a secure and verifiable proof of the entire history of the chain.

Another classic consensus model is the Practical Byzantine Fault Tolerance (pBFT) algorithm, proposed as a solution to the Byzantine Generals' problem [8]. The algorithm is known for transaction finality and attack resistance but works only with a limited number of consensus members since it was designed for leader-based systems that can reduce the quadratic communication complexity of pBFT-like protocols.

The Proof of Stake (PoS) family of algorithms [12] was inspired by the social ecosystem of humans, where the trustworthy ones (risk takers) are entitled to decide over the next state of the network. Ouroboros [19] and Algorand [15] opened the road for Casper [7] and other variations that are following a similar path. Considering the acceptable trade-off between energy efficiency and decentralization, PoS has few proven vulnerabilities [13] and it is considered one of the most balanced consensus methods.

Besides the briefly presented consensus models above, there are many other algorithms and variations that are worth mentioning: Proof of Burn, Proof of Elapsed Time, Proof of Capacity, Proof of Identity and others [3].

From an analytical point of view, several studies [4, 5, 27, 28] have reached an important conclusion: each classical consensus mechanism comes with performance trade-offs, each has its own advantages and disadvantages, and thus, it can perform better or worse, depending on the application context, business perspective and hardware infrastructure constraints.

1.2 Enterprise Perspective

If we analyze the range of blockchain models from public to enterprise, we can identify the following focus points:

- transparency was traded for on-demand permission-based access and controlled privacy [36]
- the development and assessment of benchmarks and frameworks are focused on analyzing the existing architectures in order to identify the best choice for a specific context [11, 26];
- the optimization of algorithms is targeting only certain parameters: high-performance [21], efficiency [24], decentralized storage [32]. In this case, a multi-objective optimization would be very hard to perform;
- already functional products are using domain-focused solutions, tailored for individual applications and technologies [2, 20]
- industrial international standardization initiatives are already mature enough to be implemented in the next generation of products [16, 18].

Looking over the above points, we can conclude the following: for multi-organization enterprise networks, there is no easy way to establish a sole consensus mechanism because the need for flexibility is higher than the need for cross-organization communication.

1.3 Practical Solutions

Pioneers of enterprise-ready blockchain solutions are already using permissioned consensus methods that are flexible enough to match a large variety of use-cases. For example, Corda R3 is not using block-based ordering, instead, it uses notaries for transaction ordering and timestamping services to reach consensus (validity and uniqueness) [6]. The Hyperledger family clearly states that there is no magic key that opens all doors, that is why various products are using different consensus mechanisms, most of them based on permissioned voting systems: Kafka for Fabric, Proof of Elapsed Time for Sawtooth, Sumergi in Iroha and RBFT in Indy [34].

There are many use-cases, pilot applications or even commercial solutions developed using Hyperledger [29] or Corda [25]. Most of these solutions require a custom adaptation effort, particularly because of these two aspects:

- the virtual machine model needs more components in order to be able to process smart contracts written in high-level programming languages required for a large variety of constraints, rules and integration needs [1, 33];
- the consensus model needs on-demand difficulty control [22] to allow fast adaptation. Such an engine would be able to offer more flexibility for cross-domain, cross-shard [1], multi-organization rules or consortiums [10].

2 Dynamic Consensus

2.1 Starting Point

Based on the summarized analysis presented above (classic consensus models, enterprise perspective, practical solutions), this section introduces the concept of Dynamic Consensus, a meta-mechanism meant to increase the flexibility and adaptability of blockchain solutions by leveraging the usage of multiple consensus rules at the same time.

An enterprise network is a private permissioned network with thousands of cross-domain or cross-organization processes that are functioning in parallel. As depicted in Fig. 1, the data exchange is intense and for a realistic scenario, it has to be governed by a continuously changing set of rules.

The needs for such a business ecosystem are very hard to reproduce using fixed logic components. Such a network would be cost-ineffective and underperforming with a classic consensus mechanism where all nodes have to be aware in real-time of the entire state space. Also, a single consensus mechanism would probably reduce the parallelism capacity of the system.

2.2 Hierarchical Structure

An enterprise network is usually built across a consortium of organizations (companies) aiming for interoperability and data exchange. Each organization has its own domains (departments or sub-organizations) and each domain controls multiple shards (data processing clusters). Each shard is responsible for a specific industrial process and

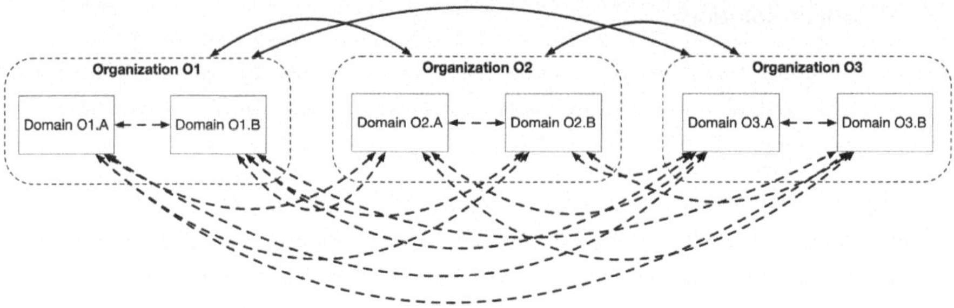

Fig. 1. Data exchange in a multi-organization environment

contains several nodes (processing agents). The nodes perform the actual consensus computations. Figure 2 shows the hierarchical structure of an enterprise network that is designed for a generic consortium.

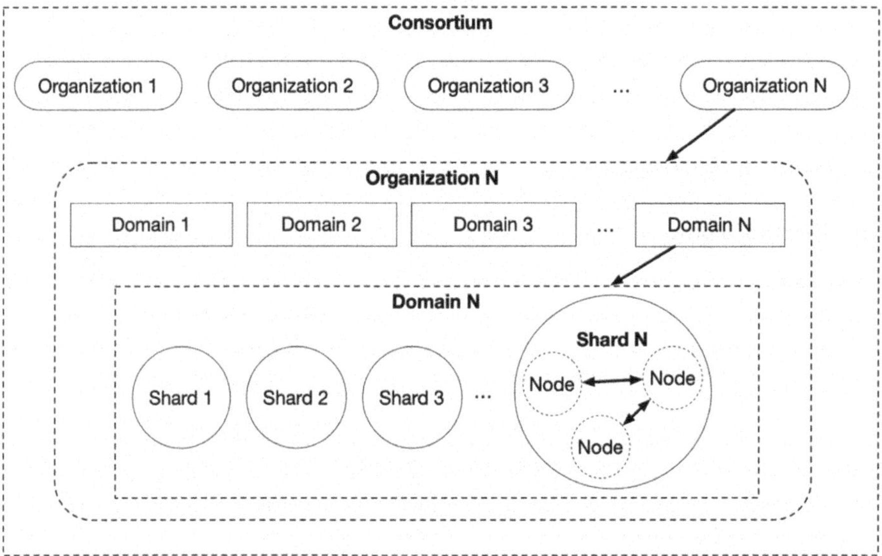

Fig. 2. Hierarchical design of an enterprise network

In a blockchain architecture, any data-exchange that affects the decentralized ledger is formally considered a transaction. In the above-presented design, once a transaction is issued, it is fairly easy to identify its origin and target by using the system hierarchy. This is the first step that allows the grouping of transactions based on their source of emittance and their source of impact. In this case, local transactions are processed only at a lower level with a reduced number of nodes, while cross-entity transactions require more decision nodes. This approach saves processing and propagation time and allows the use of an ad-hoc consensus at a lower scale and the use of an overall consensus of all nodes over some global (system-wide) transactions.

2.3 Channels and Virtual Domains

Side channels, plasma, state channels and other similar mechanisms [9, 35], are solutions that keep only relevant information on-chain while the processing is performed off-chain.

In the context of the above-presented consortium design, we propose to use channels as communication rooms where members (nodes) from multiple organizations can subscribe to specific topics. The members of a channel are not dependent on the hierarchical localization and channels are turning into virtual cross-organization domains. Each such domain will be created with the approval of the involved organizations. All operations inside the channel can be processed with a topic-driven consensus algorithm chosen by the participating members (nodes). Figure 3 presents an example of a channel that acts as a virtual domain across organizations.

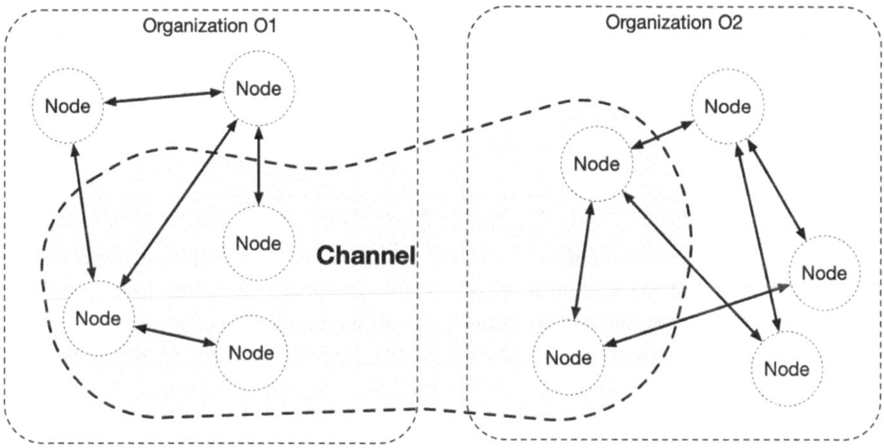

Fig. 3. Channel acting as a virtual domain across organizations

2.4 Indexing Functions

In the configuration stage, each organization defines the following parameters for its nodes:

- contextual power (C_p). The higher the value, the larger the impact of the node in the decision process. C_p can be considered a low pass filter, where a node with a maximum C_p can be involved in all the decisions across the consortium, while a node with a minimum C_p will be involved only in the least important decisions;
- location interest (L_i). Each location has a unique identifier across the network. Using an array, each node specifies its interest value (boolean) for every specific location in the network (organization, domain, virtual domain)

From a generic perspective, a transaction has the following fields: type(t), asset(as), source(so), target(ta), timestamp(ts), context(c), location effect(le). Considering the system-wide constant expiration timestamp(θ) and predicate functions $e(ts)$, $d(so,ta)$ defined as:

$$e(ts) = \begin{cases} 0\, if\ ts \geq \theta \\ 1\, if\ ts < \theta \end{cases} \quad d(so, ta) = \begin{cases} 0\, if\ so, ta \notin same\, network \\ 1\, if\ so, ta \in same\, network \end{cases}$$

The importance function (IMP_{func}) is a classification function that outputs the corresponding consensus algorithm and is defined as follows:

$$IMP_{func} : < T, As, C, L, Ad, Ad, Ts > \; \rightarrow X$$

$$IMP_{func}(t, as, so, ta, ts, c, le) = classify(< t, as, le, d(so, ta), e(ts) >) \text{ where}$$

$$X = \{PoW, PoS, pBFT, \ldots\} \text{ - contains all classes of alg. supported by the system}$$

$$T = \{t | t \in N\ and\, 0 - register\, organization,\ 1 - move\, asset, \ldots\}$$

$$As = \{as | as \in N\, and\, 0 - no\, asset,\ 1 - numeric\, asset, \ldots\}$$

$$C = \{c | c \in N\, and\, 0 - critical,\ 1 - business, 2 - operational, \ldots\}$$

$$L = \{le | le \in N\, and\, 0 - consortium,\ 1 - organization, 2 - local, \ldots\}$$

$$Ts = \{ts | ts \in N\ and\ ts\, is\, the\, timestamp\}$$

$$Ad = \{a | a\, is\, an\, account\, address\}$$

The exact mathematical weights of the classification function are dependent on the configuration parameters of each system. Multiple interactive simulations and relevant training data can reveal optimal values based on hardware specifications, network size and business case. At the system level of each node, there is a shared lookup routing table (Fig. 4) that stores the routing address for all the available consensus algorithms. For each transaction the IMP_{func} determines the associated class of the consensus algorithm. The algorithm address is determined by querying the lookup table for the determined class.

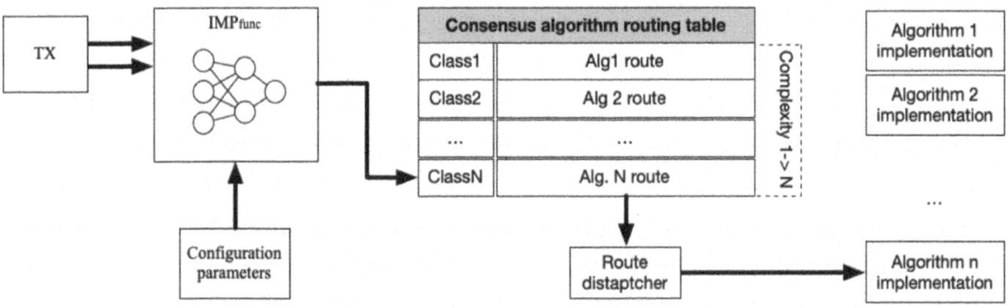

Fig. 4. Indexing table of consensus algorithms

The algorithm dispatch process is dynamically executed at the node level for every visible transaction in the pool. A node decides to participate in a consensus round for a transaction if the following conditions are met at the same time:

- the context(c) of a transaction corresponds to the contextual power of the node (C_p);
- the location effect(le) of a transaction corresponds with the location interest (L_i) of the node.

2.5 Overall Architecture

Most of the enterprise-oriented blockchain architectures [6, 17, 34] already offer a high level of abstraction. From a software engineering perspective, in order to migrate towards a dynamic consensus capable platform, the static consensus module should be replaced with a meta-consensus component. This component acts as a dispatcher for all transactions. From the perspective of a consortium owner, Fig. 5 displays an overall functional architecture.

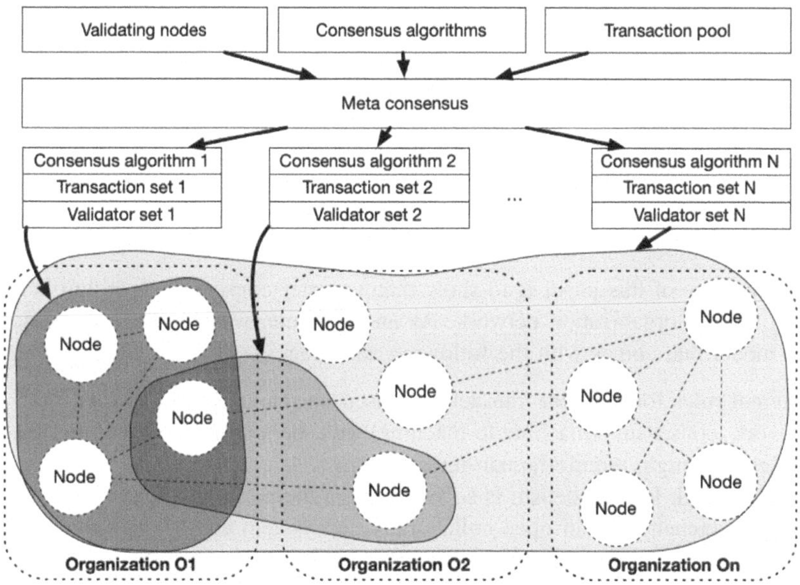

Fig. 5. Dynamic consensus architecture

The level of consensus parallelism can be observed in the context of multiple heterogeneous groups that are processing transactions with different consensus algorithms. The most important nodes should run on a hardware configuration with higher security requirements, thus, they can participate in multiple groups at the same time.

2.6 Prerequisite Actions for Integration

In order to integrate the dynamic consensus model in an existing multi-organization infrastructure, the following guidelines need to be followed:

- the previously described indexing functions are based on a common set of rules that has to be defined at the consortium level. All existing rules will balance the importance, location and context according to application-specific needs. A new rule added to the set has to be validated by the entire network with a top-level voting-based consensus algorithm;
- for reducing the complexity of simulation, the initial state can be: all the nodes receive the same power and the meta consensus is fed with only one entry. Gradually, adding more algorithms and adjusting the power of the nodes should be a context-driven decision;
- the existing static consensus has to be generalized and abstracted in order to be able to swap the modules;
- the host network should activate the support for state decoupling procedures like state sharding [23], layering, etc.

3 Conclusions

The performance of enterprise blockchains is a topic of great interest. Some of the existing platforms are highly adapted to specific needs but are unable to scale and others are highly scalable but not flexible enough for cross-organizational collaboration.

The purpose of this paper is to show that a master consensus algorithm is suboptimal in a multi-organization network. As an substitute, we propose a dynamic consensus model that comes with the following advantages:

- different rules for different transactions based on relevance and impact;
- network consensus (machine-to-machine) can be separated from the business consensus (might require human intervention);
- the bottleneck for parallelism is solved using a hierarchical structure;
- the data exchange in an open collaborative ecosystem can be regulated to match enterprise privacy and security policies;
- from an architectural perspective, the presented model covers the decentralization, consensus and security properties described by the DCS Theorem [30];

Dynamic Consensus is currently being integrated and tested on the Insolar Assured Ledger Platform [17] on a sharded multi-organization infrastructure. Future publications will include comparison sheets between classic static implementations and the proposed dynamic approach.

References

1. Al-Bassam, M., Sonnino, A., Bano, S., Hrycyszyn, D., Danezis, G.: Chainspace: A Sharded Smart Contracts Platform, arXiv:1708.03778v (2017)
2. Ali, A.A., El-Dessouky, I., Abdallah, M., Nabih, A.: The quest for fully smart autonomous business networks in IoT platforms. In: Proceedings of the 3rd Africa and Middle East Conference on Software Engineering, pp. 13–18. ACM, New York (2017)
3. Anwar, H.: 101Blockchain: Consensus Algorithms Blockchain. https://101blockchains.com/consensus-algorithms-blockchain/ (2018)
4. Baliga, A.: Understanding Blockchain Consensus Models, Persistent Systems. https://www.persistent.com/ (2017)
5. Ballandies, M.C., Dapp, M.M., Pournaras, E.: Decrypting distributed ledger design-taxonomy, classification and blockchain community evaluation. arXiv preprint arXiv:1811.03419 (2020)
6. Brown, R.G.: The Corda Platform: An Introduction, Corda R3 Documents (2018)
7. Buterin, V., Griffith, V.: Casper the Friendly Finality Gadget, arXiv:1710.09437v4 (2019)
8. Castro, M., Liskov, B.: Practical Byzantine fault tolerance. In: Proceedings of the Third Symposium on Operating Systems Design and Implementation (OSDI 1999), pp. 173–186. USENIX Association, USA (1999)
9. Coleman, J., Horne, L., Xuanji, L.: Counterfactual: Generalized State Channels. https://l4.ventures/ (2018)
10. Dib, O., Brousmiche, K.-L., Durand, A., Thea, E., Hamida, E.B.: Consortium blockchains: overview, applications and challenges. Int. J. Adv. Telecommun. **11**(1&2), 51–64 (2018)
11. Dinh, T.T.A.D, Wang, J., Chen, G., Liu, R., Chin, O.B., Tan, K.: BLOCKBENCH: a framework for analyzing private blockchains. In: Proceedings of the 2017 ACM International Conference on Management of Data (SIGMOD 2017), pp. 1085–1100. ACM, New York (2017)
12. Garcia Ribera, E.: Design and Implementation of a Proof-of-Stake Consensus Algorithm for Blockchain, Ph.D. Thesis at Universitat Politècnica de Catalunya (2018)
13. Gazi, P., Kiayias, A., Russell, A.: Stake-bleeding attacks on proof-of-stake blockchains. In: IEEE Crypto Valley Conference on Blockchain Technology (CVCBT), pp. 85–92 (2018)
14. Gervais, A., Karame, O.K., Wüst, K., Glykantzis, V., Ritzdorf, H., Capkun, S.: On the security and performance of proof of work blockchains. In: Proceedings of the 2016, ACM SIGSAC Conference on Computer and Communications Security (CCS 2016), pp. 3–16. Association for Computing Machinery, New York (2016)
15. Gilad, Y., Hemo, R., Micali, S., Vlachos, G., Zeldovich, N.: Algorand: scaling byzantine agreements for cryptocurrencies. In: Proceedings of the 26th Symposium on Operating Systems Principles, pp. 51–68 (2017)
16. Gramoli, V., Staples, M.: Blockchain standard: can we reach consensus? IEEE Commun. Stan. Mag. **2**(3), 16–21 (2018)
17. Insolar Team: Insolar Technical Paper (2019). https://insolar.io/uploads/Insolar%20Tech%20Paper.pdf
18. International Organization for Standardization, ISO/TC 307 - Blockchain and distributed ledger technologies (2016). https://www.iso.org/committee/6266604.html
19. Kiayias, A., Russell, A., David, B., Oliynykov, R.: Ouroboros: a provably secure proof-of-stake blockchain protocol. In: Katz, J., Shacham, H. (eds.) Advances in Cryptology. Lecture Notes in Computer Science, vol. 10401. Springer, Cham (2017)
20. Kim, H., Laskowski, M.: Towards an ontology-driven blockchain design for supply chain provenance. Intell. Syst. Acc. Financ. Manage. **25**(1), 18–27 (2018)

21. Khan, N.: FAST: a MapReduce consensus for high performance blockchains. In: Proceedings of the 1st Workshop on Blockchain-enabled Networked Sensor Systems (BlockSys 2018), pp. 1–6. ACM, New York (2018)
22. Kraft, D.: Difficulty control for blockchain-based consensus systems. Peer-to-Peer Netw. Appl. **9**(2), 397–413 (2015). https://doi.org/10.1007/s12083-015-0347-x
23. Luu, L., Narayanan, V., Zheng, C., Baweja, K., Gilbert, S., Saxena, P.: A secure sharding protocol For open Blockchains. In: ACM SIGSAC Conference on Computer and Communications Security, pp. 17–30, ACM, New York (2016)
24. Milutinovic, M., He, W., Wu, H., Kanwal, M.: Proof of luck: an efficient blockchain consensus protocol. In: Proceedings of the 1st Workshop on System Software for Trusted Execution (SysTEX 2016), pp. 1–6 ACM, New York (2016). Article 2
25. Mohanty, D.: R3 Corda for Architects and Developers. Apress (2019). ISBN-13: 978-1484245316
26. Moorsel, A.V.: Benchmarks and models for blockchain: consensus algorithms. SIGMETRICS Perform. Eval. Rev. **46**(3), 113 (2018)
27. Nguyen, G., Kim, K.: A survey about consensus algorithms used in blockchain. J. Inf. Process. Syst. **14**(1), 101–128 (2018)
28. Pîrlea, G., Sergey, I.: Mechanising blockchain consensus. In: Proceedings of the 7th ACM SIGPLAN International Conference on Certified Programs and Proofs, pp. 78–90. Association for ACM, New York (2018)
29. Shah, N.: Blockchain for Business with Hyperledger Fabric. BPB Publications (2019). ISBN-13: 978-9388511650
30. Slepak, G., Petrova, A.: The DCS Theorem, arXiv:1801.04335v1 (2017)
31. Stifter, N., Judmayer, A., Schindler, P., Zamyatin, A., Weippl, E.R.: Agreement with Satoshi - On the Formalization of Nakamoto Consensus. IACR Cryptology ePrint Archive (2018)
32. Tang, Y., Zou, Q., Chen, J., Li, K., Kamhoua, C.A., Kwiat, K., Njilla, L.: ChainFS: blockchain-secured cloud storage. In: IEEE 11th International Conference on Cloud Computing, pp. 987–990 (2018)
33. Tara, A., Ivkushkin, K., Butean, A., Turesson, H.: The evolution of blockchain virtual machine architecture towards an enterprise usage perspective. In: Silhavy, R. (eds.) Software Engineering Methods in Intelligent Algorithms. CSOC 2019. Advances in Intelligent Systems and Computing, vol. 984. Springer (2019)
34. The Linux Foundation, Hyperledger Architecture, Volume 1, Introduction to Hyperledger Business Blockchain Design Philosophy and Consensus (2017)
35. Vasa, V.: Difference Between SideChains and State Channels, Hackernoon (2018). https://hackernoon.com/difference-between-sidechains-and-state-channels-2f5dfbd10707
36. Zhang, J.: Kaleido - Permissions & Privacy: Core Elements of an Enterprise Blockchain (2019). https://kaleido.io/permissions-privacy-in-enterprise-blockchain/

The Ground Objects Monitoring by UAV Using a Search Entropy Evaluation

N. E. Bodunkov, N. V. Kim, and Nikita A. Mikhaylov(✉)

Moscow Aviation Institute, National Research University, Moscow, Russia
nkim2011@list.ru, n.mikhaylov.mai@mail.ru

Abstract. The paper presents an approach of entropy search ground objects by unmanned aerial vehicles. The UAV path is planned by optimization proposed criterion of throughput. This criterion takes into account a priori distribution of probability of presence search object in search area, the distance between UAV and next possible waypoint and search object observability evaluated with known distribution density of observed mark. The article shows that resulting search time can be significantly decreased by clarifying the initial estimates of the search object presence distribution in search area, that's leads to decreasing an initial entropy of object position. The authors offer use the situation analyze approach to reduce number of possible regions of search. The situation analyze approach is based on a description of search situation evaluated with: description of the search object, description of the search area, description of an environment and the relationships between those factors. The article presents the results of computer modeling of the proposed approach #CSOC1120.

Keywords: Unmanned aerial vehicle · Searching for mobile ground objects · Route planning · Search entropy

1 Introduction

Searching for ground objects with unmanned aerial vehicles (UAVs) is among the most important parts of the modern ground surface monitoring. Using UAVs in emergencies such as searching for people and various objects of interest in fires or natural disasters is often more efficient than using manned aircraft or ground vehicles [1].

Secondary search is different in that an a priori probability of an object position is known during UAV route planning. This information helps dismiss random or continuous search [2] and direct UAV towards regions of maximum probable presence of objects involved. Object finding is done using computer vision methods [3, 4]. Setting up object search and tracking is discussed in [5–10]. Articles [11–15] offer search algorithms including those based on using an object's visual signs. Articles [16–18] review options of UAV route planning before ground objects search.

Meanwhile, the known articles describing search route planning do not consider observation conditions such as precipitations, smoke, fog or obstructing objects like buildings, trees etc. That is why the search in such a region may be not enough efficient. Besides, in case of various objects and significant search area dimensions it is preferable to consider distances from UAV to individual objects.

R. Silhavy (Ed.): CSOC 2020, AISC 1226, pp. 443–452, 2020.
https://doi.org/10.1007/978-3-030-51974-2_42

Articles [19, 20] have shown that search route planning may involve information theory methods [21], which enable to consider the above-mentioned factors. In particular, a single UAV route planning involves maximum capacity criterion calculated with entropy assessment. Here every next waypoint chosen is a point (region) on the map, monitoring of which results in maximum reduction of search entropy per time unit. Article [19] has shown that the use of maximum capacity criterion helps significantly improve searching efficiency as against using a known maximum a priori probability criterion.

Further search efficiency improvement can be done by reducing initial (a priori) entropy. Meanwhile, the known works did not review issues of direct reduction of initial search entropy.

The objective of this article is improving search efficiency by means of developing initial entropy reduction methods.

These planning methods can be applied for different search tasks such as searching for suffering distress; military reconnaissance missions etc. The usefulness of the methods includes increasing search efficiency and reduction of resources involved in search operations.

2 Methods

In general, the informational description of the process can be presented as

$$H_f = H_b - I = H_b - C \cdot T \tag{1}$$

where H_f is end entropy of search, H_b is initial (a priori) entropy of search, I is useful information of an object of interest obtained during search time T, C is amount of useful information received by a searching system per time unit (capacity).

$$H_b = -\sum_{n=1}^{N} P_n \log_2 P_n \tag{2}$$

$$H_f = -\sum_{m=1}^{M} P_m \log_2 P_m \tag{3}$$

where $n = 1 \ldots N$ are possible outcomes before the searching begins, $m = 1 \ldots M$ are possible outcomes after the searching completed, $P_{n\,\mathrm{or}\,m}$ is probability of n or m outcomes.

For example, for the case of equal distribution of probabilities $P_n = const = \frac{1}{N}$ entropy is calculated as:

$$H_b = \log_2 N \tag{4}$$

In case of given requirements for search results such as probability of correct detection of an object searched the limits (H_{fr}) shall be set for end entropy as follows:

$$H_f \leq H_{fr} \tag{5}$$

from which, with H_b and C known, we may obtain the search time required:

$$T_r = \frac{H_b - H_{fr}}{C}. \tag{6}$$

The reduction of a priori entropy H_b shall obviously lead to reduction of search time T, which often is a minimized efficiency criterion.

As it is seen from (2), (4) value H_b is defined by a value of possible outcomes N and distribution of probabilities. In particular, N defines possible (a priori) number of positions of objects involved. In worst case (at equiprobable distribution) and given that the search area S is available, as per (4), we obtain:

$$H_b = \log_2 \frac{S}{\Delta s} \tag{5}$$

where Δs is an object surface.

3 UAV Flight Path Planning

In a real-time secondary search UAV defines next waypoint (region) at every control stage. In particular, such waypoint or region may be a point with a maximum a priori object presence probability. Upon arrival at a destination region UAV is to inspect it. At visual search UAV equipment receives an image of an area surveyed, where the object marks are searched. According to values of the marks a decision is made as to whether the object involved is present or not. The search process can be followed by choosing the next waypoint. The search is to be continued till the whole given search area is surveyed, the given search period is over or other limits are executed.

This work offers using an approach based on maximizing capacity (useful information received) at every k stage of the search [19, 20]

$$\max C_k(x,y) = \frac{H_k(O(x,y)) - H_k(O(x,y|u_r))}{T_k}$$

where $u_r \in U_r = u_{r,1}, u_{r,2}\ldots, u_{r,i}, j = 1..M$ is a mark, the value of which is used to make a decision, T_k is a forecasted UAV flight time at k stage of path planning, $O(x,y)$ is a event in point x, y: the search object presence or not.

The difference of entropies in a numerator of formula (6) corresponds to amount of useful information of object position in point x, y, with mark u_r obtained.

Entropy $H_k[O(x,y)]$ is an initial entropy of the search k stage and calculated by formula (2).

Posterior entropy helps consider various particularities of a search situation and can be calculated [22] as follows

$$\begin{cases} H_k(O(x,y)|u_r) = -\sum_{m=1}^{N} P(O_m(x,y)) \sum_{l=1}^{N}\sum_{j=1}^{M} P(u_{r,j}|O_m(x,y)) \cdot H_{l,r,j} \\ H_{l,r,j} = P(O_l(x,y)|u_{r,j}) \cdot \log_2 P(O_l(x,y)|u_{r,j}) \end{cases}$$

where N is a total number of possible outcomes with a probability of P_m, $P(u_{r,j}|O_m(x,y))$ is a probability of obtaining of k value of mark $u_{r,i}$, given that the event $O_m(x,y)$ has happened, $P(O_l(x,y)|u_{r,j})$ is a probability of $O_l(x,y)$ event, given that j value of mark u_r is obtained. This probability can be obtained using Bayes' formula:

$$P(O_l(x,y)|u_{r,j}) = \frac{P(O_l(x,y)) \cdot P(u_{r,j}|O_l(x,y))}{\sum_{n=1}^{N} P(O_n(x,y)) \cdot P(u_{r,j}|O_n(x,y))}$$

Probabilities $P(u_{r,j}|O(x,y))$ are calculated through integration of given dencities of mark u_r distribution within the limits defined by corresponding detection thresholds [22]. Selection of a next waypoint x^*, y^* corresponds to the fulfillment of the condition (3).

Planning methods include the following procedures:

1. Defining initial conditions: number of objects M for searching, prior distribution of $P(O_m(x,y))$, map of observability (as u_r distribution dencities) number of UAVs and their initial position.
2. Evaluating information capacity map I for each object of M (4),
3. Calculating capacity $C(x,y)$ based on a current position of every UAV.
4. Choosing a waypoint by each UAV in turns according to (6), provided that a waypoint was before chosen by another UAV its $C(x,y) = 0$.
5. Performing flight and inspection of points chosen. Make a decision of presence the searching object.
6. Recalculating $P(O_m(x,y))$ (if object presence $P(O_m(x,y)) = 1$, otherwise Bayes' formula is using) and repeating procedures pp 2–5 as one of UAVs has surveyed its point.
7. The process is to continue till all objects M are found.

4 Method of Describing the Search Situation

It is possible to reduce the initial entropy by clarifying the initial estimates of the reliability distribution $P(O_m(x,y))$ of the appearance of an object in different places in the search area. Clearly, the following factors need to be analyzed in order to obtain these estimates:

- The interaction of a mobile object and the environment ("*Where* the object can *pass*?");
- The target of the search object ("*Where* and *how* is the search object moving?").

The combination of these factors will be called a description of the search situation – Q. In turn, the reliability of finding an object in a certain search area will be a function of this description – $P(O_m(x, y)) = F(Q(x, y))$, and the description itself will take the form:

$$Q = \langle SO, Env \rangle$$

where SO – is a description of an searching object, Env – is a description of the environment.

To generate descriptions and obtain estimates of $P(O_m(x, y))$ we suggest using the situation analysis approach. During the analysis, fuzzy rules are checked sequentially from a pre-prepared knowledge base, for example:

IF *landscape.roughness* = *"high"* ***and*** *object.type* = *"passenger",* ***then*** P_n= *"low»*

IF *landscape.roughness* = *"high"* ***and*** *object.type* = *"tractor",* ***then*** P_n= *"high»*

IF *landscape.ground_state* = *"solid"* ***and*** *object.type* = *"passenger",* ***then*** P_n= *" high»*

The basic rules link the environment in a certain area (for example, landscape) and the search object to the reliability of its presence in it. However, as shown in the example above, rule parameters can be fuzzy. The result of the rules is also fuzzy (it takes the values "*Very Low*", "*Low*", " *Average*", "*High*", "*Very High*"). The conversion to confidence values occurs according to known fuzzy inference rules.

Parameter values are entered in the rules from the situation description. Some parameters (for example, descriptions of the search object, surface types) are entered in advance as a prior information, some are obtained from external sources or sensors (for example, information about weather conditions), others are formed by additional fuzzy rules, for example:

If *landscape.type* = *"ground"* ***and*** *fallout.rain* = *"strong",* ***then*** *landscape.-ground_state* = *"viscous»*

5 Results

Let us review an area, where a ground vehicle is searched (Fig. 1).

On the picture paved roads (highway) are shown in red; service roads, which are impassable by passenger vehicles in particular conditions, are shown in yellow; cross-country areas passable only by tracked vehicles are not marked.

The total area of the spot is around $2 \cdot 10^6$ m^2. The total area of motor ways in the region is $4.81 \cdot 10^4$ m^2, where $2.98 \cdot 10^4$ m^2 is a highway; $1.81 \cdot 10^4$ m^2 are service roads.

It is known that search performance W_P is calculated as follows:

$$W_P = V_{UAV} \cdot W_{obs}$$

where V_{UAV} is UAV ground speed; W_{obs} is observation width. Let us make an assumption that the UAV moves in a uniform manner at a constant altitude h_{UAV} at a

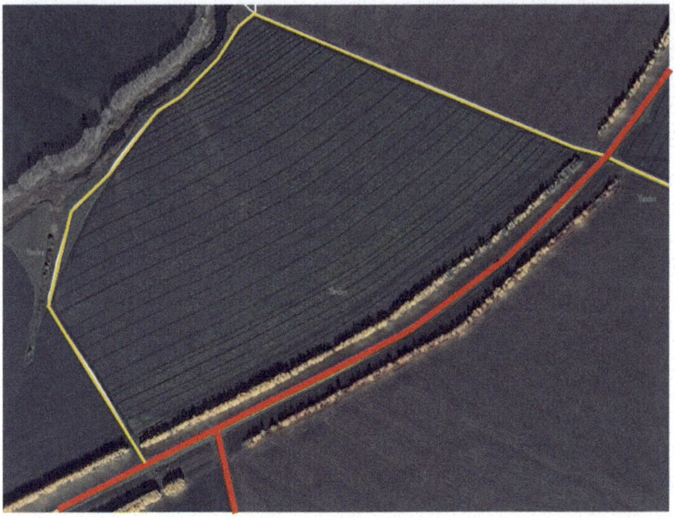

Fig. 1. Search area

speed of $V_{UAV} = 33\text{m/s}$. We also assume that the vision angle of the computer vision system is 60°, then $W_{obs} \cong h_{UAV}$. At $h_{UAV} = 100$ m $W_P = 3300\,\text{m}^2/\text{s}$. Let us now divide the search area into equal plots 100 × 100 m, then time to fly over and inspect a single plot $T_{obs} = 3$ s. Let us consider the search for three possible object types, which are passenger car, off-road vehicle and tracked vehicle (TV).

Let us assume that the external conditions are such that the passenger car cannot pass along service roads and cross-country terrain, then the distribution of probabilities of possible presence shall appear as shown in Fig. 2.

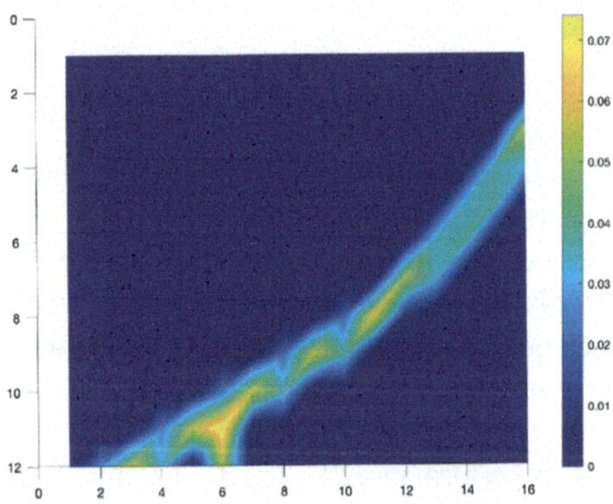

Fig. 2. Probability distribution of presence the passenger car on the search area

We assume the search operating area is an area, where the object presence probability is as follows $P(O(x, y)) \geq 0.005$.

The off-road vehicle can pass along service roads but it is not capable of moving along cross-country terrain. In this case the distribution of probabilities of possible presence shall appear as shown in Fig. 3.

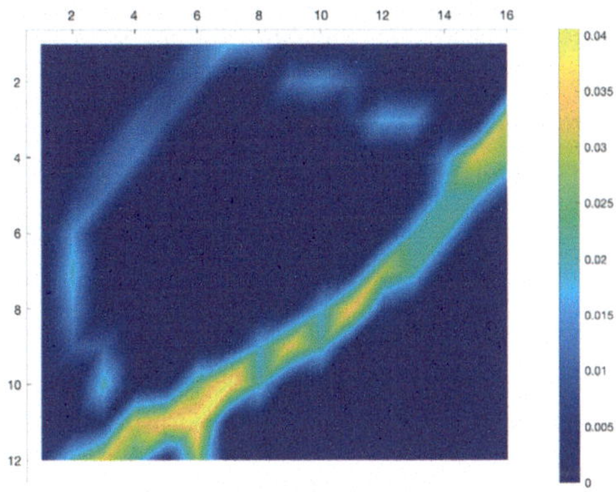

Fig. 3. Probability distribution of presence the off-road vehicle search area on the search area

TV can appear in any region with equal probability. In this case the search initial entropy is calculated using formula (2).

Using the complete search method provided in [19] we shall obtain results shown in Fig. 4. Here the cases are reviewed when all regions of the search area are inspected, where the objects may be found.

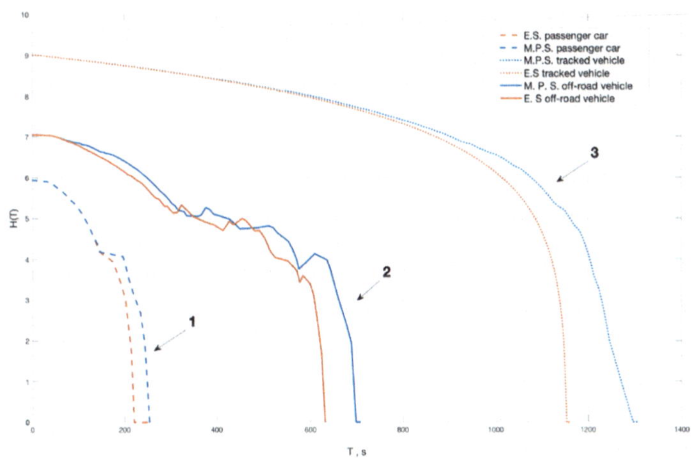

Fig. 4. Results of modeling of search process as to different vehicles: 1) passenger car; 2) off-road vehicle; 3) tracked vehicle. The red proposed approach, the blue search by maximum probability

Table 1 contains summary data of initial entropy, relative useful information obtained through case analysis as well as end time of search.

Table 1. Summary data of initial entropy.

Object searched	H_0, bit	ΔH_0, bit	T, s	ΔT, s
Tracked vehicle	9	–	1163	–
Off-road vehicle	7	2	631	532
Passenger car	5,9	3,1	220	943

6 Discussion

As shown by diagrams (Fig. 4), application of case methods helps reduce initial search entropy, which in fact means the reduction of the operating are (Table 1) and decrease search time.

The 3 type objects search modeling results show that in every case the maximum capacity search is more efficient than maximum a priori probability search. This happens because the latter case does not involve observability and distance to the next object searched.

The mentioned options of objects possible position evaluation did not consider possible starting positions, speed and moving behavior. Taking into account these factors helps reduce initial entropy but requires making corresponding data bases and knowledge bases.

7 Conclusion

- An offer has been made as to the technique of using of case methods to reduce initial (a priori) search entropy, which ensures significant decrease of operating areas inspection time.
- Initial entropy reduction corresponds to operating areas of search thanks to specifying possible object presence positions defined using production rules.
- The offered technique helps, in a quantity-related manner as evaluation of a priori entropy, assess case analysis results, which ensure capability of matching various algorithms as to ultimate search efficiency.

Acknowledgments. Financial support was provided by the Russian Foundation for Basic Research (project 19-08-00613-a).

References

1. Evdokimenkov, V.N., Kim, N.V., Kozorez, D.A., Mokrova, M.I.: Control of unmanned aerial vehicles during fire situation monitoring. INCAS Bull. **11**, 66–73 (2019)
2. Abchuk, V.A., Suzdal', V.G.: Search for Objects. Sovetskoe radio, Moscow (1977)
3. Bernd, J.: Digital Image Processing. Springer, Heidelberg (2005)
4. Lucas, B.D., Kanade, T.: An iterative image registration technique with an application to stereo vision. In: 7th International Joint Conference on Artificial Intelligence, pp. 121–130 (1981)
5. Fan, J., Wu., Y., Dai, S.: Discriminative spatial attention for robust tracking. In: Daniilidis, K., Maragos, P., Paragios,N. (eds.) Computer Vision – ECCV 2010. Lecture Notes in Computer Science, vol 6311. Springer, Heidelberg (2010)
6. Yan, X., Wu, X., Kakadiaris, I.A., Shah, S.K.: To track or to detect? an ensemble framework for optimal selection. In: Fitzgibbon, A., Lazebnik, S., Perona, P., Sato, Y., Schmid, C. (eds.) Computer Vision – ECCV 2012. LNCS, vol 7576. Springer, Heidelberg (2012)
7. Karasuru, B.: Review and evaluation of well-known methods for moving object detection and tracking in videos. J. Aeronaut. Space Technol. **4**(4), 11–22 (2010)
8. Yilmaz, A., Javed, O., Shah, M.: Object tracking: a survey. ACM Comput. Surv. **38**(4), 45 (2006)
9. Lienhart, R., Maydt, J.: An extended set of haar-like features for rapid object detection. In: 2002 Proceedings International Conference on Image Processing, vol. 1, pp. 1. IEEE, Rochester, NY, USA (2002)
10. Türmer, S., Leitloff, J., Reinartz, P., et al.: Evaluation of selected features for car detection in aerial images. In: 2011 ISPRS Hannover Workshop, pp. 14–17, Hannover (2011)
11. Lin, Y., Yu, Q., Medioni, G.: Efficient detection and tracking of moving objects in geo-coordinates. Mach. Vis. Appl. **22**, 502–520 (2010)
12. Lin, F., Lum, K.Y., Chen, B.M., et al.: Development of a vision-based ground target detection and tracking system for a small unmanned helicopter. Sci. China Ser. F: Inf. Sci. **52**, 2201 (2009)
13. Farnebäck, G.: Very high accuracy velocity estimation using orientation tensors, parametric motion, and simultaneous segmentation of the motion field. In: Proceedings of the Eighth IEEE International Conference on Computer Vision, vol. 1, pp. 171–177, Vancouver, Canada. IEEE (2001)
14. Obermeyer, K.: Path planning for a UAV performing reconnaissance of static ground targets in Terrain. In: AIAA Guidance, Navigation, and Control Conference, Guidance, Navigation, and Control and Co-located Conferences, p. 11. Chicago, USA (2009)
15. Zhang, J., Liu, L., Wang, B., et al.: High speed automatic power line detection and tracking for a UAV-based inspection. In: International Conference on Industrial Control and Electronics Engineering (ICICEE), pp. 266–269, IEEE, Xi'an, China (2012)
16. Kim, N., Chervonenkis, M.: Situational control unmanned aerial vehicles for traffic monitoring. Mod. Appl. Scie. **9**(5), Special Issue, Canadian Center of Science and Education. ISSN (printed) 1913-1844. ISSN (electronic): 1913-1852 (2015)
17. Kim, N.V., Bodunkov, N.E., Mihaylov, N.A.: Automatic decision making by the onboard system of an unmanned aerial vehicle during traffic monitoring. Vestn. Moskovsk. Aviatsionnogo Inst. **25**(1), 99–108 (2018)
18. Amelin, K.S., Antal, E.I., Vasil'ev, V.I., et al.: Adaptive control of an autonomous group of unmanned aerial vehicles. Stokhasticheskaya Optim. Inf. **5**(1), 157–166 (2009)

19. Mikhaylov, N.A.: Route planning of search for UAV using entropy approach. In: IV Vserossi'skii nauchno-prakticheskii seminar Bespilotnie transportnie sredstva s elementami iskusstvennogo intelekta, pp. 126–135, Kazan, Russia (2017)
20. Kim, N.V., Mikhaylov, N.A.: An Entropy approach in solving the search problem by a UAV group. Stanki i Instrument 9, pp. 28–31 (2019)
21. Kogan, I.M.: Applied Information Theory. Radio i svyaz', Moscow (1981)
22. Gorelik, A.L., Skripkin, V.A.: Recognition Methods. Nauka, Moscow (2004)

Optimization of Classification Thresholds for States of Transionospheric Radio Links Described by the Normal Distribution for Ensuring the Accuracy of UAV Positioning

Gennadiy Ivanovich Linets, Sergey Vladimirovich Melnikov(✉), and Alexander Mikhailovich Isaev

North-Caucasus Federal University, Stavropol, Russia
kbytw@mail.ru

Abstract. Nowadays the development of satellite navigation systems is largely integrated into modern society. The GPS (USA) and GLONASS (Russia) navigation systems are most commonly used. However, all the satellite navigation systems are negatively affected by the artificial irregularities arising in the ionosphere, which introduce the greatest error in the accuracy of an unmanned aerial vehicle (UAV) positioning. In such cases, radio link perturbations are defined by one of the random distributions: normal distribution, Rayleigh distribution, Rice distribution, Nakagami distribution. Existing methods of counteracting the perturbations cannot adequately suit the emerging practical needs since they are based on the predictive models. One of the promising ways to solve this problem is using of the methods of automated monitoring and control of satellite links state. The control and monitoring of such systems is determined by solving a problem of recognition and classification of an object, but such approach leads to inevitable occurrence of type I (false alarm) and type II (anomaly undetection) errors. In order to minimize the anomaly undetection errors when crossing classes, it is necessary to determine the classification threshold.. In this case, it is possible to solve the problem mathematically using the available values of the system parameters. The solution to the problem is expressed with simultaneous equations which are put in a convenient form for solving by three quasi-Newtonian methods that allow to simplify the algorithm (Powell, Broyden, and Krylov-Newton). #CSOC1120.

Keywords: GPS · GLONASS · UAV · Type I errors · Type II errors · Normal distribution law

1 Introduction

One of the factors affecting the positioning accuracy of robotic unmanned aerial vehicles (UAV) is the state of satellite links, the accuracy of which is described by the selected classification thresholds. It is known [1] that the possible link states of the satellite navigation systems are described by the following random distributions: normal, Rice, Nakagami, Rayleigh. The choice of one of these random distributions is

R. Silhavy (Ed.): CSOC 2020, AISC 1226, pp. 453–469, 2020.
https://doi.org/10.1007/978-3-030-51974-2_43

determined by the current state of the ionosphere and depends on a number of natural and artificial ionospheric irregularities having different physical nature. It should be noted that the change in the ionosphere state is a random stochastic process which involves monitoring ionosphere parameters with the required frequency in real time. In [2], it was shown that the UAV positioning error depends on the correct operation of the GPS/GLONASS system in determining pseudorange and on using the current constellation of ephemeris which are characterized by the parameters that describe the geometric relative position of the satellites. It has been established that the selection of the current ionosphere state model (random distribution) and processing the transionospheric link data should be performed using the ground control centers which have large power and information resources in comparison with the UAV. In the update of the current state of the links having a significant impact on the UAV positioning accuracy, the problem of determining the optimal classification thresholds in order to minimize the probability of a significant ionospheric change undetection arises. There are several papers in which the quality of satellite links is studied [3], however, the problem of determining the optimal classification thresholds in case of significant changes in the ionosphere properties was inadequately treated. In contrast to the aforementioned works, this article solves the problem of determining the optimal classification thresholds for a transionospheric radio link with the normal distributed state.

The scope of the UAV has expanded significantly over the past few years. Alongside with the military industry, UAVs are widely used in solving civilian problems [4]: from agricultural needs to infrastructure health examination of various objects (buildings, power lines, etc.). In economic terms, the use of light-class UAV with a weight of up to 10 kg is the most optimal.

The positioning problem of a quadcopter is solved using GPS/GLONASS global satellite navigation systems. It is known that the artificial ionospheric irregularities (AII), which have different origins, have a significant effect on the short-wave signal propagation [1, 5–7].

Due to the GPS/GLONASS signal distortion, it becomes necessary to use methods increasing the accuracy of the UAV positioning. A number of effective methods to reduce the AII impact on the accuracy of a GPS/GLONASS receiver positioning, such as DGPS or the method proposed in [2, 3], involve the use of stationary information systems. The functions of these systems are to monitor the ionosphere state and to compute the correction components of the GPS/GLONASS signal. However, these solutions do not provide the necessary update rate of the relevant data. Computation of the GPS/GLONASS signal correction components directly in the area of use of an UAV is the most effective solution taking into account the ionosphere inhomogeneity and its AIIs which is not currently applied in practice.

During the operation of the UAV, a flight control center (FCC) is deployed in the landing/launch area. Due to the stationarity of the FCC during the operation of the UAV, it is reasonable to implement computation of the correction components on its basis. The advantage of this approach is the provision of the necessary update rate of latest information which will allow to maintain the mobility and positioning accuracy of the FCC and the UAV. In this case, it is reasonable to change the parameters of the satellite link in order to achieve its optimal state. This will reduce the ionospheric impact on the accuracy of the FCC positioning. It should be noted that the use of such a system on the UAV board is cost-inefficient due to the load capacity restrictions of UAV.

It was shown [8] that modern space communication systems operate under great uncertainty, and the information about a number of their parameters may be inaccurate. The stochastic nature of the ionosphere development reduces the effectiveness of manual adjustment of the radio link parameters. The most natural solution to this problem is to use an adaptive system for monitoring and control of the state of transionospheric radio links. Similar solutions are used in the radio links operating in other frequency bands [10]. The adaptive monitoring system allows to change the radio link properties in an optimal way with the required rate to keep the radio links in the working condition when exposed to ionospheric perturbations.

The monitoring and control system should have the following functions: identification of the control target state; development of the control actions according to the control goals and taking into account the current environment state; application of the control action to the control target. In order to identify the state of complex objects, pattern recognition methods should be used including machine learning methods.

It was shown in [11] that when solving the identification problem, the occurrence of the type I and type II errors are inevitable. The article proposes a solution to the problem of optimizing thresholds for classification of the type I and type II errors of the system monitoring and controlling the transionospheric radio link state in order to minimize the anomaly undetection probability described by the normal distribution.

2 Methods

Satellite links are non-stationary. By nonstationarity of satellite links the change in the link state over time is meant. Two devices are used to identify the radio link state, the first one is a detector which allows to detect the anomaly data, the second one is a recognizer which allows to classify the radio link states.

A change in the radio link properties affects a change in the radio link state. The common signal state is characterized by one of the random distributions depending on reflection, diffraction, and scattering of the signal. Some types of signal fluctuations can lead the signal to a normal distribution.

The identification system recognizes the radio link state using the properties given. The detection device draws a sample on a complex attribute x, and the recognition device determines belonging to the corresponding class.

In this case, errors in the classification of the radio link state recognition and the anomaly detection are possible along the change in the ionosphere properties.

It is required to optimize the probability of type I and type II errors. The anomaly undetection is the most undesirable error, which means a change in the radio link performance when the GPS/GLONASS signals' error probability is greater than usual.

For this, in the general case, the recognition problem will be reduced to the testing of a number of hypotheses $B_1, B_2, \ldots, B_i, \ldots B_k$ where B_i is the hypothesis assuming that the object belongs to the A_i class. We assume that the a priori probability distributions of these hypotheses, i.e. the probability values of belonging an object $P(B_i)$ to a

A_i class, are given. Assuming $\sum_{i=1}^{k} P(B_i) = 1$ since the object must belong to a certain class. Under this condition, the distribution density is $p_i(x) = p(x_i/B_i)$.

The identification system uses two hypotheses $B_1 = N$ and $B_2 = \overline{N}$ with the corresponding a priori probabilities of the occurrence of a normal $p_1 = p(B_1) = p(N)$ and anomalous $p_2 = p(B_2) = p(\overline{N})$ situations in the system. Also $p_1 + p_2 = 1$.

We use the Neumann-Pearson criterion as a decision function ensuring the highest accuracy of the identification system. We define the false alarm probability at a constant level C and require a minimum error of undetection of the system operation mode violation [1, 10]. Then

$$P_{np}^{\min} = \min p_2 \beta(x_0) \tag{1}$$

with the constraint of:

$$P_{n.m.} = p_1 \alpha(x_0) = C = const \tag{2}$$

where $\alpha(x_0)$ are type I errors; $\beta(x_0)$ are type II errors.

The distribution densities of the x and y features are set in the normal distribution form Fig. 1a and b.

The type I errors ("false alarm" error, Fig. 1a) and the type II errors ("anomaly undetection", Fig. 1, in the bottom):

$$\alpha(x_0) = \int_{x_0}^{\infty} f(x/N)dx, \tag{3}$$

$$\beta(x_0) = \int_{-\infty}^{x_0} f(x/\overline{N})dx. \tag{4}$$

The expressions of distribution densities of the attribute x of proper $f(x/N) = f_1(x)$ and faulty $f(x/\overline{N}) = f_2(x)$ system functioning are in the following form. Then the type I and type II errors of a detector (first stage) will be respectively equal [1]:

$$\alpha_{dtr} = \int_{x_0}^{\infty} f_1(x)dx; \beta_{dtr} = \int_{-\infty}^{x_0} f_2(x)dx. \tag{5}$$

Type I and type II errors for the recognizer are determined in a similar way (second stage)

$$\alpha_{rec} = \int_{y_0}^{\infty} f_1(y)dy; \beta_{rec} = \int_{-\infty}^{y_0} f_2(y)dy \tag{6}$$

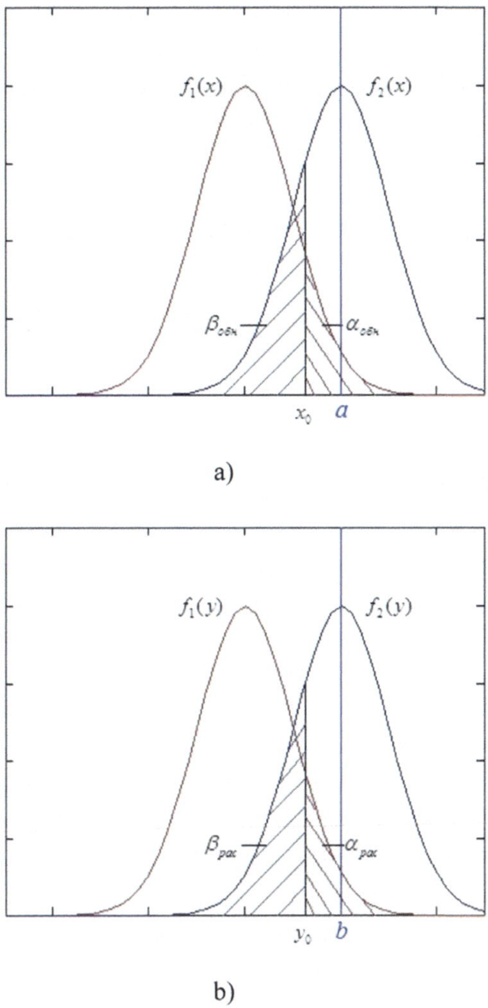

Fig. 1. Probability density functions of x and y attributes.

The distribution densities of attributes x and y for both detection and recognition devices are:

$$f_1(x) = \frac{1}{\sqrt{2\pi}} \cdot e^{-x^2/2}; f_2(x) = \frac{1}{\sqrt{2\pi}} \cdot e^{-(x-a)^2/2};$$

$$f_1(y) = \frac{1}{\sqrt{2\pi}} \cdot e^{-y^2/2}; f_2(y) = \frac{1}{\sqrt{2\pi}} \cdot e^{-(y-b)^2/2}.$$

It is necessary to find the optimal thresholds for the x_0 and y_0 classification.

Taking into account formulas (5) and (6), formulas (1) and (2):

$$P_{f.a.} = p_1 \int_{x_0}^{\infty} f_1(x)dx \int_{y_0}^{\infty} f_1(y)dy = C = const; \tag{7}$$

$$P_{om}^{\min} = \min p_2(1 - \int_{x_0}^{\infty} f_2(x)dx \int_{y_0}^{\infty} f_2(y)dy). \tag{8}$$

Since the thresholds x_0 and y_0 in the expression (7) are related by one functional relationship $x_0 = \varphi(y_0)$, after differentiating (8) with respect to y_0 and setting it to zero, we obtain

$$\frac{dx_0}{dy_0} \cdot f_2(x_0) \cdot \int_{y_0}^{\infty} f_2(y)dy - f_2(y_0) \cdot \int_{x_0}^{\infty} f_2(x)dx = 0. \tag{9}$$

The general statement of the problem of optimizing the (x_0, y_0) classification thresholds is reduced to solving the simultaneous equations:

$$\begin{cases} \int_{x_0}^{\infty} f_1(x)dx \cdot \int_{y_0}^{\infty} f_1(y)dy = C/p_1 \\ \frac{dx_0}{dy_0} \cdot f_2(x_0) \cdot \int_{y_0}^{\infty} f_2(y)dy - f_2(y_0) \cdot \int_{x_0}^{\infty} f_2(x)dx = 0 \end{cases}. \tag{10}$$

with the constraints of: $p_1 = const$ at $p_1 + p_2 = 1$.

It is required to determine the optimal thresholds of the x_0 and y_0 classifications from the solution of the simultaneous Eq. (10).

To solve the simultaneous Eq. (10), we use the normalized distribution function in the following form [7]:

$$\varphi(x) = \frac{1}{\sqrt{2\pi}} \cdot \int_{x}^{\infty} e^{-u^2/2} du$$

The relations (7) for this function:

$$\varphi(x) = \frac{1}{2} - \varphi_0(x), \tag{11}$$

where

$$\varphi_0(x) = \frac{1}{\sqrt{2\pi}} \cdot \int\limits_0^x e^{-u^2/2} du$$

is the Laplace function.

The following relations are determined for this function:

$$\int\limits_x^\infty f_2(u)du = \int\limits_{x-a}^\infty f_1(u)du = \varphi(x-a), \tag{12}$$

$$(\varphi(x))'_x = -\frac{1}{\sqrt{2\pi}} \cdot e^{-x^2/2}. \tag{13}$$

For a further solution based on (12), we present the simultaneous Eq. (10) in a convenient form:

$$\begin{cases} \varphi(x_0) \cdot \varphi(y_0) = C/p_1 \\ \varphi(x_0 - a) \cdot \varphi(y_0 - b) \to \max \end{cases}. \tag{14}$$

Applying equality: $\frac{x+y}{2} \geq \sqrt{x \cdot y}$, where $x > 0$, $y > 0$ and $x + y = const$, and $\max(x, y)$ is achieved at $x = y$. Then the simultaneous Eq. (14) takes the form:

$$\varphi(x_0) \cdot \varphi(y_0) = C^*, \tag{15}$$

$$x_0 - a = y_0 - b. \tag{16}$$

Where we get:

$$\Delta = a - b$$

$$x_0 = y_0 + \Delta, \tag{17}$$

$$\varphi(y_0 + \Delta) \cdot \varphi(y_0) = C* = const. \tag{18}$$

We proceed with the approximate solution of Eqs. (10) by the Newton method for simultaneous nonlinear equations.

Differentiating the Eq. (11) with respect to x, we obtain:

$$\varphi'(x) = (\frac{1}{2} - \varphi_0(x))'_x = -(\frac{1}{\sqrt{2\pi}} \cdot \int_0^{y_0} e^{\frac{-u^2}{2}} du)'_x = -\frac{1}{\sqrt{2\pi}} e^{\frac{-u^2}{2}} u.$$

According to the formulas (11), (12) and (13), the Eqs. (10) take the form:

$$\begin{cases} \varphi(x_0) \cdot \varphi(y_0) = \frac{C}{p_1} \\ \frac{dx_0}{dy_0} \cdot \frac{1}{\sqrt{2\pi}} \cdot e^{\frac{-(x_0-a)^2}{2}} \cdot \varphi(y_0 - b) - \frac{1}{\sqrt{2\pi}} \cdot e^{\frac{-(y_0-a)^2}{2}} \cdot \varphi(x_0 - a) = 0 \end{cases} \tag{19}$$

Differentiating the first of equation of the (14) with respect to y_0, we obtain:

$$(\varphi(x_0) \cdot \varphi(y_0))'_{y_0} = (\frac{C}{p})'_{y_0},$$

or

$$(\varphi(x_0))'_{y_0} \cdot \varphi(y_0) + \varphi(x_0) \cdot \varphi(y_0))'_{y_0} = 0.$$

Given that $x_0 = x_0(y_0)$, we have

$$(\varphi(x_0))_{x_0} \cdot (x_0)_{y_0} \cdot \varphi(y_0) + \varphi(x_0) \cdot (\varphi(y_0))_{y_0} = 0 \tag{20}$$

According to (13), equation transformed to:

$$-\frac{1}{\sqrt{2\pi}} \cdot e^{-\frac{x_0^2}{2}} \cdot \frac{dx_0}{dy_0} \cdot \varphi(y_0) - \frac{1}{\sqrt{2\pi}} \cdot e^{-\frac{y_0^2}{2}} \cdot \varphi(x_0) = 0.$$

From equations, we obtain

$$\frac{dx_0}{dy_0} = \frac{e^{-\frac{y_0^2}{2}}}{e^{-\frac{x_0^2}{2}}} \cdot \frac{\varphi(x_0)}{\varphi(y_0)} \tag{21}$$

Expressions (21) are substituted into the Eqs. (19), then the second equation is transformed into:

$$\frac{e^{-\frac{y_0^2}{2}}}{e^{-\frac{x_0^2}{2}}} \cdot \frac{\varphi(x_0)}{\varphi(y_0)} \cdot \frac{1}{\sqrt{2\pi}} \cdot e^{-\frac{(x_0-a)^2}{2}} \cdot \varphi(y_0 - b) - \frac{1}{\sqrt{2\pi}} \cdot e^{-\frac{(y_0-b)^2}{2}} \cdot \varphi(x_0 - a) = 0$$

Which is equivalent to

$$e^{-\frac{x_0^2}{2}} \cdot e^{-\frac{(x_0-a)^2}{2}} \cdot \varphi(x_0) \cdot \varphi(y_0 - b) - e^{-\frac{(y_0-b)^2}{2}} \cdot \varphi(y_0) \cdot \varphi(x_0 - a) = 0$$

Given that:

$$\frac{x_0}{2} - \frac{(x_0 - a)^2}{2} = x_0 \cdot a - \frac{a^2}{2}$$

and

$$\frac{y_0}{2} - \frac{(y_0 - b)^2}{2} = y_0 \cdot b - \frac{b^2}{2},$$

then the latter will be:

$$-e^{x_0 \cdot a - \frac{a^2}{2}} \cdot \varphi(x_0) \cdot \varphi(y_0 - b) - e^{y_0 \cdot b - \frac{b^2}{2}} \cdot \varphi(y_0) \cdot \varphi(x_0 - a) = 0$$

Then the Eqs. (19) are transformed to the following:

$$\begin{cases} \varphi(x_0) \cdot \varphi(y_0) = \frac{C}{p_1} \\ -e^{x_0 \cdot a - \frac{a^2}{2}} \cdot \varphi(x_0) \cdot \varphi(y_0 - b) - e^{y_0 \cdot b - \frac{b^2}{2}} \cdot \varphi(y_0) \cdot \varphi(x_0 - a) = 0 \end{cases} \tag{22}$$

The following designations are to be introduced:

$$\begin{cases} F_1(x_0, y_0) = \varphi(x_0) \cdot \varphi(y_0) - \frac{C}{p_1} = 0 \\ F_2(x_0, y_0) = -e^{x_0 \cdot a - \frac{a^2}{2}} \cdot \varphi(x_0) \cdot \varphi(y_0 - b) - e^{y_0 \cdot b - \frac{b^2}{2}} \cdot \varphi(y_0) \cdot \varphi(x_0 - a) = 0 \end{cases} \tag{23}$$

The solution being found should be as [11]:

$$\overline{X}_{(n+1)} = \overline{X}_{(n)} - W^{-1}(\overline{X}_{(n)}) \cdot \overline{F}(X_{(n)}) \tag{24}$$

where

$$\overline{X}_{(n)} = \begin{pmatrix} x^0_{(n)} \\ \hline y^0_{(n)} \end{pmatrix}; \; \overline{F}(X_{(n)}) = \begin{pmatrix} F_1(x^0_{(n)}, y^0_{(n)}) \\ \hline F_2(x^0_{(n)}, y^0_{(n)}) \end{pmatrix}.$$

Then

$$W^{-1}(x^0_{(n)}, y^0_{(n)}) = \frac{1}{\left| W(x^0_{(n)}, y^0_{(n)}) \right|} \cdot \begin{pmatrix} D & -B \\ -C & A \end{pmatrix}, \tag{25}$$

where $A = \left.\frac{\partial F_1}{\partial x_0}\right|_{\substack{x_0 = x^0_{(n)} \\ y_0 = y^0_{(n)}}}$, $B = \left.\frac{\partial F_1}{\partial y_0}\right|_{\substack{x_0 = x^0_{(n)} \\ y_0 = y^0_{(n)}}}$, $C = \left.\frac{\partial F_2}{\partial x_0}\right|_{\substack{x_0 = x^0_{(n)} \\ y_0 = y^0_{(n)}}}$, $D = \left.\frac{\partial F_2}{\partial y_0}\right|_{\substack{x_0 = x^0_{(n)} \\ y_0 = y^0_{(n)}}}$.

The results of differentiation of the Jacobian determinant [11]:

$$\frac{\partial F_1}{\partial x_0} = \left(-\frac{1}{\sqrt{2\pi}} \cdot e^{-x_0^2/2}\right) \cdot \left(0.5 - \frac{1}{\sqrt{2\pi}} \int\limits_0^{y_0} e^{-u^2/2} du\right);$$

$$\frac{\partial F_1}{\partial y_0} = \left(-\frac{1}{\sqrt{2\pi}} \cdot e^{-y_0^2/2}\right) \cdot \left(0.5 - \frac{1}{\sqrt{2\pi}} \int_0^{x_0} e^{-u^2/2} du\right);$$

$$\frac{\partial F_2}{\partial x_0} = a \cdot e^{x_0 \cdot a - \frac{a^2}{2}} \cdot \varphi(x_0) \cdot \varphi(y_0 - b) + \frac{1}{\sqrt{2\pi}} \cdot e^{-\frac{(x_0-a)^2}{2}} \cdot (e^{y_0 \cdot b - \frac{b^2}{2}} \cdot \varphi(y_0) \cdot \varphi(y_0 - b));$$

$$\frac{\partial F_2}{\partial y_0} = \frac{1}{\sqrt{2\pi}} \cdot e^{-\frac{(y_0-b)^2}{2}} \cdot \varphi(x_0 - a) - e^{x_0 \cdot a - \frac{a^2}{2}} \cdot \varphi(x_0) - b \cdot e^{y_0 \cdot b - \frac{b^2}{2}} \cdot \varphi(y_0) \cdot \varphi(x_0 - a))$$

Thus, the Jacobian determinant at a point (x_0, y_0) is:

$$W(x_0, y_0) = \begin{pmatrix} \frac{\partial F_1}{\partial x_0} & \frac{\partial F_1}{\partial y_0} \\ \frac{\partial F_2}{\partial x_0} & \frac{\partial F_2}{\partial y_0} \end{pmatrix},$$

where

$$\varphi(x) = \frac{1}{2} - \varphi(x_0) = \frac{1}{2} - \frac{1}{\sqrt{2\pi}} \cdot \int_0^x e^{-t^2/2} dt.$$

The matrix inverse to the Jacobian matrix is found according to [11]:

$$W^{-1}(x_0, y_0) = \frac{1}{|W(x_0, y_0)|} \cdot \begin{pmatrix} D & -B \\ -C & A \end{pmatrix},$$

$$|W(x_0, y_0)| = AD - BC \tag{26}$$

There are several solving methods to solve non-linear equations (or simultaneous non-linear equations): graphical, analytical and numerical methods.

Graphical methods are the least accurate, but they allow one to determine the most approximate values in complex equations which can be used further for finding more accurate solutions to equations.

Analytical methods (or direct methods) allow us to determine the exact solution values of the equations. This method allows to represent the roots in a ratio form (with formulas). However, the vast majority of nonlinear equations encountered in practice cannot be solved by direct methods. In such cases, numerical methods are used that make it possible to obtain an approximate root value with any given accuracy ξ.

Using a numerical methods for solving nonlinear equations is an iterative computation process which consists in sequentially refinement of an initial approximation of the root values of the equation (or the imultaneous equations).

In the numerical approach, the problem of solving nonlinear equations is divided into two stages:

1. root localization (isolating);
2. root refinement.

Root localization refers to the process of finding the approximate root value or finding such intervals within which a unique solution is contained

Root refinement refers to the process of approximation of the root values with a given accuracy by any numerical method for solving nonlinear equations.

The disadvantage of most iterative methods for finding roots is that they can find only one root of the function when used once, and it is not known which one.

Quasi-Newtonian methods are also based on an iterative formula, but these methods are simpler to implement and less demanding on computing resources. All quasi-Newtonian methods are the first-order methods [12]. Following quasi-Newton methods were used for root localization: Powell [15], Broyden [14], and Newton-Krylov [13].

3 Results

As an example, let us determine the optimal classification thresholds for a recognizer and a detector with the normal attribute distribution for a device for detecting objects by the x attribute and a recognition device by the y attribute. The input parameter values are set to a = 1, b = 1, C = 0.05.

Let us determine the dependence of the convergence of the simultaneous equations in the necessary range of intersection of two distributions on the initial x and y values. The computation results are shown in Table 1.

Table 1. Computation results.

x, y	Roots found		
	Powell	Krylov	Broyden
(0.0, 0.0)	(1.470518583974019, 1.470518583974019)	(1.4704763171150939, 1.4704763171150934)	(1.470517022694477, 1.470524675173994)
(0.5, 0.0)	(1.4705185840396138, 1.4705185839776098)	(1.4705169257014734, 1.4705168898302041)	(1.4707324232471022, 1.4706148059637696)
(1.0, 0.0)	(1.470518583979655, 1.4705185839776904)	(1.4705065249536569, 1.4705044777389271)	(1.470469610734382, 1.4704856467491376)
(1.5, 0.0)	(1.4705185835055892, 1.4705185838762278)	(1.4705183594359037, 1.4705182826873693)	(1.4705871724984407, 1.470577042306709)
(2.0, 0.0)	(1.470518583973996, 1.470518583973372)	(1.47052966133056, 1.4704499694683175)	(1.4706464911846981, 1.4706900222357329)
(0.0, 0.5)	(1.4705185839776092, 1.4705185840396056)	(1.4705168901239067, 1.470516925373911)	(1.4702520304710196, 1.470364622709813)
(0.5, 0.5)	(1.47051858397402, 1.4705185839740187)	(1.470337860380618, 1.4703378603806176)	(1.4703385070077908, 1.4704437874613079)

(continued)

Table 1. (*continued*)

x, y	Roots found		
	Powell	Krylov	Broyden
(1.0, 0.5)	(1.4705185839741133, 1.4705185839740675)	(1.4705170359475923, 1.4705165643315945)	(1.4705137026594486, 1.47051275459333)
(1.5, 0.5)	(1.4705185839742922, 1.4705185839744725)	(1.4705186002367079, 1.4705178077379661)	(1.4705943732235285, 1.470583014188955)
(2.0, 0.5)	(1.4705185839730803, 1.4705185839739285)	(1.4705500914147824, 1.4704280861370715)	(1.4708018792832904, 1.470675167979375)
(0.0, 1.0)	(1.4705185839776809, 1.4705185839796424)	(1.4705044781977994, 1.470506524620427)	(1.4705186096459988, 1.4705197196489153)
(0.5, 1.0)	(1.4705185839740658, 1.4705185839741008)	(1.4705165644962386, 1.4705170358036683)	(1.4705024663620216, 1.4705109079989174)
(1.0, 1.0)	(1.4705185839740171, 1.4705185839740171)	(1.4704813135781367, 1.4704813135781367)	(1.4703809220671016, 1.4704465720400386)
(1.5, 1.0)	(1.4705185839662946, 1.4705185839620056)	(1.4705258989203303, 1.470495129928691)	(1.470505377634064, 1.470482204526423)
(2.0, 1.0)	(1.4705185839826154, 1.470518583967009)	(1.4705183618523348, 1.4705185958883824)	(−6254514287335.554, -23581.9397199785)
(0.0, 1.5)	(1.4705185838762294, 1.4705185835055896)	(1.4705182826338574, 1.4705183594309075)	(−6254514287335.554, −23581.9397199785)
(0.5, 1.5)	(1.4705185839744628, 1.4705185839742938)	(1.4705178077842755, 1.4705186001986115)	(1.4704267191481042, 1.470511070474073)
(1.0, 1.5)	(1.4705185839620056, 1.4705185839662946)	(1.4704951388369116, 1.4705258861368944)	(1.470531797281466, 1.4704913785399094)
(1.5, 1.5)	(1.47051858397402, 1.47051858397402)	(1.4705145924744243, 1.4705145924744243)	(1.4705973920900781, 1.4705719904059384)
(2.0, 1.5)	(1.4705185839709334, 1.470518583973861)	(1.470518360958065, 1.4705185890061587)	(−1420210.8676142222, −1481607.8575498462)
(0.0, 2.0)	(1.4705185839733712, 1.470518583973992)	(1.4704500316231959, 1.4705296436505544)	(−1420210.8676142222, −1481607.8575498462)
(0.5, 2.0)	(1.47051858397393, 1.4705185839730794)	(1.4704280858573713, 1.470550091876533)	(−1420210.8676142222, −1481607.8575498462)
(1.0, 2.0)	(1.4705185839670103, 1.4705185839826134)	(1.470518596187498, 1.4705183616218973)	(1.4706584797523, 1.470711949449169)
(1.5, 2.0)	(1.4705185839738621, 1.470518583970932)	(1.4705185889369656, 1.470518361066864)	(1.4706898255284777, 1.4705509853583683)
(2.0, 2.0)	(1.47051858397394, 1.4705185839740982)	(1.470518071378803, 1.470518071378803)	(1.4704902559676507, 1.470551326112785)

As it is shown in Table 1, the Powell and Newton-Krylov methods converge irrespective of the initial x and y values which indicates the stability of the methods with respect to such input data. However, the Broyden method is more sensitive, and it does not converge in five cases. Consider the time taken to find these roots.

In this case, the average execution time for the Powell method = 0.029647189 s, for Newton-Krylov method = 0.036272392 s, and for Broyden method = 0.607344 497 s.

We will perform an another test to choose between Newton-Krylov and Powell methods according to the obtained solutions. In order to do this, we will set the input parameters in the following order (x, y), (a, b, C) and put it in Table 2.

Table 2. Computation results of test.

Initial value	Solution		
	Powell	Krylov	Broyden
((0.52, 0.52), (0.5, 0.5, 0.005))	(1.4705185839740198, 1.4705185839740196)	(1.4703745777209132, 1.4703745777209132)	(1.4691046335316795, 1.4722747019286029)
((0.77, 0.52), (0.75, 0.5, 0.005))	(1.9955499921226, 0.7807593694896523)	(1.995523152961913, 0.7808338016124815)	(1.9951613225484763, 0.7816493075157911)
((1.02, 0.52), (1.0, 0.5, 0.005))	(2.2546151965706263, 0.21739505263950856)	(2.254615675991736, 0.21739518063353588)	(2.254664420085599, 0.21728028497807436)
((1.27, 0.52), (1.25, 0.5, 0.005))	(2.394390528678955, −0.2550876791675289)	(2.3943904286953757, −0.2550794657502552)	(2.3941254050388356, −0.25531757210909395)
((1.52, 0.52), (1.5, 0.5, 0.005))	(2.474193056400394, −0.6708854247125697)	(2.47419183361394, −0.6707970383606562)	Break
((1.77, 0.52), (1.75, 0.5, 0.005))	***(4.899722034145561, -0.19357747949594462)***	Break	Break
((2.02, 0.52), (2.0, 0.5, 0.005))	***(7.8715254096811895, -4.715990752411883)***	Break	Break
((0.52, 0.77), (0.5, 0.75, 0.005,))	(0.7807593694896637, 1.995549992122594)	(0.7808338154846147, 1.9955231444120671)	(0.7807930799399725, 1.995545504460687)
((0.77, 0.77), (0.75, 0.75, 0.005))	(1.4705185839740185, 1.4705185839740187)	(1.4705159506784877, 1.4705159506784877)	(1.4707669800900258, 1.4703961755820398)
((1.02, 0.77), (1.0, 0.75, 0.005))	(1.8707688316250257, 0.9824985960599391)	(1.870768718339154, 0.9824987490551098)	(1.8708986846828362, 0.9824481371954045)
((1.27, 0.77), (1.25, 0.75, 0.005))	(2.1163828372907423, 0.5491798554486199)	(2.1163771665594173, 0.5491923351443124)	(2.116088877823689, 0.5490368673454641)
((1.52, 0.77), (1.5, 0.75, 0.005))	(2.27395205840655, 0.16271379919979195)	(2.2739613428986343, 0.16270690839775528)	(2.274042373503607, 0.16276491309751587)
((1.77, 0.77), (1.75, 0.75, 0.005))	(2.378423438943877, −0.1894945973206386)	(2.3783980110873224, −0.18943740542685447)	(2.378463418733371, −0.18956755533909453)
((2.02, 0.77), (2.0, 0.75, 0.005))	***(6.441828997028069, 6.2108800783588105)***	Break	(2.448781976550669, −0.5194639791128046)
((0.52, 1.02), (0.5, 1.0, 0.005))	(0.21739505263950837, 2.2546151965706267)	(0.2173951723845279, 2.254615677731046)	(0.21741041262496327, 2.2545801701019266)

(*continued*)

Table 2. (*continued*)

Initial value	Solution		
	Powell	Krylov	Broyden
((0.77, 1.02), (0.75, 1.0, 0.005))	(0.9824985960599385, 1.8707688316250235)	(0.9824987521537848, 1.870768716491563)	(0.9825529518335661, 1.8707611470059207)
((1.02, 1.02), (1.0, 1.0, 0.005))	(1.4705185839740178, 1.470518583974021)	(1.4704945587386775, 1.4704945587386775)	(1.4705104300456562, 1.4705273888334798)
((1.27, 1.02), (1.25, 1.0, 0.005))	(1.7966548367608342, 1.0887149619664582)	(1.796654103044827, 1.0887154495836793)	(1.7966836166183306, 1.088598123231547)
((1.52, 1.02), (1.5, 1.0, 0.005))	(2.0216626731543528, 0.7341737790204484)	(2.021665611974986, 0.7341740694987305)	(2.0218871645748426, 0.734461031252174)
((1.77, 1.02), (1.75, 1.0, 0.005))	(2.1810359726160162, 0.40512830034043673)	(2.181035943425561, 0.4051283200104612)	(2.1810228567970626, 0.405111523485289)
((2.02, 1.02), (2.0, 1.0, 0.005))	(2.296269150264971, 0.09622754708321496)	(2.2962691497211445, 0.09622754960232836)	(2.296310003855564, 0.09620360926558824)
((0.52, 1.27), (0.5, 1.25, 0.005))	***(−0.255087679167596, 2.3943905286789637)***	***(−0.2550795340808276, 2.3943904370294877)***	***(−0.2551925928853162, 2.3942938908367832)***
((0.77, 1.27), (0.75, 1.25, 0.005))	(0.5491798554486303, 2.1163828372907387)	(0.5491922997107899, 2.116377179097693)	(0.5489860411267222, 2.1161104160732274)
((1.02, 1.27), (1.0, 1.25, 0.005))	(1.0887149619664582, 1.7966548367608342)	(1.0887154484146113, 1.7966541034579746)	(1.088740084024289, 1.796660433282746)
((1.27, 1.27), (1.25, 1.25, 0.005))	(1.4705185839740198, 1.4705185839740198)	(1.4704892111736338, 1.4704892111736338)	(1.4705236870461822, 1.4703007491645033)
((1.52, 1.27), (1.5, 1.25, 0.005))	(1.7482005254446502, 1.1535966674701683)	(1.748185073135233, 1.1537396975010938)	(1.7482700941002651, 1.1535326281717475)
((1.77, 1.27), (1.75, 1.25, 0.005))	(1.954389388016497, 0.8508974591048265)	(1.9543912803756296, 0.8508987052654585)	(1.9543658937748634, 0.8508008173250531)
((2.02, 1.27), (2.0, 1.25, 0.005))	(2.110166512096933, 0.5621898866983358)	(2.110166577375479, 0.5621896284887417)	(2.110213753400425, 0.5623148210750332)
((0.52, 1.52), (0.5, 1.5, 0.005))	**(−0.6708854247125637, 2.4741930564003947)**	**(−0.6707970109199565, 2.4741918289776548)**	**(−0.6708783846456665, 2.474159234611462)**
((0.77, 1.52), (0.75, 1.5, 0.005))	**(0.16271379919983464, 2.2739520584064845)**	**(0.1627069112535348, 2.2739613421194256)**	Break
((1.02, 1.52), (1.0, 1.5, 0.005))	(0.7341737790204425, 2.021662673154366)	(0.7341740724126367, 2.0216656109304827)	(0.7342243994714774, 2.0217186953553945)
((1.27, 1.52), (1.25, 1.5, 0.005))	(1.153596667470165, 1.7482005254446518)	(1.1537396971421583, 1.7481850736445985)	Break
((1.52, 1.52), (1.5, 1.5, 0.005))	(1.4705185839740291, 1.4705185839740291)	(1.4704850401557372, 1.4704850401557372)	(1.4705201588453776, 1.4705218109630158)
((1.77, 1.52), (1.75, 1.5, 0.005))	(1.7144567429017625, 1.1968595170872651)	(1.7144587639414448, 1.1967822742010137)	(1.714575822281423, 1.1968929984032148)
((2.02, 1.52), (2.0, 1.5, 0.005))	(1.9049918486761035, 0.9302731251155038)	(1.9049825328155037, 0.9302471953112844)	(1.9049918673044004, 0.9302730119642653)
((0.52, 1.77), (0.5, 1.75, 0.005))	(1.6204881370432689, 6.107206015902219)	(297.397157726856, −24.812282921952377)	Break

(*continued*)

Table 2. *(continued)*

Initial value	Solution		
	Powell	Krylov	Broyden
((0.77, 1.77), (0.75, 1.75, 0.005))	**(−0.1894945973206076, 2.3784234389438783)**	**(−0.18943741854292642, 2.3783980112534433)**	**(−0.189496516608766, 2.378422293266525)**
((1.02, 1.77), (1.0, 1.75, 0.005))	(0.40512830034042857, 2.1810359726160184)	(0.4051283200157589, 2.1810359434238613)	(0.40542283491332753, 2.1812463103792425)
((1.27, 1.77), (1.25, 1.75, 0.005))	(0.8508974591048181, 1.9543893880165)	(0.8508987052133384, 1.9543912803245065)	Break
((1.52, 1.77), (1.5, 1.75, 0.005))	(1.1968595170872696, 1.7144567429017592)	(1.1967822776734836, 1.7144587616978957)	(1.1968790524930548, 1.7145177158673324)
((1.77, 1.77), (1.75, 1.75, 0.005))	(1.4705185839740238, 1.4705185839740151)	(1.470518373464926, 1.470518373464926)	(1.4704362147832133, 1.4703802599303348)
((2.02, 1.77), (2.0, 1.75, 0.005))	(1.6899283961485199, 1.2273803622412478)	(1.6899273884034025, 1.227379280809866)	(1.690035408411977, 1.2274135020273504)
((0.52, 2.02), (0.5, 2.0, 0.005))	**(−1.6794260466005106, 8.127533398512524)**	Break	**(−74.59609808044135, 17.016557705040103)**
((0.77, 2.02), (0.75, 2.0, 0.005))	(0.5007495809322493, 7.516200291708267)	Break	Break
((1.02, 2.02), (1.0, 2.0, 0.005))	(0.09622754708324371, 2.296269150264973)	(0.0962275495915967, 2.2962691497237344)	(0.09624109452882525, 2.29626159512336)
((1.27, 2.02), (1.25, 2.0, 0.005))	(0.5621898866983358, 2.110166512096933)	(0.5621896283021092, 2.110166577419122)	**(−241628.57313519288, −294985.4919446978)**
((1.52, 2.02), (1.5, 2.0, 0.005))	(0.9302731251155092, 1.9049918486761015)	(0.9302471955612616, 1.9049825326481844)	Break
((1.77, 2.02), (1.75, 2.0, 0.005))	(1.227380362241245, 1.6899283961485212)	(1.2273792780577126, 1.6899273905108414)	(1.227365032356153, 1.6899090344225653)
((2.02, 2.02), (2.0, 2.0, 0.005))	(1.4705185839740218, 1.4705185839740171)	(1.47047098990193, 1.47047098990193)	(1.4705306870179513, 1.4704915433079635)
Average execution time	0,023784946	0,644701596	0,618834296

As it is shown in the Table 1, the advantage of the Powell method is obvious since, in several cases, The Newton-Krylov method terminates without finding any solution with the longer execution time. However, the Powell method finds the wrong solution for such cases (highlighted in the Table 2). Additional studies are required to identify the causes of a decrease in the stability of the simultaneous Eq. (23).

4 Conclusion

In this article, the solution for determining the optimal classification thresholds is obtained for the case when the state of the transionospheric radio links is described by the normal distribution law. The solution was obtained using quasi-Newtonian methods and has sufficient accuracy. The obtained algorithm can be used to optimize the classification of the transionospheric radio link states during fading described by the normal distribution. Further, it is necessary to optimize the classification thresholds for the ionosphere states described by the generalized four-parameter law.

Acknowledgments. These research was supported by the scientific project "Development of a robotic unmanned aerial vehicle of a multi-rotor type using a strapdown inertial navigation system" of the Federal Target Program for 2014–2020 (unique identifier RFMEFI57818X0222) with financial support from the Ministry of Science and Higher Education of Russia, on the basis of the core facilities centre of NKFU using the scientific equipment "Hardware and software complex of passive ionosphere monitoring NovAtel GPStation-6".

References

1. Pashintsev, V.P., Katkov, K.A., Gahov, R.P.: Satellite navigation during ionospheric disturbances. NCSTU, Stavropol, 259 p. (2012). (in Russian)
2. Al-Salihi, N.K.: Precise positioning in real-time using GPS-RTK signal for visually impaired people navigation, system. School of Engineering and Design Brunel University, London, United Kingdom, September 2010
3. Aleksandrovich, K.K., Petrovich, P.V., Konstantinovich, K.E.: Information system for monitoring the ionosphere. Izvestiya Samarskogo nauchnogo tsentra RAN, № 2–3 (2016). (in Russian)
4. Pashintsev, V.P., Koval, S.A., Strekozov, V.I., Bessmertnyiy, M.Y.: Detection of artificial ionospheric disturbances via satellite navigation systems. Teoriya i tehnika radiosvyazi. #1, pp. 112–116 (2013). (in Russian)
5. Korolev, E.V., Drevin, K.A., Vladimirov, V.M.: Researching the effect of ionospheric delay on determination of the pseudodality in space navigation systems alnosti navigatsionnyih kosmicheskih apparatov sputnikovyih radionavigatsionnyih sistem. Aktualnyie problemyi aviatsii i kosmonavtiki. #9 (2013). (in Russian)
6. Lobanov, B.S.: Investigation of possibility creation bulk formations in the ionosphere that effectively interact with electromagnetic radiation in the ultra-wide range of frequencies. Teoriya i tehnika radiosvyazi. #3, pp. 16–24 (2009). (In Russian)
7. Pashintsev, V.P., Gamov, M.V.: Ionospheric influence on time delay measurement of signal in satellite navigation systems. Radioelektronika, T. 45, #12, pp. 3–13 (2002). (In Russian), Advances in Intelligent Systems Research, vol. 166 217
8. Ippolito Jr., L.J.: Satellite Communications Systems Engineering: Atmospheric Effects, Satellite Link Design and System Performance, 394 p., August 2008. ISBN 978-0-470-75444-3
9. Goot, R., Mahlab, U.: Group operation of frequency adaptive radio communication systems, pp. 37–40 (1996). https://doi.org/10.1109/eeis.1996.566887
10. Bouchachia, A.: Adaptation in classification systems. In: Hassanien, A.E., Abraham, A., Herrera, F. (eds.) Foundations of Computational Intelligence Volume 2. Studies in Computational Intelligence, vol. 202. Springer, Berlin (2009)
11. Tong, X., Feng, Y., Li, J.J.: Neyman-Pearson classification algorithms and NP receiver operating characteristics. Sci. Adv. **4**(2), eaao1659 (2018). https://doi.org/10.1126/sciadv.aao1659
12. Fang, X., Ni, Q., Zeng, M.: A modified quasi-Newton method for nonlinear equations. J. Comput. Appl. Math. **328**, 44–58 (2018)

13. Yildirim, A., Kenway, G.K.W., Mader, C.A., Martins, J.R.R.A.: A Jacobian-free approximate Newton–Krylov startup strategy for RANS simulations. J. Comput. Phys. **397**, paper № 108741 (2019)
14. Gondzio, J., Sobral, F.N.C.: Quasi-Newton approaches to interior point methods for quadratic problems. Comput. Optim. Appl. **74**(1), 93–120 (2019)
15. Ghosh, D.: A davidon-fletcher-powell type Quasi-Newton method to solve fuzzy optimization problems. Commun. Comput. Inf. Sci. **655**, 232–245 (2017)

Spatial Secrecy of a Low-Frequency Satellite Communication System with a Phased Antenna Array

A. F. Chipiga[1(✉)], V. P. Pashintsev[1], G. V. Slyusarev[1], A. D. Skorik[2], and I. V. Anzin[1]

[1] North Caucasus Federal University, Stavropol, Russia
afchipiga@ncfu.ru, pashintsevp@mail.ru, van356@mail.ru
[2] Russian Institute of Powerful Radio Engineering, Saint Petersburg, Russia

Abstract. A method for evaluating the spatial secrecy of a low-orbit low-frequency satellite communication system using an onboard phased array antenna has been developed. Analysis of the results obtained shows that the use of a phased array of 4 helical antennas on board of artificial Earth satellite with an orbit height of 700 km provides a spatial secrecy coefficient of more than 7 dB on the carrier frequency of 60 MHz outside of almost the entire European part of Russia and its western borders #CSOC1120.

1 Introduction

It is known [1, 2] that problems of ensuring information security of satellite communications are usually solved with using encryption. However, modern cryptographic devices do not provide information protection when the computational capabilities of radio interception tools become more powerful [3]. Therefore, significant efforts have been made to implement the physical layer (PHY) protection method for implementing secure wireless transmission with using differences in the characteristics of communication and radio interception channels. In particular, the authors [4] presented the concept of a telephone listening channel and laid the foundation for information security at the physical level, which allows to achieve an ideal secure transmission if the quality of radio interception is lower than the quality of transmission over the communication channel. Later, this method for protecting information at the physical level was used in various data transmission channels [5, 6].

It is known [7] that one of the main ways to increase the noise immunity and energy secrecy of a satellite communication system (SCS) is to increase its spatial secrecy with using a transmitting antenna with a narrow directional pattern (DP) on an artificial earth satellite (AES).

In the works devoted to the noise immunity of on-ground wireless networks, simple scenarios were considered in which all communication means are equipped with a single antenna [8, 9]. However, the results of the use of several antennas in wireless communication networks have shown significant advantages in terms of data transfer speed and energy efficiency [10–12]. In addition, the possibilities of using multipath

R. Silhavy (Ed.): CSOC 2020, AISC 1226, pp. 470–479, 2020.
https://doi.org/10.1007/978-3-030-51974-2_44

antennas with narrow DP to improve the information security of satellite communications are being widely studied [13, 14].

However, the use of a narrowly directed transmitting antenna of the AES does not allow for high energy secrecy of the SCS when the radio intercept (RI) receiver is located close (up to 10 km) from the ground receiver of the SCS. Therefore, in [15–28], a method was developed to ensure high energy secrecy of the SCS at close placement of the RI receiver (which uses one (n = 1) antenna) with using a reduced to f_0 = 30… 100 MHz carrier frequency (in which the propagation of radio waves is accompanied by scattering on the inhomogeneities of the ionosphere, the appearance of relative phase shifts of incoming rays and deep fading of received signals) and separated signal reception on several (n = 2…4) antennas.

However, when transmitting signals using a lower (to f_0 = 30…100 MHz) carrier frequency, it is difficult to implement a narrow DP of the transmitting antenna (for example, a helical one) of the AES. Therefore, with a wide DP of the on-board low-frequency SCS's transmitter, the service area can be so extensive that the RI receiver can be placed abroad and several (n = 2…4) spaced antennas can be used there. Therefore, a methodology of evaluating the energy secrecy of low-orbit low-frequency SCS when selecting the width of DP of the transmitting antenna of the AES with zero radiation level within the borders of Russia and the arbitrary removal of the radio intercept receiver was developed in [26, 27].

In accordance with the results of this method, it is possible to provide a very high coefficient of energy secrecy of the low-frequency SCS when the radio interception receiver is arbitrarily removed. Moreover, when it is placed close, the high energy secrecy of the low-frequency SCS $S(R_r) > 22$ dB is provided with spatially spaced signal reception, and at reconnaissance distance $R_r > 740$ km it is provided with the spatial secrecy of the radiation of the transmitting antenna of the AES.

It seems obvious that it is possible to increase the spatial secrecy of a low-frequency SCS with using a phased antenna array (FAR) as a transmitting antenna on the board of the AES.

The purpose of this work is to develop a method for evaluating the spatial secrecy of a low-orbit low-frequency SCS using an onboard phased antenna array.

2 Methodology for Assessing the Spatial Secrecy of the SCS

Figure 1 shows a model for placing radio assets of SCS and radio intelligence to assess the spatial secrecy of a low-frequency SCS at an arbitrary distance (R_r) from a radio intercept receiver (RIR) to a ground-based receiver of SCS (that uses single helical antennas) due to the use of a transmitter with a FAR consisting of several (for example, four) helical antennas on a low-orbit AES (H_{AES} = 700 … 1500 km).

Let's analyze the energy and spatial secrecy of a low-frequency SCS using FAR (Fig. 1).

Recall [15–23, 26, 27], that the noise immunity of the SCS is provided if the actual signal-to-noise ratio h^2 at the receiver input is $h^2 \geq h^2_{all}$ equal or higher than the allowed value h^2_{all} (which provides an error probability equal to the value $P_{err} = P_{err\,all} = 10^{-5}$ allowed in the SCS). The condition for ensuring the energy secrecy of the SCS is that

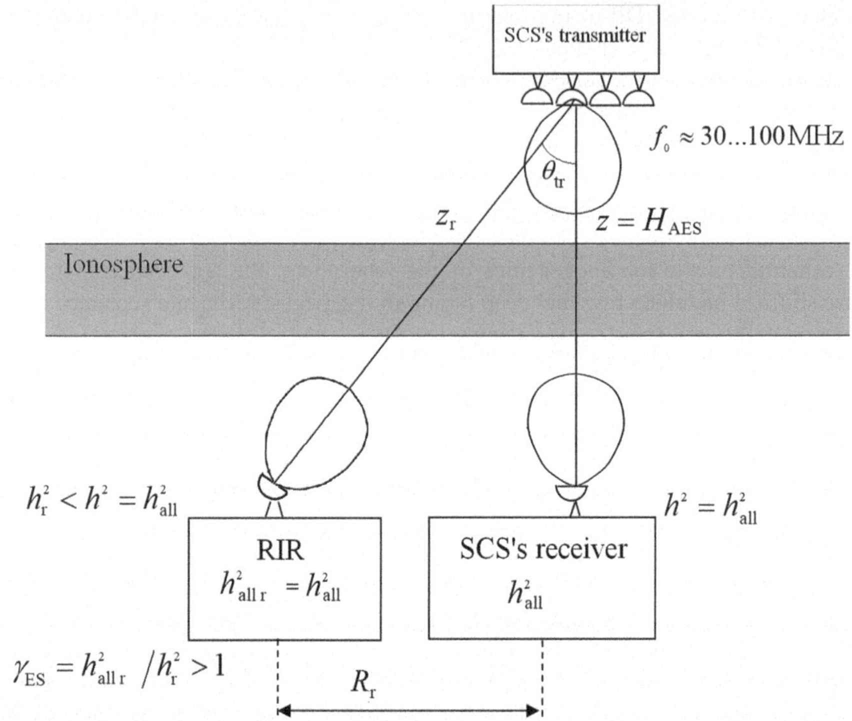

Fig. 1. Model for placing radio assets for assessing the energy secrecy of low-frequency SCS using on-board FAR

the actual signal-to-noise ratio h_r^2 at the input of the radio intercept receiver does not exceed the allowed value $h_{all\,r}^2$ necessary to achieve the allowed probability of erroneous signal reception $h_{all\,r}^2 = \psi^{-1}(P_{err\,all\,r})$. This condition $\left(h_r^2 < h_{all\,r}^2\right)$ can be written as an excess of the energy secrecy coefficient over one: $\gamma_{ES} = h_{all\,r}^2 / h_r^2 > 1$.

A detailed form of the condition for ensuring energy secrecy of SCS is given in [15–23, 26, 27]:

$$\gamma_{ES} = \frac{h_{all\,r}^2}{h_r^2} \approx \frac{1}{F_t^2(\theta_{tr})} \frac{z_r^2}{z^2} \frac{h_{all\,r}^2}{h_{all}^2} > 1, \quad (1)$$

where $F_t^2(\theta_{tr}) \le 1$ is normalized power DP of the transmitting antenna of an AES in the direction θ_{tr} to the RI receiver; z_r and z are ranges of radio lines from AES to RI receivers and SCS respectively.

Condition (1) can be written as a function of a distance from the radio intercept receiver to the receiver of SCS $(R_r \sim \theta_{tr})$ or from the reconnaissance distance as following [26, 27]:

$$\gamma_{\mathrm{ES}}(R_{\mathrm{r}}) \approx \frac{1}{F_t^2(R_{\mathrm{r}})} \frac{z_{\mathrm{r}}^2(R_{\mathrm{r}})}{z^2} \frac{h_{\mathrm{all\,r}}^2}{h_{\mathrm{all}}^2} = S(R_{\mathrm{r}}) \Delta h_{\mathrm{all}}^2 > 1, \quad (2)$$

Where

$$S(R_{\mathrm{r}}) = F_t^{-2}(R_{\mathrm{r}}) z_{\mathrm{r}}^2(R_{\mathrm{r}}) / z^2 \geq 1 \quad (3)$$

is the coefficient of spatial secrecy of the SCS (gradually increasing as the radio intercept receiver is removed to a distance R_{r} from the receiver of SCS); $\Delta h_{\mathrm{all}}^2 = h_{\mathrm{all\,r}}^2 / h_{\mathrm{all}}^2$ is a gain in energy secrecy from the use of spatially spaced signal reception in a low-frequency SCS, i.e. the component of the energy secrecy coefficient of the SCS that is due to a decrease in frequency and the use of spaced reception and independent of the reconnaissance distance R_{r}.

According to Fig. 1 RI receivers and receivers of SCS use the same signal processing scheme and ensure equalities $h_{\mathrm{all\,r}}^2 = h_{\mathrm{all}}^2$ and $\Delta h_{\mathrm{all}}^2 = h_{\mathrm{all\,r}}^2 / h_{\mathrm{all}}^2 = 1$. In this case the energy secrecy coefficient of the SCS is equal to the spatial secrecy coefficient: $\gamma_{\mathrm{ES}}(R_{\mathrm{r}}) = S(R_{\mathrm{r}})$.

According to (1–3), the spatial secrecy coefficient of the SCS $S(R_{\mathrm{r}} \sim \theta_{t\mathrm{r}})$ depends on the normalized reconnaissance distance $z_{\mathrm{r}}(R_{\mathrm{r}})/z = z_{\mathrm{r}}(R_{\mathrm{r}})/H_{AES}$, which is determined with the reconnaissance distance $R_{\mathrm{r}} \sim \theta_{t\,\mathrm{r}}$ [26, 27].

The reconnaissance distance R_{r} is related to the angle on the direction of reconnaissance $\theta_{t\mathrm{r}}$ with the following dependence:

$$\theta_{t\mathrm{r}} = 90^\circ - \frac{R_{\mathrm{r}}[\mathrm{km}]}{222,4} - \operatorname{arc\,tan}\left[\frac{H_{AES}}{(2R_E + H_{AES}) \tan(R_{\mathrm{r}}[\mathrm{km}]/222,4)}\right], \quad (4)$$

where $z = H_{AES}$ (height of low-orbit AES is H_{AES} = 700 … 1500 km).

The maximum reconnaissance distance, i.e. the distance between the AES and the location of the reconnaissance receiver (radio intercept receiver), is defined as [28]

$$z_{\mathrm{r}}(R_{\mathrm{r}}) = R_E \sin(R_{\mathrm{r}}[\mathrm{km}]/111,2) / \sin \theta_{t\mathrm{r}}, \quad (5)$$

where $R_E \approx 6370$ km is the radius of the Earth.

Analysis of expressions (4, 5) and Fig. 1 shows that as the distance R_{r} between radio intercept receivers and receivers of SCS increases, the angle on the direction of the reconnaissance receiver $\theta_{t\mathrm{r}}(R_{\mathrm{r}})$ and the reconnaissance distance $z_{\mathrm{r}}(\theta_{t\mathrm{r}})$. In addition, as the height of the AES's orbit (H_{AES}) decreases, the angle on the direction of the reconnaissance receiver $\theta_{t\mathrm{r}}(R_{\mathrm{r}}, H_{AES})$ and the reconnaissance distance $z_{\mathrm{r}}(\theta_{t\mathrm{r}}, H_{AES})$ increase.

To define a second component of the spatial secrecy of SCS (3) $S(R_{\mathrm{r}}) \sim F_t^{-2}(R_{\mathrm{r}})$, let's use the expression (4), that establishes a relationship $\theta_{t\mathrm{r}} = \psi(R_{\mathrm{r}}, H_{AES})$, and the general expression for the normalized voltage DP of the cylindrical helical antenna consisting of n_h coils [29], that with the traditional values of the angle of elevation of the spiral loop $\alpha \approx 12 \ldots 14^\circ$ is reduced to the simplified form [27–29]

$$F_t(\theta_{tr}) \approx \frac{2}{\pi n_h} J_0(\sin\theta_{t\,r}) \cos\theta_{tr} \frac{\sin \pi n_h \xi}{\xi^2 - 1}, \quad (6)$$

where $J_0(x)$ is Bessel function; $k = 2\pi/\lambda_0$ is a wave number; $\xi = 1 + 0.22(1 - \cos\theta_{tr}) \tan\alpha$ is a coefficient of the wave's deceleration. According to [29], with $\xi \approx 1$, $\sin \pi n_h \xi / (\xi^2 - 1) = \pi n_h / 2$. So with $\theta_{tr} = 0$ we will have $J_0(\sin\theta_{tr}) = J_0(0) = 1$ and $F_t(\theta_{t\,r} = 0) = 1$.

The normalized DP of a FAR consisting of 4 helical antennas is related to the DP of a single helical antenna with a general dependence [30]:

$$F_{t\,far}(\theta) = F_t(\theta)\cos(0,5kd_1 \sin\theta),$$

where d_1 is a distance between the centers of spirals in a row. So the normalized power DP of a helical FAR used on a board of AES can be expressed via the DP of a single helical antenna (6) as

$$F^2_{t\,far}(\theta_{tp}) = F_t^2(\theta_{t\,r})\cos(0,5kd_1 \sin\theta_{tr}) = F_t^2(\theta_{tr}) \cdot f^2_{far}(\theta_{t\,r}), \quad (7)$$

where $f_{far}(\theta) = \cos(0,5kd_1 \sin\theta_{tr})$ is a coefficient of FAR.

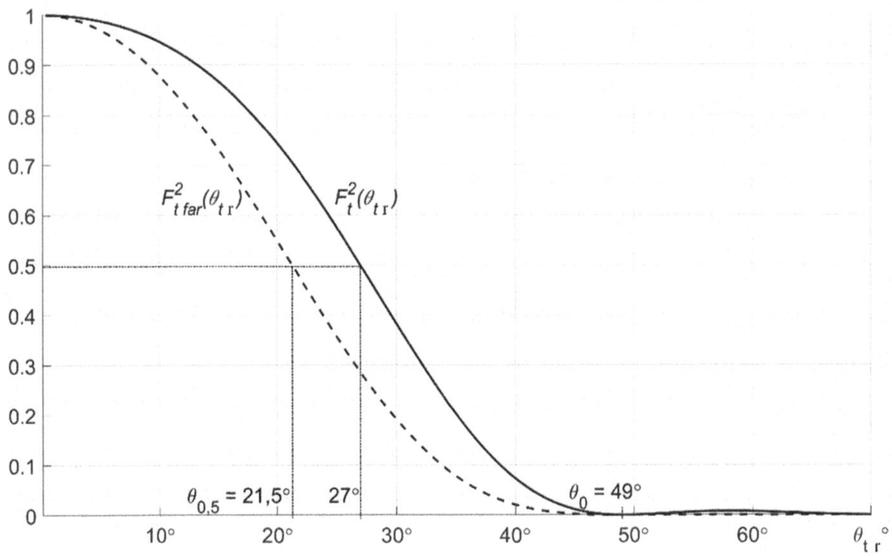

Fig. 2. Normalized directional patterns of a helical antenna and a phased antenna array of 4 helical antennas

Figure 2 shows the normalized power DP of a single helical antenna $F_t^2(\theta_{t\,\mathrm{r}})$ and a phased antenna array of 4 helical antennas $F_{t\,far}^2(\theta_{t\,\mathrm{r}})$, constructed according to (6, 7) with n_h = 13, $\alpha \approx 13°$, λ_0 = 5 m (f_0 = 60 MHz), d_1 = 2,5 m.

The analysis of Fig. 2 shows that the width of the DP of the helical antenna at the half power level is $2\theta_{0,5} \approx 27° \cdot 2 = 54°$ and at the zero radiation level it is $2\theta_0 \approx 2 \cdot 49° = 98°$. For the case of FAR, the width of the directional pattern is maintained at the level of zero radiation $2\theta_{0far} \approx 98°$, and at the level of half power is $2\theta_{0,5far} \approx 21.5 \cdot 2 = 43°$ (i.e., which is 80% of a single antenna).

In accordance with the dependence (4) $\theta_{t\mathrm{r}} = \psi(R_\mathrm{r}, H_{AES})$ and expression (7) for the DP of FAR $F_{t\,far}(\theta_{t\,\mathrm{r}})$, the normalized power DP of FAR can be represented as a function of the distance R_r from the RI receiver in the following form:

$$F_{t\,far}^2(\theta_{t\,\mathrm{r}}) = F_t^2(\theta_{t\,\mathrm{r}}) \cdot f_{far}^2(\theta_{t\,\mathrm{r}}) \equiv F_{t\,far}^2(R_\mathrm{r}) = F_t^2(R_\mathrm{r}) \cdot f_{far}^2(R_\mathrm{r}). \quad (8)$$

Therefore, the spatial secrecy coefficient of a low-frequency SCS using an on-board FAR can be expressed in terms of the spatial secrecy coefficient of the SCS using a single antenna (3) ($S(R_\mathrm{r}) = F_t^{-2}(R_\mathrm{r}) z_\mathrm{r}^2(R_\mathrm{r})/z^2 \geq 1$) as

$$\begin{aligned} S_{far}(R_\mathrm{r}) &= F_{t\,far}^{-2}(R_\mathrm{r}) z_\mathrm{r}^2(R_\mathrm{r})/z^2 \\ &= F_t^{-2}(R_\mathrm{r}) f_{far}^{-2}(R_\mathrm{r}) z_\mathrm{r}^2(R_\mathrm{r})/z^2 = S(R_\mathrm{r})/f_{far}^2(R_\mathrm{r}) \end{aligned} \quad (9)$$

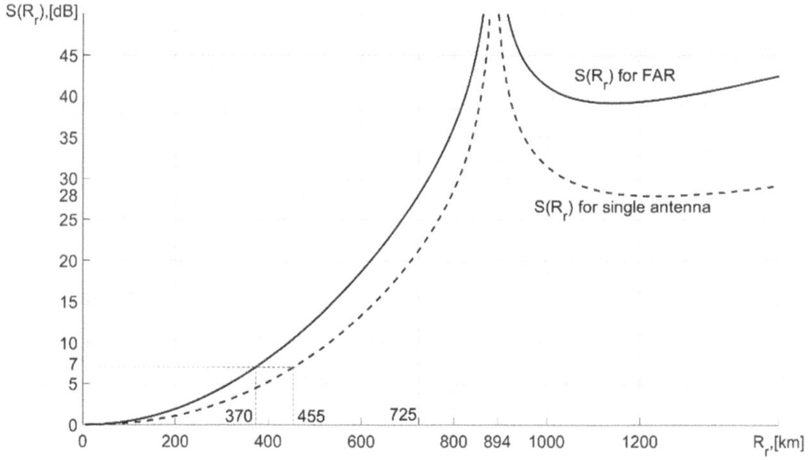

Fig. 3. Dependences of the spatial secrecy coefficient of SCS using a helical antenna and FAR on the reconnaissance distance

Figure 3 shows the dependences of the spatial secrecy coefficient of a low-orbit low-frequency SCS using a helical antenna $S(R_\mathrm{r})$ and FAR consisting of 4 helical antennas $S_{far}(R_\mathrm{r})$ on a AES on the reconnaissance distance (R_r).

These dependencies are constructed according to expressions (3–6) and (7–9) for the following source data: AES's orbit height H_{AES} = 700 km, carrier frequency f_0 = 60 MHz (λ_0 = 5 m), number of helical antenna coils n_h = 13 and helical coil lift angle $\alpha \approx 12...14°$.

3 Discussion

Analysis of Fig. 3 shows that the spatial secrecy coefficient of the SCS for both a single antenna and a FAR takes the maximum value $(S(R_r) \to \infty)$ when the angle $(\theta_{tr} \sim R_r)$ between the direction of the reconnaissance receiver and the subsatellite point is equal to half the width of the directional pattern for the zero radiation level of the AES's transmitting antenna $(\theta_{tr} = \theta_0 \approx 49°)$ and $F_t^2(\theta_{tr} = \theta_0) = 0$. This angle of zero radiation corresponds to the reconnaissance distance $(\theta_0 \sim R_r)$, which at the height of AES H_{AES} = 700 km is R = 894 km. A high level of spatial secrecy of the low-frequency SCS $S(R_r) > 28$ dB is provided when the RIR is removed at a distance R_r > 800 km for a single helical antenna and R_r > 725 km for a FAR consisting of 4 antennas. This is due to the fact that at a reconnaissance distance R_r = 725 km the spatial secrecy

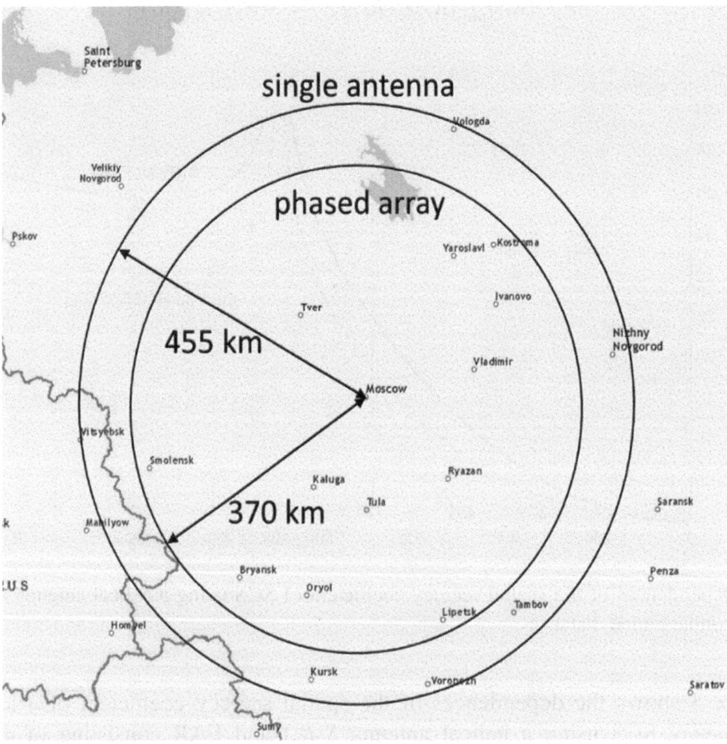

Fig. 4. The area of providing spatial secrecy $S(R_r) \geq 7$ dB in a low-frequency SCS when using a helical and FAR consisting of 4 antennas on an AES.

coefficient of the SCS using on-board helical antenna on the AES, which is approximately equal $S(R_r) \approx$ 22 dB, is significantly lower than when using FAR (28 dB).

If we consider $S(R_r) > 7$ dB to be acceptable level of spatial secrecy, it is provided when the RIR is removed at a distance $R_r > 455$ km when using a single helical antenna on board of AES, and for FAR the same level is provided at a distance $R_r > 370$ km. For clarity, the indicated reconnaissance distances are displayed on the map of Russia, assuming that the SCS's AES's subsatellite point is located in Moscow.

Analysis of Fig. 4 shows that the use of FAR consisting of 4 antennas on board of a low-orbit AES (H_{AES} = 700 km) provides an acceptable coefficient of spatial secrecy of the low-frequency (f_0 = 60 MHz) SCS $S(R_r) \geq 7$ dB outside of almost the entire European part of Russia and its western borders.

4 Conclusion

A method for evaluating the spatial secrecy of a low-orbit low-frequency SCS using an on-board phased antenna array has been developed. It includes the following stages: 1) development of a model for placing the SCS's radio assets and radio intelligence (Fig. 1) to assess the spatial secrecy of a low-frequency SCS when the radio intercept receiver (RIR) is arbitrarily removed from the ground receiver of the SCS using an on-board FAR consisting of 4 helical antennas; 2) getting the dependence (3) $S(R_r) = F_t^{-2}(R_r) z_r^2(R_r)/z^2$ of the spatial secrecy coefficient of low-frequency SCS on normalized directional pattern of the transmitting antenna of the AES $F_t^2(R_r)$ to a reconnaissance distance R_r from the radio intercept receiver and normalized reconnaissance distance $z_r(R_r)/H_{AES}$; 3) determination of the relationship (4) $R_r \sim \theta_{tr}$ between the distance R_r and the direction θ_{tr} of reconnaissance; 4) determination of range of reconnaissance $z_r(R_r)$ according to (4); 5) determination of normalized directional patterns of the on-board helical antenna (6) $F_t(\theta_{tr})$ and FAR consisting of 4 antennas (7) $F_{t\,far}^2(\theta_{t\,r}) = F_t^2(\theta_{tr}) \cdot f_{far}^2(\theta_{tr})$ for low-orbit low-frequency SCS (Fig. 2); 7) estimation of the spatial secrecy coefficient of a low-orbit low-frequency SCS using an on-board FAR according to (9) $\Pi_{far}(R_r) = \Pi(R_r)/f_{far}^2(R_r)$ and Fig. 3.

Analysis of the results shows (Fig. 4) that the use of FAR consisting of 4 helical antennas on board a low-orbit AES (H_{AES} = 700 km) provides an acceptable coefficient of spatial secrecy of the low-frequency (f_0 = 60 MHz) SCS $S(R_r) \geq 7$ dB outside of almost the entire European part of Russia and its western borders.

Acknowledgements. The work was supported by the Russian Foundation for Basic Research, project No. 18-07-01020.

References

1. Roy-Chowdhury, A., Baras, J., Hadjitheodosiou, M., Papademetriou, S.: Security issues in hybrid networks with a satellite component. IEEE Wirel. Commun. Mag. **12**(6), 50–61 (2005)

2. Murtaza, A., Pirzada, J., Xu, T., Liu, J.: A lightweight authentication and key sharing protocol for satellite communication. Int. J. Comput. Commun. Control (IJCCC) (2020). https://doi.org/10.17706/ijcce.2020.9.1.46-53
3. Cruickshank, H., Howarth, M., Iyengar, S., Sun, Z., Claverotte, L.: Securing multicast in DVB-RCS satellite systems. IEEE Wirel. Commun. Mag. **12**(5), 38–45 (2005)
4. Wyner, A.: The wire-tap channel. Bell Syst. Technol. J. **54**(8), 1355–1387 (1975)
5. Leung-Yan-Cheong, S.K., Hellman, M.E.: The Gaussian wire-tap channel. IEEE Trans. Inf. Theory **24**(4), 451–456 (1978)
6. Csiszár, I., Körner, J.: Broadcast channels with confidential messages. IEEE Trans. Inf. Theor. **24**(3), 339–348 (1978)
7. Tuzov, G.I, Sivov, V.A., Prytkov, V.I., et al.: Noise immunity of systems with complex signals. Sov. Radio, Moscow (1985)
8. Bloch, M., Barros, J., Rodrigues, M., McLaughlin, S.: Wireless information-theoretic security. IEEE Trans. Inf. Theor. **54**(6), 2515–2534 (2008)
9. Barros, J., Rodrigues, M.: Secrecy capacity of wireless channels. Proc. ISIT **2006**, 356–360 (2006)
10. Alves, H., Souza, R.D., Debbah, M., Bennis, M.: Performance of transmit antenna selection physical layer security schemes. IEEE Sig. Process. Lett. **19**(6), 372–375 (2012)
11. An, K., Liang, T., Yan, X., Zheng, G.: On the secrecy performance of land mobile satellite communication systems. IEEE Access **6**, 39606–39620 (2018)
12. Li, Y., An, K., Liang, T., Yan, X.: Secrecy performance of land mobile satellite systems with imperfect channel estimation and multiple eavesdroppers. IEEE Access **7**, 31751–31761 (2019)
13. Lei, J., Han, Z., Castro, M.A.V., Hjorungnes, A.: Secure satellite communication systems design with individual secrecy rate constraints. IEEE Trans. Inf. Forens. Secur. **6**(3), 661–671 (2011)
14. Lin, M.L., Ouyang, J., Zhu, W., Panagopoulos, A.D., Alouini, M.: Robust secure beamforming for multibeam satellite communication systems. IEEE Trans. Veh. Technol. **68** (6), 6202–6206 (2019)
15. Chipiga, A.F., Senokosova, A.V.: Information protection in the space communication systems using change sin radio wave propagation conditions. Cosm. Res. **45**(1), 52–59 (2007)
16. Chipiga, A.F., Senokosova, A.V.: A method to ensure energy security of satellite communication systems. Cosm. Res. **47**(5), 393–398 (2009)
17. Pashintsev, V.P., Tsimbal, V.A., Chipiga, A.F.: Analytical dependence of the energy secrecy of satellite communications on the choice of carrier frequency. In: Proceedings of the 18th International Scientific and Technical Conference "Radiolocation, Navigation, Communication", pp. 2113–2120. VSU, Voronezh (2012)
18. Chipiga, A.F., Pashintsev, V.P., Tsymbal, V.A., Shimanov, S.N.: Procedure for calculating the dependence of the energy concealment factor on carrier frequency selection for low-frequency satellite communications system. Autom. Control Comput. Sci. **50**(6), 408–414 (2016)
19. Chipiga, A.F., Pashintsev, V.P., Anzin, I.V., Lyakhov, A.V.: Dependence of the energy secrecy of a low-frequency satellite communication system on the removal of the radio detection receiver. Spec. Equip. **3**, 19–26 (2017)
20. Pashintsev, V.P., Chipiga, A.F., Lyakhov, A.V., Anzin, I.V.: Energy secrecy of low-frequency satellite communication systems from signal detection. Spec. Equip. **3**, 10–18 (2017)

21. Pashintsev, V.P., Peskov, M.V., Smirnov, V.M., Smirnova, E.V., Tynyankin, S.I.: Procedure for extraction of small-scale variations in the total electron content of the ionosphere with the use of transionospheric sounding data. J. Commun. Technol. Electron. **62**(12), 1336–1342 (2017). https://doi.org/10.1134/S1064226917110158
22. Pashintsev, V.P., Katkov, K.A., Katkov, E.K., Gakhova, N.N., Gakhov, R.P., Titov, A.I.: Forecast accuracy of determining pseudo range in satellite navigation system through analysis of data from ionosphere monitoring. J. Fund. Appl. Sci. **9**(1S), 899–913 (2017)
23. Chipiga, A.F., Pashintsev, V.P., Tsymbal, V.A., Zelenevskiy, V.V.: Low-frequency satellite communication system technical means' parameters synthesis by the requirements for energetic concealment and noise immunity. Autom. Control Comput. Sci. **52**(3), 243–249 (2018). https://doi.org/10.3103/S0146411618030057
24. Pashintsev, V.P., Peskov, M.V., Kalmykov, I.A., Zhuk, A.P., Senokosov, M.A.: Method for the evaluation of ionospheric diffractive and dispersive properties impact on the interference immunity of satellite communication systems. Int. J. Civil Eng. Technol. **9**(13), 44–61 (2018)
25. Chipiga, A.F., Shevchenko, V.A., Pashintsev, V.P., Kostyuk, D.M.: Estimation of energy secrecy of a low-orbit low-frequency satellite communication system with an arbitrary multiplicity of spatially spaced reception. News Inst. Eng. Phys. **4**(50), 49–55 (2018)
26. Pashintsev, V., Chipiga, A., Anzin, I.: Energetic concealment of low-frequency sattelite communication system with arbitrary recession of radiointercepting receiver. In: Proceedings of the 2018 Multidisciplinary Symposium on Computer Science and ICT, CEUR Workshop Proceedings, Stavropol, Russia, pp. 237–244, 15 October 2018
27. Pashintsev, V.P., Chipiga, A.F., Anzin, I.V.: Energy secrecy of a low-orbit low-frequency satellite communication system when the radio intercept receiver is arbitrarily removed. Manag. Commun. Secur. Syst. **4**, 122–135 (2018)
28. Pashintsev, V.P., Gahova, N.N., Katkov, E.K., Zaytseva, T.V., Balabanova, T.N.: Probability of erroneous reception of navigational radio signals under ionospheric disturbances. COMPUSOFT Int. J. Adv. Comput. Technol. **8**(6), 3201–3205 (2019)
29. Yurtsev, O.A., Runov, A.V., Kazarin, A.N.: Helical Antenna. Sov. Radio, Moscow (1974)
30. Drabkin, A., Zuzenko, V., Kislov, A.: Antenna-Feeder Devices. Sov. Radio, Moscow (1974)

Direction Finding of Ionospheric Formation with Small-Scale Inhomogeneities Based on GPS Monitoring's Data Processing

V. P. Pashintsev[1], V. A. Tsimbal[2], A. F. Chipiga[1(✉)], M. V. Peskov[1], and M. A. Senokosov[1,2]

[1] North Caucasus Federal University, Stavropol, Russia
pashintsevp@mail.ru, afchipiga@ncfu.ru
[2] Institute of Engineering Physics, Serpukhov, Russia

Abstract. A method and an algorithm for determining the coordinates of the ionospheric formation's region with small-scale inhomogeneities of electronic concentration based on mathematical processing of data from the outputs of the dual-frequency receiver of GPStation-6 for monitoring the standard deviation of small-scale fluctuations in the total electronic content of the ionosphere and continuous measurement of the current values of the elevation angle and azimuth of the navigation spacecraft has been developed #CSOC1120.

Keywords: Ionosphere · Full electronic content · Small-scale inhomogeneities · GPS monitoring · Dual-frequency receiver · Time series · Subinospheric point · Geographical coordinates

1 Introduction

It is known [1–9] that there may be intense fluctuations in the phase and the amplitude of the received signals (otherwise, flickering, fading, scintillation) in transionospheric radio channels, which cause a significant (by orders of magnitude) decrease in the quality of satellite communication systems (SCS) and satellite radio navigation systems (SRNS). The reason for this is the formation of regions with small-scale inhomogeneities of electronic concentration in the ionosphere (most often in areas of Equatorial and polar latitudes) [4–11]. Therefore, it is an urgent task to monitor the spatial position of small-scale ionospheric formations in order to predict the quality indicators of SCS and SRNS in areas of equatorial and polar latitudes and to select alternative radio routes.

Analysis of the causes of ionospheric flickering in the SCS shows the following [4–11]. The effect of natural disturbances in the area of equatorial and polar latitudes can cause the appearance of specific extensive (up to 1000 km or more) ionospheric formations. They consist of small-scale inhomogeneities (fluctuations) of the electronic concentration (ΔN_i) relative to the average (background) value $(\overline{N})$. Moreover, the ratio $\Delta N_i/\overline{N}$ of the fluctuation and regular components of the electronic concentration in a small-scale ionospheric formation $N = \overline{N} + \Delta N_i$ can exceed this ratio by 1–2

R. Silhavy (Ed.): CSOC 2020, AISC 1226, pp. 480–487, 2020.
https://doi.org/10.1007/978-3-030-51974-2_45

orders of magnitude in the normal mid-latitude ionosphere (where $\Delta N_i/\overline{N} = 0, 1\ldots1\%$). The average electronic concentration $\overline{N}$ changes with increasing of the height (h) reaching a maximum value $\overline{N_m} \equiv \overline{N}(h_m)$ at the height of the maximum ionization of the ionosphere layer F$(h = h_{max} \equiv h_m)$. Measurement of the ionosphere's altitude profile $\overline{N}(h)$ up to the height of the ionization maximum $(h = h_m)$ can be performed with using data from the ionosphere vertical sensing (IVS) station. A method for measuring the altitudinal change in small-scale fluctuations in electronic concentration $\Delta N_i(h)$ including one at the height of the ionization maximum $\Delta N_i(h_m)$ with using data from IVS station is known [14]. However, it is not used because of the complexity of the technical implementation. In addition, the IVS station provides reliable data for measuring the electronic concentration $N(h) = \overline{N}(h) + \Delta N_i(h)$ in a limited area of the ionosphere above the station with a radius of approximately 500 km.

Nowadays the most advanced means are GPS monitoring of the ionosphere, i.e. passive monitoring of the ionosphere using the dual-frequency receiver of the GPStation-6 of the GPS satellite radio navigation system (SRNS) [2, 3, 15–17]. It can be used to obtain data on the total electronic content (TEC) of the ionosphere $N_T = \int N(h)dh$ on the radio wave propagation (RWP) routes from all "visible" navigation spacecrafts (SC) to the ground-based dual-frequency receiver of the GPStation-6 at a distance of up to 1000 km. It is obvious that the presence of small-scale fluctuations of the electronic concentration $N(h) = \overline{N}(h) + \Delta N_i(h)$ in the ionosphere causes the appearance of small-scale fluctuations ΔN_T in the results of TEC measurement $N_T = \overline{N}_T + \Delta N_T$ against the background of the average value $\overline{N_T} = \overline{N_m}h_e$ determined with the equivalent thickness h_e of the ionosphere [4–7, 16].

Under the conditions of ionospheric disturbances and the appearance of ionospheric formations, small-scale fluctuations of TEC $(\Delta N_{Ti} \sim \Delta N_i h_e)$ can increase by 1–2 orders of magnitude compared to the normal (undisturbed) ionosphere. Therefore, RWP in the SCS and SRNS with a carrier frequency $f_0 \sim 1$ GHz in these conditions is accompanied by scattering of radio waves on small-scale inhomogeneities, the appearance of multipath and fading of received signals, which increase in proportion to the growth of the relative time of delays of incoming rays $\left(\Delta\tau_i \sim \Delta N_{Ti}/f_0^2\right)$ and lead to a significant deterioration of the noise immunity of the SCS and the accuracy of positioning of the SRNS.

Therefore, it is an urgent task to locate the spatial coordinates of the area of small-scale ionospheric formations based on mathematical processing of GPS monitoring data for inhomogeneous ionospheric TEC.

The purpose of the report is to develop an algorithm for determining the coordinates of the ionosphere region with small-scale inhomogeneities based on mathematical processing of data of the total electronic content of the ionosphere monitoring using the dual-frequency receiver of the GPStation-6.

2 Method for Determining the Coordinates of Ionospheric Regions with Small-Scale Inhomogeneities Using GPS Monitoring's Data

To achieve the purpose of the report, let's analyze the principle of operation and construction of the complex for determining the coordinates of ionospheric regions with small-scale inhomogeneities. A method for monitoring ionosphere parameters using GPS SRNS signals is known [1–3]. The essence of the method is that the passage through the ionosphere of a radio signal emitted at two carrier frequencies f_1 and f_2 from the SRNS's spacecraft (SC) is accompanied with various delays ($\Delta\tau_1 \sim N_T/f_1^2$ and $\Delta\tau_2 \sim N_T/f_2^2$) and phase changes ($\Delta\varphi_1 \sim N_T/f_1$ and $\Delta\varphi_2 \sim N_T/f_2$). They allow to continuously determine the ionosphere's TEC $N_T \sim (\Delta\varphi_2 - \Delta\varphi_1)$ along the radio path "SC of SRNS – Dual-frequency receiver of SRNS" at any time t using navigation measurements of the dual-frequency receiver of SRNS. Thus, the time series of values of the ionosphere's TEC $N_T(t)$ are obtained. In the presence of small-scale inhomogeneities in the ionosphere, a method [18, 19] has been developed that allows (with expanding the capabilities of the receiver of the GPStation-6) to calculate the ionosphere's TEC $\left(N_T = \psi\left(f_1, f_2, z_1', z_2', \varphi_1', \varphi_2'\right)\right)$ with a sampling frequency $f_s = 50$ Hz (sampling period is $\tau_s = 0,02$ s) based on the results of dual-frequency (f_1, f_2) measurements of pseudo-distances $\left(z_1' \sim \Delta\tau_1 c;\ z_2' \sim \Delta\tau_1 c\right)$ and pseudo-phases $\left(\varphi_1', \varphi_2'\right)$, to isolate small-scale fluctuations $\Delta N_T(t)$ from the time series of the ionosphere's TEC $N_T(t) = \bar{N}_T(t) + \Delta N_T(t)$ and then use a standard procedure to evaluate their standard deviation (SD) $\sigma_{\Delta N_T}(t) \sim \left\langle \Delta N_T^2(t) \right\rangle^{0,5}$.

Figure 1 illustrates the principle construction of the complex coordinates of the regions of the ionosphere with small-scale inhomogeneities (SSI), consisting of a dual-frequency receiver of SRNS, the output of which produces results of dual-frequency (f_1, f_2) measurements of pseudo-distances $\left(z_1'(t), z_2'(t)\right)$ and pseudo-phases $\left(\varphi_1'(t), \varphi_2'(t)\right)$, the unit for determining the SD of small-scale fluctuations of the TEC of the ionosphere, which calculates the current value of the TEC of the inhomogeneous ionosphere $N_T(t) = \bar{N}_T(t) + \Delta N_T(t)$ on the RWP route from SC of SRNS to the dual-frequency receiver of the GPStation-6 and the SD $\sigma_{\Delta N_T}(t) \sim \left\langle \Delta N_T^2(t) \right\rangle^{0,5}$ of its small-scale fluctuations $\Delta N_T(t)$ and the unit for determining the coordinates (latitude φ_P and longitude l_P) of the ionosphere's region with SSI. These coordinates at any point in time $(\varphi_P(\mathrm{t}), l_P(\mathrm{t}))$ can be calculated from the known formulas [1] for calculation of geographical coordinates subionospheric point (SIP) for radio route "SC of SRNS – Dual-frequency receiver of SRNS" based on information about height of maximum ionization of the ionosphere $(h_{\max})$ and current data about the elevation's angle $\theta_S(t)$ and azimuth $\alpha_S(t)$ of the orbit of travel of the SC, obtained from the navigation messages received by the radio receiver of SRNS. A subionospheric point is a projection on the Earth's surface of the intersection point of the radio route "SC of SRNS - Dual-frequency receiver of SRNS" with the region of maximum ionization of the ionosphere at the height $h_{\max}$ that forms the main contribution to the variations of the TEC [1].

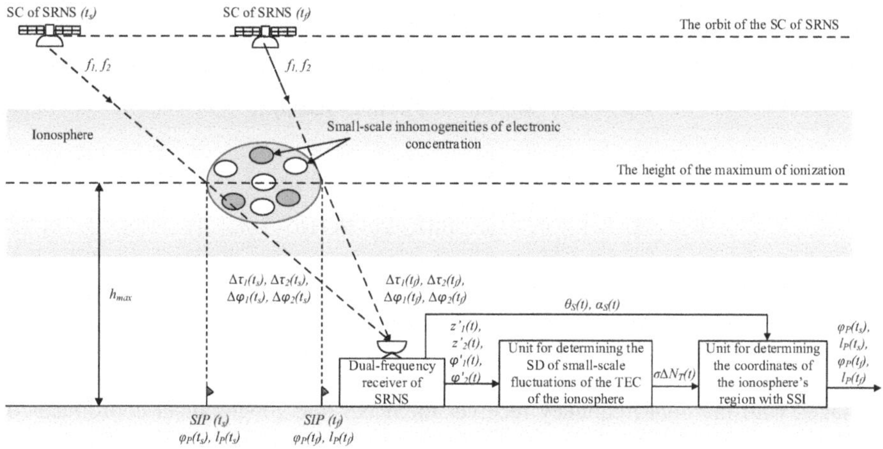

Fig. 1. Principle of operation and construction of the complex for determining the coordinates of ionospheric regions with small-scale inhomogeneities

Figure 1 shows positions of two subionospheric points (SIPs), corresponding to the moments of the start (t_s) and the finish (t_f) of the intersection of the radio route from a moving SC of SRNS to dual-frequency receiver of SRNS on the borders of the region with small-scale inhomogeneities of the ionosphere, where the SD of small-scale fluctuations of the TEC increases to a some limit value $\sigma_{\Delta N_T L}$ (i.e. $\sigma_{\Delta N_T}(t_{s,f}) \geq \sigma_{\Delta N_T L}$).

Based on the developed principles of operation and construction of a complex for determining the coordinates of ionospheric regions with small-scale inhomogeneities, it is possible to justify the method for determining the geographical coordinates of the ionospheric region with SSI.

At the first stage of the method in the unit for determining the coordinates of the ionosphere region with SSI based on data of the SD of small-scale fluctuations of the ionosphere's TEC $\sigma_{\Delta N_T}(t)$ the time interval $[t_s, t_f]$ is determined within which the small-scale fluctuations of the TEC reaches and exceeds the specified limit value $\sigma_{\Delta N_T \Pi}$ (i.e. $\sigma_{\Delta N_T}(t) \geq \sigma_{\Delta N_T L}$).

The starting moment t_s is fixed when the SD of small-scale fluctuations of the TEC exceeds the limit value $(\sigma_{\Delta N_T}(t) \geq \sigma_{\Delta N_T L})$ for the first time, and the finishing moment t_f is fixed when the SD of small-scale fluctuations of the TEC becomes less than the limit value. At the same time, the threshold value is compared with both the current $(\sigma_{\Delta N_T}(t) \leq \sigma_{\Delta N_T L})$ and previous values of the SD of small-scale fluctuations of TEC $(\sigma_{\Delta N_T}(t - \tau_S) \geq \sigma_{\Delta N_T L})$ which differ in time by the value of the sampling interval of measurements (τ_S).

At the second stage of the method the latitude $\varphi_p(t)$ and longitude $l_p(t)$ of the SIP (with an overlay on the map) is calculated at the moments of time t_s and t_f determined at the previous stage to determine the position and linear dimensions of the ionosphere region with SSI. Formulas are known [1] for restoring the geographical coordinates (latitude φ_P and longitude l_P) of the SIP based on the azimuth α_S and elevation angle θ_S

of the navigation SC of SRNS and the geographical coordinates of the receiver of SRNS (φ_B, l_B):

$$l_P = l_B + \arcsin(\sin \psi_P \sin \alpha_S \sec \varphi_P), \tag{1}$$

where

$$\psi_P = (\pi/2) - \theta_S - \arcsin(R_E \cos \theta_S / (R_E + h_{\max})) \tag{2}$$

is the central angle between the observation point (the receiver of SRNS) and the subionospheric point,R_E is the Earth's radius.

On the basis of expressions (1–3) and data on changes in the elevation angle $\theta_S(t)$ and azimuth $\alpha_S(t)$ of SC obtained from the results of processing received navigation messages with the dual-frequency receiver of SRNS, the geographical coordinates of the SIP at moments of time t_s and t_f are determined at the second stage with using the following dependencies:

$$\varphi_P(t_{s,f}) = \arcsin\left[\sin \varphi_B \cos(\psi_P(t_{s,f})) + \cos \varphi_B \sin(\psi_P(t_{s,f})) \cos(\alpha_S(t_{s,f}))\right]; \tag{3}$$

$$l_P(t_{s,f}) = l_B + \arcsin\left[\sin(\psi_P(t_{s,f})) \sin(\alpha_S(t_{s,f})) \sec(\varphi_P(t_{s,f}))\right] \tag{4}$$

where

$$\psi_P(t_{s,f}) = (\pi/2) - \theta_S(t_{s,f}) - \arcsin\left(R_E \cos \theta_S(t_{s,f}) / (R_E + h_{\max})\right). \tag{5}$$

3 Results

It follows that, in accordance with [18, 19], the method for determining the coordinates of ionospheric regions with small-scale inhomogeneities based on GPS monitoring data consists of a sequence of the following operations:

1) calculation of ionospheric TEC and formation of a time series $N_T(t)$ based on the results of dual-frequency (f_1, f_2) measurements of pseudo-distances $(z_1'(t), z_2'(t))$ and pseudo-phases $(\varphi_1'(t), \varphi_2'(t))$ using the dual-frequency receiver of GPStation-6;
2) passing a time series $N_T(t) = \bar{N}_T(t) + \Delta N_T(t)$ through a digital filter to determine small-scale fluctuations of the ionosphere's TEC $\Delta N_T(t)$;
3) calculation of the SD $\sigma_{\Delta N_T}(t)$ of small-scale fluctuations of the ionosphere TEC $\Delta N_T(t)$ obtained at the previous stage;
4) comparison of the SD of small-scale fluctuations of the ionosphere TEC $\sigma_{\Delta N_T}(t)$ with the limit value $\sigma_{\Delta N_T L}$;
5) determination of the starting t_s and the finishing t_f moments of the ionosphere's region with SSI where the SD of small-scale fluctuations of the TEC becomes equal to or exceeds the limit value: $\sigma_{\Delta N_T}(t_{s,f}) \geq \sigma_{\Delta N_T L}$.

6) determination of the latitude $\varphi_P(t_{s,f})$ and longitude $l_P(t_{s,f})$ of the SIP at moments t_s and t_f.

Taking into account these operations, the block diagram of the algorithm for determining the coordinates of the ionosphere region with small-scale inhomogeneities based on GPS monitoring data is shown on Fig. 2.

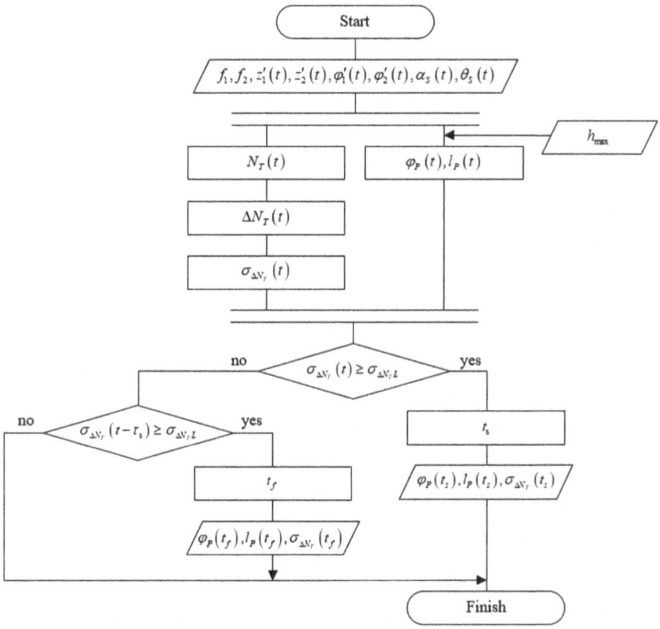

Fig. 2. Algorithm for determining the coordinates of the ionosphere's region with small-scale inhomogeneities based on GPS monitoring data

4 Conclusion

Thus, on the basis of mathematical processing of data from the GPStation-6 receiver outputs for continuous dual-frequency (f_1, f_2) measurements of pseudo-distances $(z_1'(t), z_2'(t))$ and pseudo-phases $(\varphi_1'(t), \varphi_2'(t))$ to the navigation spacecraft, as well as its current elevation angle $\theta_S(t)$ and azimuth $\alpha_S(t)$ an algorithm has been developed (Fig. 2), which allows to determine the geographical coordinates and estimate the linear dimensions of the region of ionospheric formation with small-scale inhomogeneities of electronic concentration. The essence of the algorithm is to determine the starting (t_s) and the finishing (t_f) moments, when the SD of small-scale fluctuations of the TEC becomes equal to the limit value $\sigma_{\Delta N_T}(t_{s,f}) = \sigma_{\Delta N_T L}$, and then determine the latitude $\varphi_P(t_{s,f})$ and the longitude $l_P(t_{s,f})$ of the subionospheric points at moments t_s and t_f.

The obtained data can be used to predict changes in the quality indicators of the SCS and SRNS in the regions of polar and equatorial latitudes, as well as to search for and select alternative radio routes in the conditions of ionospheric disturbances.

Acknowledgements. The work was supported by the Russian Foundation for Basic Research, project No. 18-07-01020.

References

1. Afraimovich, E.L., Perevalova, N.P.: GPS monitoring of the earth's upper atmosphere. In: Scientific Center for Reconstructive and Regenerative Surgery of the East Siberian Scientific Center of the Siberian Branch of the Russian Academy of Medical Sciences. State Institution, Irkutsk, 480 p. (2006)
2. Shanmugam, S., Jones, J., MacAulay, A., Van Dierendonck, A.J.: Evolution to modernized GNSS ionospheric scintillation and TEC monitoring. In: Proceedings of IEEE/ION PLANS 2012, Myrtle Beach, South Carolina, pp. 265–273 (2012)
3. Carrano, C., Groves, K.: The GPS segment of the AFRL-SCINDA global network and the challenges of real-time TEC estimation in the equatorial ionosphere. In: Proceedings of the 2006 National Technical Meeting of the Institute of Navigation, Monterey, pp. 1036–1047 (2006)
4. Maslov, O.N., Pashintsev, V.P.: Models of transionospheric radio channels and noise immunity of space communication systems. Suppl. J. Infocommun. Technol. (4), 357 (2006)
5. Pashintsev, V.P., Solchatov, M.E., Gakhov, R.P.: Influence of the Ionosphere on the Characteristics of Space Information Transmission Systems: Monography. Fizmatlit, p. 184 (2006)
6. Pashintsev, V.P., Sapozhnikov, A.D., Vititlov, L.L.: Analytical method for evaluating of ionosphere on the noise immunity of space communication systems. Telecommun. Radio Eng. **46**(12), 84–87 (1991)
7. Pashintsev, V.P., Kolosov, L.V., Tishkin, S.A., Smirnov, A.A.: Influence of the ionosphere on signal detection in space communications systems. J. Commun. Technol. Electr. **44**(2), 132–139 (1999)
8. Gakhova, N.N., Katkov, E.K., Pashintsev, V.P., Zaytseva, T.V., Balabanova, T.N.: Probability of erroneous reception of navigational radio signals under ionospheric disturbances. Compusoft **8**(6), 3201–3205 (2019)
9. Groves, K.: Monitoring ionospheric scintillation with GPS. In: Colloquium on Atmospheric Remote Sensing Using the Global Positioning System, Boulder, CO, pp. 1–59, 20 June–2 July 2004
10. Aarons, J.: Global morphology of ionospheric scintillations. Proc. IEEE **70**(4), 45–66 (1982)
11. Secan, J.A., Nickisch, L.J., Knepp, D.L, Snyder, A.L., Kennedy, E.J.: Investigation of Plasma Phenomena in the Ionosphere Under Natural Conditions and Under Conditions Artificially Perturbed by HAARP Air Force Research Laboratory, Final Report 31, 122 p., August 2008
12. Pashintsev, V.P., Katkov, K.A., Katkov, E.K., Gakhova, N.N., Gakhov, R.P., Titov, A.I.: Forecast accuracy of determining pseudo range in satellite navigation system through analysis of data from ionosphere monitoring. J. Fundam. Appl. Sci. **9**(1S), 899–913 (2017)
13. Davies, K.: Ionospheric Radio Waves, p. 460. Blaisdell, Waltham (1969)

14. Pashintsev, V.P., Galushko, Y.I., Koval, S.A., Senokosova, A.V., Gribanov, E.V.: Method for determining the intensity of ionospheric inhomogeneities using vertical sensing data Patent for invention of Russian Federation № 2403592. - Bul. № 31, 10 November 2010
15. Pashintsev, V.P., Peskov, M.V., Smirnov, V.M., Smirnova, E.V., Tynyankin, S.I.: Procedure for extraction of small-scale variations in the total electron content of the ionosphere with the use of transionospheric sounding data. J. Commun. Technol. Electr. **62**(12), 1336–1342 (2017)
16. Pashintsev, V.P., Peskov, M.V., Kalmykov, I.A., Zhuk, A.P., Senokosov, M.A.: Method for the evaluation of ionospheric diffractive and dispersive properties impact on the interference immunity of satellite communication systems. Int. J. Civ. Eng. Technol. (IJCIET) **9**(13), 44–61 (2018)
17. Pashintsev, V.P., Chipiga, A.F., Tsymbal, V.A., Shimanov, S.N.: Procedure for calculating the dependence of the energy concealment factor on carrier frequency selection for low-frequency satellite communications system. Autom. Control Comput. Sci. **50**(6), 408–414 (2016)
18. Pashintsev, V.P., Peskov, M.V., Solntsev, K.P.: Complex for determining statistical characteristics of fluctuations in the total electronic content of the ionosphere using the dual-frequency receiver of the GPStation-6. In: Reports of the First all-Russian Conference Modern Signal Processing Technologies (MSPT-2018), 31 October–2 November 2018. Series: Scientific All-Russian Conferences, no. VII, pp. 164–168. Bris, Moscow (2018)
19. Pashintsev, V.P., Peskov, M.V., Polezhaev, A.V., Toiskin, V.E., Kabanovich, S.G.: Expanding the capabilities of the dual-frequency receiver of the GPStation-6 for measuring small-scale fluctuations in the total electronic content of the ionosphere. In: International Conference Radio-Electronic Devices and Systems for Infocommunication Technologies, REDS - 2018. Reports. Series: Scientific Conferences Dedicated to the Radio Day, no. LXXIII , pp. 33–37. Bris-M (2018)

Adaptive IoT-Based HVAC Control System for Smart Buildings

A. V. Kychkin(✉), A. I. Deryabin, O. L. Vikentyeva, and L. V. Shestakova

National Research University Higher School of Economics, 38, Studencheskaya Street, Perm 614070, Russian Federation
avkychkin@hse.ru

Abstract. The article studies the experience of automation of heating, ventilation, and air conditioning (HVAC) systems of buildings with regard to the technical capacities of the Internet of Things (IoT). Using the data from IoT devices maintains the set quality parameters throughout the entire operation period, which is achieved with the compensatory and predictive control algorithms. The objective of the research is to increase the HVAC control efficiency in smart buildings using the control system with the adaptation circuit, which proactively compensates any disturbances. The proper operation of the circuit requires accumulation of information of the venue during the operation period, which is used for building the transfer functions of the HVAC of the building. Continuous adaptation of the control system model to reality is a way to continuously optimize the adjustments of the regulation algorithm, ensuring effective operation of the local temperature regulation circuits. The capacities of the IoT controller-based control system and the generation of a compensatory-predictive control signal with the placement of the control algorithm in a "cloud" on a server are demonstrated with the indoor temperature control model. The simulation models of the indoor temperature changing processes are studied: the indoor temperature changing process model without a control system; model with a PI-regulator and disturbance compensation; the disturbance compensation model for the IoT controller-based control system. The structural and parametric identification of the model is carried out with the active experiment method #CSOC1120.

1 Introduction

The modern buildings featuring a complex engineering infrastructure with a monitoring and control system are referred to as smart buildings (SB) [1, 2]. The number and types of SB in a modern city is continuously growing: among them, there are large office buildings, shopping malls and entertainment centres, hotel facilities, airport terminals, industrial workshops, social buildings, logistic and warehouse facilities etc. The operation of HVAC systems in such buildings requires considering a number of factors, such as resource saving, minimization of operation expenses, security improvement, provision of comfortable conditions for labour and recreation. This causes the problems

R. Silhavy (Ed.): CSOC 2020, AISC 1226, pp. 488–504, 2020.
https://doi.org/10.1007/978-3-030-51974-2_46

of storing and processing large volumes of data which at the present moment are used in an extremely inefficient way.

It should be noticed that the HVAC life cycle is long and it goes beyond the life cycle of the standards encompassing the security, comfort, energy effectiveness and other requirements. This is the reason why a large number of today's public buildings equipped with automated systems do not comply with the modern standards and therefore require a large-scale reconstruction. This brings the need to consider the rational data use aspects back at the stage of the information model design [3, 4].

HVAC operation can be optimized by means of automation of the microclimate control, communication systems and networks using the IoT (Internet of Things) technologies and respective analysis data. The implementation of an IoT-oriented control system with a flexible multilevel architecture means introduction of several control circuits including one machine learning circuit and one adaptation circuit.

The questions of controlling the smart buildings' subsystems have been studied by many authors. Thus, works [5, 6] elaborate on the questions of cyber-physical systems' control. Particularly, the cyber-physical control term is introduced. Cyber-physical control is netcentered. In the centre, there is a control core. The entire system is based on the intellectual nodes of operation control or the intellectual nodes of cyclic control.

The work [7] offers a computer model of an intellectual building, able to simulate the operation of the main utilities of the building and utilities subsystem control algorithms. The authors regard a smart building as a hybrid system. The hybrid nature of the system manifests itself in the description of the interacting elements of the system and their behaviour as both continuous and discreet processes. As a modelling tool, the Simulink system of Matlab package is used. The electric lighting control and the building microclimate control systems' models are built. The quantitative experiments were used to derive a series of logical rules that may be used to develop the smart building control algorithms.

The authors of [8] consider the problem of predictive energy consumption analysis of the building. The distinctive feature of the approach is its applicability to an object during operation of the building at a certain period of time to keep the identification discreet for the user.

In [9], the MatLab model of indoor heating processes behaviour for the automatic control of the energy saving processes in the building has been developed and implemented. The mathematical modelling was based on the differential balance equation method. As dependent variables for the balance design, the heating energy supplied to the object and dispersed in the environment was used. The quantitative experiment makes it possible to assess the heating circuit parameters' coefficients (thermal capacity of walls and floor structures, heat emission of heating devices, heating radiation coefficients), to calculate the thermal energy saving at regulating the indoor temperature for 24 h; to assess the main features of the heating modes of the building (heating and cooling time influenced by the changing parameters of the system).

The work [10] researches the influence of the solar radiation energy on the heat processes inside the building. In comes up with a comprehensive model of the building and a block of weather simulation for assessing the energy saving potential of the typical buildings and structures. The simulation model is based on the Simulink application of Matlab package. The modelling results prove the energy effectiveness of the multi-circuit control system with a distributed structure compared to a centralized system with one regulation circuit.

For controlling the energy saving processes in a public building, the work [11] offers a mathematical model that takes the radiation and convection fluxes into account. The model consists of three differential thermal balance equations.

The work [12] develops mathematical models for the heating mode of the building, suggesting different variations of control input: heating system capacity, temperature of the heat carrier at the heating system input and heat carrier flow rate. The constructed models make it possible to take the unsteadiness of the external air temperature into account.

Despite the experience accumulated in the sphere of automation of buildings and technical capacities of IoT, the approach to HVAC control by means of adaptation still remains understudied. Within the said algorithm, we shall study the compensatory and predictive algorithms that maintain control over the climate parameters in the building based on the IoT platforms.

2 Methods

The adaptive approach to the HVAC control process implies the presence of an environment, an object, a subject, and a control algorithm. The control algorithm consists of the control inputs that switch status of the controlled object from one to another. In this situation, the control over a smart building's HVAC is a process of organizing an intentional impact on the controlled object, or the SB, in order to achieve a number of targets {Ti}: resource saving (T1), reduction of operation expenses (T2), enhancement of safety (T3), provision of comfortable conditions of labour and recreation (T4). The HVAC control system can be presented as a diagram (see Fig. 1).

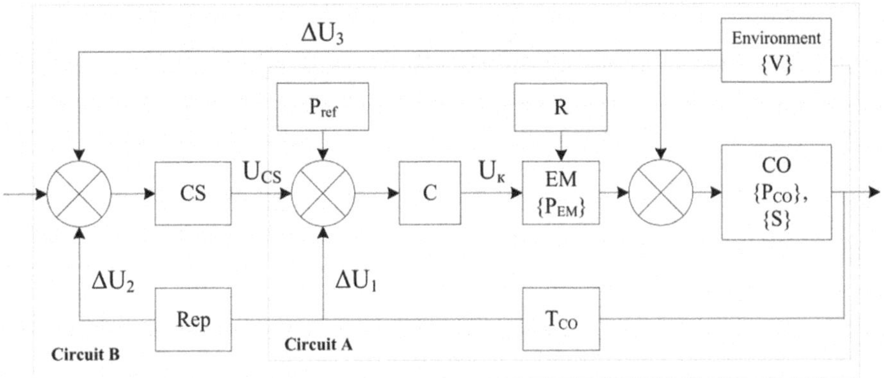

Fig. 1. HVAC control system for SB.

The legend of Fig. 1:

CS - control subject;
CO - control object described with a set of parameters Pco = {p1, p2, ... pi ... pk};
Pref - reference values of the parameters characterizing the condition of the control object.
EM - executive mechanism which uses a set of resources {R} for its work, under the influence of controlling signal Uc changing the state Si of the control object.
C - controller used for controlling the executive mechanism based on the values of the parameters acquired from the CO sensors, reference values of parameters Pref and the control input Ucs.
{V} - external environment parameters.
Tco -transducers for measuring the control object parameters.
Rep - report system the control subject uses to collect information about CO

Based on the information on the values of the CO parameters collected from the reporting system, the information on the current resources and the information on the disturbing influences of the external environment V, the control subject makes a decision on forming a control input Ucs.

Within the control system, three disturbing influences can be found:

- $\Delta U1$ - the input from the transducers of the control object, presenting its changes.
- $\Delta U2$ – the input coming to the control subject from the reporting system.
- $\Delta U3$ – input from the external environment.

These disturbing inputs are compensated by two control circuits:

- Circuit A including the IoT controller, EM, CO with transducers, providing automatic feedback on the control object using the controller;
- Circuit B including the CS, the IoT controller, EM, CO with transducers, the reporting system, providing feedback on the control object in the manual mode using the reporting system.

Between the CO parameters and the EM parameters there is a proportion of G1 Pco $\times$ Pem. Changing the Pem parameter may cause the change of one or several parameters Pco.

Between the multitude of parameters of the external environment {V} and the multitude of the parameters of the control object {Pco} there is a proportion G2 $\subset$ Pco $\times$ V, as the change of Vi yields the change of one or several parameters Pco.

Therefore, the targets T1 and T2 may be achieved by minimizing the consumed resources R, and the targets T3 and T4 may be achieved by establishing the values of the control object parameters Pco and Pem in accordance with the reference values Pref.

At the present time, the control subject in HVAC systems is the operator, a person who uses the information system to acquire information about the condition of the heaters, air conditioners and fans, and establishes the executive mechanism parameters Pem in accordance with the reference values of the indoor climate. At that, the values of the CO parameters are influenced by the disturbances V of the external environment

(Fig. 1). At the same time, it is impossible to efficiently control Pco for the following reasons:

- V, Pco, Pem have different laws of changes in time;
- the dependency of Pco on Rem and V is not always obvious;
- the system has a response rate that does not allow to change Pco quickly when V is changed;
- reference values do not always correspond to real conditions depending on V and the CO state.

The problem may be solved with a system where the control subject will be a Machine Learning technology-based smart system. Such system must search for dependencies between the executive mechanisms' parameters Pem and the control object parameters Pco, the environment disturbances V the control object parameters Pco. These dependencies may be used for forecasting the values of the parameters Pco and Pem, which may help setting the required values with respect to the HVAC response rate in advance.

At the present time, in the IoT architecture of the smart building HVAC control system, similarly to [13], three levels of components are considered (see Fig. 2):

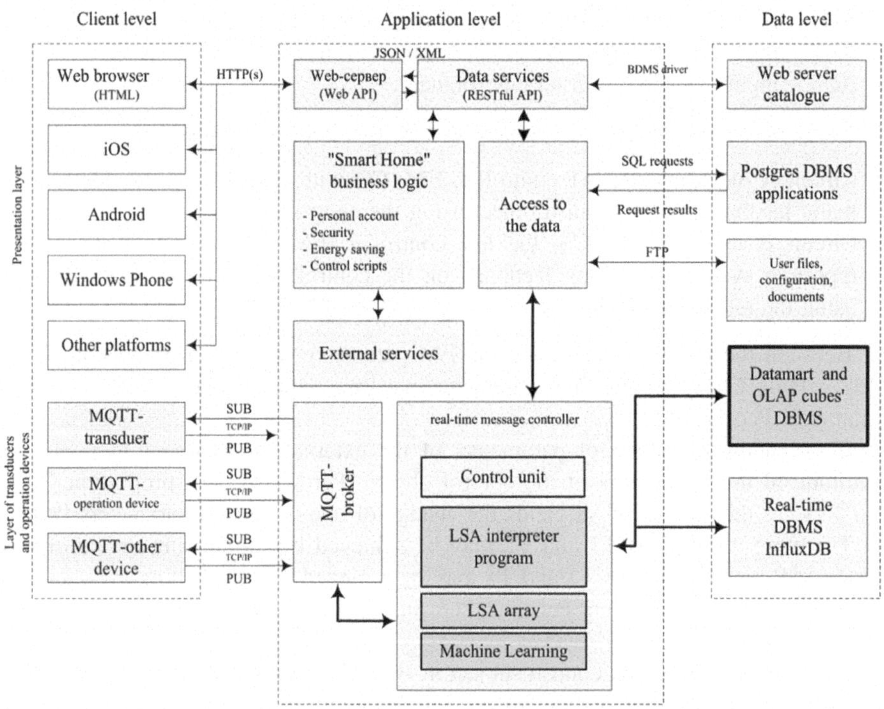

Fig. 2. Three levels of components in the IoT architecture of the HVAC control system with the adaptation model.

- client level;
- application level;
- data level.

The Machine Learning module shall include a report generating block for the CS (see Fig. 3).

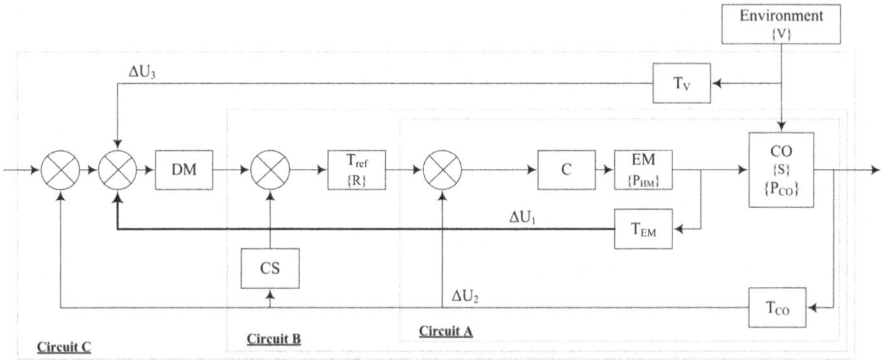

Fig. 3. HVAC control system with an additional circuit for SB.

In the suggested solution, we may use three information models (Fig. 2):

1. transducer and executive mechanism level model (transducers, controller, DB for storing the controller-processed information - circuit A;
2. SB logic model (web interface, SB business logic, DB for storing information at the CB level, communication elements) - circuit B;
3. adaptivity model (datamart, OLAP cubes, physical process modelling subprogram) - circuit C. This model is supposed to be used to accumulate the history of the environment and executive mechanisms'parameters' date during the set period of time. Machine Learning methods may be used to find the dependencies between Pem and Pco, V and Pco, and to use these dependencies for automatic setting of the threshold values of Pref used for CO control. Moreover, the need for resources required for EM may be calculated.

The datamart and OLAP cubes are used to store the parameter values of the environment, Pco, Pem. Datamart is a multidimensional space of parameters collected from the transducers. The Machine Learning and modelling program contains the algorithms of parameters' processing and dependencies' detection.

As additional control circuits are added, there occurs a problem related to the growing volume of the real-time information, as besides the CO parameters, the parameters of the environment and EM will be used. Therefore, there increases the number of transducers required for the measurement of such parameters. This problem is suggested to be solved by using transducers with adjustable actuation threshold on the CO. In this case, the signal from the transducer gets binary value Di = f(Pi) = {0.1}, where Di is a signal from the transducer corresponding to the parameter

Pi, Pi ∈ Pco. Function Di gets the value 1, if Pi>, otherwise Di gets the value 0. Transformation of the transducer values into the binary format makes it possible to use discreet automaton models (such as logical schemes of algorithms, LSA) for implementation of the control algorithms in the controller [14]. For this purpose, the LSA storage module and the LSA interpreter are added (Fig. 4). Environment and executive mechanism transducers are used for the measurement of digital parameters.

This way, there are three levels that can be distinguished in the smart building control system architecture: client level, application level, and data level; and three layers: presentation level, operation unit layer and analytics layer.

Presentation Layer. At an established moment of time, the application server transmits a signal to the controller to send the developed bulk of data from the real-time database. Through the browser, the system user reaches the application which requests the data from the application DB and returns them to the user; at that, the data may be presented in the required form (graphs, diagrams etc.). To build the parameters' space in the analytics layer, it is necessary to synchronize the transducer layer timers and the operation units with the analytics layer. For this purpose, the presentation layer timer is used.

Transducer and Operation Unit Layer. The transducer and operation unit layer timer, issues a signal to the real-time message controller for reception of information through the set intervals of time. The CO transducer returns the value 0, if the value of the respective parameter does not exceed the established threshold value and 1, otherwise. The controller returns the acquired value to the program automaton (LSA), which processes the value and, if necessary, makes up a control command for the operation unit. The controller records all acquired values into the real-time database. After the established intervals of time, the timer issues a signal for getting information from the transducers of the environment and EM transducers by the controller. EM and environment transducers provide information in the form of digital values. The acquired values are also recorded in the real-time database.

Analytic Layer. At a set moment of time, the DBMS of the datamart and OLAP cubes issues a signal to the controller to transmit a formed bulk of data from the real-time database to the datamart. After that, the Machine Learning procedure is launched to process the information of the conditions of the control object, the environment and the operation units for the given period of time, stored in the datamart. After this information is processed, the information about adjustment of the threshold values of the transducers for the next period of the smart building operation will be received for the controller.

3 Results

The controllers that directly operate the "things" of the building HVAC (IoT devices) are connected to the existing computer network of the building with the access to the Internet using wireless access points (WAP) The multitude of WAP covering the entire operation space of the IoT devices in the building participates in transmitting the data packages between themselves, as well as with the local Internet of Things server (IoT server), which may be actualized on the basis of the CSC (control system controller) or a PC work station [15, 16]. The flows of measurement information from the IoT devices are forwarded to the database server (DB server) in charge of accumulating and structuring the information.

IoT-based network complex architecture. The tasks of the control system are accumulation, processing and analysis of the internal and external parameters of the building and its utilities followed by setting new target values, fulfilled by means of the following:

- Circuit A, including the IoT device, transducers (T) and operating unit (OU) in charge of the feedback on the control object in the automatic mode using the MC program scripts;
- Circuit B, including the smart, digital transducers and meters (DT), the network of MCi, CSC and WAP infrastructure, and in charge of the feedback on the control object in the automatic/automated modes by means of the local control system scripts;
- Circuit C, including the communication means of the world wide web Internet access and the external IoT servers, and in charge of the feedback on the control object in the automatic/automated modes by means of the analytical apparatus, including intellectual data analysis on the basis of Machine Learning technologies [17, 18].

Figure 4 presents the architecture of the network control complex of the building HVAC system based on the IoT devices, built on the basis of [19–21]. With circuit C and the Machine Learning technologies applied at IoT servers, the dependencies of the internal building parameters on each other and the external parameters are determined, the target efficiency indicators are calculated to establish the deviations of the reference and actual values.

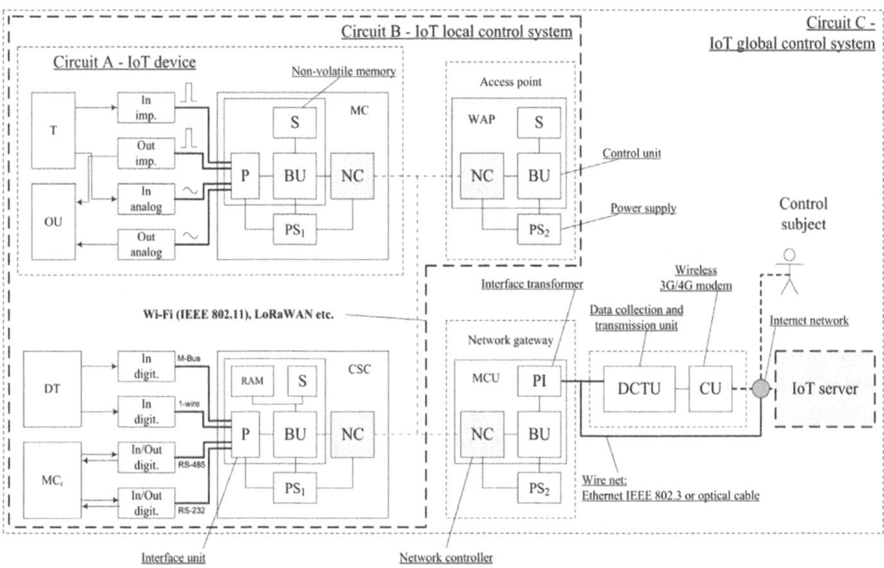

Fig. 4. IoT-based architecture of the network control system of the building.

Based on the continuous analysis of data from the IoT devices and the comparison of the data with the set building control modes as scripts, the faulty functions of the automated system may be detected. It can be done by means of continuous adaptation and optimization of the building HVAC control system parameters, by means of automatic monitoring of the large number of standard modules, through the analysis and predictive management of devices with regard to the comfortable and safe conditions for the office employees and staff, as well as the low energy consumption [22] of the described scripts of Ci (see Fig. 4).

Block Diagram of the IoT-Based Building Multicircuit HVAC Control System. Based on the suggested architecture of the IoT-based network complex of the building HVAC multicircuit control, the control system block diagram was built to monitor the changes in the building and to ensure the adaptation of the used scripts to the changing conditions, see Fig. 5. The elements of the control system diagram may be described with the typical transfer functions of the first or second order device, the lag block and require further identification (see Fig. 5).

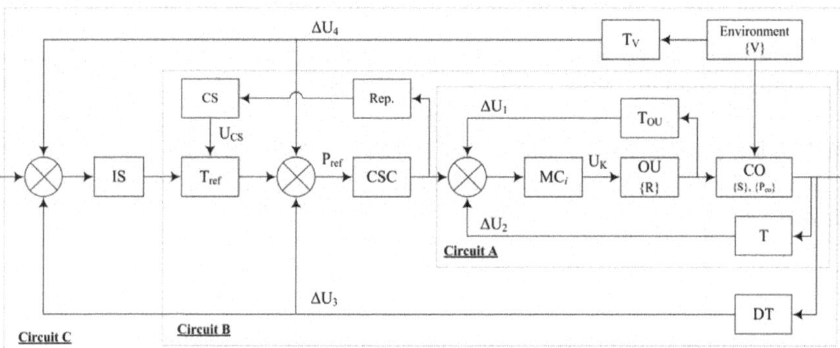

Fig. 5. Block diagram of the IoT-based building HVAC multicircuit control system with adaptation.

Circuit A of the control system operates based on the following principle. A control object CO characterized with a series of parameters $P_{co} = \{p_1, p_2, \ldots\ p_i\ \ldots\ p_k\}$ is directly affected by the executive mechanism EM, which uses a set of resources $\{R\}$ to operate. Under the impact of the control signal U_c and the control script, the object changes its state in such a way, that for each transition between the states there is a one-valued script C_i. OU signal is transmitted from the controlling MC_i of the *IoT* device regulating the output impact on the CO with the OU - T_{OU} transducer. ΔU_1 is the effect from the operation unit transducers, presenting the changes of their states. For registration of the output parameters P_{out} from the control objects, the T transducers were used. ΔU_2 is the effect conveyed to the controlling MC from the control object.

To solve the problem of increasing information processing lags, the MC and CSC implementing the discreet automaton model-based control algorithms shall be used, such as LSA, or logical schemes of algorithms. The application of LSA requires much less computing capacity of the *IoT* device compared to other controller programmer methods, described by the well-known standards, including IEC 61131-3 for PLC programming languages; IEC 61131-5 for programming the connection between the PLC and other programmable systems; IEC 61131-7 for the fuzzy control and regulation programming language.

Circuit B of HVAC control system operates as follows. To the input of the controlling MC_i of the *IoT* device a signal is conveyed for CSC which, in its turn, acquires a set of reference values of the parameters characterizing the state of the control object - P_{ref} established by the control subject, or CS, as input values. Based on P_{ref}, values of the control object parameters acquired from the digital transducers and meters DT, information of the present resources and the disturbing effects of the environment $\{V\}$, CSC makes up the control inputs, controlling the group of the *IoT* devices. ΔU_3 – effect conveyed by the digital transducers and meters installed on the control object. ΔU_4– effect from the external environment. Circuit B CSC may perform the functions of the local *IoT*, the control system server.

In order to solve these problems, the *IoT*-based building control system was introduced the circuit C actualized on the basis of the external server, where the Machine Learning algorithm using intellectual system will act as the control subject. Identification of dependencies between the parameters of the execution mechanisms, P_{co} control object parameters and V external environment disturbances helps forecasting the control inputs with regard to the lag of the system, [23]. In this situation, the system input will be conveyed the vector of the regulated control object parameters - P_{in}.

4 Discussions

The capacities of the IoT controller-based adaptive HVAC control system and the transmission of a control signal with the placement of the control algorithm in a server are demonstrated with the indoor temperature control model. This task is solved in the automated *BEMS (Building Energy Management System*) building control systems.

The research object is a standard office building with the floorspace of around 1000 sq.m. and four floors. The building is erected with the cast-in-place concrete frame technology with filling the external walls with D500 grade gas concrete blocks. The thickness of the external walls of the building is 400 mm. The building is heated with the double-pipe system with bimetallic heating units. The control object is the temperature adjustments of the control object premises. The main task of the regulation is stabilization of the comfortable temperature adjustments on the building premises at the level of 20 °C. The main disturbances for such CO include:

- external environment (atmosphere) temperature possible to forecast [24];
- heat carrier temperature.

Control input is the degree of valve opening on the line of the heat carrier supply into the indoor heating system, i.e. the indoor temperature is maintained by changing the valve position (i.e. regulation of the heat carrier flow rate by the heating devices).

Figure 6 presents a diagram of the imitation model of the indoor temperature changing process. The structural and parametric identification of the model is carried out with the active experiment method (see Fig. 6).

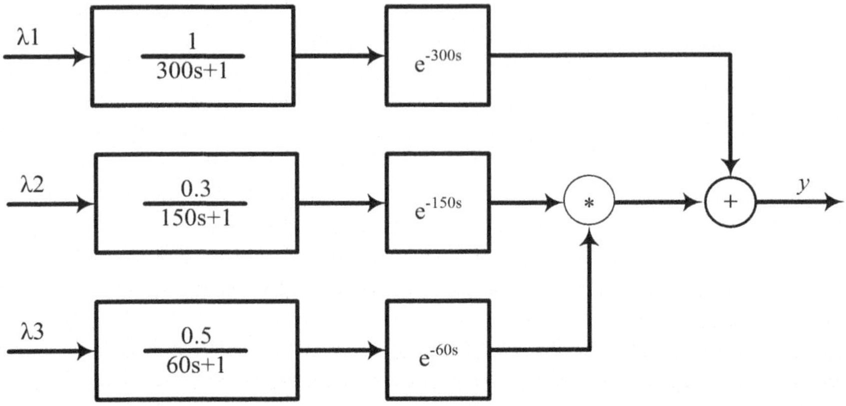

Fig. 6. Indoor changing process imitation model in OpenModelica environment, where: λ1 – ambient temperature; λ2 – heat carrier temperature; λ3 – opening degree of the heat carrier supply valve.

The control quality may be improved by using the computing capacities of the cloud server for calculating the compensatory corrections introduced to compensate the disturbance impact. Transmission function (TF) of a compensator is calculated with the following formula:

$$W_k(p) = \frac{-W_\lambda(p)}{W_{pe\varepsilon}(p)}$$

where $W_\lambda(p)$ - TF of the object in the compensated disturbance channel; $W_{pe\varepsilon}(p)$ - regulator TF.

The imitation model of the indoor temperature changing process when controlled by PI regulator with disturbance compensation has been made up (see Fig. 7).

CS structure with disturbance compensation based on the cloud server and IoT controller is presented in Fig. 8.

The model considers the following influences:

- temperature of the environment with transmission coefficient 1, time constant 300 min, lag 300 min,
- heat carrier temperature with transmission coefficient 0.3, time constant 150 min, lag 150 min,
- valve opening degree with transmission coefficient 5, time constant 60 min, lag 60 min.

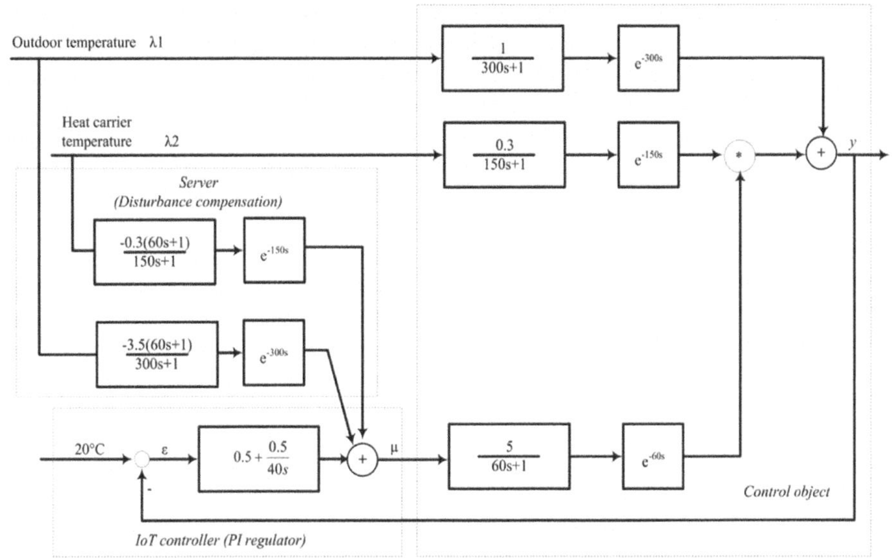

Fig. 7. Imitation model in OpenModelica environment, control system with PI regulator and disturbance compensation.

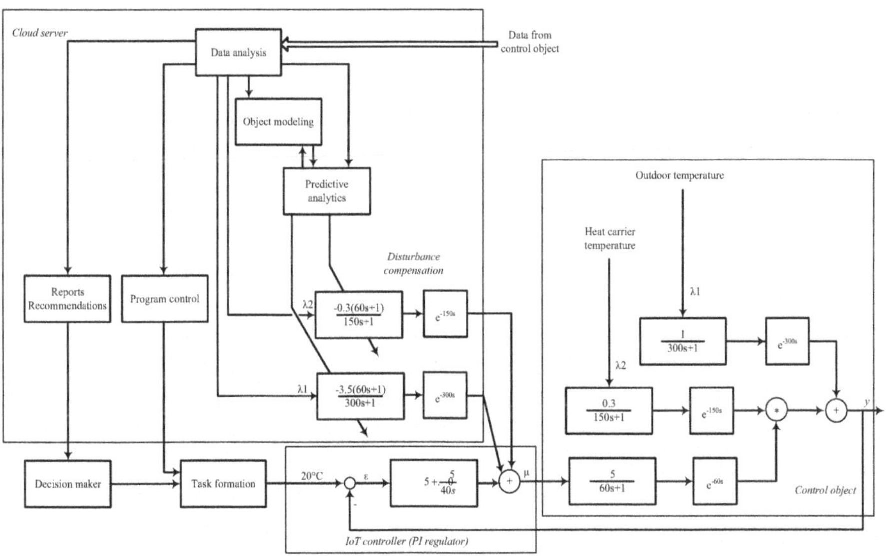

Fig. 8. CS structure with disturbance compensation for IoT controller-based CS.

Input conditions: temperature of the environment 20 °C, heat carrier temperature 20 °C, valve opening degree 50%, indoor temperature 20 °C.

The valve opening degree and the heat carrier temperature are interconnected parameters. When the valve is closed, the heat carrier temperature makes no impact on the indoor temperature, and at the heat carrier temperature of 20 °C the valve opening degree makes no impact on the indoor temperature.

Figure 9 presents the result of 10 days' modelling under disturbances, at the constant control input value - the valve opening degree (47%) (see Fig. 9).

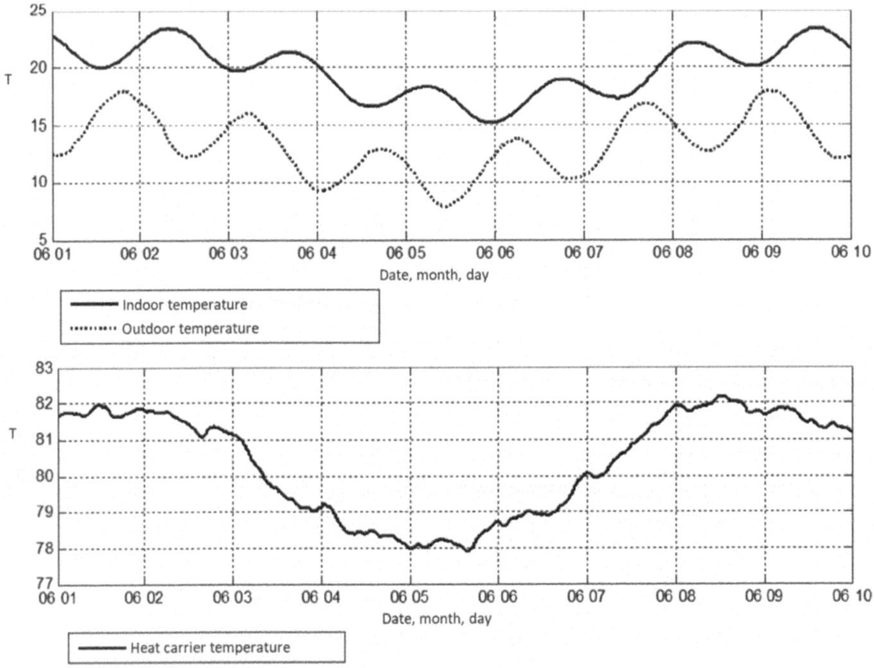

Fig. 9. Indoor temperature changing modeling result without CS.

The signals imitating the changes in the environment temperature, including some random components, daily temperature fluctuations and changes of temperature during several days have been formed. Therefore, during 24 h the temperature of the environment changes from colder values during night, increasing by 3–5 °C by midday, and the daily average temperature also changes with regard to the low-frequency noise during ten days. Heat carrier temperature varies from 78 to 82 °C throughout the entire operation period. At the set conditions, at the absence of an automatic regulation system (constant heat carrier supply valve opening degree - 47%), indoor temperature varies from 15.1 to 23.4 °C. To monitor the control process quality, we shall use the root mean square deviation (RMS deviation) value which for the present model constitutes 5.06.

Figure 10 presents the result of the imitation modelling of the indoor temperature changing process under the same disturbances when controlled by the PI regulator and under the compensatory effects conveyed by the cloud server, at the transducer request frequency of 1 min, 1 h, 3 h, 6 h.

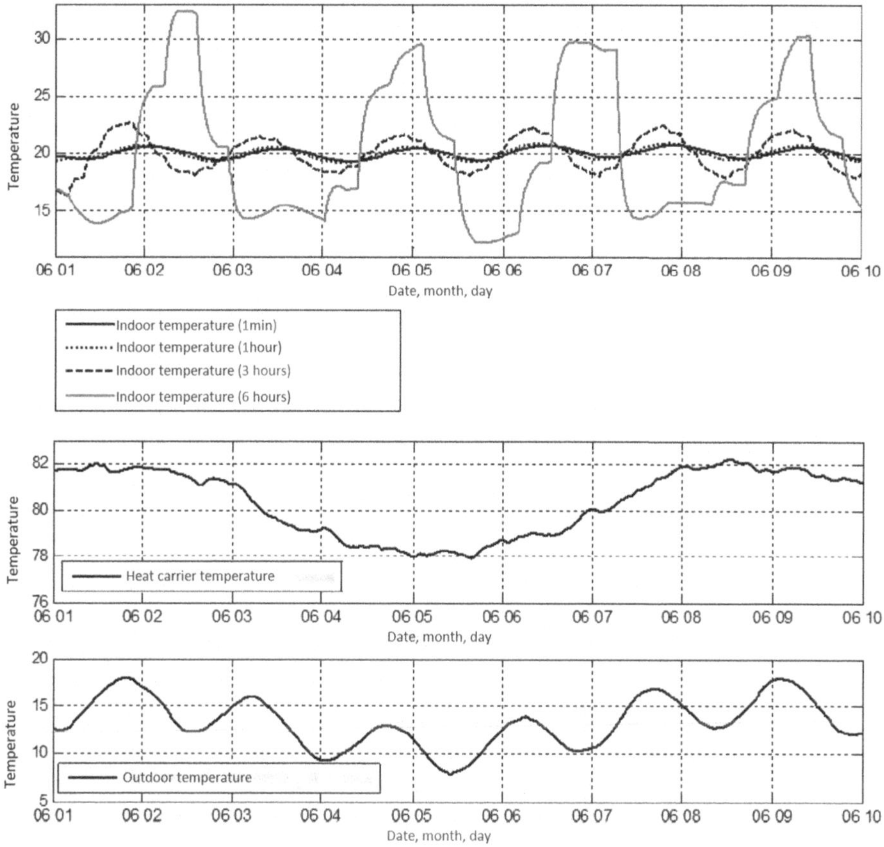

Fig. 10. Result of modelling the indoor temperature of the building when controlled with a PI regulator and disturbance compensation.

Control processes were modelled for different transducer request frequencies in order to assess the variation of the RMS deviation for the situations when the cloud server cannot be contacted for some period of time.

At the transducer request frequency of 1 min, the indoor temperature changes from 19.3 to 20.7 °C, RMS deviation - 0.15.

At the transducer request frequency of 1 h, the indoor temperature changes from 19.2 to 20.9 °C, RMS deviation - 0.23.

At the transducer request frequency of 3 h, the indoor temperature changes from 16.3 to 22.7 °C, RMS deviation - 2.17.

At the transducer request frequency of 6 h, the indoor temperature changes from 12.2 to 32.4 °C, RMS deviation - 35.4.

The increased transducer request frequency reduces the regulation quality, just like in the case of the classical regulation algorithm. However, when the controlled disturbance compensation algorithms are used, the regulation quality is equal in quality to the PI-regulator system at the transducer request frequency of 1 h (see Fig. 10).

Figure 11 presents the reaction of the CO to the step change of opening the heat carrier supply valve from 70% to 80% (see Fig. 11).

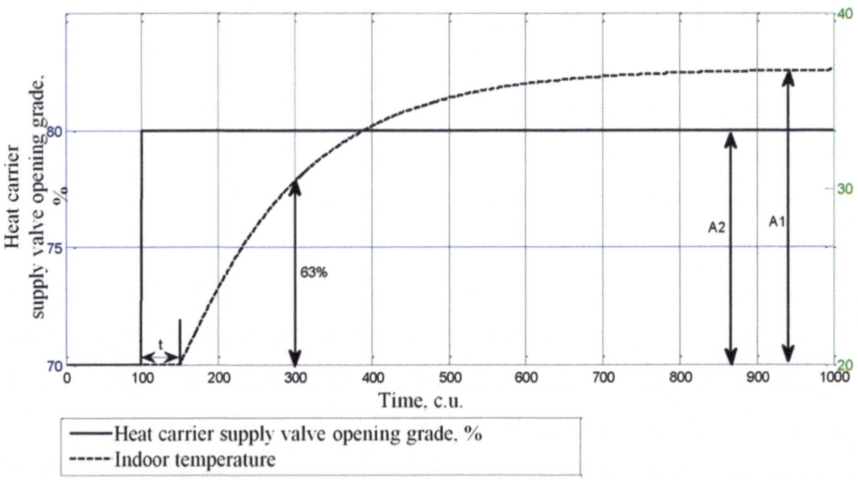

Fig. 11. Smart building HVAC CO channel identification.

Step disturbance supplied at the time moment 100. Reaction to the disturbance occurred at the time moment 150. Therefore, the lag for the channel is 50 units of time. When the input influence changed by 10%, the output changed by 16.74 °C; therefore, the transmission coefficient equals to 1.674. At the moment of time 300, the output value reached 63% of the established condition; therefore, the constant value of time equals to 150 units of time.

5 Conclusion

Controlling the building HVAC with IoT devices implies the presence of several circuits. The first circuit is used for getting automatic feedback on the control object with a series of simplified scripts on the microcontroller of the controller, directly connected to the "smart" or Internet thing. The second circuit ensures automatic/automated feedback with the group control system scripts, with the involvement of specialized local networks controllers. The third circuit provides automatic/automated feedback with an analytic apparatus of the external server.

In the multicircuit system, the control subject is the building operator, using the remote user interface for the building HVAC information output and establishes the parameter values in accordance with the reference values. Using Machine Learning mechanisms in the external building control circuit makes it possible to search for dependencies between the executive device parameters and the control object parameters, the external environment disturbances and the control object parameters in the technical subsystems of the building. The established dependencies are suggested to be used for predictive control, able to make up the controlling signals in advance with due regard to the lag of the system.

The solution described above is recommended to be implemented in the BEMS automated buildings' control systems featuring IoT devices.

References

1. White paper: "Green Intelligent buildings". Mapping of companies and activities in the US within "smart" buildings. Innovation Centre Denmark "Silicon Valley", Palo Alto, CA, USA (2014)
2. Wonga, J.K.W., Li, H., Wang, S.W.: Intelligent building research: a review. Autom. Constr. **14**(1), 143–159 (2005)
3. Zhang, J., Seet, B.-C., Lie, T.T.: Building information modelling for smart built environments. Buildings **5**(1), 100–115 (2015)
4. Hong, T., Feng, W., Lu, A., Xia, J., Yang, L., Shen, Q., Im, P., Bhandari, M.: Building Energy Monitoring and Analysis. Lawrence Berkeley National Laboratory, Berkeley (2013)
5. Kudzh, S.A., Tsvetkov, V.Y.: Netcentered control and cyper-physical systems. Educ. Res. Technol. **2**(19), 86–91 (2017)
6. Tsvetkov, V.Y.: Distributed smart control. State Counsel. **1**, 16–22 (2017)
7. Zhu, C., Leung, V.C.M., Shu, L., Ngai, E.C.-H.: Green Internet of Things for smart world. IEEE Access **3**, 2151–2162 (2015)
8. Lü, X., Lu, T., Kibert, C.J., Viljanen, M.: A novel dynamic modeling approach for predicting building energy performance. Appl. Energy **114**(C), 91–103 (2014)
9. Zhang, J., Seet, B.-C., Lie, T.T.: Building information modelling for smart built environments. Buildings **5**(1), 100–115 (2015)
10. Strizhak, P.A., Morozov, M.N.: Mathematical modelling of building heating regime considering the solar radiation heat input. Tomsk Polytech. Univ. News Georesour. Eng. **326** (8), 36–46 (2015)
11. Stepanov, V.M., Sergeeva, T.E.: Analysis of the buildings' heat-mass-exchange process mathematic models for the development of an electromechanic system control inputs. Tula State Univ. News Tech. Sci. **12**(2), 158–164 (2015)
12. Panferov, V.I., Anisimova, EYu., Nagornaya, A.N.: To the theory of mathematical modelling of the heating circuit of the buildings. South Ural State Univ. Newsl. **14**, 128–131 (2006)
13. Faizrakhmanov, R.A., Frank, T., Kychkin, A.V., Fedorov, A.B.: Sustainable energy consumption control using the MY-JEVIS energy management data system. Russ. Electr. Eng. **82**(11), 607–611 (2011)
14. Chivilikhin, D., Shalyto, A., Vyatkin, V.: Inferring automata logic from manual control scenarios: implementation in function blocks. In: Proceedings of the 13th IEEE International Symposium on Parallel and Distributed Processing with Applications (ISPA 2015), Helsinki, Finland, pp. 307–312. IEEE (2015)
15. Kychkin, A.V.: Synthesizing a system for remote energy monitoring in manufacturing. Metallurgist **59**(9–10), 752–760 (2016)

16. Lyakhomskii, A.V., Perfil'eva, E.N., Perfil'eva, E.N., Kychkin, A.V., Genrikh, N.A.: Software-hardware system of remote monitoring and analysis of the energy data. Russ. Electr. Eng. **86**, 314–319 (2015)
17. Braga, L.C., Braga, A.R., Braga, C.M.A.: On the characterization and monitoring of building energy demand using statistical process control methodologies. Energy Build. **65**, 205–219 (2013)
18. Kychkin, A.V., Mikriukov, G.P.: Applied data analysis in energy monitoring system. Probl. Energet. Reg. **2**(31), 84–92 (2016)
19. Gubbi, J., et al.: Internet of Things (IoT): a vision, architectural elements, and future directions. Future Gener. Comput. Syst. **29**(7), 1645–1660 (2013)
20. Kelly, S.D.T., Suryadevara, N.K., Mukhopadhyay, S.C.: Towards the implementation of IoT for environmental condition monitoring in homes. IEEE Sens. J. **13**(10), 3846–3853 (2013)
21. Tao, F., et al.: IoT-based intelligent perception and access of manufacturing resource toward cloud manufacturing. IEEE Trans. Industr. Inf. **10**(2), 1547–1557 (2014)
22. Menezes, A.C., Cripps, A., Bouchlaghem, D., Buswell, R.: Predicted vs. actual energy performance of non-domestic buildings: using post-occupancy evaluation data to reduce the performance gap. Appl. Energy **97**, 355–364 (2012)
23. Mahdavi, A., Schuss, M., Suter, G., Metzger, A.S., Camara, S., Dervishi, S.: Recent advances in simulation-powered building systems control. In: Proceedings of the 7th International IBPSA Conference (Building Simulation 2009), Glasgow, UK, 27–30 July 2009, pp. 267–268 (2009)
24. Guan, L.: Preparation of future weather data to study the impact of climate change on buildings. Build. Environ. **44**(4), 793–800 (2009)

Technology of Self-orientation of Aircraft Relative to External Objects

Jaffar Daeibal(✉) and Nikolai Sergeev

Institute of Computer Technology and Information Security, Southern Federal University, Taganrog, Russian Federation
jaafer19831l@gmail.com, nesergeev@sfedu.ru

Abstract. The paper considers the problem of autonomous orientation of rotorcraft. The task of orientation is an integral part of the task of autonomous piloting. From a fundamental standpoint, the vast majority of methods for solving this problem work in a strictly deterministic environment and are based on a developed mathematical apparatus. From a technological point of view, well-known approaches are based on a combination of satellite positioning technologies and sensors. The advantages of such technologies are low cost and the ability to reduce the computational complexity of algorithms by increasing the number of sensors used at various stages of solving the problem. But such technologies do not receive mass adoption because they have low positioning accuracy and are dependent on external observation conditions. The main problem that impedes the solution of this problem is the balance between the accuracy of positioning, the computational complexity of the algorithms and the stability of the system in non-deterministic environments. To solve this problem, a technology is proposed for presenting information on the position of the aircraft relative to the object of observation, based on complex linguistic variables. The techniques for representing the position of an object in two- and three-dimensional space are described. The technology of position coding in the RGB color palette, used to calculate the position of the object, as well as for the purpose of training the system by the operator in the future, is presented.

1 Introduction

There is the task of synthesizing a control system for a rotary-wing aircraft designed for automatic take-off and landing [1, 2]. This task requires solving the following problems:

1. Orientation of the aircraft relative to the terrain;
2. The formation of control actions in the coordinate system of the rotorcraft to maintain the take-off/landing trajectory with minimal deviations from the axis.

In this paper, we consider the problem of orienting a rotorcraft. Existing approaches are based on a combination of satellite positioning technology and sensors. The advantages of such technologies are low cost and low computational complexity of the

The original version of this chapter was revised: The author name has been changed from "Jaafer Daiebel" to "Jaffar Daeibal". The correction to this chapter is available at https://doi.org/10.1007/978-3-030-51974-2_59

R. Silhavy (Ed.): CSOC 2020, AISC 1226, pp. 505–517, 2020.
https://doi.org/10.1007/978-3-030-51974-2_47

algorithms. But such technologies have not received commercial or mass use. The reason for this is their main drawback - positioning accuracy.

To date, the typical accuracy of GPS positioning is 6–8 m in the horizontal plane [3]. With the use of additional technical devices it can be improved. But this is true for an open surface, where there are no natural and artificial barriers to receiving a satellite signal. Thus, with a landing pad size of 20 by 20 m, the deviation can be more than 25% and turn out to be critical.

Known systems for automatic monitoring of objects in space, based on image recognition algorithms. They solve the problem of determining the position of the observed object relative to the observation area. However, their disadvantage is computational complexity, which leads to more complex algorithms and an increase in the size of the hardware component. This in turn reduces the payload that the aircraft can carry.

Since the task of automatic take-off and landing is non-deterministic, its solution can be achieved by applying the theory of fuzzy sets [4]. For this, we will consider the task of orienting a rotorcraft as an identification of the position of the take-off and landing site relative to the location of the device. The runway is indicated by the sign H (Heliport).

Known works in this field are focused on dividing the observed situation into coordinate planes and determining the position of an object in a fuzzy coordinate system on each axis and then calculating the position by combining the obtained values. Ultimately, this does not significantly reduce the computational load.

An alternative solution may be such a representation of the position of the object in which there is no need to split the observed plane into separate coordinates [5]. Suppose that the position of an object can be determined in a form similar to that used by the operator: close, far, far left, far right, etc. In this case, it is not necessary to form separate points on the coordinate axes to represent the position of the object, which will significantly reduce the computational complexity of the algorithms [6].

The task of ensuring automatic takeoff and landing of the aircraft in this way can be divided into the following stages:

1. Identification of an object in an image from a digital video camera
2. Determination of the position of the object on the plane
3. Determining the position of an object in space
4. The formation of control actions to ensure autonomous take-off or landing.

The problems of identifying objects in images are successfully solved by existing algorithms [7] and are not the subject of this study. This paper presents an approach to identifying the position of an object on a plane using fuzzy sets that does not involve splitting into coordinate axes.

2 The Concept of Proposed Technology

In general, the task of orientation and automatic control of a rotary-wing aircraft can be represented by the following scheme (Fig. 1):

The problems of identifying objects in images are successfully solved by existing algorithms [7] and are not the subject of this study. Solving the problem of modeling the position of an object based on fuzzy logic, which reduces the computational load,

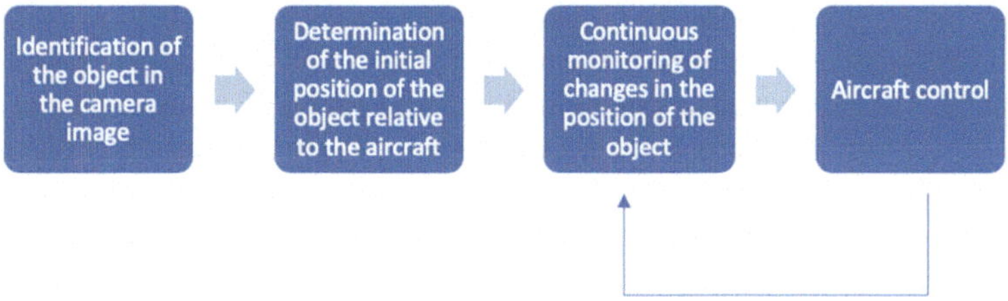

Fig. 1. The process of orientation and control of a rotary-wing aircraft

requires the definition of a new coordinate system, for which there is no need to divide the observed plane into separate coordinates [5]. Suppose that the position of an object can be determined in a form similar to that used by the operator: close, far, far left, far right, etc. In this case, it is not necessary to form separate points on the coordinate axes to represent the position of the object, which will significantly reduce the computational complexity of the algorithms [6].

3 Related Work

The main problems in the autonomous control of rotary-wing aircraft are: accurate measurements (or optimal estimates) of the location of the landing platform, as well as UAVs and a stable trajectory in the presence of interference and uncertainties. To solve these problems, several approaches to the autonomous landing of unmanned aerial vehicles have been proposed. Erginer and Altug proposed the design of a PD controller for orientation control in combination with a vision-based tracking system that allows the quadrocopter to land autonomously on a stationary landing pad [8]. Voos and Nourghassemi presented a control system consisting of controlling the orientation of the inner loop using linearization of feedback, controlling the speed and height of the outer loop using proportional control and a 2D tracking controller based on linearization of feedback for an autonomous landing UAV with four rotors on a moving platform [9]. Ahmed and Pota introduced an advanced non-linear flyback controller for landing an unmanned aerial vehicle using a leash [10].

Robust control methods were also used to land the UAV to deal with uncertain system parameters and interference. Shu and Agarwal used a mixed H2/H∞ control technique, where the H2 method is used to optimize the trajectory, and the H∞ method minimizes the influence of disturbances on the productivity output [11]. Van et al. they also used a mixed H2/H∞ technique to ensure that the UAV tracks the desired landing path under the influence of uncertainties and interference [12]. In their approach, the H2 method was formulated as a linear quadratic Gaussian problem (LQG) for an optimal dynamic response, and the H∞ method was adopted to minimize the effect of soil and atmospheric interference.

Computer vision has been used in a decisive role in many autonomous landing methods. Lee et al. [13] introduced image-based visual services (IBVS) to track the landing platform in two-dimensional image space. Serra et al. [14] also adopted dynamic IBVS along with a translational optimal flow for velocity measurement.

Borowczyk et al. [15] used the AprilTags auxiliary visual system [16] together with an IMU and GPS receiver integrated into a moving target moving at speeds up to 50 km/h. Beul et al. [17] demonstrated an autonomous landing on a golf cart operating at a speed of ~4.2 m/s, using two cameras for high-frequency pattern recognition in combination with an adaptive yaw strategy.

Learning-based management methods for autonomous landing have also been studied to achieve optimal management policies in the face of uncertainty. Polvara et al. [18] proposed an approach based on a hierarchy of deep Q-networks (DQNs), which can be used as a high-class management policy for navigation at different stages. With optimal policy, they demonstrated autonomous landing of a quadrocopter in a wide variety of simulated environments. A number of approaches based on adaptive neural networks have also been adopted to make the trajectory controller more reliable and adaptive, ensuring that the controller is able to direct the aircraft to a safe landing in the presence of various interference and uncertainties [19–23].

The Predictive Control (MPC) model is a control algorithm that uses a process model to predict conditions in the future time horizon and calculate its optimal system input by optimizing a linear or quadratic lens without feedback, taking into account linear constraints. Researchers have already introduced it to other problems. Templeton et al. [24] introduced a terrain mapping and analysis system for autonomous landing of a helicopter in an unprepared terrain based on MPC. Yu and Xiangzhu [25] introduced a predictive controller model to avoid obstacles and plan the route for launching the carrier aircraft. Samal et al. [26] presented a model of a predictive controller based on a neural network for processing external noise and changing system parameters to control the height of an unmanned helicopter. Tian et al. [27] presented a method that combined MPC with a genetic algorithm (GN) to solve the problem of joint UAV search.

Feng, Zhang et al. [28] proposed a new control method that allows micro UAVs to land autonomously on a moving platform in the presence of uncertainties and interference. The main focus of this control method is to implement such an algorithm in an inexpensive, lightweight embedded system that can be integrated into micro-UAVs. The proposed approach consists of visual target tracking, optimal target location and predictive model control, for optimal UAV control.

A review of existing approaches showed that fuzzy logic, despite its advantages, is used as an auxiliary tool or not used at all. Therefore, to solve the synthesis problem for the takeoff and landing control system of a rotorcraft, it is necessary first of all to determine the way of representing its coordinates and to perform its fuzzification.

4 Methods for Determining the Position of an Object in One-Dimensional Space

We solve the problem of identifying the position of the object relative to one observed axis. We will describe the position of the linguistic variable Bi, "POSITION OF THE OBJECT". We define Bi, on the base term-set T1 I = ("LEFT ANGLE", "CENTER", "MIDDLE", "RIGHT ANGLE"}. Such a term-set is convenient for describing the position of an object if the visual axis is perpendicular to the axis of movement of the object. Linguistic variable Bi, as follows:

$$Ti = \{ti1, ti2, ti3\}$$

As the subject scale X, you can choose to remove the object from one of the edges of the image. Information about the position of the object we will receive from a video camera located on a rotary-wing aircraft. Since the object is monitored using a digital video camera, in order to get rid of the subsequent calculation of coordinates, the address space of the memory frame can be declared the object plane.

If the object approaches the aircraft on which the video camera is installed, then the image size of the object increases. Thus, by the size of the image, one can judge the location of the object on the gutter and determine the control actions to achieve the chosen goal. In Fig. 2 shows the membership functions for the terms of the linguistic variable B_i. The unit values of the membership functions of the terms $\{t_{i1}, t_{i2}, t_{i3}\}$ correspond to images O1, O2, O3 in the upper part of the figure. Images of two arbitrary positions of the object O4, and O5 are also presented.

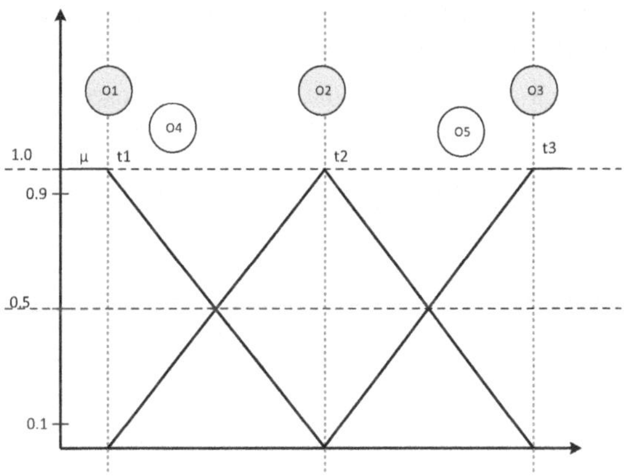

Fig. 2. Membership functions for terms of the linguistic variable B_i

The values of the linguistic variable Bi (O4) = (<t_{i1}/0.7>, <t_{i2}/0.3>, <t_{i3}/0>) and Bi (O5) = (<t_{i1}/0>, <t_{i2}/0.4>, <t_{i3}/0.6>) determine the position of these two objects on the X axis. Next, we define a way to represent the position of the object in two-dimensional space.

5 Methods for Identifying the Position of an Object in Two-Dimensional Space

We will try to get rid of the calculation of the values of quantities in standard measures, even for intermediate attributes. To justify this approach, let us conduct the following thought experiment. If you propose a person to overcome the sequence of obstacles of the same type with increasing dimensions (width or height), then, having overcome the previous obstacle, the next "too high" obstacle, he will refuse to overcome. If we propose to estimate the height of the obstacle more accurately (in meters) than "Too high", then, most likely, without (or having) an insurmountable obstacle in sight, he will imagine how it was and would mentally compare it with the measure. That is, we can conclude that the "measurements" of a person are functional, since in another situation the same obstacle may turn out to be "not high", and that a person resorts to using standards most often when it is necessary to transfer information about a certain image to another person or present on external media. To make decisions based on a digital image, a similar "functional" measurement mechanism is used, i.e. subject scales for a linguistic variable can be constructed, for example, in the number of pixels, and not translate them (the number of pixels) estimated dimensions in centimeters.

We introduce a number of definitions. A complex linguistic variable will be called a linguistic variable defined on a hierarchical set of terms, where each term is, in turn, an independent linguistic variable, which in the future can be used independently or as an element of the term set of another linguistic variable. The subject scale of such a linguistic variable can represent the result of an operation on several scales.

For example, the position of an object on a plane (Fig. 3) will be described by a complex linguistic variable:

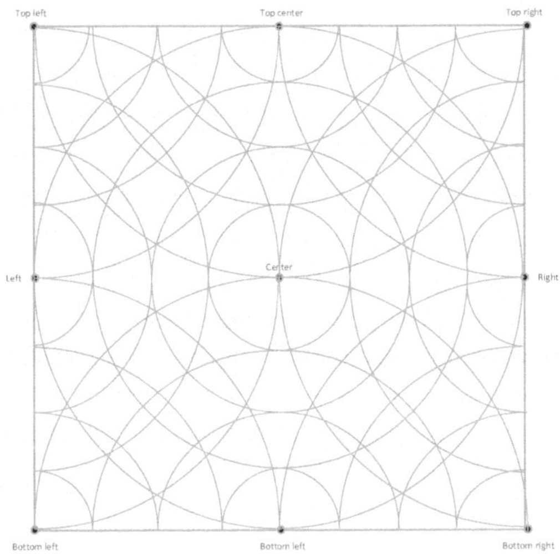

Fig. 3. Representation of a plane to determine the position of an object

$$< \beta u, T, X, G, M >$$

where βu is the name of the linguistic variable - "OBJECT POSITION ON THE PLANE"; T is a basic term set, T = {"EXTREME MIDDLE LEFT CORNER", "CENTER OF THE MIDDLE EDGE", "EXTREME MIDDLE RIGHT ANGLE", "NEAR LEFT CORNER", "NEAR RIGHT CORNER", "CENTER OF RIGHT BOUNDARY", "FAR LEFT CORNER", "FAR RIGHT ANGLE", "FAR EXTREME LEFT ANGLE", "CENTER OF FAR RED CORNER", "EXTREME FAR RIGHT CORNER, T, T, R, T},}} T. L-extreme terms,-angle terms; X is the domain of definition in the classical form-subject scale, in our case, the subject plane (the product of one-dimensional subject scales), because observation of the object is carried out using a digital video camera, the subject plane can declare the address space of the frame in memory in order to get rid of the subsequent calculation of coordinates; G is the syntax procedure; M-semantic procedure.

Complex linguistic variables have three-dimensional membership functions obtained by rotating symmetric membership functions relative to a perpendicular dropped from points for μ = 1, for central terms t € T. by angle π, for extreme t € T L and angular terms t € *T* < relative to perpendiculars to the maximum and minimum of the object scale at angles π and π/2, respectively (Fig. 4).

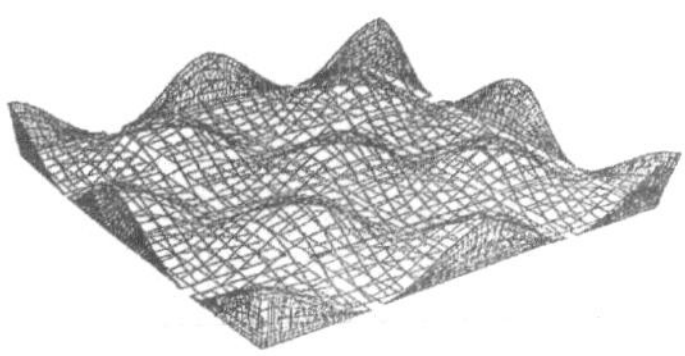

Fig. 4. Membership surfaces of complex linguistic terms

When passing to triangular/trapezoid functions (Fig. 5), the surfaces of the membership functions take the form of cones, truncated cones, half-cones, quarter-cones, as shown in Fig. 4.

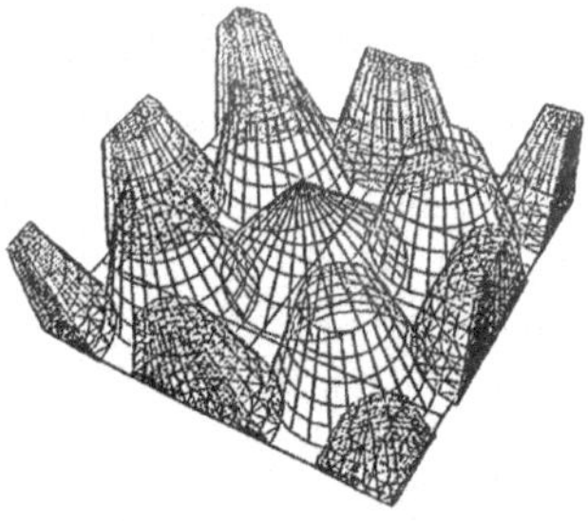

Fig. 5. Membership surfaces obtained from triangular and trapezoidal membership functions

The use of accessories for fuzzy modeling of surfaces can lead to an increase in the volume of calculations. In this regard, it is of interest to projection of sections, a variable that describes the position of an object on a platform on the term set T = {"EXTREME NEAR LEFT ANGLE", "CENTER OF THE NEAR FACE", …} parallel to the base of the surfaces of membership functions at some level μ (μ sectional projection).

They represent concentric circles (ellipses). However, when using triangular and trapezoidal membership functions to form membership surfaces, they can be represented by two arbitrary μ-sections and their projections onto the platform plane. For simplicity of further reasoning, we will define membership surfaces with a zero cross section and an uncertainty cross section—a 0.5 cross section. Terms for which the values of degrees of membership are not zero in a given observation frame are called dominant.

In the case of determining the position of an object using only two projections of μ-sections, uncertainty in the position of the object can occur. In this case, as shown in Fig. 6, finding the object on the projections of the sections is not difficult, because there are only two such points.

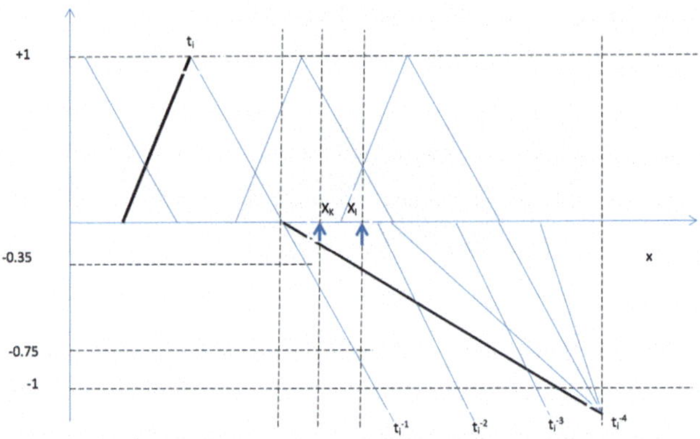

Fig. 6. Using negative continuations of membership functions

We introduce three types of linguistic variables:

The composite linguistic variable βvv. (T1. T2), (X1, X2), G, M, V>
The dual linguistic variable β β, T, (X1, X2), G, M, W>
The complex linguistic variable <βu, T, X, G, M>

This is necessary to create linguistic macrostructures that cover a certain commonality of elementary categories of the subject area, which have a complete semantic meaning. Such a combination, by virtue of the introduced restrictions and assumptions, contains built-in inference elements for decision making. A knowledge engineer who uses such structures to describe a domain may not care about the already built-in inference components. However, when designing more complex structures, it is necessary to comply with the restrictions imposed by such macro attributes. Here are some differences between the introduced linguistic variables.

The composite linguistic variable is intended to reduce to a single linguistic attribute data from several, usually identical, information sources. Each source is represented by its own term set defined on its own subject set. You can give an example with a description of the position of the object according to the information received from several cameras included in the video surveillance system.

The dual linguistic variable is described on a single term set, but some terms are described on two subject scales, i.e. such terms actually have two membership functions. Subject scales can represent different physical quantities, but they must necessarily depend on the state of some attribute, which is described by the used term set. Thus, either both channels indirectly describe some property that is not available on the day of direct measurement, or, for some reasons, it is necessary to obtain information simultaneously from two sources.

The complex linguistic variable differs from the classical definition of the term set and the form of the domain of definition. As noted above, each term, in turn, can be described as an independent linguistic variable with terms. described on a linear subject scale. Thus, we can say that in the complex linguistic variable Bu, the term set and the subject set are formed by the product of the term sets and subject scales of two (or three) linguistic variables β1, and β2, i.e. T = T 1, x T2, X1 x X2. Obviously, a linguistic variable in this definition is convenient for representing the position of an object on a plane and in space.

When modeling decision-making processes, it is necessary to provide for the resolution of the following situation. Suppose two specific values of a certain linguistic variable have zero values of the membership function of the term of interest to us. We need to find out which of the values "nevertheless" is closer to the reference value or which value has the higher degree of reliability of the term of interest. The values of adjacent (neighboring) terms can not help much, for example, when the terms are difficult to arrange, as is the case with the linguistic variable "POSITION OF THE OBJECT ON THE PLANE" (see above).

To clarify the position of the object on the plane, it is proposed to search (scan) the projections of the sections of the uncertainty of the negative extensions of the membership surfaces of neighboring non-dominant terms. After going over to reasoning according to the plan of sectional projections, one may doubt the expediency of using an excessive number of terms.

6 The Method of Representing the Position of the Object

To determine the position of an object relative to the observed plane, it is necessary to encode the values of fuzzy linguistic variables that define zones on the plane. The numerical representation of the position is not always acceptable, since the position of the object is characterized not by one but by a multitude of nonzero values of fuzzy linguistic variables (from 1 to 6) (see Fig. 7).

As can be seen from the figure, such a representation is complex and requires certain costs for perceiving and determining a position based on available data -

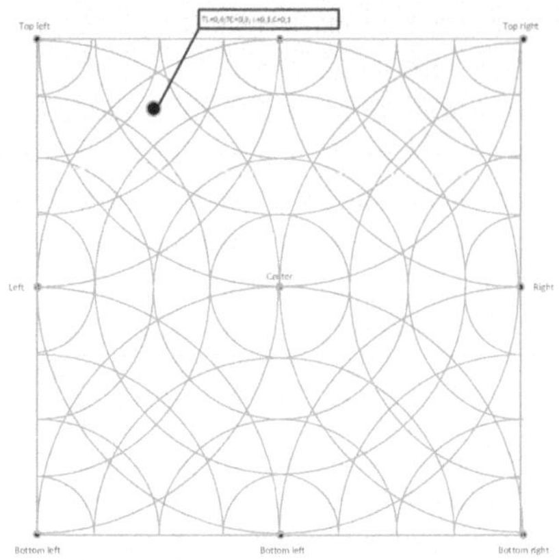

Fig. 7. Numerical representation of the position of the object

numbers vary in the same ranges, and the object's belonging to a particular area on the plane is determined by the membership of numerical values to points. It is necessary to find a way to represent the situation in which one value of one variable is enough. Such a variable may be the color of the pixel, uniquely interpreted by both the operator and the computer program. To set the color of the pixel, any color scheme can be selected, the color depth of which allows you to encode the required number of variations.

The most common color model used in digital images is RGB [28]. Thanks to the additive principle of constructing this model, one can represent the coordinates of the object as follows:

$$\begin{cases} Red(0\ldots255) : Left(0)\ldots Right(255) \\ Green(0\ldots255) : Bottom(0)\ldots Top(255) \end{cases}$$

Therefore, the extreme left lower point will correspond to black, and the extreme right upper point will be yellow (see Fig. 8). The blue component of the RGB model (Blue (0 … 255) will be used in the future for height coding.

We define the discretization step necessary for coding the coordinates of the observed plane. According to the scheme proposed in the previous section (see the figure), the position of the object is represented on the plane using nine fuzzy linguistic variables from TL to BR. Each variable can take 11 values (from 0 to 1 inclusive in increments of 0.1), which correspond to the location of the object relative to a given point on the plane. In addition, each point can potentially intersect five neighboring three times (by values of 0.5 … 0.1), which leads to the imposition of color palettes of each point and a new palette.

Therefore, the minimum palette used to encode the position of an object should include 9 * 11 colors for encoding the values of fuzzy linguistic position variables, as well as 3 * 5 colors for encoding intersection points. In total, to encode the coordinates of the plane when using a step of values of fuzzy linguistic variables of 0.1, at least 114 colors or an 8-bit RGB model palette are required.

Fig. 8. Representation of coordinates using a color palette

7 Conclusions and Further Research

The paper proposed a technique for determining the position of an object on a plane using a fuzzy coordinate system, in which there is no need to divide the plane into the coordinate axis.

The use of complex linguistic variables and coordinateless representation allows to provide low computational complexity of the algorithms, and thereby minimize the hardware requirements.

The use of color schemes when representing the position of the object is also aimed at simplifying the processing and perception of the position of the object relative to the observed plane.

Further research under this topic will be aimed at solving the following problems:

1. Determining the position of the object in three-dimensional space, taking into account the height. This task will require the development of a new complex linguistic variable of procedures for the additive addition of blue to the existing palette
2. Determination of the movement of an object in two-dimensional and three-dimensional space. This task will require the development of new linguistic variables to represent the speed, acceleration and direction of movement of the object
3. Prediction of the position of the object based on observational data
4. The synthesis of the control system and the formation of control actions for a rotorcraft.

References

1. Scherer, S., Chamberlain, L., Singh, S.: Autonomous landing at unprepared sites by a full-scale helicopter. Robot. Auton. Syst. **60**(12), 1545–1562 (2012)
2. Bondarev, V.G., Lopatkin, D.V., Smirnov, D.A.: Automatic landing of aircraft. Bulletin of Voronezh State University, Series: System Analysis and Information Technologies, no. 2, pp. 44–51 (2018). GPS.gov. GPS.gov: GPS Accuracy (2017). https://www.gps.gov/systems/gps/performance/accuracy/
3. Pshikhopov, V., Sergeev, N., Medvedev, M., Kulchenko, A.: The design of helicopter autopilot. SAE Technical Papers, 5 (2012)
4. Sergeev, N.E.: Fuzzy Models of Instrumental Motor Actions of the Operator. Publishing house Rost. University, p. 135 (2004)
5. Nomenchuk, A.Y., Sergeev, N.E.: About one of the ways of managing the takeoff and landing of the helicopter. In: Promising Systems and Control Tasks Materials of the Twelfth All-Russian Scientific and Practical Conference and the Eighth Youth School-Seminar "Information Management and Processing in Technical Systems", pp. 271–282 (2017)
6. Samoylov, A., Kucherova, M., Tchumichev, V.: Model of an intellectual information system for recognizing users of a social network using bioinspired methods. In: Advances in Intelligent Systems and Computing, vol. 985, pp. 147–155 (2019)
7. Kucherova, M.S.: The analysis of approaches to identification of individuals by digital images. In: Innovative Technologies and Didactics in Teaching Collected Papers 2017, pp. 140–144 (2017)
8. Erginer, B., Altug, E.: Modeling and PD control of a quadrotor VTOL vehicle. In: Proceedings of the 2007 IEEE Intelligent Vehicles Symposium, Istanbul, Turkey, 13–15 June 2007, pp. 894–899 (2007)
9. Voos, H., Nourghassemi, B.: Nonlinear control of stabilized flight and landing for quadrotor UAVs. In: Proceedings of the 7th Workshop on Advanced Control and Diagnosis ACD, Zielo Gora, Poland, 17–18 November 2009, pp. 1–6 (2009)
10. Ahmed, B., Pota, H.R.: Backstepping-based landing control of a RUAV using tether incorporating flapping correction dynamics. In: Proceedings of the 2008 American Control Conference, Seattle, WA, USA, 11–13 June 2008, pp. 2728–2733 (2008)
11. Shue, S.-P., Agarwal, R.K.: Design of automatic landing systems using mixed H/H control. J. Guid. Control Dyn. **22**, 103–114 (1999)
12. Wang, R., Zhou, Z., Shen, Y.: Flying-wing UAV landing control and simulation based on mixed $H2/H_{\infty}$. In: Proceedings of the 2007 IEEE International Conference on Mechatronics and Automation, ICMA 2007, Harbin, China, 5–8 August 2007, pp. 1523–1528 (2007)
13. Lee, D., Ryan, T., Kim, H.J.: Autonomous landing of a VTOL UAV on a moving platform using image-based visual servoing. In: Proceedings of the IEEE International Conference on Robotics and Automation (ICRA), Saint Paul, MN, USA, 14–18 May 2012, pp. 971–976 (2012)
14. Serra, P., Cunha, R., Hamel, T., Cabecinhas, D., Silvestre, C.: Landing of a quadrotor on a moving target using dynamic image-based visual servo control. IEEE Trans. Robot. **32**, 1524–1535 (2016)
15. Borowczyk, A., Nguyen, D.-T., Nguyen, A.P.-V., Nguyen, D.Q., Saussié, D., Ny, J.L.: Autonomous landing of a multirotor micro air vehicle on a high velocity ground vehicle. J. Guid. Dyn. **40**, 2373–2380 (2016)
16. Olson, E.: AprilTag: a robust and flexible visual fiducial system. In: Proceedings of the IEEE International Conference on Robotics and Automation, Shanghai, China, 9–13 May 2011, pp. 3400–3407 (2011)

17. Beul, M., Houben, S., Nieuwenhuisen, M., Behnke, S.: Landing on a moving target using an autonomous helicopter. In: Proceedings of the 2017 European Conference on Mobile Robots (ECMR), Paris, France, 6–8 September 2017, pp. 277–286 (2017)
18. Polvara, R., Patacchiola, M., Wan, J., Manning, A., Sutton, R., Cangelosi, A.: Toward end-to-end control for UAV autonomous landing via deep reinforcement learning. In: Proceedings of the 2018 International Conference on Unmanned Aircraft Systems (ICUAS), Dallas, TX, USA, 12–15 June 2018, pp. 115–123 (2018)
19. Juang, J., Chien, L., Lin, F.: Automatic landing control system design using adaptive neural network and its hardware realization. IEEE Syst. J. **5**, 266–277 (2011)
20. Lungu, R., Lungu, M.: Automatic landing system using neural networks and radio-technical subsystems. Chin. J. Aeronaut. **30**, 399–411 (2017)
21. Qing, Z., Zhu, M., Wu, Z.: Adaptive neural network control for a quadrotor landing on a moving vehicle. In: Proceedings of the 2018 Chinese Control and Decision Conference (CCDC), Shenyang, China, 9–11 June 2018, pp. 28–33 (2018)
22. Lee, S., Shim, T., Kim, S., Park, J., Hong, K., Bang, H.: Vision-based autonomous landing of a multi- copter unmanned aerial vehicle using reinforcement learning. In: Proceedings of the 2018 International Conference on Unmanned Aircraft Systems (ICUAS), Dallas, TX, USA, 12–15 June 2018, pp. 108–114 (2018)
23. Templeton, T., Shim, D.H., Geyer, C., Sastry, S.S.: Autonomous vision-based landing and terrain mapping using an MPC-controlled unmanned rotorcraft. In: Proceedings of the IEEE International Conference on Robotics and Automation, Roma, Italy, 10–14 April 2007, pp. 1349–1356 (2007)
24. Wu, Y., Qu, X.: Obstacle avoidance and path planning for carrier aircraft launching. Chin. J. Aeronaut. **28**, 695–703 (2015)
25. Samal, M.K., Anavatti, S., Garratt, M.: Neural network based model predictive controller for simplified heave model of an unmanned helicopter. In: Proceedings of the International Conference on Swarm, Evolutionary, and Memetic Computing, Bhubaneswar, India, 20–22 December 2012, pp. 356–363 (2012)
26. Tian, J., Zheng, Y., Zhu, H., Shen, L.A: MPC and genetic algorithm based approach for multiple UAVs cooperative search. In: Proceedings of the International Conference on Computational and Information Science, Shanghai, China, 16–18 December 2005, pp. 399–404 (2005)
27. Feng, Y., Zhang, C., Baek, S., Rawashdeh, S., Mohammadi, A.: Autonomous landing of a UAV on a moving platform using model predictive control. Drones **2**, 34 (2018)
28. The Evolution of Color Pigment Printing. Artfacts.org. Accessed 29 Apr 2013

Efficiency Estimation of Single Error Correction, Double Error Detection and Double-Adjacent-Error Correction Codes

N. D. Kustov, E. S. Lepeshkina(✉), and V. Kh. Khanov

Reshetnev Siberian State University of Science and Technology, Krasnoyarsk, Russia
klepkal111.93@mail.ru

Abstract. Failures in memory are a serious problem for ensuring the reliability of spacecraft onboard equipment. To correct errors in memory, Single Error Correction and Double Error Detection (SEC-DED) codes are traditionally used. However, they do not correct multiple errors in the code word, the probability of occurrence of which increases with the development of microelectronic technologies. More recently, a new class of codes Single Error Correction, Double Error Detection and Double-Adjacent-Error Correction (SEC-DED-DAEC) code. However, these codes are not only characterized by increased code word implementation complexity and redundancy, but also by the probability of double non-adjacent error mis-correction. The SEC-DED-DAEC currently includes several codes. In this work, an experimental effectiveness estimation of this class correction codes is carried out, for which a method for evaluating the code was proposed, combining the indicators of redundancy, complexity and DAEC mis-correction probability. Based on the estimation, recommendations are given on the use of specific SEC-DED-DAEC codes in electronic space instrument engineering. The results of the work can be used in the design of fault-tolerant memory, including cryptographic systems for spacecraft. #CSOC1120

Keywords: Multiple cell upset · Correction code · Double error correction code · Estimation of double-adjacent-error correction code

1 Introduction

Onboard equipment operating in space under the influence of cosmic radiation is sensitive to short-term failures caused by exposure to ionizing particles. This can lead to failures in electronic devices. The interaction of charged particles with transistors can cause short-term and permanent failures, depending on the position and amount of charge transferred to the material as a result of collisions of particles with silicon. Similar effects that occur under the influence of ionizing radiation are called single events (SEE – Single Event Effects). They can lead to recoverable or non-recoverable failures [1]. A short-term manifestation of SEE is called a fatal error if the device has not received permanent damage. A striking example is a single memory cell failure (SEU – Single Event Upset). This is the inversion of discharges that occurs when an

R. Silhavy (Ed.): CSOC 2020, AISC 1226, pp. 518–525, 2020.
https://doi.org/10.1007/978-3-030-51974-2_48

ionizing particle strikes a transistor of a memory cell, provided that the charge is sufficient to reverse the state. The memory cell is still working properly, performing read and write operations, but the information stored in it is corrupted.

It should be noted that the concept of a removable error does not reduce the severity of the problem. On the contrary, the short-term and random nature of single failures leads to the extreme complexity of their search and correction. The probability of an SEU making the entire array of memory cells potentially faulty. It should be noted that with the miniaturization of the component base, there is a possibility that several memory cells will fall under the influence of a charged particle. This creates multiple cell failures (MCU – Multiple Cell Upset) in the memory arrays. If the damaged bits belong to one memory word, then the MCU is called Multiple Bit Upset [2].

Correction codes are used to correct errors in memory. There are several types of correcting codes. The most widely used error correction codes in memory are codes with Single Error Correction and Double Error Detection (SEC-DED). To correct failures of two-bit MBUs, a new class of codes Single Error Correction, Double Error Detection and Double Adjacent Error Correction, SEC-DED-DAEC [3] has recently appeared. An experimental evaluation of correcting codes effectiveness of this class is carried out in this work. For this, a code evaluation method was proposed that combines the indicators of redundancy, complexity, and DAEC mis-correction probability.

2 Related Work

SEC-DED codes are capable to correct single-bit errors and detect two-bit ones using redundant bits called parity bits. The advantage of such codes is the high decoding process speed with a small number of parity bits. However, radiation-induced single effects can lead to multiple failures of MBU cells in adjacent areas where two or more error bits occur in the same word. In this case, the corrective capabilities of the SEC-DED codes are not enough to correct the errors that occur.

Bose–Chaudhuri–Hocquenghem codes (BCH) [4] can be used as an alternative (reference) solution with the ability to correct errors of higher multiplicity, however, the overhead of their implementation in terms of the parity bits number and decoder design complexity are large.

More recently, SEC-DED-DAEC has been proposed that are capable of correcting single-bit errors and adjacent two-bit errors. For this group of correction codes, the number of parity bits, as well as modulo 2 (XOR) addition operations, performed during the generation of the syndrome in the decoder, is less than that of the BCH codes that allow the correction of double errors (DEC BCH). In addition, the logic of the SEC-DED-DAEC decoder is simpler than the logic of the DEC BCH decoders, since the possibility of correcting double non-adjacent errors for these codes is not required. However, the SEC-DED-DAEC codes are characterized by mis-correction of double non-adjacent errors, since the syndromes for double non-adjacent errors can be equal to the syndromes of double adjacent errors.

There are several SEC-DED-DAEC codes based on binary linear block codes. They are shortened Hamming codes, which provide the possibility of detecting and correcting errors of a greater multiplicity than Hamming codes [5]. To ensure the functioning of the syndrome for double adjacent errors, it must be separated from the syndromes for single errors and double adjacent errors. However, the probability of mis-correction remains, unless the syndromes for double adjacent errors and double non-adjacent errors are not mutually separated.

For example, the Dutta codes [6] provides correction of the double adjacent error by moving the columns in the H-matrix (verification matrix) obtained from the SEC-DED Hsiao code [7], but their error correction coefficient is very high. The columns of even and odd weight in the H-matrix are used in the Richter [8], Datta [9] and Ming [10] codes, but despite attempts to reduce the frequency of mis-correction, this problem is still relevant in these codes. In addition, even though the Datta codes have a rather low error correction rate, they require a large number of parity bits and XOR operations to generate the syndromes. In the H-matrix for the Neale code, a repeating unit matrix of the "Fire code" [11] is used to detect successive errors and correct double adjacent errors [3]. Although the number of XOR operations in the syndrome for Neale codes is lower than in other codes, the error correction rate is slightly higher than in the above Datta codes.

The fundamental difference between such codes, determining their corrective capabilities, is directly the H-matrix on which they are based. Such matrices are constructed according to several predefined conditions, which differ slightly for each code. The conditions for constructing the H-matrix are selected on the basis of hypothetical assumptions about their effectiveness, as well as conclusions made on the basis of experimental results obtained using one or another code.

So in order to achieve the possibility of correcting a single error and detecting a double error (for SEC-DED codes), the H-matrix must comply with the following rules:

1. each column of the matrix is a nonzero vector;
2. each column of the matrix is different;
3. result of the XOR operation on any two columns is not equal to any of the other columns.

The first two rules provide a code distance of three, which allows codes to correct single errors. The third rule provides a code distance of four, which in turn allows detecting double errors.

For SEC-DED-DAEC codes, the set of conditions is usually somewhat wider. So, for example, for the Hoyoon-Yongsurk code [12], the H-matrix must meet the following conditions:

1. each column of the matrix is a nonzero vector;
2. each column of the matrix is different;
3. the result of the XOR operation on any two columns is not equal to any of the other columns;
4. columns with odd and even weight are not placed sequentially;
5. all columns with odd weight have a weight of more than two.

The fourth rule, it is claimed, allows distinguishing between syndromes of double adjacent errors from syndromes of double non-adjacent errors. The fifth rule makes it possible to distinguish between syndromes of double adjacent errors in information bits and syndromes in parity bits. Although the syndromes of double non-adjacent errors in information bits and parity bits may overlap with the syndromes of double adjacent errors, this is considered acceptable, since in reality, as a rule, most information bits are physically stored separately from parity bits.

Since mathematical proof of a particular H-matrix effectiveness is not possible, just like enumerating all possible matrices, certain algorithms based on various heuristic selection methods are compiled to construct them. As a result, from the set of matrices compiled by these algorithms, one is selected that has the highest efficiency (that is, the lowest probability of mis-correction) with the minimum possible decoder complexity (minimum number of XOR operations on the code word).

In this regard, each code ultimately has its own unique parameters. Such parameters, in turn, entail requirements for the hardware on which one or another encoding method is implemented. For example, the requirements for the number of memory cells required to store parity bits. In addition, depending on the specifics of the equipment operation, the speed of the encoding and decoding processes or the likelihood of mis-correction can be critical when choosing an encoding method. Thus, it is not possible to choose a universal coding method that could be used for any task. Therefore, there is a need for a partial estimate of the code effectiveness.

3 SEC-DED-DAEC Codes Efficiency Estimation Method

The paper proposes a methodology for evaluating the effectiveness of correcting codes. It is proposed to evaluate the effectiveness of codes in three main parameters:

1. code word redundancy (number of correction bits) – this parameter of codes determines the necessary amount of additional memory to perform the encoding task;
2. the number of units in the verification matrix (the number of XOR operations) – the parameter determines the complexity of the decoder, due to the number of XOR operations in the decoding process;
3. the probability of double non-adjacent error mis-correction – the parameter shows the probability of erroneous detection of double non-adjacent error as double adjacent and, as a consequence, mis-correction of code word bits.

If the redundancy of the code word and the number of units in the check matrix are encoding parameters known in advance, then the probability of twofold non-adjacent errors mis-correction is a parameter that must be determined by empirical research that is, modeling of encoding methods.

It is proposed to apply a transforming function to each of the studied parameters, which allows one to average its estimate with respect to other parameters. Thus, each parameter will be evaluated on a scale of [0; 10], and the sum of the results of the each parameters evaluation will determine the effectiveness of a particular code.

Converting Functions:

1. *Redundancy estimate* = *10* × *NIB*/*TNB*, where *NIB* is the number of information bits in the codeword, and *TNB* is the total number of bits of the codeword;

2. *Decoder complexity estimate* = *(200 − NXOR)/20*, where *NXOR* is the number of units in the check matrix, and *200* is selected as the maximum value of the number of units inherent in the code matrix of the DEC BCH;
3. *Error correction estimate* = *(100 − MCV)/10*, where *MCV* is the mis-correction value of a double non-adjacent error obtained experimentally for the code in percent.

A overall estimate for codes can be obtained by summing the results at the output of the functions: *overall estimation* = *(redundancy estimate* × *SF1)* + *(decoder complexity estimate* × *SF2* + *error correction estimate* × *SF3).*

Moreover, for each estimation, significance coefficients (SF1, SF2, SF3) are provided that allows to set the priorities of a particular parameter depending on the specifics of the task. In the general case, when the significance factors are equal to “1”, the total score is determined on a scale of [0; 30].

4 Experimental Results

The following codes with a 32-bit information word were chosen as the studied ones:

1. Hsiao code (39, 32);
2. BCH code with double error correction (44, 32);
3. Dutta code (39, 32);
4. Datta code (42, 32);
5. Neale code (42, 32);
6. Reviriego code (39, 32) [13];
7. Cha-Yoon code (39, 32) [14];
8. Hoyoon-Yongsurk code (41, 32).

SEC-DED-DAEC codes were modeled using a program written in C++, and DEC BCH code was modeled in MATLAB. Each code passed 10,000 iterations in a double non-adjacent error mode (bit errors are guaranteed to be added to two random different non-adjacent bits of the code word).

As a result, a summary Table 1 was obtained that takes into account all three investigated parameters.

Applying the proposed technique to the data obtained, the following results were achieved (see Table 2).

Values of the overall estimate for equal 1 significance indicators are distributed in the range from 16.895 to 22.059.

Table 1. Code parameter values.

Code	The number of code word bits	The number of units in the verification matrix	The probability of mis-correction of a double non-adjacent error (%)
Hsiao (39, 32)	39	96	37
BCH (44, 32)	44	200	0
Dutta (39, 32)	39	96	56,5
Datta (42, 32)	42	140	21,1
Neale (42, 32)	42	80	15,6
Reviriegȯ (39, 32)	39	103	61,6
Cha-Yoon (39, 32)	39	83	59,3
Hoyoon-Yongsurk (41, 32)	41	116	4,3

Table 2. Code parameter estimations.

Code	Redundancy estimate	Decoder complexity estimate	Error correction estimate	Overall code estimation
Hsiao (39, 32)	8,205	5,200	6,300	19,705
BCH (44, 32)	7,273	0,000	10,000	17,273
Dutta (39, 32)	8,205	5,200	4,350	17,755
Datta (42, 32)	7,619	3,000	7,890	18,509
Neale (42, 32)	7,619	6,000	8,440	22,059
Reviriego (39, 32)	8,205	4,850	3,840	16,895
Cha-Yoon (39, 32)	8,205	5,850	4,070	18,125
Hoyoon-Yongsurk (41, 32)	7,805	4,200	9,570	21,575

5 Evaluation

The given data and estimates allow to draw some conclusions about the effectiveness of each of the codes under study.

Hsiao code has low redundancy, as well as relatively low decoder complexity; however, the probability of a double non-adjacent error mis-correction is high. Overall estimate is 19,705.

The BCH code with double error correction faultless corrects double non-adjacent errors; however, it has the highest redundancy and complexity of the decoder. In addition, these indicators increase linearly with respect to the number of bits in the code word, which indicates its inefficiency for space memory using. The overall estimate for this code is 17.273.

The Dutta code has the same decoder redundancy and complexity indicators as the Hsiao code, but it loses significantly in terms of a double non-adjacent error mis-correction probability, which also indicates its low efficiency. Overall code estimate is 17.755.

The Datta code has relatively high redundancy and decoder complexity. Despite the fact that the probability of mis-correction is small, this code loses to the codes of Hoyoon-Yongsurk and Neale. Overall code estimate is 18,509.

The Neale code has high redundancy, however, the complexity of the decoder is the smallest, and the probability of mis-correction is relatively small. Overall code estimate is 22,059 – the maximum indicator of all studied codes.

The Reviriego code has low redundancy, like the Hsiao code. The probability of erroneous code correction is the highest; the complexity of the decoder is average. Given an optimized decoding scheme, this code is most productive, but the likelihood of erroneous encoding is not acceptable. The total code estimate is 16.895 – the worst overall indicator for the ones considered in this work.

The Cha-Yoon code also has low redundancy, but the error correction rate is very high. Despite the low complexity of the decoder, the Neale code is superior to the Cha-Yoon code in other parameters. Overall code estimate is 18.125.

The Hoyoon-Yongsurk code has an average redundancy rate as well as decoder complexity relative to other codes. Moreover, the probability of mis-correction of twofold non-adjacent errors is minimal. The total code estimate is 21.575 – the second value of all the codes studied.

Thus, only two codes had a total code estimate of over 20: these are the Neale and Hoyoon-Yongsurk codes. Obviously, the effectiveness and complexity of a particular method of redundant coding significantly depends on the required level of ensuring the reliability of error correction. What is important that the usual characteristics are: minimum code distance, redundancy, etc. are not decisive when choosing a code for the problem in question, and therefore, in practice, codes are often chosen with the worst characteristics, but which make it easier to parry the error. This means that codes that win solely in terms of redundancy are often not considered for practical implementation, since they lose in other indicators with the Neale code and the Hoyoon-Yongsurk code.

Comparing the codes of the Neale and Hoyoon-Yongsurk, the following recommendations can be given. When the spacecraft is in high orbits under conditions of intense exposure to radiation effects, the Hoyoon-Yongsurk code can be recommended as the main means of ensuring memory stability along with other circuitry solutions. This is due to its ability to detect double non-adjacent errors and errors of greater multiplicity with the highest probability. If the performance indicators for the correction algorithm are critical and/or the effect of cosmic radiation is not large (low orbits), it is recommended to use the Neale code. This code does not guarantee error-free correction of double non-adjacent errors, but the probability of error correction is relatively small, however, the complexity of its decoder is the smallest.

6 Conclusion

The paper presents an estimation of the SEC-DED-DAEC codes effectiveness. The estimation was carried out according to the parameters of code word redundancy, implementation complexity, and DAEC mis-correction probability. A method is proposed combining these indicators into a common one. Experimental studies and evaluations made it possible to draw conclusions about the effectiveness of the studied codes. The codes with the greatest efficiency were identified with their using recommendations.

Research is planned to continue in the direction of the practical use of SEC-DED-DAEC codes in the design of FPGA-based on-board equipment for spacecraft, including cryptographic processors for communication system of satellite – ground station.

Acknowledgments. The reported study was funded by RFBR, project number 19-38-90052.

References

1. Nicolaidis, M.: Soft Errors in Modern Electronic Systems. Springer, New York (2011)
2. Gaillard, R.: Single event effects: mechanisms and classification. Front. Electron. Test. **41**, 27–54 (2011)
3. Neale, A., Sachdev, M.: A new SEC-DED error correction code subclass for adjacent MBU tolerance in embedded memory. Device Mater. Reliab. IEEE Trans. **13**(1), 223–230 (2013)
4. Naseer, R.: A framework for soft error tolerant SRAM design. A Dissertation presented to the faculty of the graduate school University of Southern California, pp. 134 (2008)
5. Hamming, R.: Error detecting and error correcting codes. Bell Syst. Tech. J. **29**(2), 147–160 (1950)
6. Dutta, A., Touba, N.: Multiple bit upset tolerant memory using a selective cycle avoidance based SEC-DED-DAEC code. In: 25th IEEE VLSI Test Symposium, pp. 349–354 (2007)
7. Hsiao, M., Bossen, D., Chien, R.: Orthogonal latin square codes. IBM J. Res. Dev. **14**(4), 390–394 (1970)
8. Richter, M.: New linear SEC-DED codes with reduced triple bit error miscorrection probability. In: 14th International on-Line Testing Symposium, pp. 37–40 (2008)
9. Datta, R., Touba, N.: Exploiting unused spare columns to improve memory ECC. In: Proceedings of 27th IEEE VLSI Test Symposium Santa Cruz, CA, US, pp. 47–52 (2009)
10. Ming, Z., Yi, X., Wei, L.: New SEC-DED-DAEC codes for multiple bit upsets mitigation in memory. In: Proceeding of 19th IEEE/IFIP International Conference on VLSI and System-on-Chip, Hong Kong, pp. 254–259 (2011)
11. Adi, W.: Fast burst error-correction scheme with fire code. IEEE Trans. Comput. **7**, 613–618 (1984)
12. Hoyoon, J., Yongsurk, L.: Protection of on-chip memory systems against multiple cell upsets using double-adjacent error correction codes. Int. J. Comput. Inf. Technol. **03**, 1316–1320 (2014)
13. Reviriego, P., Martínez, J., Pontarelli, S., Maestro, J.: A method to design SEC-DED-DAEC codes with optimized decoding. IEEE Trans. Device Mater. Reliab. **14**(3), 884–889 (2014)
14. Cha, S., Yoon, H.: Single-error-correction and double-adjacent-error-correction code for simultaneous testing of data bit and check bit arrays in memories. Trans. Device Mater. Reliab. **14**(1), 529–535 (2014)

A Formal Model of the Decision-Making Process Under Uncertainty

Nikita Gorodilov, Gennadiy Chistyakov(✉), and Maria Dolzhenkova

Department of Computers, Vyatka State University,
Moskovskaya, 36, 610000 Kirov, Russia
gennadiychistyakov@gmail.com

Abstract. The paper defines the concepts of decision making and uncertainty and determines the main features of the decision-making task. The authors develops a decision-making model under uncertainty; its distinctive feature is the possibility of presenting criteria for evaluating alternatives in any known measurement scales. The author gives an example of using the developed model to solve the problem and makes conclusions regarding the developed decision-making model under uncertainty.

Keywords: Decision making · Alternative solution · Multicriterial · Uncertainty · Poorly structured problem

1 Introduction

In the modern world we often have to make decisions in a constantly increasing and poorly formalized flow of information. Situations are often complicated due to the fact that a decision needs to be taken promptly, there is practically no opportunity for a detailed analysis of options and the resulting consequences of their implementation. It is extremely difficult to consider, evaluate, compare and take into account various points of view.

We understand "solution" as the complex of the possibilities, the process of searching for the best option and the obtained answer. Making a decision means to choose the best option or several options from the set of possible ones [1]. The main elements of the decision-making process are [2, 3]:

- *The person who makes the decision*, that is, the person or group of people who is responsible for the decision.
- *The goal* as the final result.
- *Alternative solutions* are means to achieve the goal. If there is only one possible solution, there is no problem of choosing it.
- *External conditions* are a set of phenomena that affect the final decision. The decision maker does not always have complete information about the external environment.
- *Outcomes* are evaluating decision-making results. Here we use such concepts as "utility", "fine", "loss", etc.

R. Silhavy (Ed.): CSOC 2020, AISC 1226, pp. 526–535, 2020.
https://doi.org/10.1007/978-3-030-51974-2_49

- *The decision making rule* is a mechanism that allows to choose the best solution, that is, algorithm of developing a solution. The rule reflects the knowledge of the task and the attitude of the decision maker towards future outcomes.

2 A Formal Decision-Making Model

2.1 Decision-Making Process Steps

In the decision-making process, we can distinguish the following steps [1,4]:

1. *Diagnosis of the problem*
   - *Formulation of the problem.* To formalize the problem, it is necessary to establish its causes, evaluate the available data related to the solution of the problem, determine the factors that affect the solution in general and the restrictions imposed.
   - *Identification of alternatives.* In practice, it is necessary to limit the number of options due to the large amount of information and the limit of time required to formulate and evaluate alternatives. Moreover, the choice is made according to a certain standard until an alternative satisfies the decision-maker.
   - *Assessment of alternatives.* It consists in determining the advantages and disadvantages of each alternative and possible consequences.
2. *Solution of the problem.* The most acceptable alternative that received the maximum score at the previous stage is chosen.
3. *Evaluation of the result of solving the problem.* Evaluation is necessary to understand the effectiveness of the obtained decision. Depending on the type of decision, the effectiveness may have qualitative or quantitative characteristics.

2.2 Description of Uncertainty

The degree of structuring or formalization of a problem significantly influences the decision-making process [5]. This concept was introduced by G. Simon and A. Newell (1958). It is associated with a difference in the quantitative, qualitative, objective and subjective information connected with the problem. Well-structured problems involve the expression of core dependencies by objective models. Poorly structured problems require qualitative descriptions based on subjective judgments of a person.

Examples of well-structured problems are military operations, the choice of a route for transporting cargo, and the distribution of work among performers. Examples of poorly structured problems are the selection of investment or research projects.

Poorly structured problems have the following features:

- Absence, shortage or inability to obtain objective information at the time of setting the task.

- The inability to build a fully formalized model of the problem situation and the need to obtain additional data on behalf of the decision maker or expert.
- Combining quantitative and qualitative characteristics of the problem situation, and many criteria for evaluating solutions.
- Lack of a way to objectively choose the best solution and, as a result, making choices based on subjective preferences of the person.

Thus, we should note that poorly structured problems are solved under conditions of uncertainty.

The concept of uncertainty is heterogeneous in manifestation and in content. The first scientist who drew attention to the problem of uncertainty was the American economist F. Knight. According to Knight, "we live in a world subject to change, in a realm of uncertainty. Therefore, we act on the basis of an opinion that can be justified to a greater or lesser degree and be of great or small value; we are not completely ignorant, but we do not have complete and perfect information, but only partial knowledge" [6]. In addition, there are the following interpretations of uncertainty [7]:

- Lack of awareness and the need to act based on opinion, not knowledge.
- Ignorance, incomplete and inaccurate knowledge of the laws of business activity.
- Uncertainty due to limited information.
- Uncertainty associated with the conflict, with risk, as information parameters of the decision-making process.
- Unreliability, insufficiency or complete lack of information when making decisions in scientific and other studies.
- Insufficiency of information about the conditions under which economic activity will take place, a low degree of predictability and prediction of these conditions.
- Incomplete or inaccurate information on the prerequisites, conditions or consequences of the activity.
- Uncertainty caused by the action of factors that a person cannot influence (objective uncertainty). It is connected with a lack of knowledge of the subject about the external environment or inaccurate ideas about his own needs, incentives, desires (subjective uncertainty).
- Incomplete or inaccurate understanding of the values of various parameters in the future, generated by various reasons and incomplete or inaccurate information on the conditions for implementing the decision, including costs and results.

Based on this information, we can formulate the concept of uncertainty. It is incompleteness, inaccuracy or complete lack of information on the prerequisites, conditions or consequences of the implementation of the decision.

2.3 Features of Modeling Decision-Making Tasks Under Uncertainty

A characteristic feature of real-world tasks is multicriteria [8,9]. It means that the choice of the best solution depends on several criteria. Therefore, when comparing alternatives, we should take into account every indicator.

The main statements should be taken into account when building a multi-criteria model of a decision-making problem under uncertainty [10,11]:

- Several decision-making alternatives. If there is only one alternative, there is no decision-making problem.
- Many criteria for evaluating alternatives, every one is important when choosing a solution.
- The criteria for evaluating alternatives can be quantitative and qualitative. Quantitative criteria are determined on the basis of statistical data or on the basis of mathematical models. Qualitative criteria are determined on the basis of expert information.
- Restrictions imposed on the decision-making process. They are imposed on the criteria for evaluating alternatives and are used in determining the best decision.
- Ability to take into account the subjectivity of the decision maker. Subjectivity is observed when identifying alternatives, qualitative criteria for evaluating alternatives and in determining the restrictions imposed on the decision-making process.

2.4 A Formal Model of Decision-Making Under Uncertainty

Formally, the decision-making problem D can be represented as a quadruple:

$$D = <K, A, G, P>,$$

where K is criteria for evaluating alternatives, A—alternatives to the decision-making process (DM process), G—restrictions, and P—function for choosing alternatives.

The criteria for evaluating alternatives are multitude:

$$K = \{k_1, k_2, \ldots, k_n\},$$

where k_i is criterion for evaluating alternatives, moreover, $k_i \in K$, $i \in \mathbb{N}$ and values $k_i \in S$ (criterion measurement scale).

Alternatives to the DM process can be taken as:

$$A = \{a_1, a_2, \ldots, a_m\},$$

where a_i an alternative, moreover, $a_i \in A$, $i \in \mathbb{N}$ and values $a_i \in Sn$ (name scale).

Restrictions are functions that evaluate alternative criteria:

$$G = \{fg_1, fg_2, \ldots, fg_l\},$$

where fg_i function of the restriction imposed on one of the evaluation criteria, with the known alternatives, moreover, $fg_i \in G$, $i \in \mathbb{N}$.

$$fg_i = f(k_j, a_k) = \begin{cases} 0, \text{if the criteria satisfy the purpose or restriction} \\ 1, \text{otherwise} \end{cases},$$

where k_j—criteria subject to restriction, a_k—alternatives.

Function of Selecting Alternatives

$$P = f(K, G, A) = a_k,$$

where K—criteria for evaluating alternatives, G—restrictions and purposes, A—alternatives. The result of its work is a multitude a_k—alternatives that are a solution to the problem of DM, while $a_k \in A$.

Criteria Can Take any of the Measurement Scales. Metrology distinguishes five types of scales, which can be represented as follows:

$$S = \{Sn, So, Sd, Sr, Sa\},$$

where Sn—name scale, So—scale of order, Sd—scale of intervals (differences), Sr—scale of relations, Sa—absolute scale.

Each measurement scale can be formally represented as:

$$S_i = \langle X, f_i, Y \rangle,$$

where X is the set of real objects, situations, events or processes, Y—the set of elements (values) of the sign system, f_i—the function of mapping objects to the values of the sign system, which can be represented as:

$$f_i = f(X) = Y$$

The name scale is formally presented as:

$$Sn = \langle X, f_{sn}, Y \rangle,$$

where f_{sn} is name scale conversion function, which is represented as:

$$f_{sn} = f(X) = Y = \begin{cases} y_1 & \text{if } f(x_i) = y_1 \\ y_2 & \text{if } f(x_i) = y_2 \\ \dots & \\ y_m & \text{if } f(x_i) = y_m \end{cases},$$

where $f(x_i)$, $i = 1 \dots n$ is the function determining the correspondence of a real object with the value of a sign system, n—number of objects of the real world, m—the number of objects of a sign system, $n, m \in \mathbb{N}$.

On the name scale, only the equivalence operation is possible, which for any two objects A and B allows to establish the truth of one of the following statements: $A = B$ or $A \neq B$.

The order scale is formally represented as:

$$So = \langle X, f_{so}, Y \rangle,$$

where Y is the set of objects of the order scale, sorted in ascending or descending order, which is represented as:

$$Y = \begin{cases} y_1 \leq y_2 \leq \cdots \leq y_m & \text{in ascending scale} \\ y_1 \geq y_2 \geq \cdots \geq y_m & \text{in descending scale} \end{cases}$$

f_{so} is the transformation function of the order scale, which is represented as:

$$f_{so} = f(X) = Y = \begin{cases} 0 & \text{if } f(x_i, y_1) = False \\ y_1 & \text{if } f(x_i, y_1) = True \text{ and } f(x_i, y_2) = False \\ \ldots & \\ y_m & \text{if } f(x_i, y_m) = True \end{cases},$$

where $f(x_i)$, $i = 1 \ldots n$ is the function determining the correspondence of a real object with the value of a sign system, n—number of objects of the real world, m—the number of objects of a sign system, $n, m \in \mathbb{N}$. In general $x_i \in Y$.

On the order scale, an equivalence and order operation is possible, which for any A and B allows in the case $A \neq B$ to establish the truth of one of the following statements: $A > B$ or $A < B$.

The scale of intervals (differences) is formally represented as:

$$Sd = \langle X, f_{sd}, Y \rangle,$$

where f_{sd} is the interval scale conversion function, which is represented as:

$$f_{sd} = f(X) = a \cdot X + b,$$

where a is a unit of the considered quantity (scale), b—the reference point.

On the scale of order, the operation of equivalence, order and additivity is possible, which for any A and B allows to understand how much one object is larger than another: $A + c = B$ or $A = B - c$.

The scale of relations is formally presented as:

$$Sr = \langle X, f_{sr}, Y \rangle,$$

where f_{sr} is the interval scale conversion function, which is represented as:

$$f_{sr} = f(X) = a \cdot X,$$

where a is a unit of the considered quantity (scale).

There is a natural origin in the scale of relations, that is, for the function of transforming the scale of intervals $b = 0$. On the scale of order, the operation of equivalence, order, additivity, and relation is possible, which for any two numbers

A and B allows to understand how many times one object is larger than the other: $A/c = B$ or $A = B * c$.

The absolute scale is formally presented as:

$$Sa = <X, f_{sa}, Y>,$$

where f_{sa} is the absolute scale conversion function, which is represented as:

$$f_{sa} = f(X) = Y,$$

where X is the set of objects of the scale of relations, Y—set of values.

Absolute scales have all the features of relationship scales, but in addition there is a natural unambiguous definition of the unit of measurement (natural zero and unit).

To determine the alternatives of the DM process, it is necessary to pass to the input of the alternative assessment function a set of criteria characterizing the alternatives, a set of goals and limitations imposed on the criteria. At the output, we will have a set of alternatives satisfying the DM process [12,13].

3 An Example of Using a Formal Decision-Making Model

It is necessary to solve the problem of choosing a water supplier for the organization. The decision maker must identify the candidate to make the contract. It is required to identify many alternatives A, many criteria for evaluating alternatives K and many restrictions G imposed on the decision-making process.

Three organizations are alternatives: "Pure water" (a_1), "Fresh water" (a_2) and "Drinking water" (a_3).

The alternatives are evaluated on five criteria:

- Product quality (k_1). The criterion is measured on a scale of order (So) and has three values in ascending order: poor, medium and good quality.
- Cost of the product (k_2). The criterion is measured on an absolute scale (S_a). So the values—are measured in rubles per cubic meter (rub/m^3).
- Supplier history (k_3). The criterion is measured in the order scale (So) and has three values—in ascending order: bad history, lack of supplier history and good supplier history.
- Additional service (k_4). The criterion is measured in the order scale (So) and has two values—in ascending order: an additional service is absent or present. Additional services may include equipment repairs and other services.
- Availability of additional goods from the supplier (k_5). The criterion is measured in the order scale (So) and has three values in ascending order: there are no additional goods, there is a small range of additional goods and there is a wide range of additional goods. Additional products may include: various equipment for bottling water and any types of stationery if the supplier specializes in working with large organizations.
- Speed of delivery (k_6). The criterion is measured in the scale of relations (Sr). So the values are measured in days of delivery.

Table 1. Parameters for each alternative

Company/Criteria	"Pure water"	"Fresh water"	"Drinking water"
Product quality (k_1)	Medium	Medium	Good
Cost of the product (k_2)	30.60	32.60	33.60
Supplier history (k_3)	Good	No	Good
Additional service (k_4)	No	Yes	No
Availability of additional goods (k_5)	Wide range	Wide range	Small range
Speed of delivery (k_6)	0.3	0.1	0.4

After defining the criteria, it is necessary to assign the values of the assessment parameters for each alternative, which are presented in the Table 1.

The next steps are to identify the restrictions imposed on the decision-making process. In this case, the decision-maker determines the restrictions based on the following considerations: the company represented by the decision-maker is faced with the task of finding a supplier to work for a long time, and it is good if the supplier can provide not only water. Also for the decision-maker, the most important criterion is the quality of the product when it is impossible to compare alternatives. Based on these conditions, we have the following restrictions:

- Additional goods are required (criterion k_5, restriction g_1). As the criterion is measured on a scale of order, then all values of the criterion greater, or equal to the "small range", satisfy the condition.
- A supplier should have a good history (criterion k_3, restriction g_2). As the criterion is measured on a scale of order, then all values of the criterion greater, or equal to "good", satisfy the condition. In this case, only one value of the criterion satisfies the condition.

Below are all the initial data of the decision-making task:

- $K = \{k_1, k_2, k_3, k_4, k_5, k_6\}$—product quality, cost of the product, supplier history, additional service, availability of additional goods, speed of delivery.
- $A = \{a_1, a_2, a_3\}$—"Pure water", "Fresh water", "Drinking water".
- $G = \{g_1, g_2\}$—availability of additional goods, good supplier history.

After determining the necessary initial data, it is necessary to decide which alternative is the best. For this operation, the developed model uses the following function: $P = f(K, G, A)$. In the general case, the function $f(K, G, A)$ is an algorithm for determining the best solution for the known alternatives, criteria and restrictions.

To determine the best alternative, the first thing to do is to remove solutions that do not satisfy the restrictions G:

- All alternatives satisfy the restriction g_1.
- Alternatives a_1 "Pure water" and a_3 "Drinking water" satisfy the restriction g_2. The alternative a_2 does not satisfy the restriction, since the value of the criterion "There is no history" is less than "Good history" based on the restrictions imposed on the values of the order scale.

Now the decision maker is faced with the task of choosing between two alternatives a_1 and a_3. As we noted earlier, an alternative with the best criterion for k_1 product quality is the best for the organization. In our case, such an alternative is a_3, as the criterion k_1 is higher than of alternative a_1 and, although the cost of products k_2 is higher for alternative a_1 and alternative a_3 has only a small range of additional products, the chosen alternative is the best. So, $P = f(K, G, A) = \{a_3\}$.

Thus, the alternative a_3 was selected (supplier "Drinking Water") during the decision-making process of choosing a water supplier for the organization.

In the process of solving the problem, both subjective and objective data can be used. Here are some explanations. When determining alternatives A, restrictions and purposes G, the function of choosing alternatives P and criteria K, the decision maker can use information obtained on the basis of statistical methods or other formalized algorithms.

In the presented example, objective data were used to determine the values of the k_2 criterion (cost of production), and the values of the k_3 criterion (supplier history) were obtained on the basis of a subjective opinion.

4 Conclusion

This article presents the concepts of decision making and uncertainty and develops a formal model of decision making under uncertainty. As the initial data of the model, it is necessary to determine alternatives, criteria for evaluating alternatives and restrictions imposed on the decision-making process. A distinctive feature of the developed model is the ability to determine the criteria for evaluating alternatives in any measurement scale, as well as the ability to use any decision-making algorithms with the known source data.

The presented example of solving a problem allows to conclude that it is possible to use the developed model to solve decision-making problems under uncertainty.

References

1. Petrovsky, A.: Decision-Making Theory. Academy Press, Moscow (2009)
2. Boldyrev, A.S.: Basic Concepts of Decision-Making Theory. Bull. St. Petersburg Univ. Ministry Internal Aff. Russ. **57**, 87–91 (2013)
3. Ozernoy, V.M: Decision-making (review). Avtomat. Telemech. **11**, 103–121 (1971)
4. Terentyev, N.Y.: Styles and decision-making process. Probl. Acc. Finance **13–1**, 63–67 (2014)

5. Malkova, T.B., Malkov, A.V.: Practical method for risk assessment in energy under uncertainty. Econ. Ind. **12–1**, 63–69 (2018)
6. Knight, F.: Risk, Uncertainty and Profit. Business Press, Moscow (2003)
7. Aralbaeva, F.Z., Karabanova, O.G., Krutalevich-Levaeva, M.G.: Risk and uncertainty in managerial decision-making. Bull. Orenburg State Univ. **4**, 132–139 (2002)
8. Stepanov, V.V., Kucher, V.A.: Formalization of the decision-making process for managing information security in automated systems using the Bayesian approach. Sci. J. KubSAU **111**, 1746–1757 (2015)
9. Solodukhin, K.: Fuzzy strategic decision-making models based on formalized strategy maps. Advances in Economics, Business and Management Research **47**, 543–547 (2019)
10. Byatez, I.V.: The reasons for the origin and measurement of uncertainty and risk in the economy. Bull. Udmurt Univ. Ser. Econ. Law. **4**, 14–18 (2011)
11. Tychinsky, A.V.: Uncertainty in the adoption of managerial decisions. Bull. South. Fed. Univ. Tech. Sci. **52**, 118–122 (2005)
12. Milaya, A.S., Savkova, E.O.: Formalization of the tasks for decision-making on optimization of the work on customer service under uncertainty. Econ. Ind. **11–1**, 82–93 (2018)
13. Kapustin, V.F.: Uncertainty: types, interpretations, accounting in modeling and decision-making. Bull. St. Petersburg Univ. **2–5**, 108–114 (1993)

IoT in Traffic Management: Review of Existing Methods of Road Traffic Regulation

Dmitry Elkin[1(✉)] and Valeriy Vyatkin[2,3]

[1] Southern Federal University, Institute of Computer Technologies and Information Security, Chekhova. 2, 347900 Taganrog, Russia
delkin@sfedu.ru

[2] Department of Electrical Engineering and Automation, Aalto University, Otakaari 1B, 00076 Aalto, Finland
vyatkin@ieee.org

[3] Department of Computer Science, Electrical and Space Engineering, Luleå University of Technology, Luleå 97187, Sweden

Abstract. Internet of Things (IoT) is an important upcoming technology for making infrastructures of our society smart and adaptable to the users' needs. One such infrastructure is transportation. This article discusses existing methods and algorithms for automated management of traffic flows with a purpose of identifying hot spots and methods for applying IoT in this sector. Advantages and disadvantages of the existing methods are discussed, including evaluation of their effectiveness.

Keywords: IoT · Traffic management · Intelligent transportation system · ITS · Traffic · Traffic jam

1 Introduction

The relationship between roads and cities is obvious. The main features of the road infrastructure are a long service life and high cost, which means that this problem can be solved at the moment while keeping these objects in their original condition. Economic calculations [13] show that effectively manage traffic on existing roads is more profitable than building new ones, although it is also not very cheap from an economic point of view. The experience of many large megacities of the world [14] shows that the construction of new and reconstruction of existing highways and roads with a constant increase in the number of vehicles does not allow to completely reduce the difference between the carrying capacity of roads and the level of demand for road transport.

The high costs of building new road infrastructure, restricting travel, and environmental factors are pushing companies and the government to look for solutions to manage traffic flows more efficiently. We order to mitigate the negative effects of congestion and optimize the use of limited public funds.

In European countries such as Denmark, UK, Germany, and the Netherlands, the use of advanced approaches to the dynamic redistribution of flows using automated systems gave the following results [1]:

R. Silhavy (Ed.): CSOC 2020, AISC 1226, pp. 536–551, 2020.
https://doi.org/10.1007/978-3-030-51974-2_50

- an increase in the average throughput of the road network in the period of overload (during peak hours) by 3–7%;
- an increase in the overall throughput of the transport network by 3–22%;
- reduction of primary accidents by 3–30%;
- reduction of secondary accidents by 40–50%;
- alignment of the speed of the transport stream during the overload period of the road network;
- more predictable behavior of road users;
- improves safety on the road;
- reduction of traffic congestion on highways.

Thus, when using the latest means and developments in the field of traffic management, it is possible not only to relieve the road network from systematically arising jams but also to significantly reduce government spending on the organization of traffic in general.

The Internet of Things (IoT) is the current megatrend in the development of information and control technologies. The purpose of this work is to review the possibility of applying the Internet of Things technology to the organization of road infrastructure for the dynamic management of traffic flows.

In Sect. 2, we look at existing models for traffic management. These models are now used in most cities. Then, in Sect. 3, we consider the most popular algorithms for managing traffic flows and consider their advantages and disadvantages. In Sect. 4, we identify the shortcomings of existing algorithms and an overview of their effectiveness in actual use. In the fifth part, we will review the application of IoT in the management of traffic flows, the existing approaches of algorithms and methods, as well as identify promising areas of development.

2 Typical Models of Traffic Management

Traffic lights is the ultimate means to control traffic flows, and computer controllers are commonly used for switching the signals of traffic lights to deal with the ever-changing traffic intensity. To optimize the control process, controllers with several programs (peak hour, day and nighttime) are used.

Multi-program hard-coded regulation helps to reduce the delay but is not optimal. It is not able to take into account the accidental arrival of vehicles at the intersection. The solution to this problem is adaptive control with feedback to the transport stream. To do this, controllers receive continuous information about the state of the transport stream from transport detectors located in the intersection zones. Transport detectors are designed to register passing vehicles and determine the parameters of the traffic flow [21].

Further development of the idea of coordinating the work of individual traffic lights was the creation of automated traffic management systems that are able to control traffic signaling in the whole city, combining managed traffic lights with a single control center and synchronizing their work.

To describe the organization of traffic management, a layered architecture composed of 5 layers is useful, described as follows [21]:

The lowest level of automatic control system is responsible for such functions as stabilization or programmatic changes in the parameters of the object in accordance with the settings that are set by the upper level of the automated control systems. At the automatic control system level, technical means such as digital controllers and traditional continuous controllers are used.

The level of automated control systems optimizes the management of a limited set of objects that are subject to the appropriate optimizers. The tasks of traffic management at this level of the system may differ from the overall task of the entire system functioning. At least, it is necessary to take into account the features of the subsystems subordinate to the optimizers. The technical means used at the level of the automated control systems and at higher levels of the hierarchy should use newest technologies, high-speed communication and information processing facilities.

The level of coordination is responsible for coordinated management of the work of local optimizers that is carried out to achieve the overall goal of the entire system. For optimization, criteria are used that allow it to be carried out at any level of the system.

The level of operational decision-making accommodates the governing body (a team of decision makers or intelligent software) for heuristic solutions of global system-level optimization problems. At this level, the overall goals and objectives of the system are transformed into settings for the lower levels of the hierarchy. Also, management resources are distributed among the individual subsystems and decisions are made for various emergency situations.

At the heart of all levels of traffic management are specialized algorithms for optimal control of transport flows, we will consider them in more detail in the following section.

3 Existing Algorithms for Traffic Management

3.1 Management Algorithms at the Local Crossroads

Crossroads that are more than 1 km apart from each other in the established practice are considered independent and can be controlled separately from each other. Algorithms for controlling them are implemented in the road controllers. These algorithms can be divided into two groups. In both groups, traffic signaling is controlled in accordance with the criterion of the minimum of total vehicle delays at the intersection [22].

The first group is based on methods for determining control parameters: the duration of the traffic light cycle and the distribution of periods within a cycle based on the average characteristics of the transport stream.

The second group is based on methods of switching signals of traffic lights according to the instantaneous behavior of the traffic flow.

1) Management based on the length of the queue. For more efficient use of the green signal periods, it is also possible to minimize the switching time of the traffic light signal from the green signal to the red signal, but it is necessary to limit the duration

of this period to prevent the appearance of very long control cycles and guarantee the safe passage of roads by pedestrians.

2) Management based on queue length and saturation flow. The algorithm skips only the vehicles that were accumulated during the red traffic light signal. It is implemented on the basis of information on the saturation flux at each approach to the intersection. The intensity of the queue's departure is determined by the saturation flux, so if the time interval between vehicles exceeds the interval in the saturation stream, this means that all the accumulated vehicles have departed, and the time of the signal change has come.

3.2 Network Management Algorithms

Network management algorithms are used at those intersections that are located at a distance of less than 1 km from each other and connected in the same network. To calculate the control parameters, information on the transport situation from all intersections of the network is required. At the network level, it is common to determine the duration of the control cycles and the shifts of the beginning of the cycles at adjacent intersections.

1) Hard network management. Network rigid control provides coordination of the traffic lights in a certain area of control. One of the most popular and widely used products for traffic management in the world are the software packages of the Laboratory of Transport Research (TRL) of Great Britain. They are used in more than 110 countries for a wide variety of assessments and modeling work. Engineers can also optimize fixed traffic signal switching programs in existing micromodels using VISSIM and AIMSUN software products [2].
2) Adaptive Network Management. Hard control algorithms, which are based on the assumption of repeatable traffic situations in the same hours of the day or days of the week, in the case of the high amplitude of the instantaneous values of traffic flow intensity cannot cope with the situation, which will lead to skyrocketing queue and the ability to block the adjacent intersections. In such cases, the most effective use of adaptive management methods. The algorithm underlying the system is SCOOT - Split, Cycle and Offset Optimization Technique was proposed by Hunt in 1982 [3].

The algorithm underlying the OPAC system is Optimized Policies for Adaptive Control [4] was proposed by Gartner et al. in 1982. The OPAC algorithm uses the optimal sequence of a limited search (OSCO), to plan the entire horizon and use the final cost to be fined the vehicle remains in the queue on the horizon. Horizon is 60 s, 10 s of which the head is the period associated with the information coming from the traffic detectors in real time, and the remainder (tail) portion of a predicted information. OPAC tests showed a better result of 5–15% of the existing methods, with a great advantage at a high degree of saturation.

To automate the process of updating a fixed traffic signal switching plan, an AUT algorithm was developed. This algorithm allows constantly collects data from transport detectors throughout the network. To calculate typical flows for each time of day, the data is processed, and prepared to calculate new coordination plans. Advantages resulting from the application of UTOPIA, show an increase in vehicle speed by an average of 15% and a 28% increase in the speed of public transport, which is given priority [6].

The algorithm underlying the SCATS system is the Sydney Coordinated Adaptive Traffic System was proposed by Sims et al. in 1984 [5]. SCATS consist of 3 levels of control: central, regional and local. For each intersection, the algorithm distributes the calculations between the regional computers in the traffic calculation center and the road controllers. The central level is controlled by a central computer that interacts with other levels, primarily for monitoring purposes.

SCATS combine adaptive control of traffic light signals with conventional control methods. This approach allows satisfying various operational needs of the system.

The basic algorithms developed to solve transport problems allowed to collect colossal data on the principles of traffic management in large cities and on highways. These data are excellent for application in the development of new methods for traffic organization.

4 Practical Application of Existing Algorithms

The algorithms considered in the previous part are among the most widely used in practice for managing traffic flows [7]. In order to compare their performance with the same parameters of the road network, it is necessary to carry out simulation modeling or explore the real part of the road network after the implementation of a traffic management system based on one of the presented algorithms and identify the optimal criteria by which you can evaluate the effectiveness of the investigated algorithms.

After reviewing various studies to assess the effectiveness of real-time transport management [8, 10–12], the following evaluation criteria can be identified:

- total vehicle delay;
- number of stops;
- travel time;
- average car delay;
- the capacity of the road network;
- traffic saturation;
- flow rate;
- queue length.

In the works of Stevanovic et al. and Kergaye et al. [8, 9], considering the possibilities of the SCOOT and SCATS algorithms to respond to changes in the motion parameters, the authors came to the following conclusions:

- SCOOT and SCATS provide similar delays.
- Both systems seem to be able to cope with the predominantly unsaturated demand for transportation.
- The lengths of SCATS traffic light cycles are more flexible, while SCOOT tends to keep cycle lengths on average, lower, for a longer time.

In the Slavin's study [10], it was found that, in general, the conditions of movement before and after SCATS were significantly different in terms of speed and volume. SCATS showed statistically significant improvements in traffic speed at one minor intersection, even when traffic volumes showed a statistically significant increase. At the main crossroads, the results were mixed and inconclusive.

In the work of Doshi et al. [11], OPAC, SCOOT, SCATS were considered as three strategies of adaptive control, each of which has its own concept of controlling the synchronization of signals at intersections with different characteristics. The study revealed that with the growth of traffic on the road network, it is observed that the advantages achieved with the help of adaptive control prototypes decrease with a higher ratio of volume and capacity. With a large number of phases and a highly variable need for traffic, such algorithms respond slowly, affecting the priority level of the passage.

Pavleski and Koltovska, in their work [12], established that the efficiency of the adaptive control of the UTOPIA traffic signal provided the best criteria, the average queue length, the maximum queue length, and vehicle throughput at intersections compared to fixed time control.

In the field of real-time traffic management, approaches and solutions are known that are aimed at solving problems: estimating traffic congestion of transport network sections, technical solutions for redirecting traffic flow by lanes, methods and routing algorithms. The algorithms considered in the work often undergo changes, but the differences in the new ones consist of reducing the computational complexity or increasing the reliability of the decision made. At the same time, the already developed algorithms and architectures OPAC, SCATS, SCOOT and UTOPIA have sufficient parameters for their direct application.

However, during the review, we identified common problems to existing approaches:

- slow reaction to a change in traffic situation;
- the complexity of refinement and implementation;
- the high cost of existing solutions;
- small coverage area.

To solve the problems that were identified during the work, the researchers apply new approaches to traffic management, based on the concept of IoT, big data processing, the use of multi-agent systems, cloud technology, V2X strategy. Next, consider the existing approaches to cope with the problems mentioned earlier.

5 IoT in Traffic Management

5.1 Is It Possible Using Microcomputers for Traffic IoT?

At the moment, for traffic management, many different elements of the road infrastructure are used. But the main features directly controlling traffic flows are classic elements: traffic lights and road signs. Now many researchers are working to implement the traditional traffic management using the IoT paradigm. Most prototypes for testing algorithms and approaches are implemented based on microcomputers like Intel Edison Raspberry PI, and similar. Are these computers suitable for working with real traffic systems? In [35], Masek, Pavel, et al. describe the proposed framework for traffic modeling, which is capable of running on top of low-power equipment. The study was as close as possible to the real road traffic conditions. They created scenarios that follow the structure of the practical urban intersections in Brno. For the performance evaluation, they developed a unified test environment in Java programing language, where the considered scenarios were implemented. Thus, it was found that for the integration of the Internet of things to control traffic flows in cities, you can use, compact and low-power devices. For example, the Raspberry Pi 2 (Model B), can now be used.

Regarding data transmission networks in IoT systems Keertikumar, M et al. found out in their work [36]. The basic wireless standards used for the communication of IoT elements are considered. It was found that in a smart transportation system if the vehicle is moving at speed higher than 20 km/h. The GPS communication will not be the right solution because the bandwidth of GPRS/GSM is in the range of 64 to 128 kilobytes per second. Thus, for efficient transport management, communications should be based on modern standards and optimized data transfer protocols according to IoT standards. Also, in the article Kaminski, Nicholas J. et al. simulated the operation of a traffic control system based on IoT [37]. Using the agent-based modeling (ABM) method, a traffic control system based on IoT was simulated. One of the results of the study showed that the accuracy of the information that is transmitted to the subject for decision-making is not very important. Since accuracy did not have a big impact on the performance of the traffic control system. Thus, during the simulation, the importance of the protocols and technologies used to communicate IoT elements was confirmed. This is necessary to organize effective traffic flow management.

In practice, Misbahutddin et al. using the Raspberry Pi microcomputer [8], a traffic light model was created that works according to a specific algorithm. The traffic light can work both in automatic mode and in manual control mode by a police officer. The scheme in this paper can be further improved if the situational traffic information is automatically passed to the RPi. RPi is controlling the lights at an intersection so that the authorities can make and implement the quick decision. To perform this automation, RFID, or any other wireless sensor technique may be used. But prioritizing certain roads or specific directions for longer durations may likely cause traffic problems for the different routes. Therefore, it is recommended that the suggested dynamic traffic light controlling scheme should only be applied after a thorough study of traffic patterns. And these traffic patterns are first that develop during a test period.

A slightly different direction in the work Thakur, Tanvi Tushar, et al. They presented [28] an algorithm for managing traffic flows at intersections in the Internet of Things paradigm. These algorithm transport sensors transmit information to the microcontroller to calculate the traffic signal cycle. The essence of the algorithm: if the traffic density is below 60% - low, so extra time given to them will divert to the busy lane. In case if traffic density is increased by 60%, it will flow the fixed time slot. At the time of Emergency, Priority will be given to the lane with the Emergency vehicle. The developed algorithm was programmed in the Texas Instruments TM4C129E microcontroller, which is specially designed to implement IoT capabilities. Also, in the study

[9], it is proposed to configure adaptive traffic light regulation on a specific section of the road network. The collected traffic information is processed using the Intel Edison microcomputer. Further, through wireless channels (for example, WI-FI), data is transmitted to the Microsoft Azure IoT cloud service. The authors claim that the algorithm was modeled in MATHLAB under various conditions and showed better performance than the fixed TLS algorithm, by 68.06% in Average Queue Lengths and 66.67% in Average Waiting Time.

5.2 IoT and Cloud Computing

As in the study [9], scientists propose various combinations of IoT and cloud computing to increase the pooling of benefits, for example researchers He, Wu, Gongjun Yan, and Li Da Xu [18], to solve transport problems, offer a new multi-level cloud platform for transport data. The platform using cloud computing and IoT technologies. The proposed multi-level cloud platform allows you to integrate various sensors, actuators, controllers, GPS devices, mobile phones, and other equipment for accessing the Internet. Platform as well as use network technologies (wireless sensor network, cellular network, satellite network, etc.), cloud computing, IoT, and middleware. This platform supports V2V and V2I communication mechanisms and can collect and exchange data between drivers, vehicles, and roadside infrastructures, such as cameras and streetlights. Cloud-based intelligent parking services and cloud-based data mining services on roads allow you to solve parking problems in large cities. Also, the traditional model of traffic management has several significant drawbacks that can be eliminated using the approach proposed in work Wu He et al. [18]. This study makes contributions by proposing a novel software architecture for the vehicular data clouds in the IoT environment, which has the capabilities to integrate numerous devices available within vehicles and devices in the road infrastructure. However, research on integrating IoT with vehicular data clouds is still in its infancy, and existing study on this topic is highly insufficient.

New energy-efficient communication technologies are used to create cloud architecture. Nor, Ruhaizan Fazrren et al. propose a system [33] for monitoring traffic jams at intersections. Monitoring will be carried out using sensors in each lane that are connected to the cloud server using LoRaWAN global communication technology. As a sensor, the magnetometer module GY-271 is used. The developers presented a four-level system architecture. At the lower level, an information collection sensor and a transmission node, at an average level a data reception gateway and LoRa cloud server for processing received data, at the upper level an analytics application with a user interface. The authors assembled a model of the proposed system and developed a model for analyzing the intersection congestion based on the data obtained. The system was tested, and the results of the traffic congestion of the intersection were obtained. The result is used to change the mode of operation of the traffic light.

Efficient cloud computing requires quality data. Researchers are now offering various ways to collect traffic data using IoT. For example, one of the most used technologies for obtaining information about cars is RFID, let's consider why.

5.3 Ways to Collect Traffic Data Using IoT

For efficient data collection in [23], the authors propose the development of the Smart Traffic Management System (STMS) for smart cities. The system will be based on the Internet of things and big data processing. RFID, according to the authors, uses these technologies more efficiently and cheaper than existing ones. For their mathematical model for calculating the optimal TLS cycle, they use traffic densities of various streets at a specific time as information. RFID as the main tool, is also used by Choosri N. et al. Authors developed a prototype IoT application for traffic management [24]. The primary technology used in the system is the passive RFID UHF technology for collecting vehicle information. The authors propose this approach instead of the existing in Thailand, manual and semi-automatic TLS management. The proposed solution is aimed at modernizing technologies, i.e., RFID and IoT, to adapt them to current operations, instead of completely changing the operation of the system. The authors claim that IoT, like information, is automatically transmitted to other intersections. In the course of the work, a prototype was developed to verify the concept. A particular reader reads data from the RFID tag of a passing car with a distance of up to 9 meters. Data transfer for the calculation of TLS. However, according to the authors, the experiment serves only to verify our system configuration, and it needs to be additionally tested in working conditions, including in real terms. Also, vehicle identification using RFID has been proposed by Yu, Minghe et al. [26]. The practical aspects of the use of RFID technology in controlling the movement of traffic flows were examined. The authors described the anti-collision protocol, .data cleaning algorithms, and missing data filling.

In the work Al-Sakhan [20], proposed architecture employs key technologies: Internet of Things, RFID, wireless sensor network (WSN), GPS, cloud computing, agent and other advanced technologies to collect, store, manage and supervise traffic information. The proposed system can provide a new way of monitoring traffic flow that helps to improve traffic conditions and resource utilization. However, now another tool is gaining popularity as in the work of Kanungo et al. [15] tells about a system that consists of video cameras on the traffic junctions for each side as if it is a four-way junction. Therefore, four video cameras will be installed over the red lights facing the road. Cameras would be capturing video and broadcasting it to the servers where using video and image processing techniques the vehicle density on every side of the road is calculated and an algorithm is employed to switch the traffic lights accordingly.

To improve the quality of traffic control, researchers offer various concepts of IoT-based traffic flow control systems. For example, V2IoT, AoT, Internet Agents, SIoA, and the like. The proposed systems have various features, then we will consider them.

5.4 Variations of IoT Systems

In [31] because existing systems lack intelligence due to hardware and software limitations. The authors proposed the concept of “Agents of Things” (AoT). The essence of this concept is that each thing must be intelligent. Things to interact directly with each other or to interact through other systems. AoT architecture is proposed consisting of six levels, in this architecture things. They interact with each other through software agents. The main principle of AoT is intelligence, in contrast to IoT. To solve the problem of accident warning is proposed in the AoT concept device - Road accident monitor (RAM). Their drivers and special services of the accident. In the article [34], the authors work in the same direction. Researchers propose to develop an Internet of Things-based distributed intelligent system with self-optimization for the control of traffic signals and the monitoring of environmental parameters. The main feature of the proposed solution is that all entities interact among themselves and form a multi-agent system. Each agent is made responsible for the performance of the device it represents and for the connection to the IoT. The authors state that the system will be adaptable, easily expandable, and robust.

Anass, Rghioui et al. tried to apply the Kerner three-phase traffic theory to realize a synchronized system. The authors proposed establishing an Intelligent Transport System that will provide automatic management of traffic lights [40]. Communication in the system based on the concept of the Internet of Things for various traffic lights controllers to enable them to collaborate. The authors propose a new concept 'V2IoT' an improved intelligent transportation system offering the opportunity to exploit the benefits of smart mobility. The solution is to use a wireless sensor network interconnected calculating the density of a road and transfer these data to the base station placed at a traffic light controller level. The main idea it's communication between controllers to exchange information with each other, and they can prevent the condition of the road before it is congested. Based on the proposed approach, a simulation was launched. The simulation showed that when using the V2IoT principle in traffic control, the amount of harmful emissions into the atmosphere is significantly reduced. Using the proposed approach, average travel time in peak decreases by 10%, the total waiting time in peak decreases by 30%, the average waiting time in peak decreases by 5%.

Adaptive TLS is one of the main problems in traffic management. In an article by Bui and Khac - Hoai Nam [25], a new approach based on the Internet of things and optimization using game theory is proposed. The paradigm of IoV (Internet of Vehicles) is considered, in which each car is considered as an intelligent object. The article suggests the main aspects of modeling and simulation of the connected intersection, which based on IoT. Highlights, algorithms for smart traffic lights control to improve traffic flow in real-time by applying the game theory. The modeling architecture of the connected intersection, how to get data of smart objects such as vehicles and sensors is discussed. During the study, two-game models were proposed for two scenarios of traffic. An agent-based modeling environment Netlogo simulator was also simulated. The proposed approach was compared with two existing ones, and it was revealed that the method suggested in the article reduces the waiting time at intersections by an average of 20%.

Milanés et al. [16] present a V2I-based traffic management system, which proposes a solution to the problem of regulating traffic flow in urban areas, in which different bottleneck situations may coexist. The system contributes to avoiding accidents by alerting the driver in advance of potential collisions. The system includes an intelligent controller that uses a reference safety distance and the appropriate speed as fuzzy inputs. The output sent to the driver is information on how the vehicle is being driven.

Researchers suggest various ways to optimize traffic based on IoT. The primary way to optimize this, of course, is to optimize traffic signals to ensure that traffic passes without delay. Various systems, such as the "green wave" are implemented using IoT.

5.5 IoT and Traffic Light

Phan, Cao Tho, et al. in their work [29] presented an approach to organizing a green wave on a section of a road network with several intersections and using IoT to monitor the state of traffic flow and control traffic signals. In the course of work, an application for traffic control at intersections was developed. The application can work in automatic and manual modes. The IoT application uses a PLC S7-1200 at every intersection to control traffic lights system. PLC transfers parameters to the host machine and receives parameters from the host machine at the operation center using the TCP/IP protocol. The application also provides a simulation mode for checking settings. So far, the system has been tested for one arterial road. In future work, authors will propose a method to apply the system for a network of arterial roads.

The concept of intelligent traffic lights in the IoT paradigm has been actively disseminated. For example, the work of Miz Volodymyr, and Vladimir Hahanov [27] considers the process of implementing STL (Smart Traffic Light), as an element of the road infrastructure and (CTMS) Cognitive road traffic management system, as the basis for STL. STL able to change the signal time depending on the volume of traffic flows. STL control method uses the maximum Boolean derivative on the lines of motion. STL switching is delayed if all sides of the intersection are full for minimizing conflicts on all lines of traffic flows. STL switching cycle has a minimum period equals to the crossing time. In work, the authors concluded that CTMS is characterized by interrelated cloud components: infrastructure, monitoring, and traffic management. The central controlling part of CTMS is the STL, which allows us to: optimize the traffic time, cost and quality to address the social, humanitarian, economic, criminal, insurance and environmental needs and reduce existing road infrastructure and save. At the same time, Tielert, Tessa, et al. offer the concept of vehicle interaction with a traffic light [39] to optimize travel time and fuel economy based on IoT. Traffic-light-to-vehicle communication (TLVC) has received an increasing amount of attention. TLVC promises beneficial effects for individual drivers even at low penetration rates of the new communication technology. In the course of the study a simulation of the movement of a car through intersections with an adapted speed using IoT was made. The simulation showed that for a single vehicle and single traffic light, TLVC reduces fuel consumption by up to 22% and CO, NOx and particulate matter emissions by up to 80%, 35%, and 18%, respectively. At the same time, for several vehicles and intersections on the route 3 km, fuel economy is 8%.

New applications of IoT technologies in road infrastructure are continually appearing. Various sensors, traffic lights, lighting, traffic signs, and even road fences and parking lots.

5.6 IoT and Road Infrastructure

In a study [32] by Jabbarpour et al. introduced the concept of intelligent highway fences as part of the IoT paradigm. The authors suggest that often when there is a traffic jam on one part of the highway, the other part is free. Intelligent fencing will help solve this problem. Road-Side Unit (RSU) special equipment checks the flow rate. If the flow velocity is less than 8 m/s, then a jam has formed. Then, according to the algorithm developed by the authors, the intellectual fence opens an additional lane for movement on the free part of the road. The authors suggest that the developed system will reduce the number of traffic jams.

In the same time, Wang, in [17] developed a new Reservation-based Smart Parking (RSP) system to optimize parking management. They proposed a dynamic pricing scheme for satisfying the different needs of drivers and service providers, which is based on real-time parking information. The pricing scheme is integrated with the proposed RSP system in which the parking price is dynamically adjusted in response to the relationship between demand and supply and congestion level. Upon receiving parking prices, drivers make their reservations to maximize their benefits according to the utility function. Based on the obtained results from a simulation study, the authors conclude that the proposed RSP has a positive effect on traffic. For example, the system increases the revenue for service providers, provides service differentiation for users with different needs alleviates traffic congestion, caused parking searching, and reduces the amount of traffic searching for parking. Also, the work of Khanna et al. [19] system that proposed provides real-time information regarding the availability of parking slots in a parking area. Users from remote locations could book a parking slot for them by the use of our mobile application. The efforts made in this paper are intended to improve the parking facilities of a city and the quality of life of its people.

Researchers have also come up with concepts in which IoT elements of the road infrastructure come to the aid of people in vital situations. For example, it was proposed to implement a system for driving emergency vehicles to their destination without traffic congestion in a minimum amount of time.

5.7 IoT and People

Chowdhury Abdullahi considered [30] the possibility of securing the unimpeded movement of an emergency vehicle on roads using the Secured Traffic Management System. For the effective operation of the system, it is proposed to determine the type of accident, traffic congestion, determine the optimal route and the priority of the emergency vehicles. When the emergency vehicle approaches the intersection, the system calculates the parameters based on the received data and provides priority for the intersection. The system was modeled in the SUMO program. The results showed that the travel time of emergency vehicles using the proposed EPCS system is 12.7 min, versus 19.4 with normal regulation and 17.5 with a green wave. Also, researchers Venkatesh, H. et al. propose the use of IoT in traffic control to give priority to ambulances when driving through intersections [38]. The ambulance driver is supposed to send a request to the signal point using his GPS location on a cloud server using GSM technology. When the request was sent to the signal board, it automatically sends the received acknowledgment and the traffic signal switches to green. The device in the ambulance consists of: an ARM processor, a GPS module and a GSM module for transmitting data to a traffic light control system. The authors believe that GPS, GPRS, and network, with the help of the Internet of Things, is found, to construct an intelligent traffic monitoring system. The development system can serve an excellent facility to make a path to the ambulance in traffic load to reach the hospital.

6 Conclusion and Future Work

In the course of the review, we found that the Internet of Things is now a rapidly developing technology, creating a universal standard in wired and wireless communications. IoT allows people and objects to communicate anytime, anywhere, with anything and anyone, ideally using any media and any services. Internet of things has created a technological revolution in the future of computing and communications. IoT creates a mixture between the physical worlds and the information world together, giving tremendous opportunities in traffic management of urban infrastructure such as smart cities, industrial control, environmental monitoring, and similar. IoT in traffic management replaces existing approaches related to strict and adaptive traffic light regulation and is actively implemented in ITS.

The study showed that the introduction of IoT technologies in the transport structure would significantly reduce the waiting time for cars in the queue at intersections [9, 25, 41], the total travel time [25, 40, 42], save fuel [27, 29, 39], reduce harmful emissions into the atmosphere [39], reduce the travel time of emergency vehicles to their destination [30, 38], solve the parking problem [17–19] and still show many other positive effects. Thus, it significantly improves the road situation in cities.

However, there is a big problem in developing complex systems of this kind. The problem is testing such implementation systems, searching for material resources and the necessary equipment that meets the necessary quality and reliability standards. Although many research results have already been obtained, the existing road infrastructure is not ready, and the developed approaches do not allow us to utilize the potential of existing technologies fully.

Thus, in our opinion, future research should be aimed at:

- Search for ways to implement new approaches to traffic management using existing road infrastructure;
- A solution to the problem of distribution of computing load when driving, to reduce the cost of equipment;
- Increased flexibility and scalability of the traffic management system through decentralization;
- Improving the quality of traffic management by finding the best options when negotiating with agents.

Therefore, we will direct our research to the development of the approach of distributed control systems organized within the IoT paradigm, but with an implementation based on the existing road infrastructure (PLC controllers, communication lines). In the approach we consider proposing, the road infrastructure would be organized in such a way that each element of the infrastructure is an intelligent agent. The intelligent agent could be an autonomously functioning traffic light, road sign, road fence, marking element, and similar. All agents communicate with each other to organize the optimal process of traffic control.

Acknowledgments. The reported study was funded by RFBR, project number 19-37-90102.

References

1. Mirshahi, M., Obenberger, J., Fuhs, C.: Active traffic management: the next step in congestion management (2007)
2. Wallace, C.E. et al.: TRANSYT-7F user's manual. No.UF-TRC-U32 FP-06/07 (1984)
3. Hunt, P.B., et al.: SCOOT-a traffic responsive method of coordinating signals. No. LR 1014 Monograph (1981)
4. Gartner N.H., Tarnoff P.J., Andrews C.M.: Evaluation of optimized policies for adaptive control strategy. Transportation Research Record. No. 1324 (1991)
5. Sims, A.G., Dobinson, K.W.: The Sydney coordinated adaptive traffic (SCAT) system philosophy and benefits. IEEE Trans. Veh. Technol. **29**(2), 130–137 (1980)
6. Mauro, V., Di Taranto, C.: Utopia. IFAC Proc. **23**(2), 245–252 (1990)
7. Zhao, Y., Tian, Z.: An overview of the usage of adaptive signal control system in the United States of America. Appl. Mech. Mater. **178**, 2591–2598 (2012)
8. Misbahuddin, S., et al.: IoT based dynamic road traffic management for smart cities. In: 2015 12th International Conference on High-capacity Optical Networks and Enabling/Emerging Technologies (HONET). IEEE (2015)
9. Chong, H.F., Ng, D.W.K.: Development of IoT device for traffic management system. In: 2016 IEEE Student Conference on Research and Development (SCOReD). IEEE (2016)
10. Slavin, C., et al.: Statistical study of the impact of adaptive traffic signal control on traffic and transit performance. Transp. Res. Record. **2356**(1), 117–126 (2013)
11. Ozbay, K, et al.: Evaluation of adaptive control strategies for NJ highways. No. FHWA-NJ-2006-001 (2006)
12. Pavleski, D., Koltovska-Nechoska, D., Ivanjko, E.: Evaluation of adaptive traffic control system UTOPIA using microscopic simulation. In: 2017 International Symposium ELMAR. IEEE (2017)
13. Leontiadis, I., et al.: On the effectiveness of an opportunistic traffic management system for vehicular networks. IEEE Trans. Intell. Transp. Syst. **12**(4), 1537–1548 (2011)
14. An, S.-H., Lee, B.-H., Shin, D.-R.: A survey of intelligent transportation systems. In: 2011 Third International Conference on Computational Intelligence, Communication Systems and Networks. IEEE (2011)
15. Kanungo, A., Sharma, A., Singla, C.: Smart traffic lights switching and traffic density calculation using video processing. In: 2014 Recent Advances in Engineering and Computational Sciences (RAECS). IEEE (2014)
16. Milanes, V., et al.: An intelligent V2I-based traffic management system. IEEE Trans. Intell. Transp. Syst. **13**(1), 49–58 (2012)
17. Wang, H.: A reservation-based smart parking system (2011)
18. He, W., Yan, G., Da Xu, L.: Developing vehicular data cloud services in the IoT environment. IEEE Trans. Industr. Inf. **10**(2), 1587–1595 (2014)
19. Khanna, A., Anand, R.: IoT based smart parking system. In: 2016 International Conference on Internet of Things and Applications (IOTA). IEEE (2016)

20. Al-Sakran, H.O.: Intelligent traffic information system based on integration of Internet of Things and Agent technology. Int. J. Adv. Comput. Sci. Appl. (IJACSA) **6**(2), 37–43 (2015)
21. Roess, R.P., Prassas, E.S., McShane, W.R.: Traffic Engineering. Pearson/Prentice Hall, Upper Saddle River (2004)
22. De Souza, A.M., et al.: Traffic management systems: a classification, review, challenges, and future perspectives. Int. J. Distrib. Sensor Netw. **13**(4), 1550147716683612 (2017)
23. Rizwan, P., Suresh, K., Babu, M.R.: Real-time smart traffic management system for smart cities by using Internet of Things and big data. In: 2016 International Conference on Emerging Technological Trends (ICETT). IEEE (2016)
24. Choosri, N., et al.: IoT-RFID testbed for supporting traffic light control. Int. J. Inf. Electron. Eng. **5**(2), 102 (2015)
25. Bui, K.H.N., Jung, J.E., Camacho, D.: Game theoretic approach on Real-time decision making for IoT-based traffic light control. Concurrency Comput. Pract. Experience **29**(11), e4077 (2017)
26. Yu, M., et al.: An RFID electronic tag based automatic vehicle identification system for traffic IOT applications. In: 2011 Chinese Control and Decision Conference (CCDC). IEEE (2011)
27. Miz, V., Hahanov, V.: Smart traffic light in terms of the cognitive road traffic management system (CTMS) based on the Internet of Things. In: Proceedings of IEEE East-West Design & Test Symposium (EWDTS 2014). IEEE (2014)
28. Thakur, T.T., et al.: Real time traffic management using Internet of Things. In: 2016 International Conference on Communication and Signal Processing (ICCSP). IEEE (2016)
29. Phan, C.T., et al.: Applying the IoT platform and green wave theory to control intelligent traffic lights system for urban areas in Vietnam. TIIS **13**(1), 34–51 (2019)
30. Chowdhury, A.: Priority based and secured traffic management system for emergency vehicle using IoT. In: 2016 International Conference on Engineering & MIS (ICEMIS). IEEE (2016)
31. Mzahm, A.M., Ahmad, M.S., Tang, A.Y.C.: Agents of things (AoT): an intelligent operational concept of the Internet of Things (IoT). In: 2013 13th International Conference on Intellient Systems Design and Applications. IEEE (2013)
32. Jabbarpour, M.R., Nabaei, A., Zarrabi, H.: Intelligent guardrails: an IoT application for vehicle traffic congestion reduction in smart city. In: 2016 IEEE International Conference on Internet of Things (iThings) and IEEE Green Computing and Communications (GreenCom) and IEEE Cyber, Physical and Social Computing (CPSCom) and IEEE Smart Data (SmartData). IEEE (2016)
33. Nor, R.F.A.M., Zaman, F.H.K., Mubdi, S.: Smart traffic light for congestion monitoring using LoRaWAN. In: 2017 IEEE 8th Control and System Graduate Research Colloquium (ICSGRC). IEEE (2017)
34. Turcu, C.E., Găitan, V.G., Turcu, C.O.: An internet of things-based distributed intelligent system with self-optimization for controlling traffic-light intersections. In: 2012 International Conference on Applied and Theoretical Electricity (ICATE). IEEE (2012)
35. Masek, P., et al.: A harmonized perspective on transportation management in smart cities: the novel IoT-driven environment for road traffic modeling. Sensors **16**(11), 1872 (2016)
36. Keertikumar, M., Shubham, M., Banakar, R.M.: Evolution of IoT in smart vehicles: an overview. In: 2015 International Conference on Green Computing and Internet of Things (ICGCIoT). IEEE (2015)
37. Kaminski, N.J., Murphy, M., Marchetti, N.: Agent-based modeling of an IoT network. In: 2016 IEEE International Symposium on Systems Engineering (ISSE). IEEE (2016)
38. Venkatesh, H., Perur, S.D., Jagadish, M.C.: An approach to make way for intelligent ambulance using IoT. Int. J. Electr. Electron. Res. **3**(1), 218–223 (2015)

39. Tielert, T., et al.: The impact of traffic-light-to-vehicle communication on fuel consumption and emissions. In: 2010 Internet of Things (IOT). IEEE (2010)
40. Anass, R., Yassine, H., Mohammed, B.: IoT for ITS: a dynamic traffic lights control based on the Kerner three phase traffic theory. Int. J. Comput. Appl. **145**(1), 0975–8887 (2016)
41. Liu, Y., Lei, L., Chen, W.P.: Intelligent traffic light control using distributed multi-agent Q learning. In: 2017 IEEE 20th International Conference on Intelligent Transportation Systems (ITSC). IEEE (2017)
42. Bui, K.-H.N., Jung, J.J.: Internet of agents framework for connected vehicles: a case study on distributed traffic control system. J. Parallel Distrib. Comput. **116**, 89–95 (2018)

On the Development of an Expert Decision Support System Based on the ELECTRE Methods

Tatiana Kravchenko[1], Timofey Shevgunov[1,2](✉), and Alexander Petrakov[2]

[1] Higher School of Economics, National Research University, Myasnitskaya Ulitsa 20, 101000 Moscow, Russia
{tkravchenko, tshevgunov}@hse.ru

[2] Moscow Aviation Institute, National Research University, Volokolamskoe Shosse 4, 125993 Moscow, Russia
nio4@mai.ru

Abstract. Analytical justification of decision options using decision support systems (DSS) significantly improves the quality of decisions. The use of the currently existing DSS, which usually includes one or two decision-making methods, does not always lead to the desired results, since each method is based on certain assumptions and is not universal. The noticeable effect is achieved when many decision-making methods are included in one DSS knowledge base. The systems that meet these requirements belong to the class of Expert Decision Support Systems (EDSS), which currently includes more than 50 decision-making methods. Expanding the EDSS knowledge base, made by including new methods in it, allows choosing the most suitable solution method for each decision-making task. Addition of the decision table model, which is the basis of the system knowledge base, allows developing EDSS without completely processing the system program code. The ELECTRE methods were adopted for expanding the EDSS knowledge base. The basis for the selection was their key feature, which consists in the fact that they do not use the alternative valuation convolution operation given in different scales according to individual criteria. The article shows the adapted algorithms of the ELECTRE family methods ready for inclusion in EDSS. The algorithm is proposed for obtaining a criterion matrix being based on alternatives that serve as input information for the ELECTRE methods in cases where there is no objective information to fill it. The results of the study can be used to develop EDSS, that open the way to analytically substantiate solutions using methods that were not previously used in the system.

Keywords: Expert Decision Support System · ELECTRE · Knowledge base

R. Silhavy (Ed.): CSOC 2020, AISC 1226, pp. 552–561, 2020.
https://doi.org/10.1007/978-3-030-51974-2_51

1 Introduction

The decision support systems (DSS) based on the methods of the decision making theory remain particular interest among all types of information-analytical systems. A common DSS is usually understood as an interactive computer system that helps the decision maker (DM) to utilize available information with a higher performance and choose better models to solve poorly structured tasks or tasks that are difficult to formalize [1, 2].

Generally, EDSS uses more than 50 mathematical methods based on underlying decision-making models. The best known and described DSSs are such systems as Super Decisions, Decisions Lens, Expert Choice, Expert Decision Support System (EDSS) [3, 4]. The first two systems in the list use the analytic hierarchy process (AHP) and analytic network process (ANP) designed by Thomas L. Saaty [5, 6]. Such systems as Expert Choice, Imperator 3.1, Expert, OPTIMUM, Vybor, MPRIORITY, and WinEXP+ use only HAM [3].

One of the most promising contemporary group of methods is known as the ELECTRE family. The ELECTRE methods include ELECTRE I, ELECTRE Iv, ELECTRE IS, ELECTRE II, ELECTRE III, ELECTRE IV, ELECTRE TRI. The author of the basic ELECTRE method is Professor B. Roy [7]. The algorithms of these methods are described in detail in [8–10].

The paper is organized as follows. In Sect. 2, three of the algorithms (ELECTRE IS, ELECTRE II, ELECTRE III) are adapted to further usage in EDSS terms. The Sect. 3 propose the way of embedding the algorithms to the knowledgebase. The paper ends with conclusion.

2 ELECTRE Family

The adaptation of the ELECTRE methods has been done to incorporate them further into a typical EDSS. The algorithms of the methods are written in a single terminology which facilizes formulating a set of decision rules in the form ready to include the selected method in the system knowledge base.

2.1 ELECTRE IS Features

Input conventions:

- $\{X_1, X_2, \ldots, X_I\}$ is a set of alternatives;
- $\{K_1, K_2, \ldots, K_L\}$ is a set of criteria, where, for each criterion, its direction is also set: maximization or minimization;
- $\{Z_1, Z_2, \ldots, Z_L\}$ is a set of criteria weights, the weights are normalized;
- F_{il} is the evaluation of alternative X_i by criterion K_l;
- the set of the elements $\{F_{il}\}$ form a "criterion-alternative" matrix;
- X_iSX_k is a logical relationship "X_i alternative is no worse than X_k", which may take the valuse of "true" or "false";

- v_l is the veto-threshold for the criterion indexed by l, which denotes the critical difference between the evaluation of alternatives by the lth critetion, which allows us to veto the claim that one alternative is "no worse" than the other;
- q_l is the indifference threshold for the lth criterion, which denotes the difference between the evaluations of two alternatives according to the lth criterion, which is not significant for the decision maker (DM);
- p_l is the preference threshold for the lth criterion, which denotes the difference between the evaluations of two alternatives according to the lth criterion, which is essential for DM and allows us to conclude that one alternative is strictly superior to another according to this criterion;
- η is the coefficient that is applied to the indifference threshold level q_l to correct it in the process of determining the "veto situation";
- s is a concordance level, which is considered the minimum sufficient to determine the relationship X_iSX_k;
- The concordance index $\mathrm{SO}(X_i,X_k)$ is between the alternatives X_i and X_k

It is crucial to note that the p_l score should always be greater than q_l, since the preference threshold p_l denotes the difference between alternative evaluation, which is significant for DM, and indifference threshold q_l is insignificant.

A key difference of the ELECTRE IS method is the use of a pseudocriteria [8]. The concept of the pseudocriteria is based on the fact that the evaluation of alternatives do not have the absolute accuracy, and the difference between the evaluated metrics of the alternatives can be interpreted differently. In order to take this fact into account, the ELECTRE IS method uses the indifference and preference thresholds, which allow taking into account the degree of superiority of one alternative over another for each criterion.

The procedure for ranking alternatives by ELECTRE IS is based on a sequential pairwise comparison of alternatives. The evaluation of the concordance index $SO(X_i, X_k)$ for each pair (Xi, Xk) made of the compared alternatives depends on whether the criterion is the maximization or minimization direction.

For the criteria with maximization direction, the following rules hold:

1. The concordance index, which is denoted by $SO_+(X_i, X_k)$, is equal to the sum of the weights of all of those criteria by which the alternative X_i is better than X_k.
2. If the difference between the evaluated values of alternatives X_i and X_k by criterion l is less than the indifference threshold, then its weight is not taken into account.
3. If the difference between the evaluations of alternatives X_i and X_k by criterion l is greater than or equal to the indifference threshold, but less than the preference threshold, then its weight is taken into account with a reduction factor $\varphi_l(X_i, X_k)$, where l is the number of criterion.
4. If the difference between the evaluations of alternatives X_i and X_k by the lth criterion is greater than or equal to the preference threshold, then its weight is taken into account with the coefficient 1.

Having combined the above-listed rules, one can write

$$
\begin{aligned}
SO_{+}(X_i, X_k) = & \sum_{l=1}^{L} Z_l\{\, l: \; F_{il} - F_{kl} \geq p_l\} \\
& + \sum_{l=1}^{L} \varphi_l(X_i, X_k) \cdot Z_l\{\, l: \; q_l \leq F_{il} - F_{kl} < p_l\},
\end{aligned}
\tag{1}
$$

where the reduction factor is found as

$$
\varphi_l = \frac{F_{il} - F_{kl} - q_l}{p_l - q_l}. \tag{2}
$$

For the criteria with minimization direction, other rules are applied:

1. The concordance index, which is denoted by $SO_{-}(X_i, X_k)$, is equal to the sum of the weights of all the criteria by which the alternative X_k is better than X_i.
2. If the difference between the evaluations of alternatives X_k and X_i by criterion l is less than the indifference threshold, then its weight is not taken into account.
3. If the difference between the evaluations of alternatives X_k and X_i by criterion l is greater than or equal to the indifference threshold, but less than the preference threshold, then its weight is taken into account with a reduction factor $\varphi_l(X_i, X_k)$.
4. If the difference between the evaluations of alternatives X_k and X_i by criterion l is greater than or equal to preference threshold, then its weight is taken into account with the coefficient 1:

Having combined the above-listed rules, one can write

$$
\begin{aligned}
SO_{-}(X_i, X_k) = & \sum_{l=1}^{L} Z_l\{\, l: \; F_{kl} - F_{il} \geq p_l\} \\
& + \sum_{l=1}^{L} \varphi_l(X_i, X_k) \cdot Z_l\{\, l: \; q_l \leq F_{kl} - F_{il} < p_l\},
\end{aligned}
\tag{3}
$$

where the reduction factor is found as

$$
\varphi_l = \frac{F_{kl} - F_{il} - q_l}{p_l - q_l}. \tag{4}
$$

The concordance index $SO(X_i, X_k)$ for the whole set of the criteria, including those the maximization and minimization direction:

$$
SO(X_i, X_k) = SO_{+}(X_i, X_k) + SO_{-}(X_i, X_k). \tag{5}
$$

Further, for each pair of the alternatives by each criterion, the presence of a veto situation is checked. The veto situation arises in the following cases.

As long as the criteria with maximization direction are considered, the veto will be raised if the condition

$$F_{kl} - F_{il} \geq v_l - q_l \cdot \eta \tag{6}$$

is satisfied, where

$$\eta = \frac{1 - SO(X_i, X_k)}{1 - s}. \tag{7}$$

For criteria with minimization direction, the veto arises provided

$$F_{il} - F_{kl} \geq v_l - q_l \cdot \eta. \tag{8}$$

The coefficient η is applied to the indifference threshold level q_l to correct it in the stage of determining the veto situation. The greater the value by which concordance index $SO(X_i, X_k)$ exceeds the concordance level s, the less η is. Consequently, the probability of a veto situation goes down, since the product $q_l \cdot \eta$ decreases.

If the concordance index $SO(X_i, X_k)$ greater than or equal to a given concordance level s, then the $X_i S X_k$ relation arises between the alternatives X_i and X_k. However, assuming there is a veto situation by at least one criterion between the alternatives, the relation S cannot arise between them.

2.2 ELECTRE II Features

Method ELECTRE II brings additional convention, since it uses two concordance levels s_1 and s_2 to define the ranking ordering for two alternatives. Concordance level s^1 is used to determine the preference relation S of the first order, and level s^2 of the second order. In the second ranking order of alternatives, both 'weak' and 'strong' preference relationships are considered (S^2), and in the first order there are only the most 'strong' preference relations (S^1), in which experts express great confidence. The inclusion of 'weak and strong' preference relationships into the second ranking order is carried out provided one specifies the concordance level s^1 greater than s^2: the lower the concordance level is, the lower the requirements for the value of the concordance index to determine the relationship S between the alternatives are.

The ELECTRE II method allows full ranking by using the procedure known as the alternative distillation [8]. The best alternative is the one that is preferred by the largest number of alternatives, and the worst alternative is the one that is preferred by the least number of alternatives.

The procedure for ranking alternatives by ELECTRE II is based on the sequential comparison of each pair of the compared alternatives. For each pair of the alternatives, a concordance index is calculated and a veto situation is checked. The veto situation arises in the following cases:

1. For the criteria with a direction of maximization, the difference between the evaluation of alternatives X_k and Xi by criterion l is greater than or equal to its veto threshold v_l:

$$F_{kl} - F_{il} \geq v_l. \tag{9}$$

2. For the criteria with a direction of minimization: the difference between the evaluation of alternatives X_i and X_k by criterion l is greater than or equal to its veto threshold v_l:

$$F_{il} - F_{kl} \geq v_l. \tag{10}$$

If the concordance index $SO(X_i, X_k)$ is higher or equal to the given concordance level s1, then the relation $X_iS^1X_k$ arises between the alternatives X_i and X_k. If a veto situation arises between the alternatives by at least one criterion, then the relation S_1 cannot arise between them.

If the concordance index $SO(X_i, X_k)$ is higher or equal to the given concordance level s^2, then the relation $X_iS^2X_k$ arises between the alternatives X_i and X_k. If a veto situation arises between the alternatives by at least one criterion, then the relation S_2 cannot arise between them.

The final step in ranking alternatives using the ELECTRE II method is the distillation of alternatives. Let us denote with r_{X_i} the number of alternatives X_k, against which the alternative $X_i : X_iS^1X_k$, with $k \neq i$, is preferred.

First of all, let define the set D_1, which includes all the best alternatives, preferred against the greatest number of other alternatives with respect to S^1; it has been formed by experts on the basis of r_{X_i} values). If the set D_1 contains more than one alternative, the value r_{X_i} for them is to be recalculated in such a manner that only those alternatives that are included in the set D_1 are considered. Based on the recalculated values r_{X_i}, a new set of the best alternatives D_2 with respect to S_1 has been formed.

The procedure is being repeated until the only best alternative is determined or all the alternatives included in the set of the best are equal to r, which is the value calculated taking into account only those alternatives that are among the set of the best. The best alternative (or several alternatives) gets rank 1 and is removed from consideration. The distillation procedure is then repeated again to determine the alternative with rank 2, and so on, until all of the alternatives are finally ranked.

If several alternatives have rank 1 and cannot be ranked using the distillation procedure, then they form the set N, in which all alternatives are considered equivalent. If it is necessary to rank the elements of the set N, then the S^2 preference relation is determined between the alternatives in the given set, with which the distillation procedure can be repeated on the set N.

The procedure for determining the S_2 preference relation is similar to the procedure used to determine the preference S^1 relation. However, the concordance level s^2 is used to apply it. Thus, the relation S^2 arises between the alternatives, if the concordance index is higher or equal to the given concordance level s^2.

Since the concordance level $s^1 > s^2$, then the S_2 preference relation allows the DM to identify weaker preference relationships. Thus, the S^2 preference relation can help the DM to choose between alternatives that are considered equivalent based on ranking using the S^1.

2.3 ELECTRE III Features

ELECTRE III brings an additional convention:

- $SO_l(X_i, X_k)$ is the concordance index between the alternatives X_i and X_k by the lth criterion;
- PR is the overall relation of preference;
- λ is the preference relation threshold, which is the value of PR, which DM considers to be minimally sufficient to determine the relation S between two alternatives;
- $ND(X_i, X_k)$ is the overall relation of non-concordance.

The high accuracy of the ranking method ELECTRE III is achieved due to its possibility of using several pseudocriteria (similar to ELECTRE IS) and alternative distillation procedure (introduced ELECTRE II). The use of pseudocriteria allows one to take into account unstructured data when alternatives is under ranking, and the distillation procedure makes it possible to assign a different rank to each alternative. Therefore, the ranking of alternatives by method ELECTRE III is based on a consistent comparison of each pair of alternatives.

For a criterion with the "maximization" direction, the concordance index $SO_l(X_i, X_k)$ is calculated for each characteristic using the following formula:

$$SO_l(X_i, X_k) = \begin{cases} 1, \text{ if } F_{il} - F_{kl} \geq p_l, \\ \frac{F_{il} - F_{kl}}{p_l}, \text{ if } q_l < F_{il} - F_{kl} < p_l, \\ 0, \text{ if } F_{il} - F_{kl} \leq q_l. \end{cases} \quad (11)$$

Similarly, one can write the formula for a criterion with the "minimization" direction:

$$SO_l(X_i, X_k) = \begin{cases} 1, \text{ if } F_{kl} - F_{il} \geq p_l, \\ \frac{F_{kl} - F_{il}}{p_l}, \text{ if } q_l < F_{kl} - F_{il} < p_l, \\ 0, \text{ if } F_{kl} - F_{il} \leq q_l. \end{cases} \quad (12)$$

Thus, the concordance index by criterion $l\, SO_l(X_i, X_k)$ can take a value from the range [0, 1]:

- 1 means 'X_i is strictly better than X_k by criterion l';
- from 0 to 1 means 'non-strict preference';
- 0 means 'no preference'.

Next, the non-concordance index is calculated for each criterion l. As long as criteria with the "maximization" direction is considered, one can write:

$$NE_l(X_i, X_k) = \begin{cases} 1, \text{ if } F_{kl} - F_{il} \geq q_l, \\ \frac{F_{kl} - F_{il}}{v_l}, \text{ if } q_l < F_{kl} - F_{il} < v_l, \\ 0, \text{ if } F_{kl} - F_{il} \leq v_l. \end{cases} \quad (13)$$

On the other hand, for a criterion with the "minimization" direction, one obtains:

$$NE_l(X_i, X_k) = \begin{cases} 1, \text{ if } F_{il} - F_{kl} \geq q_l, \\ \frac{F_{il} - F_{kl}}{v_l}, \text{ if } q_l < F_{il} - F_{kl} < v_l, \\ 0, \text{ if } F_{il} - F_{kl} \leq v_l. \end{cases} \tag{14}$$

The non-concordance can assume a value from 0 to 1: 0 value means "there is no reason to believe that the alternative X_i is worse than the alternative X_k by criterion l". From 0 to 1—"there is reason to believe that the alternative X_i is worse than the alternative X_k» . If $NE_l(X_i, X_k) = 1$, then criterion l vetoes the claim that the alternative X_i is better than X_k.

To evaluate how one alternative is better than the other, the general concordance index:

$$C(X_i, X_k) = \sum_{i=1}^{L} Z_l \cdot SO_l(X_i, X_k), \tag{15}$$

If $C(X_i, X_k) = 1$, then alternative X_i has strict preference for alternative X_k for all the criteria. If $C(X_i, X_k) = 0$, there is no reason to believe that alternative X_i is anything better than alternative X_k.

In order to evaluate how one alternative is worse than the other, the general non-concordance relation is used for all criteria $ND(X_i, X_k)$:

$$ND(X_i, X_k) = \prod_{l:NE_l(X_i, X_k) > C(X_i, X_k)} \frac{1 - NE_l(X_i, X_k)}{1 - C(X_i, X_k)}. \tag{16}$$

The general non-concordance relation can take a value from 0 to +∞. In case, where $ND(X_i, X_k) = 0$, then at least one criterion vetoes the claim that the alternative X_i is better than X_k. If the non-concordance relation tends to +∞, then by all criteria $SO(X_i, X_k) = 1$. Accordingly, the smaller values of the general non-concordance relation $ND(X_i, X_k)$ mean less reason to say that the alternative X_i is worse than X_k.

In order to determine the relation S between two alternatives, two main aspects of the comparison must be taken into account simultaneously: relation of concordance and non-concordance. The value of PR implements the accounting for the relation of concordance and non-concordance for the corresponding pair of alternatives:

$$PR(X_i, X_k) = ND(X_i, X_k)C(X_i, X_k) \tag{17}$$

PR takes its value from 0 to +∞. It is obvious that $PR = 0$ if at least one criterion vetoes the statement that the alternative X_i is better than X_k, since $ND(X_i, X_k) = 0$. Therefore, there is no reason to believe that the alternative X_i is better than the X_k by any of the criteria, as $C(X_i, X_k) = 0$.

According to ELECTRE III, the alternative X_i is considered "not worse than" X_k, if the value $PR(X_i, X_k)$ is no less than the λ preference relation threshold. The higher λ is, the higher the "strictness" of ranking alternatives takes place. The final step in ranking alternatives using the ELECTRE III method is distilling the alternatives, in the same manner as it is carried out for ELECTRE II.

3 Embedding New Methods in the EDSS Knowledge Base

As a comparative analysis of the methods showed, the most effective method is ELECTRE III. Given that the criterion-alternative matrix can be determined by an expert approach, we will include a combined method in an EDSS that allows to obtain an averaged matrix of 'criterion-alternative' and use the Roy principle (named after the method family author) to agree on alternatives for different attributes. Let us call this method PUR_q Roy.

When one develops a module for choosing the adoption method, depending on the conditions of the problem, the EDSS used the decision table (DT) model [4]. Being the necessary conditions to select the appropriate method, DT introduced questions about the elements of any decision-making task; conditions inputs are the results of checking conditions or possible answers to asked questions. The action section of DT is formed by various decision-making methods included in the EDSS. The decision-making rules with this interpretation of DT show which decision-making methods should be chosen for various combinations of answers to asked questions.

The whole set of such decision-making rules forms the EDSS *knowledge base*. To include new methods in the system, the change may be required in the composition of the questions as well as the answers. In this case, a new decision-making rule is additionally included in the DT. The PUR_q Roy method will be used if the following conditions hold. Several problem situations with given probabilities of their occurrence. The principle of matching evaluations of alternatives in various problem situations with given probabilities of their appearance – the majority. Awareness of the consequences of the decision – complete certainty at one stage. The number of experts involved in solving the problem – several experts. The principle of harmonizing evaluations of alternatives given by individual experts – the majority. Number of signs for evaluating alternatives – several signs. The degree of comparability of the features – incomparable. The principle of harmonizing evaluations of alternatives for individual criteria – Roy. The method of setting evaluations of the relative significance of the characteristics – characteristics weights are specified. A way to set many alternatives – the finite set. A way to set preferences on many alternatives – quantitative evaluations of alternatives for each characteristic are given.

4 Conclusion

As a result of the study, the algorithms of the following methods have been adapted: ELECTRE IS, ELECTRE II, ELECTRE III. They were described in terms of EDSS and brought to a form convenient for software development. For inclusion in EDSS, the PUR_qRoy method is proposed, which is based on the ELECTRE III method and an algorithm for obtaining a criterion-alternative matrix. It is shown how the new method can be embedded in the EDSS knowledge base. The results of the study can be used to develop the Expert Decision Support System, allowing to analytically substantiate decision options using various decision methods.

References

1. Burstein, F., Holsapple, C.W.: Handbook on Decision Support Systems 1: Basic Themes. Springer-Verlag, Heidelberg (2008)
2. Trakhtengerts, E.A.: Komp'juternye sistemy podderzhki prinjatija upravlencheskih reshenij (Computer-aided control decision support systems). Problemy upravlenija (Control problems) **1**, 13–28 (2003)
3. Kravchenko, T.K., Seredenko, N.N.: Vydelenie priznakov klassifikacii sistem podderzhki prinjatija reshenij (Feature extraction for classification of decision support systems). Otkrytoe obrazovanie (Open education) **4**, 71–79 (2010)
4. Kravchenko, T.K.: Ekspertnaja sistema podderzhki prinjatija reshenij (An expert decision support system). Otkrytoe obrazovanie (Open education) **6**, 147–156 (2010)
5. Saati, T.L.: Fundamentals of Decision Making and Priority Theory, 1st edn. RWS Publications (2000)
6. Saati, T.L.: Decision Making for Leaders: The Analytic Hierarchy Process for Decisions in a Complex World Analytic Hierarchy Process Series, vol. 2. 3rd edn. RWS Publications (2001)
7. Roy, B.: Classement et choix en présence de points de vue multiples (la méthode ELECTRE). La Revue d'informatique et de recherche opérationelle (RIRO) **8**, 57–75 (1968)
8. Roy, B.: The outranking approach and the foundation of ELECTRE methods. Theory Decis. **31**, 49–73 (1991)
9. Figueira, J., Greco, S., Ehrgott, M.: Multiple Criteria Decision Analysis. State of the Art Surveys, 134–152 (2004)
10. Kangas, A., Kangas, J., Pykäläinen, J.: Outranking methods as tools in strategic natural resources planning. Silva Fennica **35**(2), 215–227 (2001)

Practical Model for Evaluating the Risk of a Person to Commit a Criminal Offence

Anca Avram(✉), Tudor Alin Lung, and Oliviu Matei

Technical University of Cluj-Napoca, University Center at Baia Mare, Baia Mare, Romania
anca.avram@ieee.org, alinlung33@gmail.com, oliviu.matei@holisun.com

Abstract. As the society develops, the volume of criminal offences and the complexity of crimes also increases. Data mining techniques can be used to help the police departments and explore already existing data to predict for a person the potential risk of committing a criminal offence. This information, known in advance by the authorities, could lead to smart preemptive actions that could lower the crime rates. The purpose of this research is to find a way of modeling the existing information about persons and test what are the machine learning algorithms that give the best results. The work is specific to the Romanian context, the results are very promising, but intensive tests on larger sets of data should be performed.

Keywords: Machine learning · Criminal offence risk · Data mining

1 Introduction

Crimes are a difficult subject in our society that cannot be ignored and needs constant attention and investigation in order to find possible ways to prevent more crimes happening.

The dynamics of the criminal phenomenon of the last years and the efforts of the state institutions and of the civil society to deal with them in a coherent and effective way, imposes the realization of a national strategy for crime prevention. This is fully achievable because there is a theoretical basis devoted to this area: internal and external normative acts, programs and recommendations of the Council of Europe and the United Nations (UN) and strategies recently developed at the level of other states. We can see how, all over the world, statistics offer significant increases in crime, but also increasing efforts to cope with this phenomenon, at the state level [1].

In the information age we are in, the information from the online environment gathered both by the administrators of the social sites and by the specialized institutions of the state, about people, offers a very big data flow.

The idea of the project came from the need of the Romanian National Police to identify as quickly and as easily as possible persons, who could potentially commit criminal offences. As a result, the scope of this project is strict for

R. Silhavy (Ed.): CSOC 2020, AISC 1226, pp. 562–570, 2020.
https://doi.org/10.1007/978-3-030-51974-2_52

providing the information needed by the Police Department, in order to reduce as much as possible the crime rate at national level.

This article wants to tackle both theoretical and practical aspects of creating a system of analysis for people, system that predicts with a certain accuracy what is the risk of a person to commit a criminal offence, based strictly on the knowledge available for that person, like the environments and entourage that he/she frequents, income, dependants and other existing information [11].

Using the RapidMiner platform [13], as well as the data available, a very useful tool in identifying potential criminals can be created, tool that could be useful to the Romanian National Police to achieve a climate of public order and security, as it was also done by Oatley and Ewart in project OVER [15] made in cooperation with the United Kingdom authorities.

The paper is structured as follows. Section 2 presents the current state of art in the area of predicting criminal risks. It is followed by the experimental overview presented in Sect. 3 - this contains details on the objectives of the research, the way the data model was constructed and the results obtained after the experiments were performed. Sect. 5 presents the conclusions and discussions for this research.

2 State of Art

According to King et al. [10], artificial intelligence in the study of crime is still a relatively young and inherently interdisciplinary field, covering socio-legal studies for formal science, but offering an analysis of potential problems related to artificial intelligence as a means of committing crimes as well as a possible way of solving them.

In 1979, Cohen and Felson in their work [4] observed that circumstances can influence the behaviour of offenders. It is supported the idea in which, human ecological theory allows infiltration into the daily life of illegal activities, commonly referred to as criminal offences, which align the legal activities of everyday life, thus giving rise to the offender.

Kiani et al. [9] demonstrate that by using data mining on a subject in a criminal database by applying certain types of equations, an accurate result is achieved, thus motivating both the necessity and efficiency of using data mining to identify potential criminogenic factors. Ryan and Hofmann [16], proved that prediction models can be used with a resultant high degree of accuracy by grouping the class labels as opposed to prediction models which treat each crime as a specific class label.

As Farsi et al. [6] observed, a lot of studies in the field of cybercriminology have a focus on what actions can be taken to reduce the impact of an already committed crime. Farsi et al. presented an overview of predictive analysis techniques that could be used within the context of cybercriminology. This is an inspiration for the current work that focuses on discovering a predictive model that could be used to identify the persons with a potential risk of committing criminal offences.

Nagamatsu in [14] says that war, crime and terrorism are all incidents caused by the specific intentions of the people. However, these events, like accidents and natural disasters, are largely influenced by the development of scientific technology and social structures of political disagreement and economic disparities.

Chen et al. [3] tried introducing a framework that implies data mining, for analysis of eight major crime types, from a data set with data collected from the Tucson and Phoenix police departments. Their work, together with the work performed by Davidson in [5] and by Kanazawa in [8] was a starting point for the current research when establishing the relevant attributes to be studied for a specific person when evaluating the criminal risks.

Although there has been made research in the area of crimes, since each country has their particularities, the current work aims at first to find a model fit to the Romanian reality for predicting criminal risks, and afterwards evolve in a more general model, that could be fit also in other countries.

3 Methods

The main focus of this project was to develop a model for analyzing the risk of committing a criminal offence for a person and to study what are the best machine learning algorithms that could be of use in predicting that risk.

The value obtained could be used in real life scenarios, by police officers or other justice related departments. It is not a categorical value, implying in any way that a person that has a very high risk of committing a criminal offence should be punished in advance, on the contrary, it could raise awareness and maybe preventive actions could be taken by the proper authorities (for example sending the person to a psychologist, if this seems suitable or taking preventive actions in a specific area, if more persons are considered).

Also, it needs to be mentioned that at this point of research, the article focuses on predicting "everyday" criminal offences and not very complex crimes that involve modes of operation, criminal groups, as well as the need to know how certain specific areas work. As a next step, the model could be further extended to provide an overview for different types of crimes.

3.1 Experimental Objectives

The entire experiment was thought considering the following main objectives:

(1) Adding a complex set of data referring to both the implemented persons and to certain specific criteria, key in establishing the desired result.
(2) Training the most favorable model discovered after analyzing several machine learning algorithms.
(3) Obtaining real data and applying them to the trained data set in order to obtain the desired result.
   - Data collection from different sources as well as from different types of human personalities.

- Diversification of the criteria in order to train the network in producing results for a broader set of crimes.
- Selecting through the Rapid Miner application, only the criteria essential to obtain the desired result

(4) Providing real-time output data and establishing the risk of committing a criminal offence.

In the current research, the first two objectives are in focus. For the third and fourth objective, that imply obtaining a larger set of real data and processing it in real time, there have been some steps performed, meaning that it was tried to obtain authentic information from the Romanian Ministry of Internal Affairs, the Center for the Coordination of Scientific Research. Although the idea was perceived as promising, it was not yet possible to obtain the necessary data, so that they can be introduced into the application, thus limiting the research to the data set created for the project development.

3.2 Experimental Data

An important note that needs to be mentioned is that the data on which the prediction experiments were performed, were obtained at random from a larger database containing records related to the criminal history of more persons and comply with the principles and rules imposed by the European Union, through the General Data Protection Regulation (GDPR).

Data was stored in csv format. At this point the data set that is the starting point for the study is a modest size, but if this would turn in an application being used at national level, the amount of data to be analyzed would exponentially increase, hence a tool that would support big data was needed. So, as in [12] where large financial data is analyzed, the tool chosen to run our experiments was Rapid Miner.

Data that was used as starting point for this research was represented in a proportion of 70%, so binomial classification [2] was performed in order to evaluate the most appropriate attributes to be of use in training the model. After the binomial classification, the following attributes for a person were considered when creating the model:

- Had relations that subsequently led to the commission of an offense;
- Instigated the commission of an offense (instigator);
- Cooperated in committing an offense (co-author);
- Helped in some way in committing a crime (accomplice);
- Committed any act provided by the criminal law (author);
- Was the victim of any criminal offence;
- Neglect of family care;
- Deviant parental behaviors and attitudes (parental offense, parental violence or tolerance of violence);
- Family separation (conflicts between spouses or divorce);
- Meets all the criteria for committing offenses;

- Is/was part of any criminal organization/group;
- He/she suffers from certain diseases that could cause him/her commit criminal offenses;
- Is a recidivist;
- Was investigated for committing an offense but the process was classified;
- Aware of legal consequences that come after committing an offense;
- Risk of committing criminal offence.

Table 1 is an example of data from the historical background of more persons, data that was already converted to binary values.

Table 1. Example of historical background information for more persons

Had relations that led to an offense	0	0	0	0	0	1
Instigator	0	0	0	0	0	0
Co-author	1	0	0	0	0	0
Accomplice	0	0	0	0	0	0
Author	1	0	0	0	0	0
Victim	0	0	0	0	0	0
Neglect of family care	1	0	0	0	0	0
Deviant parental behaviour	0	0	0	0	0	1
Family separation	0	0	0	0	0	0
Meets all criteria	1	0	0	0	0	0
Is part of a criminal group	0	0	0	0	0	0
Diseases	1	1	1	0	1	0
Is recidivist	0	0	0	0	0	0
Was prior investigated	0	0	0	0	1	0
Aware of legal consequences	1	0	0	0	1	0
Risk of criminal offence	1	0	0	0	1	0

The previously created information is about the historical records existing related to the background of a person, while we also introduced a set of information that could be filled in by the authorized personnel on the spot, when needed, in order to evaluate the risk for a person and act upon the result. This "on the spot" model has the following attributes:

- Approximate age;
- Entourage (alcoholic, narcotic, NA, normal, good, weak);
- Level of language understanding (low, medium, high);
- Tone used to communicate (high voice, normal voice, low voice, firm voice, emotional voice);
- Gender;
- Living area (urban or rural);

- Number of persons living in the same space;
- Type of home (house, flat, shelter, homeless);
- Has ID card;
- Number of family members;
- Studies (no of finished years of study);
- Military status;
- Marital status (married, unmarried);
- Religion (orthodox, catholic, Muslim etc.).

This information is more about the context a person is living and status of the person at the moment of the investigation. It is information that if filled in appropriately could lead to good predictions, because people are very much influenced by the environment and entourage and tend to act in a similar manner in similar conditions.

4 Results

The experiment was performed on a set of 1455 records. The data set was split randomly and 90% was used for training the model, while 10% was used for testing it. In a later phase of the research, the model should be used against real data.

The process for analyzing the data and obtaining a prediction on the criminal risk of a person is depicted in Fig. 1 and explained as follows.

Data Modeling. The data modeling part is responsible with the load of training data, preprocessing and obtaining of a validation model that will be evaluated later on the test data.

Data is obtained from a csv file containing both information related to historical criminal background of a person and information related to current situation.

Preprocessing is performed in order to clean data and prepare it for use in the algorithms that will be applied in order to obtain the model. These transformations refer to elimination of outliers, conversions from nominal values to numerical values to support the machine learning algorithms that will be tested [7].

Prepare Testing Data. Preparing the test data is a similar sub-process that consists of loading the prepared test data and process it to have the same format as the data used in the training sub-process, so that the obtained model can later on be validated.

Test and Process Results. The prediction that is the subject of the analysis is the value of the binary column: "Risk of committing a criminal offence". The validation process uses the operator "Vote" from Rapid Miner, that builds a classification model upon the data set and learning algorithms studied. This operator uses a majority vote (for classification) on top of the predictions of the

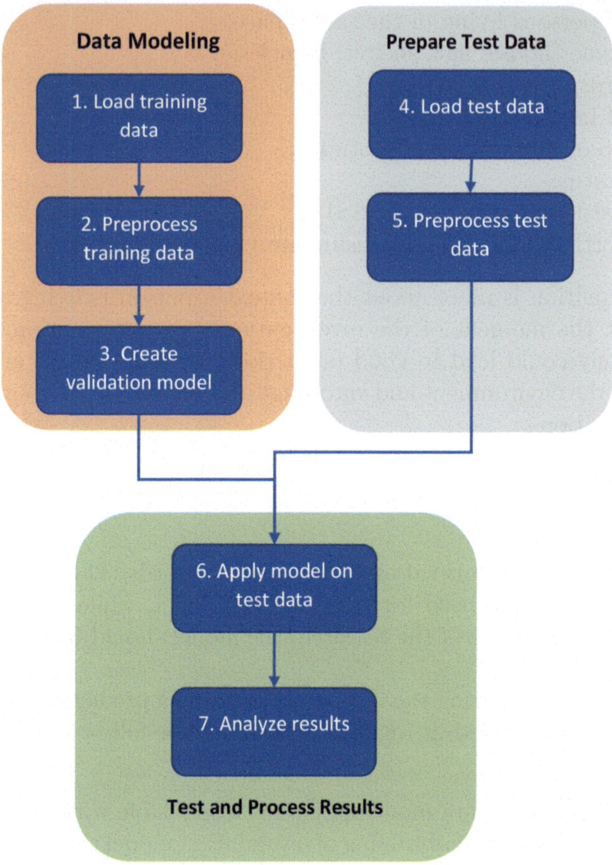

Fig. 1. Process of evaluating criminal risk for a person

base learners provided in its subprocess. Table 2 presents the most important algorithms that data was tested upon and the obtained results.

The results are promising and prove that the model implemented would be a good starting point in predicting the risk of committing a criminal offence. One of the reasons some of the results are so close to the 100% precision is the lack of more real life scenarios, so that the resulted model could be over fitting, after the machine learning algorithms are run. In terms of performance, the Random Tree and k-Nearest Neighbor seem to be the most suitable options. Naïve Bayes also seems a good candidate for our scenario. Even though it is a high-bias classifier, this could be a situation where this could be highly applicable, because it uses Gaussian probability to model the attribute data.

Table 2. Obtained results per algorithm for evaluating criminal risk

Algorithm	Accuracy of prediction	Execution time
Decision tree	99.93%	13 s
Stump decision	99.93%	2 s
Random tree	72.22%	<1 s
Random forest	81.45%	8 s
Gradient Boosted Trees	93.01%	22 s
K Nearest Neighbor	72.84%	<1 s
Naïve Bayes	99.93%	2 s
LDA	30.02%	6 s

5 Discussions

Being able to know that a person is at high risk of committing a criminal offence, in specific circumstance, could be useful and of great help mainly to the state's authorized institutions in order to identify the potential offenders. The result is not categorical, but after inserting the data and finding the outcome, these can focus the attention only on certain entities targeted by the application for the purpose of crime prevention.

The data that serves as basis for the research presented in this paper is based on information gathered from the historical background of a person, correlated with information on the current status of the person. It focuses on providing a prediction on the risk of committing an offence. The results are promising with the test data created. The natural step forward in the research would be running the same processes and understanding the results on real life scenario and adjusting the source attributes considered.

This research does not provide a result for all types of criminals, whereas it is based on the perpetrator criteria of the common offender, respectively the one who comes from environments with social problems, such as the environments and entourages they frequent, the incomes, the persons in maintenance and many other such criteria.

Another possible further development would be to move from the idea of generalization to that of customization, for example focusing on the features of the people who commit crimes with violence. These traits could be given by the environment in which he was exposed, possible traumas suffered in childhood/adolescence/maturity, health problems, mental problems etc. That would require specialised input on this matter.

As an optimization, a small adaptation of the model can be made, in that, we can establish a word library, to be searched on social networks, e.g. Facebook, Twitter, Instagram and thus identify themselves after analyzing the text, or as they say in the specialized language the analysis of feelings, potential offenders.

Also, if the model proves to be consistent after being used on more real life scenarios, the presented models and scenarios could be adapted also to the reality specific in other countries.

References

1. Aebi, M.F, Akdeniz, G, Barclay, G., Campistol, C., Caneppele, S., Harrendorf, B., Heiskanen, S, Hysi, M., Jehle, V., Jokinen, JM., et al.: European sourcebook of crime and criminal justice statistics 2014 (2014)
2. Beniwal, S., Arora, J.: Classification and feature selection techniques in data mining. Int. J. Eng. Res. Technol. **1**(6), 1–6 (2012)
3. Chen, H., Chung, W., Xu, J.J., Wang, G., Qin, Y., Chau, M.: Crime data mining: a general framework and some examples. Computer **37**(4), 50–56 (2004)
4. Cohen, L.E., Felson, M.: Social change and crime rate trends: a routine activity approach. Am. Sociol. Rev **44**, 588–608 (1979)
5. Davidson, R.N.: Crime and Environment. Routledge, Abingdon (2019)
6. Farsi, M., Daneshkhah, A., Far, A. H., Chatrabgoun, O., Montasari, R.: Crime data mining, threat analysis and prediction. In: Cyber Criminology, pp. 183–202. Springer (2018)
7. García, S., Luengo, J., Herrera, F.: Data Preprocessing in Data Mining. Springer, Heidelberg (2015)
8. Kanazawa, S., Still, M.C.: Why men commit crimes (and why they desist). Sociol. Theory **18**(3), 434–447 (2000)
9. Kiani, R., Mahdavi, S., Keshavarzi, A.: Analysis and prediction of crimes by clustering and classification. Int. J. Adv. Res. Artif. Intell. **4**(8), 11–17 (2015)
10. King, T.C., Aggarwal, N., Taddeo, M., Floridi, L.: Artificial intelligence crime: an interdisciplinary analysis of foreseeable threats and solutions. Sci. Eng. Ethics, 1–32 (2019)
11. Kolvin, I., Miller, F.J.W., Fleeting, M., Kolvin, P.A.: Social and parenting factors affecting criminal-offence rates: findings from the newcastle thousand family study (1947–1980). Br. J. Psychiatry **152**(1), 80–90 (1988)
12. Lane, J., Kim, H.J.: Big data: web-crawling and analysing financial news using rapidminer. Int. J. Bus. Inf. Syst. **19**(1), 41–57 (2015)
13. Mierswa, I., Wurst, M., Klinkenberg, R., Scholz, M., Euler, T.: Yale: rapid prototyping for complex data mining tasks. In: Proceedings of the 12th ACM SIGKDD international conference on Knowledge discovery and data mining, pp. 935–940. ACM (2006)
14. Nagamatsu, S.: War, Crimes, and Terrorism. Springer, Heidelberg (2019)
15. Oatley, G.C., Ewart, B.W.: Crimes analysis software: 'pins in maps', clustering and bayes net prediction. Expert Syst. Appl. **25**(4), 569–588 (2003)
16. Ryan, J., Hofmann, M.: Communities and crime: creating a data mining prediction model. IT T, 61 (2013)

Diagnostics of the Machines and Devices Based on the SIMATIC ProDiag

Michal Kopcek and Miroslava Fackova(✉)

Faculty of Materials Science and Technology in Trnava,
Institute of Applied Informatics, Automation and Mechatronics in Bratislava,
Slovak University of Technology, J. Bottu 52, 91724 Trnava, Slovakia
{michal.kopcek,miroslava.fackova}@stuba.sk

Abstract. The paper is focused on the diagnostics of machines and devices based on the SIMATIC ProDiag library for the machines produced by company Skartek, s. r. o. The first part is dedicated to theoretical problem description and analysis. The next part of the work describes a sample example with configuration of supervisions, which serves as a demonstration for the implementation of the ProDiag library. Testing of the example and the implemented ProDiag library functions are followed by an evaluation of the ProDiag library's utility within production of the Skartek company. The next part is comparison of alarm handling by conventional ways and by using the ProDiag library. At the end we will show the advantages and disadvantages of the shown methods of alarm handling.

Keywords: PLC · SIMATIC ProDiag · Alarm

1 Introduction

The submitted paper deals with the utilization possibilities of the SIMATIC ProDiag library within the machines and devices from the production portfolio of Skartek company.

A clear alarm system forms a prerequisite for a fast and effective service intervention in the event of a failure [6, 8]. However, handling of an alarm could be often very time consuming and exhausting procedure requiring higher skills and practical experience from the programmer (automation specialist). Therefore, pre-prepared function libraries offer a good way to optimize and speed up the development of the control systems software, while transforming the implementation of the software into a few step configurations rather than programming. This paper analyses the Siemens SIMATIC ProDiag library and introduces a proof of concept in a model solution to compare this approach with the conventional ways of alarm handling to determine the added value. The results of this experiment are summarized at the end of this paper.

2 SIMATIC ProDiag Software Library

The SIMATIC ProDiag software library is used for machines and devices diagnostics. It is fully integrated into the TIA Portal environment and allows to configure the tracking of the machine states to diagnose failures in the process. The integrated HMI

R. Silhavy (Ed.): CSOC 2020, AISC 1226, pp. 571–583, 2020.
https://doi.org/10.1007/978-3-030-51974-2_53

visualization provides support for solving these issues. Monitoring messages and alarms in the process displayed on the HMI could be generated for various types of failures. These messages provide specific information about the monitoring mode, location and cause of the failure. In addition, an information about solving these problems is provided. Thus, the operator can evaluate the failure and quickly take appropriate action in case of danger [7].

The user configures the appropriate system or process component diagnostics using the ProDiag library. These data are entered by simple editing into input data files. ProDiag provides process-independent diagnostic rules that identify unexpected process failures. Diagnostic systems perform online real-time monitoring. These systems continuously receive process signals from the device as input and provide the identity of possible component failures as output in case of failure occurrence, e.g. clogged filters, missing signal, etc. [5].

3 Model Solution

In order to be able to assess appropriateness of the ProDiag library a model solution was designed and implemented in conformance to the internal company programming standard [4]. The work cycle was designed according to the company experiences. This example consists of the application control algorithm with implemented ProDiag library functions.

The model is conceived as a simple assembly line with three workstations (WS) with palette transfer conveyor for palette transport between operators as shown in Fig. 1. The palette could be moved in the two directions by the conveyor. The palette could be also moved towards the operator and back on WS 1 and WS 3 using pneumatic actuators. Therefore, optical barriers are presented on these posts. The WS 1 serves for loading of the part to the jig on the palette by operator N°1. The WS 2 simulates the manual assembly process, which is irrelevant from the diagnostics point of view, therefore was not simulated. The WS 3 serves for unloading of the part from the line.

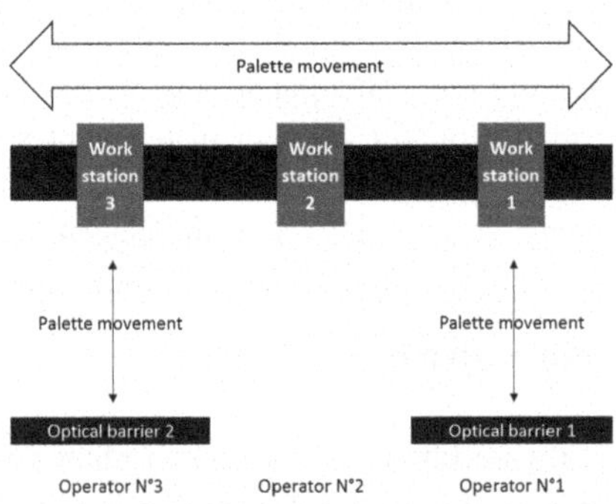

Fig. 1. Model solution

3.1 Initialization and Automatic Cycle of the Production Line

The device initialization ensures that the device meets the start position conditions which means, that both optical barriers are active, and the palette is situated on WS 1 in the line position.

The working cycle begins on WS 1, when the automatic cycle is activated, the palette moves towards the operator N°1 (perpendicular to the conveyor). After the palette reaches the end position, the operator could load the part to the jig while interrupts the optical barrier. The optical barrier is reactivated when the operator finishes the part loading and leaves the barrier space. The palette returns to the line position on WS 1. Consequently, is the palette transferred to the WS 2 and then to WS 3 after the assembly process was done. When the palette reaches the end position of the conveyor, it moves towards the operator N°3 (perpendicular to the conveyor), who could unload the part from the jig while interrupts the optical barrier. After the optical barrier is reactivated, the palette moves back to the initial position on WS 1 and a new cycle begins.

3.2 Inputs and Outputs Description

The input and output control signals of the model are described in Table 1 and Table 2. The separate signal is represented by symbolical and absolute address as it was implemented in the programable logic controller (PLC) and its purpose is explained in the comment.

Table 1. Inputs

Symbol	Address	Comment
btnStart	I0.0	Start cycle button
snOptBar1	I0.1	Optical barrier WS 1
snOptBar2	I0.2	Optical barrier WS 3
snPresencePart	I0.3	Part presence sensor
snPaletteExtended	I0.4	Palette in operator position
snPaletteInserted	I0.5	Palette in line position
snPalettePosition1	I0.6	Palette presence sensor WS 1
snPalettePosition2	I0.7	Palette presence sensor WS 2
snPalettePosition3	I1.0	Palette presence sensor WS 3
snStopper1Open	I1.1	Stopper is open sensor WS 1
snStopper1Close	I1.2	Stopper is closed sensor WS 1
snStopper2Open	I1.3	Stopper is open sensor WS 2
snStopper2Close	I1.4	Stopper is closed sensor WS 2
snStopper3Open	I1.5	Stopper is open sensor WS 3
snStopper3Close	I1.6	Stopper is closed sensor WS 3
btnJobDone	I1.7	Job confirmation button
btnJobStarted	I2.0	Job start button
snAirPressure	I2.1	Air pressure sensor
edmMotorBreaker	I2.2	Circuit breaker feedback
snJobRunning	I2.3	Job running sensor

Table 2. Outputs

Symbol	Address	Comment
yPaletteExtended	Q0.0	Palette to operator position
yPaletteInserted	Q0.1	Palette to line position
yStopper1	Q0.2	Pneumatic stopper WS 1
yStopper2	Q0.3	Pneumatic stopper WS 2
yStopper3	Q0.4	Pneumatic stopper WS 3
mMotorFW	Q0.5	Motor forward
mMotorBW	Q0.6	Motor backward

4 Configuration of the SIMATIC ProDiag Function

The functions of the SIMATIC ProDiag software library are configured in a separate function block through both local and global supervisions. Global supervision can be applied to Boolean tags in a tag table, to a global data block, or to an Array. Local supervisions are configured on the Bool type variables located in the function block.

Supervisions of the model solution are configured in the FB_ProDiag function block on Boolean tags in the tag table and on global data block variables.

The ProDiag library offers to configure five types of supervisions:

- Operand supervision.
- Interlock supervision.
- Action supervision.
- Reaction supervision.
- Position supervision.

4.1 Operand Supervision

The operand supervision checks whether the compressed air supply is active when initialization or automatic cycle is started. Otherwise, an alarm of the supervision of the type "Error" is activated.

Configuration of the operand supervision (Fig. 2):

1. Open the tag table.
2. Right-click on the "snAirPressure" tag and choose the option "Add new supervision".
3. Select "Operand" as the type of supervision.
4. The trigger value is "False".
5. Set Delay time to 0 ms.
6. Choose the tag "btnStartCycle" as the Condition 1 and set "True" as its trigger value. Up to 3 conditions could be used if needed.
7. Choose the alarm category "Error".

8. Select FB_ProDiag as destination function, the "error flag" will be automatically created in the instance data block.
9. The alarm message is written to the field "Specific text field".

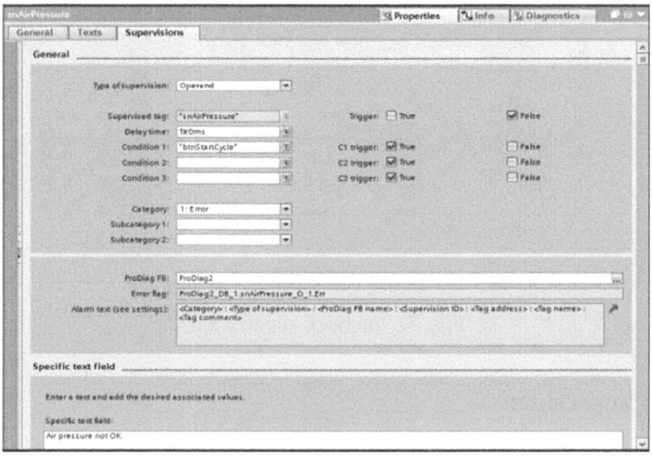

Fig. 2. Operand supervision

4.2 Interlock Supervision

The interlock supervision monitors the optical barriers. It checks the activation of the optical barrier when the palette is in the line position. An alarm of the supervision of the type "Error" is activated, if the condition is not met.

Configuration of the interlock supervision (Fig. 3):

1. Open the tag table.
2. Right-click on the "snOptBar1" tag and choose the option "Add new supervision".
3. Select "Interlock" as the type of supervision.
4. The trigger value is "False".
5. Set Delay time to 0 ms.
6. Choose the tag "yPaletteInserted" as the Condition 1 and set "True" as its trigger value. Up to 3 conditions could be used if needed.
7. Choose the alarm category "Error".
8. Select FB_ProDiag as destination function, the "error flag" will be automatically created in the instance data block.
9. The alarm message is written to the field "Specific text field".

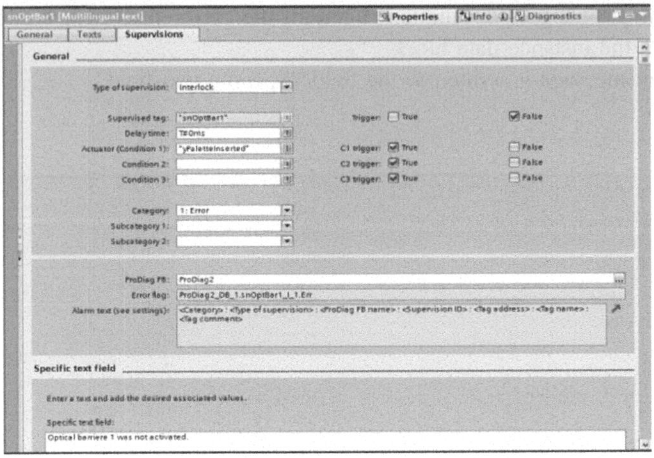

Fig. 3. Interlock supervision

4.3 Action Supervision

The action supervision monitors whether the requested movement (action) was executed in the given time. Otherwise, an alarm of the supervision of the type "Error" is activated.

This kind of supervision is used in the example e.g. for monitoring of the palette movements. These tags are monitored multiple times during the cycle, after the conveyor movements and when the "gbInitOK" auxiliary variable is activated. A monitoring of the sensor "snPaletteExtended" was chosen as an example supervision. If the sensor is not activated even after the pre-set delay, the alarm is activated.

Configuration of the action supervision (Fig. 4):

1. Open the tag table.
2. Right-click on the "snPaletteExtended" tag and choose the option "Add new supervision".
3. Select "Action" as the type of supervision.
4. The trigger value is "False".
5. Set Delay time to 1400 ms.
6. Choose the tag "gbInitOK" as the Condition 1 and set "True" as its trigger value. Up to 3 conditions could be used if needed.
7. Choose the tag "gDBCycle.Step1" as the Condition 2 and set "True" as its trigger value.
8. Choose the alarm category "Error".
9. Select FB_ProDiag as destination function, the "error flag" will be automatically created in the instance data block.
10. The alarm message is written to the field "Specific text field".

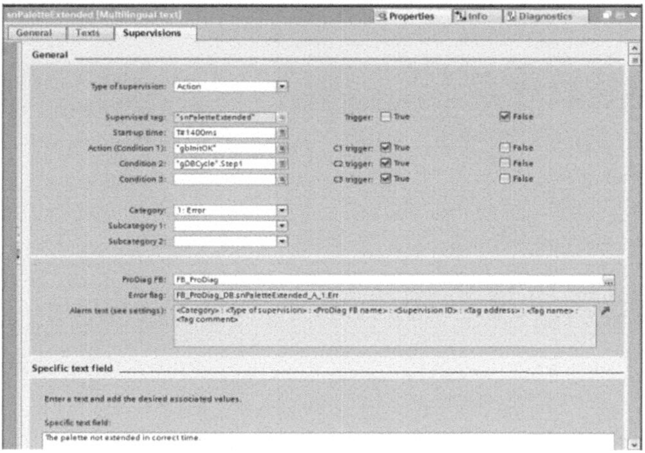

Fig. 4. Action supervision

4.4 Reaction Supervision

The reaction supervision monitors whether the requested movement (action) reached the end position in the given time, respectively whether it was not interrupted before reaching of the end position. Otherwise, an alarm of the supervision of the type "Error" is activated.

This kind of supervision is used in the example e.g. for monitoring of the palette movements to the operator position. If the sensor "snPaletteExtended" is not activated within the pre-set delay after the control output "yPaletteExtended" was activated, the alarm is present, because a failure occurred during the palette transport.

Configuration of the action supervision (Fig. 5):

1. Open the tag table.
2. Right-click on the "snPaletteExtended" tag and choose the option "Add new supervision".
3. Select "Reaction" as the type of supervision.
4. The trigger value is "False".
5. Set Delay time to 200 ms.
6. Choose the tag "yPaletteExtended" as the Condition 1 and set "True" as its trigger value. Up to 3 conditions could be used if needed.
7. Choose the alarm category "Error".
8. Select FB_ProDiag as destination function, the "error flag" will be automatically created in the instance data block.
9. The alarm message is written to the field "Specific text field".

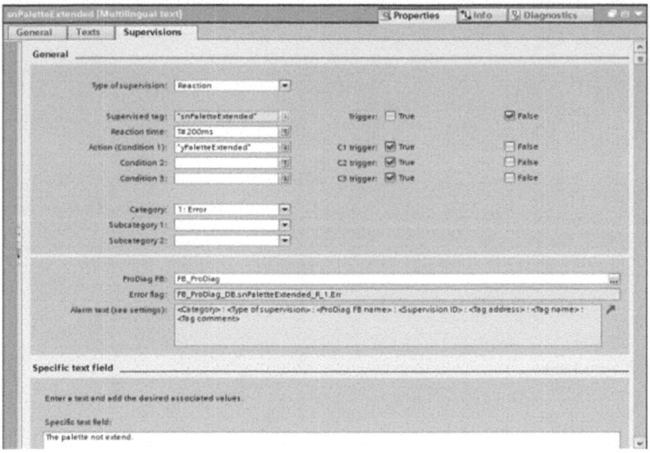

Fig. 5. Reaction supervision

5 Testing of Model

The performed tests were documented in test forms, which structure is derived from the best practice guides for factory acceptance testing widely used within the validation phase of the process control systems life cycle [2, 3].

Two tests of the interlock supervision, which monitors the optical barrier "snOptBar1" situated on WS 1 in the model line, are described as an example. The monitoring activates the alarm of the category "Error – Interlock", when the optical barrier is not active while the palette is in the line position.

The first test checks normal behaviour of the system, if the sensor works correctly, then no alarm message should be displayed, respectively no alarm is active. This situation is shown in Fig. 6.

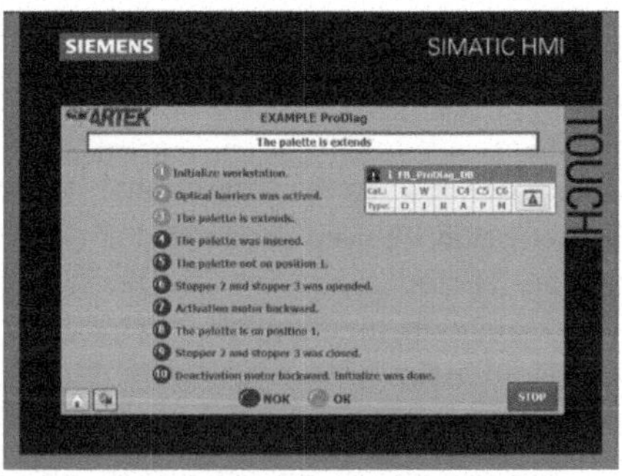

Fig. 6. Test without active failure

The second test checks the system behaviour during the off-normal conditions. The failure was activated on purpose. In such a case the alarm should be activated, the alarm message should be displayed as well as the part of the machine cycle control algorithm in which the process is stucked. The resulting displays are shown in Fig. 7 and Fig. 8 from the operator point of view. As mentioned before, all tests were documented. The form of the second described test is shown in Fig. 9 as an example.

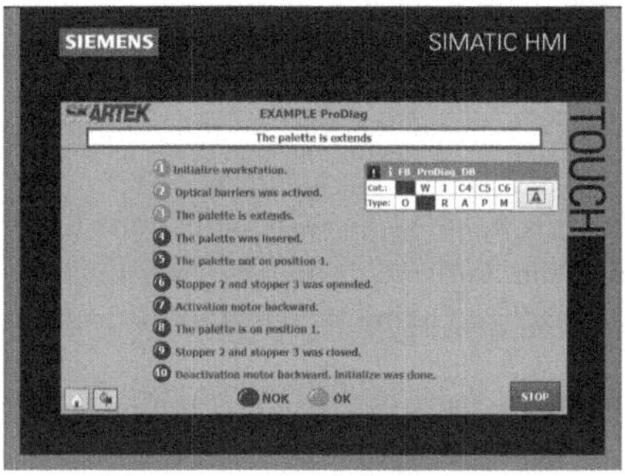

Fig. 7. Test with active failure

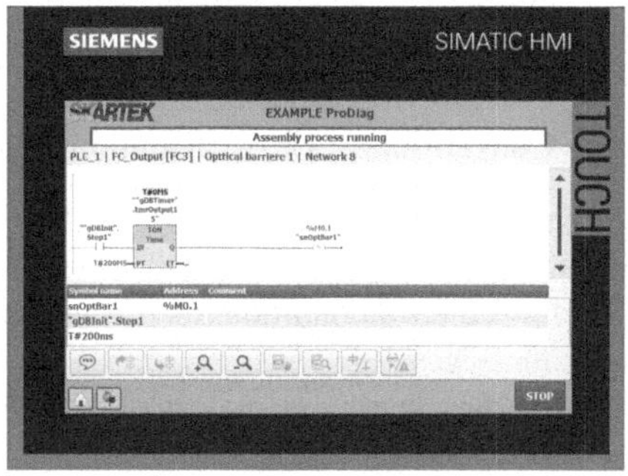

Fig. 8. PLC code display

Model solution

Number: 01-00-01

Name: Interlock supervision of the optical barrier snOptBar1 on the WS 1 – active failure

Configuration:

Supervised tag – snOptBar1

Trigger – FALSE

ProDiag FB – FB_ProDiag

Type of supervision – Interlock

Category – Error

Delay time – T#0ms

Condition – yPaletteInsered (TRUE)

Test execution:

- The palette is in the line position on the conveyor
- Sensor snPalettePosition1 is active
- Violation of condition of type Interlock at the ProDiag function block
- Force snOptBar1 value to FALSE

Expected values:

- Error notification in ProDiag graphic visualization
- Enable the PLC code button
- Display a specific error in the control program

Test result:

- Error notification in ProDiag graphic visualization
- Enable the PLC code button
- Display a specific error in the control program

Test result: POSITIVE

Surname: Fačková Name: Miroslava Date: 01.04.2019

Fig. 9. The test form

6 Comparison of the Alarm Handling Methods

After implementation and testing phase of the model solution an evaluation of the utility of the ProDiag software library within the production of the Skartek company could be done. The following text is dedicated to the comparison of alarm handling by conventional ways commonly used in the company and by using the ProDiag software library.

6.1 Conventional Alarm Handling

Currently, the alarm handling of the machines within Skartek production is done using the function "FC_Alarm", which was internally standardized. This function includes programmed error condition treatments that are visualized using the HMI in the form of alarm messages.

The programming procedure of the alarm handling using the function "FC_Alarm":

1. Click "Add new block" in the PLC tree and select a function. Rename function to "FC_Alarm".
2. Click "Add new block" in the PLC tree and select a data block. Rename this data block to "gDBError".
3. Create variable "wError1" of the data type word in the data block "gDBError". Add additional variables if needed.
4. Create algorithm for the alarm handling of the identified failure into a separate network in the "FC_Alarm".
5. Click "HMI tags" in the HMI tree and add a new tag "gDBError_wError1".
6. Link this newly created tag with the PLC tag "gDBError.wError1". For multiple PLC tags representing alarms, create additional HMI tags in the same way.
7. Click "HMI alarms" in the HMI tree and add a new alarm in the "Discrete alarms" tab.
8. In the "Trigger tag" field, select the "gDBError_wError1" HMI tag and select the "gDBError.wError1.x0" tag in the "Trigger address" field. In the "Alarm text" field, enter a specific alarm error message. Create additional alarms in the same way.
9. Open "GlobalScreen" in the HMI tree. Add "Alarm window" from the graphic symbols to the screen.
10. On the "General" tab, check "Pending alarms" and enable "Errors". Enter the name "Pending alarms" in the "Window" tab. In the "Miscellaneous" tab, enter the name "Alarm window_Pending". Set additional graphic options as needed.
11. On the "GlobalScreen" screen, add "Alarm window" from the graphic symbols.
12. On the "General" tab, check "Pending alarms" and enable "System". Enter the name "Alarm window" in the "Window" tab. Enter "System events" in the "Miscellaneous" tab. Set additional graphic options as needed.
13. After PLC and HMI compilation and subsequent download is the alarm handling functional.

An example algorithm for the alarm handling of the optical barrier failure is shown in Fig. 10.

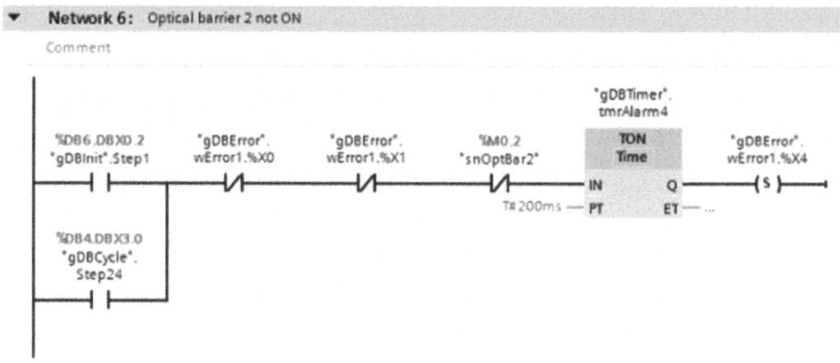

Fig. 10. Conventional ways alarm handling

6.2 Supervision Configuration Using the ProDiag Library

Monitoring of fault conditions using ProDiag library is configured through the "FB_ProDiag" function block. The function block contains all configured global variable supervisions. Alarm messages are visualized on the HMI.

The configuration procedure is as follows:

1. Click "Add new block" in the PLC tree and select the function block. Set the programming language to PRODIAG (incl. IDB) and name this function block "FB_ProDiag".
2. The instance data block "FB_ProDiag_DB" is created automatically.
3. Open the I/O tags table in the PLC tree.
4. Left click on the tag that should be monitored and select "Add new supervision". Configure the supervision over a particular tag as described before. Create all the desired supervisions.
5. After the supervision is created, its instance is automatically added in "FB_ProDiag_DB".
6. Click "HMI tags" in the HMI tree and add a new "ProDiag.State" tag.
7. Link this newly created tag with the PLC tag "FB_ProDiag_DB.State".
8. Create a new screen in the HMI tree and name it "ProDiag". Add components "ProDiag overview" and "Alarm view" to this screen.
9. In the "General" tab of the "Prodiag overview" graphical component set the tag to "ProDiag.State". In the "Miscellaneous" tab, enter the name "ProDiag overview".
10. In the "Alarm view" graphic component, check "Pending alarms" and "Unacknowledged alarms". Enable all alarm classes. Enter the name "Alarm view" in the "Miscellaneous" tab.
11. Click "Add new block" in the PLC tree and add organization block OB250. Name it "ProDiagOB".
12. After PLC and HMI compilation and subsequent download is the alarm handling fully functional.

An example configuration for the supervision of the optical barrier failure is shown in Fig. 11.

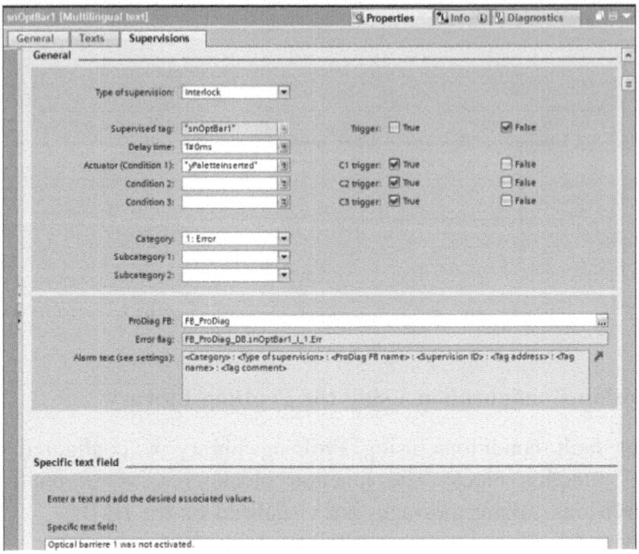

Fig. 11. Alarm handling using ProDiag

7 Discussion and Conclusion

The results of the comparison of the conventional ways of the alarm handling and the ProDiag library should be helpful within the decision during development in Skartek company. Comparison of the main advantages and disadvantages are shown in Table 3 and Table 4.

Table 3. Advantages

Conventional ways	ProDiag
No additional cost on licences	Simple supervisions configuration without changes in the original software
Implementation according to the internal company standard	Simple visualizing using the tools in TIA Portal
Hardware independent programming	Simple interconnection between PLC and HMI

Table 4. Disadvantages

Conventional ways	ProDiag
Time consuming programming with changes in the original software	Necessity to buy the license
Complicated interconnection between PLC and HMI	Works only with the CPUs S7-1500 from SIEMENS

Due to the comparison of the two presented methods for alarm handling it could be stated, that after the initial cost of using the ProDiag software library, this method is more practical and cost effective. Configuring the supervisions instead of programming separate algorithm saves time, thus reducing the cost of the final product. The ProDiag library offers HMI graphical components allowing to display the parts of control algorithm, in which is the execution stucked due to not fulfilled condition, along with an error message. Such level of diagnostics is very hard to achieve by conventional way, furthermore, this kind of development is very costly.

On the other hand, the ProDiag library is nowadays available only for the programmable logic controllers with the CPU S7-1500 series and for HMIs of the Comfort series and higher. Such hardware means higher initial costs. Furthermore, the analysis [1] of the Skartek company production shows, that high performance control systems are appropriate only in cca 10% of realized projects. For the rest 90% of machines the CPUs S7-1200 series with Basic series HMIs are used.

However, saving the programmers time and accelerating the development and commissioning processes even in the 10% of the produced machines could create additional space for increasing the production with the same human resources capacity.

Acknowledgments. This work was supported by Slovak Scientific Grant Agency VEGA within the project 1/0232/18 “Using the methods of multi-objective optimization in a production processes control”.

References

1. Fačková, M.: SIMATIC ProDiag based machine and plant diagnostics, in Slovak (Diagnostika strojov a zariadení na báze SIMATIC ProDiag), Diploma thesis, Supervisor: Kopček, M. UIAM MTF STU, Trnava. MTF-30179-73437 (2019)
2. Kopček, M.: Management of FAT for the process level control systems, in Slovak (Manažment FAT pre systémy riadenia procesnej úrovne), Habilitation thesis. UIAM MTF STU, Trnava (2014)
3. Kopček, M., et al.: Systematic approach to factory acceptance test planning. In: ESCAPE-24: 24th European Symposium on Computer Aided Process Engineering. Part A, 1st edn., Hungary, Budapest, 15–18 June 2014. Elsevier, Amsterdam, pp. 1597–1602 (2014). ISBN 978-0-444-63456-6
4. Kopček, M.: Standard programming guide v1.1., in Slovak, (Štandard programovania v1.1., internal documentation). Skartek (2019)
5. NED: Software PRODIAG, © Nuclear Engineering Division 2016. Web access (2016): https://www.ne.anl.gov/codes/prodiag/. Accessed 16 June 2016
6. Rothenberg, D.H.: Alarm Management for Process Control: A Best-Practice Guide for Design, Implementation, and Use of Industrial Alarm Systems. Momentum Press (2009). ISBN-13 978-1-60650-003-3
7. Siemens: Machine and Plant Diagnostics with ProDiag V4.1, © Siemens AG 2019. Web access (2018). https://support.industry.siemens.com/cs/ww/en/viev/109740151. Accessed 16 June 2016
8. Strémy, M., et al.: Introduction to programmable logic controllers. In: Slovak, (Úvod do programovateľných logických automatov). Qintec, Trnava (2011). ISBN 978-80-969846-9-5

Power and Frequency Scheduling Using Equal Throughput Strategy in PD-NOMA Systems

Ya. V. Kryukov[(✉)], D. A. Pokamestov, A. A. Brovkin, and E. V. Rogozhnikov

Tomsk State University of Control and Radioelectronics, Tomsk, Russia
kryukov.tusur@gmail.com

Abstract. The paper proposes a method for scheduling frequency and power resources between users to ensure equal throughput in the downlink channel when non-orthogonal multiplexing of users in power domain of orthogonal subcarriers (PD-NOMA/OFDMA). This method allows the multiplexing of several user signals with different power in a single frequency-time segment. The existing approaches to resource scheduling are based on the use of ideal signal-code sequences with perfect noise immunity characteristics. This does not allow an accurate estimate of the PD-NOMA system throughput in real operating conditions. To solve this problem, it is necessary to design a new method of power scheduling that takes into account the type and characteristics of real modulation and coding scheme (MCS). The novelty of the paper is to take into account currently used MCS's based on QAM modulation and LDPC coding in power scheduling. A designed scheduling method is using strategy of achieve equal throughput of users and maximize system throughput. To estimate the PD-NOMA system throughput using the proposed methods, a simulation model of the PD-NOMA/OFDMA system's downlink channel has been developed. The average throughput gain of the PD-NOMA/OFDMA system relative to OFDMA was 19% over 100 simulation iterations #CSOC1120.

Keywords: NOMA · PD-NOMA · Resource scheduling · Multiplexing · Equal throughput strategy

1 Introduction

The Power Domain Non-Orthogonal Multiple Access (PD-NOMA) method is one of the most promising methods for multiplexing user channels for future wireless broadband telecommunication systems [1–4]. PD-NOMA allows non-orthogonal user channel multiplexing in power domain of single time-frequency resource element. The method is non-orthogonal due to the use of controlled adjacent-channel interference when generating a group signal which is eliminated at the point of receipt. PD-NOMA increases the system spectral efficiency (SE) due to power sharing between users adaptive to the channel conditions. The effectiveness of PD-NOMA is confirmed by numerous experiments, including those described in [5].

It was shown in [6] that PD-NOMA can be effectively used in conjunction with the orthogonal frequency multiplexing method (OFDMA,) using the advantages of both

R. Silhavy (Ed.): CSOC 2020, AISC 1226, pp. 584–595, 2020.
https://doi.org/10.1007/978-3-030-51974-2_54

methods. The combination of PD-NOMA and OFDMA allows to simultaneously schedule the resource in 3 domains: time, frequency, and power.

Currently, PD-NOMA is actively being researched in many RnD groups. The review papers [1–5] show the effectiveness of PD-NOMA technology and identify problems to be addressed. The vast majority of scientific papers are devoted to the theoretical study of PD-NOMA, using Shannon theorem. It was shown [6] that the use of PD-NOMA can provide an increasing of system throughput by 1.5 or more times depending on the usage scenario. It is worth noting that the PD-NOMA ultimate SE is calculated in the papers. It can be significantly lower in real operating conditions.

The aim of this paper is to evaluate the SE of the downlink PD-NOMA communication channel using existing MCS's. To solve this problem, a simulation model has been developed for the generation and processing of a multiple communication channel where OFDMA acts as the main method of multiplexing channels, and PD-NOMA as an additional one. The work uses its own method of scheduling power and frequency resources. Its strategy is to achieve equal throughput of user channels while maximizing system throughput.

During the simulation, the throughput of OFDMA and PD-NOMA/OFDMA systems in typical operating scenarios was calculated. It is shown that the proposed method of power and frequency scheduling provides a gain in the PD-NOMA/OFDMA system throughput.

2 PD-NOMA Multiplexing

The PD-NOMA group signal S generation for M users is according to:

$$S = \sum_{m=1}^{M} \sqrt{p_m} \cdot X_m,$$

where p_m is the power coefficient and x_m is the data symbol of the m-th user [5]. In this case, signal S is divided into M layers, and each user occupies its own non-orthogonal layer. Coefficients p_m is calculated taking into account the signal-to-noise ratio SNR_m of the m-th user. Moreover, coefficients $p_1 > \ldots > p_M$ are ordered in increasing order, according to $SNR_1 < \ldots < SNR_M$. The SNR_m value can be obtained according to $SNR_m = |\alpha_m|^2 \cdot P/N_m$, where α_m is the channel fading coefficient and N_m is the power of AWGN in the m-th channel, and P is the output signal power $P = \sum_{m=1}^{M} p_m$.

The PD-NOMA signal is processed on the side of each user device using the Successive Interference Cancellation (SIC) [7]. According to the SIC, user signals are demodulated and removed from the total received signal sequentially, starting with the one of the highest power (with partial power p_1) and ending with the one of the lowest power (with partial power p_k).

Currently, non-orthogonal multiplexing of more than 2 channels is not of practical interest due to the high computational complexity and many unsolved problems [1–6]. Therefore, hereinafter, we consider the PD-NOMA multiplexing of two users ($M = 2$).

In this case, the available power P is scheduled between two layers: the upper layer with power p_1 and the lower layer with power p_2 ($p_1 > p_2$). One of the most acute problems of PD-NOMA is the obtaining of values p_1 and p_2 according to the required quality of service (QoS) and channel state of each user. Several methods have been proposed for the calculation [5, 8–11] which do not take into account the characteristics of real MCS's based on QAM and LDPC coding.

Therefore, in this paper, we use our own method of power scheduling [12] for two-layer channel multiplexing using QAM (with modulation indices 4, 16, 64, etc.) The method allows to find the ratio of layers power $\gamma = p_1/p_2$ for a given index of QAM, symbol error rate ρ_{err} (SER) and the known SNR in each user channel.

3 System Model

Let K users be located in the service area of the base station (BS). The purpose of BS is multiplexing all users and scheduling the resource between them. OFDMA is used as the main multiplexing method, and PD-NOMA is additional. All users have the same QoS request. The base station must to ensure equal bitrates in user channels while achieving maximum system throughput.

We suppose that base station knowns channel state information (CSI) of each user, presented as an estimates of the signal-to-noise ratio $SNR_{1..K}$. We assume that the channel state does not change during the transmission and the amplitude frequency response is flat during the bandwidth.

The available frequency resource consists of F_{bw} orthogonal OFDM subcarriers. Using OFDMA, it is possible to place the data of single user on each orthogonal subcarrier, and the task of the base station is to select MCS and the schedule of subcarriers between the user. On the other hand, additional PD-NOMA allows allocating data of two users on one subcarrier. In this case, the base station shall divide the users into paired groups for non-orthogonal multiplexing, determine the MCS for users within the group and distribute the power and frequency both within the same group and between all groups.

To compare the OFDMA and PD-NOMA/OFDMA systems, the system throughput is obtained (C_{sum}^{oma} and C_{sum}^{noma}, respectively). Let the coefficient G shows the throughput gain in the PD-NOMA/OFDMA system relation to OFDMA. In case of $G > 1$, the use of the PD-NOMA method provides a gain in system throughput and vice versa.

$$G = C_{sum}^{noma} / C_{sum}^{oma} \tag{1}$$

4 OFDMA System Capacity

In this scenario, BS forms downlink channel using only the OFDMA method. Each k-th ($k = \{1, \ldots K\}$) user is assigned an MCS with SE value R_{k} and a set of F_k^{oma} orthogonal subcarriers. Accordingly, the throughput of the k-th user is:

$$C_k^{oma} = R_k^{oma} \cdot F_k^{oma} \tag{2}$$

The R_k^{oma} value is determined by the SE of the used MCS selected adaptively to CSI of the k-th user. The final set of MCS is predefined, and the required value of SNR is known for each MCS. Each user channel uses MCS with maximum SE, satisfying the threshold SNR value. To ensure equal throughput of all users ($C_1^{oma} = C_2^{oma} = \ldots = C_K^{oma}$), it is necessary to allocate the available frequency resource according to:

$$F_k^{oma} = \frac{F_{bw}}{R_k \cdot \sum_{i=1}^{K} \frac{1}{R_i}} \tag{3}$$

Throughput of the OFDMA system C_{sum}^{oma} can be calculated as follows:

$$C_{sum}^{oma} = \sum_{k=1}^{K} C_k^{oma} \tag{4}$$

5 PD-NOMA/OFDMA System Capacity

In this scenario, the PD-NOMA/OFDMA combined multiple access technique is used to generate the group signal. The task of the base station is to determine the most effective combination of pairwise user associations, as well as the distribution of power and frequency resource between them.

Calculation of the maximum spectral efficiency of each combination of pairwise user associations. The base station forms all possible combinations of pairwise user associations on the "each with each" principle. According to combinatorics, U combinations are possible in total:

$$U = \frac{K!}{2(K-2)!}$$

Let $\mathbf{V} = \{V_1, V_2, \ldots, V_U\}$ be the array of all possible different combinations of users, where each element of array $V_u = \begin{pmatrix} n_u \\ m_u \end{pmatrix}$ contains a pair of user indices n_u and m_u ($n_u, m_u \in \{1, \ldots, K\}$) corresponding to the u-th ($u = 1 \ldots U$) combination. In this case, the user with the number n_u is located on the upper layer, and the user with the number m_u on the lower layer. According to the basic principle of PD-NOMA, the user with the lowest SNR in a pair is placed on the upper layer getting the largest share of power, and vice versa. Users are sorted in increasing order of SNR ($SNR_1 < \ldots < SNR_K$), therefore $n_u < m_u$.

$$V_1 = \binom{1}{2}, V_2 = \binom{1}{3}, \ldots, V_U = \binom{K-1}{K}$$

Selecting of MCS in Each Pair. With non-orthogonal multiplexing of two user channels, the available power is spited between them. Using flexible power distribution, the PD-NOMA method allows the combination of different MCS's on the non-orthogonal layers, giving the gain to the lower (m-th) or upper layer (n-th). Various MCS combinations can be used by user within pair V_u, and the type of MCS on the upper layer affects the type of MCS in the lower layer. In each u-th pair, it is required to determine the only effective combination of MCSs, taking into account the following criteria:

- Ensuring the equal SE of the users in pair $R^u_{n_u} = R^u_{m_u}$.
- Maximization of the total SE of two channels in pair

For each u-th pair, all possible MCS combinations are determined. For each combination, using expressions (2–13) from [12], the power of the upper and lower layers is obtained for known values of SNR and SER. If condition (1) from [12] is not satisfied, then the use of this combination will lead to a significant increasing SER. In this regard, this MCS combinations are excluded. The base station determines the only MCS combination for each pair of users based on the criteria described above.

Let $R^u_{n_u,m_u} = R^u_{n_u} + R^u_{m_u}$ be the total SE of the u-th pair with non-orthogonal multiplexing of the n-th and m-th users using MCS's with SE values $R^u_{n_u}$ and $R^u_{m_u}$. For example, $R^1_{1,2}$ is the total SE of the 1-th pair with PD-NOMA multiplexing of 1-th and 2-th user. Matrix **SR** contains $R^u_{n_u,m_u}$ of all pairs:

$$\mathbf{SR} = \{R^1_{1,2}, R^2_{1,3}, \ldots, R^u_{n_u,m_u}, \ldots, R^U_{K-1,K}\} \tag{5}$$

Allocation of Users into Pair Groups. The task of the base station is to select $K/2$ combinations from **V** taking into account **SR**, which use will maximize the system throughput C^{noma}_{sum}. In our work, this problem is solved by forming all possible combinations of pairs by enumeration and choosing the most effective combination. Excluding repetitions, the total number of combinations A can be obtained as:

$$A = \prod_{i=1}^{K/2} (K - 2i + 3)$$

For a large number of users, the enumeration method has high computational complexity; therefore, the optimization problem was solved in [13]. In this paper, the optimization problem is not addressed.

The combination of the pairs selected by BS is group $Z = \{Z_1, \ldots, Z_{K/2}\} \in V$, where $Z_i = \binom{n_i}{m_i}$ contains the indexes of users in i-th pair, where n_i and m_i is the user numbers on upper and lower layer, respectively. In this case, $n_{i+1} \notin \{n_1, \ldots, n_i\}$ and $m_{i+1} \notin \{m_1, \ldots, m_i\}$ at $i = 1 \ldots K/2$.

$$Z_1 = \begin{pmatrix} n_1 \\ m_1 \end{pmatrix}, \ldots, Z_i = \begin{pmatrix} n_i \\ m_i \end{pmatrix}, \ldots, Z_{K/2} = \begin{pmatrix} n_{K/2} \\ m_{K/2} \end{pmatrix}$$

The SE of the i-th pair R_i^z can be determined from **SR** taking into account the indices n_i, m_i from Z_i. The SE of the pairs selected by BS is recorded in $\mathbf{R}^z = \{R_1^z, \ldots, R_{K/2}^z\}$.

Frequency Allocation. The available frequency resource of the system is shared between pairs $Z_1, \ldots, Z_{K/2}$ with SE values $R_1^z, \ldots, R_{K/2}^z$, so that the throughput of the pairs is the equal $C_1^z = C_2^z = \ldots = C_{K/2}^z$. The throughput of the i-th pair is calculated as $C_i^z = C_{n_i}^z + C_{m_i}^z$, where $C_{n_i}^z$ and $C_{m_i}^z$ are the throughputs of users of the upper and lower levels from i-th pair. The frequency resource for the i-th pair F_i^z can be calculated as:

$$F_i^z = \frac{F_{bw}}{R_i^z \cdot \sum_{j=1}^{K/2} \frac{1}{R_j^z}} \tag{6}$$

Expression (6) in similar to (3) using the summary SE of the user pair. Further, it is required to ensure equal user throughput within the i-th pair $C_{n_i}^z = C_{m_i}^z$. For this, 2 types of resource scheduling can be used (Fig. 1):

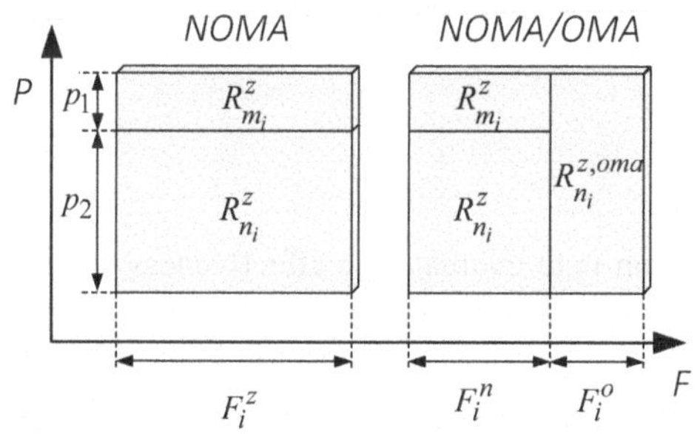

Fig. 1. Various frequency scheduling in pair.

- NOMA pair. Users is multiplexed only in power domain in case of equal SE of the layers $R_{n_i}^z = R_{m_i}^z$. In this case, the throughput of each user is obtained:

$$\begin{aligned} C_{n_i}^z &= F_i^z R_{n_i}^z \\ C_{m_i}^z &= F_i^z R_{m_i}^z \end{aligned} \tag{7}$$

- NOMA/OMA pair. In case of unequal SE of the layers $R^z_{n_i} \neq R^z_{m_i}$, the frequency resource F^z_i is divided into two subchannels $F^z_i = F^n_i + F^o_i$. The first subchannel with frequency band F^n_i is used by two non-orthogonal multiplexed users at the same time, and the second subchannel F^o_i is used by the single user with the lowest SE in the pair (n_i-th user). To ensure the equal throughput within pair $C^z_{n_i} = C^z_{m_i}$, it is necessary to correctly schedule F^n_i and F^o_i within the pair frequency resource F^z_i. Condition $C^z_{n_i} = C^z_{m_i}$ can be written as:

$$F^n_i R^z_{n_i} + F^o_i R^{z,oma}_{n_i} = F^n_i R^z_{m_i}, \tag{8}$$

where $R^{z,oma}_{n_i}$ is the SE of the n_i-th user in subchannel F^o_i determined by the MCS (like in OFDMA scenario). From (8) we derive the expression for obtaining F^n_i and F^o_i:

$$\begin{aligned} F^n_i &= \frac{-F^z_i \cdot R^{z,oma}_{n_i}}{R^z_{n_i} - R^z_{m_i} - R^{z,oma}_{n_i}} \\ F^o_i &= F^z_i - F^n_i \end{aligned} \tag{9}$$

Then, the throughput of the i-th pair is:

$$C^z_i = F^n_i (R^z_{n_i} + R^z_{m_i}) + F^o_i R^{z,oma}_{n_i} \tag{10}$$

And total PD-NOMA/OFDMA system throughput is:

$$C^{noma}_{\Sigma} = \sum_{i=1}^{K/2} C^z_i \tag{11}$$

6 Simulation

The purpose of simulation is to evaluate the effectiveness of the proposed method for scheduling frequency and power resources between users upon reaching the equal bitrate and obtaining the value of G according to (1) for ten users (K = 10).

Simulation Scenario. Ten users are located at a random distance $d_{1..K}$ in the 10–200 m from BS. Values $d_{1..K}$ obey the uniform distribution density function. The channel fading coefficients$\alpha_{1..K}$ are obtained individually for each user using the model of signal propagation in free space: $\alpha_k = 20\log_{10}(4\pi \cdot d_k \cdot f_0/c)$. Carrier frequency is f_0 = 6 GHz. We assume that additive noise is white Gaussian complex noise with zero mean. Users noise power devices is constant and equal to $N_{1..K}$ = −100 dBm. Output signal power is P = −10 dBm. The SNR_k value obtaining from $SNR_k = |\alpha_k|^2 P/N_k$. The frequency response of each channel is flat and does not change with time. The downlink bandwidth is N_{sc} = 1024 subcarriers with a frequency spacing of 15 kHz.

The quality of transmission in each user channel is estimated by obtaining the block error rate (BLER) with the total number of transmitted bits equal to 10^8. To schedule the power according to [12], the symbol error rate value $\rho_{ser} = 1e^{-2}$ was used which provides BLER < 10^{-8} in each user channel using LDPC coding. Using ρ_{ser}, we can control the distance between QAM symbols in the constellation diagram.

Available MCS's. The OFDMA system uses a set of 12 MCS's based on a QAM and LDPC combinations which are defined for 802.11ax [14]. The PD-NOMA/OFDMA system uses a set of 7 MCS's where the type of QAM modulation is changed, and the coding rate is constant $R^{ldpc} = 5/6$. By simulation, it was found that with $R^{ldpc} = 5/6$ all bit errors are corrected, which are a consequence of the close proximity of the modulation symbols on the constellation diagram when $\rho_{ser} = 1e^{-2}$. The obtaining of optimal R^{ldpc} according to ρ_{ser} is a different task and is not considered in this paper. Available MCS's for OFDMA and PD-NOMA systems are given in Table 1.

Table 1. Available MCSs for OFDMA and PD-NOMA multiplexing.

MCS Num.	OFDMA		PD-NOMA	
	MCS	SNR threshold, dB	MCS	SNR threshold
1	1/2 BPSK	2	5/6 BPSK	Not fixed. Depends on MCS's in the pair
2	1/2 QPSK	5	5/6 QPSK	
3	3/4 QPSK	9	5/6 QAM-16	
4	1/2 QAM-16	11	5/6 QAM-64	
5	3/4 QAM-16	15	5/6 QAM-256	
6	2/3 QAM-64	18	5/6 QAM-1024	
7	3/4 QAM-64	20		
8	5/6 QAM-64	25		
9	3/4 QAM-256	29		
10	5/6 QAM-256	31		
11	3/4 QAM-1024	36		
12	5/6 QAM-1024	40		

Simulation Example. To demonstrate the resource scheduling result by the proposed method, we provide an example of multiplexing eight users. Table 2 shows the user multiplexing parameters for OFDMA and PD-NOMA/OFDMA systems.

Table 2. Scheduling result.

UE	SNR, dB	OFDMA		PD-NOMA/OFDMA					
		F^{oma}, subcar.	MCS OMA	MCS NOMA	p, W	F^n, subcar	Pair	F^o, subcar	MCS OMA
1	21.9	149	7	2	0.8215	131	1+6	145	7
2	23	148	7	3	0.9765	130	2+8	97	7
3	24.8	148	7	3	0.9489	261	3+4	–	–
4	26.3	134	8	3	0.0511	261	3+4	–	–
5	27.1	134	8	3	0.9628	174	5+7	58	8
6	33.3	101	10	5	0.1785	131	1+6	–	–
7	35.7	101	10	4	0.0372	174	5+7	–	–
8	44.2	81	12	5	0.0235	130	2+8	–	–

Resource scheduling for OFDMA and PD-NOMA/OFDMA systems according to Table 1.3 is given in Fig. 2a and b. For each user, the type of QAM modulation, partial power, LDPC coding rate, and the number of allocated subcarriers are given. With this allocation, a strategy ensuring equal throughput for all users is achieved. In addition, the throughput gain in each user channel of PD-NOMA/OFDMA system is ~29%.

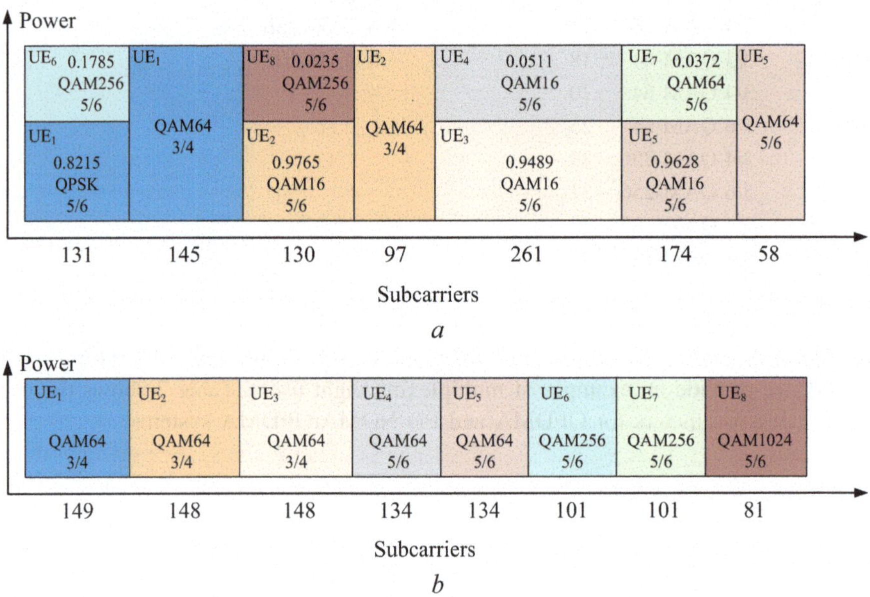

Fig. 2. Scheduling result: a – OFDMA system; b – PD-NOMA/OFDMA system.

In as an example, Fig. 3 shows the constellation of a signal with non-orthogonal multiplexing of the 1st and 6th users: a – at the BS output; b – at the 1-th user demodulator input; c – at the 6-th user SIC demodulator input; d – at the 6-th user demodulator of own signal. The method of power scheduling proposed in [12] allows (adaptively to CSI) power scheduling and adjusting the distance between the constellation points using ρ_{ser} so that the bit errors can be corrected during LDPC decoding.

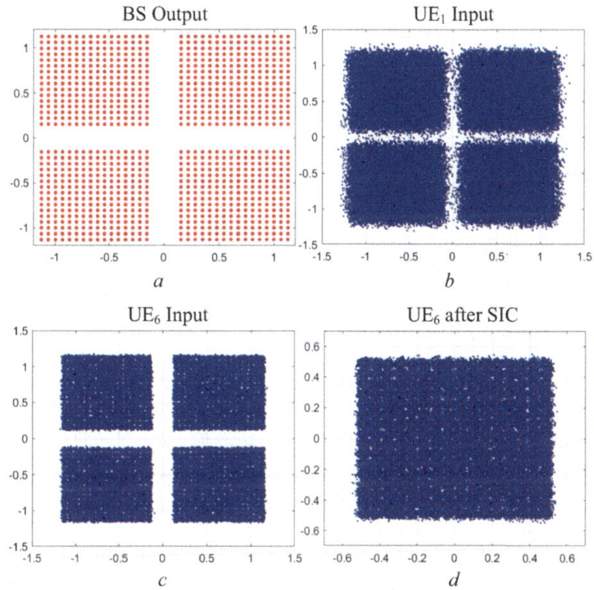

Fig. 3. PD-NOMA constellation diagram of the pair that contained 1st and 6th user

Iterative Simulation Result. Figure 4 shows the gain value G for iterative simulation with a random values of $d_{1..K}$ in each next iteration. The red markers show the value G for each iteration, and the blue line shows the average value G_{aver} over 100 iterations. According to the iterative simulation results, the average gain of the PD-NOMA/OFDMA system throughput is G_{aver} = 1.194 (or 19.4%). The minimum value is G_{min} = 1.08, the maximum is G_{max} = 1.34.

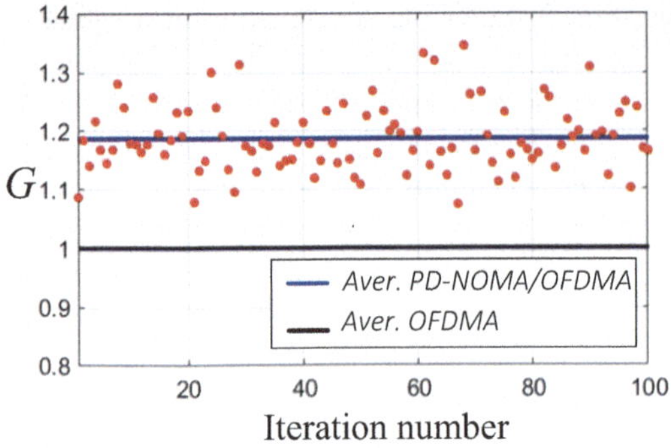

Fig. 4. Channel capacity gain of the PD-NOMA/OFDMA compared with OFDMA

7 Conclusion

The paper proposes a power and frequency resource scheduling method in the downlink channel using PD-NOMA and OFDMA simultaneously which provides equal user throughputs while maximizing the overall system throughput. The proposed method consists of following steps:

1) Organization of all possible pairs of users in and determination of the MCS's for each pair.
2) Selection of user pairs that provide maximum SE.
3) Scheduling the power and frequency resources between user pairs and users within pair.

As an example, the allocation of the communication resource by the OFDMA and PD-NOMA/OFDMA method for 8 users with the given SNR is provided.

A simulation model has been developed that implements resource allocation by the designed method, taking into account the characteristics of real MCS's based on QAM and LDPC. The simulation result is the effectiveness of the proposed method, and the average gain in throughput of the PD-NOMA/OFDMA system relative to OFDMA over 100 iterations is 19.4%.

Acknowledgments. The work is supported by the Russian Science Foundation grant. Project number 19-79-00087.

References

1. Benjebbour, A., Saito, K., Li, A.: Non-orthogonal multiple access (NOMA): concept, performance evaluation and experimental trials. In: 2015 International Conference on Wireless Networks and Mobile Communications (WINCOM), 14 January 2015, pp. 1–6, Marrakech, Morocco (2015)

2. Saito, Y., Benjebbour, A., Kishiyama, Yo., et al.: System-level performance evaluation of downlink non-orthogonal multiple access (NOMA). In: 2013 IEEE 24th International Symposium on Personal, Indoor and Mobile Radio Communications (PIMRC), pp. 611–615 (2013)
3. Saito, Y.: Non-orthogonal multiple access (NOMA) for cellular future radio access. In: 2013 IEEE 77th Vehicular Technology Conference (VTC Spring) (2013)
4. Islam, S.M.R., et al.: Power-domain non-orthogonal multiple access (NOMA) in 5G systems: potentials and challenges. IEEE Commun. Surv. Tutor. **19**(2), 721–742 (2016)
5. Kryukov, Ya.V., Pokamestov, D.A., Rogozhnikov, E.V., et. al: Experimental Research of PD/NOMA. In: 2018 19th International Conference of Young Specialists on Micro/Nanotechnologies and Electron Devices EDM, pp. 176–179, Erlagol, Russia (2018)
6. Kryukov, Ya.V.: Generation and demodulation signals of multi-channel communication systems with channel multiplexing by power. Ph.D. dissertation, Tomsk State University of Control Systems and Radioelectronics (2015)
7. Higuchi, K., Benjebbour, A.: Non-orthogonal Multiple Access (NOMA) with successive interference cancellation for future radio access. IEICE Trans. Commun. **98**(3), 403–414 (2015)
8. Kryukov, Ya.V., Demidov, A.Ya., Pokamestov, D.A.: Power calculation algorithm in non-orthogonal multiple access. Proc. TUSUR 19(4), 91–94 (2016)
9. Oviedo, J.A., Sadjadpour, H.R.: A fair power allocation approach to NOMA in multiuser SISO systems. IEEE Trans. Veh. Technol. **66**(9), 7974–7985 (2017)
10. El-Sayed, M.M., Ibrahim, A.S., Khairy, M.M.: Power allocation strategies for Non-Orthogonal Multiple Access. In: 2016 International Conference on Selected Topics in Mobile & Wireless Networking (MoWNeT), pp. 1– 6, Cairo (2016)
11. Fu, Y., Salaün, L., Sung, C.W., et. al.: Subcarrier and power allocation for the downlink of multicarrier NOMA systems. IEEE Trans. Vehic. Technol. **93**(12), 11833–11847 (2018)
12. Kryukov, Y.V., Pokamestov, D.A., Rogozhnikov, E.V.: PD-NOMA power coefficients calculation while using QAM signals. In: Vishnevskiy V., Samouylov K., Kozyrev D. Distributed Computer and Communication Networks (DCCN 2019). Communications in Computer and Information Science, vol 1141. Springer, Cham
13. Sun, Y., Ng, W., Ding, Z., et al.: Optimal joint power and subcarrier allocation for MC-NOMA systems. In: 2016 IEEE Global Communications Conference (GLOBECOM), pp. 1–6, Washington DC (2016)
14. Konstantinos, D., Bellalta, B.: Optimal Resource Allocation in IEEE 802.11ax Uplink OFDMA with Scheduled Access (2018)

Problems of Socio-Cyber-Physical Systems Development and Implementation: State-of-Art and Directs of Research

Ekaterina Shcherbakova[1] and Boris Sokolov[2(✉)]

[1] ITMO University, Princeton University, 49 Kronverksky Pr., St. Petersburg 197101, Russia
Komaro2@mail.ru

[2] St. Petersburg Institute for Informatics and Automation of the Russian Academy of Sciences, 39 14th line, St. Petersburg 199178, Russia
sokolov_boris@inbox.ru

Abstract. In this article there are the concepts of modern Russian and foreign scientists who analyze the current state of the problems of modeling socio-cyber-physical systems (SCPS). Development of complex SCPS required the creation of new approaches, models, methods and technologies. But certain problems began to arise, starting from tasks such as research, modeling and using of the model. It is possible to obtain fundamentally new knowledge as a result of the fact that a human is presented in SCPS. This knowledge was not previously entered into the SCPS knowledge base or known to its resources [16]. Also there are problems associated with communication in socio-cyber-physical systems. For example, one of the problem is in the interaction between devices at the software level [1]. To eliminate the shortcomings, multimodal interfaces are being actively developed, which are aimed at creating a way of interaction between a computer and a person, similar to the interaction between humans. In this way we use all known modalities that are accepted for human communication. In addition, this development will help to solve the main problem of SCPS which means to increase reliability by duplicating information through several channels. There is also a problem in the development of technical requirements for socio-cyber-physical systems. The difficulty in developing such requirements is that they can change throughout the life cycle of the system due to the constant development of socio-cyber-physical systems.

Keywords: Socio-cyber-physical systems · Big data · Internet of Things · Modeling of socio-cyber-physical systems · Industry 4.0

1 Introduction

In recent years, research has been widely developed related to the new model of economics (knowledge economy) and the 4th industrial revolution (Industry 4.0, Logistics 4.0 and Mobility 4.0), which is primarily due to the emergence of cyber-physical systems. This is a fundamentally new class of complex systems, the

R. Silhavy (Ed.): CSOC 2020, AISC 1226, pp. 596–606, 2020.
https://doi.org/10.1007/978-3-030-51974-2_55

development of which requires the creation of new approaches, models, methods and technologies [4].

Intelligent devices are becoming more complex and gaining more opportunities, while remaining relatively affordable [12]. Many of them use access to high-speed wireless networks, including 4G networks. Any object on the Internet of Things (IoT) can extract information from the environment, manage the received data and provide it to other devices or users [2].

The purpose of the study of this article is to consider the concepts of modern Russian and foreign scientists who analyze the current state of the problems of modeling socio-cyber-physical systems (SCPS). Development of complex SCPS required the creation of new approaches, models, methods and technologies. But certain problems began to arise, starting from tasks such as research, modeling and use of the model.

The main goal of this article is achieved in the process of considering three issues. Firstly, this is the definition of the current state of knowledge of SCPS. Secondly, the main problems of socio-cyber-physical systems have been identified. As a result of the fact that a person is presented in SCPS, it is possible to obtain fundamentally new knowledge. This knowledge was not previously entered into the SCPS knowledge base or known to its resources. Also there are problems associated with communication in socio-cyber-physical systems. For example, one of the problem is in the interaction between devices at the software level. In this case, it is difficult to coordinate, synchronize and harmonize the whole range of devices that people use in the modern world. To eliminate the shortcomings, multimodal interfaces are being actively developed, which are aimed at creating a way of interaction between a computer and a person, similar to the interaction of a person with a person, using all known modalities that are accepted for human communication. In addition, this development will help to solve the main problem of SCPS which means to increase reliability by duplicating information through several channels.

Another urgent problem that arises in the development of technical requirements for socio-cyber-physical systems was identified. The difficulty in developing such requirements is that they can change throughout the life cycle of the system due to the constant development of socio-cyber-physical systems [7]. The presence of very complex relationships, interactions and influences between the components of systems can contribute to unexpected and even unacceptable consequences in the structure and behavior of the system [13]. One of the perspective approaches supporting the adaptability and breakdown of systems is the development of continuous requirements based on maneuverability and flexibility.

2 The Problem of Knowledge Acquisition in Socio-Cyber-Physical Systems

In the modern world, characterized by an abundance of information, the cognitive abilities of a person who independently makes decisions has become lacking. A person can no longer takes into account all the necessary information when making decisions - there is too much data, and you need to look at them in general - match elements, analyze, make predictions. It can take days or even weeks for a person to process the data that a computer can handle much faster. This problem has two solutions: to transfer all the functions of a person to a computer or to increase a person's ability using the same computer. Simplification of the analysis of information by providing combined and generalized data remains for computing systems, but a person already makes a final decision on their basis in a socio-cyber-physical system.

People and organizations can be represented as objects of the socio-cyber-physical system. For example, in marketing, behavioral analysis is used to forecast demand and plan the distribution of stocks and offers. People and organizations can be involved in the socio-cyber-physical system by creating updatable and expandable profiles that describe their preferences and strategies. This model also allows you to predict the consequences of a decision and the actions which were taken. Profiles will allow you to individualize the way the system interacts with live users [16].

Socio-cyber-physical systems naturally combine machines and humans, which leads to research related to areas within the field of knowledge management. This is due to the fact that a person has his own knowledge, experience, technical intuition. The most urgent problem of SCPS today is the problem of acquiring knowledge.

The presence of a person in SCPS makes it possible to obtain fundamentally new knowledge, which means knowledge that not only was not previously entered into the SCPS knowledge base and known to its resources, but absolutely newly invented knowledge. The acquisition of knowledge is the process of enriching new knowledge of an object, for example, a software agent, information system, person, etc.

The objects that acquire knowledge are the resources of SCPS. The concept of SCPS assumes an integral connection, cybernetic and social space [8]. Physical space consists of various information-computing physical devices. These devices are making cybernetic space. Social spaces are used by people with their knowledge, mental ways and sociocultural elements. All three spaces are interconnected. Cyber space is accessible to people with physical space. The listed spaces represent the multiple resources that make up them.

The information interaction of the physical, cybernetic and social spaces of the SCPS is carried out through the interaction of the resources that form these spaces. In this article information resources and physical resources with an information component are combined into the category of cybernetic resources since physical resources cannot interact without support from information technology. Thus, cybernetic and social resources are allocated as resources with the ability to interact. Both of these types can interact with each other, as well as with resources of another type. At the same time, active cybernetic and social resources can initiate interaction.

3 Problems of Socio-Cyber-Physical Systems in Accordance with the Type of Communication

In addition to the problem of acquiring knowledge in developing socio-cyber-physical systems, there is a number of other problems. Doctor of Technical Sciences, Professor of the Russian Academy of Sciences (RAS) Andrei Leonidovich Ronzhin noted in his report at the conference "Digital Industry: State and Development Prospects of 2018" that the problems of the last decade in the modeling of socio-cyber-physical and collaborative robotic systems remain relevant at the present. We can identify five main problems:

1. Planning and navigation through previously unknown territory (obstacle avoidance, obstacle recognition, recognition of the type of obstacle based on semantic technologies, which means an understanding of what type of obstacle it is, whether it is necessary to bypass it, or to moved). It gives rise to collisions in the group behavior of robots.
2. The problem of energy. This mainly concerns robotic tools. The main difference of it is mobility, especially for ontropomorphic robots like humans, where the main idea is that they move and work in the space that is normal for human behavior. It is not only a flat surface, a homogeneous environment, but such spaces as cosmos, water, air, ordinary rooms equipped with stairs, escalators, elevators and other irregularities.
3. Security. The concept of security is the safety of the behavior of the robot when it interacts with other robots, or with humans. For this the zoning is performed. The zone for robots is separated from the zone of humans's work. In addition when a human crosses or appears in the area of the robot's action, its speed decreases, it acts more accurately when transferring, for example, some object to a person using programmed action or stops the action. If the problem is different, for example, spy, the enemy penetrated counter-control, then this is already a security problem. In this case we need to ensure cyber security in order to prevent the control of the robot which is controlled by a person from outside. This is one of the goals of the Industrial Internet.
4. Anthropomorphic manipulations. These is the design of manipulators similar the human in order to perform accurate operations, both in production and on the human body. It is one of the main applications of medical robots (for example, the Da Vinci robot and medical analogs). It is just a product of operations, teleoperations, when qualified medical personnel (qualified operator) controls various kinds of instruments for performing surgical operations inside the human body.
5. Intuitive human-machine interfaces for the implementation of interaction, communication between a human, a robot or a socio-cyber-physical system. Automation and the creation of digital twins are primarily focused on production and are developing faster in industry, where everything is logical, standardized, because there are artificial signals of a known physical form in a known bandwidth with a known sampling frequency. If a person is taken as an object of research, it is necessary to understand its variability. If it is speech, then consider acoustics, vocabulary, semantics, syntax, and so on. In order to adequately fulfill the requirements of the user, it is necessary to understand the context, give a personalized assessment of this user. It is necessary to understand and provide required interfaces and services.

The problems of interaction of socio-cyber-physical systems can be considered relative to different types of communications.

The main types of communications.

1. Human-human. When people are in a similar space, for example, in the same audience, when they have the same knowledge, they understand each other perfectly. Otherwise, when they communicate remotely or in different languages, certain problems appear. It could be worse by human-machine interaction. Such shortcomings are often found during communications in telecommunications mode, during teleconferences.
2. Human-device. Communication is carried out through the interface that the developer provides. The interface has to have the natural modality processing facilities (speech processing, gestures, facial expressions, body position) and the necessary technical support. In this case the interaction turns out to be quite stable. Otherwise, there are problems with perception and understanding.
3. Interaction between devices. In this case, problems may occur at the level of hardware software. As a result it is quite difficult to coordinate, synchronize and harmonize the whole range of devices that a person uses.

4 Multimodal Interfaces. Difficulties in Their Development

Currently, there is an active development of multimodal interfaces, which are aimed at creating a way of interaction between a computer and a human, similar to the interaction of a human with a human, using all known modalities which are accepted for human communication [6]. This development solves the main problem of socio-cyber-physical systems - increasing reliability. This is due to the fact that information data is transmitted through several channels, where the information may duplicate. It is increases the accuracy of recognition, and, accordingly, increases the efficiency of interaction.

The second advantage of multimodal interfaces is their use for people with disabilities. If a person does not have any of the modalities, he can replace it by another means (for example, if he has hearing problems, he makes up for it with gestures).

Currently, various applications of socio-cyber-physical systems are being developed: embedded, stationary, intelligent environments. The intellectual environment is a more complex socio-cyber-physical system in the image of a medical chamber, an intellectual hall (situational center), and a smart city. In an intelligent meeting room, a person sees only means of displaying information and means of inputting information, such as video cameras, microphones, and other sensors in order for the system to fully perceive and recognize the context. Huge computing and network resources are not visible in such an intellectual space. These resources are aimed at processing what is happening in the hall, personified in understanding what each of the participants is doing, writing it down and providing the person with the appropriate information service.

It is important to consider several aspects during developing multimodal interfaces. Firstly, these are the characteristics of the client device, its size, computational capabilities, network capabilities, capabilities of the sensors of multimedia devices that are available in it. Secondly, the user's parameters, his skills, capabilities, preferences, limitations, his mobility, characteristics of the interaction environment, noise level, physical parameters of the atmosphere, the number of users, the distance between the user and the device, the ability to connect to network resources, etc. Finally, these are the specifics and pragmatic conditions that are imposed by the type of application: informational, educational and applied.

As for audio localization, the main problem in speech technologies is that it is necessary to cut off the useful signal from extraneous noise. A bimodal system was developed at the Russian Academy of Sciences (RAS), when the position of the face and the position of the sound source were recognized simultaneously. Due to this, it became possible to recognize and record only those signals which are in a narrow area of space directly in front of the system itself.

5 Problems in Overcoming a Number of Technical and Social Obstacles

As a result of the introduction of socio-cyber-physical systems, certain technical and social obstacles were identified. First, the global cooperation of large corporations and small companies around the world is required. In other words, "all machines must speak the same language". For example, if a company receives a material (an unfinished product) that cannot be processed (cannot read an RFID chip programmed on a different frequency), then it will not be suitable for production. The common languages and platforms should be indicated to all. The technology of various manufacturers and companies will be based on them and called "end-to-end standardization" [9].

In 2014, the German Minister of Economy claimed that the government had concerns that "the big data which need for Industry 4.0 are collected not by national companies, but by four companies from Silicon Valley" [10]. Sigmar Gabriel meant that focusing on new technologies of Industry 4.0 would bring to power a very small group of corporations, and this would cause extremely strong inequality in the business sphere.

The report of the experts of the World Economic Forum says that in 2020 automation, robotics and new technologies will leave 5 million people out of work. Even the Chinese economy will suffer since such technologies will allow the transfer of plants from developing countries closer to consumers.

Fears are aimed at the fact that a significant reduction in jobs will not be offset by the emergence of new ones. For example, in many modern factories there have been formed departments with thousands of people who monitor production processes and equipment conditions. This can become a very serious problem against the background of a growing world population.

Cybersecurity also remains a problem: it is difficult to create completely secure networks, and the integration of physical objects with the Internet makes hacker attacks bigger and more dangerous. Hackers will be able to stop production processes and cause significant damage due to cyber-attacks.

The development of requirements is currently used mainly in the field of software development regarding the problems of the engineering approach and technical requirements for socio-cyber-physical systems. It means that the requirements are mainly defined for cyberspace [14]. However, the "Internet of Things" (IoT) has begun a new stage in the evolution of cybersystems, where cyberspace is closely integrated with physical space [18]. Not so long ago, the next stage of evolution began, when cyber and physical space are closely intertwined with social space. Thus, in the nearest future, in most cases, the development of requirements will be used not only for software (cyber) systems, but also for complex systems that combine all three dimensions - social, cyber and physical [3, 17, 18].

The engineering approach and the development of technical requirements mainly consists in creating software that is a relatively stable and predictable cyberspace. But even for such software, problems arise in the development of requirements when, for example, they change, are absent or become unsuitable for the desired product. It is becoming increasingly difficult to develop technical requirements with the advent of socio-cyber-physical systems, which are complex heterogeneous systems [11]. It is not only standard practice when requirements change, supplement and develop. Also requirements often appear during the life cycle of a system. In socio-cyber-physical systems, requirements are defined not only for cyberspace, they must also cover social and physical spaces. There are very complex interconnections, interactions, and influences between system components that can lead to unexpected and even unacceptable consequences in the structure and behavior of the system as a whole. One of the promising approaches supporting the adaptability and breakdown of systems is the development of continuous requirements based on maneuverability and flexibility [15]. Therefore, it is necessary to timely identify deficiencies in existing practices for creating requirements for socio-cyber-physical systems and develop the necessary adjustments and improvements in the process of introducing new requirements.

6 Problems of Complex Modeling and Proactive Management of SCPS

In the study of management processes for such complex objects as SCPS, the fundamental role is played by the law of necessary diversity, formulated by R. Ashby in 1952. However, the concept of centralized control with negative feedback, which underlies N. Wiener's classical cybernetics, isn't always currently adequate to modern control processes. In particular, it doesn't allow describing the behavior of complex distributed self-organizing SCPS. In contrast to the existing traditional cybernetic approaches, these studies have shown that it is necessary for SCPS and the processes of their self-organization to move to another level of organization of control processes (the level of complexity management). There are proposed two directions for implement this management concept - narrowing the variety of environmental influences on the SCPS and expanding the variety of control actions on the SCPS.

In the first area, the most promising methods are focused on a polymodel description of a specific subject area; classification and ordering of models, establishing relationships between them, based on the search for rational multicriteria solutions (compromises) in the presence of unavoidable threshold information and time constraints, as well as methods based on overcoming problems of large dimension and uncertainty when describing a subject area using decomposition (composition methods), aggregation (disaggregation), coordination, approximation, linearization, relaxation, reduction (immersion).

In turn, in the second direction, the most promising methods are based on the ideas of a self-similar recursive description and modeling of objects of study (introducing categories of macro-state, structural state, multi-structural state), dual control technology, and proactive structural control dynamic SCPS (including a flexible combination of hierarchical and network management principles). In contrast to the traditionally practiced reactive management, the proactive management of SCPS allows to prevent in advance the prerequisites (and not the consequences) of the occurrence of emergency (critical) situations due to the structurally functional redundancy purposefully formed at each specific stage of the SCPS based on the knowledge gained as a result of multivariate proactive forecasting of possible options by the functioning of the SCPS.

Most of the listed methods for managing complexity can be used to analyze possible approaches to organizing processes controlled by self-organization of SCPS.

In contrast to the well-known approaches that determine the world-wide level of modeling the processes of controlled self-organization and involve the use of finite-dimensional and recurrent dynamical systems, there are great prospects for research areas related to the implementation of a neo-cybernetic concept. This concept involving synthesis as models of targeted behavior hierarchical economic-production systems under control and disturbing influences, as well as network models of cause-effect relationships between processes of the occurrence of mass events in these systems, describing the relationship between the endogenous characteristics of the systems; as well as models of self-organization and evolution of socio-economic-production systems; models of equilibrium states of socio-economic-production systems; models of the dynamics of socio-economic-production systems when deviating from an equilibrium state under the influence of control and disturbing influences; analytical and simulation G models of multiagent systems. This will provide control over details of forecasts of SCPS behavior and the ability to select a forecast interval that is consistent with a given level of detail, and will allow us to analyze the processes of evolution and reconfiguration of socio-economic systems.

At the same time, the development of technologies for parametric adaptation of a complex of models of processes of controlled self-organization of SCPS based on mosaic statistics is advisable to be carried out on the basis of data reconstruction procedures. In the general case, the parametric and structural adaptation of the models of proactive control of the structural dynamics of SCPS should be based on combined methods of the theory of optimal control and evolutionary modeling. An important role in the indicated processes of adaptation of the models of proactive control of the structural dynamics of SCPS should be assigned to specialists in the corresponding subject areas, who provide constant updating of the models under consideration with using the capabilities of modern and promising intelligent interfaces (InI).

7 Conclusion

This article examined the main problems associated with complex modeling and proactive management of SCPS.

Currently, for SCPS, the necessary improvements are still being formulated as events of a high level of complexity. In the future, specific methods, models and artifacts which will be suitable for use will be developed. The future direction of research is to study existing continuous approaches, methods, models and tools for developing technical requirements and assessing their impact on the development of requirements specifically for socio-cyber-physical systems by submitting applications to specialized design organizations.

Also in this article were examined the main existing problems in the development of requirements that have already been identified in software development, and determined the necessary improvements in the development of requirements. In addition, socio-cyber-physical systems were presented, the main characteristics of these systems are indicated, and additional steps are listed in developing requirements to support these characteristics.

From a technical point of view, many complex problems still have to be solved, at least in an effective and applicable way in real conditions. At the moment these problems are:

1. The heterogeneity of the data. Data heterogeneity is a serious problem that can adversely affect the effectiveness of interactions and the development of communication protocols. Systems must be able to support a large number of different applications and devices.
2. Reliability. Socio-cyber-physical systems can be used in such critical areas as healthcare, infrastructure, transportation and many others. The basic requirements are reliability and safety, as actuators have an impact on the environment. In fact, the influence of actuators can be irreversible, so the likelihood of their unexpected behavior should be minimized. In addition, the environment is unpredictable, so systems must be able to continue to operate in unforeseen circumstances and adapt in case of failures.
3. Data management. It is necessary to store and analyze big data coming from various network devices, process them and display the results in real time. Data can be managed using delayed or online streaming processing, depending on the purpose of the system. Information can often change, and processing is based on adaptive and constant requests while using streams in real time.
4. Confidentiality. The problem is to maintain a balance between maintaining confidentiality, protecting personal data and the availability of data to provide better services. Serious data confidentiality problems arise since cyber-physical systems manage significant amounts of data, including confidential information such as health, gender, religion, and many other personal information. Privacy policies are needed for cyber-physical systems, therefore, a data depersonalization tool is needed [5]. It allows you to delete personal information before the system processes the data.

5. Security. Socio-cyber-physical systems must ensure the safety of communications, since all actions are coordinated between devices in real time. Such systems expand the scale and scope of the interaction between physical and computing systems, which complicates security tasks. Traditional security infrastructures are not enough to solve this problem, and new solutions must be found. It is necessary to protect both incoming data and stored data collected for future use. And finally, cyber-physical systems are based on heterogeneous applications and wireless communications, what often complicates security.
6. Real time. Socio-cyber-physical systems control a significant amount of data received from sensors. Computational processing should be efficient and timely, as physical processes continue regardless of the results of the calculations. They must have the capacity or necessary power to support immediate processing to meet this requirement. Because failure to take timely action can result in substantial damage.

The research described in this paper is partially supported by the Russian Foundation for Basic Research (grants: 17-29-07073-ofi-i, section № 1-2: 18-07-01272, section № 3, 19-08-00989 section № 4-5), state research 0073–2019–0004 section № 6-7.

References

1. Antsaklis, P.: Goals and challenges in cyber-physical systems research. IEEE Trans. Autom. Control **59**(9), 3117–3119 (2014)
2. Awais, A., Muhammad, B., Sadia, D., Shehzad, K., Muhammad, M.U., Anand, P., Alavalapati, G.R., Nasro, M.-A.: Socio-cyber network: the potential of cyber-physical system to define human behaviors using big data analytics. Future Gener. Comput. Syst. **92**, 868–878 (2018)
3. Balandin, S., Andreev, S., Koucheryavy, Y.: OLD Internet of Things, smart spaces, and next generation networks and systems. In: 15th International Conference, NEW2AN 2015 and 8th Conference, ruSMART 2015, St. Petersburg, Russia, 26–28 August 2015. Lecture Notes in Computer Science (including Subseries Lecture Notes in Artificial Intelligence, Lecture Notes in Bioinformatics), vol. 9247, pp. 80–94 (2015)
4. Bennaceur, A., Ghezzi, C., Tei, K.: Modelling and analysing resilient cyber-physical systems. In: 14th IEEE/ACM International Symposium on Software Engineering for Adaptive and SelfManaging Systems (SEAMS) (2019)
5. Bindu, P.V., Thilagam, P.S., Ahuja, D.: Discovering suspicious behavior in multilayer social networks. Comput. Hum. Behav. **73**, 568–582 (2017)
6. Calinescu, R., Camara, J., Paterson, C.: Socio-cyber-physical systems: models, opportunities, open challenges. In: 5th International Workshop on Software Engineering for Smart Cyber-Physical Systems (2019)
7. Camara, J., Garlan, D., Moreno, G., Schmerl, B.: Evaluating trade-offs of human involvement in self-adaptive systems. In: Mistrik, I., Ali, N., Kazman, R., Grundy, J., Schmerl, B. (eds.) Managing TradeOffs in Adaptable Software Architectures, pp. 155–180. Morgan Kaufmann, Boston (2017)
8. Dang-Pham, D., Pittayachawan, S.: Applications of social network analysis in behavioural information security research: concepts and empirical analysis. Comput. Secur. **68**(July), 1–15 (2017)

9. Fleischmann, H., Kohl, J., Franke, J.: A reference architecture for the development of socio-cyber-physical condition monitoring systems. In: 11th System of Systems Engineering Conference (SoSE) 2016, pp. 1–6 (2016)
10. Ioan, B., Tecaru Berekmeri, C.V., Blebea, I.: Considerations regarding the integration-intrication processing the nature and technology. ACTA Univ. Cibiniensis **64**(1), 82 (2014)
11. Lace, K., Kirikova, M.: Required changes in requirements engineering approaches for socio-cyber-physical systems. In: CEUR Workshop Proceedings, vol. 2075 (2018)
12. Moldovan, D., Copil, G., Dustdar, S.: Elastic systems: towards cyber-physical ecosystems of people, processes, and things. Comput. Stand. Interfaces **57**, 76–82 (2018)
13. Mueller, E., Chen, X.-L., Riedel, R.: Challenges and requirements for the application of industry 4.0: a special insight with the usage of cyber-physical system. Chin. J. Mech. Eng. **30**(5), 1050–1057 (2017)
14. Rho, S., Vasilakos, A.V., Chen, W.: Cyber physical systems technologies and applications. Future Gener. Comput. Syst. **56**, 436–437 (2016)
15. Roche, R., Suryanarayanan, S., Hansen, T.M.: Cyber-Physical-Social Systems and Constructs in Electric Power Engineering. The Institution of Engineering and Technology, London (2016)
16. Smirnov, A.V., Levashova, T.V.: Acquisition of knowledge in socio-cyberphysical systems in the process of informational interaction of resources. In: Information and Control Systems, vol. 6, pp. 113–122 (2017)
17. Smirnov, A., Shilov, N., Fedotov, Y., Krotov, K.: Context-aware self-configuration of flexible combined transportation chains: theoretical and technological foundations. In: Proceedings of mobil.TUM 2015 Conference, Munich, Germany (2015)
18. Sokolov, B., Yusupov, R., Verzilin, D., Sokolova, I., Ignatjev, M.: Methodological basis of socio-cyber-physical systems structure-dynamics control and management. In: Digital Transformation and Global Society, pp. 610–617 (2016)
19. Zhuge, H.: Cyber-physical society-the science and engineering for future society. Future Gener. Comput. Syst. **32**(C), 180–186 (2014)

Modeling and Analysis of Survival to Patients of Pancreatic Cancer in Krasnoyarsk Territory

Artem Kazarin[1], Natalia Lukyanova[1,2](✉), and Olga Melnikova[3]

[1] Voino-Yasenetsky Krasnoyarsk State Medical University,
1 Zheleznyak Partizan Street, 660022 Krasnoyarsk, Russia
artem.kazarin.96@bk.ru

[2] Institute of Mathematics and Computer Science of Siberian Federal University,
79 Svobodny Avenue, 660041 Krasnoyarsk, Russia
nalukyanovakrsk@gmail.com

[3] Reshetnev Siberian State University of Science and Technology,
31 Krasnoyarsky Rabochy Avenue, 660037 Krasnoyarsk, Russia
melok325@gmail.com

Abstract. Any serious medical research involves the use of information technologies of statistical analysis in processing the results of clinical observations, experimental data and measurements. In the paper the survival analysis patients by the medical database of pancreatic cancer in Krasnoyarsk Territory is represented. Survival analysis is based on the survival function. Survival function modeling was performed on the basis of three approaches: parametric, semi-parametric and non-parametric. The survival model was fitted on the base of the selection of distributions. The closest distribution is the Weibull distribution. Semiparametric modeling made it possible to construct a survival curve for a combination of prognostic factors based on the Cox proportional risk regression model. Two parameters were identified as prognostic for survival: distant metastases, number of chemotherapy courses taken. An assessment was made of the overall survival of patients with pancreatic cancer and an assessment of the survival of patients after surgical and chemotherapeutic treatment in various subgroups by age and gender using the non-parametric Kaplan-Meier method #CSOC1120.

Keywords: Survival analysis · Censored data · Kaplan-Meier analysis · Cox regression · Pancreatic cancer · Krasnoyarsk territory

1 Introduction

Pancreatic cancer (PCa) are one of the urgent medical and social problems of healthcare throughout the World. PCa is a formidable and difficult to diagnose and treat cancer with low survival rate. In the Russian Federation in 2018 in the structure of the incidence of the population of malignant diseases, pancreatic

R. Silhavy (Ed.): CSOC 2020, AISC 1226, pp. 607–616, 2020.
https://doi.org/10.1007/978-3-030-51974-2_56

cancer was 11th (3.1%). The average Russia indicator of untimely diagnosis (IV stage) with PCa is 58.9%. There is a high mortality rate in the first year since the diagnosis was established (average Russian indicator – 66.9%, Krasnoyarsk Territory – 62.2%) [1]. A last comprehensive assessment of hospital morbidity in Krasnoyarsk Territory was carried out in work [2].

The mathematical apparatus of statistics in medicine is one of the tools for analyzing experimental data and clinical observations. Survival analysis methods are intensively using in medicine [3]. One of the important characteristics describing the course of the disease is the life expectancy of patients from the moment of diagnosis or after surgery/chemotherapy. Any statistical research involves processing of information using specialized software.

2 Methods and Materials

Goal of Research: To analyze the survival of patients with PCa in the Krasnoyarsk Territory. To achieve the goal, the following tasks were set and solved: (1) to assess the overall survival of patients with this pathology, to build survival curves, to select the distribution that most accurately describes the data, (2) to assess the survival of patients after surgical and chemotherapeutic treatment in subgroups by age and gender, (3) to identify factors affecting the increase or decrease in the chances of survival of patients with PCa.

Survival analysis is a combination of statistical methods most often used in medical research, for example, as in [4,5]. Survival assessment methods are understood as the study of the patterns of occurrence of the expected event in representatives of the observed sample in time [6,7]. In paper, an event is understood as a fatal outcome, representatives (objects of research) – patients with an established diagnosis of PCa. The study included 400 observations taken from the Cancer Register of Krasnoyarsk Regional Clinical Oncology Center named after A.I. Kryzhanovsky. Both full and censored observations were taken into account. Censored observation is the patient's medical history, which at the time of the last examination does not end with the onset of death. Censorship right-hand: the date of beginning of observation of each patient is the date of diagnosis. The time interval was considered in days. In this research, in 332 (83%) patients, the outcome is fatal, in the remaining 68 (17%), surgery or chemotherapy. Distribution of patients by gender: male – 204 (51%), female – 196 (49%).

At the data preprocessing stage determining the structure and types of input data, filling the database in MS Excel based on the Cancer Register. Patients are writing in rows, indications – in the columns:

1. Gender of a patient (M-Male; F-Female).
2. Date of birth (in DD.MM.YYYY format).
3. Date of diagnosis (in DD.MM.YYYY format).
4. Date of clinical outcome (date of death for complete observations, date of last operation / chemotherapy for censored observations in DD.MM.YYYY format).

5. Code of observation is the binary indicator: 0 - the patient dropped out of the study for some reason, but was alive at the time of dropping out (censored); 1 - the patient died (complete).
6. ICD-10 diagnosis (International Classification of Diseases). C25 – malignant neoplasm of the pancreas includes the following localizations:
   - C25.0 – of the pancreatic head;
   - C25.1 – of the pancreatic body;
   - C25.2 – of the pancreatic tail;
   - C25.7 – of other parts of the pancreas;
   - C25.8 – malignant neoplasm of the pancreas which is beyond the above mentioned one or more locations;
   - C25.9 – malignant neoplasm of pancreas, unspecified.
7. Clinical stage cancer (I, II, III, IV).
8. Primary tumor volume according to the Tumor Node Metastasis (TNM) classification (-1 – TX: cannot be assessed, 0 – T0: absence of tumor, 1, 2, 3, 4 – T1, T2, T3, T4: degree of increase in tumor size).
9. Metastasis to regional lymph nodes according to TNM classification (-1 – NX: cannot be assessed, 0 – N0: absence of metastases, 1, 2, 3 – N1, N2, N3: degree of metastases).
10. Distant metastases according to TNM classification (-1 – MX: cannot be assessed, 0 – M0: absence of metastases, 1 – M1: presence of metastases).
11. Surgical treatment (0 – no, 1 – yes).
12. Chemotherapy (0 – no, 1 – yes).
13. Number of chemotherapy courses taken.

Indicators calculated based on the input data:

1. Age (number of years at the date of clinical outcome).
2. Period of cases in days (calculated as the difference between the end date of the observation (date of clinical outcome) and the start date of the observation (date of diagnosis)).

Data analysis was performed in SPSS, STATISTICA, MS Excel. Survival analysis is a class of statistical models to estimate the probability of occurrence of an event. Kaplan-Meyer method and life tables are using to construct survival curves [3,8]. The two methods have similarities in basic principles of application. These methods take into account not the actual dates of the beginning of the observation or the outcome (since they are different), but the period of time from the beginning of the observation to the occurrence of the outcome or the last contact with the observed. To compare survival between groups of patients formed by different filters, taking into account gender, age classification, cancer stage, presence/absence of surgical treatment and/or chemotherapy, Log Rank criterion was used, as well as Breslow criterion and Tarone-Ware criterion. Cox regression, or the model of proportional risks, is a graphical construction and mathematical representation in the form of coefficients of a regression equation, exponentials of the risk coefficients of an event occurring as a function of time, and an assessment of the influence of each of the independent variables on this risk [9]. This method involves predicting the risk of an event occurring for the object in question and evaluates the effect of independent variables (predictors) on this risk. Only in the Cox regression to assess the impact of the set of predictors on the outcome, taking into account the observation period.

3 Results and Discussions

A medical database containing information on 400 patients with PCa was research. The mean age of patients was 64.24$\pm$ 11.6 years (format $M \pm \sigma$, where M is the mean, σ is the standard deviation). The age category of 60–69 years is found most often (35.75% of cases of analyzed cohort) in the distribution of patients with PCa throughout the sample, in complete and censored observations (see Fig. 1).

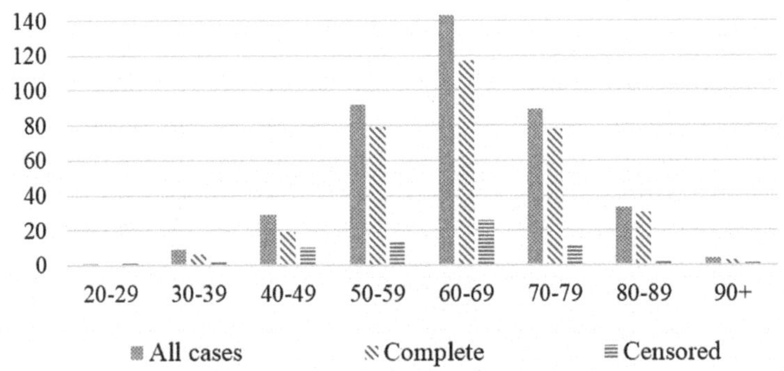

Fig. 1. Distribution of observations over time by age

The prognosis of the disease depends on what stage the diagnosis was made. When studying patients with PCa at the stages of the malignant process, it was found that for the first time at 365 patients (91.25% of cases) were registered the III and IV stages (see Table 1). Patients with stage I and stage II of the tumor process were registered in a cohort at 35 people (8.75% of cases). This once again suggests that early diagnosis of PCa is a difficult task. This is mainly due to the latent course of the disease (in our case, 96 observations were noted in the extracts of the cancer register) and untimely appeal to a medical institution (57 observations). The most susceptible to the malignant process (215 people) is the head of the pancreas (5.0).

Prediction based on survival analysis allows us to determine the probability of the outcome studied for any moment in time, i.e. to evaluate the survival function $S(t) = P\{T > t\} = 1 - F(t)$, where T – non-negative random variable representing the period of time before the patient's fatal outcome. The cumulative distribution function is respectively $F(T) = P\{T \leq t\}$.

Modeling $S(t)$ is possible based on three approaches. The parametric approach proceeds from the assumption of a certain shape of the curve $S(t)$ and

Table 1. Distribution of patients by the severity of the cancerous process

Stage cancer	Total obs.	Number of complete obs.	Number of Censoreds obs.(%)	Me, days	95% CI for Me
I	8	4	4 (50%)	161	0.0–345.2
II	27	14	13 (48.1%)	255	120.6–389.4
III	87	65	22 (25.3%)	93	63.3–122.7
IV	278	249	29 (10/4%)	47	38.3–55.7
Sum total	400	332	68 (17%)	57	47.0–67.0

Note: obs. – observations; Me – median of expected life-time

performs the fitting of the survival model based on the selection of distributions. Semiparametric modeling allows the construction of a survival curve for a combination of prognostic factors based on the Cox proportional risk regression model. With a nonparametric approach, even weak assumptions for $S(t)$ are absent and the data set speaks for itself. In the case when covariates are naturally determined (for example, the patient's gender) or absent, the non-parametric Kaplan-Meier method is used in the survival analysis.

For prediction it is often necessary to know the form of the survival function. For this purpose, various distribution families are used. During the procedure of fitting the theoretical distribution to the nonparametric life expectancy data, the following types of distributions depending on the parameters were investigated:

- Exponential distribution: $S(t) = exp\,(-\lambda t)\,;$
- Weibull distribution: $S(t) = exp\,(-\lambda t^{\gamma})\,;$
- Gompertz distribution: $S(t) = exp\left(-\lambda\gamma^{-1}(exp(\gamma t) - 1)\right).$

Data for analysis was the period of cases in days and the censoring variable. Parameter estimation is implemented in STATISTICA using three methods: the least squares method (Weight1) and two weighted least squares methods with different weights (Weight2, Weight3) [10]. The results of estimations for the parameters of the above distributions are presented in the Table 2. The closest distribution is the Weibull distribution with parameters estimated by the second method of weighted least squares (Weight 3), the hypothesis of the proposed distribution is accepted (p = 0.051). The smaller the resulting chi-square value, the more accurate the parameter estimation. Thus, parametric family of distributions that best describes the data, which are random values of the patient's lifetime, was found.

$$S(t) = exp\left(-0.024971 \cdot t^{0.811376}\right)$$

The following Fig. 2 shows the Kaplan-Meyer overall survival curve for all observations. The median life expectancy was 57 days (95% confidence interval (CI): 47.03–66.97). Only 6 patients had one-year survival rate (1.5% of total observed).

Surgical treatment was performed at 247 patients (62% of the analyzed cohort). Only 64 patients (16%) underwent chemotherapy. For comparative analysis we form two subgroups: the first subgroup has a complex of surgical and

Table 2. Parameter estimates for different distributions

Distribution	Weight	λ	γ	χ^2	Log likelihood
Exponential	Weight1	0.010206	–	27.75212	−564.706
Exponential	Weight2	0.008769	–	29.22160	−565.441
Exponential	Weight3	0.009557	–	26.57072	−564.115
Weibull	Weight1	0.005995	1.069200	40.54692	−571.103
Weibull	Weight2	0.017774	0.885789	18.15752	−559.909
Weibull	Weight3	0.024971	0.811376	17.06511	−559.362
Gompertz	Weight1	−4.91575	0.000829	43.59258	−572.626
Gompertz	Weight2	−4.58651	−0.000313	25.28178	−563.471
Gompertz	Weight3	−4.57884	−0.001099	24.05901	−562.859

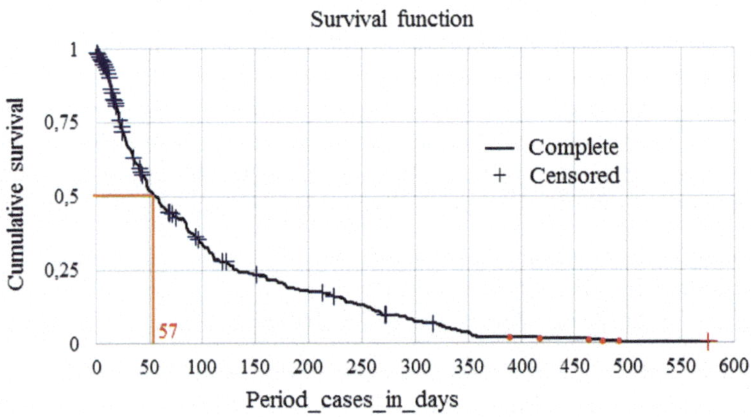

Fig. 2. Graph of overall survival for the entire sample population (red dots and red cross denote complete and censored observations (respectively) which survived the one-year rate)

chemotherapeutic treatment, the second subgroup does not. For the first subgroup, 56 cases were selected, of which 30 were complete observations, 26 (46.4%) were censored. The median survival for the first group was 171 days (95% CI: 80.63–261.37). The second subgroup includes 145 observations (of which 143 cases were with III and IV stages), all were complete. The median survival of the second subgroup was 36 days (95% CI: 26.29–45.71). On Fig. 3 shows the Kaplan-Meyer overall survival functions in the subgroups under consideration. Differences in survival in groups of patients are statistically significant according to the criterias: Log Rank (χ^2 = 10.76; $p < 0.001$), Breslow (χ^2 = 23.34; $p < 0.001$), Tarone-Ware (χ^2 = 17.79; $p < 0.001$). The effect of surgical treatment is targeted, and chemotherapy is aimed at combating cancer cells and metastases that are widespread throughout the body. Thus, surgical treatment in combination with chemotherapy has a positive effect on the clinical outcome of pancreatic cancer, prolonging the life of the cancer patient.

Comparison of survival in subgroups of patients by gender was performed. The distribution of patients is as follows: women – 196 (49%), of which 160 are complete, 36 are censored observations; men – 204 (51%), of which 172 are complete, 32 are censored observations. The median life expectancy for men was 48 days (95% CI: 33.69–62.31), for women 67 days (95% CI: 48.90–85.10). Differences in survival between gender groups in the studied cohort of patients are statistically insignificant according to the criterias: Log Rank ($\chi^2 = 1.01$; $p = 0.32$), Breslow ($\chi^2 = 2.30$; $p = 0.13$), Tarone-Ware ($\chi^2 = 1.87$; $p = 0.17$).

Comparison of the overall survival in subgroups of patients on age is carried out: over 60 years old (≥60) and under 60 years old (<60). Distribution of patients in these subgroups the following: (≥60) – 269 people, from them 228 – complete observations, 41 – censored; (<60) – 131 patient, of which 104 – complete observations, 27 – censored. The median of the expected life time for patients (≥60) made 52 days (95% of CI: 42.53–61.47), for patients (<60) – 65 days (95% CI: 39.64–90,36). Differences in survival in these subgroups are statistically insignificant according to the criterias: Log Rank ($\chi^2 = 1.45$; $p = 0.23$), Breslow ($\chi^2 = 1.74$; $p = 0.19$), Tarone-Ware ($\chi^2 = 1.66$; $p = 0,20$).

The age group of older persons (60–74 years) is considered. The median overall survival for this group was 59 days (95% CI: 41.93–76.07). Using the SPSS filter in this age group, we select those who are diagnosed with IV stage and form two subgroups: the first subgroup has surgical and chemotherapeutic treatment, the second – absence of treatment. The median of the expected life time for the first subgroup was 107 days (95% CI: 87.86–126.14), for the second subgroup 34 days (95% CI: 21.79–46.21). The behavior of the graphs of survival functions for the studied subgroups is similar to the graphs in Fig. 3. The median survival of a subgroup of elderly patients who received a combination of surgery and chemotherapy is 1.8 times higher than the median overall survival for elderly people, and 3.15 times higher than the survival of elderly patients, not receiving no therapy at all. Despite the elderly age of patients with the last stage of cancer, in most cases, surgical treatment in combination with chemotherapy prolongs the life of cancer patients.

At the time of establishment of the diagnosis at 65.25% of patients of the considered sample have distant metastases, and at 25.25% of patients regional lymph nodes are involved (see Table 3). The tumor in the pancreas quickly reaches large sizes, gives metastases to other organs (liver, lungs, pleura, peritoneum), it is impossible to estimate regional lymph nodes in 225 patients in the research.

When analyzing life times, it is particularly relevant to find out whether some continuous variables are related to observed life times. If there is such a dependence, it is necessary to choose a suitable mathematical model and evaluate the model parameters. It is problematic to build such a model using classical multiple regression for two reasons. First, life times are usually not normally distributed, which is a serious violation of assumptions for estimating multiple regression by least squares method. Times of life usually have an exponential

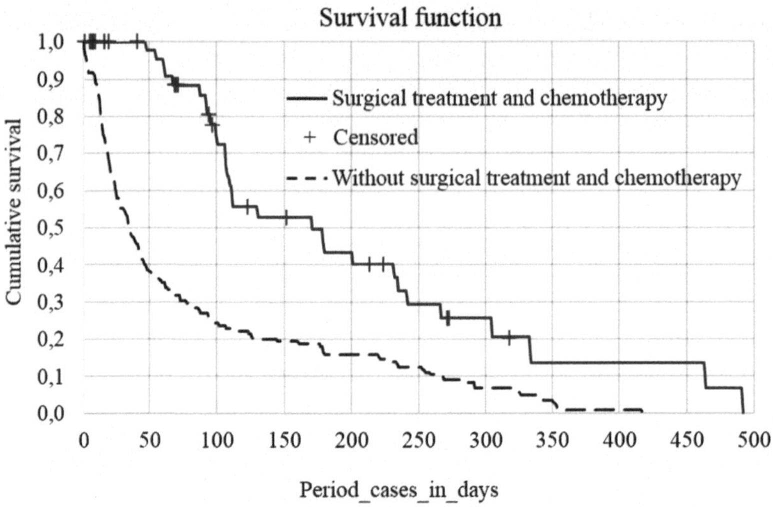

Fig. 3. Graphs of overall survival functions in groups with and without surgical treatment and chemotherapy

Table 3. Distribution of patients (abs) according to the TNM system

TNM	TX	T0	T1	T2	T3	T4	NX	N0	N1	N2	N3	MX	M0	M1
Cohort	6	0	3	37	212	142	225	74	80	17	4	18	121	261
Male	2	0	2	20	103	77	116	37	39	10	2	12	55	137
Female	4	0	1	17	109	65	109	37	41	7	2	6	66	124

distribution or Weibull distribution. Second, there is a problem with censored observations.

An analysis was made according to the Cox model, which considers three types of indicators: result (complete/ censored observation), period of cases, predictors, and allows you to evaluate the independent contribution of several predictors at the same time, namely, the stage and type of localization of the tumor process, the primary tumor volume, metastasis to regional lymph nodes, distant metastases, the presence of surgical treatment or chemotherapy, the number of completed chemotherapy courses. Two parameters were identified as prognostic for survival: distant metastases, number of chemotherapy courses taken (see Table 4). Others parameters had not independent prognostic significance, including metastasis to regional lymph nodes ($p = 0.057$). The constructed Cox regression model is statistically significant. The maximum-likelihood method (−2LL in SPSS) for the base model without predictors is 3273.742. The value of this indicator for the model with predictors is significantly lower (−2LL=3169.869; $\chi^2 = 84.952$; $p < 0.001$), therefore, the model has greater predictive power compared to the base. The Exp(B) indicator reflects the Hazard ratio (HR) [6,9].

If HR>1, then this independent variable will affect the clinical outcome of PCa and thereby reduce life expectancy, if HR<1, then an independent predictor acts as the so-called protective factor, to some extent prolonging life.

Table 4. Significant predictors in Cox's proportional risk model

	B (SE)	Wald	p-value	Exp(B)	95% CI Exp(B)
Distant metastases	0.63 (0.14)	19.06	<0.001	1.88	1.42–2.50
Courses of chemotherapy	−0.25 (0.07)	13.89	<0.001	0.78	0.68–0.89

Note: B is the regression coefficient, SE is the standard error.

Distant metastases (HR=1.88, 95% CI: 1.42–2.50) affect the risk of death in PCa, reduce the life expectancy of patients. The number of chemotherapy courses taken is a protective factor in PCa (HR = 0.78, 95% CI: 0.68–0.89).

4 Conclusion

The obtained results on up-to-date data (2017–2019) of the Cancer Register of Krasnoyarsk Territory are consistent with what is written in the literature on this problem. Pancreatic cancer affects both men and women equally. No statistically significant differences were revealed by gender and age structure. High prevalence of PCa is observed in the elderly. The largest share among the representatives of the observed sample in time is occupied by the age category 60–69 years. According to the localization of the tumor process in the pancreas, the head of the gland is most often affected. Unfortunately, most patients were with the last stage of the oncological process and metastases to various organs. This is mainly due to the latent course of the disease and untimely treatment in a medical institution. The median overall survival was 57 days (95% CI: 47.03–66.97). The probability of survival with a malignant neoplasm of the pancreas decreases every day. Despite the modern methods of diagnosis and therapy, unfortunately, not all patients survive to one year. Distant metastases have the greatest effect on the early development of death in PCa, and reduce the life expectancy of cancer patients. Parametric model of survival functions are obtained as a result of fitting the theoretical distribution to the data of a nonparametric study of life expectancy. Weibull distribution was chosen as the distribution that most accurately describes the data.

References

1. Kaprin, A.D., Starinskiy, V.V., Petrova, G.V.: Malignant neoplasms in Russia in 2018 (morbidity and mortality). In: Hertsen, P.A. (ed.) Moscow Oncology Research Center – Branch of FSBI NMRRC of the Ministry of Health of Russia, Moscow (2019). (in Russian)

2. Okladnikova, E.V., Ruksha, T.G.: Analysis of hospital morbidity for pancreatic cancer in Krasnoyarsk krai. Siberian J. Oncol. **6**, 61–67 (2015). (in Russian)
3. Glantz, S.A.: Primer of Biostatistics, 7th edn. McGraw-Hill Medical, New York (2012). (in Russian)
4. Slinin, A.S., Bydanov, O.I., Karachunskiy, A.I.: Analysis of survival and possibility of certain events in patients with acute leucosis. Pediatr. Haematol. Oncol. Immunopathol. **15**(3), 34–39 (2016). https://doi.org/10.20953/1726-1708-2016-3-34-39. (in Russian)
5. Zulkarnaev, A.B.: Features of survival analysis on patients on the «waiting list» for kidney transplantation. Bull. Siberian Med. **18**(2), 215–222 (2019). https://doi.org/10.20538/1682-0363-2019-2-215-222. (in Russian)
6. Ohno-Machado, L.: Modeling medical prognosis: survival analysis techniques. J. Biomed. Inform. **34**, 428–439 (2002). https://doi.org/10.1006/jbin.2002.1038
7. Clark, T., Bradburn, M., Love, S., et al.: Survival analysis Part I: basic concepts and first analyses. Br. J. Cancer **89**, 232–238 (2003). https://doi.org/10.1038/sj.bjc.6601118
8. Kaplan, E.L., Meier, P.: Non-parametric estimation for incomplete observations. J. Am. Stat. Assoc. **53**, 457–481 (1958)
9. Bradburn, M., Clark, T., Love, S., et al.: Survival analysis Part II: multivariate data analysis - an introduction to concepts and methods. Br. J. Cancer **89**, 431–436 (2003). https://doi.org/10.1038/sj.bjc.6601119
10. Borovikov, V.P.: Popular introduction to modern data analysis in the STATISTICA system: methodology and technology of modern data analysis: tutorial. Goryachaya linia - Telekom, Moscow (in Russian) (2014)

Approximation a Reachability Area in the State Space for a Discrete Task

Vitaly Ushakov[1,2(✉)]

[1] St. Petersburg Institute for Informatics and Automation of the Russian Academy of Science, 14th Line 39, St. Petersburg 199178, Russia
mr.vitaly.ushakov@yandex.ru

[2] Saint-Petersburg State University of Aerospace Instrumentation, Bolshaya Morskaya Street, 67, St. Petersburg 190000, Russia

Abstract. To solve the problem estimating and analyzing the quality indicators an control automated system for moving objects, it is proposed to apply an original approach based on the use reachability areas to evaluate the quality indicators of control systems for similar objects and use reachability areas formed in the space of significant system-technical indicators as an information basis. Because the analytical solution such problems even on modern computers requires significant (large) computational resources, it is proposed to use analytical geometry (and, in particular, projective geometry) to solve them. Thus, in order to build reachability area, as well as construction the target set at reachability area, the article discusses the use JavaScript with the Ganja.js package. The article will give specific recommendations for solving the investigated problem.

Keywords: Optimal control · Reachability area · Approximation reachability areas · Pareto set · Orthogonal projection · Projection operator · Complex dynamic objects · Interdisciplinary approach · Ganja.js · JavaScript

1 About Reachability Areas

At the end of the 20th century, it seemed that the control theory in large dimension problems, uncertainty, nonlinearity, stationarity went beyond its capabilities, and, in fact, came to a standstill. And, it seemed that there would be no new scientific discoveries in this area, but it was necessary to go somewhere in the heuristic.

But in the theory of optimal control, there is a concept - reachability area [1–4]. That is, in the framework of this scientific study, an attempt is made to get away from the logical-dynamic description for complex objects to a multidimensional geometric description; to distract from nature, from functional spaces, from piecewise-continuous controls. Since, performing work with multidimensional geometric images, which are behind the managerial processes, and also performing the replacement of managerial processes that occur in time for areas that do not change in time, we get a completely new and unique solution to the problem facing us. Using multidimensional geometry, the theory of project operators, those problems are solved that the optimal control theory could not solve, since it does not have the corresponding mathematical

R. Silhavy (Ed.): CSOC 2020, AISC 1226, pp. 617–624, 2020.
https://doi.org/10.1007/978-3-030-51974-2_57

apparatus. But the obvious advantage of mathematics is that it works in different spaces, and leaving the main space in the conjugate space it is possible to solve the same problems, but in a completely original way.

Also, one of the advantages of the proposed approach is the availability of free software for implementing the appropriate methods and algorithms using modern PCs, which will automate and speed up the calculation. In particular, in open source software there are constructive tools that make it possible to perform multidimensional projection of the target set onto the reachability area. And, working with multidimensional geometric images, it is possible by designing the target set on the reachability area to get the Pareto set [5, 6]. But the most important thing is that it is possible to obtain results in geometric images due to the presence of a direct connection between dynamic models with geometric images and return to the control space and solve problems such as the redistribution task, the modernization task, and the schedule task (control theory in scheduling theory [7–9]).

2 Approximation Reachability Areas

To date, two concepts have been developed for determining or evaluating the reachability areas possible phase states for system at different points in time: construct or approximate. The task which is considered in the article is one of the fundamental problems for optimal control theory in complex dynamic systems [10]. Reachability areas play an important role in control, observation, and forecasting tasks. Based on the construction and approximation for reachability areas many problems in the optimal control theory are solved [3, 10–12].

As examples we can cite approaches to constructing approximations reachability area, based on solving the corresponding optimal control problems in constructing various classes of ellipsoids [3], parallelepipeds, polygons or triangles. The approximation reachability area [13, 14] can be performed by various methods for example polyhedra or ellipsoids. To solve the task (development an algorithm for complex planning of operations based on the Krylov-Chernousko method) set in [15] a simplified method for approximation reachability areas was used.

All these problems are reduced to constructing or evaluating reachability areas in which the phase vector for system can be located and to operations with these reachability areas, which subsequently serves as the basis for the development various numerical algorithms for finding a solution to a boundary or optimization problem for the considered dynamical system. However, the practical construction of reachable areas [16–18] especially in complex dynamic systems of large dimension is a very difficult task even when using modern computers. Therefore, in practice methods are often used to approximate these reachability areas with the required accuracy instead of directly constructing them.

It is well known that information about the reachability area and its main characteristics essentially replace all the information necessary for solving the assessing tasks the capabilities for any dynamic system, the stability for its functioning, the options synthesis for creating and developing these systems. As already mentioned, knowledge about reachability areas makes it easier to solve optimal control problems. On the other hand, the optimal control methods and firstly the Pontryagin maximum principle make it possible in some cases to research and construct reachability areas.

3 Pareto Set

The Pareto set as you know is the set of unimprovable alternatives. Obviously, points that do not belong to the Pareto set are not the best alternative (the best solution in our case). The selection the Pareto set is the first step in comparing alternatives (solutions).

In accordance with the Edgeworth-Pareto principle, the selected variants cannot be outside the Pareto set, and any Pareto-optimal variant can be selected. Having only the set and the vector criterion one can construct the Pareto set, although serious computational difficulties may arise. Next one should choose within the Pareto set or in other words narrow [19, 20] this set to an acceptable size.

It is well known (for example [21]) that in the general case the Pareto set has a rather complicated structure, and therefore its construction often encounters insurmountable computational difficulties.

It is assumed that the choice is made by a decisionmaker. In order for the choice made to be most consistent with the achievement for goal (that is, it was "the best" or "optimal" for the given decisionmaker), it is necessary to take into account the "tastes" and "preferences" of the decisionmaker which to some extent revealed would be the meaning of a "best" solution.

As stated in [22], the main advantage using the Pareto set is that when it is realized simultaneously (in parallel), all the main tasks - the integrated planning task, modernization management task and the automated system functioning task for controlling moving objects [1] within the framework of a single logical-dynamic description are solved processes under consideration. And the analysis for all cases of the relative position for the target set and the reachability area carried out in [2] provides a constructive way to construct the Pareto set in optimal control problems. When solving our problem, obtaining a Pareto set is a solution to this problem.

4 Construction Pareto Set in the State Space Using the Projective Geometry

Construction Pareto set in the state space [23, 24] using the projective geometry.

To construct the Pareto set in the state space, it is proposed to use the following algorithm (using JavaScript and Ganja.js [25, 26]):

Step 1. Formation, analysis and preparation of source data.
Step 2. Planning the process of solving the problem of constructing a Pareto set. Determination the relative position for the target set and reachability area [1]. Determining the time required to execute the algorithm.
Step 3. Formation the target set (dynamic system reachability area) in JavaScript. The result is passed to Ganja.js.
Step 4. Building the target set and approximation the reachability area in Ganja.js.

Step 5. Orthogonal projection the target set onto the reachability area [1, 22] in Ganja.js using projective geometry (and projecting operators [11, 27, 28], in particular). As a result (according to [1, 3]), the formation the Pareto set occurs.
Step 6. Formation and interpretation the output results, presenting them in a form convenient for subsequent use (for example, for developing adaptive plans for the development complex technical systems [22, 29]).

After completing all the steps the algorithm, one can consider various service situations that are based on the results modeling the reachability area.

Then it is possible to select a specific variant a multi-structure macrostate for a complex technical system with simultaneous adaptive synthesis of both control technologies for the use these systems and specific control plans (programs) that implement these technologies under given conditions.

5 Using Ganja.js to Build Pareto Set

Ganja.js [25, 26] (Geometric Algebra - Not Just Algebra) is a Geometric Algebra code generator for JavaScript. It generates Clifford algebras and sub-algebras of any signature and implements operator overloading and algebraic constants.

Mathematically, an algebra generated by ganja.js is a graded exterior (Grassmann) algebra (or one of its subalgebras) with a non-metric outer product, extended (Clifford).

Technically, ganja.js is a code generator producing classes that reificate algebraic literals and expressions by using reflection to rewrite functions containing algebraic constructs to their procedural counterparts.

Practically, ganja.js enables real math syntax inside JavaScript, with element, vector and matrix operations over reals, complex numbers, dual numbers, vectors, spacetime events.

The algorithm considered in the previous section served as the basis for the developed prototype software package that performs the construction the Pareto set (Fig. 1, the Pareto set is highlighted in green, target set – blue, reachability area – in red).

As an example, we take the task (differential equations for a dynamical system) from [30] and construct a reachability area for this dynamical system.

The perturbed roll motion is determined by the following differential equations:

$$\frac{d\gamma}{dt} = \omega_x; \quad \frac{d\omega_x}{dt} = c_1\omega_x + c_2\delta_e, \tag{1}$$

where γ is the heel angle, ω_x is the heel angular velocity, and δ is the laying angle the ailerons rudders.

Control satisfies the restriction:

$$|\delta_e(t)| \le \delta_{eM}. \tag{2}$$

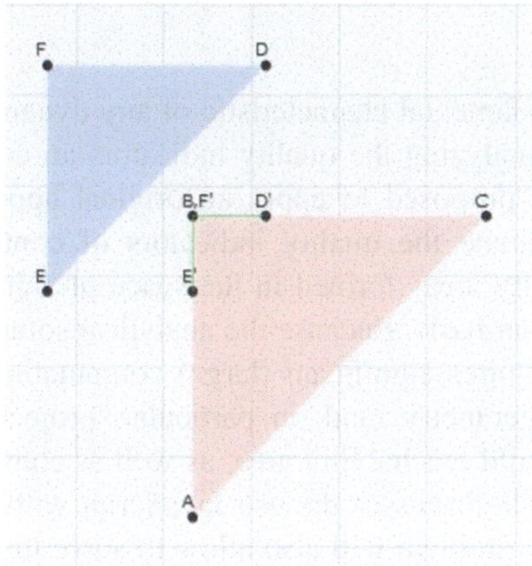

Fig. 1. Example in Ganja.js

The initial conditions are set:

$$t_0 = 0, \quad \gamma(0) = \gamma_0, \quad \omega_x(0) = \omega_{x0}. \tag{3}$$

In Fig. 2, the reachability area for system (1) is constructed in the $0\gamma\omega_x$ plane for $\theta = 1.7$ s under the initial conditions: $\gamma(0) = 3$, $\omega_x = -3$ 1/s. System parameters had the following meanings: $c_1 = -2$ 1/s; $c_2 = 2$ 1/s^2; $\delta_{eM} = 1$.

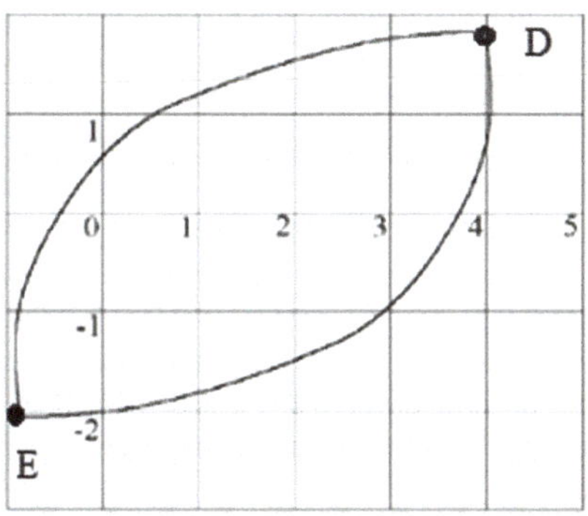

Fig. 2. Dynamic system (1) reachability area

Now there is a software implementation based on orthogonal projection, projection operator in JavaScript and Ganja.js.

6 Conclusion

Reachability area is a fundamental characteristic of any dynamic system. To solve the problem estimating and analyzing the quality indicators an control automated system for moving objects, it is proposed to apply an original approach based on the use reachability areas to evaluate the quality indicators of control systems for similar objects and use reachability areas formed in the space of significant system-technical indicators as an information basis. Because the analytical solution such problems even on modern computers requires significant (large) computational resources, it is proposed to use analytical geometry (and, in particular, projective geometry) to solve them. Thus, in order to build reachability area, as well as construction the target set at reachability area, the article discusses the use JavaScript with the Ganja.js package.

Also, the solution this problem will also allow to solve the rescheduling task from [31], which is based on an analysis of structure-dynamics control in a complex dynamic object [32].

Acknowledgements. The reported study was funded by RFBR, project numbers (**19-38-90221**, **17-29-07073**, **18-07-01272**, **18-08-01505**, **19-08-00989**, **20-08-01046**), state research **0073-2019-0004**, project «Innovative information technologies for analysis of negative impact on the cross-border region forests (KS1309 InnoForestView)» of the South-East Finland – Russia CBC 2014–2020 Programme.

References

1. Okhtilev, M.Y., Sokolov, B.V., Yusupov, R.M.: Intellectual Technologies of Monitoring and Controlling the Dynamics of Complex Technical Objects. Nauka, Moskva (2006). (in Russian)
2. Gubanov, V.A., Zakharov, V.V., Kovalenko, A.N.: Introduction to systems analysis. LGU, Leningrad (1988). (in Russian)
3. Chernousko, F.L.: State Estimation of Dynamic Systems. CRC Press, Boca Raton (1994)
4. Clarke, F.H., Ledyaev, Y., Stern, R.J., Wolenskii, P.R.: Qualitative properties of trajectories of control systems: a survey. J. Dyn. Control Syst. **1**(1), 1–48 (1995)
5. Engau, A.: Pareto solutions in multicriteria optimization under uncertainty. Eur. J. Oper. Res. **281**(2), 357–368 (2020). https://doi.org/10.1016/j.ejor.2019.08.040
6. Nogin, V.D.: Narrowing the Pareto set: an axiomatic approach. FizMatLit, Moskva (2016). (in Russian)
7. Khmelnitsky, E., Kogan, K., Maimom, O.: Maximum principle-based methods for production scheduling with partially sequence-dependent setups. Int. J. Prod. Res. **35**(10), 2701–2712 (1997)
8. Ivanov, D., Sokolov, B., Kaeschel, J.: A multi-structural framework for adaptive supply chain planning and operations with structure dynamics considerations. Eur. J. Oper. Res. **200** (2), 409–420 (2010). https://doi.org/10.1016/j.ejor.2009.01.002
9. Chauhan, S.S., Gordon, V., Proth, J.-M.: Scheduling in supply chain environment. Eur. J. Oper. Res. **183**(3), 961–970 (2007). https://doi.org/10.1016/j.ejor.2005.06.078
10. Sokolov, B., Dolgui, A., Ivanov, D.: Optimal control algorithms and their analysis for short-term scheduling in manufacturing systems. Algorithms **11**(5), 99–110 (2018). https://doi.org/10.3390/a11050057

11. Sokolov, B., Ushakov, V.: Model-algorithmic support for abilities calculating of control system based on projection operators. In: Silhavy, R. (eds.) Computer Science Online Conference 2019, AISC, vol. 986, pp. 342–348. Springer, Cham (2019). https://doi.org/10.1007/978-3-030-19813-8_35
12. Krasovsky, N.N.: Motion Control Theory (Linear Systems). Nauka, Moskva (1968). (in Russian)
13. Shorikov, A., Bulaev, V., Goranov, A., Kalev, V.: Approximation of reachable domains of nonlinear discrete controlled dynamical systems. BSU Bull. Math. Inform. **1**, 52–65 (2018). https://doi.org/10.18101/2304-5728-2018-1-52-65. (in Russian)
14. Shorikov, A., Goranov, A.: A technique for approximating the reachability region of a nonlinear controlled dynamic system. Appl. Math. Control Sci. **2**, 112–121 (2017). (in Russian)
15. Potryasaev, S.: A dynamic model and algorithms for integrated planning of operations and resource allocation in a corporate information system (2009). (in Russian)
16. Ovseevich, A.: Reachability areas for controlled systems, their properties, approximations and applications (1996). (in Russian)
17. Chernousko, F., Ovseevich, A.: Some properties of optimal ellipsoids approximating reachability sets. Dokl. Math. **67**(1), 123–126 (2003)
18. Filippova, T.: Ellipsoidal estimates of reachable sets for nonlinear control systems with bilinear uncertainty. Recent Adv. Comput. Optim. **838**, 121–131 (2020). https://doi.org/10.1007/978-3-030-22723-4_8
19. Nogin, V., Protodyakonov, I., Evlampiev, I.: Fundamentals of Optimization Theory. Higher school, Moskva (1986). (in Russian)
20. Nogin, V.: The problem of narrowing the Pareto set: approaches to the solution. Artif. Intell. Decis. Making **1**, 98–112 (2008). (in Russian)
21. Petrovsky, A.: Decision Theory. Academy, Moskva (2009). (in Russian)
22. Sokolov, B., Mironov, A., Shestopalova, O., Yusupov, R.: Methodological and Methodical Foundations of Proactive Development Control of Complex Objects and their Use in Modernization the Information System for the Plesetsk Cosmodrome. Modern Problems of the Creation and Operation of Arms, Military and Special Equipment, St. Petersburg (2018). (in Russian)
23. Polyak, B., Nazin, S., Durieu, C., Walter, E.: Ellipsoidal parameter or state estimation under model uncertainty. Automatica **40**(7), 1171–1179 (2004). https://doi.org/10.1016/j.automatica.2004.02.014
24. DeRusso, P.: State Variables for Engineers. Wiley, New York (1998)
25. Keninck, S., Wieser, E., Song, U., Hadfield, H.: enkimute/ganja.js. https://github.com/enkimute/ganja.js/tree/v1.0.167. https://doi.org/10.5281/zenodo.3635774
26. enkimute/ganja.js. https://github.com/enkimute/ganja.js
27. Kantorovich, L., Akilov, G.: Functional Analysis. Pergamon Press, Oxford (1982). https://doi.org/10.1016/C2013-0-03044-7
28. Poznyak, A.: Advanced Mathematical Tools for Automatic Control Engineers: Deterministic Techniques. Elsevier Science, Amsterdam (2008). https://doi.org/10.1016/B978-0-08-044674-5.X5001-9
29. Ivanov, D., Sokolov, B.: Adaptive Supply Chain Management. Springer, London (2010). https://doi.org/10.1007/978-1-84882-952-7
30. Tolpegin, O.: Areas of Reachability of Aircraft. BSTU "Voenmekh", St. Petersburg (2002). (in Russian)

31. Teslya, N., Potryasaev, S.: Execution plan control in dynamic coalition of robots with smart contracts and blockchain. Information **11**(1), 28 (2020). https://doi.org/10.3390/info11010028
32. Sokolov, B., Zelentsov, V., Mustafin, N., Burakov, V.: Methodology of complex objects structural dynamics proactive management and control theory and its application. In: Arseniev, D., Overmeyer, L., Kälviäinen, H., Katalinić, B. (eds.) International Conference Cyber-Physical Systems and Control 2019. LNNS, vol. 95, pp. 169–177. Springer, Cham (2020). https://doi.org/10.1007/978-3-030-34983-7_17

Open Data Quality Management Based on ISO/IEC SQuaRE Series Standards in Intelligent Systems

Václav Vostrovský[1], Jan Tyrychtr[1(✉)], and Roman Kvasnička[2]

[1] Department of Information Engineering, Faculty of Economics and Management, Czech University of Life Sciences Prague, Prague, Czech Republic
tyrychtr@pef.czu.cz

[2] Department of Systems Engineering, Faculty of Economics and Management, Czech University of Life Sciences Prague, Prague, Czech Republic

Abstract. The complexity of current data needs and their analysis requires increasingly complex methods of both database and standards, and the involvement of several other disciplines. In this article, we introduced the idea of using a Data Quality Management (SQuaRE standards) in the relation to open data. However, the issue of Data Quality Management in the relation to open data is defined in this standardization too generally. For this reason, we need to address this issue as whole. This article defines the basics of Data Quality Management in the relation to open data for using in intelligent systems. Within our specification, the roles of individual subjects of Data Quality Management are defined, including their functions and responsibilities. The article then also identifies possible specific representatives of these entities and identifies the umbrella institution of the whole Data Quality Management for the agricultural sector case study.

Keywords: Data Quality Management · ISO/IEC SQuaRE · Open data · Quality evaluation · Quality Model Division

1 Introduction

Data quality is one of the most important problems in data management (Fan and Geerts 2012). Data quality management (DQM) represents a critical business discipline that requires a combination of corresponding guidelines, processes, techniques and technologies to preserve the correctness, completeness, timeliness, and consistency of enterprise data and information. DQM is defined as the processes that ensure the integrity of an organization's data during their collection, aggregation, application, warehousing, and own analysis. The businesses and institutions that want to deal with data quality management should include the data profiling, cleansing and exploitation into their strategy, and effectively motivate all data "owners" who are fully accountable for keeping data quality procedures and rules. Most common problems identified by (Vetro et al. 2016) are lack of metadata, incomplete data, lack of information on data updates.

R. Silhavy (Ed.): CSOC 2020, AISC 1226, pp. 625–631, 2020.
https://doi.org/10.1007/978-3-030-51974-2_58

Open data can be a greatly beneficial not only for the public sector but also for the private sector (Cappgiemini Consulting 2016). These benefits of open data may relate to revenue growth, cost savings and improved efficiency (see Table 1).

Table 1. Benefit of open data (Luna-Reyes and Bertot 2014).

	Drive revenue through multiple areas	Cut costs and drive efficiency	Generate employment and develop future-proof skills
Benefit to private sector	Drive new business opportunities	Reduced cost by not having to invest in conversion of raw government data	Gain skilled workforce
		Better decision making based on accurate information	

The above-mentioned economic benefits of using open data can be expected even in connection with the private/public sector. These benefits can become a reality only if the open data will be up to date, correct and complete, which must be the main task of DQM. DQM for the needs of private sector must provide the basic for the data and information management through the following key principles:

- *Accountability:* Designation or identification of a supplier responsible for the development and oversight of open data.
- *Integrity:* The quality of open data that indicates their authentication, timeliness, accuracy.
- *Availability:* The structure and availability of open data enabling their timely and effective using.
- *Disposition:* Processes and measures to ensure the legal and ethical storage and transmission of open data.

1.1 Data Quality Model

These key principles are included in the individual data quality characteristics within the framework of international standardization of software and data quality. The general data quality model defined in the standard ISO/IEC 25012 is composed of the following characteristics (Three rivers district council 2008):

- *Integrity* ensures that data is preserved, from source to consumer, without distortion from accidental or malicious intervention.
- *Validity* ensures that data conforms to business rules such as format, range, and logic.
- *Accuracy* refers to the degree of data correctness and conformity to a given standard.
- *Precision* refers to the level of granularity of the data.
- *Credibility* refers to the reasonableness of data or how believable it is.

- *Timeliness* means that although some data might be perfect in every other way, unless it is current or comes at the appropriate time, it might be of no value.
- *Completeness* suffers when users choose to use fewer fields or enter less data, either to save time or cost, but in a way that compromises decision-making capability.
- *Conciseness and higher quality data* imply greater brevity and being more succinct in entering data A competence in maintaining quality (i.e., consistency and completeness) of corporate data positively affects the firm's adoption intention for big data analytics (Kwon et al. 2014).
- *Redundancy*, making two or more entries of the same data, is a common problem with a large database.
- *Consistency* refers to the level of data repeatability - would someone enter the same data under the same circumstances in the future.
- *Objectivity* refers to the lack of biases as humans enter, approve or manipulate computerized maintenance management system (CMMS) data. Lack of objectivity may produce distortions that might be recognizable.
- *Utility* refers to the data's applicability and usefulness.
- *Accessibility* means that data should be available and easy to obtain for those authorized to enter or retrieve it.
- *Usability*, one of the most critical aspects of data quality, refers to the intuitiveness and ease of use of data, including ease of learning and remembering. Traceability is important when you need to know the source of the data to determine its credibility and to obtain further information.
- *Flexibility* means higher-quality data won't be compromised when the process or the database supporting it are changed.

1.2 SQuaRE Divisions

The basis for the effective functioning of DQM must be international standardization of SW and data quality, i.e. a set of standards SQuaRE. This set of standards consists of five main divisions:

- *ISO/IEC 2500n - Quality Management Division*. The standards that form this division define all common models, terms and definitions referred further by all other standards from SQuaRE series. The division also provides requirements and guidance for the planning and management of a project.
- *ISO/IEC 2501n - Quality Model Division*. The standards that form this division present quality models for system/software products, quality in use and data. A service quality is under development. Practical guidance on the use of the quality model is also provided.
- *ISO/IEC 2502n - Quality Measurement Division*. The standards that form this division include a system/software product quality measurement reference model, definitions of quality measures, and practical guidance for their application. This division presents internal measures of software quality, external measures of software quality, quality in use measures, data quality measures from "Inherent" and "System dependent" point of view. Quality measure elements forming foundations for the quality measures are defined and presented.

- *ISO/IEC 2503n - Quality Requirements Division*. The standards that form this division help specifying quality requirements. These quality requirements can be used in the process of quality requirements elicitation for a system/software product to be developed, designing a process for achieving necessary quality, or as inputs for an evaluation process.
- *ISO/IEC 2504n - Quality Evaluation Division*. The standards that form this division provide requirements, recommendations and guidelines for system/software product evaluation, whether performed by independent evaluators, acquirers or developers. The support for documenting a quality measure as an Evaluation Module is also presented.

2 Methods

For the design of our solution to definition of Data Quality Management in the relation to open data, we use a set of standards SQuaRE (international standardization of SW and data quality). In this article, we adopt the Quality Model Division (ISO/IEC 2501n). The Quality Model Division (ISO/IEC 2501n) defines the following subjects of DQM:

- *Acquirer* (an individual or organization that acquires data from a supplier).
- *Evaluator* (an individual or organization that performs an evaluation. An evaluator can, for example, be a testing laboratory, the quality department of an organization, a government organization or a user).
- *Developer* (an individual or organization that performs development activities, including requirements, analysis, design, implementation and testing data during the data-life cycle).
- *Maintainer* (an individual or organization that performs operation and maintenance activities of data).
- *Supplier* (an individual or organization that provides of data).
- *User* (an individual or organization that uses data to perform a specific function).
- *Quality manager* (an individual or organization that performs a systematic examination of the data).
- *Owner* (an individual or organization that takes responsibility for the management and financial value of the data with the legal authority and responsibility to establish for them evaluation, collections, access, dissemination, storage, security and cancellation).

3 Results

Now, we propose new procedure of Data Quality Management in the relation to open data usable in the agricultural sector as case study. Consequently, we show the proposal of roles of individual subjects of Data Quality Management including their functions and responsibilities.

The set of the international standards is intended to be used by people and institutions who must follow the data quality measures during their activities.

First, we design DQM subjects in relation to the use of open data in the agricultural sector:

- **Suppliers**

Suppliers are the subjects who supply open data, although this activity may not necessarily be their primary objective. Publishing the open data should be a part of their broader strategy to increase the trust of their customers. This business model should include primarily the companies who provide open data for better decisions of customers on the market. For example, this role should belong to the Ministry of Agriculture (MACR) and the State Veterinary Administration in the Czech agribusiness. The main objective of such DQM subject is the transparency of the whole sector. An acceptable platform for publishing of open data could be the existing Farmer's portal eAGRI.

- **Aggregators**

These subjects (institutions) should collect and process the open data. The core of their business will be in creating an added value by aggregating the open data and storing them in appropriate databases. These functions should be provided by Department of Information and Communication Technologies (ICT department) at the MACR or some of the departments of the agricultural colleges or universities.

- **Developers**

These subjects should create mobile, web or information terminal applications for end users. The functions of these DQM subjects should be provided by agricultural colleges and universities as part of their scientific research activities.

- **Enrichers**

These DQM subjects should add some expert evaluations to the open data and then offer them as a service to the clients. This way the original open data will be enriched by some relevant added value that will be useful for their users. The functions of these subjects should be also provided by agricultural colleges or universities as part of their scientific research activities

- **Enablers**

This category of the participating DQM subjects should offer the tools, methods and technologies that make the open data available for further processing. Such handling of the open data requires certain skill, whether on the side of suppliers of open data, who should publish these data, or the side of the users who want to process them. These functions should be implemented by the ICT department at the MACR, by the relevant departments of agricultural colleges or universities or by professional software companies.

- **Users**

Users are individual farmers or farms that obtain open data from suppliers and use them to support their business activities.

Now we design a new conceptual model of the roles, the necessary skills and the responsibilities of the DQM subjects to support agriculture sector. Figure 1 represents the procedure of DQM in agriculture sector. However, this procedure can be generalized to other sectors.

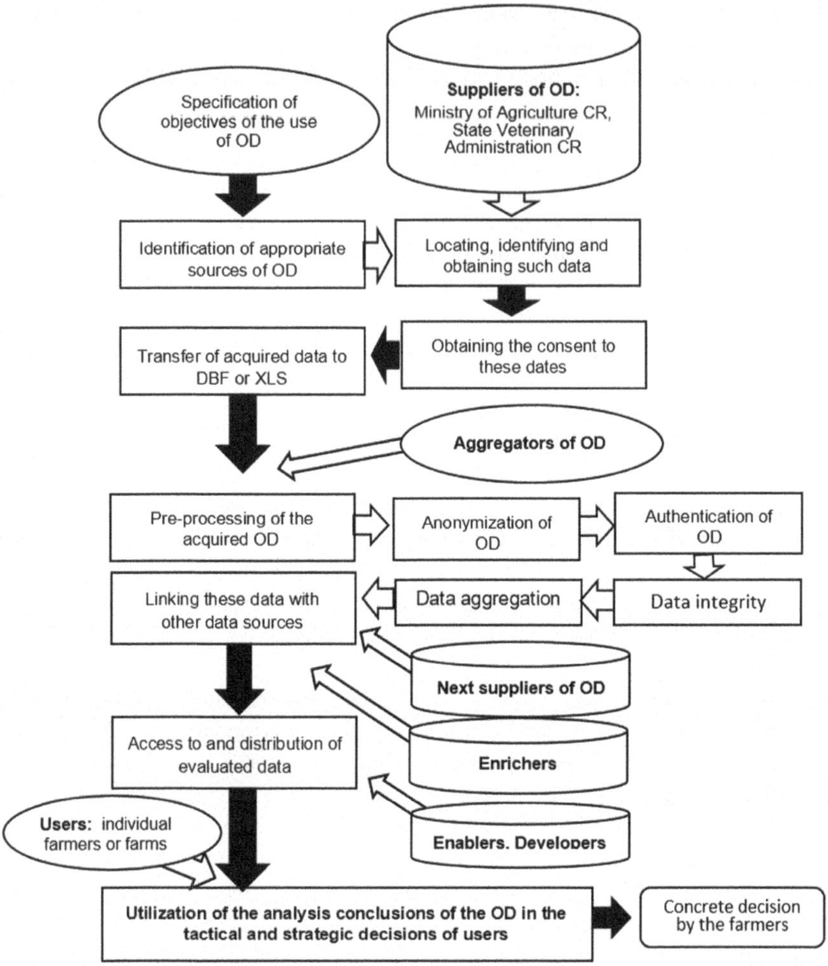

Fig. 1. A graphical representation of the roles, the necessary skills and the responsibilities of the DQM subjects.

The above mentioned and defined roles of individual subjects of DQM are a prerequisite for the effective use of open data in the agricultural sector. The MACR must have a key role to play in this matter. These roles must also include the responsibility for maintaining the necessary quality of these data. An integral part of these roles must be the unambiguous responsibility for maintaining the necessary quality of these data. This quality is determined by the degree of compliance with the individual quality characteristics defined in the international standardization of SW and data quality. Only such open data can meet the expectations associated with them in terms of Czech agriculture or other private/public sector.

4 Conclusion

In this article we presented new design of Data Quality Management in the relation to open data according to a set of standards SQuaRE. We adopted the Quality Model Division (ISO/IEC 2501n) and we proposed new procedure of DQM in the relation to open data usable in the agricultural sector as case study. Consequently, we showed the design of roles of individual subjects of DQM including their functions and responsibilities.

The results we presented in this article can be further exploited for development of the intelligent system that uses open data aggregation. Such standardized database design can be developed for several real-life problems.

Acknowledgment. This work was conducted within the project "Ambient Intelligence in Decision-Making Problems in Uncertain Conditions" (2019B0008) funded through the IGA Foundation of the Faculty of Economics and Management, Czech University of Life Sciences Prague, Czech Republic.

References

Cappgiemini Consulting: The Open Data Economy Unlocking Economic Value by Opening Government and Public Data. Capgemini-consulting (2016). https://www.capgemini.com/resources/the-open-data-economy-unlocking-economic-value-by-opening-government-and-public-data/. Accessed 4 Nov 2016

Fan, W., Geerts, F.: Foundations of data quality management. Synth. Lect. Data Manag. **4**(5), 1–217 (2012)

ISO/IEC: ISO/IEC IS 9126, Software product evaluation – quality characteristics and guidelines for their use. International Organization for Standardization, Geneva (1991)

ISO/IEC: ISO/IEC FDIS 9126-1, Information technology – Software product quality, Part 1: Quality model. ANSI, ISO/IEC JTC 1 (2000)

ISO/IEC 15939: ISO/IEC 15939, Systems and software engineering – Measurement process. ISO (2007)

ISO/IEC: ISO/IEC 12119, Information technology – Software packages – Quality requirements and testing. ISO/IEC JTC 1/SC 7 (1994)

ISO/IEC: ISO/IEC 25012, Software engineering – software product quality requirements and evaluation (SQuaRE) – data quality model, ISO/IEC JTC 1/SC 7 (2008)

ISO/IEC: ISO/IEC 25010. Systems and software engineering - systems and software quality requirements and evaluation (SQuaRE) - System and software quality models (2011)

Kwon, O., Lee, N., Shin, B.: Data quality management, data usage experience and acquisition intention of big data analytics. Int. J. Inf. Manag. **34**(3), 387–394 (2014)

Luna-Reyes, L.F., Bertot, J.C., Mellouli, S.: Open government, open data and digital government. Gov. Inf. Q. **31**(1), 4–5 (2014). https://doi.org/10.1016/j.giq.2013.09.001

Three Rivers District Council: Data quality strategy (2008)

Vetrò, A., Canova, L., Torchiano, M., Minotas, C.O., Iemma, R., Morando, F.: Open data quality measurement framework: definition and application to open government data. Gov. Inf. Q. **33**(2), 325–337 (2016)

Correction to: Technology of Self-orientation of Aircraft Relative to External Objects

Jaffar Daeibal and Nikolai Sergeev

Correction to:
Chapter "Technology of Self-orientation of Aircraft Relative to External Objects" in: R. Silhavy (Ed.): *Applied Informatics and Cybernetics in Intelligent Systems*, AISC 1226, https://doi.org/10.1007/978-3-030-51974-2_47

In the original version of the book, the following belated correction has been incorporated: The author name has been changed from "Jaafer Daiebel" to "Jaffar Daeibal" in the Frontmatter, Backmatter and in Chapter 47. The book and the chapter have been updated with the change.

The updated version of this chapter can be found at
https://doi.org/10.1007/978-3-030-51974-2_47

R. Silhavy (Ed.): CSOC 2020, AISC 1226, p. C1, 2022.
https://doi.org/10.1007/978-3-030-51974-2_59

Author Index

R. Silhavy (Ed.): CSOC 2020, AISC 1226, pp. 633–635, 2020.
https://doi.org/10.1007/978-3-030-51974-2